Springer Series in Wood Science

Editor: T. E. Timell

M. H. Zimmermann
Xylem Structure and the Ascent of Sap (1983)

J. F. Siau
Transport Processes in Wood (1984)

R. R. Archer
Growth Stresses and Strains in Trees (1986)

W. E. Hillis
Heartwood and Tree Exudates (1987)

S. Carlquist
Comparative Wood Anatomy (1988)

L. W. Roberts/P. B. Gahan/R. Aloni
Vascular Differentiation and Plant Growth Regulators (1988)

C. Skaar
Wood-Water Relations (1988)

J. M. Harris
Spiral Grain and Wave Phenomena in Wood Formation (1989)

B. J. Zobel/J. P. van Buijtenen
Wood Variation (1989)

P. Hakkila
Utilization of Residual Forest Biomass (1989)

J. W. Rowe (Ed.)
Natural Products of Woody Plants (1989)

K.-E.L. Eriksson/R. A. Blanchette/P. Ander
Microbial and Enzymatic Degradation of Wood and Wood Components (1990)

R. A. Blanchette/A. R. Biggs (Eds.)
Defense Mechanisms of Woody Plants Against Fungi (1992)

S. Y. Lin/C. W. Dence (Eds.)
Methods in Lignin Chemistry (1992)

G. Torgovnikov
Dielectric Properties of Wood and Wood-Based Materials (1993)

F. H. Schweingruber
Trees and Wood in Dendrochronology (1993)

P. R. Larson
The Vascular Cambium: Development and Structure (1994)

M.-S. Ilvessalo-Pfäffli
Fiber Atlas: Identification of Papermaking Fibers (1994)

Philip R. Larson

The Vascular Cambium

Development and Structure

With 340 Figures

Springer-Verlag
Berlin Heidelberg New York London Paris
Tokyo Hong Kong Barcelona Budapest

Dr. PHILIP R. LARSON
Chief Plant Physiologist (Retired)
North Central Forest Experiment Station
USDA Forest Service
Rhinelander, WI 54501, USA

Series Editor:
T. E. TIMELL
State University of New York
College of Environmental
Science and Forestry
Syracuse, NY 13210, USA

Cover: Transverse section of *Pinus lambertiana* wood. Courtesy of Dr. Carl de Zeeuw, SUNY College of Environmental Science and Forestry, Syracuse, New York

ISBN-13:978-3-642-78468-2 e-ISBN-13:978-3-642-78466-8
DOI: 10.1007/978-3-642-78466-8

Library of Congress Cataloging-in-Publication Data. Larson, Philip R. The vascular cambium: development and structure / Philip R. Larson. p. cm. – (Springer series in wood science) Includes bibliographical references (p.) and index. ISBN-13:978-3-642-78468-2
1. Cambium. I. Title. II. Series. QK725.L29 1994 581.4'1-dc20 94-7963 CIP

Softcover reprint of the hardcover 1st edition 1994

Typesetting: Data conversion by Springer-Verlag
SPIN: 10063432 31/3130 - 5 4 3 2 1 0 - Printed on acid-free paper

To my wife

Yvonne

Preface and Acknowledgements

A treatise on the vascular cambium must of necessity represent a compromise. This necessity for compromise becomes obvious when one recognizes the pervasiveness of the vascular cambium and the multiplicity of its derivative elements as found in a large tree. No single volume can begin to adequately cover what is known about this fascinating subject. The present volume, even though confined to cambial development and structure, also represents a compromise. The decision as to precisely what topics to include and how to discuss them was mine to make, and I am solely responsible for the contents.

My interest in the cambium began many years ago as a graduate student at Yale University. Beginning with early investigations of wood formation in trees, my research interests progressed through various stages of experimentation, finally culminating in the structure/function relations of the cambium and its products. These investigations, involving both anatomical and physiological studies, made me even more acutely aware of the pervasiveness of the vascular cambium and the difficulty of presenting it in a single volume. Not only is the vascular cambium the progenitor of all xylem and phloem, but subsumed within these two broad categories are numerous derivative anatomical elements of widely varying nature. Moreover, both the cambium and its derivative elements are subject to structural and functional modifications as the plant ages and its environment evolves.

Further contributing to the pervasiveness of the vascular cambium is its origin from the procambium as I have stressed in the Introduction. When Professor Timell first approached me to undertake the task of writing a book on the cambium, it was my intent to include the procambium. Interestingly enough, a comprehensive coverage of the procambium-cambium relationship which places the two parts of this continuum in proper perspective has never been attempted. Nonetheless, despite my strong interest in the subject, it soon became obvious that the procambium and cambium comprised too large a topic for a single volume.

I have introduced my treatise with an historical review of the cambium quite unlike those of other reviewers. Whereas others have highlighted specific events, I have attempted to trace in considerable detail the tortuous pathways by which the cambium eventually became recognized as a cellular, dividing tissue. The work of Mirbel, who not only devoted his professional career to study of the cambium but also contributed significantly to its elucidation, has been accorded special attention. Here, and elsewhere in the text, the apology offered by De Bary (1877) might well apply: "From the older literature, I have perhaps cited too much for many, and for others too little."

This historical approach has been adhered to in introducing other chapters and sections in the text. It has been my experience over the years, both in speaking with colleagues and in reading the literature, that most biologists profess knowledge of the present status of a certain biological event or process but very few are aware of how this knowledge was discovered. This oversight often leads to later dubious claims of priority. As Esau (1969) has so aptly stated, "The pace of modern research leaves little time for contemplative studies of historical documents." Consequently, I have provided the historical background in an attempt to clarify the priority for certain observations and discoveries.

In many cases, the original terminology has been retained when citing early papers, not only to preserve the author's intended meaning but also to avoid injecting unintended interpretations. For similar reasons, the reader might encounter apparent contradictions when comparing different sections of the text. This results in part because of uncertainties and contradictions in the original papers and in part because it was not possible to describe the subtle caveats, details, etc. brought forth by each author.

Although I am the author of this book and bear sole responsibility for all errors of omission and interpretation, I owe its existence to many others. I am indebted to Professor Tore Timell, who not only suggested that I write the book, but also offered valuable advice and encouragement during the years of preparation. Two institutions provided invaluable services without which this work could never have been completed. Personnel of the North Central Forest Experiment Station, USDA Forest Service, my former employer, have graciously granted me office space and full use of facilities and services since my retirement. I am particularly indebted to Drs. R. D. Lindmark, Director of the North Central Forest Experiment Station, St. Paul, Minnesota and J. G. Isebrands, Director's Representative and Project Leader, Rhinelander, Wisconsin.

Among the invaluable services provided by the Forest Service was a cooperative arrangement between the Experiment Station and the University of Minnesota Agricultural Library. Every reference cited in the text has been read either in the original volume or a photoprint therefrom. The vast majority of these papers has been provided by the dedicated librarians of the Experiment Station and the University of Minnesota's Interlibrary Loan Department. Many additional papers have been made available by the US Department of Agriculture Library in Beltsville, Maryland.

My special thanks to those who offered invaluable comments and suggestions. The manuscript was read in its entirety by Tore Timell, and selected chapters by J. R. Barnett, G. P. Berlyn, B. G. Cumbie, R. A. Gregory, L. Kucera, C. M. Larson, N. R. Lersten, C. H. A. Little, R. T. Riding, W. C. Shortle, F. W. Taylor, F. W. Telewski, E. A. Wheeler and B. F. Wilson.

Contents

Chapter 1

Introduction

The cambium has been variously defined as follows: "The actively dividing layer of cells that lies between, and gives rise to, secondary xylem and phloem (vascular cambium)" (IAWA 1964); "A meristem with products of periclinal divisions commonly contributed in two directions and arranged in radial files. Term preferably applied only to the two lateral meristems, the vascular cambium and cork cambium, or phellogen" (Esau 1977); and, "Lateral meristem in vascular plants which produces secondary xylem, secondary phloem, and parenchyma, usually in radial rows; it consists of one layer of initials and their undifferentiated derivatives" (Little and Jones 1980).

Clearly, the cambium is a diverse and extensive meristem, and no one definition will encompass all manifestations of what anatomists consider cambium. Its diversity and extent are further exemplified by a single plant, such as a temperate-zone tree, in which procambium is initiated in the embryo and perpetuated throughout every lateral, primary meristem before giving rise to cambium in the secondary body. The cambium thereafter performs its meristematic task of producing daughter cells that differentiate to specialized tissue systems. The cambium, however, does not remain static. Its derivatives vary either in form, or function, or rate of production at different positions on the tree, with age of the tree, and with season of the year. Moreover, the cambium responds both to internal signals and to external stimuli such as environment or wounding. Some plants possess no true cambium, others possess a cambium of limited extent or duration, while still others possess anomalous cambia. Obviously, cambium cannot be easily defined. It is best comprehended as a concept, a generalized idea of what cambium should be. As a concept, the cambium must be treated either as developmental states or in some specific way in which it is manifested.

Ideally, a thorough discussion of the procambium should precede that of the cambium. The procambium, a primary meristem, and the cambium, a secondary meristem, exist as a developmental continuum (Larson 1975, 1980b, 1982, 1984[1]). The procambium originates during early embryogeny. In most species, it

---

[1] In this discussion of the procambium, I have liberally cited my own publications, not because I consider them authoritative, but because the references cited therein constitute a

is first perceived during the late globular or early heart-shaped stage, usually at the level of the incipient cotyledonary node. Subsequent development of the central core of procambium is basipetal toward the radicle pole and acropetal in the elongating cotyledons and presumptive epicotyl. Once established in the embryo, the procambium progresses acropetally and continuously in the developing shoot. That is, the procambium continues its invariable acropetal advance within the apical meristem and the lateral meristems emanating from it. Protophloem, like the procambium, develops acropetally and continuously but it usually lags the procambium by several plastochrons. Protoxylem, in contrast, develops bidirectionally from an inception point situated at a level below the advancing protophloem within the bud tissues (Larson 1976a, 1979).

The cambium arises in a region of the stem designated as the primary-secondary vascular transition zone. This zone is generally associated with the stem internode that has just ceased elongating. However, because the procambium and cambium develop as a continuum, the actual transition plane within the transition internode cannot be precisely established. Various authors have attempted to establish criteria for distinguishing cambium from procambium (Esau 1954, Gustin and De Sloover 1955, Fahn et al. 1972, Larson and Isebrands 1974, Larson 1976b, Goffinet and Larson 1981). These criteria are usually qualitative, and criteria that apply reasonably well to one species might not apply to another species. Lack of consistency is undoubtedly traceable to the fact that cambium does not differentiate directly from procambium. Rather, cells characteristic of procambium grade imperceptibly to those characteristic of cambium.

Although knowledge of how the procambium-cambium continuum develops might not be a necessary prerequisite for evaluating cambial development in older trees, it is essential for understanding the early stages of cambial development. Procambium is associated with leaf traces serving newly formed leaves, and these leaf traces are organized in a precise arrangement within the stem conforming to phyllotaxy (Larson 1980a,b). Consequently, cambium, which also arises in association with the leaf traces, must conform to the same phyllotactic arrangement. These events are not confined to seedling plants. They continue throughout the life of the plant as new leaves are produced by the apical and lateral bud meristems. Each leaf has a complement of leaf traces serving it, and each leaf trace progresses through a primary-secondary vascular transition as described above. Thus, many developmental patterns established at the time of the primary-secondary vascular transition are perpetuated during secondary development. A number of these patterns, although not specifically discussed, are alluded to in subsequent chapters.

The cambium arising at the primary-secondary vascular transition zone consists of two types of cells - fusiform initials and ray cell initials. The fusiform initials are vertically elongated cells, the derivatives of which function mainly in long-

---

fairly comprehensive coverage of the pertinent literature on procambium and the procambium-cambium continuum. References pertaining to other topics mentioned in the Introduction can be found in the chapters that follow.

distance conduction, storage of reserves, and strengthening of organs. The ray cell initials are horizontally oriented strands or aggregations of cells, the derivatives of which function mainly in conduction between xylem and phloem and the storage of reserves.

It has often been said that the main function of the cambium is to divide. Both the fusiform and the ray cell initials do so in two different ways. Anticlinal divisions contribute to tangential or circumferential increase of the cambial surface and periclinal divisions contribute to radial increase of the plant organ. Authors disagree in recognizing either a uniseriate cambium consisting of an initial cell and mother cells, or a multiseriate cambium consisting of equivalent cells and mother cells. In recent years, the two viewpoints have become polarized. However, a reconciliation of the two opposing interpretations appears possible if greater emphasis is placed on the obvious similarities and less on the philosophical and terminological differences.

Anticlinal divisions, often referred to as multiplicative divisions, increase the number of cells in the tangential direction by partition walls oriented either perpendicular or oblique to the stem surface. Anticlinal divisions of cambial cells can be prolific, but not all daughter cells survive to produce new cell files. Some daughter cells are eliminated outright from the cambium, whereas others might be segmented to parenchyma cells or transformed to ray cell initials. The ultimate fate of a daughter cell is often traceable to some attribute relating to the anticlinal division. For example, daughter cells are less likely to survive as fusiform initials if they are short, if they have poor ray contacts, or if they were derived from asymmetric divisions. The cambium is a dynamic meristem and its response to different situations can be judged by how frequently its cells divide anticlinally and how many survive to divide again. Many unusual grain patterns in wood such as the whirled grain in branch crotches or the interlocked and spiral grains in tree stems can be traced to the nature and frequency of anticlinal divisions. The fusiform initials are not subject to either haphazard or uncontrolled anticlinal divisions. Rather, they are induced to divide in specific ways by stimuli, believed to be hormonal in nature, that originate in the vegetative organs of the plant.

Periclinal divisions, often referred to as additive divisions, contribute to radial growth of a plant organ by partition walls oriented parallel to the stem surface. The products of periclinal divisions differentiate as xylem and phloem derivatives. Periclinal divisions faithfully reflect the changes in cambial pattern brought about by anticlinal divisions and they pass on these changes to their immediate derivatives. Thus, by utilizing appropriate histological techniques, investigators have recorded the series of events resulting from periclinal divisions and thereby deduced prior events attributable to anticlinal divisions. Such retrospective analyses have permitted researchers to evaluate special features such as the grain patterns in trees, and to estimate changes in cambial cell size and shape due to both inherent ontogenetic processes and environmental influences. These retrospective analyses provide an efficient means of evaluating characteristics of the tender cells comprising the narrow cambial zone and events occurring therein. Many derivative elements, however, undergo post-cambial differentiation that makes them unsuitable for retrospective analyses. This obvious source of error can be easily eliminated, or at least minimized, by confining examinations to derivative elements in which the desired trait or characteristic is subject to little or no post-cambial differentiation.

The second cell type comprising the cambium, ray cell initials, gives rise to rays of the secondary plant body. Thus, rays originate as a part of the cambium and they are not considered continuations of the so-called medullary rays originating in the pith. Like the fusiform initials, ray cell initials are cytologically homogeneous although the initials within a cambial ray can differ in size and shape. The derivative elements of ray cell initials can be either upright or procumbent in form and tracheidal or parenchymatous in function. Also, like the fusiform initials, ray cell initials are dynamic and they are constantly undergoing subtle changes in size and structure. New ray cell initials are added by conversion from fusiform initials, while other ray cell initials are lost either by reversion to fusiform initials or by elimination from the cambium.

Rays of most conifers are uniseriate, occasionally biseriate. Nonetheless, they, too, are dynamic in the sense that their numbers and sizes are constantly changing. Rays of most dicotyledons, on the other hand, are multiseriate, often attaining enormous sizes. In addition to the constant addition and deletion of ray cell initials, these multiseriate rays are subject to frequent processes of merging and splitting that drastically increase and decrease their sizes. Derivatives of ray cell initials, like their fusiform counterparts, reflect the dynamic changes of the ray cell initials. Thus, the unusual or aberrant patterns often displayed by rays can be traced to similar patterns induced in the ray cell initials.

Among the many cell and tissue types that differentiate from derivatives of cambial initials are those subsumed within the category of intercellular spaces. Nonsecretory intercellular spaces are actually intercellular channels or passages that extend across the cambium between xylem and phloem. Secretory intercellular spaces are the gum and resin canals found in many conifers and dicotyledons. In all conifers and many dicotyledons these secretory canals differentiate in close proximity to the cambial initial cells. Resin canals are also developmentally dynamic. They vary markedly in distribution, in vertical extent, and in their proclivity to meander and anastomose. In addition to the gum ducts and resin canals produced in the normal course of development, many species produce traumatic canals in response to wounding. The exuding gums and resins presumably serve both to inhibit invading organisms and to protectively seal the wound surface.

Wounding of the cambium invokes cambial responses other than the differentiation of traumatic resin canals. Cambial wounding can result from a wide variety of external causes and it can be manifested in many different ways. Nonetheless, wound closure conforms to a more or less predictable pattern irrespective of the type and severity of the wound. The first response to wounding is invariably formation of a callus mass that tends to shield the exposed meristematic tissues. Callus might arise from any living tissue bordering the wound, but usually from the rays, vascular parenchyma, and undifferentiated xylem and phloem derivatives. Seldom does the cambium contribute significantly to the initial formation of callus. As the callus pad progresses centripetally inward to occlude the wound, a phellogen develops in the outer layers of callus and a cambium in the deeper layers to the inside. Phellogen formation always precedes that of cambium. However, both the phellogen and the cambium that develop within the callus pad do so in complete continuity with pre-existing phellogen and cambium in the uninjured border tissues.

The activity of the cambium, both during a season and over a period of years, has evoked the greatest interest among anatomists and others. As a consequence, various aspects of cambial activity have been researched and reviewed by authors far too numerous to mention. Yet a comprehensive review that unifies all phases of cambial activity, both intrinsic and extrinsic, has never been attempted. In the final chapter, *Cambial Zone Characteristics*, several attributes have been briefly discussed that one might logically assign to either cambial development or cambial activity. These topics have been included not only because they relate more or less directly to previous chapters, but also because they have received relatively little attention by other reviewers. Although a more thorough discussion of these topics would be desirable, it is obvious that a complete coverage of the cambium and its various manifestations is not possible in a single volume.

Chapter 2

Fossil Evidence

"The appearance of secondary wood and bast, with the power of indefinite increase by means of cambium, was the most important event in the anatomical evolution of vascular plants." This statement, made by Scott (1902), typified his attempt to trace some of the structural changes induced by the introduction of a vascular cambium and secondary growth in plants. In a later paper, Bower (1930) emphasized the role of the conducting system. In speculating as to how seed plants might meet the increased demands of this system he stated: "The more complicated and indirect, but at the same time more efficient system, is by cambial increase and the production of secondary tissues." Bailey (1953), on the other hand, stressed the role of strength tissue, noting that in the fossil gymnosperms the xylem exhibited an increasingly dominant role in strengthening stems. He maintained: "These plants with few exceptions are arboreal forms with active and sustained cambial initials."

Evidence for the origin of cambium in the fossil record was reviewed by Barghoorn (1964). Among the five major evolutionary events in the geologic history of plant life he included: "Origin and development of cambium and the capacity for theoretically unlimited secondary growth and increase in size of the plant body." Barghoorn then attempted to trace the occurrence of cambial activity in the major plant groups as they evolved in time. This work was re-examined and greatly expanded by Cichan and Taylor (1990).

Fragile meristematic cells, such as cambia, are rarely preserved in fossil plants. Special criteria must therefore be established for assessing the presence of cambia. Cichan and Taylor (1990) listed three criteria which can be summarized as follows: (1) the tissue should exhibit circumferential continuity, with the cells arranged in well-ordered files that extend radially over a significant distance; (2) the tissue should be comprised of cells that form an axial and a radial system; and (3) the cells of the tissue should exhibit structural features that are indicative of a conductive function. Based on these criteria, the authors demonstrated that a vascular cambium evolved several times in the early history of vascular plants, although the mechanism by which it arose is still unknown. Processes recognized as being important in cambial development of extant plants, such as multiplicative division of fusiform and ray cell initials and the conversion of ray cell initials to fusiform initials, were also found to be necessary for circumferential expansion of primitive plants. The bifacial, indeterminate pattern now found in extant gymnosperms and dicotyledons was apparently present only in the progymnosperms and primitive seed plants.

Most research relating to the cambium of fossil plants has been conducted on the xylem (Barghoorn 1964, Creber and Chaloner 1984, Cichan 1985a,b, 1986a,b). However, significant contributions are now being made by analyses of phloem tissues (Smoot 1984a,b, Taylor 1988), and to a lesser extent on anomalous vascular systems with included phloem (Hueber et al. 1991).

Chapter 3

Historical

3.1 Derivation of the Word Cambium

The word cambium apparently originated in the Arabic language, where it was used in medicine and surgery. Möbius (1934) attempted to trace derivation of the term from the Arabic to its first appearance in early French texts and dictionaries. However, he found it difficult to discover just how the term cambium was derived from its Arabic roots and when it was first applied to the "building sap" of plants. Schopfer (1947) encountered similar difficulties in his attempt to trace the term. Both authors cited the 1863 edition of Littré's dictionary in which cambium was defined as a "suc nutritif," a nutritive sap or substance, elaborated to furnish material for growth of the plant. In this work Littré[1] mentioned a 14th century text by Arnault de Villeneuve[2] (Schopfer) or Arnold von Villanova (Möbius) who was knowledgeable in Arabic and who referred to the cambium as a "humour." The term was also used in gardening to refer to tissues which in the process of formation appeared gelatinous. However, Littré was unclear as to how the term cambium was derived from the Latin words "cambire," or "cambio," which mean change.

The most detailed attempt to trace derivation of the word cambium was that of Lorch (1967). He, too, attributed first use of the term cambium to Arnold de Villanova although he suggested that it might have undergone a "scribal corruption" during the original translation from Latin to Arabic. Lorch recounted several medical texts wherein cambium was considered a nutritious, animal "humour" - a sort of ubiquitous, liquid "ether" that somehow helped explain the unexplainable.

---

[1] Authors mentioned in the text, but not followed by a date, are not included in the Reference list. Information concerning these authors was obtained from the appropriately cited reviewers.

[2] The names of early workers were often spelled differently, usually depending on the language of the reviewer. Also, the extended names of early workers were conveniently abbreviated. For example, Duhamel du Monceau to Duhamel, and Brisseau de Mirbel to Mirbel. I have followed the latter convention.

All three reviewers (Möbius 1934, Schopfer 1947, Lorch 1967) credited Nehemiah Grew (1682) with being first to use the term cambium in a botanical sense in his well-known edition of *The Anatomy of Plants.*[3] But, as Lorch pointed out, Grew did not clearly differentiate between plants and animals as to the nature of cambium. Nevertheless, he did refer to cambium as the sap-like formation appearing between bark and wood when they could be easily separated during early spring (Schultz 1831, Trécul 1852). About this same time in 1675, Malpighi published a discourse on plants in which he correctly located the growing zone of wood within the inner bark. An excerpt cited by Lorch made it clear that Malpighi also equated the conduction system of plants with the circulatory system of animals. Malpighi's original idea that new layers of wood in stems and branches resulted from the periodic transformation of the innermost layers of bark was adopted by many later anatomists (Sachs 1890). Yet neither Möbius nor Lorch could find any mention of the word cambium in Malpighi's text.

3.2 The Concept of Cambium as a Tissue

In the ensuing years, the word cambium appeared sporadically in botanical texts and a number of reviewers attempted to trace its tortuous development (Trécul 1852, 1853b, 1854, Sachs 1890, Lundegårdh 1922, Möbius 1934, Schopfer 1947, Baker 1948-55, Studhalter 1955, Lorch 1967). The term was seldom used in a strictly botanical sense until revived by Duhamel du Monceau in his 1758 *La Physique des Arbres,* an early precursor to modern plant physiology texts. In one experiment described by Lundegårdh (1922), Duhamel threaded fine silver wires between bark and wood and observed that the wires became embedded mostly in the wood. In contrast, when he threaded the wires in the outer bark they remained there (Fig. 3.1). Although this experiment demonstrated secondary thickening growth, Duhamel was unable to explain how it occurred. According to Trécul (1852), Duhamel referred to this gelatinous or jelly-like layer as cambium. Yet Schopfer (1947) maintained that Duhamel had difficulty in explaining how this inner layer, being in a liquid state, could be transformed into cells and tissues. Although Duhamel was apparently aware of this dilemma, his microscopic preparations did not permit him to discover the true nature of cambium.

The dilemma that faced Duhamel was not readily resolved by his immediate successors. For example, Sprengel in 1802 applied the term cambium to the German term "Bildungssaft,"[4] literally building sap. He claimed that it descended

---

[3] Grew compared the cellular structure he observed to woven lace, and from this comparison the word "tissue" was derived (Sachs 1890).

[4] The term "Bildungssaft," referring here to the cambium, should not be confused with later use of the same term. For example, Hartig (1859b) referred to transport in the phloem as the

in the plant between wood and bark, thus causing the swelling that formed above a stem girdle (Lorch 1967). Sprengel attributed the origin of new cells to small granules or vesicles contained within the mother cell. These small bodies presumably expanded and pressed together thereby contributing to the gelatinous structure of cambium (Lundegårdh 1922). Bernhardi in 1804 also assumed the cambium to be a sap with the fluid part giving rise to vessels and cellular tissues. He recognized three kinds of cambium - *Cambium lacteum* (latex), *Cambium resinosum* (resin), and *Cambium aquosum* (water) (Möbius 1934). Many others at that time, such as Link in 1806, Treviranus in 1806, Rudolphi in 1809, and Roemer in 1816, all accepted the general view that the cambium consisted either of gel-like vesicles or a sap flowing between wood and bark. For the most part, they correctly located the cambium between wood and bast (inner bark), but they incorrectly assumed that all new wood cells arose from the bast. This was essentially the same view that Hartig (1853) attributed to Grew, Malpighi, Hales, and Duhamel[5]. Hartig was of the opinion that the original "regeneration theory" or

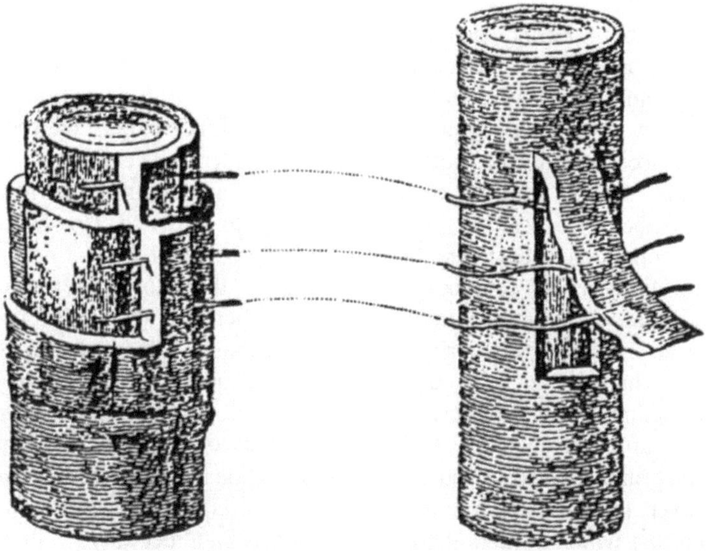

Fig. 3.1 Duhamel du Monceau's 1758 demonstration of secondary thickening growth in a dicotyledon stem. (After Lundegårdh 1922).

---

descending "primäre Bildungssaft" and return of reserve foods in the xylem as the ascending "sekundäre Bildungssaft."

[5] Trécul (1852) published an in-depth review of the history of cambium and secondary growth in plants in which he made liberal use of direct quotes from many early workers such as Grew, Duhamel, Mirbel, Knight, Kieser, and others. Hartig (1853) cited Trécul's paper and borrowed heavily from it, including the entire quote from Richard which appears later.

"transformation theory" was first formulated by Knight in 1808 and later by Kieser in 1814. According to this theory, the outer rind, along with its inner bast layer, separated from the wood during the spring growth period creating a gap. A fluid "Bildungsstoff," or building substance, was exuded into this gap and the "Bildungsstoff" was then transformed to wood layers and a new "Bastkörper," or bast body. The latter tissue then regenerated as cambium. Many years earlier, Pliny had referred to the soft inner bast as "alburnum," a tissue which presumably included the cambium along with the recent xylem and phloem (Makkonen 1967).

The concept of cambium began to change with the work of Mirbel, who recognized the cambium as a "building tissue" rather than as a "building sap." His first paper, appearing in 1808 (Lundegårdh 1922), promoted the older view of a transformation of bast to sapwood and the latter to wood. However, he did not envision a direct transformation. Rather, during the vegetation season, a delicate tissue with large vessels developed on the wood side and an open, cellular tissue on the bast side; to him, bast consisted of the entire phloem and inner cortex. Between wood and bast the cambium formed a gelatinous layer, a more or less fluid, cellular tissue. A thin layer of new bast then developed from this cambium and in this very unclear way Mirbel formulated his concept of cambium as a building tissue.

In a later paper, Mirbel (1827) responded to his critics, particularly Decandolle, who had said that Mirbel simply repeated earlier work of Duhamel. Mirbel countercharged that Decandolle had attributed to Dutrochet a definition of cambium that Mirbel claimed was borrowed from one of his own earlier papers (Mirbel 1816). In defense, Mirbel appended a transcription of this paper in its entirety.

In this 1816 paper, Mirbel admitted to having accepted at one time the view that small layers of liber, or inner bark, were transformed to sapwood. However, despite the contrary opinions of his peers, he now strongly believed that *"... never does the liber develop as wood"* (author's italics). Rather: "there forms, between liber and wood, a layer which is a continuation of both liber and wood. This regenerating layer (the couche régénératrice) is called cambium. The cambium is therefore not a fluid which moves from one place to another; it is a very young tissue that is a continuation of older tissues. It is nourished and developed by a very elaborate sap. The cambium develops during two seasons of the year between wood and bark, in springtime and in autumn. Its organization appears identical in all respects; however, the part which borders the sapwood is changed imperceptibly to wood, and that which borders the liber is changed imperceptibly to liber. This transformation is perceptible to the eye of the observer." Decandolle had attributed excerpts from the foregoing paragraph to both Duhamel and Dutrochet.

Mirbel, in this 1816 note, then raised a question which "embarasses the physiologists." That is, "... to know how the cambium, a substance with mucilagenous consistency, has enough force to repel the bark, and how, in its repulsion, it is not totally disorganized." To explain this phenomenon, Mirbel suggested that the bark expanded permitting a new "couche régénératrice" to form. The "couche régénératrice" formed a bond between wood and liber, and it therefore preceded and gave rise to the cambium. Accordingly, the cambium was renewed in this way each season (Fig. 3.2). Mirbel claimed to have conducted thorough research on the subject, having dissected and sketched specimens from 11 species of trees and noted "many very curious differences."

In yet another paper, Mirbel (1837) stated as a premise that the cambium was a tissue, or rather a cellular mucilage, of extreme delicacy. On the basis of this premise, he claimed it was not easy to demonstrate the transformation of cambium to utricles (literally, little bags or cells) and then to vessels. When he examined this delicate tissue at 400x he observed traces of white and stains of gray, which he poetically compared to "... the network that one sees in a tired eye, fixed on a white field or a luminous sky." He then increased the magnification to 700x and "... finally recognized in this material that which heretofore had never been defined." He observed "... a cellular mucilage having all the characteristics of a new organization; in short, the state of cambium."

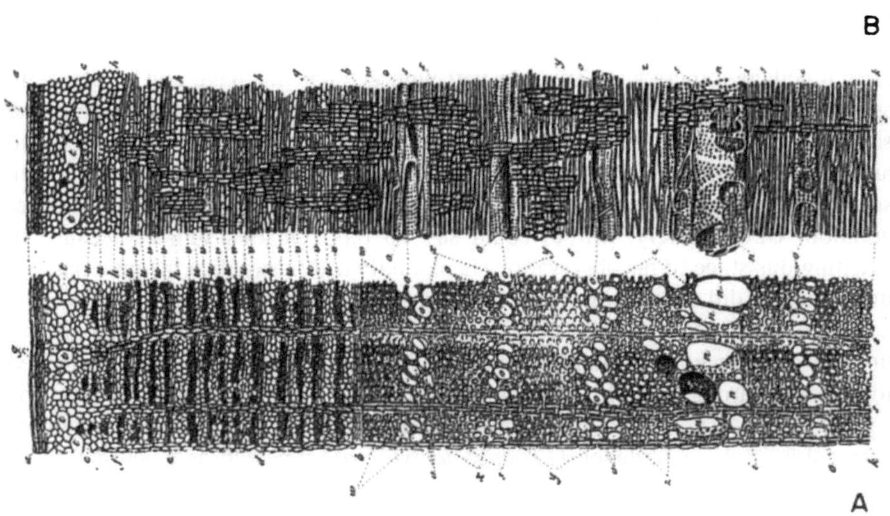

Fig. 3.2A,B. Transverse (**A**) and radial (**B**) sections of *Ulmus campestris* showing four annual increments of liber (phloem), and the fourth, along with part of the third, increment of wood. Items of interest are: *a* bark; *b* "couches régénératrice (cambium); *g* phloem medullary rays; *o* small vessels; *n*, large vessels; *s* xylem medullary rays; *v* "tubes grêles (phloem fibers). (Mirbel 1827, based on a plate engraved by M. F. Bein in 1817).

Mirbel repeated these observations on many species, and he obtained results that were ever more curious and mysterious. When he examined a very young cambium he found a "veritable minature cambium" contained within it. In fact, three generations were visible, the first being the large utricle of the series, the second arose within the first, and the third arose within the second. That is, new utricles formed successively one within the other to produce "utricle complexes." In explaining how this series developed, he noted that the large utricles and the "cambium" within them did not at first appear to change. Then, with unexpected suddeness, he observed that the large utricles became arranged in a vertical series, the walls between them disappeared, and they formed a long tube that continued to stretch and enlarge with further growth. Mirbel considered this to be a new confirmation of the theory of metamorphosis[6] of utricles to vessels. While these changes were occurring, the cambium, which previously totally filled the large utricles, was now subjected to an "inexplicable condensation". The cambium contracted to the utricle center where it was transformed to a new utricle with a new cambium in its center. On reaching this stage, the transformation process could be repeated.

Mirbel next set out to learn the origin of the spheroids or small vesicles that were liberated into the utricles, or fixed in their walls, and often contained the "immediate principles" of transformation. He pointed out that very young utricles never contained spheroids; therefore, formation of the utricles must precede that of the spheroids. In fact, he found that some utricles were entirely filled with cellular mucilage, others were filled in part with cellular mucilage and in part with spheroids, and still others contained only spheroids. Furthermore, intermediate stages occurred confirming that utricle development progressed from one stage to another. The cellular mucilage, which to him was synonymous with cambium, in some way gave rise to both the utricle wall and the new cambium for the next generation. Unfortunately, Mirbel could say no more on this subject because in reaching this stage of his observations he not only had to sacrifice his material, but that which he did observe disappeared when immersed in water.

Mirbel spent most of his career in pursuing the intricacies of cambial formation. In a major contribution published in 1839 (Schopfer 1947), he recognized four stages through which the cambium progressed from the amorphous to the cellular state. A synopsis of this lengthy paper, conducted on tissues in the roots of date palm, a monocotyledon (*Phoenix dactylifera*), appeared as a journal paper (Mirbel 1839). Mirbel had no way of knowing that monocotyledons do not possess a vascular cambium, and he therefore assumed that the date palm was no different from the dicotyledonous trees he had previously examined. Despite this shortcoming, his contribution is important in revealing the confused state of knowledge regarding the cambium at that time.

By using better optics, Mirbel (1839) was now convinced that the cambium was not a simple mucilage but a cellular tissue. Condensation of the amorphous

---

[6] For detailed reviews of the theory of metamorphosis see Sachs (1890), Green (1909), and Arber (1946).

mass to form cellular partitions occurred without an appreciable change in volume, a phenomenon he attributed to metamorphosis. During this process, the amorphous material increased in density and accumulated on the cell circumference. This state was short-lived, as the cell walls soon began to stretch and became covered with elevated "papilla" in a checkerboard pattern. Cells at this stage of development, although containing more water than formerly, were said to be gelatinous. Later, the same cells, which at first exhibited no definite form, now appeared more or less hexagonal on transverse sections. Their partitions expanded, they became thinner and drier, and as the papilla disappeared they were replaced by parallel horizontal lines resembling slender striae. When viewed on longitudinal sections, the horizontal lines appeared to extend vertically, and he suggested that a certain relation existed between the lines and the segmentation that later appeared in the vessels. However, he admitted to being very much confused regarding this idea.

Mirbel recognized three regions in the roots of date palm; peripheral, intermediate, and central. A thick layer of cambium separated the peripheral from the intermediate, and a second layer separated the intermediate from the central. He made only slight reference to the peripheral cambium and it is not clear how he envisioned the two cambia. Nonetheless, he was most excited by the intermediate zone. Utricles (cells) in the middle of this zone were oldest, and those approaching the borders of the two cambia were less developed. By older, he meant physiologically, with age dependent on the number of successive modifications each utricle underwent. The two cambia functioned simultaneously to increase the intermediate region, one adding to the interior and the other to the exterior.

Mirbel did not clearly distinguish cambium from amorphous material despite his previous assertions to the contrary. The utricles comprised the cellular material. Cambium could contribute material either directly to new utricles or to what he referred to as "méats," or intercellular spaces. By a series of transformations, utricles formed within the "méats" to eventually become the radial series of vessels. He attempted, with some success, to account for all cells of the root and their dispositions in time and space. Nonetheless, his concept of cambium was little changed from that espoused in his 1837 paper. Cambium was said to "flow" into or toward the new utricles, and new utricles formed one within another. Each of these utricles was "filled" with cambium, which eventually condensed to form the cellular partitions or walls.

A contemporary of Mirbel, the physiologist Payen (1839), had proposed a chemical theory for the nutrition of plants. He demonstrated that a contractile, granulose substance first appeared in the cambium, or the liquid phase, preceding new growth. This substance was nitrogenous. Mirbel and Payen (1843) were intrigued by the similarities of this substance with that found in animal cells; but vegetable cells were primarily carbon compounds, not nitrogenous. They speculated as to the differences between plant and animal cells, finally suggesting that the cellulose walls of plants were secreted by the cambium. Other early papers having to do with secretions producing cambium were reviewed by Trécul (1853b) in his work on wound healing.

Mirbel was not without his critics. Unger (1842), for example, presented a lengthy critique of Mirbel's (1839) paper. In particular, he objected to the way Mirbel employed the term cambium. That is, Mirbel considered cambium to be both the "organizable sap of the regenerative layer" and the cellular material so organized by this sap. Unger believed that the term cambium should be confined to the young "elementary organs"[7] that carried this formative sap as he had proposed in a previous paper in 1841.

In a rather remarkable paper, another critic of Mirbel's, le compte de Tristan (1840), undertook the daunting task of clarifying both the concept of cambium and the nomenclature of tissues produced by it. Rather than clarifying, however, his paper contributed to the existing confusion. He began with a philosophical discussion of the expression "nature d'une chose," or nature of a thing, and applied this philosophy to his conceptions of the cambium. Specifically, he objected to Mirbel's idea that all elementary organs arose in a homogeneous tissue, the cambium, and that any variation these organs underwent also occurred in the same homogeneous tissue. That is, Mirbel regarded these diverse changes in the elementary organs as simple modifications and not changes in nature. Thus, le compte de Tristan posed the question, "Are elementary organs of plants of the same nature but diversely modified, or do they originate with diverse natures"?

Le compte de Tristan conducted preliminary observations on the vascularization of *Papaver somniferum* and *Cucurbita maxima*, but his principal investigations relating to the cambium were conducted on *Rubus fruticosus* and confirmed on *Castanea* and *Quercus*. He disregarded all prior work and "... began his research essentially without antecedents and certainly without prejudice." That he did so without antecedents soon became obvious. He was of the opinion that all elementary organs had different origins which, in turn, required different cambia. Before discussing these different cambia, it is first necessary to attempt clarification of his convoluted conception of stelar anatomy and his totally original nomenclature.

Le compte de Tristan visualized the stem as consisting of three fundamental tissues. The first was called "aphrostase," meaning foam or fixed. This tissue was essentially parenchymatous, but he rejected the term parenchyma because he believed that botanists were too free in the way they comprehended it. The second tissue was called "hegémon," meaning guide or conductor. This term was applied to the longitudinal xylem bundles that either penetrated or were "guided" through the "aphrostase" tissue. He deplored the term vascular bundle, and variations thereof, because such terms drew undue attention to the vessels rather than to the true nature of the bundles. He preferred the term "filet," or "filament," hence the term "filet hegémien." The third tissue was called "proxyle," meaning before

[7] The term "elementary organ" referred to any individual cell, or sometimes a tissue, of unspecified nature.

wood. In effect, it was not yet wood but it formed the principal structure of wood when it became consolidated. Thus he also recognized "filets de proxyle."

The fundamental tissues contributed to the structure of the stem. All stem tissue inside the bark was called "endophyte." Between the bark and "endophyte" was the "cunice," which was not a tissue but simply a gap, hence an abstraction. The "endophyte," in turn, was subdivided into pith and "endostère," the solid, lignified part. The pith was derived from "aphrostase" tissue and the "endostère" from "hegémon" tissue, at least in its youth when it consisted of an ensemble of "filet hegémien." Each year a tree acquired a new layer of "endostère." The intervening tissue between "filets" and extending from pith to cortex was called "isthmes." It was of the "nature aphrostasienne." Stellar projections radiating outward from the pith were called "irradiations."

The foregoing nomenclature was offered to replace what le compte de Tristan considered the "older, defective" system. His new system was based on the three tissue systems, each of which presumably had different "origins," consequently different "natures." However, because the tissues had different origins, they necessarily arose from different cambia. He regarded the "aphrostase" and "hegémon" cambia as being separate, but debated whether the latter might have arisen from the former. He also admitted to the possibility that the "cambium proxylaire" might have arisen as a "secretion" from the other two, but again he preferred considering it a separate cambium as it was a distinctly new tissue that was neither "aphrostase" nor "hegémon" in nature. "It is, itself, a state of cambium," he wrote.

As visualized by le comte de Tristan, the "aphrostase" cambium would contribute to the interfascicular "isthmes" and related parenchymatous tissues. The "hegémon" cambium would contribute to the isolated "filets hegémien." Thus, we might regard these two sets of "cambia" as producers of primary tissues. The "cambium proxylaire," on the other hand, was definitely a producer of secondary tissues. This "cambium" arose in the outer parts of the "isthmes," or interfascicular regions. It produced new "elementary organs" that "overflowed" to encrust the "cunice," or gap region, between bark and "endostère." This new tissue, called "adélônes" was "hegémon" in nature, but the "elementary organs" were completely different from those produced by the "hegémon" cambium. Activity of the "cambium proxylaire" eventually formed a complete ring, but it did so differently in *Pinus pinaster* than it did in *Castanea*, *Quercus*, and *Rubus*. In conclusion, le compte de Tristan claimed that the three fundamental tissues, "aphrostase," "hegémon," and "proxyle," were "radically different", and they did indeed originate with diverse natures, contrary to the view of Mirbel. However, the resulting tissues became unrecognizable from one another by the diverse "adélônes" which tended to obliterate their different characteristics.

Another contemporary of Mirbel, Moldenhawer in 1812 (Lundegårdh 1922), compared one- and several-year-old woody branches and found the same cellular characteristics in both. By analogy, he deduced that a similar thickening process occurred in both herbaceous and woody plants with the exception that bundles were dispersed in the former whereas they formed a ring in the latter. The thin cambium produced separate tissues, either bast, parenchyma, or wood. Because Moldenhawer could not recognize phloem, he could not separate phloem from cambium. Nonetheless, he described the thin cambial layer as a delicate, small-celled "substance" situated between the bast bundles and the wood mass. This cambial layer produced new homologous parts, bast cells to the outside and wood cells to the inside. Moldenhawer then attempted to differentiate between monocotyledonous and dicotyledonous stem development.

A thought-provoking paper, overlooked by most reviewers, was that of Schultz (1831). In addressing the question "What is the origin of cambium"?, he listed four reasons why it could not arise from a woody sap, and four reasons why it could arise from the "Rindensäften," or rind sap. Among the latter, he noted that cambial formation began in the twigs and progressed gradually downward. Thus the rind of the twigs "dissolved" (i.e., to form cambial sap) before that of the stem. Schultz then asked, "What are the characteristics of cambium"? He pointed out that both Mirbel and Sprengle believed the cambium to be a building sap that sank downward between wood and bark to the roots. Du Petit Thouars had advanced a different theory, suggesting that the descending sap was the equivalent of roots growing downward from the twigs. This idea arose from observations that the vessels, formed in young twigs, continued in a downward course in the outermost wood layers of the stem below. Schultz disagreed with the latter theory by pointing out that the cambium was derived from rind sap, i.e., it arose from dissolution of, or secretion from, the inner part of the rind. Once formed, it was no longer a sap but an organized structure. When viewed through a magnifying glass, Schultz could see that the cambium was a tissue consisting of delicate cells and vessels which initially appeared as a gelatinous, weakly fluid mass. As development proceeded, the fluid part congealed, producing strong, organic tissues in the form of wood and rind.

Hartig (1837b), in quoting from one of his 1835 papers, also considered the cambium to be a sticky, sweetish-tasting fluid of clear consistency appearing between wood and rind. He, too, disagreed with du Petit Thouars' bud-root theory, noting that although the theory yielded a "pretty picture," it should not be taken literally. Hartig agreed that cambial activity began in the growing buds in spring, but he disagreed with the idea that cambium flowed downward from the buds. Rather, he observed a similarity between the fluid cambium and "Milchsafte," or the milky sap of trees. He viewed cambium as the fluid that not only filled the intercellular spaces between wood and rind but also the tender wood cells themselves. Thus, starch, stored as "Stärkemehl" in wood cells during winter, was gradually transformed to cambium in spring.

The view that prevailed at this time is perhaps best expressed by another contemporary of Mirbel, the French botanist Richard, who published an elementary botany text in 1846 that was most certainly based on Mirbel's work. Schultz (1831) cited a 1825 text by Richard that was also based directly on the work of

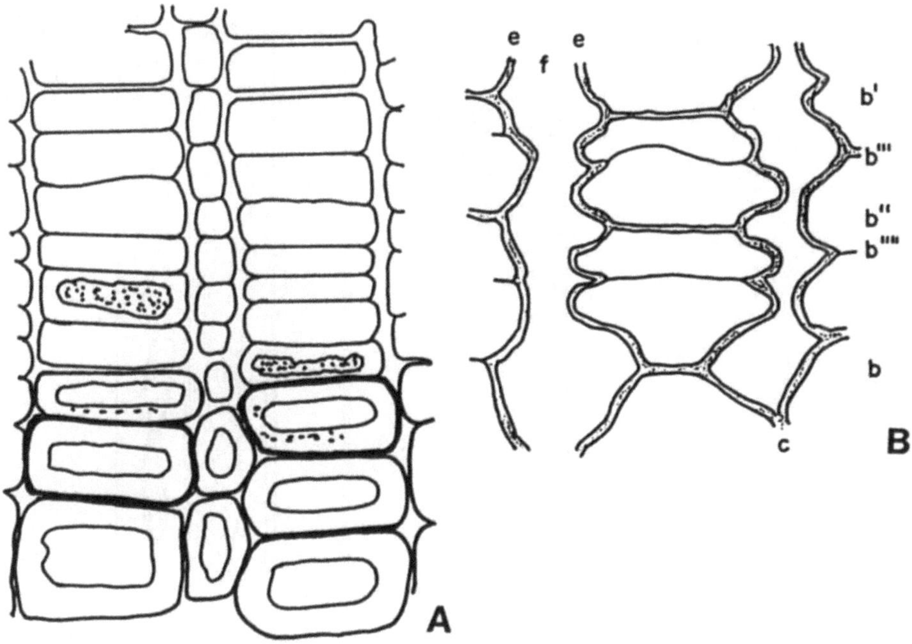

Fig. 3.3A,B. Cambial region of *Pinus sylvestris* stems in transverse section. **A** At the beginning of cambial activity on April 21 showing six to seven cambial zone cells situated between *a* and *b*. 550x. **B** Four youngest cambial zone cells from an actively growing tree. Cells swollen with concentrated sulfuric acid to emphasize wall layers. *b* Tangential walls of different ages; *c* radial walls; *e* primary cell wall; *f* intercellular substance. (After Unger 1847).

Mirbel. Trécul (1852), and later Hartig (1853), included a long quote in French from the 1846 text by Richard which I have translated as follows:

"The cambium is a nutritive fluid that in springtime flows in abundance in the celluleuse (cellular layer) which we have named the zone génératrice. It (this layer) is not transformed on the one side to a layer of bast and on the other to a layer of wood. The cambium is a fluid that essentially nourishes the plant as the blood does the animal.

At the time this phenonemon occurs in the plant there exists between wood and bark a layer of utricle tissue which unites one with the other. It is in the innermost part of this celluleuse, bordering the wood body, that one finds this fluid in abundance, the suc nutrifs (nutritive sap). This sap, by its presence, determines the formation of a large number of new utricles, either by the appearance of divisions within the existing utricles or by new utricle cells formed between them. This utricle mass separates without delay in two parts; the one, applied against the external face of the wood body, is transformed little by little to wood; the other,

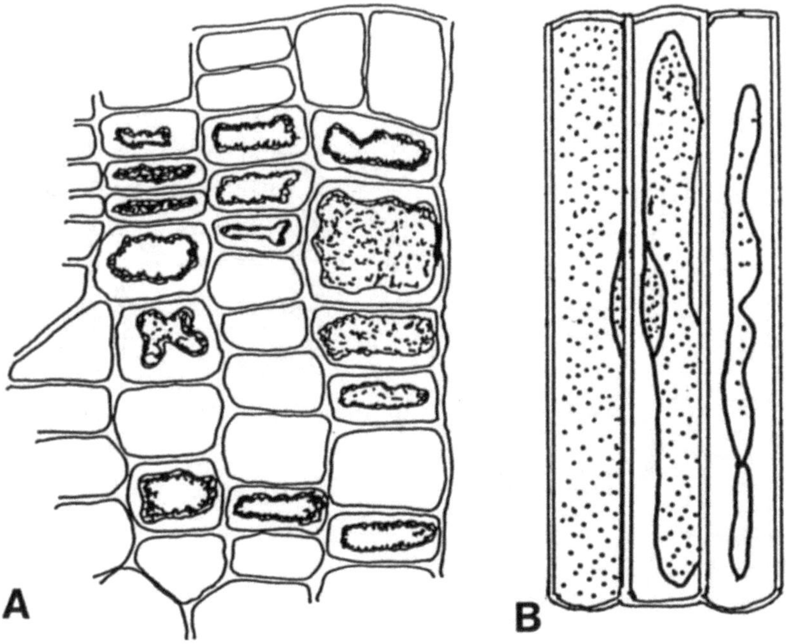

Fig. 3.4A,B. A Transverse section of cambial zone showing condition of "Primordialschläuche" detached from cell wall due to fixation in ethyl alcohol. 460x. **B** Cambial cells of *Sambucus edulus* in radial section showing presence of "Primordialschläuche" and nuclei embedded within it at left and center. 290x. (After von Mohl 1844).

which approaches the internal face of the bast, is organized into new fibers and bast; they (wood and bast) remain separated by a zone of tissue utriculaire in which the formation of new utricles persists until the time when it is arrested by phenomena within the plant."

In subsequent seasons the "zone génératrice" became the "couche régénératrice," that is, the generative layer of the first season became the regenerative layer of the following season. Hartig (1853)[8] examined the "couche régénératrice" in a number of woody species. He was unable to locate such a cell zone in dormant conifers, but he did observe a more or less broad zone of delicate fiber-like cells in the winter wood of several hardwoods. This fiber-like layer appeared to be identical with the over-wintering "couche régénératrice." His observations were based almost exclusively on differences in cell structure, which he admitted were difficult to make because of the delicate tissues. Nonetheless, he

[8] In this paper, Hartig noted that he had described the cambium in terms corresponding to the "couche régénératrice" in 1837. However, he complained that because his works appeared in forestry publications they were completely unnoticed by botanists.

was convinced that this tissue was nothing more than bast fiber tissue in which metamorphosis to cell fibers (wood) was for the most part complete but that to bast bundle fibers and sieve elements incomplete. Hartig concluded that "The so-called cambial or reproductive layer is actually bast fiber tissue that overwinters in an incomplete condition." Several years earlier, he (Hartig 1845) had observed that during bark separation in spring the two mother cells, one for bast and one for wood, remained on the bast side. He went on to point out that separation occurred "when cambium is none other than the layer of young, unhardened organ of the wood and bast body."

To account for the growth in thickness of trees, Hartig (1853) examined transverse stem sections in spring when only two mother cells were present between wood and bast. He then observed how these "Siamese twins" grew (Chap. 4.1.1). The daughter cell on the wood side was believed to be sterile and it therefore matured as a wood cell. However, the one on the bast side could divide from time to time, by constriction, to give birth to a new mother cell. Hartig (1855b) elaborated on the process by which wood and bast increased from this twin pair of permanent mother cells. The daughter cells on both wood and bast sides were at first indistinguishable from the mother cells; these together were called the "Schichtungscomplex" (stratified complex) or cambium. The first indication of differentiation was in the wall pitting of the daughter cells, at which time the wood cells were joined and the bast cells formed sieve groups.

About a decade earlier, both Unger (1847) and von Mohl (1844) had demonstrated conclusively not only that the cambial layer consisted of cells but also that the cells exhibited different developmental stages. For example, Unger produced evidence that the walls of cambial cells parallel to rind cells, as in Fig. 3.3A, were thinner than those parallel to ray cells. Von Mohl (1844) claimed to have shown the same relation in Fig. 3.4A, although the purported differences in wall thickness are by no means evident in his drawing. The cellular nature of the cambium, however, was depicted clearly by both authors. Von Mohl defined the cambial layer as a place where the number of elementary organs could increase without breaking the continuity of the tissues; i.e., new elementary organs could be inserted between previously formed ones.

This more enlightened view, unfortunately, was not always carried over to textbooks of the day. For example, Trécul, in his seventh edition of *Nouveaux Éléments de Botanique* and his *Précis de Botanique*, both published in 1852, espoused essentially the same view as the French botanist Richard. Gaudichaud (1852), in his critique of Trécul's work, made numerous quotations from these texts. Like Richard, Trécul used the animal analogy, suggesting that the cambium was the essential fluid nourishing the plant as blood did the animal. It contained all the elements required to form tissues and the different constituents that should enter into the plant body. However, just as the blood was not transformed directly into muscle, so the cambium was not transformed directly into wood. It only furnished the necessary materials for the formation of new "liber," or inner bark, and new layers of wood. Thus, the cambium was the equivalent of a fluid, nutritive sap, and the cambium itself played no role as a tissue organizer. Actual cell formation occurred in the "zone génératrice" where the "tissu utriculaire" was produced. As summarized by Gaudichaud, Trécul believed that the sap formed the cambium, the cambium formed the "tissu génératrice," and the latter formed the cellular tissues.

Gaudichaud (1852) disagreed vehemently with this view, particularly the role of the "tissu génératrice" which he considered irrelevant. He suggested that a fluid sap emanating from the buds nourished the cambium which was in a gelatinous state in the spring. In some unexplainable way, the cambium was transformed to "tissu utriculaire" and this tissue was in turn converted to wood cells on the inside and "liber" on the outside. Although Gaudichaud, like Hartig and a few others, essentially were correct in how the cambium functioned, they failed to differentiate cambium as a sap from cambium as a tissue.

A strong rivalry had existed between Trécul and Gaudichaud for several years. Gaudichaud's (1852) remarks were primarily directed at a paper read by Trécul before the French Academy of Sciences on February 16, 1852. As a consequence of the controversy that erupted between these two eminent authorities, the botanical section of the Academy established a commission to resolve the differences. Commission members were de Jussieu, Brongniart, and Richard. The Commission report, prepared by Richard (1852), constituted a strong endorsement of the views of Trécul.

In his memoire, Trécul (1852) had described the healing of a stem of *Nyssa angulisans* completely stripped of bark to a height of 45 cm. The tree was growing in a protected, humid environment and the wound healed completely under natural conditions. These results were compared with those of Duhamel du Monceau, who earlier demonstrated that debarked stems could heal in this way only if the wounds were covered and protected from desiccation. The Commission members thoroughly reviewed Trécul's conclusions and confirmed them completely. In attempting to explain Trécul's results, Richard (1852) invoked many of his own views regarding the "couche génératrice," or the ensemble of tissues that united wood and bark. According to Richard, the "couche génératrice" to the inside consisted of "utricles," more or less elongated, that resembled fibrous elements. To the outside, these "utricles" merged imperceptibly with "utriculaire" tissue of the inner bark. He then described how these "utricles" were successively transformed by "diverse metamorphoses" to their respective mature tissues. Based on this concept of the "couche génératrice," Richard then attempted to describe and explain how isolated "... plates of wood and bark developed at certain points on the debarked wood body." Somehow, this process led to complete healing of the debarked stem.

Brongniart (1852) stepped into this fray in defense of Trécul with a separate report that was more philosophical than scientific. He first attempted to clarify the distinction between the "couche génératrice" and the "cambium" which, according to French anatomists, involved more than terminology. Based on his "delicate microscopic observations," he found that the cambial zone consisted of a young tissue, thin and transparent, bathed in an abundant liquid. These observations were obviously lifted directly from Mirbel's writings. He noted further, that most "modern anatomists" designated this thin layer of young tissue which produced new wood and bark as the "couche génératrice," and that they either rejected the word "cambium" or applied it only to the nourishing liquid that bathed the young tissue.

Rejection of the word "cambium" was a direct attack on both the old "theorie du cambium" and on Gaudichaud's theory of phytons by which the latter had attempted to demonstrate continuity between vessels of the wood and young vegetative shoots. Brongniart admitted that vessel continuity was possible, but countered that it did not necessarily prove how the vessels were initiated. He favored the idea of independent initiation and subsequent joining of the vessel elements. As proof, he offered several observations of wound healing following decortication and girdling, all of which conformed with the "facts" in Trécul's memoires. In conclusion, Brongniart asked: "Why not agree with the vast majority of botanists, both old and present, that the fluid elaborated by the leaves and other living plant parts constitutes a descending sap, often mixed with an ascending sap, which together are the cause of new tissues and diverse organs, and which at all times (this fluid) penetrates young tissues making them susceptible to growth and multiplication"? Here again, Brongniart reinforced the distinction between "cambium" the fluid nutrient, and the "couche génératrice," the meristematic tissue.

In all fairness to Trécul, it should be noted that the texts Gaudichaud criticized were undoubtedly written several years before this argument erupted. Consequently, they did not necessarily express his current thoughts. In a series of papers (Trécul 1853a, 1854) concerned with the development of wood elements and the diameter growth of plants, he also professed to a different view of the cambium. He (Trécul 1853a) now subscribed to the theory that new cells were produced by "fractionation" of pre-existing "utricles" in the "couche régénératrice." For example, he examined the "utricle" layer of *Paulownia imperialis* in early spring. A cell immediately adjacent to the previous season's sapwood first expanded horizontally and then divided longitudinally by walls oriented parallel to the tree's circumference. From this division, a wood fiber and a new mother "utricle" were produced. After 12 or so similar divisions, the "utricles génératrices" stopped producing wood cells and began producing cortical cells. However, Trécul still clung to one outmoded view, noting that some young cells were susceptible to metamorphosis when necessary to meet requirements of the plant.

Trécul's (1854) next observations were made on members of the Cactaceae. As he described the "couche génératrice" in *Echinocactus couranti*, it consisted of oblong cells with thin, smooth, transparent walls, deprived of nuclei, and arranged in a horizontal, radial series. In this paper, Trécul presented an exceptionally interesting review of previous research on the cambial cell wall. Of particular interest was a lengthy quote from a paper by de Jussieu (no date) which Trécul considered "... an exceptional résumé of the state of the science."

Regnault (1860) was also a follower of Mirbel, adopting his terminology and many of his ideas. He recognized seven cellular layers or tissue systems in the stem: (1) the epidermis, (2) the "couche subéruse" and (3) the "couche herbacée," corresponding to the outer and inner cortex, respectively, (4) the "couche génératrice," consisting of two or more less distinct zones; cells of the outer zone were analogous to those of the "couche herbacée" and those of the inner zone were regularly disposed as wood fibers and rays, (5) wood cells, (6) a part internal to the wood, obviously primary xylem, and (7) the pith. The "zone génératrice" was described as consisting of delicate, transparent, fibroid elements closely pressed together. These cells were eventually transformed to lignified fibers.

I have dwelled at great length on the contributions of Mirbel because his work obviously was central to the formulation of a concept of cambium as a tissue. Hartig (1853) credited Mirbel with being the first anatomist to propose a cambial theory. Schacht's (1856) interpretation of Mirbel was essentially in agreement with that of Hartig. Yet despite Mirbel's acknowledged contributions, both Schopfer (1947) and Lorch (1967) concluded that even though he espoused the cellular concept of cambium, he failed to grasp its true cellular nature, a view with which I concur. On a contrary note, Cusset (1982) credited Mirbel with contributing the first observations of cellular divisons.

3.2.1 The Role of Cell Division

Formulation of a true concept of the cambium could not proceed further without the contributions from two closely allied areas of investigation: namely, cell division and cell protoplasm. It was now accepted that new cells formed between bast and wood, but just how they formed was not clear. Schwann, together with Schleiden, originated the cell theory in about 1839, but Schleiden assumed the lead in applying the theory to plants (Baker 1948-55, Mendelsohn 1963). According to Schacht (1856), Schleiden designated as cambium not only the building tissue between bast and wood but also that which occurred beneath the vegetative apex, the "Urparenchym," out of which vascular bundles presumably developed. According to the interpretations of both Lundegårdh (1922) and Möbius (1934), Schleiden considered the shoot apex to be the primary site of cambial formation. The cambium was believed to be a residue of the embryonal tissue mass of the upper shoot. During the differentiation process, cambium remained limited to the upper shoot and the outer side of the vascular bundle. Schleiden strongly emphasized the fact that vessels arose in a cell series, but he erroneously included phloem in the tissue he called cambium.

Although Schleiden and others at this time recognized cell division, they were unclear how it might apply to the cambium. In his early work in 1840-41, Unger, like Schleiden, considered the cambium to include what we now call procambium, since he found that this region of the shoot was also capable of producing vessels. Also, like Schleiden, he provided a fairly good description of cambial cells

(Möbius 1934). Unger argued forcefully in favor of cambial cell division, but he lacked the observations necessary to formulate a theory (Sachs 1890). Nonetheless, he described the shape of cambial cells and recognized the nature of their walls as being similar to the intercellular material of wood cells. He was also aware of the continuity of the cambium with both wood and bast, and he concluded that the wood body depended on division of cambial cells. Yet he failed to discover the true nature of cambium for he still believed that bast was produced by the parenchymatous layer enveloping cambium to the outside (Lorch 1967).

In subsequent research on cell division in *Pinus sylvestris*, Unger (1847) showed that new cambial cells formed neither at the boundary of the outermost wood cells nor at the boundary of the parenchymatous rind, but approximately in the middle between them. He noted that the radial diameter of the two smallest cells in the cambial region, taken together, was about equal to that of an adjacent wider cell (Fig. 3.3A). Therefore, it appeared obvious to him that the two smaller cells must have formed by the division of a larger one. This fact could be demonstrated by treating the tissues with concentrated sulfuric acid. In the swollen tissue, the tangential walls were always very thin relative to the radial walls (Fig. 3.3B). Also, since walls b''' and b'''' were thinner and more delicate than walls b, b' and b'', he concluded that they must be younger.

It was known at this time that the inner walls of all cambial cells were lined by an extraordinarily tender primordial membrane, which enclosed the "Primordialschlauch," or cell protoplasm (Fig. 3.4A,B). This membrane, which adhered closely to the cell wall but could be separated by chemical means, enclosed the granular cell contents. Von Mohl (1844) conducted an extensive investigation of the cambial layer in *Asclepias syrica, Pinus sylvestris* and *Sambucus edulis*. He observed that the "Primordialschlauch" divided to reproduce itself before a partition wall formed. Thus, division of the "Primordialschlauch" preceded formation of the common wall separating the new cells, a developmental phase he referred to as "gerechtfertig," literally preparation phase. He admitted to the possibility that a tender partition wall might have preceded division of the "Primordialschaluch." If so, he was unable to detect it because of its delicate nature and possible disruption during sectioning. Unger (1844) did, however, observe early development of a partition wall in dividing cells of *Tradescantia zanonia*, suggesting that it resulted from infolding of the parent cell's membrane. Both von Mohl and Unger maintained that the cell nucleus exerted no influence on the division process. Although the nucleus was intimately bound to the "Primordialschlauch" of the mother cell, both workers thought it disappeared during cell division and a new one later appeared in each daughter cell. In an 1850 publication, von Mohl still recognized a "Primordialschlauch" around which a primary membrane formed; Schacht adopted this view in his 1852 textbook (Trécul 1854).

Von Mohl (1844) never observed "Primordialschlauch" in older cells, and he therefore believed that this material was resorbed when the cell began to expand. He attributed many properties to the "Primordialschlauch," even suggesting that it might be considered "a special organ." It contained nitrogenous materials, and, in addition to its active role in cell division, it was also involved in the assimilation of nutrients into the cell. In cambial cells that still contained a nucleus, this body was situated within the "Primordialschlauch" (Fig. 3.4B) as were all cell inclusions in older cells such as chlorophyll bodies and starch granules.

In a later paper, Unger (1847) elaborated his idea that the delicate membrane enclosing the "Primordialschlauch" folded inward from both sides, thereby dividing the cell contents in two parts. He suggested that this infolding might be regarded as the first stage of new cell formation, although he now entertained the possibility that nuclear division might precede cellular division. Unger's thoughts on nuclear division were probably in reference to earlier work by Schleiden who had presented the view that during cell division two or more nuclei arose from a slimy, granulose substance, followed by each nucleus developing a membrane by sap absorption to form a new daughter cell. As pointed out by von Mohl (1844), Schleiden believed that these new cell membranes were formed by the nucleus and assumed the shape of an hour-glass. Hartig (1854c) strongly objected to this view, stating that in 15 years of observations he had never seen new cell nuclei precede formation of daughter cells, nor had he seen evidence that nuclei arose by aggregation of slime bodies.

Hartig (1853, 1854b,c) recognized what he called the "Ptychodeschlauch" and equated it with the "Primordialschlauch" of Unger (1844), von Mohl (1844), and other workers. Hartig, however, was sometimes unclear in distinguishing cell contents from the membranes enclosing them. As visualized by Hartig, the "Ptychodeschlauch" was not simple in structure. Rather, it consisted of a "Ptychodesaft," presumably protoplasm, enclosed by an inner "Ptychode" and an outer "Ptychoide"; these membranes presumably represented the tonoplast[9] and primary wall, respectively. Sanio (1863a) later equated Hartig's "Ptychode" with the extremely tender, inner lining, the "Innenauskleidung," or tertiary layer, of libriform fibers. Hartig (1853) believed that all young wood and bast cells were originally indistinguishable, but "... in the different kinds of metamorphosis, the Primordialschlauch played an important role." For example, he considered the "Ptychode" of a young cell to be a "*permanent* cell part" (author's italics). During cell maturation, the "Ptychode" was transformed to the permanent cell membrane that lined the inner cell wall, the pit canals and pit chambers through which older cells communicated, a view later advanced by Russow (1882c). Thus Hartig regarded "... the young Ptychode (Primordialschlauch) and the inner lining of mature cells as one and the same." Here, Hartig confusingly equated the wall membranes with the cell contents. In contrast to Hartig, other workers believed that the Primordialschlauch was confined to young cells, and the inner lining of mature cells had an entirely different origin. This inner cell wall had been observed by others. For example, the spiral striations appearing on the inner wall of *Pinus syl-*

[9] According to Strasburger et al. (1908), de Vries introduced the word "tonoplast" in 1885.

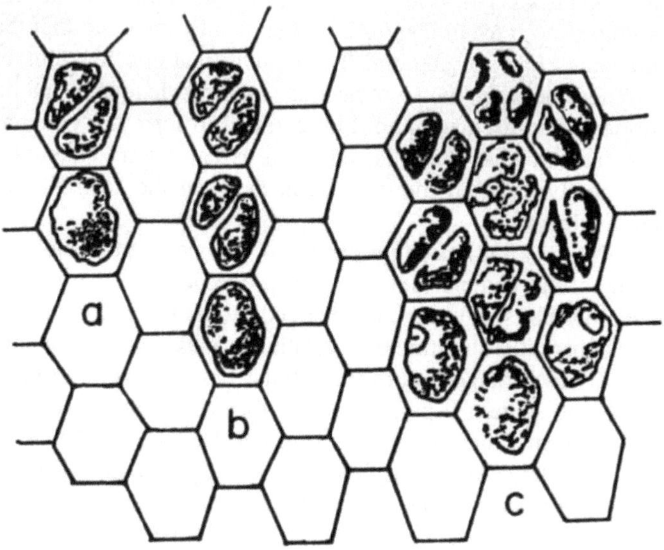

Fig. 3.5. Division of "cambial fiber" by "diagonal segmentation" (**a-c**), and condition of the "Zellschlauch" (protoplasm) during division. (After Hartig 1878).

vestris tracheids had led Meyen, many years earlier, to theorize that the cell wall was formed by the cohesion of isolated fibers (von Mohl 1844).

According to Hartig (1854c), cell division began by constriction of the inner "Ptychode" to form a dividing membrane within the mother cell followed by constriction of the outer "Ptychoide" in the same division plane. The "Ptychodeschlauch" theory formed the basis for much of Hartig's later research on the cambium and in these (Hartig 1853, 1854b,c) and other papers he proceeded to describe the fates of the daughter cells. However, as with many other early workers, the true nature of the cambium escaped him. Years later, Robert Hartig (1891) reproduced an illustration from one his father's newer texts showing how he (T. Hartig) visualized origin of a "procambial fiber" from a parenchymatous cell of the Urmeristem. However, in that text (Hartig 1878), T. Hartig considered the meristem to be cambium, not procambium as suggested by R. Hartig. As visualized by T. Hartig (1878), cambial cells within a vascular bundle that divided tangentially by diagonal segmentation produced fibers or prosenchymatous cells as in Fig. 3.5. Those that divided by radial or tangential segmentation produced parenchymatous cells. This description of cambial cell division was far more advanced than in his earlier work. Nonetheless, in that text Hartig (1878) still clung to the outmoded idea that the cell nucleus arose as a transformation of "Kernkörperchen," or nuclear particles, derived from the protoplasm.

About this time, von Mohl in 1845 observed a characteristic tissue that occurred between the bast and wood cells in both monocotyledons and dicotyledons. This tissue called "Vasa propria" (proper or special vessels) or "vaisseaux propres" (van Tieghem 1871), was believed by some earlier workers to be similar to or equated with cambium. Von Mohl disposed of this idea by showing that the cells were part of the inner bast or phloem. These investigations led him to conclude that the differences between monocotyledons and dicotyledons rested on the fact that in the former the entire cambium differentiated to form the vascular bundle, whereas in the latter a layer of cells remained as cambium. Thus, von Mohl claimed to have finally clarified the distinction between the "cambia" of monocotyledons and dicotyledons.

In a later paper, von Mohl (1858) strongly criticized the views of Schleiden, particularly his assumption that cambial development proceeded in three stages. According to Schleiden, the first stage was characterized by the formation of new cells in the form of an apparently structureless, gelatinous mass. Stem structure was determined exclusively by arrangement of the cellular tissue mass during this first stage. During the second stage, the process of cell formation had ceased, the existing cells expanded uniformly, and the volume, but not the form, of the cellular tissue mass was altered. The cambium at this stage was evident as a distinct, but delicate, cellular tissue with homogeneous contents completely saturated with sap. In the third stage, the existing cells underwent further expansion, and this process was the sole determinant of cambial form. The cambial zone now assumed a dark color as the cells were emptied of sap and replaced by air.

Von Mohl (1858) reserved his strongest criticism for Schleiden's belief that cell division was confined exclusively to the first stage. He pointed out that it was decidedly incorrect to assume that subsequent development of an organ occurred simply by the growth and expansion of pre-existing cells. On the basis of his own research, he, too, distinguished three stages of cambial development, but he defined them quite differently than Schleiden did. For the first stage, he adopted Schacht's term "Urparenchyma," or primordial parenchyma, composed of closely packed, homogeneous cells forming a uniform, structureless tissue. He noted that the term cambium could be appropriately restricted to this tissue if it had not been used by others to apply to later stages. In the second stage, the cambium separated into different classes of tissues. Part of the cells continued dividing to form parenchymatous tissues, while other groups of cells formed vessels and elongated cells by longitudinal divisions. All cells were filled with sap and they had thin, delicate walls. Consequently, the unstained tissue appeared transparent, although when stained with a carmine solution the delicate walls turned bright red. The third stage was characterized by the emergence of this tissue from the condition of cambium. It began by the appearance of intercellular, air-filled passages (vacuoles) within the parenchyma cells, in the elongation of cells such as vessels, and in the deposition of secondary walls. Transverse and longitudinal divisions continued in parts of the tissue, thus permitting recurrence of the cambial condition. Von Mohl then graciously recognized the "great merit" of Schleiden's contribution in showing that the vascular bundles in both monocotyledons and dicotyledons occurred in a cambial layer beneath the stem surface. This cambial layer produced not only the vascular bundles but concurrently the parenchymatous tissue destined for further development of the stem. However, as noted earlier, it remained for von Mohl to distinguish between these two types of cambia.

The late 1850s was a period of prolific investigation relating to the cambium. Sanio's (1857a) contribution at this time resulted from an investigation of the crystals that occurred in the cortex and bark of many species. In some species, the crystals appeared as a longitudinal series of cuboidal cells. These were not distinct, independent cells but simply the daughter cells of a larger bast cell that corresponded in size and shape to the original cambial cell (Fig. 6.28A). To account for these cells, Sanio suggested that each cambial cell divided tangentially to produce two daughter cells. If the cambial cell divided toward the outside, the inner cell became either a wood cell (fiber), a vessel, or a wood parenchyma cell through further subdivision by horizontal walls. If the cambial cell divided toward the inside, the outer cell became either a thick-walled bast cell, a thin-walled "Gitterzelle" (sieve element), or a bast parenchyma cell. The latter cell could then divide further to produce the series of short crystal-containing cells. In both of the above cases, the remaining daughter cell continued functioning as a cambial mother cell. This generalized concept of cambium was accepted readily and espoused as fact in textbooks of the day, such as Schacht (1860).

Radlkofer (1858) was another contributor at this time. He investigated *Cocculus laurifolius*, a species exhibiting anomalous secondary thickening, and examined the origin of the new cambium that gave rise to the crescentic arcs of anomalous growth. Like his peers, he assumed that new cambial cells arose from the innermost cells of the primary rind. However, he was not entirely clear when he concluded that "... formation of a new cambium appeared in the inner half of the primary bast, therefore, it was a product of the cambium itself and not a product of the Urparenchym from which the rind tissue arose." Nägeli (1858) commented on what he perceived as contradictions therein, and pointed out that this view could easily lead the reader to an impossible assumption.

Radlkofer's observations on the origin and subsequent development of the crescentic arcs of tissue were more straightforward. He observed that cambial cells were formed by an active division process in the radial direction through partition walls parallel to the stem surface. Within a crescentic arc, wood and rind cells differentiated as independent tissues, but the cells in between remained thin-walled and began producing new cells by tangentially oriented divisions. Later, the cambial cells lengthened and the wood cells that comprised the anastomosing strands elongated eight or more times. Radlkofer's main objective was to discover how these long cambial cells could form from the short cells of the rind parenchyma.

Cohn's (1857) research was oriented more toward the differentiation of wood elements. However, he did note that young wood cells derived from the cambium were originally simple. During the course of growth, they were first stretched and then divided according to the usual laws, at first in two, then in four, and finally in eight daughter cells.

The concept of cambium as a cellular tissue was now accepted by many anatomists. Yet it lacked a coherent nomenclature. Several erroneous views expressed by previous investigators were criticized and clarified by Nägeli (1858). He strenuously objected to the ideas of Schleiden, Schacht, and others, who regarded all building tissues as cambium, suggesting that such a generalization might lead to distinguishing different kinds of cambium. He then attempted to classify various tissues on a morphological basis. He first differentiated between dividing tissue and permanent tissue. Dividing tissue could exist in two forms. One, present at the beginning stage of an organ and often remaining active throughout its life, he called meristem. The other he called cambium. Meristem and cambial tissues differed with regard to form and growth of their cells, their division patterns, and the nature of the permanent tissues they produced. He recognized a "Strahlenmeristem" that produced the rays, and a "Korkmeristem" that he called "Phellogen." Cambial cells were described as being elongated, having pointed ends, and growing in the direction of the plant axis. They divided by walls parallel to their axes, and they gave rise to wood and bast. With regard to the Urmeristem, he observed that all tissues of the Urmeristem were united but it was not homogeneous. Both the cambium and the "Folgemeristem," a parenchymatous tissue that enclosed the cambium, were derived from the Urmeristem. With publication of Nägeli's paper, the concept of cambium as a dividing and reproducing tissue was now well established. However, many details of cambial formation and organization requiring further clarification still remained.

3.2.2 The Role of Cell Protoplasm

The other area of investigation that contributed to the recognition of cambium as a tissue, cell protoplasm, developed simultaneous with and in complete accord with knowledge of cell division. The building sap, together with the other gelatinous and mucilageneous substances that early workers called cambium, was largely protoplasm exuded by cambial zone cells during the crude sectioning procedures. For example, Mirbel, in his 1815 paper (Baker 1948-55), used Grew's word cambium to describe the colorless mucilage that appeared wherever new growth was about to take place. Although Mirbel considered the cambium to be a fluid, he believed that it contained the "linéamens" of a new structure. Early workers regarded the cambium as a conduit for these "life substances". They had observed bubbles, vesicles, granules, and other cell-like structures in the cambial sap; and, although they noted that new cells appeared to arise from this sap, they had difficulty observing wall formation during the division process. Even Schleiden, who is often credited along with Schwann with formulating the cell theory, believed that new cells arose by a process of condensation or coagulation of a mother substance he called "Cytoblastem" (Lundegårdh 1922). This substance, consisting of a mixture of sugar, dextrin, and slime, was not regarded as a living substance but as an organic mixture transformed to cells by a kind of fermentation process.

Nägeli in 1844 (Baker 1948-55) made a sizeable advance by discovering a "Schleimschicht," a slime layer, under the entire inner-wall surface of fully developed green algae cells. In describing this layer, Nägeli demonstrated that the cell sap was quite different from the "Cytoblastem," or slime layer described by Schleiden. A similar proteinaceous layer lining the inner cell walls of higher plants was described by von Mohl in 1844 (Baker 1948-55). There is little doubt that the layer von Mohl referred to as a "Primordialschlauch" (primordial utricle) was identical to Nägeli's "Schleimschicht." In fact, R. Hartig (1891) equated protoplasm with the "Primordialschlauch". T. Hartig (1853) had used the term "Ptychodesaftes" as a substitute term for protoplasm, and in his later work he (Hartig 1878) equated the terms "Primordialschlauch," "Ptychodeschlauch," and "Zellschlauch" with protoplasm.

First botanical use of the term "Protoplasm" was by Purkinje in 1840 (Baker 1948-55). Purkinje observed that the fluid and solid elements of cells were completely separated in space, a condition that was best expressed in tissues at the earliest stages of development such as the cambia of plants and embryos of animals. Cambium was used here in the same broad sense as used by Mirbel to include not only the cambial zone but the phloem as well. Purkinje's idea of protoplasm was also expressed in a very broad sense. For example, as a plant cell matured, Purkinje believed that "Protoplasm" separated out as a fluid part (protoplasm) and a solid part (cell wall).

The term "Protoplasm" was reintroduced by von Mohl (1846). He wrote that a cell does not contain a clear, watery sap but a viscous, colorless substance containing cell granules or particles dispersed in a slimy mass. With reference to the physiological function of this substance, he proposed the term "Protoplasm." Von Mohl was apparently unaware of Purkinje's earlier work. Möbius (1934) thought it strange and highly unlikely, however, that both workers would unknowingly arrive at the same word for the same cellular substance. Nevertheless, as Möbius pointed out, it was now clear that the gelatinous substance early workers referred to as cambium was in fact protoplasm. For a thorough review of subsequent research on plant protoplasm and the protoplasmic connections between cells, the papers by Russow (1883a), Schenck (1885), and Kienitz-Gerloff (1891), and the chapter on *The Cell* in Green (1909) are recommended.

It is obvious that the concept of cambium, as we recognize it today, gradually emerged through the contributions of many investigators. According to Lundegårdh (1922), a concept of cambium that correctly conformed with modern theory was first proposed by Unger in an 1855 paper. Lorch (1967), however, recognized Schacht as being the first to describe the cambium in modern terms in his 1853 text *Beiträge zur Anatomie und Physiologie der Gewächse*; but a few technicalities still escaped Schacht. For example, in his later text, *Lehrbuch der Anatomie und Physiologie*," Schacht (1856) still believed that a cambial cell divided by two cells forming within each mother cell rather than one cell dividing into two. Moreover, he, too, considered the cambium to extend from the thickening ring (the true cambium) into the vegetative apex (the procambium). In fact, he stated that although the cambium could increase the size of vascular bundles, it could not give rise to new ones. This role was reserved for the Urmeristem, where the cambium was thought to arise. However, in partial vindication of Schacht, it should be pointed out that Wigand (1861) considered Schacht's (1856) text to mark a turning point. The previous decade had been a prolific period of investigation and a source of many new theories. Schacht attempted to synthesize this accumulated knowledge and to re-interpret it with a liberal dose of his own theories. As a consequence, in subsequent years, his views were not only widely discussed but freely and strongly criticized.

As to my own view regarding the awarding of priority for a correct concept of cambium, I withhold judgement. Because of the conflicting terminology and vague descriptions still prevalent during the 1850s, it is difficult to assign priority to any one individual as having first assembled all the facts and clearly proclaimed them as a cambial theory.

Chapter 4

Defining the Cambium

4.1 Cambial Cell Theories

The terms cambial cell, cambial fiber, mother cell, and tissue mother cell were often employed in a nonspecific sense with reference to the cambium by early investigators. At times, one of these terms might be confined to the dividing cell(s), and at other times it might include derivative cells. The nonspecificity of the terms was due mainly to difficulties in isolating the cambium for microscopic observation, consequently, to difficulties in formulating a precise definition.

In later years, the terms cambial initials, cambium, cambial zone, and variants thereof were also used either nonspecifically, confusingly, or even erroneously. The distinction between cambium and cambial zone was not clear in the early literature, nor is it entirely clear in much of the literature today. As generally interpreted today, the cambial zone includes the cambial initials and mother cell derivatives that might or might not undergo further division. The term cambium, on the other hand, is sometimes applied to the initial cell and sometimes to the entire cambial zone as defined above. That is, in the latter case the terms cambium and cambial zone are considered synonymous.

The main source of controversy over the years has related to the nature of the cambial initial(s). Some authors recognized a single cambial initial cell in each radial file, others two or more initial cells, and still others questioned the initial cell concept itself. Several early investigators recognized initial cells, or what they referred to as mother cells, in the cambial zone but they were unsure as to their number (von Mohl 1844, Unger 1847). Schmid (1976), in a commendable contribution, attempted to clarify the nomenclature relating to the cambium. A summary of this revised terminology, along with others, will be delayed until later in this chapter. In the review that follows, the terminology employed will attempt to convey each author's intended meaning.

4.1.1 "Siamese Twin" Theory of Hartig

Hartig (1853) postulated that radial expansion of the stem occurred through the activity of a "Mutterzellen - Zwillingspaar," or a twin pair of mother cells within each radial file that he referred to as "Siamese twins." The mother cells were considered permanent cells and they were presumed to be identical in size, form, and structure. However, one mother cell contributed daughter cells exclusively to the wood side and the other exclusively to the bast side; thus, the cambium would be considered biseriate. Daughter cells on the wood side were believed to be completely sterile. That is, the mother cell on the wood side produced daughter cells throughout the growing season but then matured in the fall. The mother cell on the bast side, on the contrary, overwintered in a immature state. In the spring, it again divided by constriction to produce a new wood mother cell, thereby restoring the biseriate cambium.

To demonstrate the existence of these mother cells, Hartig cut a deep channel or groove radially in a stem, filled it with soap water, and covered it with a glass plate. He then blew air through a double-tipped glass tube, thereby forcing bubbles to arise from one tube on the wood side and from the other on the bast side. This demonstration supposedly represented the production of sterile daughter cells from the two permanent mother cells.

Hartig (1878) restated these views in his textbook, *Anatomie und Physiologie der Holzpflanzen*. He also described how the permanent mother cells matured in the winter and how, on the wood side, the inner membrane of the mother cells formed secondary walls. Hartig obviously misinterpreted the concept of cambium, for he stated that at no time could he recognize "Procambium" between the mature wood cells and the thin-walled bast fibers in the winter condition. In spring, these cells in some magical way regained their ability to redivide by the reformation of new membranes from secondary walls in the wood mother cells and by the reactivation of membranes in the bast mother cells. Spring growth in his experimental tree (*Populus serotina*) was so vigorous, however, that he could no longer identify the permanent mother cells. As his Text Fig. 63 showed, the cambial zone consisted of at least eight cells. Because of the difficulty in identifying the mother cells, Hartig admitted to the possibility of the youngest daughter cells redividing.

4.1.2 Uniseriate Initial Theory of Sanio

Sanio (1873) is often credited with formulating the cambial initial cell theory, and indeed he was first to establish a set of rules whereby initial cells could be identified with a degree of certainty. Yet as with all firsts, a sizeable body of knowledge preceded the formulation of Sanio's rules (Sect. 4.2.1). Two papers stand out with particular prominence because of the clarity of their descriptions of cambial division. The first, by Trécul (1853a), retained the older terminology by referring to cambial initials as "utricles." When he examined the "utriculaire" layer at the end of winter after the plants (*Paulownia imperialis* and *Ulmus rubra*) had resumed activity, the innermost cell in immediate contact with the sapwood was found to be radially enlarged or dilated. It then divided by a longitudinal wall parallel to the circumference of the tree. Of the two utricles that formed, the internal one became a wood fiber and the external one again divided, producing a second wood fiber and a new mother cell which enlarged like the preceding one. This process continued during the entire course of the annual increment, with each radial series producing the same "primitive cell" which elongated and divided in a continuous manner. There were times of "arrest" of the "mother utricles." After the "utricles génératrices" (on the xylem side) had produced 12 or more utricles, they ceased to divide. A cortical cell then began to multiply, but it ceased after a few divisions. In the following spring, the cell most internal to the bark, i.e., adjacent to the sapwood, revived this mode of growth, which was repeated each season.

At about this time, a very interesting paper appeared by Cohn (1857) in a rather obscure publication. Cohn investigated vessel formation in *Vitis vinifera, Ampelopis hederifolia*, and several other vines. In discussing the origin of these cells, the following statement appeared: "The developmental history shows that young wood cells in the cambium are at first simple; however, in the course of their growth they increase in length, and they divide according to the usual laws, first in two, then in four, and in individual cases also in eight daughter cells; these divisions occur before the wood cell is completely thickened (lignified): however, after it has assumed a prosenchymatous form." As written, Cohn assumed the process to be common knowledge. No references were cited by Cohn, but Cohn's paper was cited by Sanio (1863b).

In a study of several deciduous species, Sanio (1857a) assumed the cambium to consist of a single mother cell that divided tangentially to produce either a wood cell to the inside or a bast cell to the outside; thus the cambium would be considered uniseriate. He confirmed this view in *Tilia parvifolia* (Sanio 1860), at which time he concluded that each bast cell arose from a single "cambial fiber cell" and that in no way could several cambial cells be involved. However, a daughter cell might again divide by transverse walls to form bast parenchyma cells. In a later study of *Pinus sylvestris* (Sanio 1863a), he observed three cells in the cambial layer that had divided tangentially. This finding was contrary to both his earlier view of a single mother cell and Hartig's (1853) conception of twin mother cells.

In an attempt to resolve the aforementioned discrepancy, Sanio (1873) undertook an extremely detailed study of the cambium of *P. sylvestris*. When he examined a series of daughter cells, he found that the radial wall of the mother cell was curved as shown in Fig. 4.1. The lower bow of cell 3 in radial file b corresponded to that of cell 9, suggesting that the entire series was the progeny of a single cell. In this paper, he categorically affirmed the presence of a single mother cell and offered the following counterarguments to Hartig: (1) the radial series extended from the wood through the cambium to the bast. If there were two mother cells as Hartig postulated, they would be separated by a very thick wall,

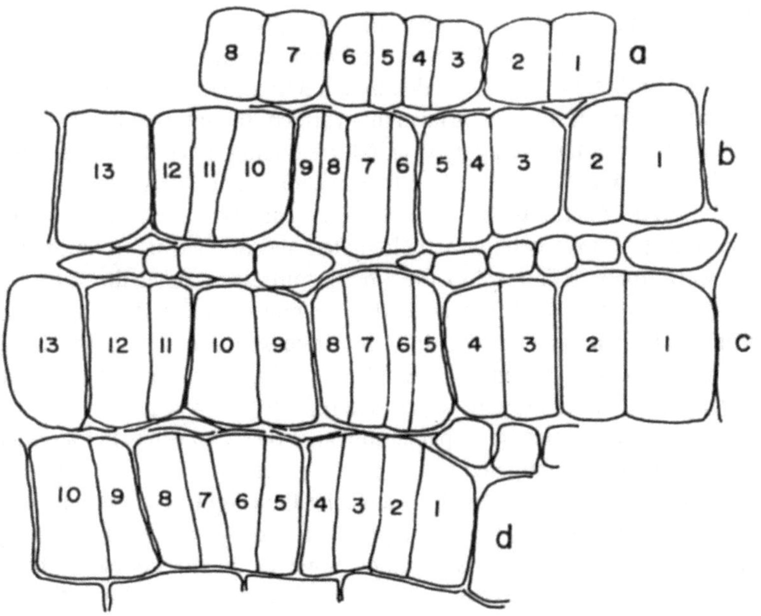

Fig. 4.1. Transverse section through the cambial zone of a 100-year-old *Pinus sylvestris* stem in August. 600x. (After Sanio 1873)

which was not the case. This argument was based on Sanio's finding that during each division a new wall was deposited within the mother cell. With a single mother cell, dividing alternately to wood and bast, tangential wall thickness would be limited (Fig. 4.1, File c). (2) When a cambial cell divided radially (anticlinally), the new cell series on both the wood and bast sides would also double. Such a sequence would be perfectly understandable with a single mother cell, but highly unlikely with two mother cells. (3) He sometimes observed "stäbchenförmige Körper," or rod-shaped bodies (trabeculae), extending radially from wall to wall through the lumens of a series of cells. These structures, which he presumed to consist of cellulose, had their origin in the cambial mother cell; and since they extended into the bast as well as the wood, they could only arise from a single mother cell (Fig. 4.27).

Sanio also spoke of "Zwillings," or twins, but in a quite different sense than Hartig. He demonstrated, in *P. sylvestris*, that each time a cambial mother cell divided, it formed a twin pair. One of the daughter cells always remained as a cambial mother cell, while the other might redivide [Fig. 4.1, File b, cells 3 (4+5)]. Frequently, the cambial cell divided twice resulting in a group of four cells, or two twin pairs (Sect. 4.5.1). Sanio assumed that mother-cell divisions alternated between wood and bast, and so formation of the twin pairs also alternated. According to this view, each radial series possessed one cell, the initial cell, which retained its cambial character. These cells were arranged around the periphery in a ring which as a whole might be regarded as an "Initialkreise," or ring of initial cells. Sanio considered these cells as more or less permanent initial cells that were invariable both during a growing season and between growing seasons.

Sanio's concept of an initial cell was not immediately accepted by later investigators. The greatest difficulty experienced by others was in attempting to identify the initial cell, and indeed the cambium itself as evidenced by the work of Müller (1875-76). This author, in his studies of *Picea* and *Quercus*, found that a strict boundary between wood and bast existed only during the winter. In spring, when new growth began, it was impossible to establish a boundary within the zone of new cells. Moreover, he could not morphologically distinguish cells of this tissue layer from the differentiating conducting cells. Not only did he question the existence of an initial cell but also the existence of twin initials in the sense of Hartig, which he referred to as two "protene." Müller finally concluded that "a cambium in the sense of a special tissue with definite morphological dignity does not exist, and that wood growth in a morphological sense is a secretion of the rind."

Velten (1875), who had studied under Nägeli in Münich in 1871, responded to Müller's criticism. He claimed that young cambial cells only appeared identical when they were examined cursorily. Detailed examination always showed a slight variation in new daughter cells from the very beginning. However, he admitted that the region where new cells formed, the "Urcambium," was difficult to discern microscopically, and that it was therefore technically difficult to establish the true origin of the cambium with certainty. Nonetheless, he proceeded to describe in minute detail the differences between "Urcambiumzellen," or cambial initial cells, and "Cambiumzellen," or daughter cells. He stated that the failure of most investigators, meaning Müller, to identify cambial initials on tangential sections was due to distortion of the tender cells, a problem which he also experienced but largely ignored.

Wieler (1886a) also questioned the concept of an initial cell because of uncertainty in attempts to identify it. The winter cambium of many woody species he examined consisted of from two to several cells, suggesting to him that the first earlywood cell(s) must be laid down in the previous season. Moreover, the xylem boundary often exhibited a zig-zag or wavy outline, further suggesting that the number of wood cells cut off by each radial file might not be uniform.

A considerably more enlightened view was taken by R. Hartig (1891). In the winter resting condition, he usually found the cambium to consist of a single cell, although it might consist of two to three cells. Division of a cambial initial in spring gave rise to a "Gewebemutterzelle," or a tissue mother cell. The latter cell, in turn, could again divide, giving rise to two to four cells. However, in most cases, he could not tell which cell in the cambial zone was the initial cell. Hartig's interpretation was a more or less direct confirmation of Sanio's concept.

While conducting an intensive investigation of pit formation, Russow (1881) had occasion to examine the cambial region of *Pinus sylvestris* and several other conifers. He believed that pit formation must naturally occur in cambial cells, and he did indeed observe that the walls of cambial cells possessed shallow indentations which later developed as bordered pits. He also criticized Sanio's (1873) concept of cambium. Although Russow agreed with Sanio's idea that the cambium began with an initial cell, he considered it to be without practical value. For example, he found it impossible to locate the initial with any degree of certainty because of the frequency with which the cell division sequence varied from the normal mode. As an alternative, he proposed that the term cambium be used with reference to all cells situated between wood and rind possessing the characteristics of cambium and dividing tangentially by longitudinal walls. Those cells that were no longer dividing but still in the process of development should be designated as either young wood or bast cells.

Sanio (1882) responded to Russow's criticism by citing 18 specific points, most of which dealt with cell wall development and pit formation. He reaffirmed his belief in the existence of a single cambial initial by stating "this distinction is not an assumption, but a result." However, he admitted that on one occasion he did observe a radial series of two contiguous cells, one of which formed bast to the outside and the other wood to the inside. These cells were separated by a conspicuously thick wall, suggesting that they could correspond with twin initial cells. He also reaffirmed his belief that daughter cells usually divided one or two times before being transformed to permanent wood or bast elements. Exceptions occurred in very slow-growing wood such as the stems of suppressed spruces or the branches of old pines.

In his rebuttal, Russow (1882c) accused Sanio of disregarding new observations and simply holding to his old opinions. Russow maintained that he did not specifically deny the existence of a cambial initial. He allowed that while the existence of such an initial, as chanced upon by Sanio, might indeed be perfectly correct, it was without practical value because of the impossibility of locating the initial with any degree of certainty. Russow nonetheless conceded that his criticism in no way lessened the meritorious value of Sanio's work.

While investigating de Vries' (1876) theory of bark pressure as a causal factor in growth ring formation, Krabbe (1884) thought it necessary to re-examine Sanio's concept of a cambial initial cell. He noted that Sanio had not considered ray cell initials in his investigation nor had he extended the theory to deciduous trees; Krabbe apparently overlooked Sanio's (1860) work on *Tilia parvifolia*. He found that Sanio's criteria for identifying initial cells applied equally well to both fusiform and ray cell initials, but they were easier to detect in the latter. For example, the tangential walls of daughter cells became more rounded as they retreated from the cambial initial, and this feature was more readily observed in ray cells because they divided less frequently. By invoking this line of reasoning, he assumed that cell 2 in ray m of a *Populus* stem (Fig. 4.2A) was the initial cell and tangential wall b was the youngest. Accordingly, cells 3+4 and 4+5 of radial files I and II, respectively, were assumed to be fusiform initial cells. However, at this stage of development, he could not say which of the two in each series would eventually remain as functioning initials. Similar conclusions were drawn for cambial zone samples of *Alnus, Clematis, Salix,* and *Tilia,* as well as *Pinus sylvestris, P. strobus,* and *Picea abies.* In all species, Sanio's rule applied. That is, a cambial initial cell divided tangentially, producing two mother cells, one of which would remain as a new initial cell while the other redivided before differentiating either as a pair of xylem or phloem cells. Ray cell initials divided in a similar manner, but the daughter cells differentiated to permanent ray cells without undergoing further division. That is, daughter ray cells never redivided, a conclusion later subscribed to by Haberlandt (1886, 1914) and Strasburger et al. (1908).

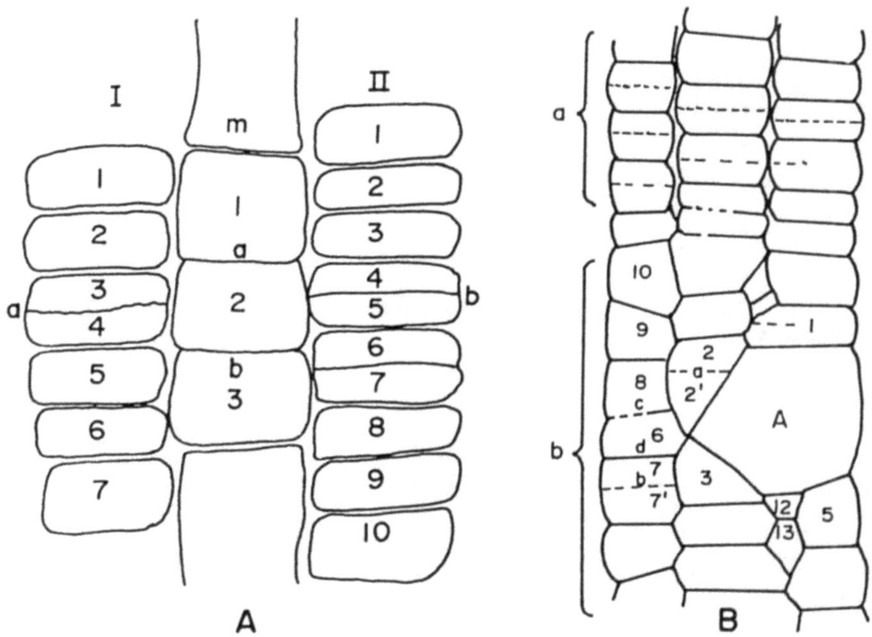

Fig. 4.2A,B. Transverse sections through the cambial zones of *Populus alba* stems. **A** Presumed identification of fusiform initial cells *Ia* and *IIb* based on position of ray cell initial *m2*. **B** Periclinal divisions of differentiating xylem cells (2 + 2', 7 + 7') in "subcambial zone" *b*; cambial zone *a*. Both 580x. (**A** After Krabbe 1884. **B** Krabbe 1886)

Krabbe (1886) reaffirmed his belief in the initial cell concept in an intensive examination of the phenomenon of sliding growth (Chap. 6.7.1). In fact, he claimed that Sanio (1873) had been first and he (Krabbe 1884) second to assume a cambial initial cell for reproducing the "thickening ring." However, during his investigations of vigorously growing *Populus, Salix,* and *Tilia,* he observed cell divisions occurring deeper in the "Splint," or the developing sapwood tissues, which he also referred to as the "subcambialen zone." He noted that no one, to his knowledge, had previously described such divisions. He considered them to be of

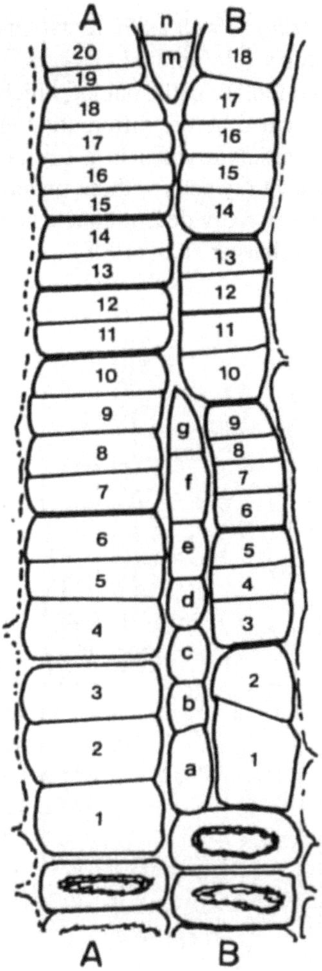

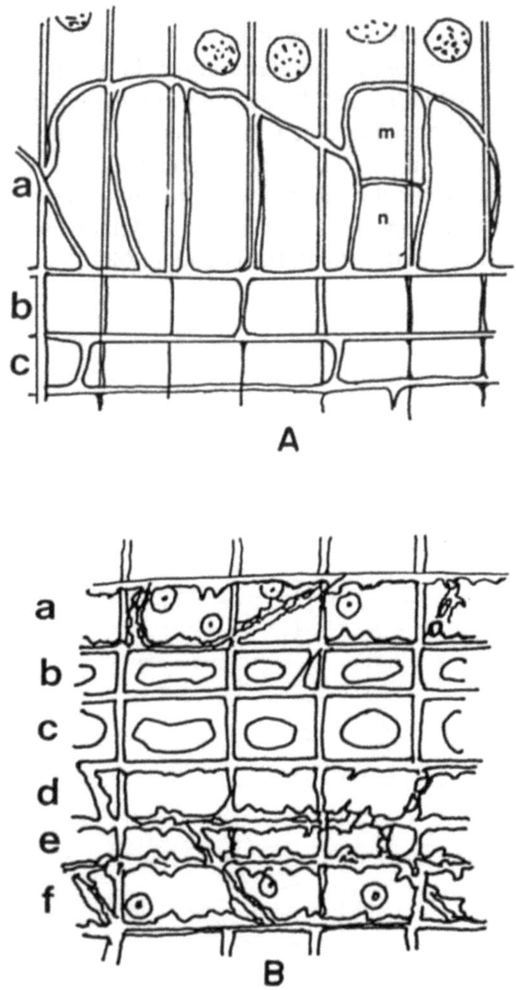

Fig. 4.3. Transverse section through active cambial zone of a *Pinus sylvestris* stem illustrating how cell divisions produced "cell complexes." 345x. (After Mischke 1890)

Fig. 4.4A,B. Radial sections through ray cambial zones of *Pinus sylvestris* stems comparing the dimensions of border cells (*a,e,f*) and interior cells (*b,c,d*) to those of adjacent tracheids and cells. **A** Phloem side. **B** Xylem side. Both 345x. (After Mischke 1890)

importance only in relation to vessel formation and expansion, and only in vigorously growing deciduous trees. Because these divisions did not appear to follow any formal rules, he tentatively regarded them as nothing more than a direct continuation of the cell divisions taking place in the thickening ring. The divisions sometimes occurred in "relatively old sapwood cells," such as cells 2+2' and cells

7+7' in Fig. 4.2B, while younger cells, such as cells 6, 8, 9, and 10, remained undivided.

Mischke (1890) was totally sympathetic toward Sanio's theory as it applied to a single initial cell. Although Sanio had emphasized the fact that each daughter cell might again divide, he had not ruled out the possibility that they might divide twice. Mischke's objective was to investigate this possibility in rapidly grown *Pinus sylvestris* trees. Interestingly, he made no mention of Krabbe's (1886) work. Neither did he mention Russow's (1881) research nor the 1882 debate between Russow (1882c) and Sanio (1882) even though these papers appeared in the same journal as his own paper. By applying Sanio's rules for distinguishing newly formed tangential walls from older ones, he attempted to identify initial cells in the cambial region. For example, in file A of Fig. 4.3, he reasoned that "Zellcomplexe," or cell complexes, 4-6, 7-10, 11-14, and 15-18 each arose by one daughter cell dividing twice; i.e., daughter cell 7 divided to yield cell 9 followed by cells 7 and 9, redividing to yield cells 8 and 10, respectively. He further rationalized, on the basis of wall structure, that the initial cell was either cell 15 or 16 in file A and, correspondingly, either cell 14 or 15 in file B.

Mischke had demonstrated that a daughter cell could divide either once or twice, but he now asked "... what is the rule and what the exception"? He examined an exceptionally fast-growing *Pinus sylvestris* tree and could find no basis for assuming that daughter cells might divide three times. He next examined a slow-growing *P. sylvestris* tree and confirmed Sanio's conclusion that, in this case, daughter cells divided only once. On the basis of these and other observations, Mischke restated Sanio's rules to show that daughter cells could divide either once or twice depending on growth intensity. Under favorable growth conditions, a daughter cell could divide twice giving rise to a four-celled complex. Under less favorable conditions, such as the beginning of the growth period, the second division of the daughter cell often failed to occur. Moreover, under extremely restrictive conditions, both divisions might be absent, in which case the daughter cell would differentiate directly to a wood or phloem element. The winter cambium usually consisted of several thin-walled cells, and the initial cell was judged to be two to three cells removed from the last-formed tracheid. Mischke interpreted this displacement to mean that the first one to several earlywood tracheids were initiated the previous fall. He predicted a similar precocious origin for the first-formed phloem cells of the following season (Chap. 11.5).

Mischke also noted that ray cell initials gave rise to daughter cells on either the xylem or phloem side at the same time as the adjacent fusiform initials did. Again, no mention was made of Krabbe's (1884) prior work on ray cell initials. Mischke made detailed measurements of ray cells in *Pinus sylvestris* and assembled these data in several tables. He concluded that the ray border cells divided more frequently, but elongated to a lesser extent than the interior cells. On the phloem side, the border cells were about the same width, while the interior cells averaged 1.5 to 3.5 times longer than the radial width of adjacent parenchyma cells (Fig. 4.4A). This meant that the border initial cells must have divided and expanded at about the same rate as the fusiform initial cells, whereas the interior cells divided less frequently but expanded to a greater extent. On the xylem side, the average border cell extended radially across two tracheids whereas the interior cells extended across from four to seven tracheids (Fig. 4.4B). This meant that all xylem ray initials, particularly the interior ones, divided less frequently but elongated radially more than the adjacent fusiform initials. The interior ray cells extended over the same average number of tracheids in both early and latewood. However, because earlywood tracheids were radially wider than latewood tracheids, the earlywood ray cells were longer. Unlike Krabbe (1884), Mischke could establish no hard and fast rules for the division of ray cell initials and daughter cells. These processes depended on the growth vigor of the tree and the respective activities of the adjacent fusiform initials.

4.1.3 Multiseriate Theory of Raatz

Raatz (1892) is credited by all subsequent writers with initiating the multiseriate cambial cell concept, and it is true that his work brought it into prominence. However, it should be recalled that years earlier Russow (1881) had abandoned Sanio's view of an initial layer by suggesting that the entire meristem lying between wood and bast be designated cambium. In addition, Krabbe (1886) observed divisions in "relatively old sapwood cells" some of which were presumably xylem daughter cells, but unrecognized at the time. Consequently, Raatz' views were not without precedence.

Raatz' (1892) polemic against Sanio's hypothesis was cautiously guarded. He graciously conceded that Sanio's conclusions were based on old, slow-growing trees and that nowhere had Sanio said that they would apply equally well to young, vigorous trees. This was an important point. Krabbe (1884) had deliberately selected old, slow-growing trees for his investigation because he found that curvature of the tangential walls was expressed more clearly in such specimens. Raatz, on the contrary, studied rapidly growing specimens of *Pinus* and *Picea* and at no time could he identify specific initial cells as described by Sanio. Rather than a favored position within each radial file, the initial cell might be displaced radially as a result of its unlimited ability to divide.

Because of this division pattern, each radial series possessed a "Wendepunkt," or turning point, where cell production shifted uninterruptedly from wood to bast and vice versa. The imaginary line connecting the Wendepunkt of adjacent radial series around the periphery was referred to as the "Wendekreise." The Wendekreise might vacillate according to the rate of cell production. For example, if growth was rapid and many wood cells were produced, the Wendekreise would shift toward the bast side. In his analyses, Raatz considered the cambium as a layer of "equivalent" cells. That is, one cell might function for a time as an initial cell on the xylem side. Then, its activity would be either suppressed or lost and a cell on the other side of the Wendekreise would become active as an initial cell.

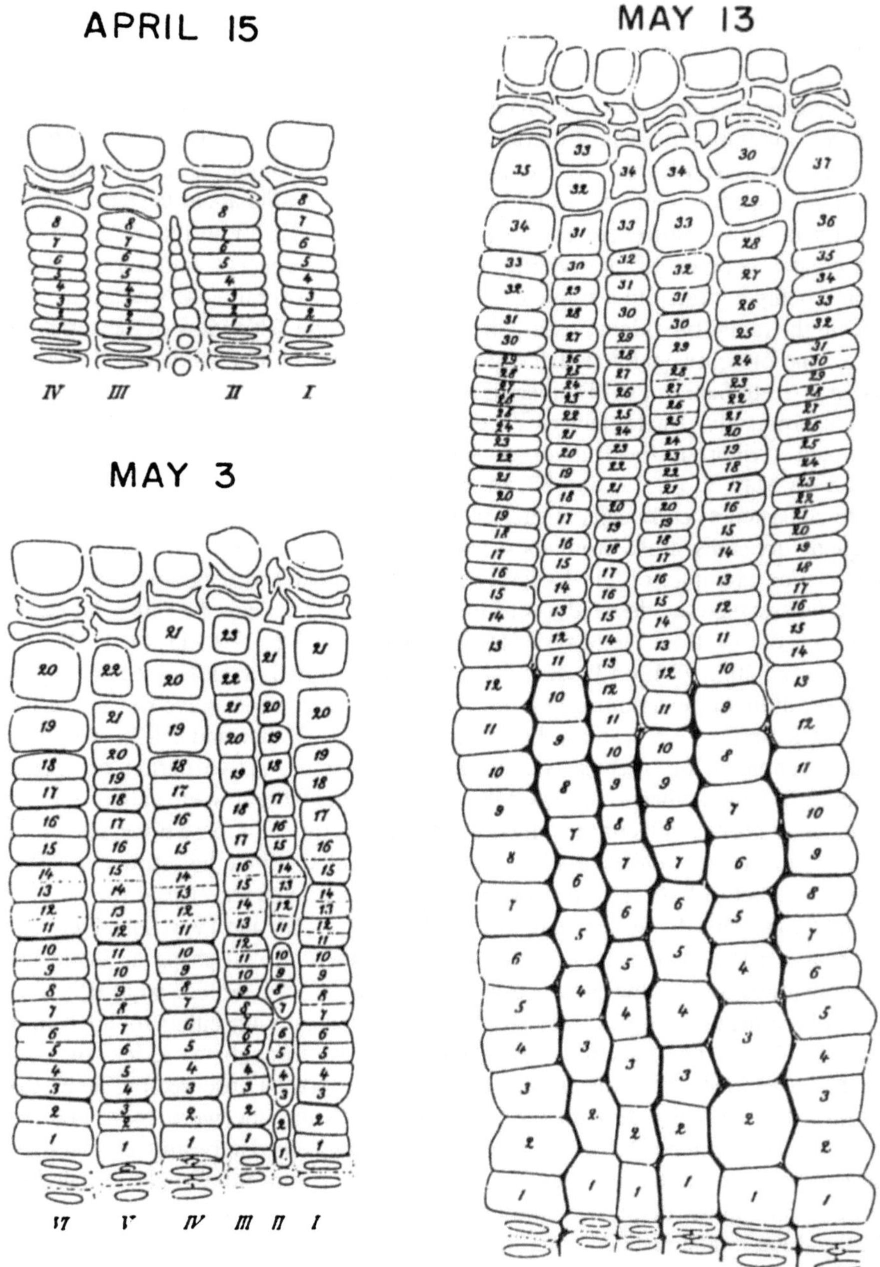

Fig. 4.5. Transverse sections of the cambial zone of a *Picea abies* stem during dormancy (*April 15*) and active cambial growth (*May 3, 13*). (Raatz 1892)

Raatz also entertained the possibility of double initial cells as visualized by Hartig. His use of the term initial cell was for convenience only. He considered all attempts to designate an initial cell as being arbitrary and inexact. Therefore, he suggested replacing Sanio's initial cell with the concept of a "dauernd theilungsfähigen Cambiummutterzelle," literally a cambial mother cell with the ability to divide continuously. By this he meant that rather than a single initial cell there existed an unspecified number of initial cells each possessing about the same potential for dividing. However, the closer one of these cells lay to the Wendekreise, the greater its potential for dividing. According to this concept, rather than cell division being confined to a single cambial initial cell, several closely adjacent cells within the radial file might possess the ability to divide. This view of the cambium as a zone of equivalent cells later became known as the multiseriate cell concept as opposed to Sanio's (1873) uniseriate cell concept.

Unfortunately, Raatz' analyses were more arbitrary than those of Sanio. He sampled a vigorous *Picea abies* on April 15 before diameter growth began and again on May 3, 13, 19, and 27. In the winter condition, April 15, the cambial zone consisted of eight cells in each radial file (Fig 4.5). By assuming that all subsequent divisions arose from these eight cells, he reconstructed the remaining series. For example, on the basis of cell arrangement and tangential wall thickness, he assigned the descendents of the original eight cells to the following series in radial file IV from the May 3 sampling date: 1-2, 3-6, 7-10, 11-14, 15-16, 17-18, 19, and 20. By May 13, some daughter cells had redivided. His reconstruction of file I at this time assumed that the original eight cells now gave rise to the following series: 1-2, 3-6, 7-10, 11-15, 16-23, 24-31, 32-34, and 35-36. Although Raatz continued his reconstructions through the most active period of growth (May 27), he nonetheless assumed that all new cells were descendants of the original eight mother cells of the winter cambium. He spoke of these divisions within cell groups as "intercalary divisions of cambial cells." Since he estimated a 10:1 ratio of wood to bast cell production, he placed the Wendekreise in the May 13 collection at cells 33, 26, 31, and 30 for cell files I, II, III, and IV, respectively.

Raatz' criteria for grouping cells was based both on cell shape as Sanio had demonstrated and on the presence of "Kurzstäbe." These "Stäbe" were described as "balkenförmige Körper," literally beam-like bodies, with either round, elliptical, or linear cross-sections extending radially through a series of cells (trabeculae) (Sect. 4.5.2). However, unlike the Langstäben observed by Sanio, that traversed many cells through the cambium, the Kurzstäbe were short, rod-like structures that extended through only one to several cells. He postulated that these structures were limited in extent because they were transmitted only to daughter cells of the mother cell in which they originated. The limited, but variable, length of these Kurzstäbe resulted from their origin either in the initial cell, a cell in the immediate vicinity of the Wendekreise, or a cell on the outer border of the dividing zone. Finally, because these Kurzstäbe occurred in various positions within a radial file and not necessarily in the same relative position in adjacent radial files, Raatz concluded that the concept of an initial cell could never be verified and was therefore invalid. A contemporary of Raatz, Perrot (1898) also found it impossible to identify a cambial initial in *Gentiana saxosa*. In this species, which possessed internal phloem, the secondary meristem functioned very slowly and it appeared that cell division occurred throughout the cambial zone.

Nordhausen (1898) was impressed with Raatz' work on conifers and decided to apply the same approach to a study of dicotyledons. He found both Langstäbe and Kurzstäbe in many woody species such as *Juglans, Populus, Salix,* and *Tilia,* but the greatest number were found in 1-year-old twigs of *Sambucus*. Some Kurzstäbe conformed to the two-celled and four-celled series observed by Sanio (1873) and Mischke (1890), respectively. However, of greatest significance were those Langstäbe that extended through a large number of cells, some through a radial series of 25 cells. Nordhausen next examined Sanio's ideas relative to "Doppelreihen," or anticlinal divisions, and the tangential wall thickness of cambial cells. In all three points - Stäbe, Doppelreihen, and tangential wall thickness - he could find no confirmation of a permanent initial cell in Sanio's sense. Many of Nordhausen's arguments were based on assumptions and suppositions rather than observations, and at least one of his supporting figures (his Fig. 3) was admittedly constructed. Moreover, he borrowed copiously from the observations of Raatz. His conclusions also conformed to those of Raatz; namely, that the cambium of dicotyledons consisted of a layer of cells all capable of dividing. No cell, by itself, possessed a greater or lesser ability to divide than other cambial cells. Thus, the cambium of both dicotyledons and conifers functioned according to the same laws.

The relation between ray cell and cambial (fusiform) initials was also investigated by Nordhausen. He was particularly critical of Krabbe's (1884) conclusion that ray cell initials adhered to essentially the same rules of division as fusiform initials did. Curiously, Nordhausen cited Mischke's (1890) work on fusiform initials but made no mention of his extensive research on ray cell initials. Rather than recognizing Mischke's border and interior cells, he referred to Kny's (1890) paper in which this author differentiated between upright "Pallisadenzelle'" and recumbent "Merenchymzelle." Nordhausen pointed out, as Mischke had before him, that the Pallisaden and Merenchym cells divided at different rates. For example, in *Salix fragilis*, he observed that a procumbent Merenchym cell was approximately three to four times longer than an upright Pallisaden cell, in which case a daughter Merenchym cell was transformed directly to a ray tracheid. He therefore reasoned that a contiguous Pallisaden cell must divide three to four times while the Merenchym cell was elongating. In the latter case, it would be impossible to determine which of the Pallisaden cells was the permanent initial cell. Nordhausen concluded that although a strong relation existed between the rates at which the ray cell and adjacent fusiform initials divided, divisions within the ray meristem depended on the type of ray cell and its degree of radial expansion.

In his introductory review, Schoute (1902) noted that the theory presented by Sanio (1873) had prevailed in a generalized form for many years. Despite the fact that new data had accumulated, authors of plant anatomy and physiology texts, seven of which Schoute cited, still adhered to the outdated view. Because of these circumstances, he thought it advantageous not only to clarify Sanio's theory but to prove that it required revision. He first addressed the question of whether the cambium consisted of one or two initials. After reviewing Sanio's arguments, Schoute concluded that they were so convincing that the question could be considered completely solved; each radial file possessed only one initial cell. He next addressed the question of how the daughter cells divided. Here, he re-analyzed Raatz' (1892) arguments in extreme detail. Schoute accepted Raatz' observations of Kurzstäbe as opposed to Sanio's Langstäbe as evidence that division of daughter cells were not confined to the immediate derivatives. He also accepted Raatz' observation that a small zone of cells in the middle of the cambium possessed more or less equal potentials for dividing. In essence, Schoute accepted all the facts in Raatz' revised theory and considered his (Raatz) presentation to be a clear and plausible explanation in support of these facts.

One major point in Raatz' theory, his definition of the Wendekreise, was questioned by Schoute. Raatz claimed that the nearer a mother cell lay to the Wendekreise, the longer it would remain in the cambial zone and the more frequently it would divide. Conversely, the farther removed the mother cell from the Wendekreise, i.e., to the outer border of the cambial zone, the less frequently it would divide, and the earlier it would be transformed to either a wood or bast element. Raatz supported this conclusion with observations on the relative thickness of tangential mother-cell walls and the length of the Kurzstäbe. Schoute accepted these conclusions, but noted that Raatz' definition of the Wendekreise was entirely mathematical, that it contained contradictions, and it was therefore incorrect.

According to Raatz, the Wendekreise was nothing more than a circle of points in the cambium. Although the Wendekreise might vacillate as the rate of cell production waxed and waned, it would nonetheless remain as a fixed point within the cambial zone. Schoute pointed out that no matter how one selected a fixed point in the cambial zone and then related it to the relative shifts of xylem and phloem cell production during the season, the resulting Wendekreise would always be highly irregular and fluctuating. Thus the Wendekreise which today might lie over a given cell in a radial series might tomorrow be several cell breadths removed either to the inside or outside. Schoute's criticism essentially duplicated that of Nordhausen (1898) in his investigation of dicotyledons.

Several minor points in Raatz' theory were also questioned by Schoute. For example, he considered Raatz' idea of a "dauernd theilungsfähigen Cambiummutterzelle" to be simply quibbling and not a matter of principle. He also questioned Raatz' idea that the cambium consisted of a layer of equivalent cells. According to Raatz, initial cells differed from daughter cells only by their special position relative to the Wendekreise, whereas Sanio had attributed special internal characteristics to the initials. Schoute agreed that Sanio was too restrictive in saying that daughter cells could divide only once, or possibly twice, but he believed Raatz went too far in saying that all cambial cells should be accorded equal value; his own opinion was that the question remained unresolved.

Schoute's paper, although strong in detail and convincing arguments, was mostly confirmatory and added little of substance. He confirmed Sanio's work supporting the idea that each radial series had one and only one initial cell. However, he also confirmed Raatz' work supporting the idea that daughter cells, situated in a small zone in the middle of the cambium, were able to divide more or less equally, but with the added caveat that this ability gradually decreased toward the outer boundary of the cambial zone. Schoute considered these questions completely settled insofar as they applied to the cambia of conifers. In his monograph *Meristeme*, Schüepp (1966) presented several possible schemes of cambial division based mainly on Schoute's work.

Schoute next turned his attention to dicotyledons. He acknowledged how previous workers, primarily Nordhausen (1898), had demonstrated that the cambia of dicotyledons developed in a manner similar to those of conifers. However, he expressed one reservation. What was the situation at the beginning of cambial activity? That is, the situation at the transition of procambium to cambium. Two possibilities existed: either the formation began in a single location and an initial cell existed from the very beginning, or several initials arose in different locations but only one survived to assume permanence. The first possibility he called "Initialencambium" in which cells were arranged in a radial series, as in conifer cambia, and all cells arose from a single initial. As an example of this type of origin, he chose the interfascicular cambium of *Ricinus communis*. The second possibility he called "Etagencambium" in which several primary cell layers participated in cambial structure but no continuous radial arrangement existed (Chap. 6.2.1). Here, he chose the interfascicular cambium of *Vitis labrusca* as an example (Fig. 4.33). Many other dicotyledons showed a clear Etagencambium during early development. In every case, however, it was very quickly transformed to an Initialencambium when the daughter cells of one cell layer suppressed the remaining ones. Unfortunately, Schoute's analyses of dicotyledon cambia rested on two false premises. First, he confined his observations to cursory examinations of interfascicular meristems and ignored the role of fascicular meristems, and secondly, like so many others at that time, he assumed that radial seriation of cells was an indication of cambium and secondary development.

Klinken (1914) discussed the cambial initial cell concept as it related to sliding growth (Chap. 6.7.1). He accepted as a first premise that each radial series in a conifer cambium arose from a single initial cell in the sense of Sanio. Differentiating cells were referred to as cambial cells. However, he also spoke of "intracambiales"[1] in a plural sense when referring to those cells in the cambial zone capable of further division and "extracambiales" when referring to the differentiating products of cambial cells; the latter cells were comparable to Krabbe's (1886) "subcambiale" cells. Although it is evident that Klinken recognized that daughter cells might divide, it is not clear from his writing how many times they might do so.

---

[1] Haberlandt (1914), in a quite different sense, considered secondary xylem to be the "intracambial" and phloem the "extracambial" increment.

Kleinmann (1923) re-examined Sanio's initial cell theory from the points of view presented by Raatz (1892) and Schoute (1902). Although he was inclined toward Raatz' concept of a cambium consisting of a layer of equal-valued cells, he did not agree with all of Raatz' ideas. Kleinmann's main argument was based on the fact that more cells were produced on the xylem side than on the phloem side, which meant that a cambial initial cell could not divide alternately with regularity. He examined radial longitudinal sections of *Sambucus nigra, Phytolacca decandra*, and *Raphanus sativus*, and in each species he observed that nuclear divisions in adjacent cell series did not necessarily correspond (Fig. 4.6). That is, the nuclear divisions did not always occur in the middle of the cambial zone. Such a condition would not be possible according to the rules of Sanio (1873) and Schoute (1902), but entirely possible according to the schemes of Raatz (1892) and Nordhausen (1898). Kleinmann noted that the cambial zone retained its boundaries despite these division patterns. He did not reject outright the initial cell theory, but instead suggested that a possible solution to the controversy might be sought in Raatz' theory of the Wendekreise. On this basis, Kleinmann defined the cambium as follows: "The cambium consists of a layer of meristematic cells in which no differences exist in their ability to divide. When cells leave the cambial zone boundary, they lose their cambial characteristics and differentiate as either wood or bast cells."

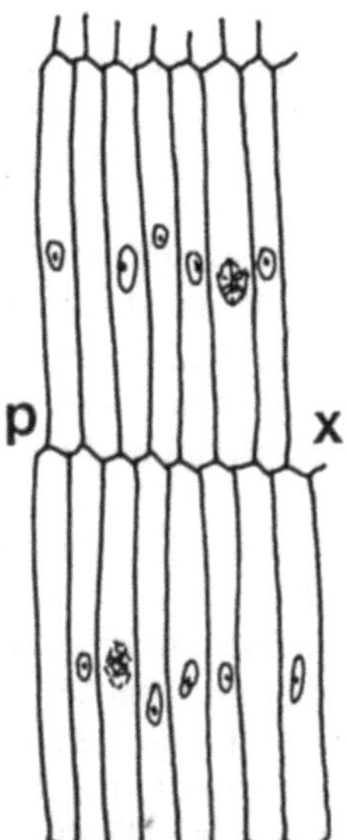

Fig. 4.6. Radial section of cambial zone of *Raphanus sativus* between phloem (*p*) and xylem (*x*) showing nuclear divisions in nonaligned cell files. (After Kleinmann 1923)

4.1.4 Uniseriate vs. Multiseriate Theories

By the mid-1920s, two general conceptions of the cambium as an initiating layer were recognized (Eames and MacDaniels 1925). Either the cambium consisted of a single layer of permanent initiating cells (Fig. 4.7A,B), or of a zone of dividing cells. According to Eames and MacDaniels, only a single cell layer could have permanent status and they specifically confined the term cambium to this cell layer. They also acknowledged that during active growth, xylem and phloem mother cells derived from the initial cells could redivide, sometimes more than once. However, these mother cells did not qualify as cambium because both the mother cells and their derivatives eventually differentiated and matured as permanent tissue cells.

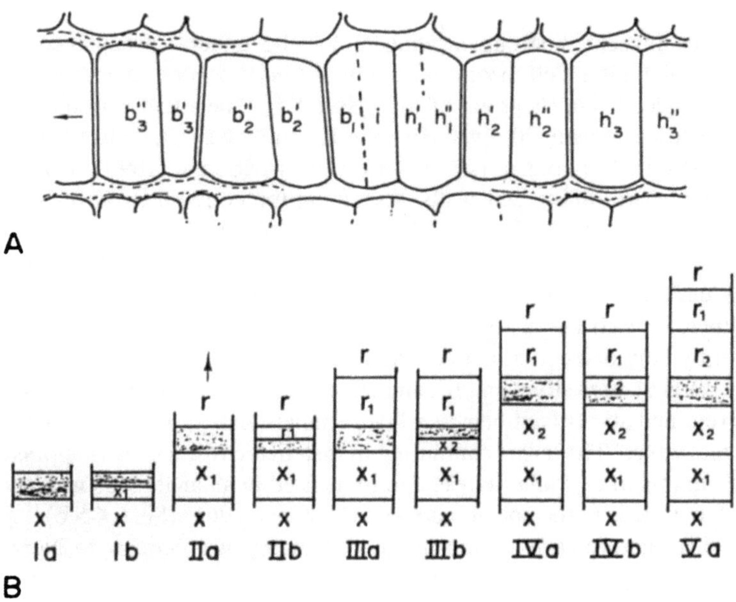

Fig. 4.7. A Rothert and Jost's (1934) conception of Sanio's Fig. 4.1. *i* Initial cell; *b* and *h* phloem and xylem derivatives, respectively. **B** Cambial division scheme based on **A**. Initial cells *shaded*; *r* and *x* phloem and xylem derivatives, respectively. *arrow* toward the bark. (After Rothert and Jost 1934)

The view of the cambium presented by Eames and MacDaniels was, in a sense, a blend of the more extreme views presented by both Sanio (1873) and Raatz (1892). This more realistic view of the cambium was accepted slowly, particularly among many European authors (see Gustin and De Sloover 1955, Catesson 1964). For example, Gustin and De Sloover listed nine textbooks which adhered to the classical concept of the cambium as a single layer of cells. Gustin and De Sloover regarded this concept as being insufficient. Although they acccepted the idea of an initial layer, they believed that derivatives produced by this layer did not necessarily differentiate directly to permanent cells. Consequently, the cambium must consist of several layers of dividing cells. However, because they were unable to distinguish among these dividing cells, they found it difficult to identify the initial layer.

Gustin and De Sloover also cited an unpublished 1947 study by Hardy on squash in which 884 cambial cell files were examined. In 71.1% of the files, the most recent division involved only one cell, and in 9.6% of the files at least two cells were in the process of dividing. However, in the latter case, the dividing cells were always separated by at least one nondividing cell. Gustin and De Sloover interpreted the second pattern to mean that the cell nearest the initial cell was not involved in division. It also meant that cells other than the initial cell were capable of dividing.

Newman (1956) presented strong arguments for the general acceptance of Sanio's (1873) cambial initial cell theory in a study of *Pinus radiata*. In addition to verifying the existence of Sanio's group of four (Sect.4.5.1), he attempted to identify initial cells in adjacent cambial rows. His criteria were the short, densely cytoplasmic cells in the initial regions of rays as observed by Mischke (1890), and the relative thickness of partitioning cell walls. Assuming the correctness of his cellular identifications, he found that cambial division patterns in adjacent rows were not necessarily synchronized. He therefore avoided the implication that any one initial cell was endowed "... with a continuing individuality from which tissue emanated." Rather, in each radial row there existed a function, the "initial function," carried on by only one cell at a time, and this function was inherited by one of the daughter cells at each division. A cambial initial, as defined by Newman, was "... the current and short-lived heir in a line of inheritance of the initial function in the cambial zone." The term "continuing meristematic residue" was applied to the initial cell in this line of inheritance, a term he used in an analogous way for the impermanence of cells in the apical meristem (Newman 1961) (Sect. 4.5.6).

Bannan's (1955) interpretation of the initial cell theory was similar to Newman's (1956), but far more thorough. He accepted Sanio's concept of an initial cell, which he called the initiating layer. In his study of *Thuja occidentalis*, Bannan followed the progressive development of the initiating layer and its derivatives throughout an entire season of growth. Dormant cambia sampled prior to bud break usually consisted of one to four cells. If two cells, the one next to the phloem would be the initial cell and the inner one a xylem mother cell, and if three cells, the inner two would be xylem mother cells. In 70% of the cases, the first division initiating spring growth occurred in the oldest xylem mother cell. In the remaining cases, the first division occurred either in a younger xylem mother cell or in the cambial initial cell. Irrespective of the site of the first periclinal division, the rapid production of earlywood cells during the vernal surge of growth was due to the division and redivision of xylem mother cells. This division pattern

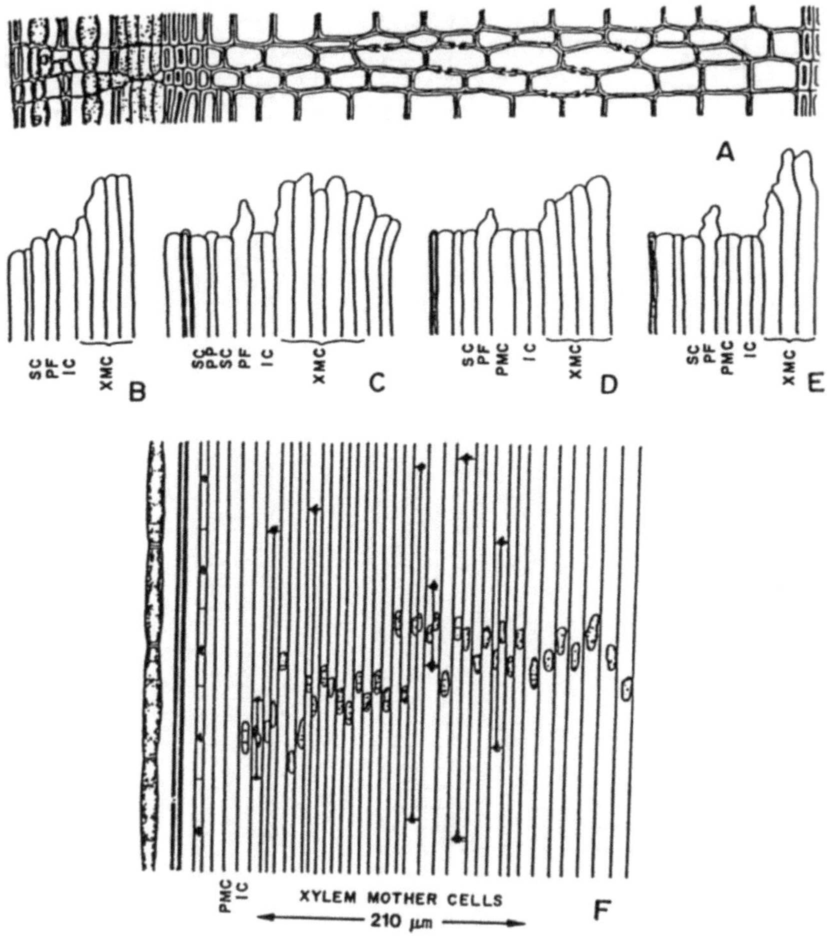

Fig. 4.8A-F. *Thuja occidentalis.* **A** Transverse section showing replication of cellular pattern following loss of an initial cell. **B-E** Drawings from radial sections depicting differences in length of cambial cell derivatives. **F** Drawing of radial section through the cambial region showing wide zone of periclinal divisions in a fast-growing tree. *SC* Sieve cell; *PF* phloem fiber; *IC* fusiform initial cell; *XMC* xylem mother cell; *PMC* phloem mother cell. (Bannan 1955)

resulted in a broad zone of xylem mother cells in various states of division (Fig. 4.8F). As the season progressed, the xylem mother cell zone gradually contracted and ceased activity in late summer.

The observations of Bannan suggested that the xylem mother cells were primarily responsible for the production of new xylem elements. The most active xylem mother cell was usually situated either immediately or closely adjacent to the initial cell (Figs. 4.9, 4.10) (for contrary patterns, see Chap. 11.4.2). Bannan

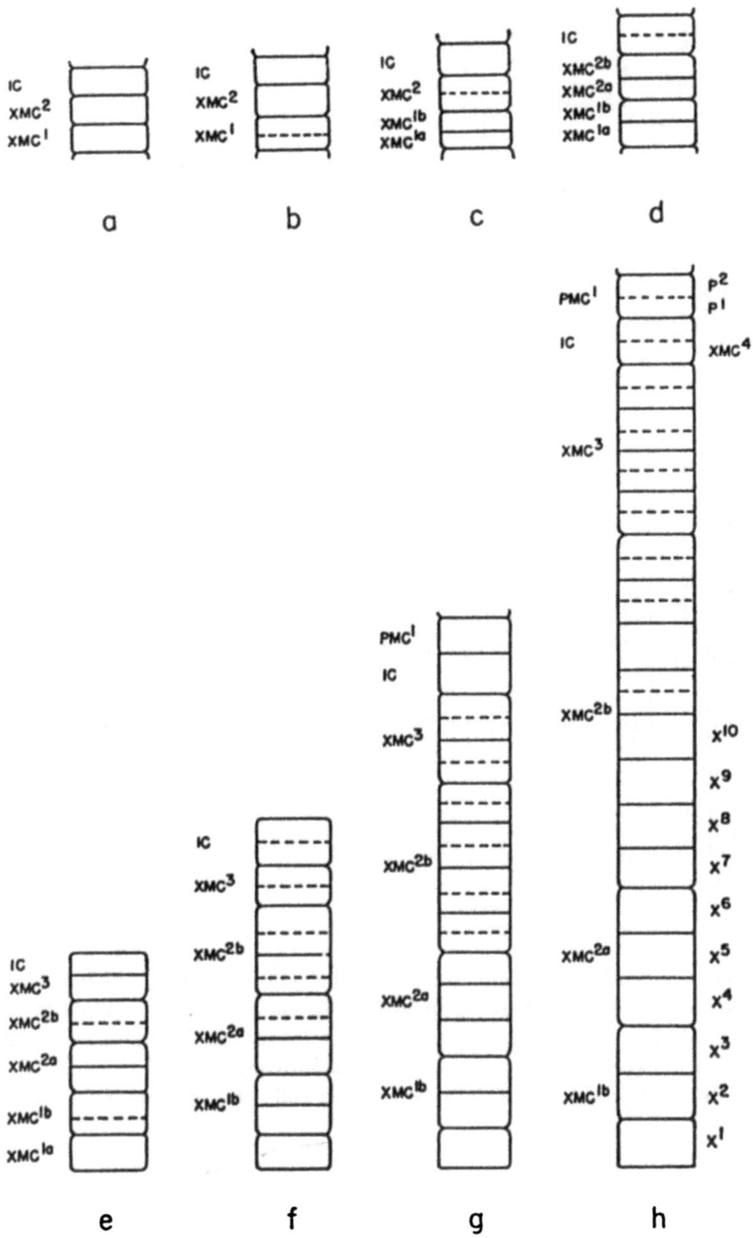

Fig. 4.9a-h. Series of diagrams showing sequence of periclinal cell divisions during cambial reactivation in *Thuja occidentalis*. Divisions began in xylem mother cells (*XMC*) which repeatedly redivided. Less frequent divisions in the tier of initial cells (*IC*) yielded new xylem and phloem (*PMC*) mother cells. *X* Differentiating xylem element; *P* phloem element. (Bannan 1962b)

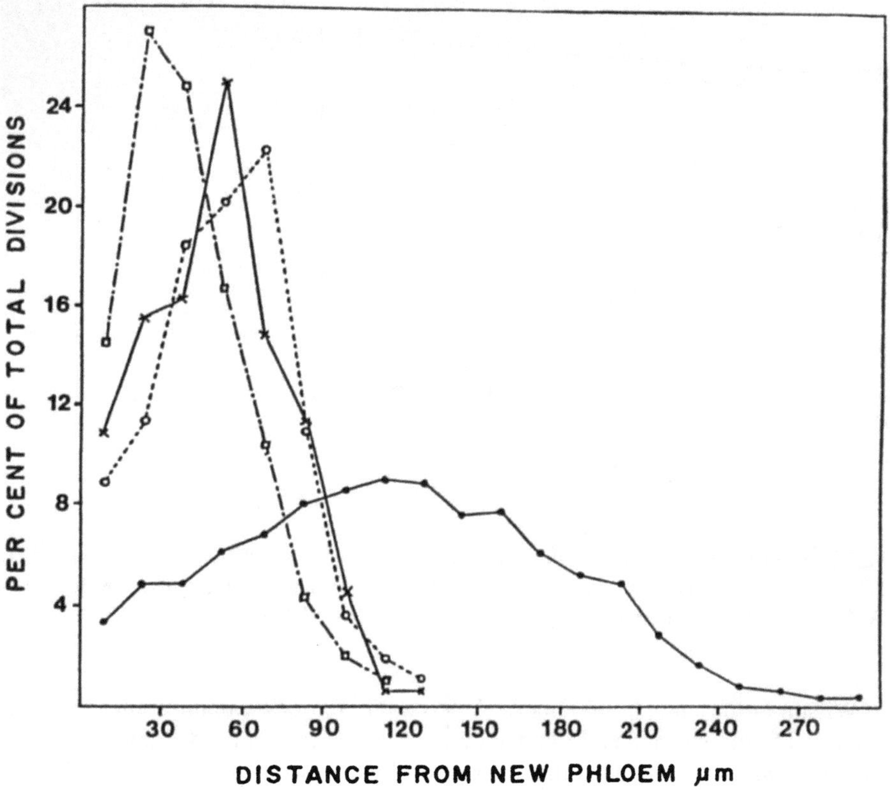

Fig. 4.10. Relative frequency of periclinal divisions across the cambial zone of four fast-growing *Thuja occidentalis* trees. Positions of the initial cells are to the *extreme left*, within the first 20 μm. Most active divisions occurred in the approximate center of the dividing and redividing xylem mother cells. Data based on 1200 division figures. (Bannan 1962b)

recognized an initial cell that intermittently divided during the season to produce new xylem mother cells. Periodically, the initial cells also produced phloem mother cells, but much less frequently than mother cells on the xylem side. Initial cells were tentatively identified on transverse sections either by the presence of an anticlinal division that resulted in the addition of a radial file on both sides of the cambium (Fig. 6.49), or by loss of an initial cell that resulted in deletion of a radial file on both sides of the cambium (Fig.4.8A). On radial microsections, xylem mother cells and their immediate derivatives were identified by the distinctive elongation of these cells (Fig. 4.8B-E). Phloem mother cells could not be directly identified other than by position because they did not elongate, although phloem fibers did. Based on the foregoing observations, Bannan (1957b) claimed adherence to the multiseriate concept. However, like so many others, multiseriate simply meant to him the recognition of mother cells in addition to the cambial initial cell.

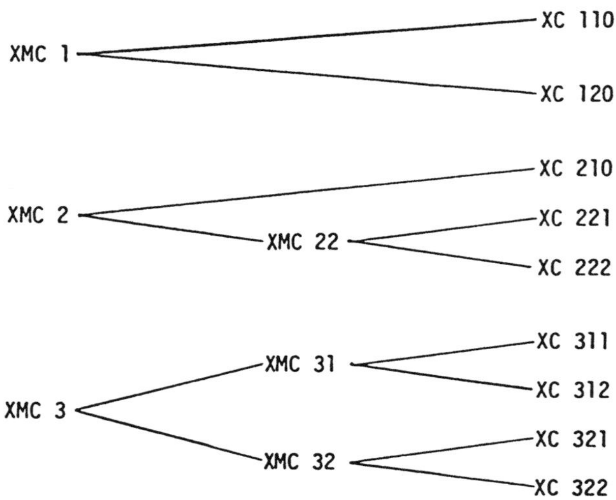

Fig. 4.11. Division sequence on the xylem side during cambial reactivation in *Picea abies*. The first 9-13 xylem cells were derived from three over-wintering xylem mother cells (*XMC 1-3*). *XMC 1* bordered on the previous year's xylem. (Höster 1974)

Wilson (1963) adopted Bannan's (1955) definition of a cambium that included xylem and phloem mother cells. He (Wilson 1964) recalculated Bannan's (1955) data on *Thuja* and concluded that the rate of division seemed to be independent of the number of cells in the cambium. Data obtained from his own studies on *Pinus strobus* agreed with those of Bannan for *Thuja*. That is, the highest frequency of division occurred in the middle of the cambial region and dropped rapidly to almost zero in the quarter of the cambium nearest the xylem, where derivatives began to differentiate. However, he interpreted this pattern differently, suggesting that "a population" of cells might perform the functions of initials. Yet in later work the concept of an initial cell was adopted (Wilson and Howard 1968). Phillips (1976) also accepted the initial cell concept in his review of the cambium.

Höster's (1974) investigation of the cambial zone in *Picea abies* also revealed the presence of an initial cell. The dormant cambium typically consisted of a phloem mother cell, the cambial initial, and three xylem mother cells. Upon resumption of growth in the spring, the phloem mother cell differentiated into a sieve cell, or two sieve cells if it divided. The three xylem mother cells usually divided according to the scheme shown in Fig. 4.11. In this way, the first 9-13 xylem cells were derived from the overwintering mother cells and all subsequent tracheids originated from the cambial initial and derivatives thereof.

One of the strongest proponents of the multiseriate concept of cambium was Catesson (1962a, 1964), who argued for a cambium in which several cells might divide simultaneously. In an extremely detailed study of *Acer pseudoplatanus*, she presented convincing evidence supporting this view. During the early part of the

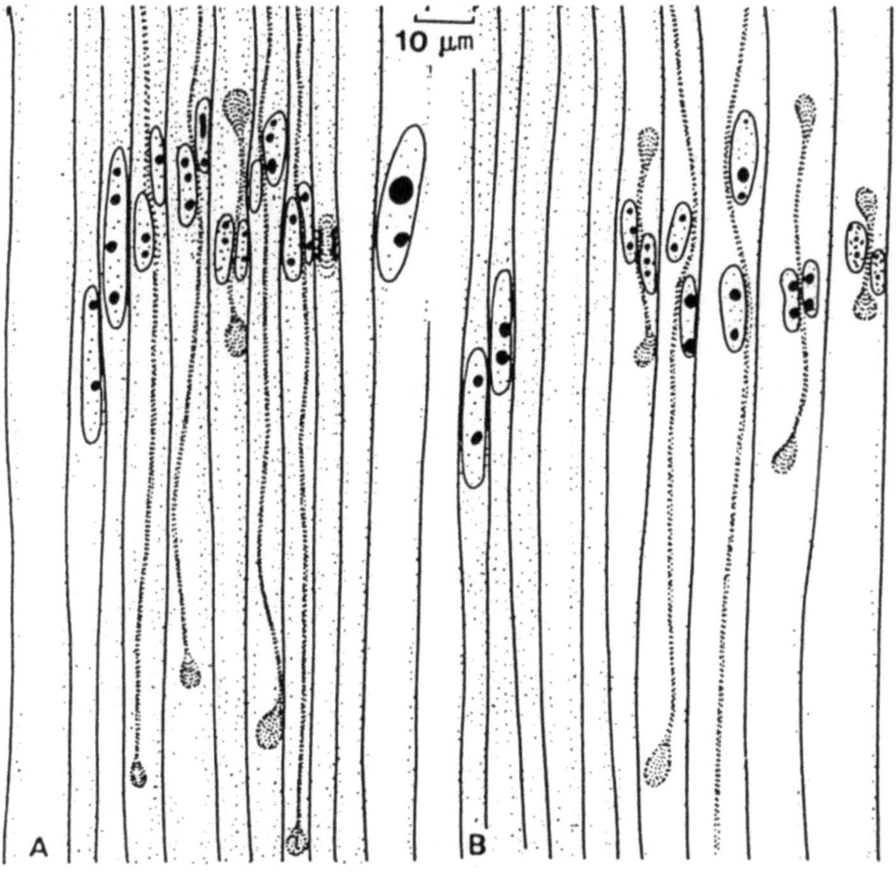

Fig. 4.12A,B. Radial sections from a *Acer pseudoplatanus* stem showing simultaneous periclinal mitoses in the cambial zone (**A**) and xylem derivatives (**B**). Cambial cells are *more densely dotted.* (Catesson 1964)

growing season (May), periclinal mitoses were found to occur simultaneously in cells of the cambium (Fig. 4.12A) and in xylem derivatives (Fig. 4.12B). No differences could be detected either in the process of mitoses, in development of the phragmoplasts, or in histology of the cambial zone cells. Consequently, she concluded that it was impossible, during the season of growth, to determine a unique initial cell for each file of cambial cells. On the contrary, the cambium was believed to consist of several similar cells endowed with equivalent abilities to divide and with each cell performing the role of an initial. That is, there is no "assise initiale privilégiée," or no privileged initial layer.

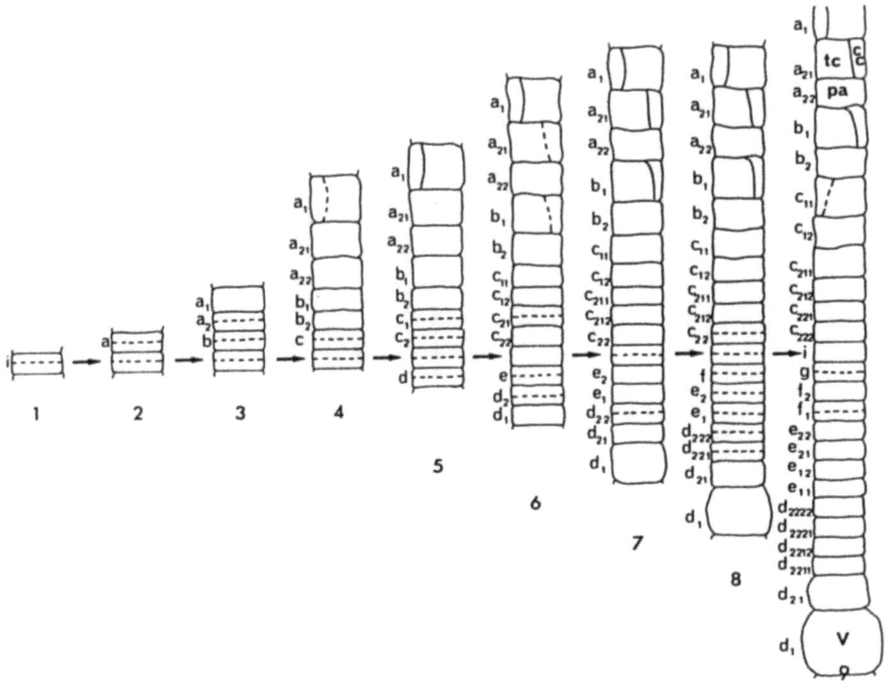

Fig. 4.13. Series of diagrams showing sequence of periclinal cell divisions during cambial reactivation in *Acer pseudoplatanus*. First divisions produced phloem derivatives (*a,b,c*) and then xylem derivatives (*d-g*). Derivatives might again divide before differentiating into vessels (*v*), parenchyma (*pa*), sieve elements (*tc*), and companion cells (*cc*). (Catesson 1984)

In a later paper, Catesson (1984) not only reaffirmed her earlier views but fortified them with additional observations and data. She deplored the fact that over the years Sanio's (1873) work not only had been misinterpreted but that the resulting interpretations had become rigidly entrenched in the literature. She favored the interpretation shown in Fig. 4.13. According to this view, the cambium might be one cell wide at the beginning of the growing season. However, as the season progressed, daughter cells assumed this function and at that time a unique cambial initial could no longer be recognized. Catesson referred to this process as "le fonctionnement alternatif," or alternative function, whereby the cambium produced xylem and phloem alternately, but not necessarily uniformly, depending on the species and growth conditions of the plant. These results and later observations (Catesson 1987) clearly indicated that at any given time there was only one cell position in each radial file capable of producing both xylem and phloem derivatives. Yet, this initial cell could not be identified on the basis of either its ultrastructure or its mitotic activity.

After many years, investigating cambial activity, Bannan (1968b) re-appraised his earlier (Bannan 1955) view of the cambium. His numerous studies of anticlinal division patterns had conclusively demonstrated that these divisions were essentially confined to an initial cell (Chap. 6.1.1). For example, in more than 20 coniferous species examined, approximately 98% of the anticlinal divisions were permanent. Since derivatives occurred in both xylem and phloem, they could only originate in the initiating layer of cambial cells. Thus, each radial file contained one initial cell, although these cells were not necessarily tangentially aligned in neighboring files, as often implied by the expression uniseriate layer. Bannan could not envision an initiating layer more than one cell wide. An initiating layer two or three cells in depth would require an exact duplication of all anticlinal divisions, a highly improbable event. He subscribed to a concept of cambium consisting of a zone of periclinal divisions several cells wide. However, the several layers of periclinally dividing cells were clearly not equivalent, and only one of them constituted the initial layer. He concluded that Sanio's (1873) original hypothesis still provided a useful concept of cambial organization.

Ultramicrotome sectioning was used by Itoh et al. (1968) in preparing a beautiful set of photomicrographs of the cambial zone of *Cryptomeria japonica*. The dormant cambium usually consisted of four cells which the authors labeled C_1 - C_4, with cell C_1 being the initial cell situated next to the phloem (Figs. 4.20, 11.17). The first division of initial cell C_1 produced a phloem mother cell which proceeded to redivide; no phloem mother cells were observed in the dormant cambium. All phloem cells produced during the growing season were derivatives of initial cell C_1. On the xylem side cambial cell C_4, nearest the mature xylem, generally matured without dividing. Occasionally, it divided once to produce two cells and rarely three times to produce four cells. Cell C_3 usually differentiated after only one or two divisions, and more rarely after three divisions producing four cells. Cell C_2 produced a total of three to seven cells by a series of two to six divisions, and initial cell C_1 produced a total of four to ten cells by a series of three to nine divisions. By an analysis of these data, the authors concluded that initial cell C_1 produced about 80% of the derivatives contributing to seasonal growth of the xylem while cells C_2 - C_4 produced the remaining 20%. A number of criteria were used to deduce division patterns, but primary emphasis was placed on the relative thickness of parental and daughter tangential cell walls.

In contrast to the apparent absence of phloem mother cells in the dormant cambium of *Cryptomeria japonica* (Itoh et al. 1968), Barnett (1992) could detect no xylem mother cells in the dormant cambium of *Aesculus hippocastanum*. The cell bordering the previous season's xylem, the boundary-layer cell, differentiated to either a vessel element or a parenchyma cell (Fig. 11.15A). It was presumably committed to differentiate as a xylem element and it did so without dividing. The cell adjacent to the boundary-layer cell was deduced to be the fusiform initial. It was the last cell in the cambial zone to enlarge and to divide. The remaining cambial zone cells were found to be committed to the phloem. They were the first to enlarge and to divide (Chap. 11.4.2), with the first division preceding that of the initial cell by at least 2 weeks. The fusiform initial was therefore recognized by a process of elimination, because the fates of all other cambial zone cells appeared to be predetermined.

In a recent study, Gahan (1989) attempted to identify subpopulations of cambial cells in stems of *Lycopersicon esculentum* by the incorporation of [3H] thymidine in nuclei of cambial initials followed by microautoradiography. The observed labeling patterns lent support to the concept of a population of true cambial initials that divided at a very slow rate. These initial cells, in turn, gave rise to a second population of daughter cells that divided rapidly, eventually differentiating as xylem and phloem elements. Results of this study suggested that the difficulty in identifying cambial cells might be explained on the one hand by the slow turnover rates of the initial cells and on the other hand by the loss of daughter cells to the rapidly dividing population. Additional suggestions for distinguishing cambial zone cells on the basis of cytochemical differences have been made by Catesson (1992) (Chap. 5.4).

4.2 Special Criteria

4.2.1 Sanio's Rules

Sanio (1873) offered several criteria for identifying cambial initial cells. These criteria, often referred to as Sanio's "rules"[2], were used by subsequent investigators in attempts to either prove or disprove the existence of initial cells (see Timell 1980, Mahmood 1990). Although the criteria were modified in various ways, the pursuit of each one added some measure of verification to the initial cell concept. Because the criteria involve both periclinal and anticlinal cambial fusiform cell divisions, they cannot be discussed consecutively. Appropriate cross-referencing will guide the interested reader to sections of the text where the criteria are discussed more thoroughly.

Not all of the following criteria are observable at any one time, and some are evident only under special circumstances or in certain species. No single criterion can be used to identify the cambial initial cell. Moreover, it must be recognized that every rule is subject to numerous qualifications and exceptions.

Radial Cell Diameter. Although not a definite criterion, radial cell diameter can be of value in identifying the initial cell in special situations. One example might

---

[2] Sanio's "rules" should not be confused with Sanio's "laws" regarding tracheid length variation in conifers discussed in Chapter 7.1.2.1.

be dormant cambia where the initial cells are sometimes especially narrow in radial diameter (Chap. 11.3.1).

Angular Cell Corners. This criterion was used by Sanio, as illustrated in Fig. 4.1. It assumed that the radial cell corners of the initial cell would be curved or bowed in the direction of division, either toward the xylem or the phloem. It was adopted by others such as Krabbe (1884) and Raatz (1892).

Tangential Wall Thickness. This criterion was also used by Sanio, as illustrated in Fig. 4.1. It assumed that a new tangential wall dividing an initial cell would be thin and tenuous relative to the radial walls (Sect. 4.5.1; Chap. 11.2).

Sanio's Four. Sanio postulated that each cambial cell would usually give rise to four cells, and proper analysis of the sequence could be used to identify the initial cell (Fig. 4.1). The two foregoing criteria were essential in determining Sanio's four. This criterion has been verified and employed by many later investigators (Sect. 4.5.1).

Anticlinal Divisions. Perhaps the most useful rule proposed by Sanio was based on the fact that new cell files derived from anticlinal cambial cell divisions would be perpetuated in both xylem and phloem (Chap. 6.1.1). In contrast, cell files derived from mother cells dividing anticlinally would be short-lived and terminate with maturation of the mother cell (Fig. 6.2).

Trabeculae. Sanio proposed that trabeculae, his Langstäben, originating in the cambium, would be perpetuated through a limited number of cells in both the xylem and phloem in much the same way as the products of anticlinal divisions. On the contrary, the Kurzstäben, described by Raatz (1892) and others, did not necessarily extend through the cambium (Sect. 4.5.2).

Alternation of Divisions. Sanio erroneously proposed that divisions of the cambial initial alternated between xylem and phloem and this sequence could be used to identify the initial. Cambial initials do alternate divisions between xylem and phloem during the growing season, but not in perfect sequence (Chap. 11.5). Several later investigators attempted to identify the cambial initial on the basis of events associated with this alternating sequence. For example, both Bannan (1955) and Srivastava and O'Brien (1966a) pointed out that the lengths of derivatives undergoing intrusive growth often exceeded that of the initial cell (Fig. 4.8B-E). Kaussmann (1963) referred to a cambium that divided toward both sides as "dipleurisch," or dipleuristic.

4.2.2 Additional Rules

Relation to Ray Cells. This criterion, suggested by the work of Krabbe (1884), Mischke (1890), and others, is based on the fact that ray cell initials divide less frequently than fusiform initials do and they often elongate radially before dividing (Chap. 8.3.3). Thus, by identifying the ray cell initial, it is sometimes possible to identify the presumptive fusiform initial by other criteria such as its narrow radial diameter and position relative to that of the ray cell initial (Wodzicki and Brown 1973).

Dormant Cambia. Although providing neither a rule nor an identifiable criterion, cambial dormancy has proven to be the growth phase during which most investigators have attempted to describe cambial initial cells. The number of cells in the cambial zone at this time is at a minimum (Chap. 11.3.1) and it is sometimes possible to identify a potential initial cell by the process of eliminating presumptive mother cells. The analyses of Itoh et al. (1968) and Barnett (1992) provide exceptional examples of how dormant cambia can be used to evaluate division processes in active cambia (Figs. 4.20, 11.15A).

4.2.3 Role of Mother Cells

The terms mother cell and daughter cell are synonyms in most cases and they differ only in how they are used. The immediate derivative of a fusiform initial dividing periclinally is a daughter cell; but this daughter cell becomes a mother cell when it in turn divides periclinally. Similarly, the immediate derivative of a fusiform initial dividing anticlinally is a daughter cell; but in this case the daughter cell becomes a new fusiform initial that can divide periclinally to produce new daughter cells. To avoid confusion, the derivatives of an anticlinal division are often called sister cells (Chap. 6).

The distinction between the uniseriate and multiseriate cambial cell concepts rests to a large degree on how one interprets the cambial mother cells. According to the uniseriate concept, the initial cell divides to produce daughter cells which might or might not redivide. One of the daughter cells assumes the function of a new initial cell and the remaining one(s) either become xylem or phloem mother cells. Mother cells are not permanent initials and, after dividing one or more times, they mature along with their daughter cells. The multiseriate concept, on the other hand, denies the existence of initial cells, primarily because in active cambia it is seldom possible to determine which of several dividing cells in a radial file might be the initial cell. Consequently, initial cells and recent mother cells are accorded equal value.

As noted, the uniseriate concept recognizes divisions by mother cells. Bannan (1955), for example, while adhering to the uniseriate concept, nonetheless recognized that mother cells might repeatedly redivide. During the vernal surge of growth, the first division was not in the initiating layer but in a xylem mother cell (Bannan 1962b) (Fig. 4.9). The highest frequency of division was in the central part of the cambial zone comprising the dividing and redividing xylem mother cells (Fig. 4.10). Yet such divisions do not invalidate the uniseriate concept.

One point that lends some credence to the multiseriate concept has been overlooked by previous investigators. Certain types of cambial injury, such as frost damage (Chap. 10.6.1) or the formation of partially missing growth rings (Chap. 11.8), result in either destruction or quiescence of the cambium, respectively. Although it is not known how the cambium becomes reestablished following these events, it has been suggested that new cambial initials might be derived from residual phloem mother cells. In this case, cambial zone cells other than the initial cells would function as initial cells in reestablishing the cambium. However, in many types of wound closure (Chap. 10.9), the cambium covering a wound site advances tangentially through the callus mass in continuity with pre-existing cambium at the wound borders.

An interesting analysis of cell division in the cambial zone of *Fraxinus excelsior* was made by Burggraaf (1974). Cells of adjacent radial files on the xylem side of the cambial zone were numbered according to their position relative to the youngest differentiating phloem cell. Cell divisions occurred over a broad zone extending up to cell number 21. Major peaks of mitotic activity were found for cell numbers 6 and 10, both with 15-17% dividing cells in adjacent files. Minor peaks occurred at cell number 0 (youngest phloem cell) with 5% dividing, and cell numbers 16 to 18 with 7% dividing. Although divisions obviously occurred throughout the cambial zone, the major and minor peaks in adjacent files suggested a surprising degree of tangential continuity.

4.2.4 Tangential Cell Alignment

A tacit assumption of Sanio's (1873) theory related to tangential alignment of the initial cells in adjacent radial files. This assumption was based on his belief that cambial cell division alternated regularly between xylem and phloem. One of the objections raised by Wieler (1886a) to Sanio's initial theory was his observation that the winter cambium often exhibited a wavy or zig-zag outline. Raatz (1892), too, observed that adjacent cell files were not in "perfect harmony," implying that cell division must be occurring in cells other than the initial layer. Moreover, he believed that one radial file must remain relatively inactive while the neighboring file was active to avoid shifting of the daughter cell walls. Therefore, according to Raatz, each radial file must have its own "Wendepunkt," or turning point.

Bannan made reference to this displacement of the initial layer in his numerous publications. For example, he (Bannan 1957b) noted that location of the functioning initial occasionally deviated from its usual, approximately tangential alignment. In a later paper (Bannan 1968b), he reaffirmed his belief that only one initial cell was present in each radial file although they were not necessarily tangentially aligned in neighboring files. This fact was emphasized in the computer model developed by Wilson and Howard (1968), where results suggested that each radial file in the sample had a different distribution of mitotic frequency. It was also evident in the model of Pulawska (1973), where an attempt was made to demonstrate the effect of cambial cell shape on periclinal division patterns (Fig. 5.5).

Newman (1956) also recognized that cambial division and differentiation patterns in adjacent radial files were not necessarily synchronized, as have many

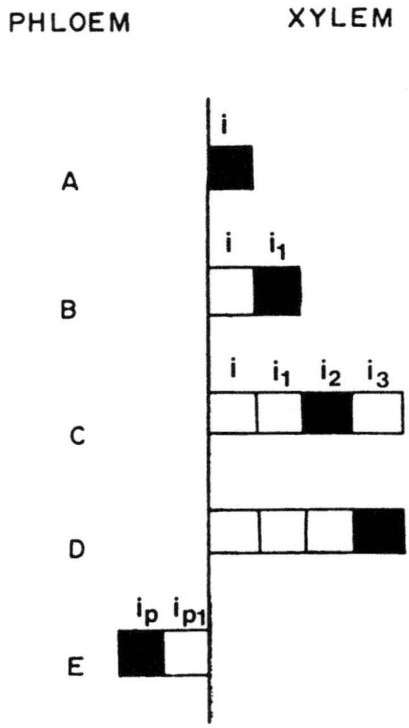

Fig. 4.14. Drawing showing several ways in which tangential misalignment of cambial cells might occur. See text for details

other investigators (Evert 1963b, Mahmood 1968, Timell 1986). Not only are both fusiform and ray cell initials radially offset from their respective neighbors, but also from each other (Wodzicki and Brown 1973). This pattern can be clearly seen in various planes of sectioning through tiers of ray cell initials (Fig. 8.6).

Wloch (1981) investigated this so-called non-parallelism of cambial cells in both normal and tumorous cambia of *Picea abies*. He found that a fusiform cell in one row of undifferentiated cells between xylem and phloem might be in contact with different cells of the neighboring rows at different levels in the cambial zone. Consequently, initial cells of two neighboring rows might be in lateral contact at one level but completely out of contact at another level. He attributed this pattern of cell alignment to the coordinated growth of contiguous cells following non-parallel, or asynchronous, periclinal divisions of the neighboring cambial initials. This pattern of nonparallelism was more accentuated in tumorous than in normal growth, and also more so during vigorous than slow growth. For example, in a normal resting cambium a cell could usually be distinguished in each row that was in contact along its entire length with one cell of the neighboring row. These cells in tangentially contiguous rows formed a continuous layer corresponding in position to those of initial cells. However, in tumorous cambia the initial cells formed a net in which great fragments of these cells protruded beyond the layer of mutual contact. Although Wloch accepted the idea of an initial cell, he was of the opinion that the cambium should be considered as an "anatomical-morphological notion." That is, "A layer of initial cells need exist only in the ideal or functional sense - but need not be a geometric reality."

Tangential misalignment might arise in one of several different ways (Fig. 4.14). According to the initial cell concept, either one of the two daughter cells might become the new initial. Consequently, if, after division, cell i_1 in File B became the new initial cell it would be offset one cell on the xylem side from the initial in File A. According to the concept of Sanio's four (Sect. 4.5.1), which many adherents of the initial cell concept accept, a daughter of the initial cell might again divide, particularly during very active growth. Consequently, cell i_2 (File C), or conceivably cell i_3 (File D), might become the new initial cell in which case the new initial cell would be offset either two or three cells from the initial cell in File A. It is unlikely that cells farther removed from cell files A or B would assume the roles of new initial cells. Tangential misalignment might also occur if, by chance, the initial function in one radial file is transferred to the phloem side in advance of adjacent files, as in File E. Each of these possibilities might be accepted as permissible under the broad interpretation of the initial cell concept.

The presence of asynchronous division patterns in neighboring cell files of the cambial zone are now widely accepted. However, based on existing evidence these patterns neither prove nor disprove the initial cell theory.

4.3 Reconciliation

It appears that Sanio's (1873) theory of a cambial initial cell has withstood the test of time, albeit with certain important modifications. The most significant modification pertains to the number of times a cambial cell might divide. Sanio was too restrictive in believing that the derivative of an initial cell could divide no more than once or twice, and he was certainly in error in stating that cambial divisions alternated uniformly between wood and bast. On the contrary, workers such as Raatz (1892) were too liberal in assuming that all cambial cells possessed equivalent potentials for dividing, the cambium here being interpreted in a broad sense. The truth lay somewhere in between.

Several reviewers (Schoute 1902, Bannan 1955, Gustin and De Sloover 1955, Catesson 1964, 1984) criticized other workers, mainly authors of textbooks, for adhering too closely to Sanio's original definition of the initial cell while not recognizing the predominant role of the daughter cells during the active growth phase. Most of these criticisms are justified. Whereas some authors simply adopted definitions published by their peers, others offered abbreviated definitions without comment. Such definitions are of course inexcusable and they have quite naturally led to both misinterpretation and confusion. As examples of how difficult it is to either interpret or categorize these varied definitions, Catesson (1984) concluded that most English speaking authors correctly interpreted the ideas of Sanio whereas French authors interpreted his ideas with much more restraint, i.e., conservatively. On the contrary, Schmid (1976) concluded that the multiseriate concept was more popular among European authors whereas the uniseriate, i.e., the more conservative, concept was more commonly accepted by American authors.

Although some readers might consider the status of cambial initial cells to be still unresolved, an attempt will be made to briefly summarize presentday interpretations of the concept. Basically, differences reside in whether the cambium should be regarded as a uniseriate layer of initial cells with a more or less permanent status and function, or as a multiseriate zone of cells in which each cell has a more or less equivalent potential for dividing. It is of interest that most proponents of the multiseriste concept recognize the presence of initial cells (Raatz 1892, Wilson 1964, Catesson 1984). That is, even a multiseriate cambium must have a starting point from the dormant state and a midpoint during the active state. It would appear, therefore, that much of the controversy resides in the inability to identify an initial cell, not in denying its presence.

Catesson (1964, 1980, 1984) advanced persuasive arguments for the multiseriate concept and Schmid (1976) proposed equally persuasive arguments favoring the uniseriate concept. This reviewer is strongly inclined toward the latter concept. After reviewing the extensive literature on the subject, it is clear that the differences are mainly a matter of interpretation, or perhaps even philosophy. As Schmid (1976) so convincingly pointed out, "The real difference between the uniseriate concept and the so-called multiseriate concept is, therefore, essentially nomenclatural, terminological, or semantic, *not* conceptual." (author's italics). Yet, to counter this argument, Catesson (1984) pointed out that the differences cannot be resolved simply on the basis of definitions. She stated: "In fact the controversy is more profound for it represents confirmation between rigid concepts and the reality of meristematic function." Thus, it is the rigidity with which both concepts are interpreted that contributes most to the controversy.

The multiseriate concept is based on a premise that, in effect, denies the existence of an initial cell while attributing to all apparently identical cells the properties and functions of initial cells. Thus, the cambium is regarded as more or less encompassing the entire zone of dividing cells. By inference, the uniseriate concept conveys the impression of a single layer of permanent initial cells. Unfortunately, this impression has all too often been accepted as fact. While theoretically true, it is not the case in practice. The initial cell refers to a position from which cell divisions intermittently alternate between xylem and phloem and from which new radial files are initiated. Although specific initial cells cannot be precisely identified, the consequences of their divisions can be easily observed. Careful analyses have demonstrated that these divisions are not necessarily synchronized in adjacent radial files, emphasizing the impermanence stressed by Newman (1956). Nonetheless, even in the most active cambia, a coherent band of cells can be recognized that many workers have referred to as the "middle of the cambium." This zone often enlarges rapidly during the vernal surge of growth, but then contracts as the rate of cell differentiation surpasses that of periclinal division. At the end of the growth period, and during the dormant period, the cambial zone is usually reduced to a narrow band of cells (Chap. 11.3.1).

One of the consequences of the uniseriate cambium is the production of xylem and phloem mother cells. Because these cells also retain the potential for division and because they are often cytologically and histologically indistinguishable from initial cells, proponents of the multiseriate concept equate them with cambium. On the contrary, proponents of the uniseriate concept also recognize that mother cells might indeed redivide, but they have no permanence as dividing cells and they eventually mature as tissue cells. For further justification of the uniseriate concept, the interested reader is referred to Schmid (1976) who not only presented forceful arguments favoring the concept but supported them with an abundant literature; and to Catesson (1980, 1984), who presented equally compelling arguments favoring the multiseriate concept.

Reconciliation of these conflicting viewpoints can easily lead to further controversy. Nonetheless, since this is a review of the cambium, I will attempt to do this. As a starting point, I will assume that most anatomists accept the presence of an initial cell in the dormant cambium although it cannot always be recognized. As discussed in Chapter 11.3.1, a number of workers in their analyses of dormant cambia have eliminated certain cells as presumptive xylem and/or phloem mother cells leaving one to several of the remaining cells as potential initial cells. Itoh et al. (1968) and Barnett (1992) have taken these analyses one step further by attempting to demonstrate that all cells of the dormant cambium are committed to a prospective fate and they can in no way be regarded as equivalent.

Major disagreement arises as to the designation of potential initial cells once the cambium begins active division. As noted in the previous discussion, adherents of both concepts accept the fact that more than one cell in a radial file is capable of division, whether these cells are called initials plus mother cells, equivalent cells, or whatever. Adherents of both concepts also accept the fact that cambial cells in adjacent files need not be tangentially aligned as depicted in Fig. 4.14.

Irrespective of how many cells within a radial file are undergoing periclinal division, only one cell can divide anticlinally and be perpetuated in both xylem and phloem. Similarly, only one cell can transfer its function from dividing xylemward to phloemward and vice versa. However, just like the initial function, which might fluctuate in its tangential alignment, so too might these latter two functions. There is no justifiable reason why a cambial layer consisting of initial cells cannot function in a coordinated manner even though the cells in the individual radial files are dividing and differentiating asychronously.

The similarities between the uniseriate and multiseriate concepts are far greater than the differences. As stressed by Schmid (1976) and so many others, the differences appear to be more terminological than conceptual. Investigators are still unable to identify initial cells either ultrastructurally (Chap. 5.3) or cytochemically (Chap. 5.4). Like elementary particles in a cloud chamber, whose presence and identities are predicted by the tracks they leave in the mist, the positions and identities of fusiform cambial initials are predicted by the derivatives they leave in the xylem and phloem. That is, by the process of division cambial initials leave "tracks" that reveal their presence, positions, and functions. In the following chapters these "tracks" will be discussed with the intent of shedding more light on the role of initial cells.

In preparing this review, I have accumulated several hundred references in which the cambium has been defined either implicitly or explicitly, many of them treating the subject in depth. These papers contain many interpretations of the cambium as well as usage of the terms cambium, cambial initials, and cambial zone. A comprehensive coverage of such a vast literature, while not impossible, is certainly impractical. My apologies to those whose work and perhaps enlightened contributions have been omitted but certainly not overlooked. Many of these papers will be cited and discussed in appropriate sections elsewhere in the text.

4.4 Terminology

As one might expect, the conflicting concepts of cambium have resulted in a conflicting terminology which a number of authors have attempted to clarify. Wilson et al. (1966) proposed a scheme in which cambium was equated with the cambial initials, and the cambial zone included the cambial initials plus the xylem and phloem mother cells (Fig. 4.15). Such a scheme would be in keeping with the concept of a uniseriate cambium. The scheme proposed by Butterfield (1975), on the other hand, considered the cambium to be an inclusive term comparable to the cambial zone of Wilson et al. (Fig. 4.16). Although Butterfield professed to being a follower of the multiseriate concept, his scheme recognized both cambial initials and mother cells. Catesson's (1964, 1984) definitions conformed more closely to those of Butterfield; she considered the terms cambium and cambial zone to be synonymous.

Schmid (1976), after reviewing the extensive literature on the cambium, compiled a table in which he compared the terms used to describe the tissues associated with both the uniseriate and multiseriate concepts of cambium. His proposed scheme for the uniseriate concept, which he favored, closely conformed with that of Wilson et al. (1966) (Fig. 4.15). That is, he suggested that the term cambium should be confined to the cambial initial cells, whereas the term cambial zone should be applied to the entire zone of dividing cells. According to this scheme, the terms cambium and cambial initials would be synonymous. This reviewer agrees completely with this particular aspect of Schmid's proposed terminology.

Timell (1979, 1980), while agreeing with Schmid's (1976) use of the terms cambium and cambial zone, disagreed with the way he segregated differentiating

MATURE PHLOEM		
DIFFERENTIATING PHLOEM	MATURING PHLOEM	
	RADIALLY ENLARGING PHLOEM	
	DIVIDING PHLOEM (Phloem mother cells)	CAMBIAL ZONE
CAMBIUM	CAMBIAL INITIAL (dividing)	
DIFFERENTIATING XYLEM	DIVIDING XYLEM (Xylem mother cells)	
	RADIALLY ENLARGING XYLEM	
	MATURING XYLEM	
MATURE XYLEM		

Fig. 4.15. Proposed terminology for the cambium and its derivatives (Wilson et al. 1966)

tissues (cell enlargement) from maturing tissues (secondary wall deposition). Timell noted that this distinction was contrary to common usage, and therefore suggested that Butterfield's (1975) criteria for differentiating phloem and xylem be accepted as shown in Fig. 4.16. As Timell pointed out, this is a more acceptable terminology.

Steeves and Sussex (1989) argued for a broader definition of cambium for a quite different reason. They suggested that the term cambial initials be retained and the term cambium be applied to the entire zone of dividing cells. Their reasoning was based both on prior, common usage of the term cambium and on the ease with which nonspecialists might understand and use the terms.

While I concur with the terminology of Schmid (1976), as modified by Timell (1979, 1980), I found it impractical to strictly adhere to the proposed terminology in the sections that follow. In most cases it was necessary to retain the terms used by various authors to avoid either misinterpretation or the unintended imposition of my interpretation. In many cases where the meaning was obvious, I have substituted the more commonly accepted terminology. Moreover, in keeping with common practice, I have used the term cambium in a nonspecific way when use of the more specific terms might be either questionable or simply less convenient.

SECONDARY PHLOEM		Mature tissue
DIFFERENTIATING PHLOEM		Secondary walls deposited (a) Cells enlarging Limited cell division (b, c)
CAMBIUM	PHLOEM MOTHER CELLS	Periclinal division
	CAMBIAL INITIAL	Periclinal and Anticlinal division
	XYLEM MOTHER CELLS	Periclinal division
DIFFERENTIATING XYLEM		Limited cell division (c) Cells enlarging Secondary walls deposited
SECONDARY XYLEM		Mature tissue

Notes:

(a) Fibers and sclereids only.

(b) Some transverse and longitudinal divisions associated with sieve tube and companion cell development.

(c) Some transverse divisions associated with the development of axial parenchyma strands.

Fig. 4.16. Proposed terminology for the cambium and its derivatives. (Butterfield 1975)

4.5 Special Features

4.5.1 Sanio's Four

4.5.1.1 Xylem

The division pattern whereby a cambial cell divides in two and then in four, giving rise to a radial tetrad of cells, has often been referred to as "Sanio's four" (Mahmood 1968). When discussing this phenomenon, most reviewers have referred only to Sanio's (1873) work on *Pinus sylvestris*, but the groundwork for his theory was laid much earlier, perhaps in the papers by Trécul (1853a) and Cohn (1857) (Sect. 4.1). In his study of *Ulmus campestris* and *Cytisus laburnum*, Sanio (1863a) mentioned that it was not so remarkable that cells were arranged in a radial series, since each cambial cell gave rise to four cells by longitudinal divisions. Later, he (Sanio 1863b) noted that if one gently macerated a series of cambial cells of *Carpinus betulus* they remained loosely united by a membrane, which he deduced to be the membrane of the cambial cell.

The foregoing events were described more directly and forcefully in his study of *Pinus sylvestris* (Sanio 1873). Here, he observed that following longitudinal division each group of four cells was enclosed by radial walls that more or less separated the mother and daughter cells from radially adjacent cells in the series (Fig. 4.1, File c, cells 5-8). That is, not only was a new tangential wall layer (cell plate) deposited during each division, but the entire contents of the daughter cells were surrounded by a new cell membrane (primary wall). Thus, the radial walls became thicker because not only was a new membrane deposited to the inside, but the original wall of the mother cell was never "resorbed." Many years earlier, Unger (1847) had commented on the fact that, under high magnification, the tangential wall between two adjacent cells appeared to be a common, simple membrane whereas the radial walls were always thicker. His drawing (Fig.3.3B) shows a remarkable resemblance to Sanio's group of four cells.

Mischke (1890), a close follower of Sanio, also found that cambial cells of *Pinus sylvestris* tended to divide in groups of four, and according to his illustrations, radial walls were thicker than tangential walls (Fig. 4.3). Raatz (1892), on the contrary, was a strong critic of Sanio's cambial initial theory and the rules that supported it. He was of the opinion that Sanio's conclusions applied only to old, slow-growing trees. That is, he agreed that the presence of one- to four-celled "Kurzstäbe," or short rods (trabeculae), were in harmony with the initial cell theory for some young vigorous trees, but that the same pattern did not prevail in older growth rings. When he examined a very fast-growing specimen of *Pinus austrica*, he found from 10 to 15 cells enclosed within a radial series. The radial walls of these cells were not conspicuously thicker than the tangential walls in any visibly systematic pattern. His criterion for grouping the cell series was Sanio's "Abrundung der Ecken," or rounding of the radial wall corners, with the first cell in the series forming a "cambial girdle." Despite his objections, Raatz' illustrations all suggested that cambial derivatives tended toward groups of two and four (Fig. 4.5). However, because these groups were not confined to a cambial initial layer, they might occur throughout the cambial zone, a point also alluded to by Schoute (1902). Nordhausen (1898), who normally adhered closely to the views of Raatz, deviated on this point. In his investigations of *Sambucus* stems, he frequently observed that the Kurzstäbe were often confined to two-celled or four-celled series as reported by Sanio (1873) and Mischke (1890).

Priestley and Scott (1939), borrowing on a previous suggestion offered by Giltay (1882), presented a drawing showing a file of parenchyma cells from a young internode of *Brassica* (Fig. 4.17A). The drawing purportedly showed a radial series of eight cells still enclosed within the wall of the parent cell. Elliot (1951) confirmed these observations in macerations of *Lupinus albus* stems. The foregoing observations, while demonstrating the presence of an enclosing wall membrane, did not confirm the existence of Sanio's four. The idea was extended to the cambium of *Pinus pinaster* by Wardrop (1952). He demonstrated the presence of "thin, feebly birefringent membranes, considered to be the remains of the parent cambial cell wall." Such membranes were observed traversing the intercellular spaces between ray parenchyma cells in *Grevillea robusta* and cambial fusiform derivatives in *Pinus pinaster*. Photomicrographic evidence of these membranes in the latter species was presented by Wardrop and Dadswell (1952).

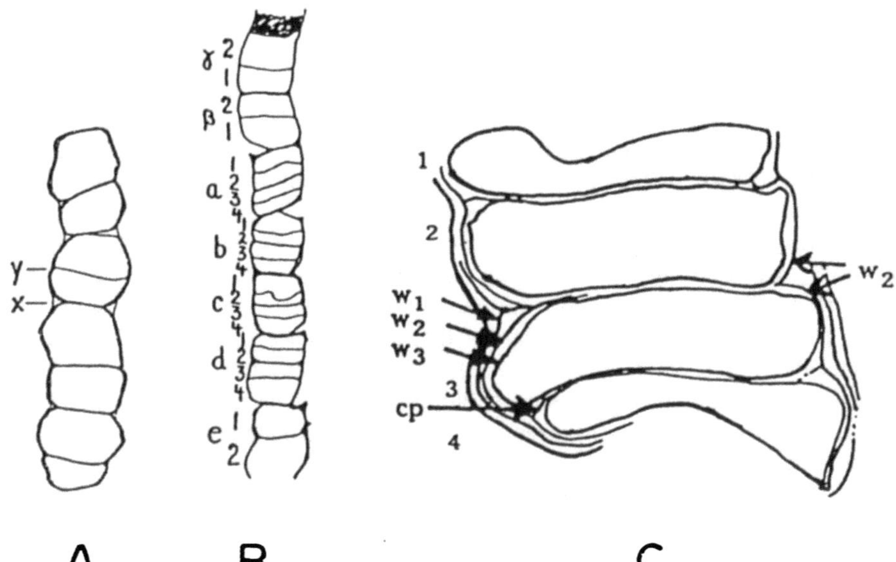

Fig. 4.17A,C. Drawings from transverse microsections showing groups of four cells newly derived from a cambial initial(s) and the enclosing parent cell wall(s). **A** File of parenchyma cells from a young internode of *Brassica*. The last division in the eight-cell file is indicated by *y* and the parent cell wall by *x*. 300x. (Priestley and Scott 1939). **B** Four groups of four xylem cells (*a-d*) and two groups of two phloem cells (β and γ) in *Pinus radiata*. As shown, a_1 is the presumed initial cell. 160x. (Newman 1956). **C** A group of four cambial zone cells from *P. radiata*. w_1 Wall around a group of four cells; w_2 wall around one pair of cells; w_3 wall around an individual cell; *cp* cell plate between two primary walls of cells 3 and 4. Approx. 1455x. (Mahmood 1968)

The first worker to seriously test this aspect of Sanio's theory was Newman (1956). He supported the contentions of Sanio (1873) and Mischke (1890) that the cambial initials and their immediate derivatives occurred as either twin pairs or groups of four. Like these workers, Newman also found that cell groups could be separated by the relative thickness of their tangential walls and the rounded appearance of their radial walls. In one such series, shown in Fig. 4.17B, groups a-e occurred in the xylem and groups β and γ in the phloem. The latter two groups were interpreted as immature phloem which invariably occurred in pairs rather than groups of four. The direction of cell division, whether toward the xylem or the phloem, could be determined by the position of the thickest tangential wall in radial series.

Newman also followed up on Sanio's suggestion that a new membrane (primary wall) was deposited around each daughter protoplast in addition to the partitioning wall or cell plate. Sanio had concluded that the greatest observable wall thickness represented at most two or three cell divisions, thus corresponding to the groups of four. Newman demonstrated this pattern in *P. radiata*, but not conclusively.

The origin and extent of pseudotransverse divisions in "xylem mother cells" of several conifers led Bannan (1957b) to conclude that the concept of Sanio's four, as described by Newman (1956), was invalid. For example, following the pseudotransverse division of a xylem mother cell, periclinal divisions continued for a variable number of successive cells and then ceased when the mother cell and its derivatives matured. The number of tracheids produced by a single xylem mother cell ranged from 2 to 30 in *Tsuga occidentalis* and up to 30+ in fast-growing *Cedrus deodara*. Bannan pointed out that periclinal divisions to at least the third order and greater would be required to produce the longer series. This greatly exceeded the second order level as suggested by Newman to achieve Sanio's four. In some cases, the first and last tracheids were either appreciably shorter or longer. This meant that the centrally located tracheids in the series were products of the last periclinal divisions before the mother cell file matured. Such a division pattern was believed to be incompatible with formation of Sanio's four.

More conclusive proof came from a study by Mahmood (1968, 1971), working under Newman, on *Pinus patula* and *P. radiata*. Mahmood introduced the term Sanio's four with reference to the last-formed xylem or phloem pair together with the divided initial cell. However, the term was rarely applied to the phloem because, like Newman, he found that the phloem derivatives usually occurred in pairs; Sanio (1873), however, believed that they did occur in groups of four (Fig.4.1, File d, cells 5-8). Mahmood also introduced the term enlarging four with reference to the group of four xylem cells radially contiguous to Sanio's four. Mahmood's conception of the cell division process in the cambial zone is shown in Fig. 4.18B. The initial cell i (column a), divided periclinally to yield a new initial i_1 and a mother cell m (column b). Following enlargement of these cells (column c), the new initial i_1 divided to form i_2 and m_1, and mother cell m divided to form daughter cells d (column d). The cells shown in column d comprised Sanio's four. The process was repeated giving rise to a new generation of Sanio's

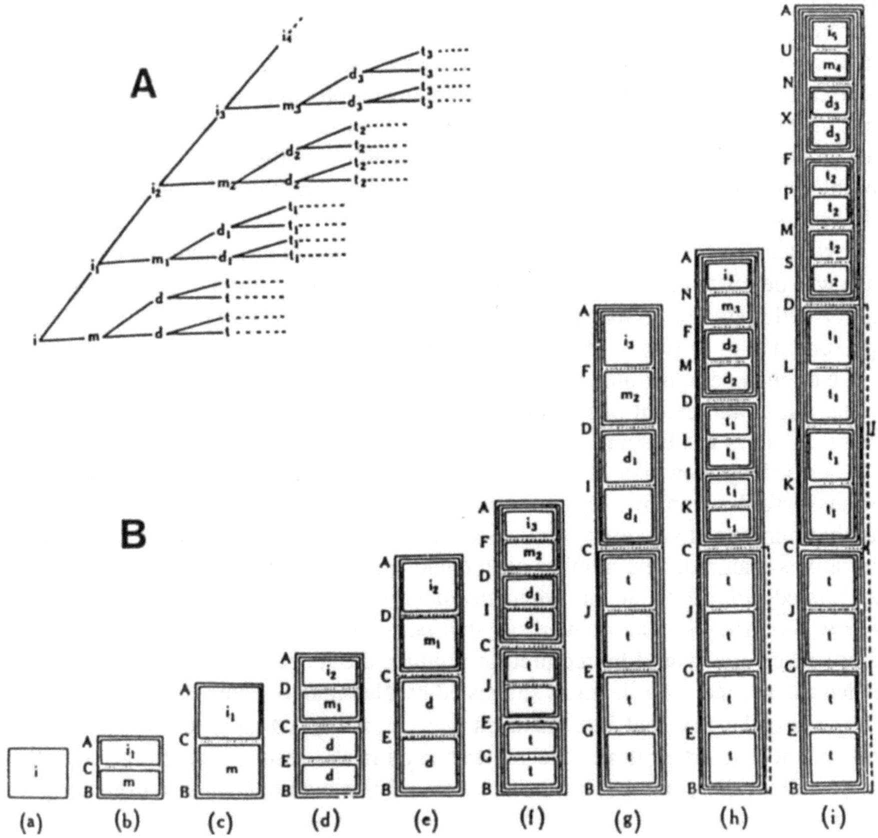

Fig. 4.18A,B. Theoretical interpretation of cambial activity and Sanio's four. **A** Sequence of cells in lineages from successive initial cells, i_1 - i_3. **B** Xylem production from a sequence of initial cells demonstrating the numbers of parental primary walls. *i* Initial cells; *m* mother cells; *d* daughter cells; *t* differentiating xylem cells. (Mahmood 1968)

four (cells i_3, m_2, d_1 and d_1, column f). Meanwhile, the two daughter cells d (column e) divided, giving rise to four xylem derivatives (column f).

Mahmood's papers contained extensive justification for the existence of Sanio's four and detailed rules for division of cambial cells and "emboxing" of the derivatives. The latter term, "Einschachtelung," as used by Russow (1881) and "emboîtement," as used by Giltay (1882), referred to the process by which a new primary wall enclosed each protoplast after cell division. In this regard, it would be of interest to know precisely what Mirbel (1839) meant by his description of cells within cells and vascular utricles within utricles (Chap 3.2). Unger (1842) referred to the phenomenon reported by Mirbel as emboîtement, or emboxing, the same term, incidentally, that Catesson (1984) used to refer to Mahmood's concept of Sanio's four.

As theoretically interpreted by Mahmood, new primary walls were added successively during each division of the cambial initial. Thus, as shown in column i of Figure 4.18B, the radially facing walls of group of four I would be four-layered, those of group of four II would be five-layered, and those of the enlarging four and Sanio's four would be six-layered. The number of tangential walls separating groups I and II at C and group II from the enlarging four at D would be eight-layered. The wall separating the enlarging four and Sanio's four at F would be seven-layered, as would the walls at corresponding positions in columns f and h.

Photomicrographs purportedly showing these wall layers were appended to Mahmood's text. An interpretative drawing of one group of four cells is shown in Fig. 4.17C. Following the lead of Newman, Mahmood found it convenient to use the thickness of tangential walls between groups of four together with the rounded corners of radial walls as criteria for identifying the cell groups. In this way, the "initial of the moment" could tentatively be identified because of its relatively thicker wall on the phloem side (Fig. 4.18B, cell A, column i). The term "initial of the moment" denoted the fact that the cambial initial cell was not a permanent cell but a position in a constant state of change (Fig. 4.18A), e.g., "a continuing meristematic residue" (Newman 1956).

Mahmood's observations, which he recently resummarized (Mahmood 1990), were all made on conventionally prepared microsections using light microscopy. Murmanis (1970, 1971) continued this line of investigation with an electron microscopy study of the vascular cambium of *Pinus strobus*. She, too, used the rounded corners of radial cell walls and the relative thickness of tangential cell walls as criteria for identifying the cambial initial cell and the grouping of cell series. The fact that ray cell initials were most often exceptionally narrow was also of value in identifying the adjacent fusiform initials; she adopted Mahmood's terminology for her analyses. Like Mahmood, Murmanis found that tangential walls between successive groups of four cells were thicker than those separating cell pairs, and the walls between pairs were thicker than those between members of a pair. However, unlike Mahmood, she could not observe individual wall layers because of a different method of specimen preparation. In some collections, the number of undifferentiated cells in the dormant cambium corresponded with the number of cells having the capacity for further division. For example, in one sample the dormant cambium consisted of the initial cell plus seven undifferentiated cells on the xylem side. One of these, which Murmanis considered to be a xylem mother cell, was first to divide upon the resumption of spring growth. The xylem mother cell divided twice and each of its three pairs of daughter cells divided

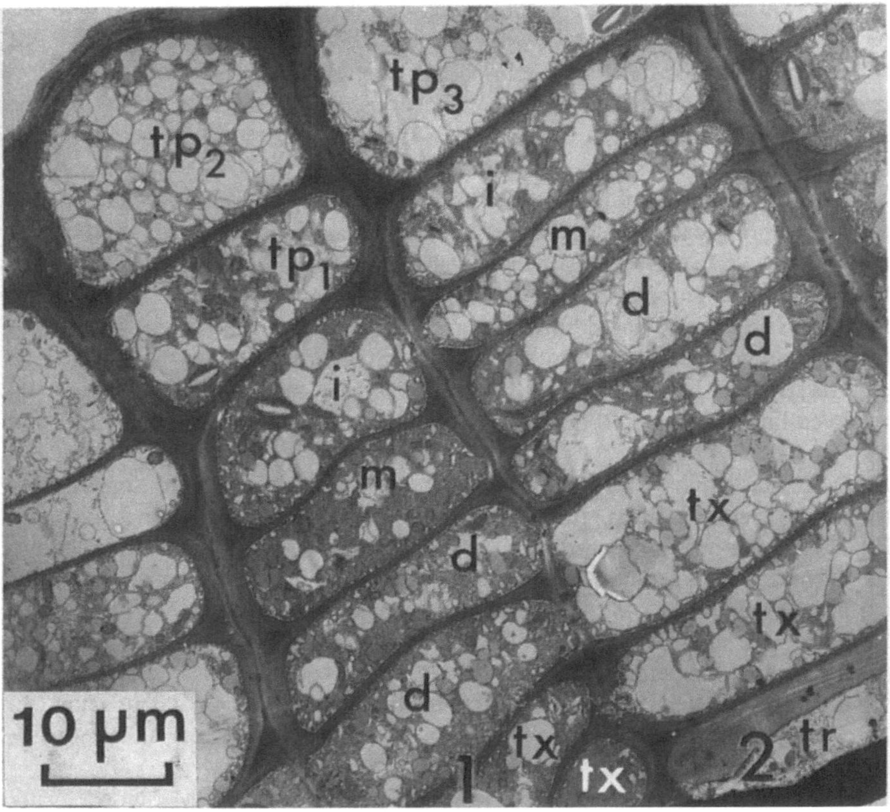

Fig. 4.19. Dormant cambial zone in *Picea abies* containing two radial files of fusiform cells with numerous small vacuoles. Transverse EM photomicrograph. *i* Initial cell; *m* mother cell; *d* daughter cell; *tx* xylem mother cell; *tp* phloem tissue cell; *tr* tracheid. (Timell 1980)

once, yielding four groups of four cells. Such groups of four cells occurred regularly in the xylem but only cell pairs were found in the phloem. A daughter cell stage comparable to that in the xylem was lacking in the phloem because periclinal divisions were confined to mother cells. Although Murmanis accepted the fact that both mother and daughter cells on the xylem side might redivide, she also recognized a cambial initial cell. The initial cell could not be identified cytologically in active cambia, but by following cell sequences the "past initial" could be identified by the divisions that gave rise to cell groups.

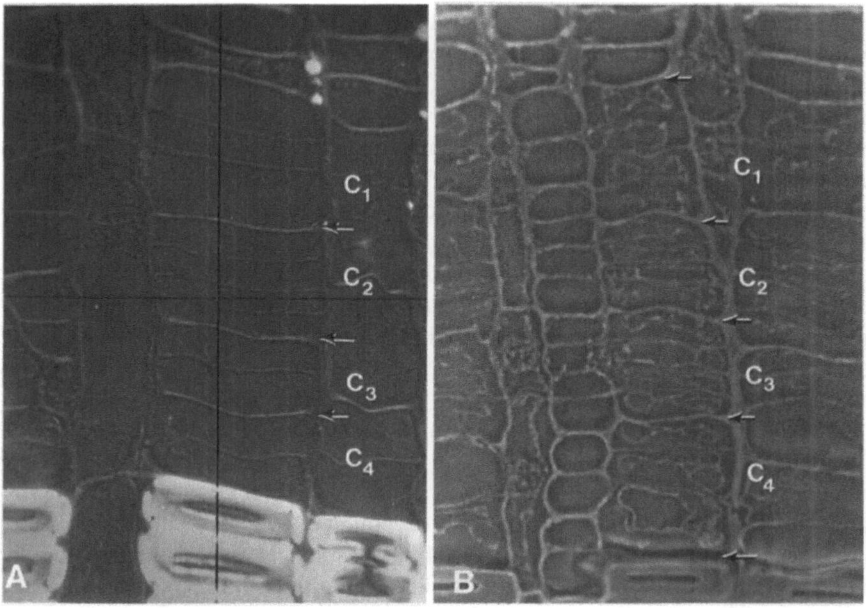

Fig. 4.20A,B. Active cambial zones of *Cryptomeria japonica* in transverse sections showing Sanio's four (C_1) and the enlarging four (C_2 - C_4). **A** C_1 - C_4 derivatives can be distinguished from each other by their highly birefringent walls (*arrows*). Only three cells are present in initial group C_1 and two cells in enlarging groups C_3 and C_4. **B** C_4 consists of only three cells after two divisions of a xylem mother cell. Polarization photomicrographs. **A** 430x. **B** 343x. (Itoh et al. 1968)

Timell (1973, 1979, 1980) followed up on the work of Murmanis in an ultrastructural study of the cambium of *Picea abies*. He, too, used the criteria established by Mahmood for detecting cell groupings and adopted his cell nomenclature and division scheme. However, he used the term expanding four for Mahmood's enlarging four.

Timell found that dormant cambia were best both for observing Sanio's four and identifying initial cells. Although the initial cell could seldom be identified with certainty, it could be detected on some sections. For example, as shown in Fig. 4.19, Sanio's four in each row consisted of an initial cell, a mother cell, and two daughter cells lying between potential xylem and phloem tissue cells. In this photomicrograph, the primary wall around each pair and the wall of the original initial enclosing the group of four can be clearly distinguished. It appeared that the cells comprising Sanio's four overwintered intact in *P. abies*. This was not true of the enlarging cells; these often overwintered as undifferentiated pairs. The arrangement of individual cells in dormant cambia, whether in pairs or groups of four, followed no specific pattern. Rather, the arrangement appeared to be largely accidental and to depend on the stage at which cambial activity ceased in the fall. No differences could be detected, either ultrastructurally or cytologically, between cambia collected from normal and compression wood parts of the tree during the dormant period. This lack of contrast also proved to be the case from collections made during active growth stages (Timell 1979). Like Timell, Höster (1974) also observed that the compression wood cambium of *P. abies* tended to occur in groups of two and four cells. Meier (1973) could recognize cell groupings in resting cambia of *P. abies*, but he had difficulty doing so in active cambia.

The publications of Timell contained many clear and convincing examples of the existence of Sanio's four. In view of these examples, he (Timell 1980) commented on the fact that several other workers had found it impossible to distinguish either Sanio's four or initial cells in their preparations. He pointed out, to the contrary, that such identifications often could be made if one carefully scrutinized the illustrations that appeared in many publications.

Itoh et al. (1968) cut thin sections with an ultramicrotome and observed them with both bright light and polarization optics. Although they neither cited Mahmood's work nor mentioned Sanio's four, their cambial-zone photomicrographs of *Cryptomeria japonica* reveal exceptionally clear examples of both Sanio's four and the enlarging four (Fig 4.20). Barnett (1972) also presented an excellent photomicrograph of Sanio's four in *Pinus radiata*, and Yoshimura et al. (1981b) used Mahmood's groups of four to identify the cambial initial layer in their studies of the "pinning" method (Chap. 10.7.1). Kuroda (1986) used the "group of four cells" to distinguish the cambial initials in her wound response study of *Pinus taeda*. Lo (1985) did not observe groups of four in his study of *Cryptomeria japonica* and *Cunninghamii lanceolata*, but he did note that twin derivatives frequently remained in contact after maceration.

Some of the most convincing illustrations of Sanio's four appeared in the papers by Imagawa (1981) and Imagawa and Ishida (1981a) on *Larix leptolepis*. Most overwintering cambial cells were separated into groups of four with three thin tangential walls between two thicker ones. During spring reactivation, each of the four cells divided (Fig. 4.21A). Four of these cells differentiated to tracheids and the remaining four continued dividing to form groups of seven or eight (Fig. 4.21B). This cell division pattern appeared to be basic to *L. leptolepis* during the active period of growth as well as to *L. kaempferi* (Imagawa 1985b). Each group of four was separated by a thick tangential wall derived from the original mother cell as described by Mahmood (1968, 1971). Occasionally, a cell with an extremely thick tangential wall was observed far removed from the cambium. Such walls presumably belonged to mother cells that had overwintered in the cambial zone.

The foregoing investigations were essentially all conducted on gymnosperms, primarily *Pinus* spp. The presence of Sanio's four has been much more difficult to detect in dicotyledons because the cell files are often distorted due to the presence of large vessels. Giltay (1882), without reference to Sanio's work, described how

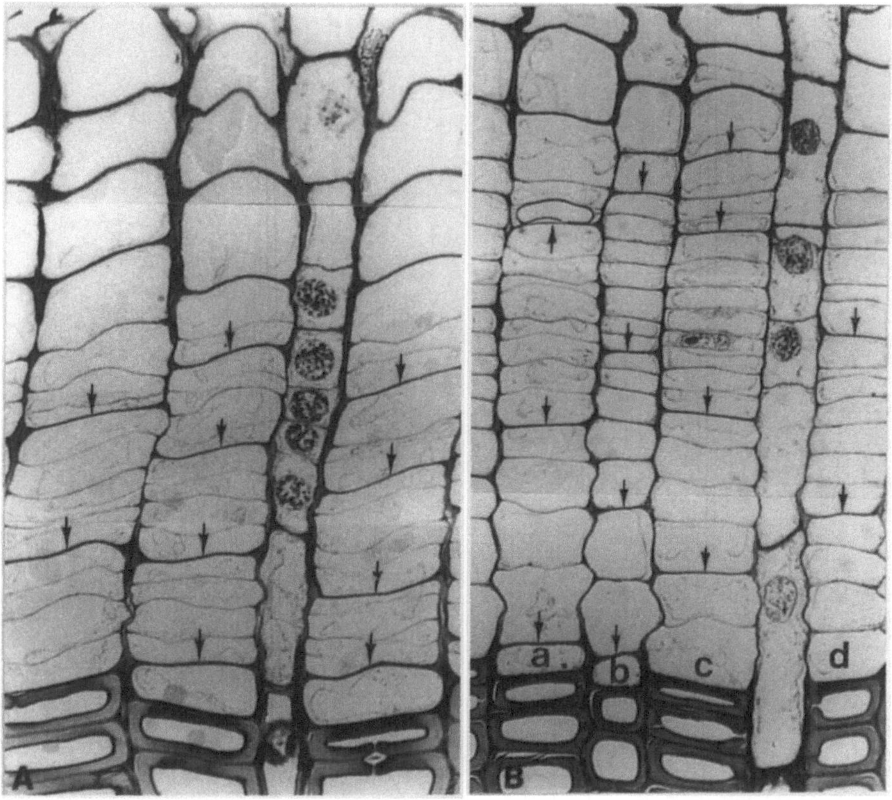

Fig. 4.21A,B. Cambial zone of *Larix leptolepis* shortly after initiation of periclinal divisions in spring (**A**) and 10 days later (**B**). The thick walls (*arrows*) shown in transverse sections are the residual walls of dormant cambial zone cells and the thin walls the results of recent divisions. **A** One cell in file to the *left* has been partitioned in six derivatives. **B** Two cells in files *a* and *c* have been partitioned in eight, and one cell in file *d* in seven derivatives. (Imagawa and Ishida 1981a)

collenchyma cells of *Hedera helix* divided by transverse partitions. When he macerated young tissue with chromic acid he observed that the new cells were enveloped by walls of the older cell, a condition he referred to as "emboîtement," or emboxing. He observed this phenomenon not only in collenchyma but in other tissues as well, and he regarded it as being of common occurrence. Giltay suggested that these "generations of emboxed cells" formed when the nucleus, dividing in the ordinary manner, secreted a homogeneous new wall that was simply disposed against the longitudinal walls which it then united transversally.

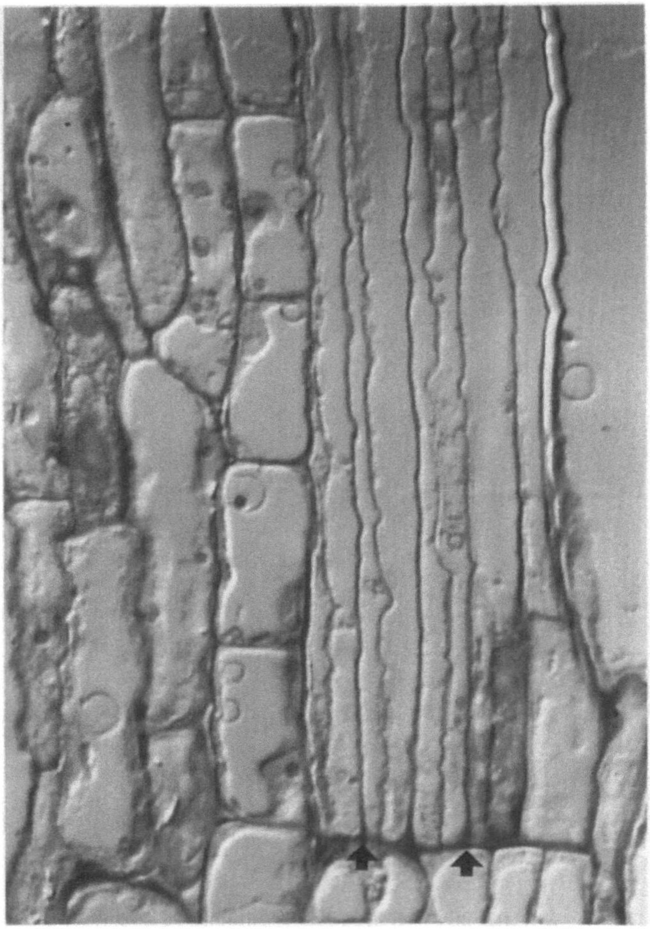

Fig. 4.22. Two groups of four cells in cambial zone of *Populus deltoides*, each presumably derived from two successive divisions of a cambial initial or immediate derivative thereof. Radial section, Nomarski 100x. (Isebrands and Larson 1973)

Murmanis (1977), noting the paucity of information on cambia of dicotyledonous species, conducted an ultrastructural study of *Quercus rubra* branchwood. Like *Fraxinus americana* (Srivastava 1966), the cambial initials remained at three to four cells at all times of the year. Murmanis used the term "cambial initials" to refer to the cells lying between the immature xylem and phloem, and "initial proper" for the one cell possessing the initial function. As in *Pinus strobus* (Murmanis 1970, 1971), groups of four cells could be determined by the relative thickness of their tangential walls. Moreover, the thickest distal tangential wall of the initial proper indicated the direction of its function, whether toward the xylem or the phloem. Derivatives of the cambial initials did not exist in recognizable groups in either xylem or phloem, probably because the cells did not divide in regular succession. The derivatives also underwent extensive intrusive growth, which resulted in the radial rows becoming distorted once they left the initial group.

Isebrands and Larson (1973) published a photomicrograph (Fig. 4.22) showing two tetrads of recently divided cells in the cambial zone of *Populus deltoides*. Although they did not pursue the problem, the cells exhibited many characteristics in common with Sanio's four.

An interesting example of Sanio's four appeared in a paper on interlocked grain in the storied cambia of *Tilia cordata* (Wloch and Zagórska-Marek 1982). Although the authors made no mention of Sanio's four, they clearly identified groups of two and four cells as shown in Fig. 6.68. They found groups of two and four sister cells in which the forked ends were practically identical. In adjacent radial groups in the xylem, presumably arising either from a previous division of the initial cell or a mother cell, the forked ends appeared different.

In Wloch's (1988) Habilitation Dissertation and Wloch (1992 personal communication), these events were discussed as they related to Sanio's four. The forked ends of cells were especially useful in studying the rearrangement of cells in the cambial zone. The cell packets on the xylem side were usually four-celled, in which case the cambial mother cell divided twice. The arrangement of the forked ends of cells within a packet was identical, but it changed progressively in successively formed packets. Wloch concluded that the position of a forked end occurred only in a fusiform initial cell. A given state was then transmitted to the mother cell which, in turn, transmitted it to the four-celled packet of xylem or to the two-celled packet of phloem. Therefore, only the initial cell underwent intrusive growth on the ray side. When the arrangement of cell ends within a packet was similar, then it was impossible to identify the initial cell. When, however, the arrangement of cell ends was dissimilar, then the initial cell was considered the most advanced cell in the packet.

The so-called emboxing of daughter protoplasts, which presumably permitted identification of the cambial initial cells, was questioned by Catesson and Roland (1981) in their study of several deciduous trees (*Acer pseudoplatanus, Populus x euramericana* var. I 214, *Quercus macrocarpa, Sambucus nigra*, and *Tilia platyphyllos*). They never observed a complete primary wall deposited around each daughter cell following the laying down of a new tangential septum. Moreover, they discounted the idea that wall layering was directly related to mitoses. Instead, they found it to be related to the simultaneous deposition and lysis of polysacccharides that occurred locally within the walls. The author's conception of these events is presented diagrammatically in (1) and (2) in Fig. 4.23, and their interpretation of the emboxing concept in (3). Catesson (1984) again addressed this question noting that through the years the emboxing concept had contributed to a "fixed image" of the relation between cambial cells and their derivatives. She reaf-

1
2
3
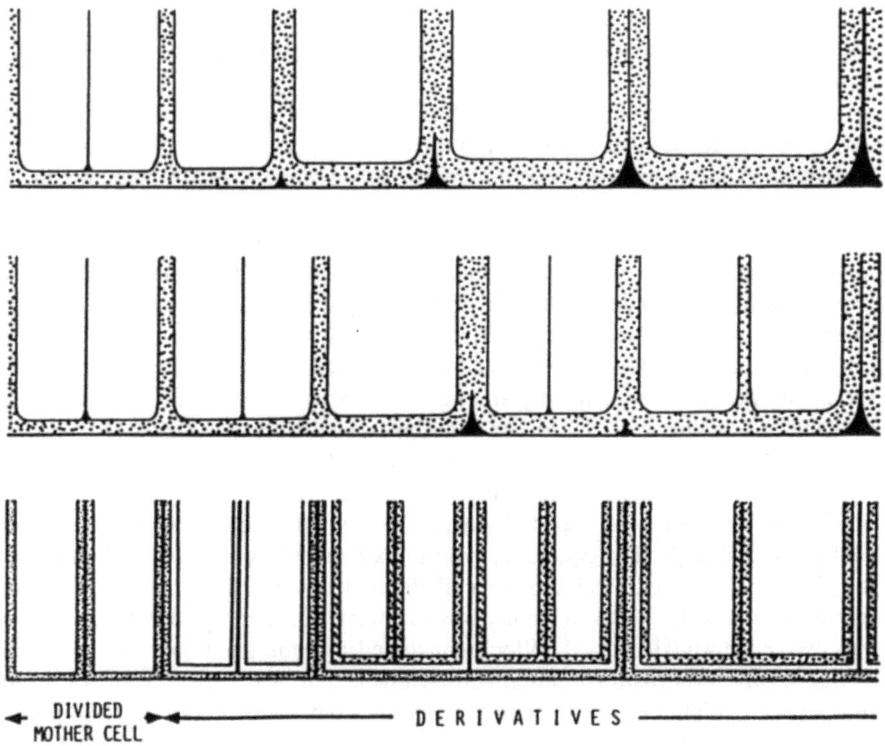

Fig. 4.23, 1-3. Sequential events of wall formation and fusion in cambial derivatives. **1** Theoretical development with no subsequent division. **2** Sequence with divisions occurring in the derivatives. **3** Succession of wall layers following divisions of cambial derivatives as postulated by Sanio's four. (Catesson and Roland 1981)

firmed her view and presented further evidence for "heterogeneity" of the cambial wall, a view that was not in harmony with the phenomenon of emboxing.

4.5.1.2 Phloem

The occurrence of Sanio's four in the phloem has never been mentioned to my knowledge other than in the theoretical scheme in Mahmood (1968). There is no reason to doubt, however, why such groups should not be present on the phloem side just as they are on the xylem side. The difficulties would be in identifying the cell groupings in the phloem because fewer cells are produced, the derivatives are more varied, and they are easily crushed. Although Sanio (1873) made little

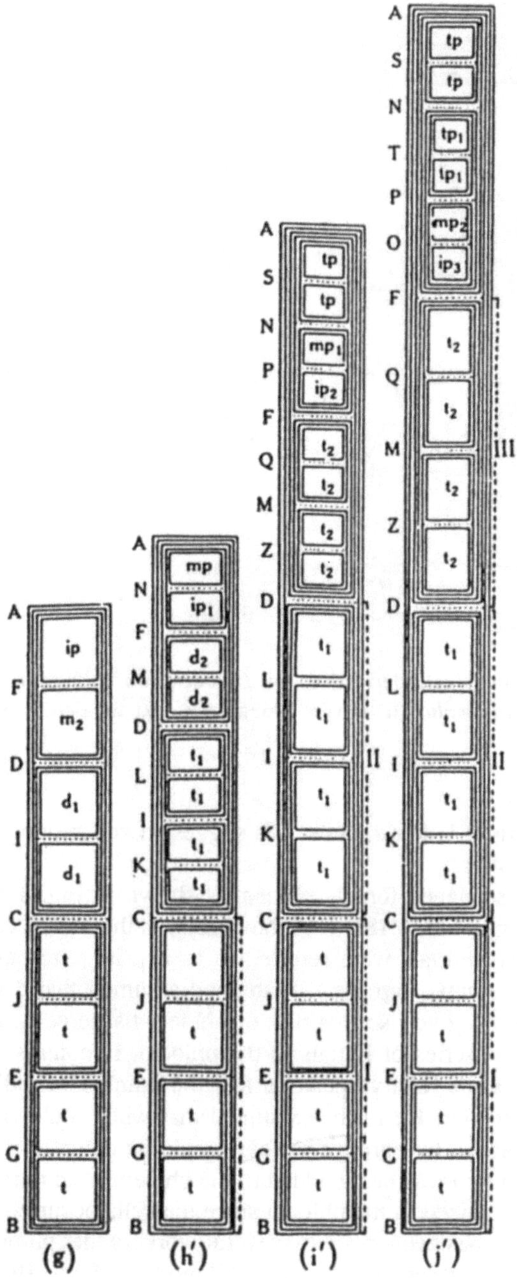

Fig. 4.24. Theoretical interpretation of the changeover from xylem to phloem production. The change began at the stage of column (*g*) in Fig. 4.18 in initial cell (*ip*). Subsequent initial cells (ip_1 - ip_3) produced pairs of phloem cells. (Mahmood 1968)

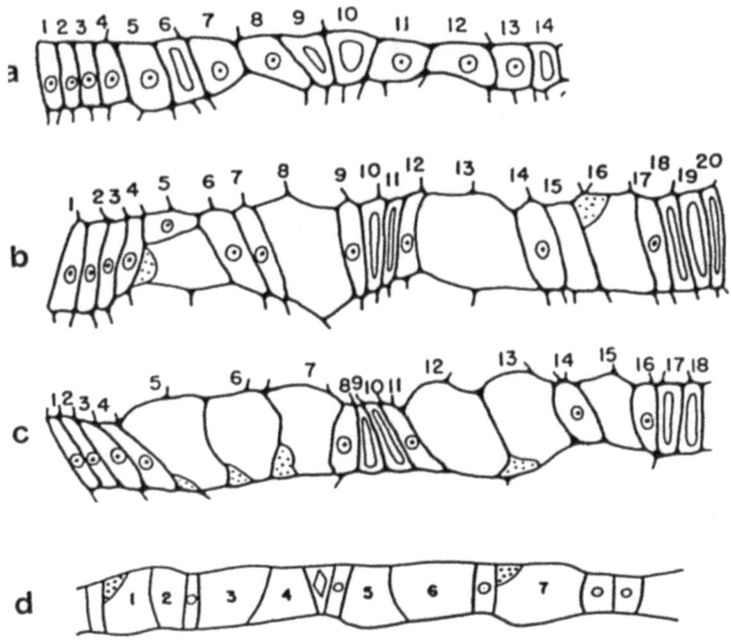

Fig. 4.25a-d. Assemblages of phloem cell derivatives in *Liriodendron tulipifera* (**a-c**)(Cheadle and Esau 1964) and *Pyrus malus* (**d**) (Evert 1963a). See text for details. **a-c** 338x. **d** 393x

mention of it, his illustration (Fig. 4.1. Ser.d, cells 5-8) suggested cell groups of four in the phloem of *Pinus sylvestris*.

Mahmood's (1968) proposed sequence for the phloem is shown in Fig. 4.24. Column g is identical to column g in Fig. 4.18B with the exception that initial cell i_3 has been labeled ip. Successive initials were referred to as ip_1, ip_2, etc., and successive phloem mother cells as mp_1, mp_2, etc. Mahmood assumed that division of a phloem mother cell would give rise to a pair of phloem tissue cells, tp, without further division. Thus, the series of initials in the phloem, in contrast to the xylem, would remain in the same relative position and the number of radial and tangential walls would be smaller. He further assumed that when a change-over from xylem to phloem occurred, two groups of four would be completed in the xylem while the first group of two was being added to the phloem. Accordingly, Sanio's four would not be recognized in the phloem since the cells occurred in pairs (Fig. 4.17B). The suggestion has been made by several workers that phloem derivatives occur in pairs both in dicotyledons (Wloch and Zagórska-Marek 1982) and in conifers (Bannan 1957c, Srivastava 1963a, Murmanis 1970, 1971).

Some indication of cell groupings in the phloem of dicotyledons can be seen in the photomicrographs of Esau and Cheadle (1955) and Cheadle and Esau (1964). The authors recognized "assemblages" of phloem cells derived from common

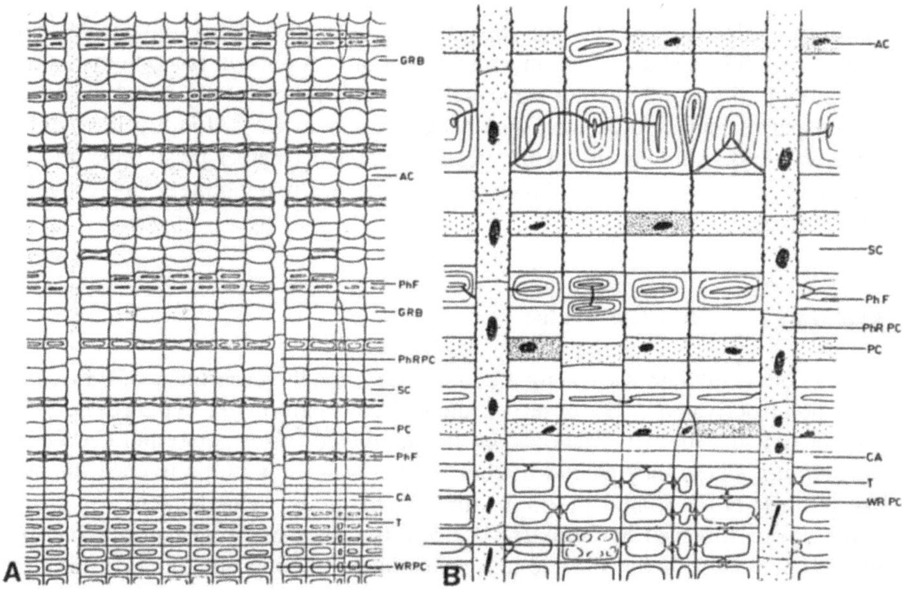

Fig. 4.26A,B. Drawings of transverse sections of the xylem, cambial zone, and secondary phloem of *Chamaecyparis pisifera* (**A**) and *Taxodium distichum* (**B**) showing repetitive sequences of phloem derivatives. *GRB* Growth ring boundary; *AC* albuminous cell; *PhF* phloem fiber; *PhRPC* phloem ray parenchyma cell; *SC* sieve cell; *PC* parenchyma cell; *CA* cambial zone; *T* tracheid; *WRPC* wood ray parenchyma cell. (den Outer 1967)

phloem mother cells. Three radial files representative of the cambial zone and phloem of *Liriodendron tulipifera* are shown in Fig. 4.25a-c. Although the presumptive groupings of cell derivatives from phloem mother cells by Cheadle and Esau (1964) did not necessarily conform to Sanio's four, one might perceive a pattern in these radial files.

A similar radial file from *Pyrus malus* is shown in Fig. 4.25d . According to Evert's (1963a) analyses, sieve elements 1 and 2, their companion cells and the parenchyma strand on the right comprise an assemblage derived from one phloem mother cell. Other assemblages, derived from subsequent mother cells, consist of sieve elements 3 and 4 and associated companion cells; the parenchyma strands between sieve elements 4 and 5; sieve elements 5 and 6, their companion cells and the parenchyma strand to the right; sieve element 7 and its companion cell; and the final two parenchyma cells to the right. An asymmetric division of a sieve element giving rise to a companion cell is shown in Fig. 11.15A.

The standard sequence for phloem cell production in *Cryptomeria japonica* was found by Itoh et al. (1968) to be a sieve cell, parenchyma cell, sieve cell, and fiber. This regular sequence was repeated throughout the growing season. This same pattern was reported by Lo (1985) in *C. japonica* and also in *Cunninghamii lanceolata*. In other conifers, such as *Chamaecyparis pisifera* and *Taxodium distichum*, the phloem elements also appear to be ordered in fairly repeatable units (Fig. 4.26). For example, each unit might consist of a phloem parenchyma, sieve cell, phloem fiber, and another sieve cell. This phloem structure, found in certain gymnosperms and dicotyledons, has been referred to as a "four-cycle rhythm" (Ziegler 1964). Although it has not been determined whether these units are derived from common phloem mother cells, the pattern is temptingly suggestive of Sanio's four.

4.5.2 Bars of Sanio - Trabeculae

Sanio is often credited with being the first to describe the occurrence of trabeculae in cambial derivatives. He observed these "stäbchenförmige, cylindrische Körper," or rod-shaped, cylindrical bodies, in *Hippophaë rhamnoides* and *Casuarina equisetifolia* where they extended from wall to wall through the tracheid lumens like a cross-beam (Sanio (1863a). Similar structures, "Querbalken," or cross-beams, were observed by Winkler (1872) in tracheids of *Araucaria brasiliensis*. Sanio (1873) later included these rod- or bar-shaped bodies among his criteria for identifying cambial initial cells of *Pinus sylvestris*. These cellulose bodies projected into the cavities of one cell after another in radial series, and because they extended into the bast as well as the wood, Sanio assumed they could only arise from a "cambial mother cell," i.e., a cambial initial cell (Fig. 4.27).

Russow (1882c) confirmed the existence of these "Cellulose stäbe" in *Abies pichta* and *Pinus sylvestris*, where he claimed to have observed them extending consecutively through the cambium in both bast and wood. On the bast side, they could be followed through three growth rings and on the wood side they continued through at least ten growth rings. The presence of these bars and their radial extent was also verified by De Bary (1877) in the wood of *Drimys winteri*; he classified them as "trachea with transverse bars."

A particularly good review of the occurrence of these structures was presented by Müller (1890) who referred to them as "Sanio'sche Balken;" this term was later translated to English as "Sanio's bars" or more commonly as "bars of Sanio." Müller credited De Bary (1877) with being the first to refer to the cells in which these structures occurred as "Trachea trabeculatae;" De Bary used the term "Trachea" as an all-inclusive term for Sanio's (1863a) "Trachea" (vessel) and "Tracheide" (tracheid or fiber). Unlike De Bary, who considered the "Balken" to be anomalous phenomena, Müller believed them to be a "decisive character for all conifers." Although he examined only 28 species, the conifers he examined represented all known families.

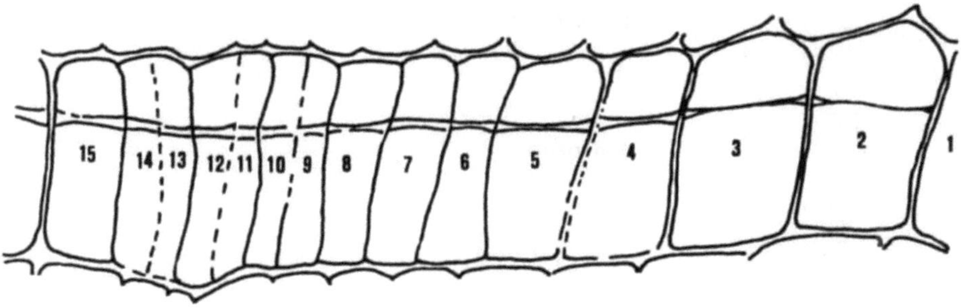

Fig. 4.27. "Querbalken," or trabeculae, extending through the cambial zone of *Pinus sylvestris*. Bast cells *9-14*; wood cells *8-1*. Transverse section 630x. (After Sanio 1873)

The occurrence of Balken structures in conifers was confined neither to particular regions of stems nor to specific stem tissues. For example, Müller observed them in the first growth ring of *Pinus strobus* and the 98th growth ring of *Pinus sylvestris*, in old root-wood or *Larix europaea*, and even in strongly hyponastic branches of *Ginkgo biloba*. He concluded that Sanio's Balken occurred in all regions and at all heights in all axial organs. They were, therefore, characteristic of all coniferous woods. Moreover, they were not a characteristic exclusive to tracheids. They were also found in "neutral" cambial cells and in phloem cells, including the sieve cells of conifers.

Müller agreed with Sanio and Russow that Balken formation began in the cambium and extended outward in the rind as well as inward in the wood. Accordingly, each segment in the wood body must possess a "conjugirt" in the phloem. He regarded the two "conjugirte" in the xylem and phloem as being equivalent topographically, but because the sequence extended across the cambium it had no beginning. However, the uninterrupted cell series exhibited a gap which always became greater with continued cambial activity. This gap represented a discontinuity, but just where the discontinuity began was impossible to determine. Presumably, Müller was referring here to the fact that because he could not precisely determine which cell within the cambial region might be the initial cell he had no way of knowing where the series began.

According to Müller, the 10-year sequence reported by Russow (1882c) was the longest, continuous series of cells with Balken formation on record. His own observations never exceeded 100 to 150 cells, but even these were difficult to follow. In some cases, as in *Araucaria*, a Balken might be appressed so close to the wall that it "appeared to grow with the radial wall." Yet when followed closely, it was "emancipated," and the Balken became clearly distinguishable in subsequent cells. Müller also described how each Balken appeared to widen or

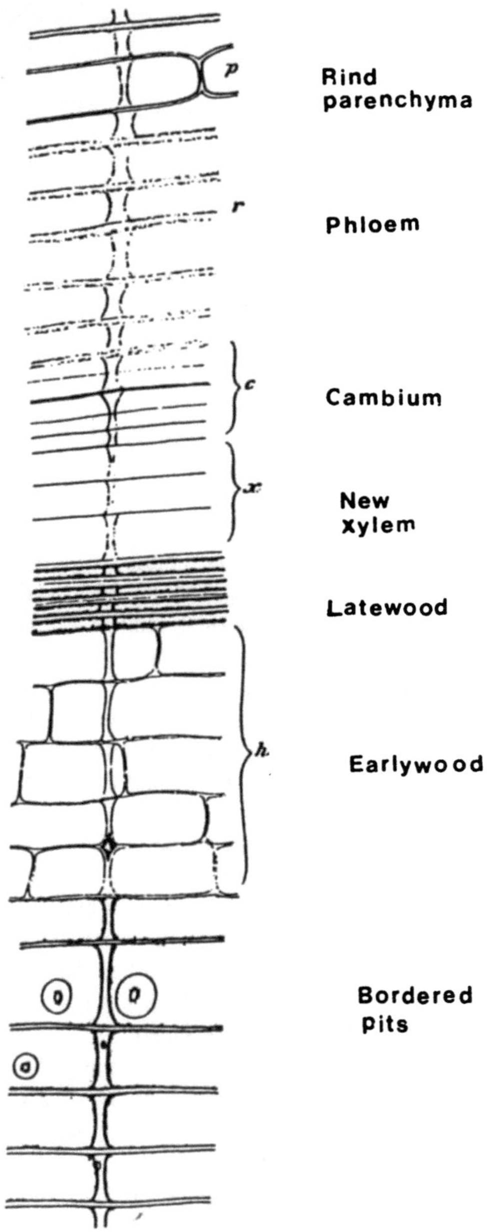

Fig. 4.28. Radial section of *Pinus sylvestris* illustrating how the "Kurzstäbes," short rods or trabeculae, would appear in different tissues. Continuity of trabeculae through adjoining tissues is strictly for purposes of illustration. 260x. (Raatz 1892)

flare where it contacted the tangential wall, and how individual Balken varied in size, shape, and distribution from cell to cell in radial series.

The occurrence of Balken formations in cambial cells was thought to be a mechanical response. The young cells would at first be under radial tension and later under compression. As to origin, Müller suggested three possibilities. The first two, that the Balken might result either from a partial resorption of the cell wall or from a secretion product of a "plasmapfropfen," or plasma plug, he rejected. The third possibility involved a detailed hypothesis by which the radial wall of a cambial cell was subjected to an advancing fold. This fold became stronger with repeated tangential divisions, and it created an obstacle to the cellular plasma stream. Termination of Balken formation in a radial cell series occurred when the Balken site in a cambial mother cell was totally resorbed.

The "Stäbe" described by Sanio (1863a, 1873) and Russow (1882c) were considered Langstäben, or long bars because they traversed many cells extending across the cambium. In addition, Raatz (1892) also observed Kurzstäbe, or short bars, that traversed only one to several cells in *Picea abies* (Sect. 4.1.3). Unlike the Langstäbe, the Kurzstäbe need not originate in a cambial initial cell, but they might arise in any dividing cell within the cambial zone. The Kurzstäbe were similar to the Langstäbe in structure except that they were broader in cross-section and extended through only one to several cells.

In agreement with Müller (1890), Raatz detected Stäbe formation in a wide variety of conifers as well as a few deciduous species, e.g., *Casuarina equisetifolia*, *Hippophaë rhamnoides*, and *Salix fragilis*. Also like Müller, he observed these structures in all parts of the tree including the secondary wood and rind of stems, branches, and roots. The segment extending through earlywood was thin and thread-like, whereas that extending through latewood was thick with broad ends where it contacted the tangential wall (Fig. 4.28). The Stäbe described and illustrated by Raatz occurred in a wide variety of lengths and configurations, some of which he referred to as "monstrosities." In tangential sections, they appeared either round, elliptical, or linear. Their structure was similar to that of the cell wall, consisting of a middle lamella and a secondary thickening lamella which was thicker in latewood than in earlywood cells. Sometimes two or more Stäbe formed within a single cell and at other times bordered pits appeared on the Stäbe walls. Although the basic structures of the Stäbe were similar among the species examined, notable differences were found in their prevalence and external forms. Raatz considered them all to be "abnormalities" despite their widespread occurrence among woody plants.

Causally, Raatz attempted to relate Stäbe formation to rudimentary pits, but he did not pursue the idea. He also entertained the three suggestions advanced by Müller, and immediately rejected the first two. After a discussion of the pros and cons, he also rejected Müller's cell wall folding hypothesis, favoring instead a "contact theorie." According to this theory, Stäbe arose from growth tensions resulting from cellulose accumulations at the point of contact between two tangential walls of a cambial mother cell.

Nordhausen (1898) confirmed the presence of both Langstäbe and Kurzstäbe in many woody plants such as *Juglans, Populus, Salix,* and *Tilia,* but they were more prevalent in 1-year-old twigs of *Sambucus*. He found Langstäbe, arising from a single cambial mother cell, extending through as many as 25 cells. They were difficult to follow in dicotyledonous woods because of irregularities in the radial rows of cells. As to origin, Nordhausen suggested that newly dividing cells were under reduced turgor pressure, somewhat similar to the process described by Müller (1890). The tangential walls were therefore in contact and somehow grew together. With increased turgor and cell expansion, the walls spread apart and the region of contact formed a Stäb. These Stäbe consisted of cellulose strands that not only grew in length but in thickness. Because they originated in a cambial mother cell, they were transmitted by repeated divisons to a large number of cells in radial series, including ray cells. Nordhausen, and Priestley (1930b) at a later date, used the presence of Langstäbe in arguments against Krabbe's (1886) theory of "sliding growth," noting that these structures could not exist if sliding growth occurred. However, Grossenbacher (1914) and others pointed out that this line of reasoning was erroneous because sliding growth was confined to the cell tips (Chap. 6.7).

In the ensuing years, numerous investigations of trabeculae appeared in the literature. Most of these dealt with the occurrence and causality of trabeculae formation and less so with their development from cambial cells. One of the more interesting studies on this subject was conducted by Müller-Stoll (1965). He set out to determine the possible influence of climatic factors on the incidence of trabeculae formation by sampling seven coniferous and four deciduous species from three different climatic zones. His results revealed a rich variety of trabeculae ranging from individual cells to extensive series and many abnormalities. In addition to observing trabeculae in prosenchymatous tissues, he also found them to be present in ground parenchyma and epidermal cells. He tentatively attributed formation of trabeculae to cold effects on the cambium, most probably late spring and early fall frosts. More importantly, however, he confirmed Sanio's (1873) conclusion that these structures originated in the cambial initials. Once formed, they were conserved in the initials for a long time, and in this way long rows of bars could extend over several growth rings.

The anatomical aspects of trabeculae formation were examined by Keith (1971). He noted that knowledge of trabeculae had advanced little since the publications of Hale (1923) and Prince (1934), although he overlooked the paper by Müller-Stoll (1965) with its extensive review of the literature. Keith (1971) accepted the fact that trabeculae originated in the cambium, but questioned Hale's theory of fungal origin. In his examination of several conifers (*Picea glauca, P. sitchensis, Pinus banksiana, P. strobus*), Keith observed that rather than being hollow, trabeculae consisted of a shell of cell wall substance enclosing a central core of middle lamella. Further verification for this structure was provided by the observation that the abrupt increase in width at the ends of individual trabeculae was not funnel-like but consisted of two or more root-like projections confluent

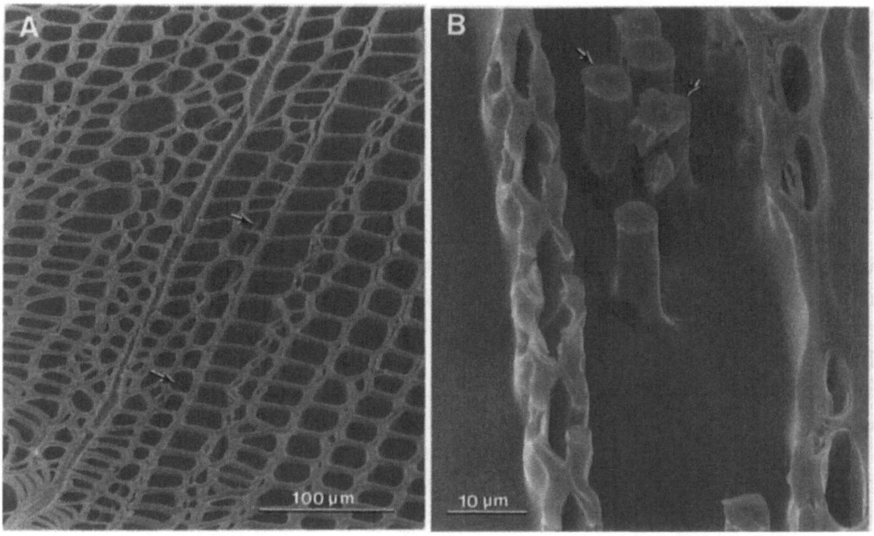

Fig. 4.29A,B. Trabeculae in *Picea sitchensis* as revealed by SEM. **A** Transverse view showing radial traverse of trabeculae (*arrows*). **B** Tangential face showing the severed ends of crowded trabeculae in a tracheid. *Arrows* indicate anastomosed trabeculae. (Ohtani et al. 1987)

with the cell wall. Consequently, the compositions of trabeculae were essentially the same as the cell walls of which they were a part, confirming the earlier, but uncited, view of Raatz (1892).

According to Butterfield and Meylan (1972), most textbooks and recent publications conveyed the impression that trabeculae were confined to conifers. Apparently they were unaware of earlier work on dicotyledons by Sanio (1863a), Raatz (1892), and Nordhausen (1898), or the extensive research by Müller-Stoll (1965). Nonetheless, they did describe trabeculae in *Knightia excelsa* and *Fuchsia excorticata* (Meylan and Butterfield 1973). In the latter species, one trabecula was found to contain a vestured pit; Raatz (1892) many years earlier had reported bordered pits on trabeculae in *Pinus sylvestris*. Other recent reports of trabeculae in dicotyledonous woods were made by Ohtani (1977) and Werker and Baas (1981).

The fact that the formation and occurrence of trabeculae did indeed incite considerable interest over the years was evidenced by the extensive literature review of Yumoto (1984). This author's own research on *Picea glauca* is the most comprehensive and richly illustrated to date on the subject. He described five basic types and two varieties which he referred to as "trabeculae and their homologous structures." These structures were found in all gradations in a wide variety of intermediate forms. Because trabeculae often occurred close together in a common row, they were believed to be intimately related ontogenetically. Although most of these structures had been documented by earlier workers, e.g., Raatz'(1892) "monstrosities," they had never been systematically described and categorized.

The notes by Ohtani (1985) and Ohtani et al. (1987) contained exceptionally clear SEM photomicrographs of abnormal trabeculae in *Abies sachalinensis* and in indented growth rings of *Picea sitchensis*, respectively (Fig. 4.29). The paper by Kucera (1978) confirmed their presence in pith parenchyma in the vicinity of vascular nodules in *Taxus baccata*. Patel (1974) reported trabeculae in tracheids and axial parenchyma of the vesselless *Pseudowintera*.

Yumoto (1984) noted that illustrations in standard anatomical textbooks usually showed long rows of trabeculae extending through many cells as first described by early investigators. Since that time, however, most workers reported trabeculae as being present either in short rows or solitary, with the notable exceptions of Müller-Stoll (1965), Grosser et al. (1985), and Süss (1989). In his work on *Picea glauca*, Yumoto observed trabeculae in both normal and compression wood where they were particularly abundant as either twins or solitary, although he occasionally found them in long rows. He could find no previous reports of trabeculae in compression wood cells, apparently missing Müller's (1890) observation of "Balken" in both the upper and lower sides of hyponastic branches of *Ginkgo biloba*. No differences were detected in the structure or occurrence of trabeculae in the two types of wood. Like others before him, Yumoto concluded that the walls of trabeculae consisted of the same layers as those of the host cell.

According to Yumoto, there was general agreement in the literature that trabeculae and their homologous structures originated in the cambium, although few developmental studies had been conducted to verify the fact. In his own work, if a trabecula primordium originated in a cambial initial, then a long row of trabeculae would form, extending from the cambium towards both xylem and phloem until either the initial cell matured or the trabecula primordium for some reason or another was eliminated. However, if the trabecula primordium originated in a xylem mother cell, then either a solitary or a short row of trabecula(e) would form only in the xylem. As to causality, the many abnormal occurrences of trabeculae suggested a genetical tendency of a particular cambial initial to form the trabecula primordium. Previous authors had advanced eight different hypotheses to account for such an initial impetus. Circumstantial evidence from these studies strongly indicated certain causal relations between abnormal cell division and the formation of trabeculae. That is, although certain conditions within the cells undoubtedly initiated trabeculae primordia, these conditions could be induced by a variety of external factors. Süss (1989) also attributed the induction of new trabeculae to some external factor. As evidence, he noted that 20% of the new trabeculae occurred in the first-formed cells of the growth ring following their induction in either cambial fusiform initials or xylem mother cells.

One of the most commonly accepted of these hypotheses is frost damage. Grosser et al. (1985) examined many diseased *Picea abies* and *Pinus sylvestris*

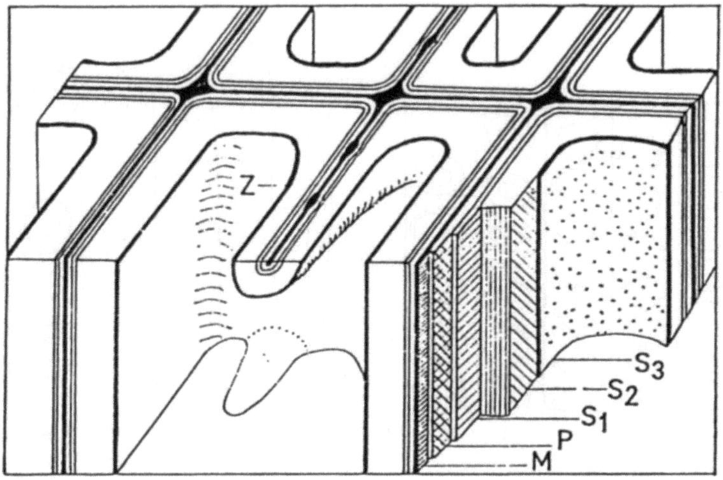

Fig. 4.30. Schematic drawing of a trabecula in *Juniperus communis* spp. *nana*. *M* Middle lamella; *P* primary wall; S_1, S_2, and S_3, secondary wall layers. (Müller-Stoll and Casperson 1988)

trees and found that the only consistent anatomical change was the pronounced increase in the occurrence of trabeculae. Whereas healthy trees had relatively few trabeculae, as many as 100 to 157/cm^2 were recorded on radial microsections of affected trees. In *P. abies* these trabeculae originated in cambial initials and often extended through many radial files of tracheids. The stimulus for their formation was obviously external. As to causality, the authors suggested that diseased trees experienced reduced resistance to frost damage.

In a later, more detailed, study of *Picea abies*, supplemented by other data, Grosser (1986) re-evaluated the frost-induction hypothesis of trabeculae formation. In the wood of 13 affected trees he found exceptionally high frequencies ranging from 1870 to 17847 trabeculae/cm^3 of xylem. Results of this study showed that trabeculae and related structures were characteristic of irregularly formed and arranged cells such as those found in wound tissue, compression wood, and "hasel growth" (Chap. 8.4.1.4). The occurrence of trabeculae in normal wood was attributed to irregularities in the normal periclinal and anticlinal divisions of fusiform initials and differentiation of their derivatives. Trabeculae that occurred in long files were induced in cambial initials. Solitary and short files were induced either in cambial initials that had lost the ability to form trabeculae after one to several divisions or in xylem mother cells. Grosser concluded that trabeculae development was related to partial collapse of cambial cells caused by irregularities in cell division. He further suggested that the structures might function as bracing elements between tangential walls to prevent collapse during local compression or tensile stress. The frost-induction hypothesis was rejected because

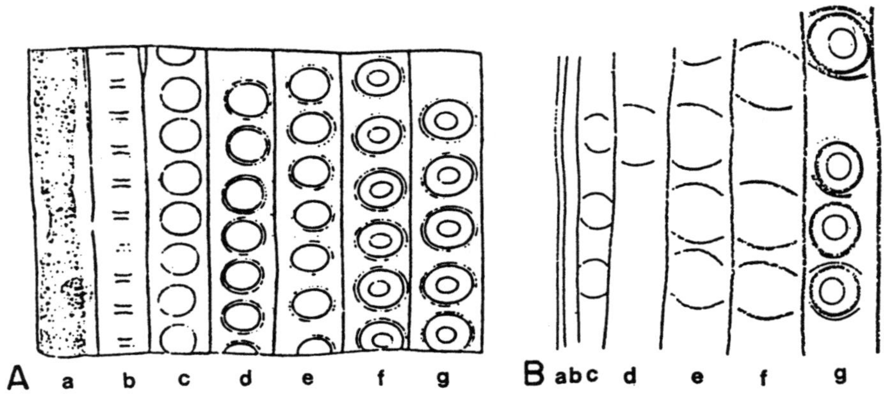

Fig. 4.31A,B. Development of bordered pits and the "double transverse lines," or crassulae, in *Pinus sylvestris*. **A** 400x. (Unger 1847). **B** 650x. (Sanio 1873)

trabeculae might be induced by any type of environmental stress or irregularity in cambial division.

Müller-Stoll and Casperson (1988) re-examined the origin and structure of trabeculae in *Larix decidua* and *Juniperus siberica*. They again found that the predisposition for trabecula formation was induced in a cambial cell and this induction then acted through a long series of cells. By employing electron microscopy, they were able to demonstrate that trabeculae were established during very early stages of cell wall formation. At this time, a central core arose from protoplasm of the cambial cell that differed in structure from the middle lamella of the host cell. On this central core was deposited a very thin primary wall, a three-layered secondary wall with a strongly thickened S_2, and an S_3 layer corresponding to a "tertiary lamella" (Fig 4.30). Thus, other than the central core, trabeculae wall structures were similar to those of normal cells. In a radial cell series, only the central core extended horizontally through the tangential walls from one cell to another. As to causality, Yumoto (1984) had listed eight hypotheses gleaned from the literature, all of which could be explained by disturbances of the cell division process. Müller-Stoll and Casperson agreed with this assessment.

4.5.3 Rims of Sanio - Crassulae

Although it is not my intent to delve into the subject of bordered pit formation, a brief survey of the early literature is necessary to clarify a problem of nomenclature. Buried in one of his papers, Sanio (1873) included a detailed account of early investigations concerning bordered pits and their formation. Most of these reports dealt with relatively late stages of xylem cell development. However, one paper by Unger (1847) described pit formation in *Pinus sylvestris* beginning with the cambial initial. As shown in Figure 4.31A, the wall of cell a appeared homogeneous. On the wall of cell b there appeared a series of "double transverse lines" indicating the beginning of pit border formation. In cell c, the pit border assumed a simple round or oval contour, and in cell d it assumed a "double contour." The pit aperature first became apparent in cell f.

Sanio (1873), while acknowledging Unger's contribution, noted that Unger was unsure of the significance of the "double transverse lines." In his own investigation of *Pinus sylvestris*, he removed all traces of protoplasm from the cells in radial sections by heating them in acetic acid. The walls of cambial and enlarging cells, revealed by this procedure, appeared smooth and of uniform thickness. During cell enlargement, localized areas of the cell wall became thinner (Fig. 4.31B), and whereas the thin areas merged gradually with the thicker wall horizontally (c,d), the thickness of the upper and lower borders became more accentuated as "doppelten Umrissen," or double rims (e,f). As visualized by Sanio, these thin spots on the radial walls of young wood cells were not themselves the beginnings of bordered pits, but simply the locations at which the pits as independent formations would arise. A "Primordialtüpfel," or primordial bordered pit, formed when the pit border surrounded the thin area and both "Umrisse" were clearly visible (g). Occasionally, as for example in earlywood cells of old stems, he observed two pits laterally adjacent in neighboring cells. These pits were united by "Querleisten," or transverse bridges, formed by a common "Umriss," or rim, that extended horizontally above and below two primordial bordered pits. As proof of this formative process, he demonstrated that not only were both Umriss and Querleisten present in mature tracheids but they appeared identical to those in cambial zone cells.

Russow (1881) questioned Sanio's assertion that cambial cells were entirely free of pit formations, attributing the lack of evidence to Sanio's clearing technique (Chap. 6.7.4.1). When radial sections were treated with chlorozinc iodide, shallow indentations appeared on the cell walls which later gave rise to bordered pits on the wood side and sieve pores on the bast side. It is possible that Sanio and Russow were viewing the cambium in different ways, Sanio as a single mother cell and Russow as a zone of dividing cells. Statements to this effect appear in several later papers (Russow 1883a,b). Russow also observed the bow-shaped borders above and below the primordial pit locations, but he referred to them as "weakly double-contoured, not strong, as Sanio presented." This statement led to a spirited controversy. Sanio (1882) first reaffirmed his view that cambial mother cell walls were smooth and showed no evidence of future pitting. As to the upper and lower primordial pit borders, he maintained they were "not strongly, but sharply separated," and attributed the discrepancy to the lithography by asking "how else should the lithographs be made"?

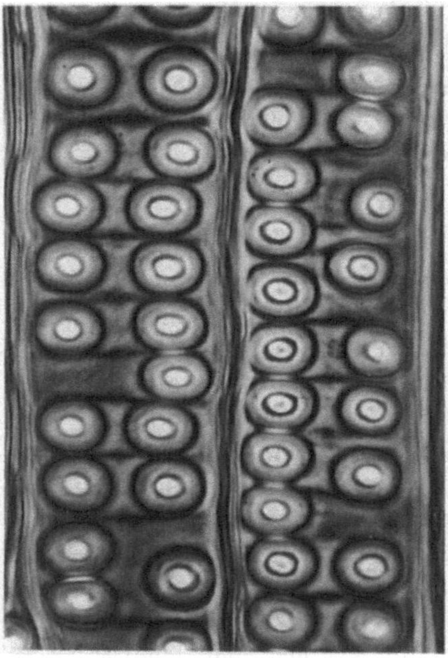

Fig. 4.32. Crassulae present above and below the radial intertracheal pitting in *Larix occidentalis*. 560x. (Core et al. 1976)

In his rebuttal, Russow (1882a,b,c) firmly stated that in all his preparations the cambial cell wall exhibited evidence of pit formation and Sanio's claims to the contrary were "groundless." This exchange, incidentally, included detailed descriptions by both authors of pit origin and development in cambial zone cells. Russow did not deny the existence of the upper and lower borders of primordial pits, but chastened Sanio for not shading the lines more heavily in his lithograph. It is of interest that throughout this exchange, both Sanio and Russow referred to these incipient pit borders as "Grenze" and not "Umrisse" as in Sanio's (1873) earlier paper.

In later papers by English-speaking authors, these structures, which were actually crassulae, were referred to as "Sanio's rims" (Gerry 1910, Groom and Rushton 1913, Grossenbacher 1914, Sifton 1915, 1922, Bailey 1919a). They appear as dark bars above and below intertracheid pitting on the radial walls of tracheids, caused by optical density of the cell wall material as illustrated by Core et al. (1976) (Fig. 4.32). For more information on pit formation in cambial cells, see Chap. 6.7.4.2.

4.5.4 Bars vs. Rims of Sanio

The term "bars of Sanio" has been confusingly applied to both trabeculae and crassulae. In the American literature, in particular, the term has been erroneously applied to crassulae. Sanio (1863b, 1873) called these structures "stäbchenförmige Körper," rod or bar-shaped bodies, as did Raatz (1892), Nordhausen (1898), and others. Müller (1890) referred to them as "Sanio'sche Balken," and it was this term that was later translated to Sanio's bars or bars of Sanio. The terms "Stäbe" and "Balken" both obviously refer to trabeculae (Müller-Stoll 1965).

One of the first references to bars of Sanio in the English literature was by Gerry (1910), but it is not clear from her paper whether the term was her translation or derived elsewhere. Nonetheless, she used it with reference to Sanio's (1873) "scheibenförmige Verdickung der Scheidewand," corresponding to the rims of cellulose above and below primordial pit locations, thus to crassulae. Most subsequent English-speaking authors adopted this interpretation.

Groom and Rushton (1913) pointed out Gerry's error and corrected it noting that the term "bars of Sanio" should refer to trabeculae and the term "Sanio's rims" to crassulae. Unfortunately, however, their advice was not heeded even by authors who later cited their paper. Bailey (1919a) acknowledged this unfortunate discrepancy and noted, quite rightly, that it would lead to considerable confusion. In his own work, he simply referred to crassulae in Sanio's original term as "Querleisten." Priestley (1930b) also called attention to this discrepancy as did Yumoto (1984) in his review. To avoid confusion, Eames and MacDaniels (1947) suggested that the terms "bars of Sanio" and "rims of Sanio" be dropped and replaced with the anatomically correct terms "trabeculae" and "crassulae," respectively. Despite these many suggestions over the years, Gerry's (1910) paper continued to serve as a primary reference and the terms "bars of Sanio" and "rims of Sanio" both became associated with crassulae (Sifton 1915, 1922, Jeffrey 1917, Hale 1923, Greguss 1955, Panshin and de Zeeuw 1980, Carlquist 1988a); Grossenbacher (1914) used "bars of Sanio" correctly with reference to trabeculae.

In botanical dictionaries, Jackson (1928) correctly defined "Sanio's bars" and "Sanio's rims" as being synonymous with trabeculae and crassulae, respectively. Swartz (1971) differentiated between "Sanio's trabeculae" and "bar of Sanio," the latter referring to crassula. Little and Jones (1980) defined "bars of Sanio" as being synonymous with crassulae.

4.5.5 Monogene vs. Polygene Cambia

Schoute (1902) had defined "Etagencambium", literally layered or storied cambium, as one in which several primary cell layers participated in cambial structure but no continuous radial arrangement existed (Fig. 4.33). As noted in Chapter 6.2.1, Beijer (1927) objected to this terminology. To replace Schoute's term Etagencambium, Beijer suggested the term "polygen." Thus, a polygen cambium or tissue would be derived from different primary cells with the result that cells in adjacent radial files would not be sister cells. A "monogen" cambium, in contrast, would be a secondary tissue in which cells in adjacent radial files were sister cells, i.e., derived from anticlinal divisions. Monogen cambia were found in all conifers and most dicotyledons. Rothert and Jost (1934) interpreted polygen to mean a radial cell series in which cells neither were in strict radial alignment nor did they possess an easily identifiable initial cell. Sárkány (1936) and Kaussmann (1963) used the term polygen cambium in slightly different ways. Kaussmann considered a monogen cambium to be one with a single fusiform initial cell, i.e., a uniseriate cambium.

Fig. 4.33. Polygene interfascicular cambium in *Vitis labrusca* showing lack of tangential alignment of division planes. 164x. (Schoute 1902)

4.5.6 Cambial vs. Apical Initials

Over the years, attempts were made to draw parallels between the division patterns of the cambium and those of the apex. In fact, the essential thesis of Priestley's (1930a,b) work was to demonstrate "... that the cambium cell is potentially equivalent to a cell of the apical meristem, and that its different form and behaviour are the result of the different conditions under which it is placed in the tissues." One of the main conditions that affected both the form and arrangement of the cambial cells, according to Priestley, was the pressure exerted on these cells by the expanding cortex.

Bannan (1955) also related cellular activity in the cambial region to that of the stem tip, noting that in the latter a small number of initials produced mother cells which in turn gave rise to tissues of the stem. Newman (1956, 1961, 1965), in keeping with his concept of "a continuing meristematic residue," also drew attention to similarities in the division patterns of apical cells and those of the cambium. Furthermore, to maintain parallelism, he wondered whether cambial initial cells should be called cambial cells since the initial cells of the apex were often called apical cells. A similar argument had been used earlier by Raatz (1892), who noted that if one accepted the initial cell concept of Sanio (1873) for the cambium, then one should also accept the "Scheitelzelle," or apical cell, concept for the shoot apex. Years later, Steeves and Sussex (1989) again noted the parallelism in the use of the terms cambium, which often includes the cambial initials and their derivatives, and apical meristem, which often includes the apical initials and their derivatives.

The fact that the cambium shares many characteristics in common with the apical meristem was also commented on by Evert (1963b) in his study of *Pyrus malus*. He noted that in both the apical meristem and the cambium, which functions like a meristem, it is extremely difficult to delimit the initials from their recent derivatives, and in both meristems the initials are continually shifting positions and being displaced. Wodzicki and Brown (1973), while accepting the fact that the cambium consists of meristematic cells, nonetheless cautioned against indiscriminate use of the term lateral meristem when referring to the cambium. They pointed out that the lateral meristems of woody plants are composed of cells other than cambial initials, just as the apical meristems of shoots and roots consist of meristematic cells other than apical initials. This line of reasoning was later elaborated on by Schmid (1976).

As a part of her intensive study of *Acer pseudoplatanus*, Catesson (1964) presented a comparison of the seasonal transformations of cambial cells with those of the shoot apex, primarily of the "anneau initial" region. The data, consisting of many ultrastructural observations and cytochemical determinations, were assembled for each month of the year in tabular form.

A slightly different approach to this question was made by Olesen (1982) in his study of cyclophysis, which he defined as the "... process of maturation of the apical meristem." He suggested that the vascular cambium is subject to two types of maturation processes, namely, those transmitted to the cambium at the time of formation from the apical meristem, and those which the cambium undergoes after its formation. Consequently, when evaluating age changes in the cambial initials and their derivatives, as in Chapter 7, it is necessary to distinguish clearly between these two processes.

Chapter 5

Cambial Cell Characteristics

5.1 Cell Shape

The shapes of cambial cells have been a matter of concern from the time they were first recognized. Braun (1854) described cambial cells simply as elongated elements with sloping walls and pointed ends. Nägeli (1858) described them similarly, but added that the pointed ends had a tendency to grow in the direction of the axis. Radlkofer (1858) considered cambial cells to be long rectangles approaching the form of a "whetstone" with oblique end walls. Most dissension among workers concerned configuration of the end wall. Müller (1875-76), for example, argued that cambial cells took the form of a pointed fiber, a view contested by Velten (1875), who claimed that Müller's cambial cells were actually differentiating fibers. Velten suggested that the cambial cells assumed the form of a

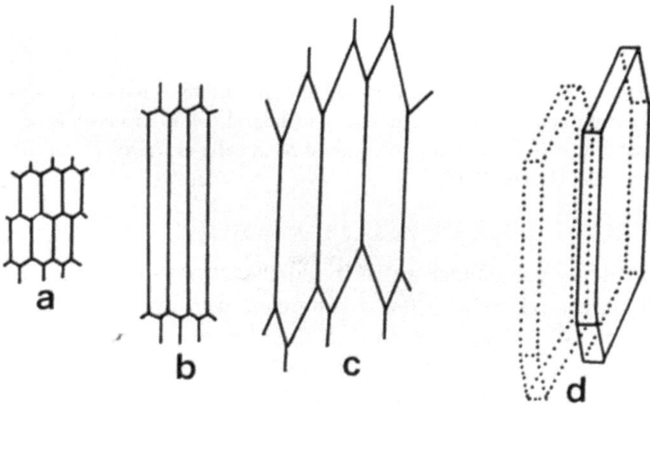

Fig. 5.1a-d. Cambial cells in *Raphanus sativus* root conforming to the shape of a hexagonal prism. **a** Transverse. **b** Radial. **c** Tangential. **d** Three-dimensional. (After Kleinmann 1923)

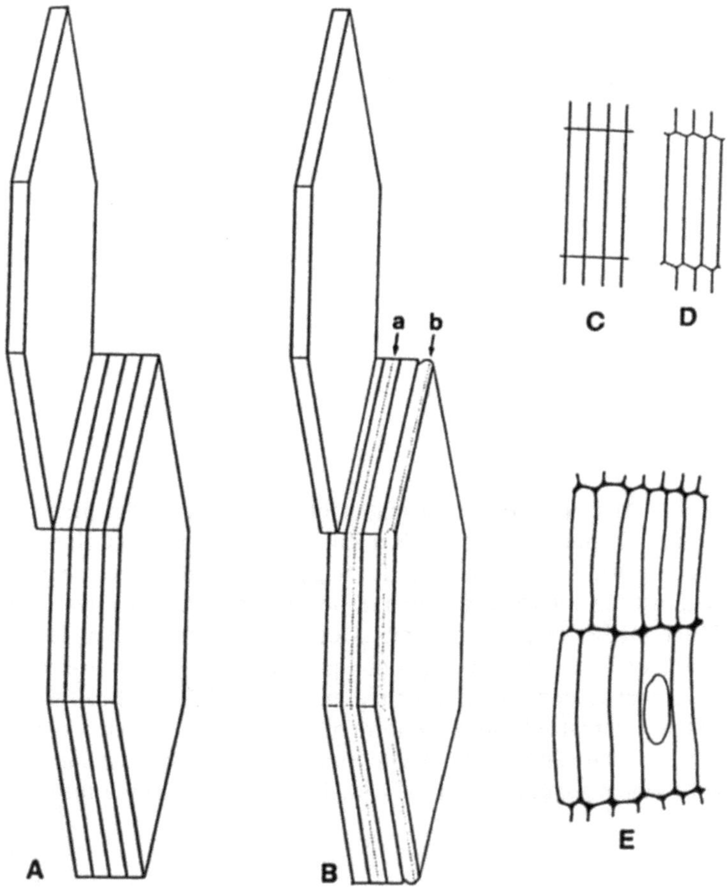

Fig. 5.2A-E. Shapes of cambial cells in *Pinus*. **A** Octahedral cells and their requisite alignment. **B** Lateral shift converting octahedral (*a*) to tetrakaidecahedral (*b*). **C** Transverse section of **A**. **D** Transverse section of **B**. **E** Drawing of cambial zone cells in *Pinus*. Diameters 694x, lengths not drawn to scale. (Lewis 1935)

rectangular/prism with slightly canted end walls. In radial sections they displayed square ends whereas in tangential sections the sloping ends were inclined toward the radial plane. Krabbe's (1886) interpretation of Velten's cambial cell is shown in Fig. 6.78B. It is essentially a hexahedron. Although he agreed with Velten, Krabbe questioned whether the wedge-shaped end walls, e.g., A, B, G, and H in Fig. 6.78B, could be maintained without the intervention of sliding growth.

The controversy continued with Mischke (1890) confirming the prismatic, sharply inclined end wall proposed by Velten, and with Raatz (1892) raising objections. Raatz did not necessarily object to the cell shape proposed by Velten, but with the belief that the shape was more or less maintained through the process of

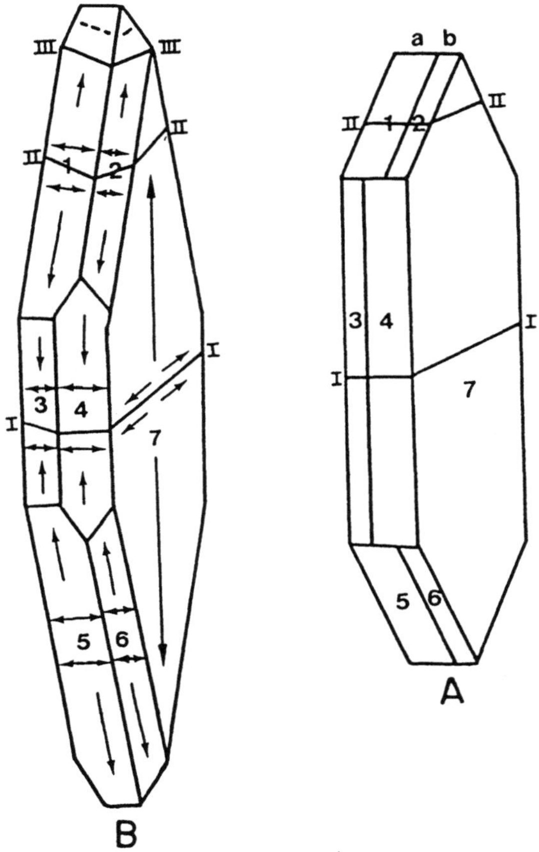

Fig. 5.3A,B. Cambial zone cells of *Pinus sylvestris* showing facets and directions of expansion. Radial walls (tangential plane) face forward. **A** Fusiform cambial initial. **B** Differentiating tracheid. Cell widths expanded about 50x relative to lengths. (After Meeuse 1942)

radial-longitudinal division. Raatz, on the contrary, demonstrated that dividing walls were sometimes obliquely oriented (pseudotransverse). Consequently, cambial cells would not always be of uniform shape. The generalized shape proposed by Velten was reaffirmed by Brown (1912) and Klinken (1914). However, like Raatz, Klinken objected to the idea that all cambial cells retained the same prismatic cross-sectional form. This objection applied particularly to dicotyledonous trees, the cambial cells of which differed in shape from those of coniferous trees. Klinken's idealized cambial cell for conifers had only four walls, two linear-lanceolate tangential walls, and two rectangular radial walls.

The generalized shape of a cambial cell as a hexagonal prism was accepted by most workers and appeared as such in many botanical texts. A typical representation from Kleinmann (1923) is shown in Fig. 5.1. However, the shape as viewed

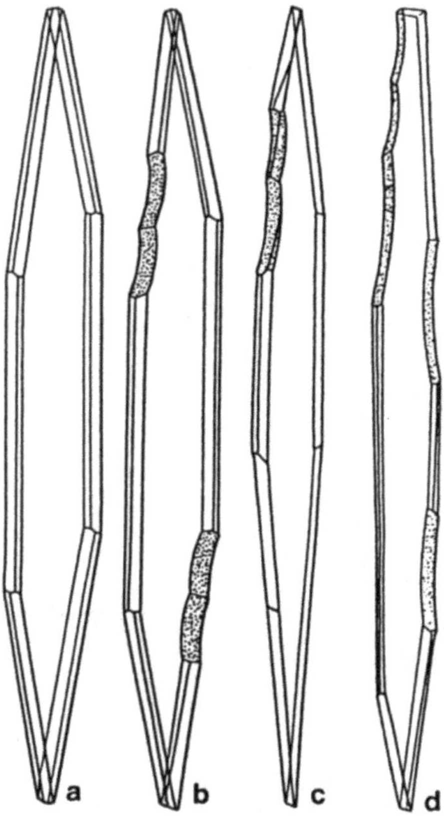

Fig. 5.4a-d. Fusiform cambial initials of *Pinus sylvestris*. **a** Hex-octahedron modified from the orthic tetrakaidecahedron of Lewis (1935). **b** Same cell with two ray contacts each with two faces. **c-d** Models of initials showing variations in shape and ray contacts. (Dodd 1948)

in the tangential plane was believed to exhibit wide variation. Priestley (1930a) attempted to modify this model. His main contribution was the suggestion that the cambial cell represented an imperfect 12- or 14-sided figure whose walls met at approximately 120° angles in symmetrical systems.

The first attempt to actually reconstruct the shape of a cambial cell from serial microsections was made by Lewis (1935) on an unspecified species of *Pinus*. Most previous authors had described the cambial cell as a octahedron, with two hexagonal tangential walls and six quadralateral radial walls (Fig. 5.2A). Since the cells must meet at four-rayed intersections, they would yield the radial cross-sectional pattern shown in (Fig. 5.2C). Lewis claimed that such intersections would be unstable under tension, and lateral sliding growth would tend to shift them to the pattern shown in (Fig. 5.2D) with the cells overlapping by thirds in alternate layers. This shift would convert the octahedron to a 14-sided tetrakaidecahedron with eight hexagonal and six quadralateral facets (Fig. 5.2B). A cell of this shape was shown to be stable (Fig. 5.2E). Meeuse (1942) arrived at a similar conclusion in his depiction of a fusiform initial cell in *Pinus sylvestris* (Fig. 5.3). He also attempted to demonstrate how such a cambial initial would elongate and expand by symplastic growth during differentiation to a tracheid (Chap. 6.7.2).

Dodd (1948) expanded on the work of Lewis (1935) by serial-sectioning a shoot of *Pinus sylvestris* transversely from a point 1 cm below the apex downward through 400, 15 μm sections. Based on these sections, 50 wax models of cambial cells were constructed (Fig. 5.4a). Measurement data from the models showed cambial cell shape to be extremely variable. The number of tangential faces ranged from 3 to 27, lateral cambial faces from 2 to 11, and ray faces from 2 to 9. Ray faces are not the same as ray contacts, because a ray contact might result in one or more ray faces on the cell surface as shown in Fig. 5.4b. Variation in the shapes of cambial initials was due mainly to the adjustments during and after divisions of both fusiform and ray cell initials that occurred within the cambial zone. The average fusiform initial had 18 faces, 14 in contact with other fusiform initials and four with ray cell initials. Thus, a typical cambial initial cell could be approximated by superimposing ray contacts on the flattened orthic tetrakaidecahedron postulated by Lewis (1935) as shown in Fig. 5.4b-d.

Three-dimensional shapes of cambial cells were reconstructed by Fujita et al. (1984) from analyses of 500 to 1000 transverse microsections cut at 2 μm. The sections were stained, photographed, and the cambial cells and their derivatives reconstructed from the photographic series. Cambial cells of *Robinia pseudoacacia* had a square, tubular body approximately 110 μm long, 18 μm high, and 4-7 μm wide, with conical "tips" at both ends approximately 35 μm long. In tangential view, the cambial cells appeared as slender hexagons. Cambial cells of *Juglans sieboldiana* had a body approximately 400 μm long, 20 μm high, and 4-8 μm wide, with tips about 100 μm long. The distinction between body and tips was less clear in the latter species. Despite the fact that cambial derivatives in *R. pseudoacacia* elongated about 300% during their differentiation to wood fibers, elongation was confined to the cell tips and length of the cell body remained essentially unchanged. Additional data on cell tips in many species can be found in Section 5.2. The relation of cell shape to cell wall mechanics was recently investigated by Gao and Pitt (1991).

5.1.1 Effect of Cell Shape on Division

Cell shape apparently can exert an influence on the division patterns of cambial initials. Or, worded differently, cambial initials that have achieved certain sizes and/or shapes appear to divide preferentially. Pulawska (1973) demonstrated, in a profusely illustrated publication, that cell shape can be a determining factor in the order of cell divisions in *Bougainvillaea glabra*, a species with storied structure. In this species, vascular bundles serving new leaves are either embedded within or lie adjacent to the peripheral meristematic cylinder. As the cylinder expands, the older bundles appear to move successively inward as they become embedded in "conjunctive tissue," consisting of lignified fibers and parenchyma (Pulawska 1972).

The peripheral meristem giving rise to secondary tissues was referred to as the "parenchymo-vascular cambium" by Pulawska (1973). In contradistinction to the vascular cambium of most plants, this meristem in *B. glabra* consists exclusively of fusiform initials that produce thin-walled parenchyma to the outside and new vascular bundles embedded in conjunctive tissue to the inside. The term cambium was applied to the meristematic layer of cells and not necessarily to the initial layer. When viewed on transverse sections, cambial cells typically assumed the shape of rectangular hexagons. When one of the cambial cells divided periclinally, the newly formed walls shifted so that the two daughter cells became either five-sided (Fig. 5.5A, cells y) or one four-and the other six-sided (Fig. 5.5B, cells x,z). In both cases, the adjacent cells (N) were transformed to heptagons. The heptagonal cells then divided periclinally so that the five-sided daughter cells from the first division again became six-sided, and one daughter cell of the new division became six-sided (A', cells z) and the other five-sided (cells y). Cells adjacent to the newly divided pair (N) were now transformed to heptagons. Similar changes in cell shape occurred following cell division in B and B'.

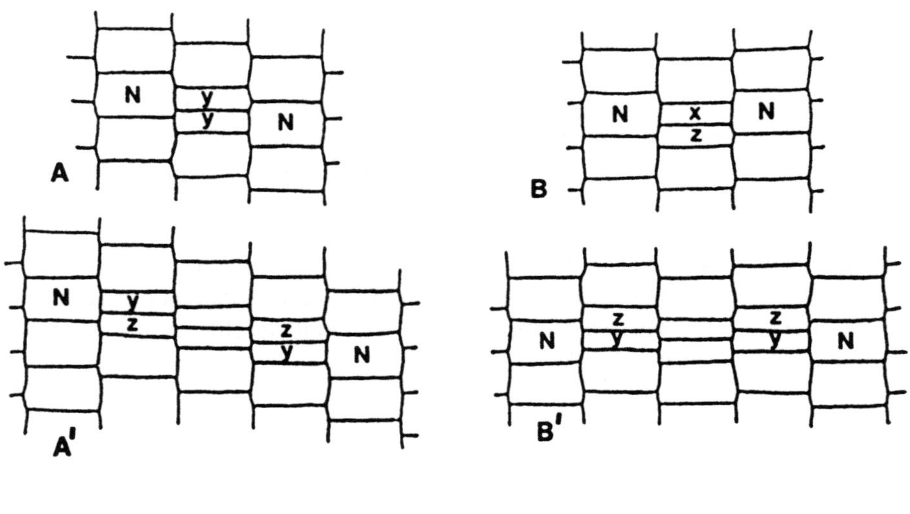

Fig. 5.5A,B. Cambial cells of *Bougainvillaea glabra* on transverse sections illustrating the "principle of disturbance" and return to their basic shapes following periclinal divisions. See text for details. (Pulawska 1973)

Periclinal division of a single cambial cell in one radial file was always followed by a periclinal division in one or both of the adjacent files, the so-called "principle of disturbance." The new division occurred in the heptagonal cell whose shape was altered by the intersecting walls of the previous division. Only one cambial cell in each radial file became the initial, not because of special properties but because of its situation. Periclinal divisions spread rapidly to the sides, although as shown in Fig. 5.5A' and B' the divisions were neither identical nor in perfect tangential alignment. After dividing, each daughter cell returned to its basic shape and number of cell contacts, i.e., a hexagonal cell with 14 contact faces. This process resulted in a peripherally continuous meristematic layer two to four cells wide within which new vascular bundles were embedded. Occasionally, new periclinal divisions occurred in parenchymatous cells external to the parenchymo-vascular cambium. Thus, the initial cells for each new cambium were derived from those of the preceding cambium. The foregoing process was then repeated and in this way successive cambia were produced. Each new parenchymo-vascular cambium was initiated concurrently with the development of vascular bundles serving newly initiated leaves.

Anticlinal divisions were also largely determined by the shapes of cambial initial cells according to Pulawska. Dividing cells conformed to the scheme described by Beijer (1927) (Chap. 6.2.1). An anticlinal division in *B. glabra* usually transformed a cell above and below the dividing cell to a heptagon in tangential view. The heptagonal cells divided, in turn, thereby initiating what appeared to be a chain reaction of anticlinal divisions as observed in several other species with storied cambia (Beijer 1927, Derr and Evert 1967). Thus, in *B. glabra* periclinal divisions were propagated laterally and anticlinal divisions vertically through series of transformed heptagonal cambial cells. Pulawska was of the opinion that the cambial initial cell in a radial file could often be determined by its size and heptagonal shape.

5.2 Fusiform Initial Size

The sizes of fusiform initials, specifically lengths and tangential widths, vary considerably among different tree species. Variability is due primarily to genetic factors related not only to inherent fusiform initial sizes and shapes but also to the frequency with which the initials divide anticlinally and elongate intrusively. Of lesser consideration is the frequency with which initials might either be lost from the cambial layer or transformed to ray cell initials.

As discussed in Chapter 7, the variability in fusiform initial lengths can be extensive both among trees of a single species and positionally within individual trees. Because of these so-called between-tree and within-tree sources of variation, comparisons of among-species differences in fusiform initial lengths must be regarded as best estimates. Proposed sampling procedures, several of which were well conceived, have seldom been adhered to; i.e., with regard to size and age of the tree, position within the tree, and whether stem or branch. Moreover, because of the difficulties in sampling and measuring the cambial initial cells, samples often consist of an insufficient number of measurements from a single sample.

Despite the foregoing qualifications many interesting data on fusiform initial sizes have accumulated. For example, average length values for cambial initials in *Pyrus communis* (Evert 1960) and *Pyrus malus* (Evert 1963a) were 551 μm and 475 μm, respectively (Table 7.11). For *Liriodendron tulipifera*, the average length was found to be 482 μm (Cheadle and Esau 1964) (Table 7.9), and for *Fraxinus excelsior* the average ranged from 250 to 300 μm (Goosen-de Roo and Spronsen 1978). Those in 3-year-old branches of *Acer pseudoplatanus* averaged 320 μm long, 2 to 4 μm in radial diameter, and 15 to 20 μm in tangential diameter (Catesson 1974). In leaning stems of *Populus deltoides*, lengths of fusiform initials on the upper, tension wood sides averaged 556 μm compared to 521 μm on the lower sides (Kaeiser 1964).

The most extensive survey of fusiform initial lengths was made by Bailey (1920a) consisting of data on 13 gymnosperms (Table 7.2) and 54 species of dicotyledons in 14 families (Table 7.4). Another extensive survey has been conducted on subtropical and tropical species by workers in India. These data represent a broad spectrum of sampling procedures and intensities ranging from a few isolated measurements to the means of 1000+ measurements made at random. Many of these species exhibit wide seasonal variations in fusiform initial size. As a general rule, fusiform initial lengths and tangential widths increase from the beginning to end of the growing season and then decrease to a minimum at the approach of dormancy (Iqbal and Ghouse 1985b, Ajmal and Iqbal 1987b, Siddiqi 1991). Because these data reside in numerous publications, many of which are difficult to obtain, a summary of mean values has been compiled in Table 5.1. Additional data have been compiled for many species in the following genera: *Acacia* (Ghouse and Iqbal 1977b, 1979, *Bauhinia* (Ghouse et al. 1976), *Cassia* (Jamal and Ghouse 1981), *Citrus* (Bartwal et al. 1983), *Dalbergia* (Ghouse and Yunus 1974b), and *Eucalyptus* (Ajmal et al. 1985, 1986b).

The designation fusiform initials in the data of Bailey (1920a), as well as in Table 5.1, does not imply elements of the initial layer, which are almost impossible to identify on tangential sections. Rather, it represents either initials or undifferentiated mother cells in the cambial zone. As long as these cells remain undifferentiated, measurements of lengths and tangential widths should be fairly representative of the true initials.

Many investigators in India included data on lengths of the tapering end walls of fusiform initials (Table 5.1). These data provide estimates of the extent of intrusive tip growth following anticlinal division of initial cells and the progressive elongation of these cell tips as the cambium expands tangentially (Chap. 6.4). Ghouse et al. (1980) suggested that the longest fusiform initials are those capable of undergoing extensive intrusive growth by producing tapering end walls.

Table 5.1. Mean lengths, tangential widths, and lengths of tapering end walls of fusiform cambial initials of subtropical and tropical tree species in India

Species	Length µm	Width µm	End wall µm	Reference
Acacia nilotica	237	54		Iqbal and Ghouse 1987
Aesculus indica	313	17		Paliwal et al. 1984
Albizia lebbeck	296	25	52	Yunus et al. 1978
Aleuosmia macrophylla	400			Paliwal and Srivastava 1969
Annona squamosa	300	21	62	Ghouse and Yunus 1974a
Bauhinia alba	263	20	58	Ghouse et al. 1976
Cassia grandis	308	18	60	Jamal and Ghouse 1981
Citrus indica	252	18		Bartwal et al. 1983
Clerodendrum inerme	254	25	53	Ghouse et al. 1980
Dalbergia sissoo[a]	203	16	20	Ghouse and Yunus 1974b
Delonix regia	306	30	99	Ghouse and Hashmi 1977b
Duranta repens	328	17	113	Ghouse et al. 1980
Erythrina indica[a]	200	38	20	Yunus et al. 1978
Eucalyptus camaldulensis	350	18	41	Ajmal et al. 1985
Ficus rumphii	312	21	40	Ajmal and Iqbal 1987b
Gmelina hystrix	363	24	166	Ghouse et al. 1980
Grewei optiva[a]	244	15		Paliwal et al. 1984
Holmskoildia sanguinea	356	25	141	Ghouse et al. 1980
Holoptelia integrifolia[a]	268	19	24	Ajmal et al. 1986b
Jacaranda minosifiolia	288	17		Anand et al. 1981
Juglans regia	535	30		Bartwal et al. 1983
Lantana camara	264	23	56	Ghouse et al. 1980
Mangifera indica	388	21		Dave and Rao 1982b
Mimusops elengi	442	24	64	Ghouse and Hashmi 1977b
Nyctanthes arbor-tristis	401	28	143	Ghouse et al. 1980
Peltophorum ferrugineum	400	24	75	Ghouse and Hashmi 1977b
Petrea volubilis	399	27	133	Ghouse et al. 1980
Polyalthia longifolia	436	23	70	Ghouse and Hashmi 1980
Pongamia pinnata[a]	181	16	18	Yunus et al. 1978
Prosopis spicigera	288	17	54	Iqbal and Ghouse 1985b
Prunus persica	258	17		Bartwal et al. 1983
Pterygota aiata[a]	298	21	37	Yunus et al. 1978
Putranjiva roxburghii	1084	18		Anand et al. 1981
Pyrus communis	477	21		Bartwal et al. 1983
Quercus incana	356	17		Paliwal et al. 1984
Rhodendron arboreum	346	17		Paliwal et al. 1984
Salmalia malabarical[a]	392	35	43	Yunus et al. 1978
Salvadora persica	138	17		Paliwal et al. 1976
Shorea robusta	292	15		Paliwal et al. 1984
Streblus asper	327	20	34	Ajmal and Iqbal 1987a

Tectona grandis	370	22	57	Ghouse et al. 1980
Terminalia arjuna	300	26	78	Ghouse and Hashmi 1977b
Vitex negundo	228	20	52	Ghouse et al. 1980

[a] Species with storied cambia.

Data on radial diameters of cambial cells are seldom provided by investigators. In most cases they are meaningless because radial enlargement often begins almost immediately after periclinal division. Moreover, radial cell diameters of cambial fusiform initials vary widely both during the division process and seasonally. Bailey (1952), in a general statement, noted that cambial initials fluctuated between 4 µm and 8 µm in radial diameter. Several other workers provided interesting data relative to the division process. For example, Bannan (1955) and Wilson (1963) claimed that cambial initials of *Thuja occidentalis* and *Abies concolor*, respectively, enlarged radially to no more than 10 µm to 15 µm before dividing. Likewise, Alfieri and Evert (1968) found that cambial initials of several *Pinus* species expanded radially to about 10 µm before periclinal divisions. Doley and Leyton (1968) placed the limiting radial diameter before periclinal divisions at 6 µm for *Fraxinus excelsior*.

5.2.1 Fusiform Initial Volume

The area occupied by fusiform initial cells on tangential sections is often referred to as fusiform initial volume. This value is simply the reciprocal of ray volume. Because ray volumes are easily measured, values have been obtained for a large number of species (Chap. 8.3.11). As shown in Fig. 8.36, fusiform initial volumes might vary from about 20 to 100%, the latter in rayless species.

Data supplied by Clowes and Juniper (1968) indicated that fusiform cambial initials of conifers are often 5 mm long and over 1 000 000 μm^3 in volume. The latter value applied to the volume of individual cells.

5.3 Ultrastructure

Despite the fact that most investigators accept the concept of a uniseriate cambium, it has been found impossible to identify a fusiform cambial initial ultrastructurally. That is, the ultrastructure of a prospective fusiform initial cell is no different from that of the immediately adjacent cambial zone cells (Srivastava 1966, Srivastava and O'Brien 1966a,b, Robards and Kidwai 1969, Evert and Deshpande 1970, Murmanis 1971, Timell 1973, 1979, Tsuda 1975a,b, Phillips

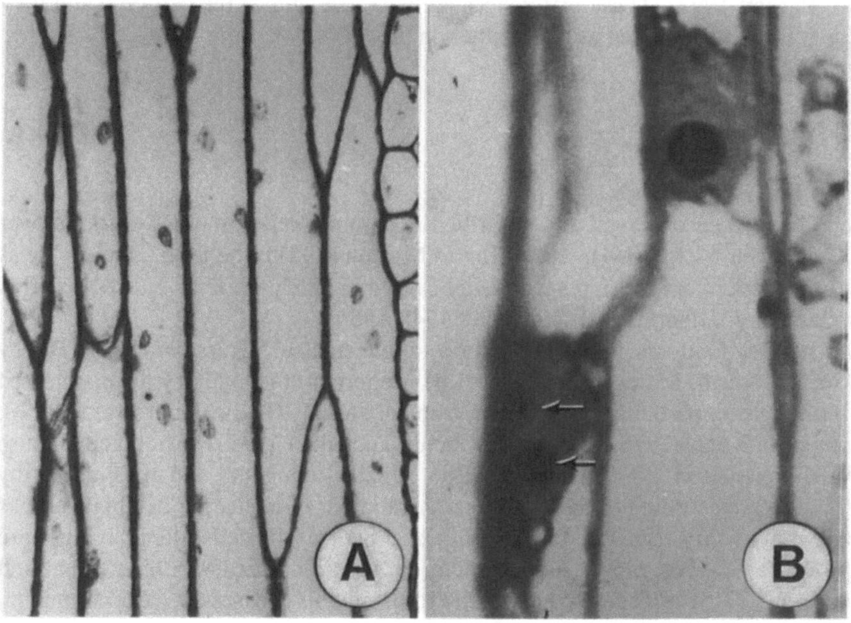

Fig. 5.6A,B. Tangential sections of *Tectona grandis* cambia. **A** Multinucleate fusiform initials. 1153x. **B** Nucleus with two nucleoli (*arrows*) in a fusiform initial. 1488x. (Dave and Rao 1981)

1980, Barnett 1981). For example, in *Pinus strobus* both the youngest sieve cells (Srivastava and O'Brien 1966b) and the youngest tracheids (Murmanis and Sachs 1969) showed essentially the same fine structures as the fusiform initials and it was almost impossible to distinguish between them. This difficulty also applies to the identification of ray cell initials and their immediate derivatives (Srivastava 1966, Srivastava and O'Brien 1966a). One of the reasons the cambial initial is difficult to identify in dormant material is the presence of one or more phloem, and occasionally xylem, derivatives that divide at season's end and overwinter in an undifferentiated state (Chap. 11.4.2). In some cases, it has been possible to recognize these potential derivatives and thereby delimit the cambium to three or four cells in both winter and summer collections (Itoh et al. 1968, Srivastava 1973, Barnett 1992).

The first indication of differentiation at the resumption of growth is the pronounced radial expansion of cambial derivatives. This expansion is due to an increase in vacuolar volume without any conspicuous change in cell fine structure (Catesson 1974). Ultrastructural differences appear once overt differentiation has begun in either a prospective xylem or phloem derivative (Cronshaw 1965, Srivastava and O'Brien 1966b, Zee 1968, Murmanis and Sachs 1969, Czaninski 1970, 1972, O'Brien 1974, Tsuda 1975b, Berlyn 1979, Hirakawa et al. 1979, Roberts

1976, Barnett 1992). These differences are characteristic of the cell type and they become more accentuated as differentiation proceeds.

5.3.1 Nucleus

5.3.1.1 Fusiform Initials

Nuclei have been observed and described in cambial cells by many workers over the years (von Mohl 1844). According to Morton (1981), the term "nucleus" was first used with reference to plant cells by Brown in 1833; the term "nucleolus" was introduced by Valentine in 1839 (Baker 1948-1955).

As a rule, fusiform initials contain a single nucleus (Russow 1882a, Bailey 1919b, 1920b,c,d, Meier 1973) (Fig. 11.4). Reports to the contrary have appeared in the literature, particularly in certain tropical species (Paliwal and Prasad 1970, Ghouse and Yunus 1973, Patel 1975, Ghouse and Khan 1977, Hashmi and Ghouse 1978, Ghouse et al. 1979, Dave and Rao 1982b, Venugopal and Krishnamurthy 1988, 1989). According to these authors, the number of nuclei per cambial cell has been found to vary from one to five, with over 80% of the cells often having more than one nuclei (Fig. 5.6A). For example, in *Acacia nilotica* var. *telia*, Iqbal and Ghouse (1987) claimed that most fusiform initials possessed several nuclei, "sometimes as many as eight."

According to Ghouse and Khan (1977), the number of cambial cells in *Psidium guajava* having the polynucleate condition tended to be higher during winter months in India (Table 5.2); but according to Venugopal and Krishnamurthy (1988), the maximum number of nuclei in several other species occurred during active growth. In seven species examined by Venugopal and Krishnamurthy (1989), in only one (*Calophyllum inophyllum*) did the fusiform initials remain uninucleate during active growth. However, in *Terminalia crenulata* 10 to 15% of the fusiform initials retained the multinucleate condition into dormancy (Venugopal and Krishnamurthy 1988). Degenerating nuclei with abnormal shapes and elongated "tails" were observed during all months of the year in *Psidium guajava* (Ghouse and Khan 1977) and during differentiation of fusiform initials to vascular elements in *Solanum melongena* (Patel 1975).

The existence of multinucleate cambial cells has been questioned by a number of workers. Barnett (1992, personal communication), for example, suggested that a cambial cell might appear to be multinucleate if two or more cells are superimposed in a tangential, longitudinal microsection. That is, because cambial cells are so narrow in radial diameter and their tangential walls so thin, several superimposed cells might appear as a single cell. Another possible source of error was mentioned by Riding (1992 personal communication). He sometimes observed what appeared to be binucleate cells in *Abies balsamea*. However, more critical examination showed these to be dividing cells with two widely separated phragmoplasts and barely distinguishable phragmosomes. If further research should prove that the multinucleate condition does exist in some species, what might be the role of these multiple nuclei during cambial cell division? This question remains unanswered (Catesson 1990). A somewhat related question, endopolyploidy, a condition in which chromosome material is replicated without nuclear division of either the nucleus or the cell, has been reported as occurring rarely in gymnosperms (Barlow 1978).

The dimensions and volumes of nuclei vary considerably in fusiform initials of different shapes and sizes. As Bailey (1920b,d) found in *Pinus strobus*, large fusiform initials often had larger and more elongated nuclei than smaller, roughly isodiametric ray cell initials. Furthermore, the nuclei in large cambial initials of mature trees were larger than those in smaller initials of young shoots. Yet there was no consistent relation between nuclear size and fusiform initial length in mature trees. The most striking variations were seasonal. Bailey (1920b) found that resting nuclei tended to be longer and narrower during the fall and winter than active nuclei during spring and summer. Similar seasonal variations in nuclear size and structure have been observed in fusiform initials of tropical species (Dave and Rao 1981, Iqbal and Ghouse 1985a). According to Catesson (1964), fusiform initials of dicotyledons contain smaller nuclei and fewer nucleoli than the longer initials of gymnosperms. An interesting comparison of the nucleocytoplasmic ratios in *Pinus strobus* and *Acer pseudoplatanus* can be found in Table 10 of Catesson (1990).

Table 5.2. Percentage of fusiform initials of *Psidium guajava* with varying numbers of nuclei in different months of the year (Ghouse and Khan 1977)

Month	Percentage of fusiform cambial initials[a]				
	1-nucleate	2-nucleate	3-nucleate	4-nucleate	5-nucleate
January	16	36	32	12	4
February	32	36	28	4	-
March	36	32	28	4	-
April	28	52	20	-	-
May	36	44	12	8	-
June	32	48	12	8	-
July	36	32	24	8	-
August	40	36	20	4	-
September	36	34	26	4	-
October	36	44	20	-	-
November	12	20	40	20	8
December	12	24	48	12	4

[a] Based on observations of 1000 fusiform initials obtained from five 20-year-old trees.

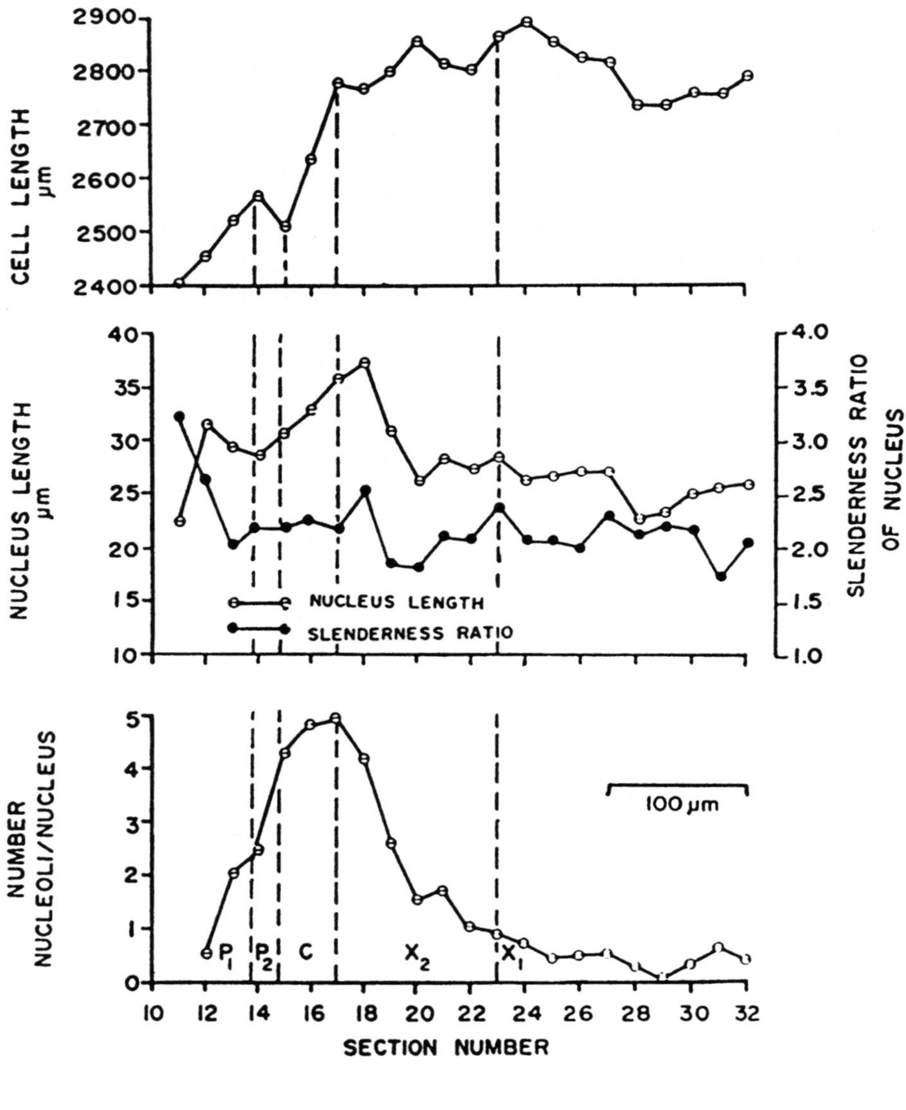

Fig. 5.7. Changes in dimensions of fusiform cells across the cambial zone and differentiating tissues within a growth ring sampled at the 0.4 m height in a mature *Picea abies* stem. P_1, P_2 Differentiated and differentiating phloem, respectively; *C* cambium; X_2, X_1 differentiating and differentiated xylem, respectively. (After Meier 1973)

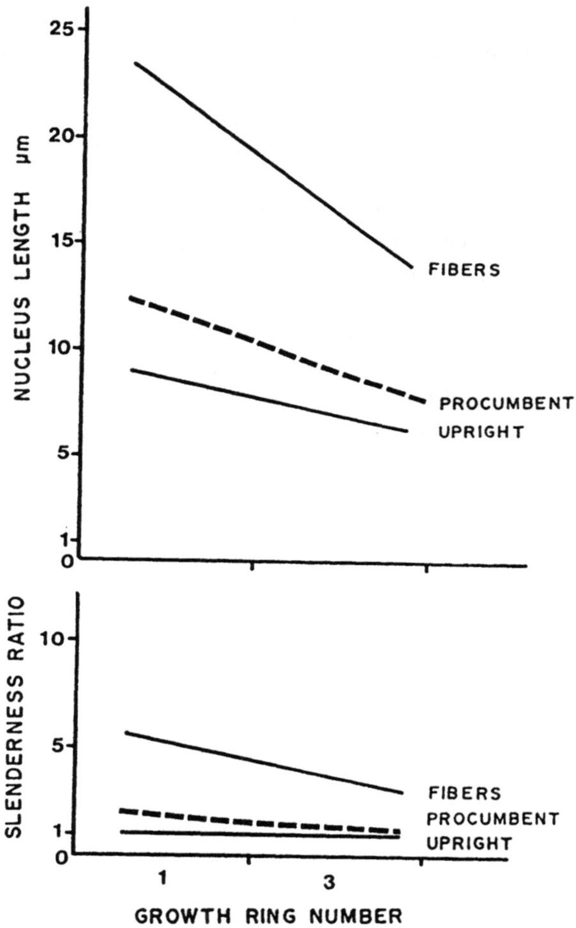

Fig. 5.8. Variations in nuclei lengths and slenderness ratios of xylem fibers and the procumbent and upright ray cells in *Sambucus racemosa*. Sampling extended from cambial region on the *extreme left*, through the outermost three growth rings, to the heartwood boundary on the *extreme right*. (After Wolkinger 1969)

Meier's (1973) results more or less confirmed those of Bailey. He examined three conifers, *Larix decidua, Picea abies,* and *Pinus sylvestris*, growing at the upper alpine tree zone. One of his criteria for comparing nuclei was the "slenderness ratio" (length/width ratio) introduced by Bosshard and Stahel (1969). Both nuclear length and slenderness ratio varied with fusiform cell length (Fig. 5.7). Fusiform cell lengths in the xylem and phloem were average values of several cell types and simply indicated trends. In the phloem, cell length decreased outward from the cambium because of post-cambial, transverse divisions, whereas in the xylem cell length first increased because of post-cambial tip elongation and then decreased in keeping with the seasonal trend of tracheid length (Chap. 7.1.2.1). The nuclear quotient was calculated by the ratio cell/nuclear length. Measurements on *P. abies* made in September at the onset of dormancy yielded a nuclear quotient of 52 for the shortest cambial cells, 80 for the longest, and an average value of 67. Nuclear quotients for *L. decidua* ranged from 29 to 61 for measurements made in October.

Slenderness ratios for fusiform initials of *Tectona grandis* were also calculated by Dave and Rao (1981). They found a high ratio during dormancy that peaked in March just before the spring resumption of cambial activity. The slenderness ratio approximately corresponded inversely with mean length of the fusiform initials which was low during dormancy and increased during active growth. In *Sambucus racemosa*, Wolkinger (1969) found that mean lengths of nuclei decreased from 23.4 μm in cells of the cambial zone to 14.3 μm in living wood fibers of the third growth ring. The slenderness ratio of cambial cell nuclei averaged 5.7 (Fig. 5.8).

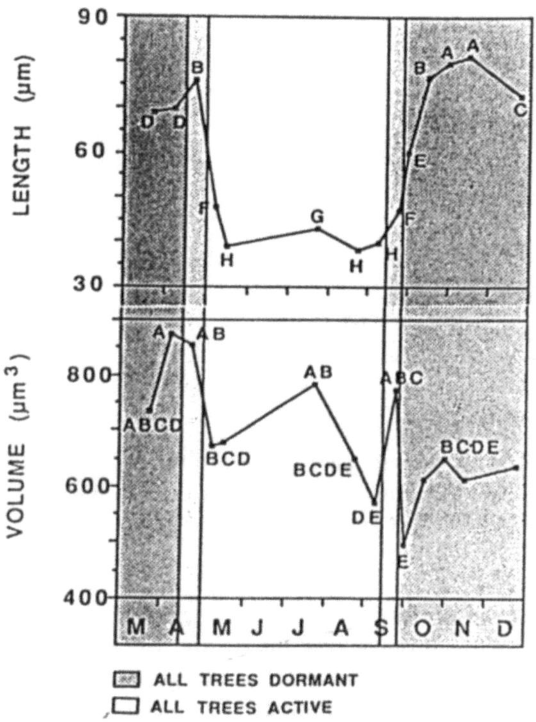

Fig. 5.9. Seasonal changes in lengths and volumes of interphase nuclei of fusiform initials at the stem base of 6-year-old *Abies balsamea* trees. Averages denoted by the *same letter* are not statistically different. (Mellerowicz et al. 1990)

A thorough study of the seasonal variation in nuclear size in fusiform initials of *Abies balsamea* was conducted by Mellerowicz et al. (1990). Nuclear shape was described as "... approximating a three-axial ellipsoid, elongated vertically with a cross section that varied from round to radially flattened." Like Bailey (1920b,d), they found that nuclei were largest during the dormant period. Lengths of interphase nuclei decreased rapidly after the initiation of cambial cell division, remained constant at a low level during the growing season, and then increased at the onset of dormancy (Fig. 5.9). Nuclear volume followed a somewhat similar declining trend during the season, although it was more erratic due to fluctuations in dimensions other than length. The slenderness ratio of *A. balsamea* nuclei increased from about 6 during the growing season to 14 during dormancy.

The nucleus of a fusiform cambial cell of *Aesculus hippocastanum* is shown in Fig. 5.10 from Barnett (1981). Interphase nuclei of cambial cells in this species

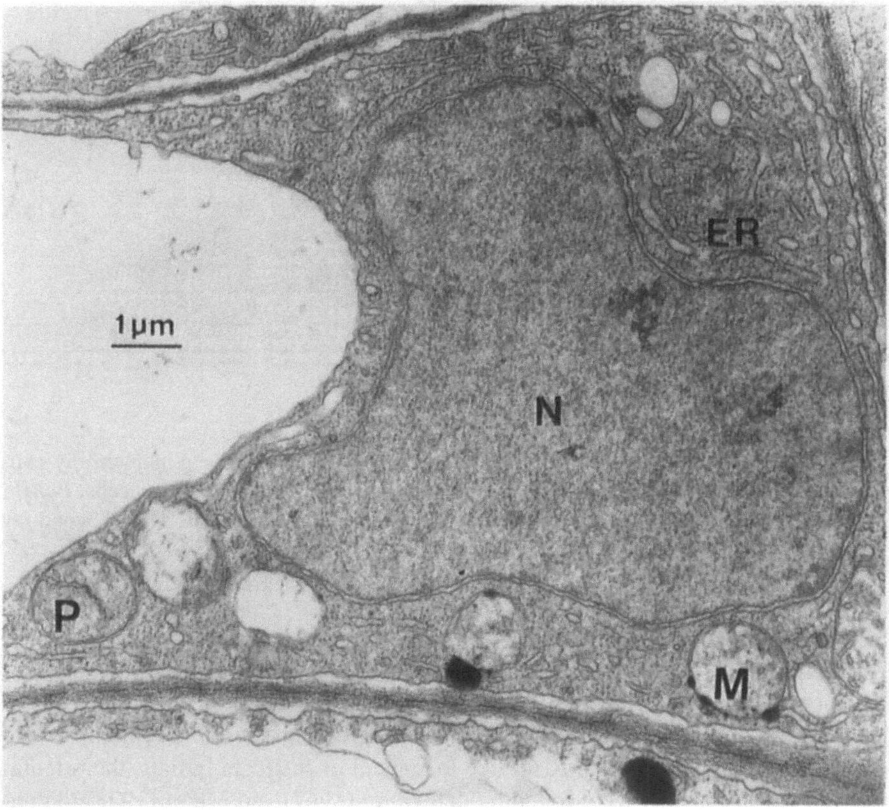

Fig. 5.10. EM photomicrograph of *Aesculus hippocastanum* during active growth showing nucleus (*N*) of a fusiform cambial initial in transverse view. *M* Mitochondrion; *P* proplastid; *ER* endoplasmic reticulum. (Barnett 1981)

have been found to contain bundles of microtubules (Barnett 1991). Each bundle was composed of between 10 and 50 microtubules arranged parallel to one another, often in a paracrystalline hexagonal array. Electron micrographs showed that one end of the bundle was associated with the nucleolus and merged with it. Although no distinct terminal structures were found, the end distal from the nucleolus lay close to the nuclear membrane. Barnett could suggest no definite role for these structures. Their presence in interphase nuclei, coupled with their association with the nucleolus, was indicative of involvement in the normal activity of the nucleolus. Moreover, the microtubules were also present in nuclei of differentiating elements of both the xylem and phloem. Thus, these microtubules appeared unrelated to those involved in the cell division process (Sect. 5.5).

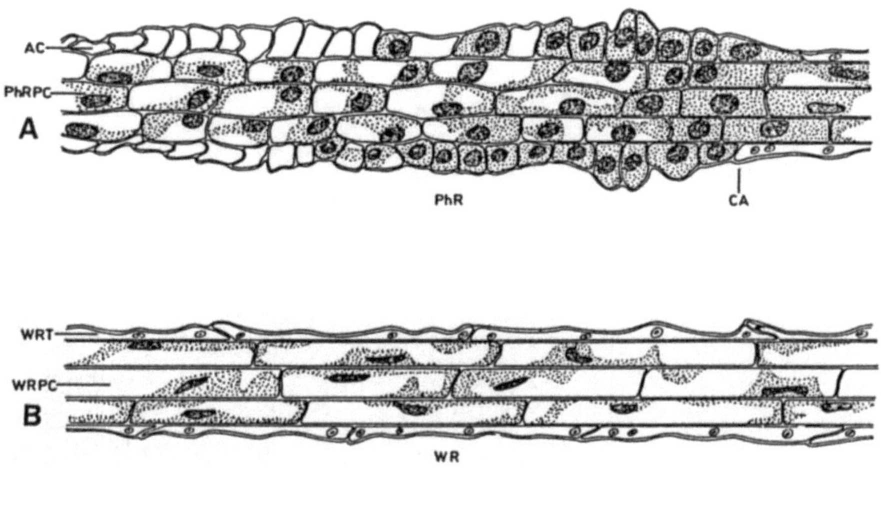

Fig. 5.11A,B. Distribution of cytoplasm and shapes of nuclei in the **A** phloem ray cells (*PhR*) and **B** xylem ray cells (*WR*) of *Pseudotsuga menziesii*. *AC* Albuminous cells; *PhRPC* phloem-ray parenchyma cells; *CA* cambium; *WRT* wood-ray tracheids; *WRPC* wood-ray parenchyma cells. (den Outer 1967)

5.3.1.2 Ray Cell Initials

Ray cell initials of most woody species are uninucleate. However, as in the case of fusiform initials, claims have been made that the ray cell initials of some tropical species are multinucleate during the active growing period (Venugopal and Krishnamurthy 1988, 1989). Also, like the nuclei of fusiform initials, the orientations of ray cell nuclei were found by den Outer (1967) and Meier (1973) to conform to the shapes and orientations of the parent ray cell initials (Fig. 5.11). Nuclei were vertically oriented in upright ray cells, horizontally oriented in procumbent ray cells, and spherical in isodiametric ray cells. As in fusiform cells,

Meier (1973) found that nuclear size varied with ray cell size in *Larix decidua* (Fig. 5.12). Ray cell heights varied only slightly, primarily by post-cambial elongation of the phloem-ray border cells. Ray cell lengths, however, varied seasonally as these cells elongated to keep pace with the more rapidly dividing fusiform cambial cells. Like Meier, Stahel (1968) observed that short, quadratic ray cells in *Populus euramericana* contained spherical nuclei. Nuclear sizes and shapes in this species varied not only seasonally but also with age of the derivatives from the cambium, or with their states of differentiation.

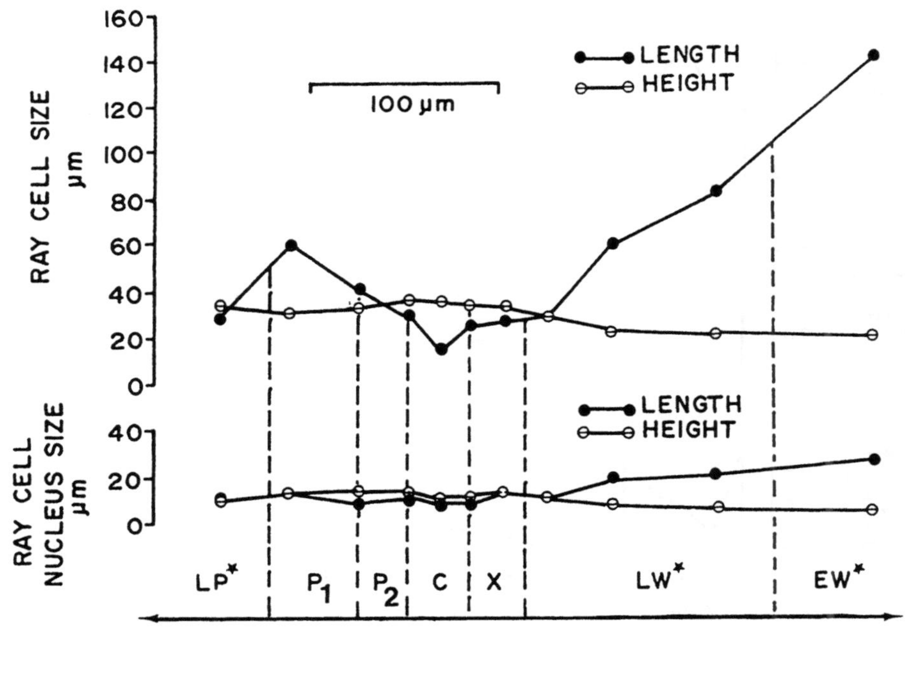

Fig. 5.12. Changes in dimensions of ray cells across the cambial zone and differentiating tissues within a growth ring sampled at the 5.8 m height in a mature *Larix decidua* stem. *P1*, *P2* Differentiated and differentiating phloem, respectively; *C* cambium; *X* differentiating xylem; *LP*, *LW*, *EW* differentiated phloem, latewood and earlywood, respectively, of previous season. (After Meier 1973)

Wolkinger (1969) showed that the largest nuclei in *Sambucus racemosa* occurred in the cambial ray cells. For example, lengths of nuclei in procumbent ray cells ranged from 12.7 µm in the cambium to 7.9 µm in the third growth ring. In the upright ray cells lengths of nuclei ranged from 9.1 µm in the cambium to 6.7 µm in the third growth ring (Fig. 5.8).

In conjunction with his study of cytology of heartwood formation, Hugentobler (1965) conducted a thorough examination of ray cell nuclei. He reported that although nuclei varied in size in different types of ray cells, a relationship existed between nuclear size and ray cell size in functionally similar elements. As shown in Fig. 5.13 for *Quercus robur*, nuclear volumes and surface areas changed seasonally with the largest values occurring in the cambial region during the summer months.

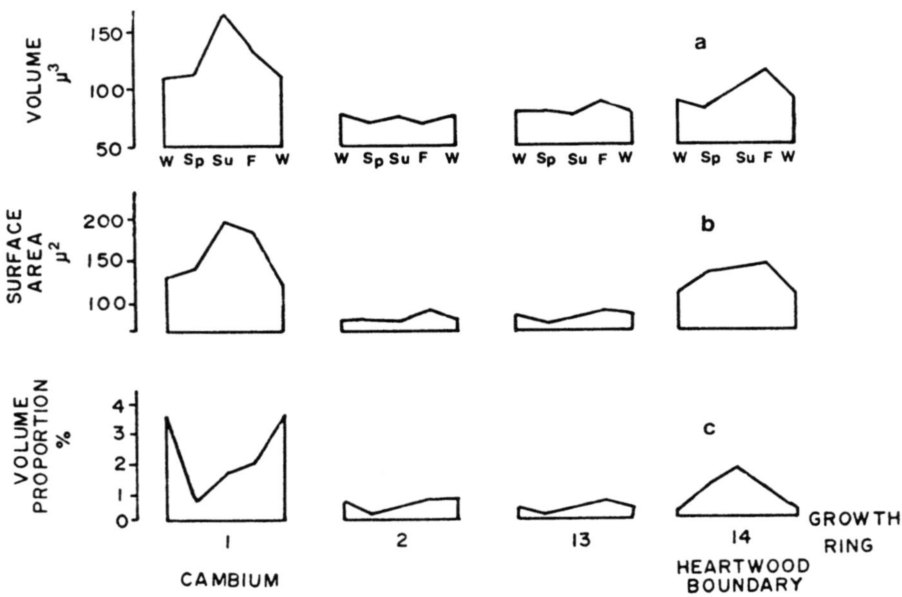

Fig. 5.13a-c. Seasonal variations in dimensions of nuclei and nucleoli in ray cells of *Quercus robur*. **a** and **b** Volume and surface area of nuclei, respectively. **c** Volume percentage of nucleoli to nuclei. *W* Winter; *Sp* spring; *Su* summer; *F* fall. (After Hugentobler 1965)

The "slenderness ratio," introduced by Bosshard and Stahel (1969), was used by the authors as a measure of the level of ray cell activity in several dicotyledonous species. The degree of slenderness in ray cell nuclei increased in both phloem and phloem derivatives for a distance of 0.5 mm outward from the cambium. The highest cellular activity was found in those ray cell tissues situated nearest the cambium. Wolkinger (1969) found practically no change in the slenderness ratio of ray cells in *Sambucus racemosa* on the xylem side of the cambium (Fig. 5.8).

5.3.1.3 Nuclear Migration

The nucleus in a resting or nondividing fusiform initial is usually situated at or near the middle of the elongated cell (Sanio 1873, Bailey 1919b, 1920b, Abbe and Crafts 1939, Itoh 1971, Catesson 1974, Farooqui and Robards 1979, Rao and Dave 1983a, Fujita et al. 1984, Rao 1985). Several hours before cell division, the nucleus migrates to the position of mitosis and becomes fixed there by strands of cytoplasm, particularly by the phragmosomes (Evert and Deshpande 1970, Itoh 1971, Rao and Dave 1983c, Furuya 1984). In the case of asymmetric divisions, as for example the cutting off of a ray cell initial, the nucleus migrates to the prospective position of the new anticlinal wall (Barghoorn 1940a) (Chap. 6.3.2.5).

The nuclei of ray cell initials also migrate to new positions prior to cell division. For example, in the dicotyledonous species examined by Hugentobler (1965), nuclei of ray cell initials were centrally located during the winter months but wall-oriented during active cell division in summer. In the ray cell initials of *Pseudotsuga menziesii* examined by Wodzicki and Brown (1973), the nuclei appeared to migrate during interphase to that part of the cell in contact with either a fusiform initial or a dividing mother cell. After a ray cell initial divided asymmetrically, the daughter cell residing nearest the fusiform initial assumed the role of a new ray cell initial. Meier (1973) also commented on the tendency for nuclei of young ray cells in both xylem and phloem derivatives to reside preferentially on the cambial sides of the cells.

5.3.1.4 DNA Content

Nuclei of dividing cells exhibit a cyclic behavior. The two key events in the cycle are mitosis and synthesis of DNA in preparation for the following mitosis (Dodds 1981). Mellerowicz et al. (1989, 1990) measured DNA contents cytophotometrically in Feulgen-stained nuclei of fusiform cambial initials from 1- (twig) and 19-year-old (stem base) cambia of *Abies balsamea* trees throughout the season. Cell and nuclear volumes of the 19-year-old exceeded those of the 1-year-old cambium ten and twofold, respectively. Telophase (2C) and prophase (4C) DNA contents varied both with cambial age and time of year. Maximum DNA contents were observed at the beginning of cambial activity in April. They were lower in the 1-year-old cambium than in the 19-year-old cambium (Fig. 5.14). DNA levels decreased during the growing season, attaining minimum levels in September in cambia of both ages. In autumn, cells accumulated in the G_1 phase of the cell cycle at which time DNA synthesis increased the 2C levels to their maximum values; blockage of the cambial cell nucleus at the G_1 stage of the cell cycle during dormancy is now a well-documented phenomenon (Catesson 1980). Mellerowicz

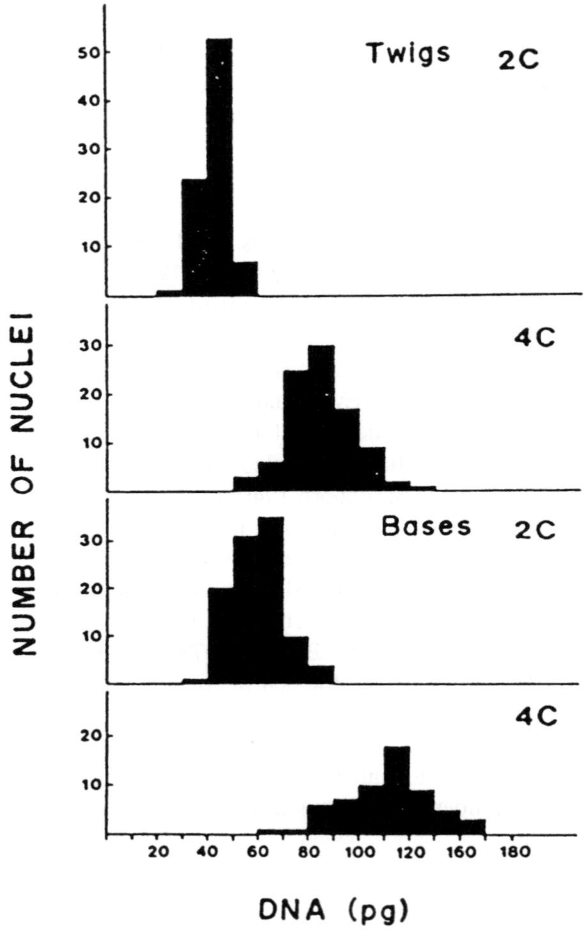

Fig. 5.14. Distribution of *2C* (telophase) and *4C* (prophase) DNA contents in fusiform cells of 1-year-old (twigs) and 19-year-old (bases) cambia of *Abies balsamea* analyzed in April. (After Mellerowicz et al. 1989)

et al. (1989) speculated that the observed differences were due either to genomic instability in the cambial tissues or to stainable properties of chromatin that changed with cambial age and seasonal activity. Similar seasonal patterns in DNA content were obtained in cambial cells of *Larix decidua* (Riding et al. 1989).

In a follow-up study conducted by Mellerowicz and Riding (1992), 1-year-old shoots from the tops of six 21-year-old *Abies balsamea* trees were sampled on May 21 and June 5, during and after, respectively, the period of high mitotic activity of the cambium. At both sampling dates, most interphase fusiform initials had an unreplicated (2C) DNA level, indicating that the majority of them were in the

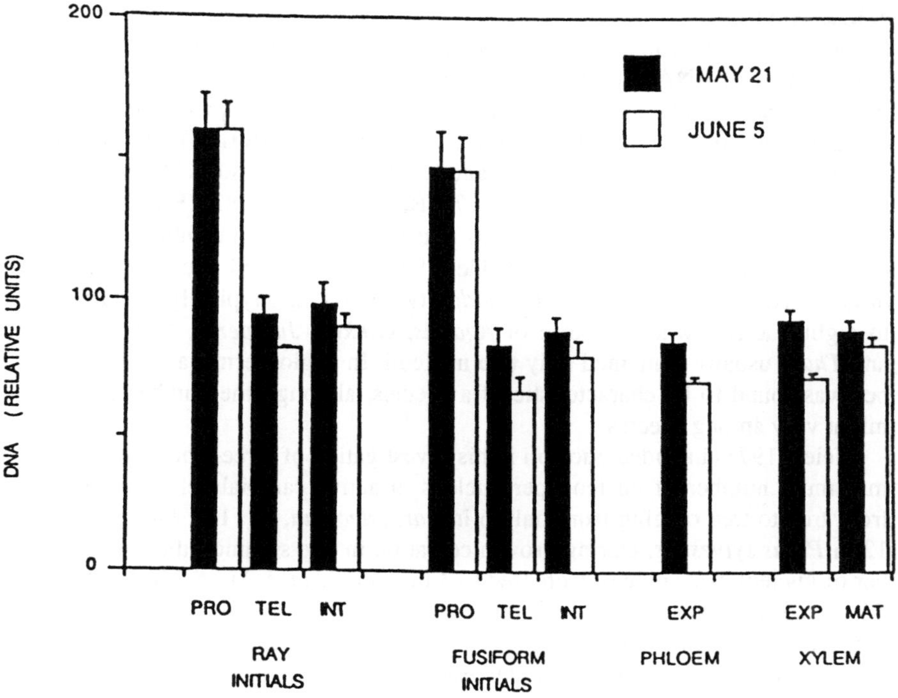

Fig. 5.15. DNA levels in the cambium and derivative cells of *Abies balsamea*. *Bars* indicate standard errors. *EXP* Expanding; *INT* interphase; *MAT* maturing; *PRO* prophase; *TEL* telophase. (Mellerowicz and Riding 1992)

G_1 phase of the cell cycle (Fig. 5.15). According to the authors, this finding indicated that fusiform initials of *A. balsamea* have a uniform ploidy level and that G_1 is the longest phase of the cell cycle. DNA contents of fusiform and ray cell initials were about the same. Thus, the greater size of fusiform initials was not paralleled by genome multiplication. DNA contents in differentiating xylem and phloem elements were also similar to those in fusiform initials during interphase and telophase, indicating a lack of endoreduplication during differentiation. On the other hand, the authors reaffirmed their previous observation that DNA levels of fusiform initials decreased during the growing season and increased to a maximum during dormancy. These observations were confirmed by Mellerowicz et al. (1992a,b), who found increasing DNA levels toward the autumn. However, the nuclei were in the G_1 phase of the cell cycle when the additional DNA was synthesized. Their experiments, involving variable temperature and photoperiodic regimes, suggested that the additional DNA was extrachromosomal.

5.3.1.5 Nucleoli

Nucleoli are spherically shaped bodies present in the nuclei of eukaryotic cells. They consist of RNA and protein and are considered sites of ribosome synthesis. Although each cambial cell usually contains a single nucleus, each nucleus might contain from one to many nucleoli (Kienitz-Gerloff 1891, Bailey 1920 b,d, Kleinmann 1923, Elliott 1935, Abbe and Crafts 1939, Catesson 1964, 1980, 1981, Itoh 1971, Patel 1975, Paliwal et al. 1976, Rao and Dave 1981) (Fig. 5.6B). Sanio (1873), for example, observed three to four nucleoli in each cambial cell nucleus of *Pinus sylvestris*. According to Zirkle (1931), cambial cell nuclei of *Abies, Cedrus, Picea, Pinus, Pseudotsuga, Sciadopitys,* and *Tsuga* typically contained four to eight nucleoli whereas those of *Agathis, Ginkgo, Juniperus, Sequoia, Taxus,* and *Thuja* usually contained only two nucleoli. In angiosperms, a particular number was found to be characteristic of a species, although the number of nucleoli might vary among species.

Meier (1973) included nucleoli in his investigation of three alpine conifers. The maximum number of nucleoli per nucleus in active cambial cells varied widely from tree to tree ranging from only 6 in *Larix decidua*, to 11 in *Picea abies*, and 12 in *Pinus sylvestris*. During growth cessation in late summer, the average number of nucleoli per nucleus decreased to 3 in *L. decidua*, 5.0 to 5.7 in *P. abies*, and 5.9 to 6.3 in *P. sylvestris*. Mellerowicz et al. (1992c), in contrast, found that the number of nucleoli per nucleus in *Abies balsamea* increased from 4.2 to 5.5 with the onset of dormancy and again decreased to 4.2 with growth reactivation in the spring. Total nucleolar volume per nucleus followed a different pattern. It reached a minimum at the onset of dormancy, increased 1.5-fold during the rest-quiesence transition, 7-fold at the time of cambial reactivation, and then decreased to a low level in June.

According to Meier (1973), the sizes of the nucleoli varied widely both among species and also within individual nuclei. In *P. abies* and *P. sylvestris*, nucleoli seldom exceeded 5 μm in diameter, whereas those in *L. decidua* sometimes attained a diameter of 7.5 μm. With progressive states of differentiation, nucleoli tended to concentrate in the center of the nucleus and to decrease in number and mass (Fig. 5.7). These changes were associated with an overall trend toward nuclear deterioration. In a similar way, Stahel (1968) found that the nucleoli in ray cells of *Populus euramericana* decreased in size with increasing distance from the cambium. The largest nucleoli occurred in newly differentiated phloem ray cells. These trends in the xylem are evident in Fig. 5.13 from Hugentobler (1965). The proportion of the ray cell nucleus occupied by nucleoli not only varied during the annual growth cycle, but it also declined rapidly inward from the cambium. A secondary peak occurred at the sapwood-heartwood boundary associated with the processes of nuclear deterioration.

The nucleoli in *Acer pseudoplatanus* cambial cells exhibited a similar pattern of seasonal activity (Catesson 1964, 1980, 1981) (Table 5.3). During winter, the nucleoli appeared dense with a well-developed fibrillar zone and a poorly developed granular zone. The granular zone expanded shortly before the onset of

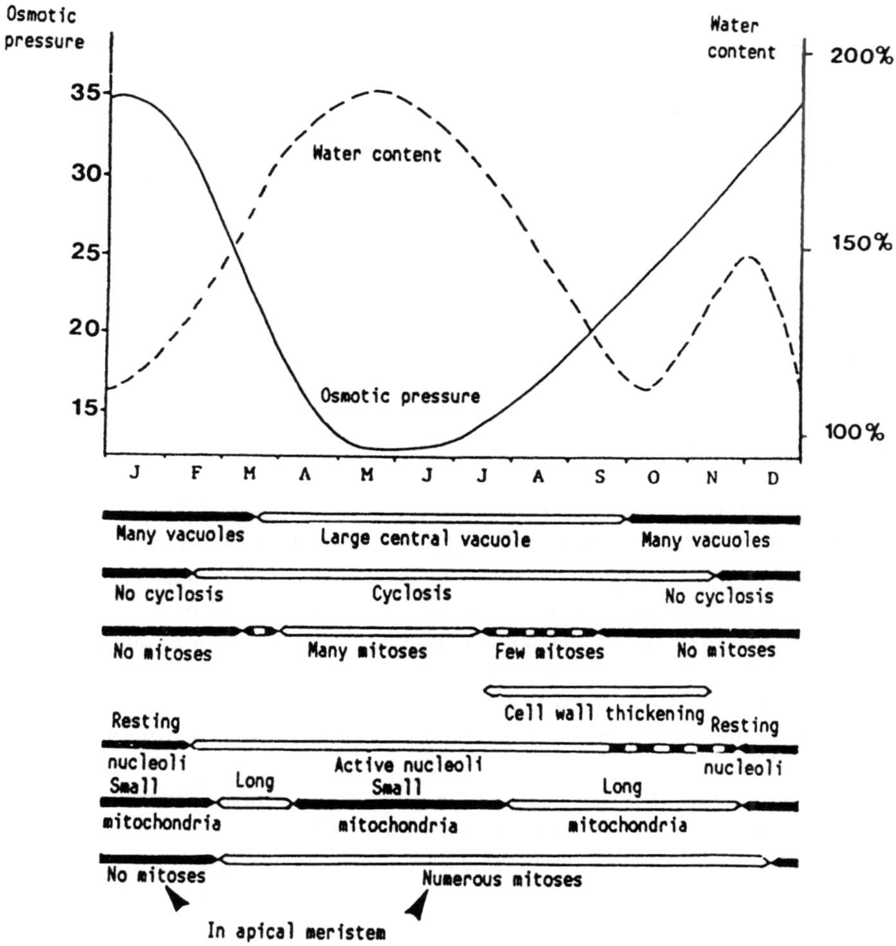

Fig. 5.16. Schematic drawing of seasonal variations observed in the cambium and apical meristems of 2-year-old *Acer pseudoplatanus* branches. Osmotic pressure is expressed in atmospheres, water content in percent dry weight. (Catesson 1980)

cambial activity, and it persisted long after mitoses ceased. In the fall, the granular zone again gave way to the dense fibrillar zone as the nucleoli contracted in size. Minimum nucleolar size occurred when osmotic pressure was maximum during cambial dormancy and vice versa during cambial activity (Fig. 5.16). A similar trend was reported by Dave and Rao (1981) for *Tectona grandis*. The area occupied by nucleoli was at a minimum in winter and reached a maximum in May, preceding the resumption of cambial activity.

Table 5.3. Seasonal variation of nucleolar volumes in ray cell initials of *Acer pseudoplatanus*[a] (Catesson 1980)

January	February	March	April	May	August	October	November
0.5	2	2	8	8	2	0.5	0.5

[a] Nucleolar volumes in μm^3.

5.3.2 The Vacuole

Vacuoles are membrane-bound cavities within the cytoplasm filled with a watery liquid, the cell sap, rather than with cytoplasm. They function in the maintenance of water balance within the cell and in the accumulation of various substances either in true solution or the colloidal state. In contrast to most meristematic cells, fusiform and ray cell initials are highly vacuolate. As pointed out by Raven (1987), 90% of the cell volume of highly vacuolate cells might consist of vacuoles. It was this vacuolation that led Krabbe (1886) to suggest that the associated increase in turgor would permit cambial cells to expand by "sliding growth" (Chap. 6.7.1).

Misconceptions as to the nature of the vacuole in cambial and other meristematic cells by early botanists were pointed out by Bailey (1930). In his own research on ten conifers and ten dicotyledons, he observed a large centrally located vacuole in fusiform initials during the active growing season. Bailey observed that the tonoplast, the term introduced by de Vries (1885) for the tenuous membrane surrounding the vacuole, appeared to move with the streaming protoplasm. During winter and periods of cambial inactivity when protoplasmic streaming ceased, the vacuoles underwent constant changes in size and shape. Ray cell initials were also conspicuously vacuolated but their vacuolar volumes were proportionately less than in fusiform initials. Two distinct categories of vacuoles were distinguished by their reactions to acidic and alkaline stains. The alkaline type stained orange or reddish-orange in aqueous solutions of neutral red, whereas the acidic type stained an intense magenta in the same solutions. Both types of vacuole often occurred within a single cell (Bailey and Zirkle 1931). By utilizing similar tests, Prat (1948) considered the xylem to be acidic, the phloem alkaline, and the cambium neutral.

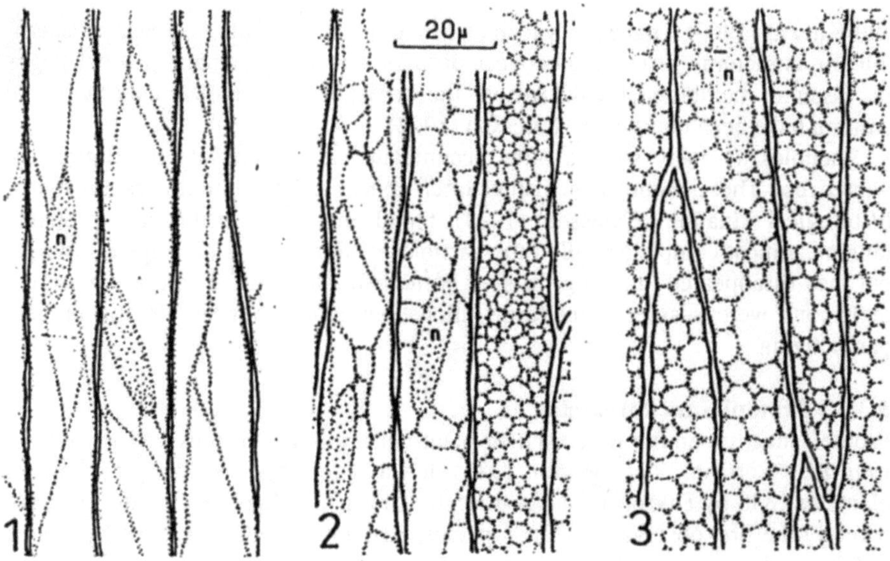

Fig. 5.17, 1-3. Drawings of cambial cells of *Acer pseudoplatanus* in the tangential plane showing condition of the vacuoles and position of the nucleus (*n*) in May (*1*), October (*2*), and December (*3*). (Catesson 1962b)

Seasonal changes in size and structure of the vacuoles in *Robinia pseudoacacia* were studied by Buvat (1956a,b). He, too, found the vacuoles of cambial cells to be disaggregated into many small spherical bodies pressed closely together during winter rest and to reaggregate as a single large vacuole during the resumption of growth. The many small vacuoles that formed around the nucleus, the phragmoplasts, and at the cell tips of *Fraxinus excelsior* had a "foamy appearance" during active cambial division in the spring (Goosen-de Roo and Spronsen 1978).

In *Holoptelea integrifolia*, it appeared that vacuolar fragmentation at the onset of dormancy began with dilation of the endoplasm (Rao and Dave 1983a). Rao and Catesson (1987) demonstrated how protrusion of a tonoplast into a vacuole and its later isolation could account for the numerous membrane remnants often found within vacuoles of cambial cells. This so-called "autumnal endocytotic membrane flow" contributed to the destruction of many cytoplasmic membranes and their sequestration in the vacuoles.

Fragmentation of the large central vacuole into many smaller ones was called "microvacuolation" by Srivastava and O'Brien (1966a). Similar seasonal changes in the vacuoles of cambial cells have been recorded in other species and the phenomenon appears to be ubiquitous in both gymnosperms (Bannan 1962b, Murmanis 1971, Timell 1973, Tsuda 1975a,b, Hirakawa et al. 1979) and dicotyledons (Catesson 1962b, Srivastava 1966, Kidwai and Robards 1969, Robards and Kidwai 1969, Evert and Deshpande 1970, Mia 1970, Farooqui and Robards 1979, Rao 1985, Barnett 1992) (Figs. 5.16, 5.17). According to Tsuda (1975a) and Robards and Kidwai (1969), change in vacuolar structure is one of the few easily discernible ultrastructural differences between dormant and active cambia. Several stages in vascular membrane breakdown at the cessation of cambial activity in the autumn have been outlined by Catesson (1992).

Vacuoles of many species often accumulate various constituents during the winter rest. These vacuolar inclusions often appear as "precipitates" as a result of histological fixation. Catesson (1974) found that such inclusions, probably tannins, almost completely filled the vacuoles of *Robinia pseudoacacia* in winter but they disappeared in spring. In *Acer pseudoplatanus*, on the contrary, vacuolar inclusions were scarce in fusiform initials but more abundant in the phloem parenchyma.

5.3.2.1 Protoplasmic Streaming

Protoplasmic streaming (cyclosis) in plant cells was first discovered in cells of *Chara* by Corti in 1774 (Morton 1981). It was later observed in cells of higher plants in 1838 by Dutrochet and in 1855 by Unger (de Vries 1885). However, apparently the first observation of this process in cambial cells was made by Velten (1872) studying under the direction of Nägeli. Velten recognized six principal kinds of "Bewegungen des Protoplasmas," or motion of protoplasm, which were either variants of the same process or artifacts. By motion or rotation he meant the movement of cell plasma in a broad circulating stream along the cell wall. Previously, this streaming motion had been demonstrated in relatively few plant cells, and Velten's objective was to show that the process was widespread in plants. He first detected it in cells of *Sida napae*, particularly in the cambial cells of tall rays. Cambial cells were found to be especially favorable material, and the process was observed in many species. The cambia of trees did not respond well when sectioned in water. However, when the cells were placed in a solution of gum arabic, rotation was observed in cambial cells of *Carpinus betulus, Pinus pumila, Quercus sessiflora*, and other species. Although Velten could not detect protoplasmic streaming in dividing cambial cells, he did observe it in those cells about to divide (Velten 1875). These cells could be distinguished from their neighbors not only by their greater volumes, but also by their more active protoplasmic streaming.

Velten's observations were confirmed by Russow (1881) in *Pinus sylvestris*. Like Velten, he demonstrated "plasma rotation" not only in cambial cells but also in young wood, bast, and ray cells. In wood cells, rotation ceased with the appearance of the secondary walls, and in phloem cells with formation of the callose plate. In developing ray cells immersed in distilled water, rotation continued for about 36 h; Bailey and Zirkle (1931) kept cambial cells alive and the protoplasm actively streaming for 1710 h in tap water. Russow (1882b), and later Wieler (1887), demonstrated protoplasmic streaming in cambial cells of several other conifers as well as a number of dicotyledonous trees. Priestley et al. (1933) also reported that they observed the "cytoplasm streaming" in newly formed cambial cells scraped from the bark surface and floated in water.

The so-called "streaming of protoplasm" was believed to be the process by which vast quantities of substances were moved within plants. In fact, de Vries (1885) suggested protoplasmic streaming as a more rapid and efficient means of transporting organic substances than Sach's theory of diffusion. Janse (1914) observed this "movement of protoplasm" in vascular tissues of *Caulerpa*, and suggested that it was intimately involved in the "impulsion basipétale," the impetus that presumably facilitated the descent of sap in trees.

The fusiform initials of conifers are characterized by large, elongated vacuoles during the active growing season (Bailey 1930). The parietal protoplasm, with its included organelles, rotates rapidly about this vacuole in a single ascending and descending stream. According to Bailey, cambial cells exhibit the most spectacular case of protoplasmic streaming, or cyclosis, of any internal tissue of higher plants. Contrary to the reports of Velten (1872, 1875), Bailey found that protoplasmic streaming was not arrested during the cell division process. It was, however, arrested in both conifers and dicotyledons during winter months when the temperatures dropped to near or below the freezing point. Unlike the rapid streaming in fusiform initials, that in ray cell initials appeared to be relatively sluggish.

Abbe and Crafts (1939) described protoplasmic streaming in the immature sieve elements of *Pinus strobus* that resumed their development in the spring (Chap. 11.4). At this time, the parietal layer of cytoplasm became progressively thinner and "It streams around the cell in a steep spiral until it reaches the tip, where it mills around and returns down the opposite side." Goosen-de Roo et al. (1983) observed the most prominent cytoplasmic streaming in fusiform cells that were differentiating to fibers in *Fraxinus excelsior*. The streaming cytoplasm occurred in longitudinal paths either within or against the parietal cytoplasm. Cytoplasmic streaming was also present in the small ray cell initials but it was slower and more varied in direction.

Catesson (1962b, 1981) related the rate of cyclosis in cambial cells of *Acer pseudoplatanus* to changes in osmotic pressure of the cytoplasm (Fig. 5.16). During active growth, the vacuole was large and more or less solitary, osmotic pressure was relatively low, and cyclosis was extremely active. As cambial dormancy set in the large vacuole was reduced to numerous small ones, osmotic pressure increased gradually to a maximum, and cyclosis completely ceased.

The rate of cytoplasmic streaming in cambial cells of *Pinus strobus* was measured under controlled conditions by Thimann and Kaufman (1958). In this species, the streaming rate increased sharply in late February, attained a maximum at the end of March, and then declined slowly until reaching a steady low value during the last 3 months of the year (Fig. 5.18). The streaming rate increased almost linearly with temperature from 5 to 34° C. It stopped completely at -1° C and showed irreversible damage above 42° C. The authors suggested that even brief exposure of cambial cells to 40° C temperature might act as a major ecological factor limiting the range of *P. strobus*.

Thimann and Kaufman demonstrated that cyclosis was at a maximum during the peak of mitotic activity. Because these events also coincided with that phase of the mitochondrial cycle when mitochondria were shortest, Catesson (1974, 1981) suggested the possibilty of a causal relation between them. Tsuda (1975b) and Parthasarathy and Pesacreta (1980) found strong evidence for the involvement of microfilaments in cytoplasmic streaming in cambial cells. For a more thorough discussion of cytoplasmic streaming and related processes in cambial cells see Catesson (1990).

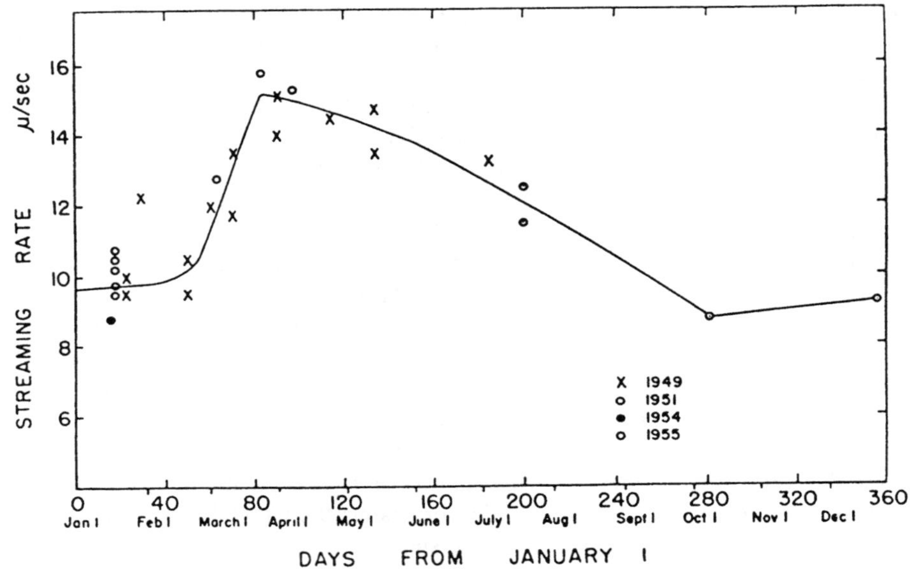

Fig. 5.18. Seasonal variation in the rate of protoplasmic streaming in cambial cells of *Pinus strobus*. Average of 4 years' data. (Thimann and Kaufman 1958)

5.3.3 Organelles

Cambial cells are rich in organelles, cellular structures that perform specialized functions. For the most part, these organelles are the same as those found in other meristematic cells (Hohl 1960). Differences occur seasonally within a species and also among different species, but these differences are usually organizational and quantitative rather than qualitative. Most workers have emphasized the fact that qualitative differences among species are relatively minor (Fig. 5.19).

Plastids are the organelles found in cambial cells that vary most both seasonally and among species. It is therefore not unexpected that plastid inclusions are the most variable cytoplasmic constituents. These inclusions consist of storage pro-

Cytological Features of Dormant Cambium

Author	Species	Dictyosomes	Endoplasmic Reticulum	Polysomes/Ribosomes
Srivastava & O'Brien, 1966	*Pinus strobus*	Numerous	Mostly smooth vesicles	Polysomes or free ribosomes
Kidwai & Robards, 1969	*Fagus sylvatica*	Few, Golgi vesicles	Smooth vesicles, some rough cisternae	Polysomes and free ribosomes
Robards & Kidwai, 1969	*Salix fragilis*	Few, no vesicles	Mainly smooth vesicles	Mostly free ribosomes
Itoh, 1971	*Cryptomeria japonica* *Pinus thunbergii*	Many vesicles and cisternae	Mostly smooth	Polysomes or free ribosomes
Murmanis, 1971	*Pinus strobus*	Few, many vesicles	Rough, sparsely distributed	Polysomes or free ribosomes
Timell, 1973	*Picea abies*	Frequent, several vesicles	Mostly smooth, few rough cisternae	Free ribosomes
Catesson, 1974	*Acer pseudoplatanus* *Robinia pseudoacacia*	Few, no vesicles	Mostly rough, few smooth cisternae Smooth	Free ribosomes Mostly free ribosomes
Tsuda, 1975	*Abies firma* *Chamaecyparis obtusa* *Cryptomeria japonica* *Pinus densiflora*	Less abundant than in active cells	Smooth (at least in fusiform cells)	Free ribosomes
Rao & Dave, 1983a	*Holoptelea integrifolia*	—	Smooth, rarely rough	Mostly free ribosomes
Rao & Dave, 1983b	*Tectona grandis*	Few, inactive	Smooth	Free ribosomes
Rao, 1985	*Aesculus hippocastanum*	Few, no vesicles	Smooth	Mostly free ribosomes
Sennerby-Forsse, 1986	*Salix viminalis*	Few, no vesicles	Mostly smooth	—

Fig. 5.19. Cytological features of dormant cambia in different species of trees. (Fahn and Werker 1990)

ducts in one form or another. Plastid lamellae, however, are usually poorly developed, perhaps because of the almost complete absence of light reaching the cambium.

Catesson (1974, 1980) summarized the various reports on plastids in cambial cells and classified them as belonging to either the "dense inclusion" or the "classical type." In the former type the lamellae contained dense material and in the latter type they were characterized by the presence of electron translucent vesicles and thylakoids. Other ultrastructural differences among species were found to be relatively minor such as the presence or absence of coated vesicles, the quantities of osmiophilic granules, or the existence of peroxisomes.

5.3.3.1 Gymnosperms

One of the more thorough ultrastructural studies was that of Timell (1973) on the cambium of *Picea abies*. Dormant cambial cells contained the usual complement of cell organelles, including mitochondria and a few, small Golgi bodies (Fig.

11.4). Only a few plastids contained starch, but lipid droplets were common. Distinguishing features of the plasmalemma were numerous invaginations that often contained granular or fibrillar materials. The most conspicuous change at the beginning of cambial activity was the disappearance of many small vacuoles and formation of a large vacuole in differentiating cells. A concomitant change occurred in distribution of the cytoplasm and its organelles, which were now located in a thin layer between the plasmalemma and the tonoplast. Other fine structural changes were relatively minor. Golgi bodies were more numerous in the active cambial cells and rough endoplasmic reticulum replaced much of the smooth. Almost identical seasonal changes were found in *Abies balsamea* (Timell 1979).

A comparison of resting and active cambia in 28 species of gymnosperms was made by Tsuda (1975a). He found that cambial cells of all species examined, whether resting or active, contained the same cytoplasmic components, although their distributions varied both seasonally and among species. As Timell (1973) found, the main differences between resting and active cambial cells were associated with the structure of the vacuole and distribution of cytoplasm and its organelles. Lipid droplets were common in resting cambia of the Pinaceae and Sciadopityaceae, but less so in cambia of Cupressaceae, Podocarpaceae, Taxaceae, and Taxodiaceae. These lipid droplets, being storage products, disappeared at the onset of cambial activity. Protein bodies were rare in all species examined and starch-containing proplastids were never observed. Other distinguishing features were the presence of smooth versus rough endoplasmic reticulum in resting and active cambia, respectively, and an increase in Golgi bodies and microtubules in the latter.

Investigators examining other coniferous species have encountered similar differences between resting and active cambia (Srivastava and O'Brien 1966a, Itoh 1971, Murmanis 1971, Barnett 1972, 1973). Although the differences appear to be relatively minor, the ultrastructural changes that occur at the beginning of cambial activity are of critical importance. They are preparatory to the rather drastic changes associated with the vital functions of cambial cell division and differentiation of the derivatives.

Species differences in cambial ultrastructure also appear to be minor and essentially insignificant. No differences of consequence have been reported in cambia of *Abies balsamea* (Timell 1979, Riding and Little 1984), *Cryptomeria japonica* and *Pinus thunbergii* (Itoh 1971), *Metasequoia glyptostroboides* (Kollmann and Schumacher 1961), *Pinus radiata* (Barnett 1973, 1975), and *Pinus strobus* (Srivastava and O'Brien 1966a, Murmanis 1971). Neither have any significant differences been recorded in cambial cell structures of normal and compression wood (Timell 1973, 1979, 1980, 1986).

5.3.3.2 Dicotyledons

Many of the same conclusions can be applied to cambial cells of dicotyledonous trees, although species differences are more common. A particularly detailed comparison of active and dormant cambial cells in *Tectona grandis* was provided by Rao and Dave (1983c). Active cambial cells were rich in all types of organelles. Golgi bodies were abundant in active cells, with hypertrophied cisternae producing numerous vesicles. Rough endoplasmic reticulum was also abundant and it occurred in the form of typical ribosome-studded cisternae. The ground meristem was rich in free ribosomes and aggregates, often appearing in circular or spiral patterns of polyribosomes. Plastids were highly variable in shape and size. They were often clustered around the nucleus, and their interiors consisted of electron-dense inclusions such as starch, plastoglobuli, and a system of tubules. Mitochondria, which are common organelles in all cambial cells, were wide in active and narrow in resting fusiform initials.

Dormant cambial cells contained all the organelles found in active cells, although they differed in amount and structure. Golgi bodies were rare and appeared to be inactive with no hypertrophied cisternae. Golgi vesicles were also infrequently encountered in the ground meristem. Endoplasmic reticulum was far less abundant and it occurred in the form of smooth, elongated cisternae with numerous separate vesicles. Ribosomes were isolated and cytoplasmic, with no polyribosome formation. Plastids also occurred in groups around the nucleus, but the membrane-bound inclusions appeared different from those in active cells. The main difference between resting and active cells was the complete absence of starch in the fusiform initials and its rare presence in ray cell initials.

Salix fragilis also provides a typical example of a dicotyledon (Robards and Kidwai 1969). Like the conifers, the resting cambium contained numerous small vacuoles and the plasmalemma displayed prominent invaginations. The cytoplasm was rich in smooth endoplasmic reticulum and numerous ribosomes. Dictyosomes of the Golgi apparatus were present but not abundant. The most striking feature of the resting cambium was the large amount of space occupied by lipid droplets and protein bodies. The latter, by contrast, are rare to nonexistent in resting conifer cambia. In the active cambium, rough endoplasmic reticulum replaced the smooth, free ribosomes were fewer, and those present were in aggregations. The cambium of *Fagus sylvatica* was quite similar in ultrastructure to that of *Salix* (Kidwai and Robards 1969, Farooqui and Robards 1979). Except for a few organelles, cytoplasmic contents remained more or less constant throughout the year, differing in degree rather than in nature.

Relatively minor differences have been reported in other dicotyledonous species such as *Acer pseudoplatanus* (Catesson 1964, 1974), *Aesculus hippocastanum* (Rao 1985, Rao and Catesson 1987, Barnett 1992), *Fraxinus americana* (Srivastava 1966), *Robinia pseudoacacia* (Buvat 1956a, Pomeroy and Siminovitch 1971), *Salix* spp. (Sennerby-Forsee et al. 1984), and *Holoptelia integrifolia* (Rao and Dave 1983a), a species with storied cambium. Mia (1970) noted that organelles of *Tilia americana* differed in certain details from other species, but Evert and Deshpande (1970) found few differences either in *T. americana* or *Ulmus americana*. The main difference they encountered was the lack of protein bodies in these species. A more detailed discussion of the ultrastructural differences among species can be found in Catesson (1974).

A recent review by Lachaud (1989) on the role of hormones in cambial activity included a rather detailed summary of the ultrastructural and cytochemical differences that occur in cambial fusiform initials during the active and resting states. Figure 5.20 is a hypothetical rendition of these states based on information gleaned from numerous investigations of both conifers and dicotyledons.

As in the case of fusiform initials, the main difference between resting and active ray cell initials appears to be in the size and structure of the vacuoles and in the plastids and their inclusions (Kidwai and Robards 1969, Robards and Kidwai 1969, Itoh 1971, Dave and Rao 1982c, Rao 1985). Fine-structure differences between fusiform and ray cell initials are also relatively minor in both conifers (Srivastava and O'Brien 1966a, Itoh 1971, Timell 1973, 1979, Tsuda 1975a,b, Riding and Little 1984) and dicotyledons (Kidwai and Robards 1969, Robards and Kidwai 1969).

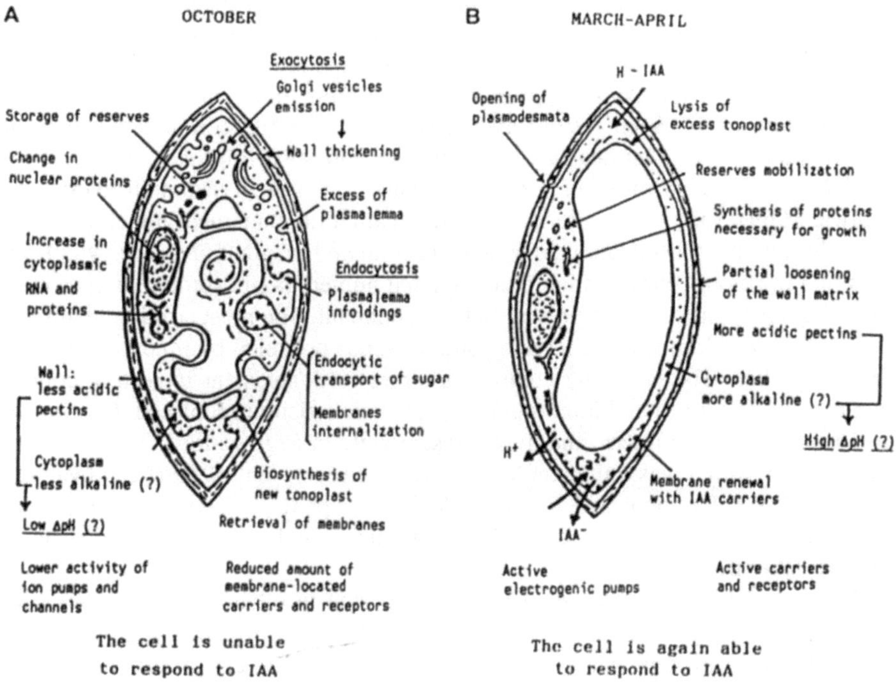

Fig. 5.20. Hypothetical scheme of events during the transition from rest (October) to activity (March-April) in cambial cells. (Lachaud 1989)

5.4 Cytochemistry

In an actively dividing cambial zone, it is almost impossible to distinguish initial cells from their immediate derivatives on the basis of ultrastructural differences. However, according to a recent report by Catesson (1992), parallel biochemical and cytochemical studies have disclosed progressive changes in radial cell wall composition across the cambial zone. She found that phloem derivatives were characterized by the early deposition of a cellulose microfibrillar skeleton and a low amount of xylans. In contrast, cellulose microfibrils were scarce in the radial walls of xylem derivatives, while xylans were rather abundant. Concomitantly, the content of acidic pectins and calcium ions in cell walls appeared to increase from the phloem to the xylem side of the cambial zone. Catesson suggested that the presence of cellulose microfibrils or the activity of certain enzymes, such as pectin methylesterase and xylan synthases, might be useful clues in recognizing the first steps of xylem and phloem differentiation in very young cambial derivatives. For more on cambial cell walls see Timell (1986) and Chapter 11.2.

In most cytochemical studies of the cambial zone, either the absence, presence, or relative abundance of organelles has been accorded greatest attention. As noted in Section 5.3.3, plastids are the organelles found in cambial cells that vary most both seasonally and among species. Plastids perform storage functions, either in the form of starch, lipids, proteins, or phytoferritin. Other cytochemical constituents found in cambial cells in smaller concentrations are tannins, crystals, and a large number of miscellaneous products.

5.4.1 Starch

One of the earliest studies of cambial chemistry was that of Hartig (1863d) who analyzed the "cambial sap" of several coniferous species and made quantitative analyses of protein, sugar, and several other constituents.[1] Years earlier, in 1834, Hartig had been first to demonstrate the presence of starch in wood cells (Sanio 1863a). Sanio (1873), however, was unable to detect either starch or chlorophyll in cambial cells of *Pinus sylvestris.*

Hartig (1863c) considered the cambial ray cells to be starch-free during the active growing season. This observation was confirmed by Russow (1882b) on several conifers and dicotyledons. In spring, before bud-break, Russow observed that the ray cells were rich in starch, and only in the region of the cambium were they starch-free. A similar conclusion was reached by Nicoloff (1911) in his investigation of a number of dicotyledonous species.

---

[1] For a review of the literature on the constituents in cambial sap of trees see Stewart (1957, 1966).

5.4.1.1 Gymnosperms

More recent reports of seasonal starch deposition patterns in fusiform cambial initials are by no means consistent. Conifer cambia as a rule are free of starch during dormancy and rich in starch during periods of active growth. Amyloplasts, or starch-containing plastids, were not observed in resting fusiform initials of either *Pinus radiata* (Barnett 1973) or *Pinus strobus* (Srivastava and O'Brien 1966a, Murmanis 1971). They were rarely observed in *Cryptomeria japonica, Pinus thunbergii* (Itoh 1971), *Metasequoia glyptostroboides* (Kollmann and Schumacher 1961), and most conifers examined by Tsuda (1975a). Plastids were abundant in *Picea abies,* and while most of them did not contain starch, a considerable number did so (Timell 1973).

Dormant ray cell initials of conifers generally contain more amyloplasts than fusiform initials do. Yet, those in some species such as *Abies balsamea* (Timell 1979) were found to be relatively starch-free during the dormant period whereas others such as *Picea abies* (Timell 1973) contained some starch, and still others such as *Cryptomeria japonica, Pinus thunbergii* (Itoh 1971), and *Metasequoia glyptostroboides* (Kollmann and Schumacher 1961) were rich in starch. These inconsistencies might be attributable, in part, to differences in the time of year during which sampling occurred (Catesson 1980), or to technical problems involved in investigating these thin and fragile meristems (Catesson 1990).

5.4.1.2 Dicotyledons

Unlike conifers, cambial cells of many dicotyledons are rich in starch during winter rest and relatively starch-free during periods of active growth (Sundvor 1956, Essiamah and Eschrich 1985). However, as low winter temperatures set in, much of the starch is converted to sugars and then back to starch when spring temperatures increase (Pomeroy and Siminovitch 1971). Starch grains were observed in resting fusiform initials of *Acer pseudoplatanus* (Catesson 1974), *Fagus sylvatica* (Kidwai and Robards 1969, Farooqui and Robards 1979), *Fraxinus americana* (Srivastava 1966), *Robinia pseudoacacia* (Buvat 1956a), and *Salix fragilis* (Robards and Kidwai 1969). Ray cell initials display similar seasonal patterns although they usually contain more starch during winter dormancy.

An entirely contrary pattern was observed in *Aesculus hippocastanum* by Rao (1985). The resting cambium in this species was characterized by the complete absence of starch during the winter. Starch reappeared in the plastids in March, and starch dissolution began in August. Carbohydrates were stored mainly as soluble sugars in vacuoles of the cambial cells. Further research on this same species by Barnett (1992) showed that winter storage occurred in the form of lipid bodies. In March, the lipid bodies disappeared from the cambial boundary-layer cells and this loss was accompanied by the accumulation of starch grains in the amyloplasts.

Starch was also completely absent from cambial initials of *Salix* spp. although it was present in ray cell initials during winter (Sennerby-Forsee et al. 1984). No mention was made of the cambium by Essiamah and Eschrich (1985). However, their photomicrographs of *Acer pseudoplatanus*, *Fagus sylvatica*, and *Fraxinus excelsior* showed no evidence of starch in dormant cambia.

Starch patterns in cambial cells of tropical species also coincide with seasonal cycles of cambial activity and leaf development. Both fusiform and ray cell initials of *Tectona grandis* were found to be essentially starch-free during the dormant period although the adjacent xylem elements were filled with starch (Dave and Rao 1982c, Rao and Dave 1983b,c). Starch grains began to appear in cambial cells with the appearance of leaves and they increased in number during the season. The number of starch grains per plastid ranged from one to many. Starch accumulation was greater in ray cell than in fusiform initial plastids, and the starch grains more or less completely filled the plastid interiors. Starch grains began to disappear at season's end, coincident with cessation of cambial activity and aging of the foliage.

A somewhat different pattern was found in *Holoptelia integrifolia* (Rao and Dave 1983a, Rao and Menon 1989). Starch was present in ray cells throughout the year but its distribution varied. Maximum starch content occurred in ray cell initials near season's end in November-December and it began to decline in February when leaves were shed. During the winter period the plastid stroma was densely packed with from one to four large starch grains. Slightly different patterns were observed in *Gmelina arborea* (Rao and Dave 1983b) and *Mangifera indica* (Rao and Menon 1989), but in each case they could be related to specific growth cycles of the species.

5.4.2 Lipids

Lipid droplets are often present in more or less large quantities in both fusiform and ray cell initials in resting cambia of conifers, as for example in *Abies balsamea* (Timell 1979, Riding and Little 1984), *Cryptomeria japonica* (Itoh 1971), *Picea abies* (Timell 1973), *Pinus radiata* (Barnett 1981), *P. strobus* (Srivastava and O'Brian 1966a, Murmaris 1971), and *P. thunbergii* (Itoh 1971). Lipid droplets were particulary numerous in the Pinaceae and Sciadopityaceae examined by Tsuda (1975a). The fact that some species, such as members of the Pinaceae, preferentially store lipids prompted Fischer (1891) to refer to them as "fat" trees and those that mainly stored starch as "starch" trees.

Lipid droplets are more variable in occurrence in the cambial cells of dicotyledons during winter rest and they tend to disappear at the onset of cambial activity. They were present, although not numerous, in resting cambia of *Aesculus hippocastanum* (Rao 1985), *Fagus sylvatica* (Kidwai and Robards 1969, Robards and Kidwai 1969), *Fraxinus americana* (Srivastava 1966), and *Salix fragilis* (Robards and Kidwai 1969). Barnett (1992) found that lipid droplets were numerous in the boundary-layer cells of *Aesculus hippocastanum* during dormancy, but less so in other cells of the cambial zone.

Lipids were considered to be abundant in fall and winter cambia of *Acer pseudoplatanus* (Catesson 1964) and *Salix* spp. (Sennerby-Forsee et al. 1984). Cambial cells of *Tectona grandis* were rich with lipid globules during the dormant season (Rao and Dave 1983b). Distribution and size of the globules decreased in the latter species at the time of bud break, reaching a minimum when cambial activity was at a maximum. The number of lipid globules again increased concomitant with the cessation of cambial activity and onset of dormancy. A similar pattern was observed in *Gmelina arborea*. Lipid globules were also present, but to a somewhat lesser extent, in both fusiform and ray cell initials of *Holoptelia integrifolia* and *Mangifera indica* during the dormant season (Rao and Menon 1989). These, too, essentially disappeared during the active growth period.

5.4.3 Proteins

Protein bodies are rare to nonexistent in the resting cambia of conifers (Itoh 1971, Timell 1980). Tsuda (1975a) made no mention of protein bodies in his cytochemical analyses of 28 gymnosperm species. Plastids with osmophilic inclusions, thought to be proteinaceous, were observed in winter cambia of *Pinus radiata* by Barnett (1973). Although he was uncertain as to their true identity in cambial tissues, Barnett did recognize protein bodies in differentiating phloem cells in summer collections. High levels of total protein, analyzed histochemically, were recorded in fusiform initials of *Abies balsamea* during winter rest (Riding and Little 1984). Protein levels then decreased during active growth.

Protein bodies are usually present, but in varying amounts, in resting cambial cells of dicotyledonous trees, and they tend to decline during active growth (Srivastava 1966, Kidwai and Robards 1969, Robards and Kidwai 1969, Farooqui and Robards 1979). In 11 dicotyledonous trees examined by Wetzel et al. (1989), the winter cambium as well as xylem ray cells were found to be packed with varying numbers of protein bodies. Cambial cells of *Betula papyrifera* contained smaller and more scattered protein bodies than those of other species, suggesting to the authors that lipid storage might be more important than protein storage in this species. Xylem ray cells of *Salix caprea* were also found by Sauter and Wellenkamp (1988) to contain numerous protein bodies.

No protein bodies were observed in fusiform initials of active cambia in *Tilia americana* and *Ulmus americana* (Evert and Deshpande 1970); these authors did not examine dormant cambia but they suggested that protein bodies might have been present had they done so. Protein bodies were present in dormant fusiform and ray cell initials of *Tectona grandis* and *Gmelina arborea* (Rao and Dave 1983b). No mention was made of their presence in cells of active cambia. In *Holoptelia integrifolia* and *Mangifera indica*, protein bodies were poorly distributed in fusiform and ray cell initials during both the active and dormant periods (Rao and Menon 1989).

The protein-rich, spherical organelles in cambial cells of *Salix microstachya* were referred to as "protein-storage vacuoles" rather than protein bodies by

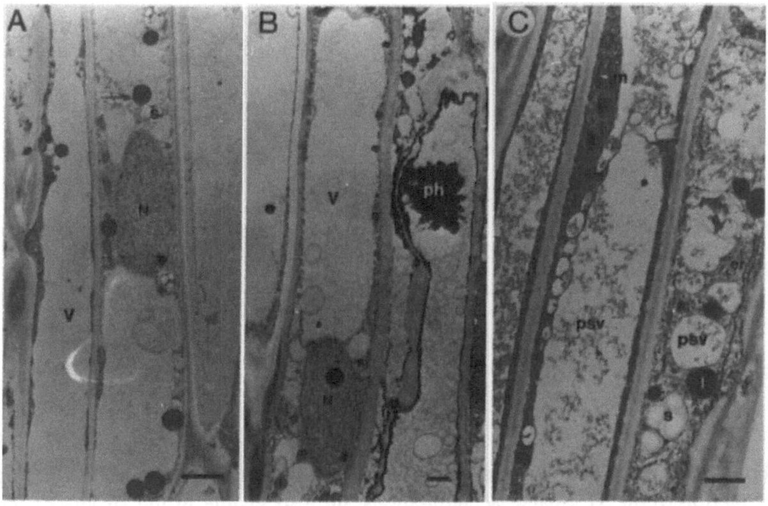

Fig. 5.21A,B. Seasonal variation in protein content in fusiform cambial cells of *Salix microstachya* shoots during mid-summer (**A**), late summer (**B**), and fall (**C**). *er* Rough endoplasmic reticulum; *l* lipid body; *m* mitochondrion; *N* nucleus; *ph* phenolic deposit; *psv* protein-storage vacuole; *s* starch grain; *V* vacuole. Tangential EM photomicrographs. **A, B** *bars* 2 µm; **C** *bar* 1 µm. (Greenwood et al. 1990)

Greenwood et al. (1990). Deposition of protein, visible as floccular material in the vacuoles of fusiform cambial cells, began appearing in mid-September and was essentially complete by mid-October in northern Canada (Fig. 5.21). At this time, the large central vacuole had been replaced by numerous protein-storage vacuoles in many of the cambial cells. Starch-containing amyloplasts, lipid bodies, and active mitochondria were also prominent features in all cells. Stacks of rough endoplasmic reticulum cisternae were present in the peripheral cytoplasm, and invaginations of the plasmalemma were common in autumn collections. However, in cambial cells where protein-storage vacuoles were abundant, the endomembrane system, primarily the rough endoplasmic reticulum, lost its organization and the cytoplasm developed a finely vesiculated appearance. By mid-January starch had been mobilized from the plastids of cambial cells, and lipid bodies were common cellular components.

5.4.4 Phytoferritin

Phytoferritin was first reported as being present in plastids of cambial cells of *Acer pseudoplatanus* by Catesson (1966) (Fig. 5.22) and a year later in those of *Salix fragilis* by Robards and Humpherson (1967). It is considered a storage product in cells such as cambial initials that are incapable of performing photosynthe-

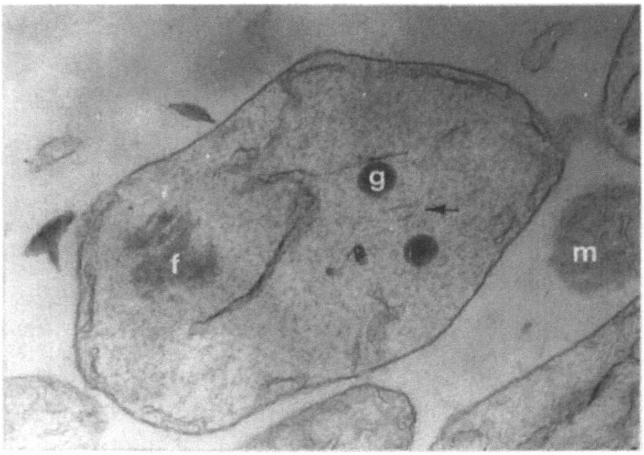

Fig. 5.22. Proplastids in a cambial fusiform initial of *Acer pseudoplatanus* containing inclusions of phytoferretin (*f*). Sparse accumulations are also visible in the stroma of the proplastids (*arrow*). *g* Osmophilic globules; *m* mitochrondrion. 48 000x (Catesson 1966)

sis. Moreover, because phytoferritin is usually present as an iron-protein complex, it is possibly a means of storing iron in a form that is nontoxic to the cells (Robards and Humpherson 1967).

Phytoferritin was found to be present in plastids of cambial cells of *Cryptomeria japonica* in both the active and dormant states, but absent in those of *Pinus thunbergii* at both times of the year (Itoh 1971). No phytoferritin was found in cambial cells of *Abies balsamea* (Timell 1973). Tsuda (1972, 1975a) reported phytoferritin to be present in proplastids of ray cell initials of the Cupressaceae, Podocarpaceae, and Taxodiaceae, but absent in fusiform initials of the same families. It was also absent in all cambial cells of the Pinaceae, Taxaceae, and Sciadopityaceae.

Reports on the presence of phytoferritin in cambial cells of dicotyledonous trees are variable. Robards and Kidwai (1969) mentioned it as being present in active cambial cells of *Salix fragilis*, but they did not discuss its presence or absence in resting cells. Czaninski (1968) and Catesson (1974) reported phytoferritin as being present in the differentiating vessel associated cells and cambial initial cells, respectively, in *Robinia pseudoacacia*. Evert and Deshpande (1970) did not detect phytoferritin in cambial cells of *Tilia americana* and *Ulmus americana*, although it was present in young axial and ray parenchyma cells near the cambium. Barnett (1992) reported the presence of phytoferritin in ray cell initials of *Aesculus hippocastanum*.

In *Tectona grandis*, phytoferritin was present in rather large quantities during the dormant period (Rao and Dave 1983b,c). It disappeared during the early stages of cambial activity and reappeared late in the season. The authors suggested that it might substitute as a storage material for starch, which is essentially absent during the dormant period in this species. Nwankiti (1982) reported the presence of phytoferritin in cambial zone cells of *Bougainvillea spectabilis*. He agreed with Robards and Humpherson (1967) that it probably served as a nontoxic storage form for iron in the cambium.

5.4.5 Tannins

Tannins were detected by Rao (1985) in cambial fusiform initials of *Aesculus hippocastanum* by their reaction with ferrous sulfate. The tannins accumulated in the vacuoles of resting cells in the form of dense bodies closely appressed against the tonoplast. Further tests with periodic acid-thiocarbohydrazide-silver proteinate suggested a linkage with cell polysaccharides.

Tannin deposition in cambial cells of *A. hippocastanum* coincided with starch accumulation. In late autumn, tannin was found in both the cytoplasm and vacuoles whereas in late winter it was confined to the vacuoles (Rao 1988b). On the contrary, autumnal tannin deposition in *Populus euramericana* was accompanied by starch disappearance. Tannin accumulated in both the cytoplasm and vacuoles, but particularly large deposits were made in the adjacent phloem ray and axial parenchyma cells. Tannins and lipids were degraded simultaneously during cambial reactivation in *A. hippocastanum*, suggesting to the author that tannins might play a nutritional role in early cambial development. Tannins were also reported in resting ray cell initials of *Holoptelea integrifolia* and *Mangifera indica* (Rao and Menon 1989).

5.4.6 Chlorophyll

Plastids of cambial cells do not as a rule contain chlorophyll, perhaps because light does not normally penetrate to the cambium. As Catesson (1974) pointed out, prolamellar bodies are infrequent in cambial plastids and when they do occur they do not form grana. The presence of chlorophyll and active photosynthesis have been recorded in the inner bark of thin-barked trees such as *Populus tremuloides* (Strain and Johnson 1963). Chlorophyll has also been detected in cambial and medullary ray cells of *Vitis vinifera* canes (Kriedemann and Buttrose 1971). However, it was not determined whether photosynthesis actually occurred in the cambial tissues. Proplastids containing rudimentary granal lamella were observed in the cambial ray cells of young, thin-barked stems of *Aesculus hippocastanum* (Barnett 1992). These putative chloroplasts were never present in fusiform cambial cells. Chlorophyll production commonly occurs in cambial tissue cultures cultivated in vitro (Vincent 1945).

5.4.7 Miscellaneous Constituents

Crystal-containing bodies were observed in resting fusiform initials of *Fraxinus americana* by Srivastava (1966) and crystals of calcium oxalate were reported in ray cell initials of *Gmelina arborea* and *Tectona grandis* by Rao and Dave (1984) and Deshpande and Vishwakarma (1992). A host of cytochemicals have been reported as being present in cambial cells such as minerals (Sieber 1912), flavonals (Feucht et al. 1986), leuco-anthocyanins (Hillis 1955), polyphenols (Hillis and Carle 1958), phospholipids (Vol and Levin 1984), triglycerides (Levin and Rubchevskaya 1985), prostaglandins (Levin et al. 1988), enzymes (Fukada et al. 1969, Kimmerer and Stringer 1988), isoenzymes (Santamour 1982), polysaccharides (Simson and Timell 1978, Timell 1986), and various hormones (Roberts et al. 1988). The foregoing list is simply indicative of the many constituents found in cambial cells, and it is by no means an exhaustive survey of the vast number of papers on the subject.

5.5 Cambial Cell Division

Although the general process by which cambial cells divide and the role of the nucleus in this process were known (Russow 1883a, Kienitz-Gerloff 1891), it was the work of Bailey that finally brought all the details into clear focus. In a series of papers, he (Bailey 1919b, 1920b,c,d) described the general process in *Pinus strobus* as follows: each fusiform initial was found to contain a single, ellipsoidal nucleus, centrally located with its long axis approximately parallel to the long axis of the cell. During mitosis, the mitotic figure was situated diagonally across the cell at an angle of 20 to 40° to the long axis of the cell (Fig. 5.23 A,D). Formation of the cell plate began when the obliquely placed spindles extended laterally by the addition of peripheral fibers (Fig. 5.23E). As more peripheral fibers were added, remains of the central fibers disappeared, leaving two separate aggregations of fibers, the so-called kinoplasmasomes, or phragmoplasts, which were connected to the first-formed part of the cell plate (Fig. 5.23B,C,F). The phragmoplasts then moved in opposite directions, extending the cell plate to the ends of the cell. This type of division resulted in two similar daughter cells each with its own nucleus, but with the two daughter nuclei lying close to one another (Fig. 6.80).

Later workers (Barnett 1972, Timell 1979) found that periclinal divisions of cambial cells in other conifers proceeded in much the same way as Bailey described. Moreover, the general features of the cell division process in cambial cells can be shown to be comparable to those in many other meristematic plant cells (Gunning and Wick 1985, Lloyd 1987, Venverloo and Libbenga 1987, Baskin and Cande 1990, Barlow 1991, Esau and Gill 1991). However, cambial cell division involves one major distinction. Cambial cells do not conform to Errera's (1888)

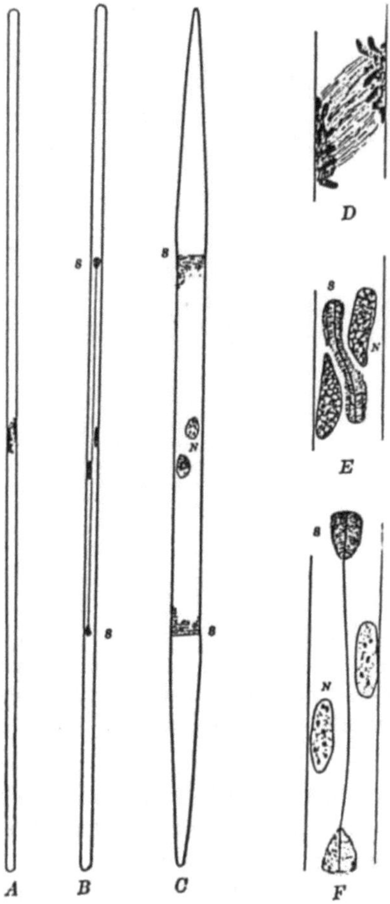

Fig. 5.23A-F. Schematic drawings of dividing cambial initials of *Pinus strobus*. **A** Radial view showing oblique position of karyokinetic figure. **B** Same, showing daughter nuclei and process of cell plate formation. **C**. Tangential view of **B**. **D** Detail of karyokinetic figure in **B**. **E** Radial view of beginning and **F** of a later stage of cell-plate formation. **N** Daughter nuclei; **S** aggregations of kinoplasmic fibers (phragmoplasts). (Bailey 1919b)

law, which states that cells tend to divide by walls of minimum area, but rather to walls of maximum area when they divide periclinally.

In one of his papers, Bailey (1920c) demonstrated that the same process of cytokinesis occurred in both conifers and dicotyledons. Pseudotransverse divisions of fusiform initials as well as divisions of ray cell initials in dicotyledons were found to be simply variations of the above process (Bailey 1920b). Figure 5.24 from Isebrands and Larson (1973), for example, shows progression of the phragmoplast and cell plate in dividing fusiform initials of *Populus deltoides*. The entire division process, as it occurred in cambial cells in the root of *Beta vulgaris*, can be easily followed in a series of photomicrographs prepared by Esau and Gill

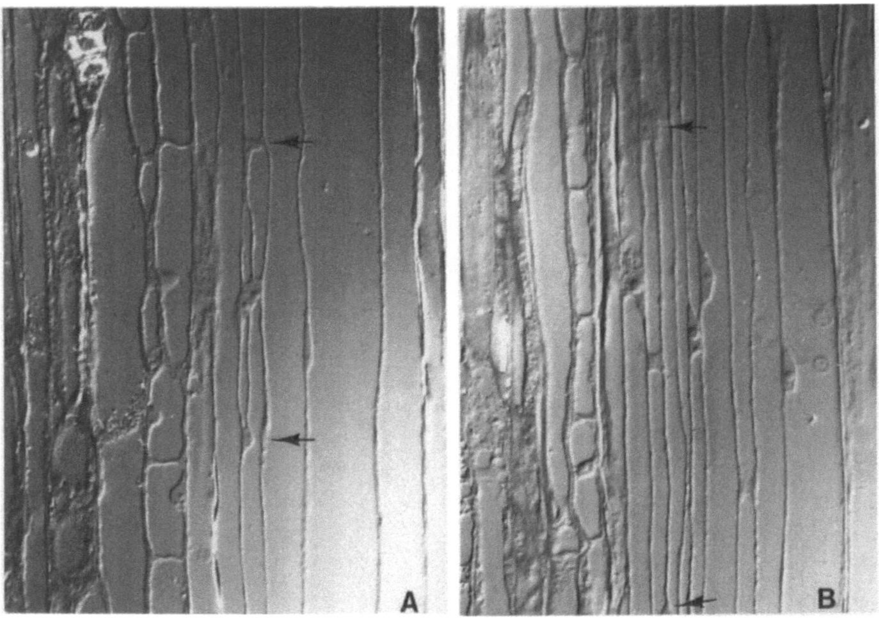

Fig. 5.24A,B. Dividing cambial cells at telophase in *Populus deltoides*. Phragmoplasts (*arrows*) are shown extending longitudinally from the divided nucleus at an early (**A**) and later (**B**) stage of cell-plate formation. Nomarski 460x. (Isebrands and Larson 1973)

(1965). A contemporary of Bailey, Kleinmann (1923), described nuclear division patterns in the cambium of several dicotyledonous trees. He found that the best time for examining nuclear division was between 03:30 and 07:30 h.

In a study of wound tissue of *Ricinus communis* and the cambia of several woody herbs, Sinnott and Bloch (1940, 1941) proposed the term "phragmosome" for the plate of cytoplasm indicative of the future wall site where the nucleus becomes located and divides. A phragmoplast, in contrast, is the spindle-shaped array of fibrils that forms between the daughter nuclei. It arises during telophase when the mitotic spindle disappears, and it is involved in construction of the cell plate. By utilizing electron microscopy, Frey-Wyssling et al. (1965) described and illustrated how Golgi bodies gathered at the cell periphery and within the phragmoplast. Later workers demonstrated that orientation of the Golgi bodies in the plane of cell division was facilitated by phragmoplast microtubules (Catesson and Roland 1981, Bakhuizen et al. 1985). The phragmoplast microtubules, in turn, have been implicated in guiding vesicles containing cell wall materials to the developing cell plate (Fosket and Morejohn 1992).

The division process was clearly illustrated in an ultrastructural study of dividing cambial cells of *Tilia americana* and *Ulmus americana* conducted by Evert and Deshpande (1970). They showed that during prophase numerous microtubules appeared in association with the centrally located mitotic spindle. These events preceded arrangement of the chromosomes along an equatorial plane oriented at right angles to the radial walls. Cell plate formation was initiated during anaphase. At this stage, the spindle microtubules traversing the future plate region composed the first microtubules of the developing phragmoplast. In both species, a broad cytoplasmic plate, or phragmosome, preceded the developing phragmoplast and cell plate through the cell. Numerous microtubules and smooth vesicles were associated with early stages of cell plate formation (Fig. 5.25A). The vesicles were derived from dictyosomes situated along the periphery of the phragmoplast and they migrated to the equatorial plane of the dividing cell, where they fused with other vesicles to form the new cell plate. Microtubules began to disappear in areas of more advanced plate formation where fusion of vesicles had resulted in formation of continuous portions of new cell wall (Fig. 5.25B-C). Concomitantly, va-

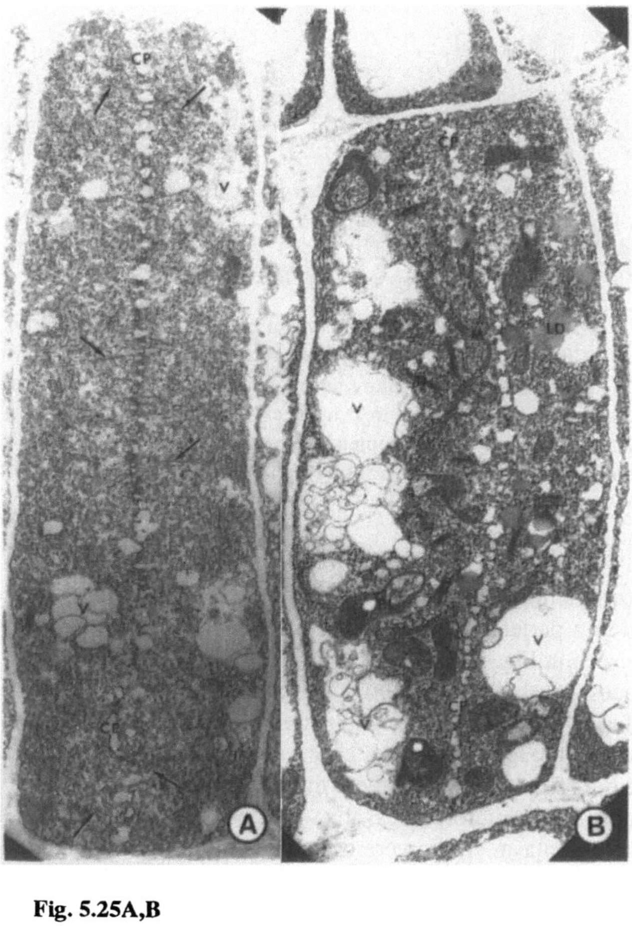

Fig. 5.25A,B

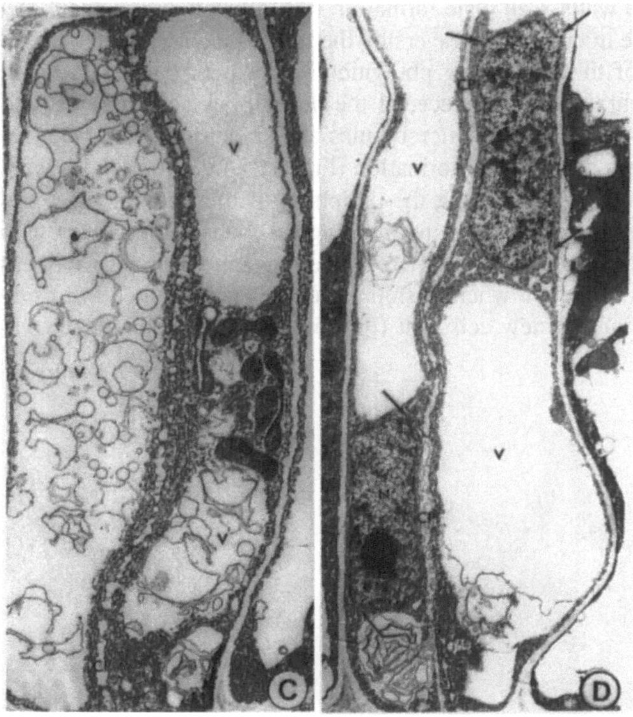

Fig. 5.25C,D. Transverse views of *Ulmus americana* fusiform cambial cells showing development of the cell plate (*CP*). **A** Cell plate associated with numerous phragmoplast microtubules (*arrows*). **B**, **C**, and **D** Sequentially more advanced stages showing increased vacuolation of cytoplasm, and connections *arrows* in **D** between nuclear envelopes and endoplasmic reticulum. *D* Dictyosomes; *LD* lipid droplet; *M* mitochondrion; *N* nucleus; *PL* plastid; *V* vacuole. EM photomicrographs **A** 5250x; **B** 7125x; **C** 7500x; **D** 6375x. (Evert and Deshpande 1970)

cuolation of the cytoplasm increased, and in late stages of cell plate formation the cytoplasm was confined to a parietal position in the cell (Fig. 5.25D). A recently formed cell plate can be seen in Fig. 5.26.

The cell division process in actively dividing cambial fusiform cells of *Fraxinus excelsior* was also followed by Goosen-de Roo et al. (1980, 1983, 1984). During pre-mitotic division, no phragmosome was observed, but a band of microtubules was present along the radial cell walls where the phragmosome would arise. The microfilament bundles and microtubules always appeared associated with each other in the parietal cytoplasm. They were oriented parallel to the longitudinal cell axis, presumably in some way related to the strong longitudinal cytoplasmic streaming (Fig. 5.27). The phragmoplast first appeared in the shape of a ring,

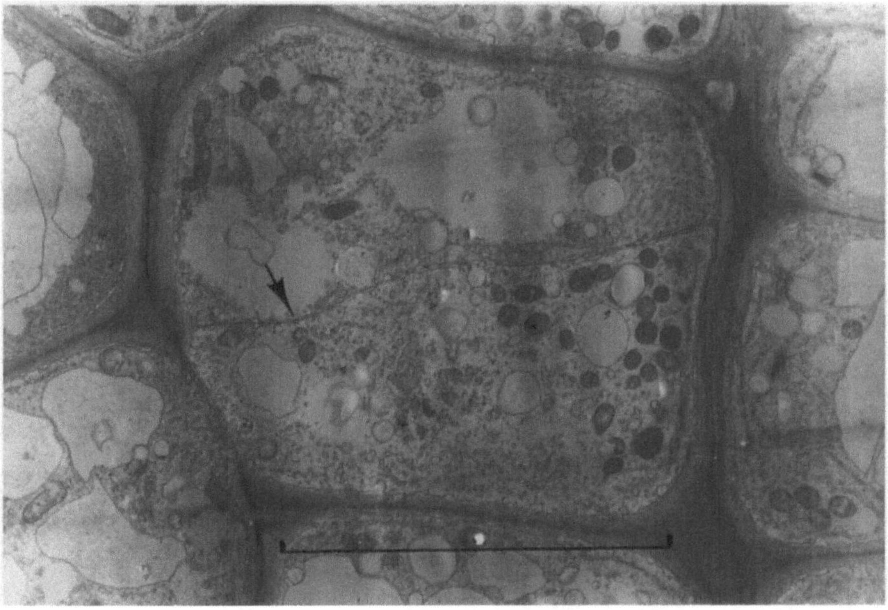

Fig. 5.26. Transverse section through a cell at the phloem end of the cambial zone of *Aesculus hippocastanum* sampled in mid-March. The cell underwent a recent periclinal division as indicated by the newly formed cell plate. *Bar* 10 μm. (Barnett 1992)

which later split in two parts. During post-mitotic division, each phragmoplast was preceded by a more or less continuous, thin strand of phragmosomes. The phragmosomes marked the plane in which the cell plate would form by the phragmoplast. According to Barnett (1981), in addition to microtubules, the phragmoplast in dividing cambial cells of *Aesculus hippocastanum* developed in association with abundant osmophilic material, ribosomes, and some evidence of dictyosomes.

Division patterns in three tropical species, *Gmelina arborea, Mangifera indica,* and *Tectona grandis,* were examined by Rao and Dave (1985a). An atypical division pattern was occasionally observed in *M. indica* in which tangential wall development was initiated at one end of exceptionally long dividing cells well in advance of the other end. The two opposing parts of the new wall met near the cell end rather than in the middle. Divisions of this type often resulted in two unequally sized daughter cells.

An interesting observation was made by Jacquiot (1978b) regarding his studies of in vitro cultures of cambial tissues. Mitoses appeared to occur normally in cambial tissue of *Fagus sylvatica* that had undergone 48 transfers over a period of about 10 years. Despite their age, cells in these cultures divided and produced typical cell types when buds of the same species were grafted to them (Jacquiot 1978a, 1984).

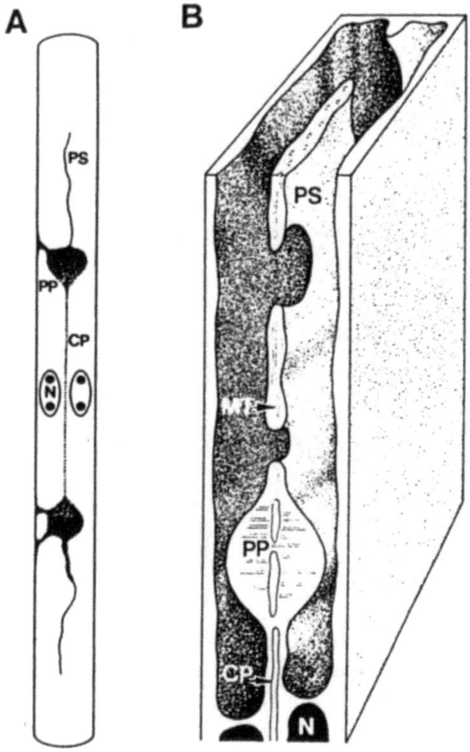

Fig. 5.27A,B. Schematic radial views of a dividing fusiform cambial initial of *Fraxinus excelsior*. **A** Phragmosomes (*PS*) proceeding in opposite directions from the daughter nuclei (*N*) in advance of the phragmoplasts (*PP*) during formation of the cell plate (*CP*). (Goosen-de Roo et al. 1980). **B** Three-dimensional representation of **A** showing orientation of microtubules (*MT*) relative to the phragmosomes and phragmoplasts. (Goosen-de Roo et al. 1984)

5.5.1 Mitotic Index

Mitotic index is defined as the fraction of a cell population undergoing mitosis at a given sampling time. When applied to the cambium, the population generally includes mother cells as well as cambial initials in a radial file. The index is calculated as a percentage value, but the percent sign is usually deleted. Although many investigators have made cursory observations of mitotic frequencies in the cambial zones of trees, relatively few have compiled sufficient data for calculating mitotic indices.

Mitotic activity in the cambial zone of 20-year-old *Pinus strobus* trees was analyzed quantitatively by Wilson (1966). Mitotic counts were made on serial tangential sections cut from 1620 increment cores collected at different positions

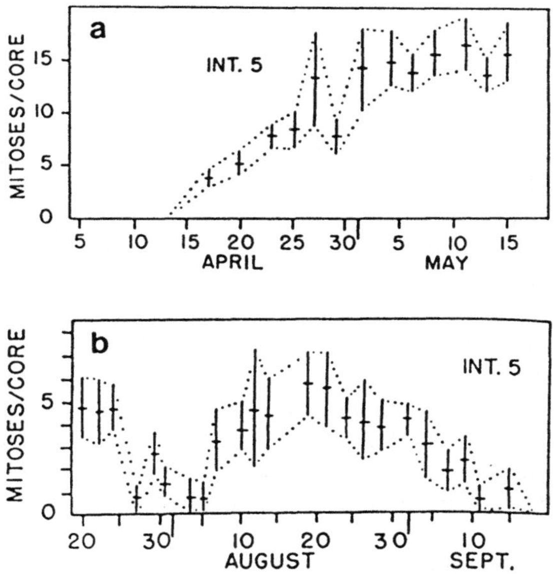

Fig. 5.28a-b. Seasonal change in the number of mitoses per increment core in the fifth internode of five *Pinus strobus* trees. **a** At the initiation of cambial activity in spring. **b** At the cessation of cambial activity in the fall. *Dotted lines* indicate the range of sampling error. (Wilson 1966)

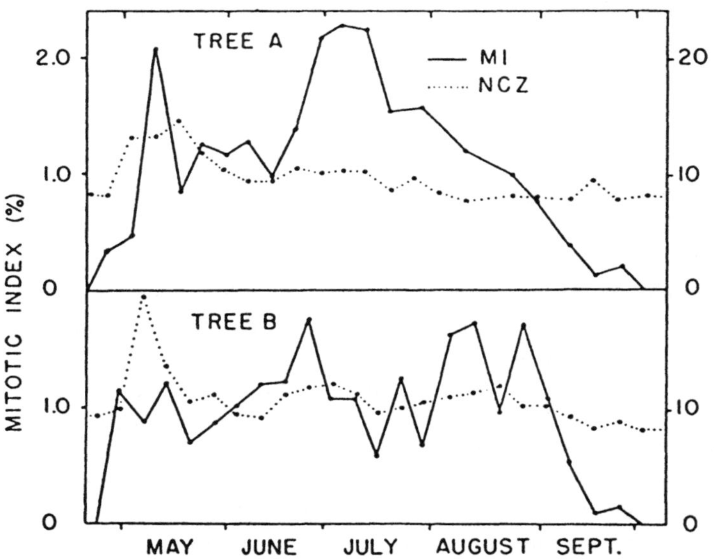

Fig. 5.29 Seasonal change in the mitotic index (*MI*) and radial number of cells in the cambial zone (*NCZ*) of two *Pinus strobus* trees. (Wilson 1968)

on the stems and at different times during the growing season. The distribution of mitoses both within an internode and throughout a tree were surprisingly uniform. Seasonally, the number of mitoses increased rapidly with the initiation of cambial activity in the spring (Fig. 5.28a), more or less stabilized during midseason, and then decreased in the fall (Fig. 5.28b). Because the cambial cell population also varied seasonally, the mitotic index declined from a high of 4.2 in May to a low of 1.9 in July, then rose to 2.6 in September before declining to zero with the onset of cambial dormancy. These values are comparable to the average 2.3 value calculated for the same species by Wilson (1964). The seasonal change in mitotic index for two *P.strobus* trees grown in New England are shown in Fig. 5.29. The mitotic index also increased for a distance of 310 cm above a girdle in *P. strobus* stems (Wilson 1968). Wilson suggested that in this case wounding had extended the normal season of cambial activity.

Gregory and Wilson (1968) calculated mitotic indices of 2.31 and 1.13 for *Picea glauca* trees growing in Alaska and New England, respectively. In a later study, Gregory (1971) confirmed these results with a mitotic index of 2.1 for *P. glauca* in Alaska. Sisson (1968) also arrived at a mitotic index of 2.3 for *Pseudotsuga menziesii*.

The seasonal variation in the frequency of mitoses in fusiform and ray cell initials of *Acer pseudoplatanus* is shown in Fig. 5.30 from Catesson (1964). Similar data are shown in Fig. 5.31 from Imagawa and Ishida (1970) for fusiform initials in *Larix leptolepis*. These data illustrate the range of variability encountered when computing mitotic indices.

The foregoing, although admittedly not a exhaustive coverage of the literature, is nonetheless suggestive of how little is known of the mitotic activities in cambia of trees.

5.5.2 Cambial Cell Cycle

The cambial cell cycle is the average time between divisions, or the length of time required for a cambial cell to complete the entire process of division. Interpretation of the cambial cell cycle is therefore at the cellular level rather than the nuclear level as usually defined (Clowes and Juniper 1968, Roberts 1976, Dodds 1981). Methods of estimating the cambial cell cycle and the limitations of these estimates have been discussed in detail by Wilson (1964).

The cell cycle for *Thuja occidentalis* during the initiation of cambial activity in the spring was estimated to be 7 days (Bannan 1955), or to vary seasonally from 4 to 14 days (Bannan 1962b). In *Pinus strobus* it was found to be about 10 days and it remained at this level not only throughout the grand period of growth but also in both slow- and fast-growing trees (Wilson 1964) (Table 5.4). Wilson interpreted this response to mean that the rate of cell division was independent of the rate of tracheid differentiation. Gregory and Wilson (1968) arrived at a similar conclusion for *Picea glauca*. They found that even though the season for cambial activity

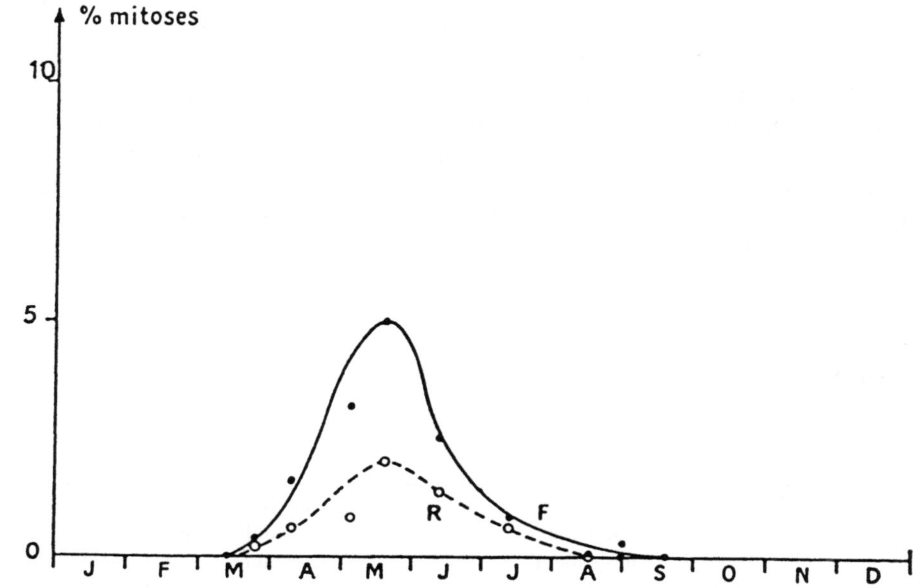

Fig. 5.30. Seasonal variation in the frequency of mitoses in fusiform (*F*) and ray cell (*R*) initials of *Acer pseudoplatanus* as a percentage of the number of initial cells. (Catesson 1964)

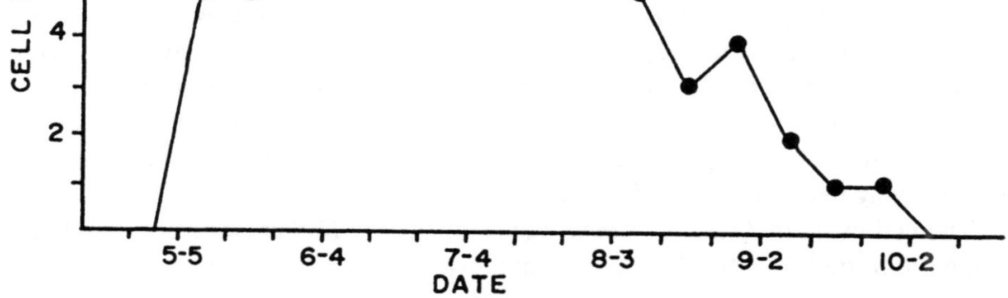

Fig. 5.31. Seasonal variation in the frequency of cell divisions in the cambial zone of *Larix leptolepis* calculated on the basis of number of cells produced at 10-day intervals. (Imagawa and Ishida 1970)

in Alaska was only about one-half as long as in New England, the rate of cell division was about twice as fast in Alaskan trees of comparable vigor; i.e., mitotic indices of 2.31 and 1.13 for Alaskan and New England trees, respectively. Additional confirmation for the independence of these processes was provided by Sisson (1968), Wilson (1968), and Gregory (1971). The cambial cell division cycle for seedlings of *Larix laricina* was estimated to be 10 days by Kennedy and Farrar (1965). In *Acer pseudoplatanus*, it was estimated to be 2 to 3 days for fusiform cells and 3 to 7 days for ray cells (Catesson 1990). Calculations of Berlyn (1982) reduced the time to about 24 h for a fusiform initial to produce a new derivative.

Table 5.4. Estimated times and rates in the cell division cycle of three *Pinus strobus* trees (Wilson 1964)

Tree	Time for cycle (days)	Time for division movement (h)	Time for mitosis (h)	Time for phragmoplast movement (h)	Rate of phragmoplast movement (μm/h)
1	10.4	26.4	5.8	20.6	57
2	9.3	26.4	4.8	21.6	47
3	10.4	21.6	5.2	16.4	105

Skene (1969), who calculated the cell cycle on the basis of the average time required for all cells of the cambial zone to divide once, arrived at about 28 days for *Pinus radiata*. Using the same method of calculation, Kutscha et al. (1975) found the cambial cell cycle for *Abies balsamea* to vary from 7 to 25 days, or an average of 13 days. In contrast to Gregory and Wilson (1968) and others cited above, who found the rate of cell division to remain essentially constant, Skene (1972) observed a variable rate in *Tsuga canadensis*. In this species, the cell division cycle decreased from 28 days at the beginning of the season to 10 days at midseason for fast-growing trees, and from 35 to 20 days for slow-growing trees.

The cell cycle in *Abies balsamea* was estimated by Mellerowicz et al. (1992a) using the formula $d = tnc^{-1}$, where t = number of days between the beginning and end of the experiment (63 days), n = average number of new cells per radial file produced during time t, and c = average of the number of fusiform cambial cells per radial file present at the beginning and end of the experiment. In 6-year-old trees grown in a natural environment, length of the cell cycle varied from 11 days in April, to 97 days in June, to 477 days in September. The authors suggested that the increase in cell cycle duration over the season might be attributed to the accumulation of cells in the G_1 phase of the cell cycle at the approach of dormancy.

Stated another way, cell production by the cambium during active growth has been estimated to be about 0.29 cells/day for *Abies pectinata* (Wodzicki and Peda 1963), 1 cell/day for *Larix laricina* seedlings (Kennedy and Farrar 1965), 3.38 cells/day for *Picea sitchensis* (Ford et al. 1978), and 0.15 cells/day for *Pinus sylvestris* (Wodzicki and Zajaczkowski 1983).

In contrast to the cambial cell division cycle, some workers have calculated the time required for the entire cell cycle including radial enlargement and differentiation as well as division (Whitmore and Zahner 1966, Wilson and Howard 1968, Shimaji and Nagatsuka 1971, Wodzicki 1971, Skene 1972, Wilson 1973, Denne and Dodd 1981). On this basis, the cell cycle of *Abies pectinata* was estimated to range from 25 to 50 days from the beginning to the end of the growing season in Poland (Wodzicki and Peda 1963). For *Larix laricina*, the cell cycle was estimated to be 20 days (Kennedy and Farrar 1965). For *Abies balsamea*, tracheid development required 14 to 37 days for normal wood and 5 to 49 days for compression wood (Kutscha et al. 1975).

Each of the foregoing approaches suffers from shortcomings such as the difficulties in identifying and quantifying the cambial zone (Catesson 1964, Swamy and Krishnamurthy 1978), in addition to sampling variations both within and among trees of a species. Moreover, short-term fluctuations in cambial activity together with the overall lack of good systems for quantitatively measuring these fluctuations detract from their general applicability (Ford et al. 1978, Ford 1981). The cited data must therefore be considered no more than approximations and useful guides for further study.

A number of workers have speculated as to the number of cycles that a meristematic cell might undergo, i.e., its longevity. As Clowes and Juniper (1968) pointed out, cambial cells do not experience a simple exponential increase. Individual cambial initials stop dividing periclinally and are replaced in the cambial cylinder by new initials derived by anticlinal divisions. Moreover, as shown in Chapter 4, the position of an initial cell is not necessarily constant following periclinal division, as its function might be taken over by a daughter cell.

Several workers have also attempted to model various aspects of the cell division and differentiation processes. For example, Wilson and Howard (1968) proposed a model that mimicked the changes in radial cell dimensions during seasonal development of a single radial file in a conifer cambium. The model was later expanded to simulate the processes of cell division, enlargement, and wall thickening during cambial activity (Howard and Wilson 1972). Burggraaf (1974) proposed a model for cell division sequences in *Fraxinus excelsior* based on assumed values for the relative rates of cell division and growth, relative duration of phases of the division cycle, and on a uniseriate layer of initial cells. A biometrical model developed by Brister (1973) was based on three growth phases, cell proliferation, cell expansion, and cell wall thickening, plus correlations in radial

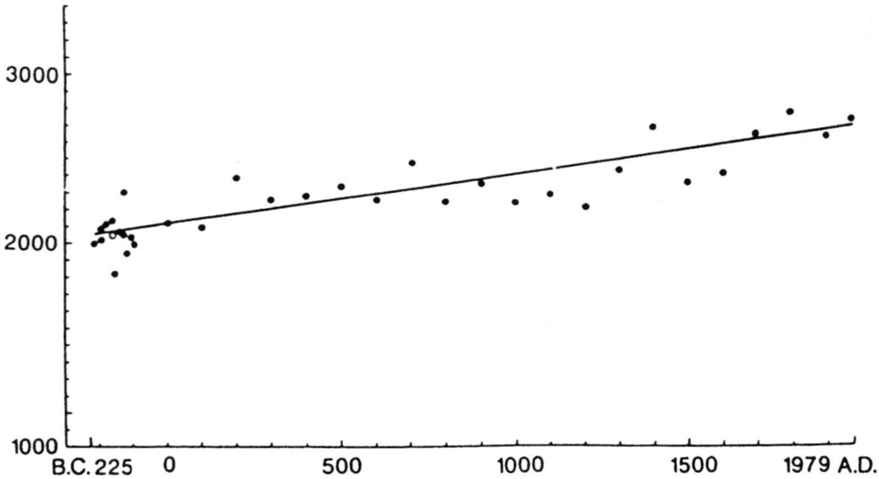

Fig. 5.32. Average tracheid length (in μm) at different age intervals determined from macerations of an increment core from a 2200-year-old *Pinus longaeva* tree. (Baas et al. 1986)

diameters of cell files across the growth rings. The model was used to simulate weekly cell production for a period of 20 weeks.

5.6 Cambial Aging

Trees are known to be long-lived organisms. Documented ages of *Pinus aristida (longaeva)* growing in the western United States exceed 4000 years, with some on record being as old as 4900 years (Currey 1965).

Continuous radial expansion requires an active vascular cambium, and various authors have speculated as to the longevity of this meristem. It must be recognized that the cambium consists of cambial initial cells and each initial has a finite lifetime. It either divides anticlinally to give rise to two daughter cells or it eventually disappears from the cambium. As pointed out by Sax (1962), plant tissues are continually renewed by meristematic growth. Consequently, aging effects are unlikely to be associated with the aging of individual cells. The question remains as to how the cambium as a tissue remains active for so long a time. Over the years, authors have speculated as to causality. Nonetheless, the answer remains elusive (Meyer 1961, Moorby and Wareing 1963, Westing 1964, Zelawski 1980).

A related question concerns cambial dormancy. The cambium can of course remain dormant during periods of unfavorable growth; but can it remain com-

Aging Process in Cambium and Xylem

Tissue		Unit time dt	Time t
Fusiform initial layer	Quantitative aging	- Decleration of division process	- Length growth of initials to maxium value - Growth in number of initial cells
Ray initial layer	Quantitative aging	- Deceleration of division process	- Growth in number of initial cells - Increase of ray matrix
Mother cell tissue	Quantitative aging	- Cell differentiation - Formation of secondary and tertiary walls	
Mechanical tissue	Qualitative aging	- Loss of protoplasm - Complete lignification	
Conducting tissue	Qualitative aging	- Loss of protoplasm - Complete lignification	- Pit closure in longitudinal tracheids - Closure of vessels by tyloses
Storage tissue	Qualitative aging	- Cells lose ability to divide	- Formation of heartwood substances - Formation of tyloses - Gradual necrobiosis

Fig. 5.33. Aging process in the cambium and xylem. (Bosshard 1965)

pletely dormant over a long period of years? This question is addressed in Chapter 11.8.3.

The question of cambial aging has been approached indirectly by examining cambial derivatives over time, the so-called age-related patterns (Chap. 7.1.2.1). Using this method, Baas et al. (1986) found that tracheid length in *Pinus longaeva* steadily increased over the last 2200 years with no sign of leveling off (Fig. 5.32). In this particular species, the "juvenile" phase of steep increase in tracheid length lasted several centuries.

Specimens from the same species (*P. longaeva*) were examined by Connor and Lanner (1990). They measured radial tracheid diameters. Despite intermittent fluctuations, mean tracheid diameter decreased only about 5% over a 4000-year time span. None of the other variables examined showed any evidence of cambial aging, suggesting to the authors that eventual decline and death of this species was due to external rather than internal causes. The results of these two studies do not lend support to the conclusion of Ghouse and Iqbal (1977a) that the decline in tracheid length with age in *Prosopsis spicigeria* was due to the "exhausted" age of the cambium.

Bosshard (1965) examined the cambium and its xylem derivatives in terms of aging processes (Fig. 5.33). His analyses showed that the cambium exhibits quantitative aging processes which are in a sense repetitive and can be recognized either over a single time unit or several to many time units. Xylem derivatives, on the contrary, exhibit qualitative aging processes that result in maturation of individual elements. Nonetheless, the xylem derivatives indirectly reflect the quantitative changes accumulated by the cambial initials over time.

Chapter 6

Anticlinal Cambial Divisions

6.1 Pseudotransverse and Transverse Divisions

6.1.1 Occurrence

Sachs (1878) introduced the terms anticlinal and periclinal with reference to the division planes of cells in the shoot apex. The terms were later adopted for the division patterns in cells of the lateral meristems. Anticlinal division is the process by which new cell files are added to the cambial layer (Figs. 6.1, 6.2, 6.11, 6.49, 9.6B). These divisions are sometimes referred to as pseudotransverse and multiplicative, and in the older literature as radial. Radial as opposed to tangential cell division was one of the criteria Sanio (1873) used to identify cambial initial cells in *Pinus sylvestris*. According to Sanio, each time a cambial initial cell divided radially, a new file of initial cells was produced that in turn gave rise to a new file of wood and bast daughter cells. The initial cell could then be located retrospectively by the doubling of a cambial initial (Fig. 6.2A). Because doubling of the radial file occurred in both wood and bast, Sanio concluded that the radial division must have taken place in the initial cell. Mischke (1890) verified this fact in *Picea abies* as well as in *P. sylvestris* by agreeing that radial divisions were confined to cambial initial cells. Hartig (1855a, 1859a, 1878), however, held fast to his view of a biseriate cambium (Chap. 4.1.1). He believed that radial increase of the cambium occurred by "diagonal segmentation" of each member of the "Muterfasernpaare," or mother-fiber pairs (Fig. 3.5), but tangential increase occurred by "radial segmentation" of the same mother fibers. According to this scheme, a continuous double file from wood to bast across the cambium would require simultaneous radial division of both members of the mother-fiber pair, a rather unlikely event.

It was implied in the work of Raatz (1892) that radial divisions were not confined to the initial cell but might occur elsewhere in the cambial zone. Interestingly, however, Raatz either failed to observe or to identify any anticlinal divisions in his seasonal analyses of *Pinus sylvestris* because none appear in his annual ring

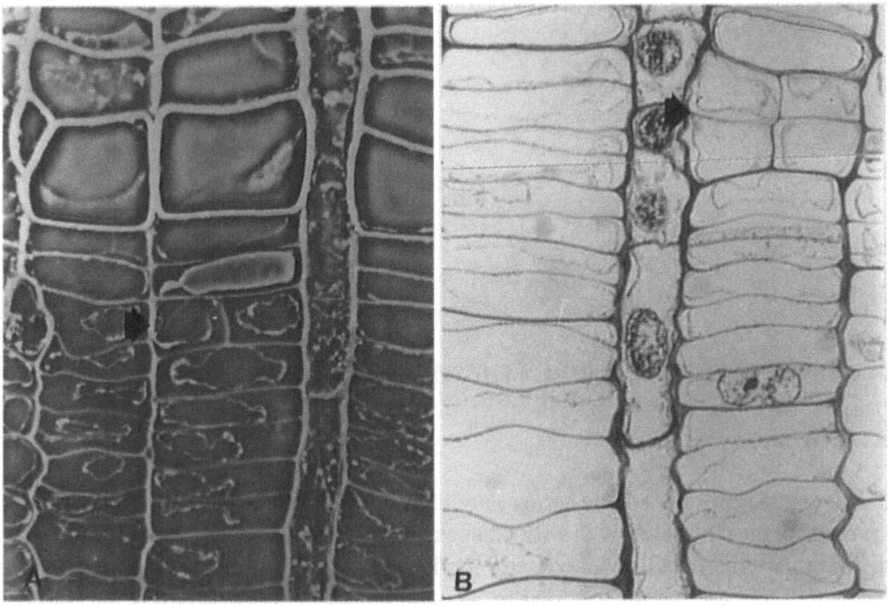

Fig. 6.1A,B. Transverse sections showing recent anticlinal divisions of fusiform initial cells (*arrows*) in *Cryptomeria japonica* (**A**) (Itoh et al. 1968), and *Larix leptolepis* (**B**) (Imagawa and Ishida 1981a). **A** Polarization 329x. **B** Light microscopy, no magnification provided.

diagrams (Fig. 4.5). Nordhausen (1898) was more specific, stating that these double files of cells not only formed series extending through the cambium but also independently in either wood or bast. On this basis, he rejected Sanio's idea of a cambial initial, suggesting instead that mother cells situated on Raatz' "Wendekreise" could serve as initial cells.

Claims that anticlinal divisions were not confined to the initial cells, such as those of Raatz (1892) and Nordhausen (1898), were not without foundation. In fact, Sanio (1873) also mentioned that a daughter bast cell sometimes divided to produce a double radial file independent of the initial cell, but he regarded these as rare occurrences, as did Neeff (1920).

A comprehensive study of pseudotransverse cambial divisions in 12 coniferous species was conducted by Bannan (1957b). His survey of over 6000 radial cell files revealed that about 98% of the pseudotransverse divisions originated in a cambial initial cell and only 2% in mother cells. The paucity of these divisions led Bannan to suggest that when a mother cell divided anticlinally it did so as an immediate derivative of an initial cell. Evidence for anticlinal division in a cambial initial was the fact that daughter cells of the division occurred in both xylem and

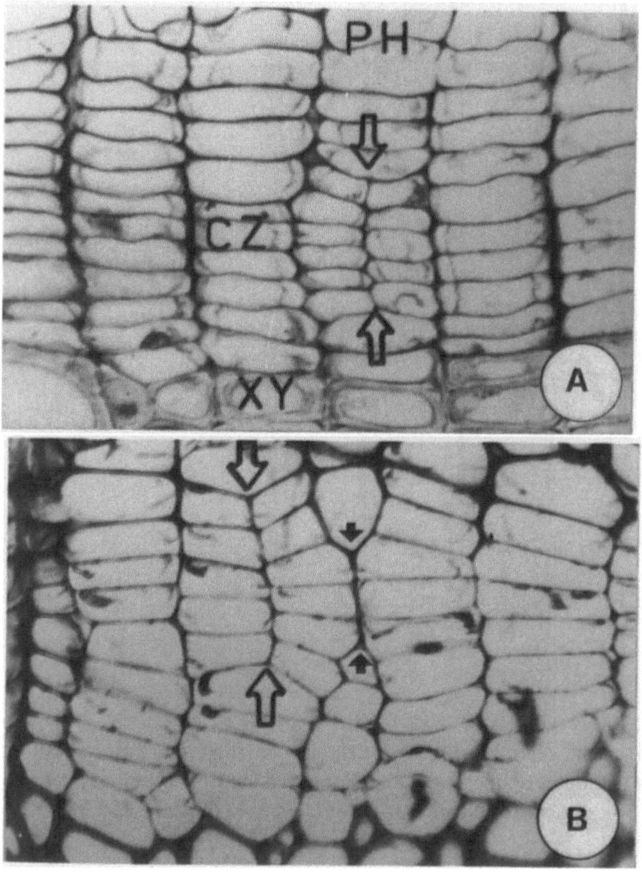

Fig. 6.2A,B. Transverse sections of cambial zones in *Tectona grandis* (A) and *Gmelina arborea* (B). **A** Dormant cambium showing doubling of a radial file following an anticlinal division of a fusiform initial the previous season (*arrows*). *PH* Phloem; *CZ* cambium; *XY* xylem. 545x. (Rao and Dave 1981). **B** Doubling of a radial file (*arrows*) and reduction by loss of a fusiform initial (*darts*). 159x. (Rao and Dave 1985a)

phloem and they were perpetuated as a radial file by a new fusiform initial cell (Figs. 6.2A,B, 6.49). In contrast, origin from a xylem mother cell resulted in a temporary file that disappeared when the mother cell and its derivatives matured (Figs. 6.2B, 6.3).

The aforementioned percentages were mean values. Bannan also showed that the proportion of pseudotransverse divisions occurring in xylem mother cells varied with species, rate of growth, time of year, and occasionally with sampling position in the tree. For example, in *Thuja occidentalis* the number of divisions occurring in xylem mother cells increased from 0.3% in growth rings less than 1 mm

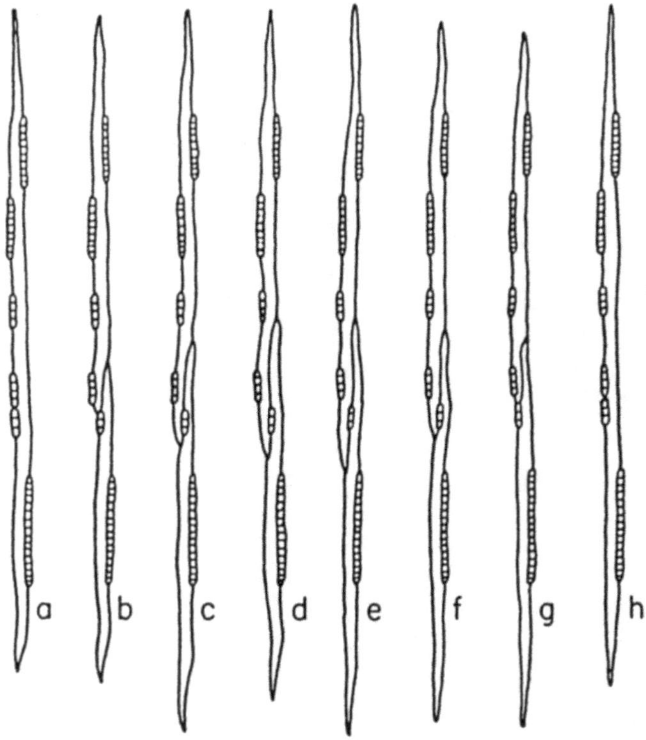

Fig. 6.3a-h. Drawings from serial sections of *Cedrus deodara* showing temporary doubling in a radial file of tracheids presumably due to the anticlinal division of a xylem mother cell rather than the initial cell. The division at **b** is no longer evident at **h**. (Bannan 1957b)

in width to 12% in rings greater than 3.0 mm in width. In the same species, 60% of the divisions originating in xylem mother cells had their inception in the first one-quarter of the growth ring and 25% in the last one-quarter. The radial extent of these temporary files was 2 to 24 cells in the earlywood and 2 to 4 cells in the latewood. In fast-growing trees of *Cedrus deodara*, 30+ tracheids were produced by a single xylem mother cell. To produce these long series, it was necessary that a mother cell and its derivatives undergo periclinal divisions of third order and greater. These periclinal divisions appeared to be concentrated in the median part of a long series; that is, the median cells were products of the last periclinal divisions before the mother cell file matured. In short series, the mother cell acted more like an initial cell; it produced xylem derivatives that underwent a limited number of redivisions as shown in Fig. 6.3.

Similar results were obtained in *Pseudotsuga menziesii* (Bannan 1964a). The frequency of pseudotransverse divisions occurring in xylem mother cells in this species ranged from a low of 2% in slow-growing to 6% in fast-growing trees (Tab. 6.1). As in *Thuja occidentalis*, the radial extent of the temporary files was greatest when a division occurred early in development of the growth ring, and it was greater in absolute distance in wide than in narrow rings. The maximum radial distance recorded was 1.5 mm in a 5.5-mm growth ring. In *Sequoia sempervirens* the proportion of pseudotransverse divisions was more closely confined to the cambial initials. Only about 0.3% occurred in derived mother cells for a ratio of 300:1 (Bannan 1966a). These values agree closely with data obtained by Hejnowicz (1968). He found that divisions in xylem mother cells in old stems of *Picea abies* and *Pinus sylvestris* comprised less than 0.5% of the total anticlinal divisions. Meier (1973) could detect no post-cambial anticlinal divisions in the three conifers he investigated.

Table 6.1. Relative frequency of anticlinal division in the vascular cambium of *Pseudotsuga menziesii* trees (After Bannan 1964a)

	Pseudotransverse divisions		
Ring width	Fusiform initials	Xylem mother cells	
mm	Number	Number	Percent
0.3-1.0[a]	894	19	2
1.0-1.9[a]	1030	40	4
2.0-3.9[a]	954	38	4
4.0-7.0[a]	498	31	6
0.3-2.0[b]	2608	47	2

[a] Trees 25-30 cm in diameter.
[b] Tree 90-210 cm in diameter.

Bannan (1957b) rarely observed anticlinal divisions in phloem mother cells of *Thuja occidentalis*. This finding might relate, in part, to the fact that Bannan (1957c) rarely observed phloem mother cells dividing periclinally more than once. However, it must be recognized that Bannan seldom examined the phloem. Those who did study the phloem in other species frequently observed anticlinal divisions occurring in the mother cells. In *Pinus pinea*, for example, Srivastava (1963a) reported that either phloem mother cells or their daughter initials in many radial files divided anticlinally once and sometimes twice.

Phloem mother cells of dicotyledons are more likely to undergo repeated anticlinal divisions than those of conifers, which might be associated with the fact that they also are more likely to undergo periclinal division. Cheadle and Esau (1964), for example, found that both periclinal and anticlinal divisions took place in phloem mother cells of *Liriodendron tulipifera*. In a more extensive study of 91 species in 44 families of dicotyledons, Esau and Cheadle (1955) found evidence of anticlinal divisions in phloem mother cells of 35 species. Anticlinal divisions in these cases were in reference to the origin of sieve elements, i.e., whether or not a phloem mother cell divided anticlinally before giving rise to a sieve element. Derivatives of anticlinal divisions in radial phloem files from *Liriodendron tulipifera* and *Pyrus communis* are shown in Fig. 4.25a-c and d, respectively, and a dividing sieve element in *Aesculus hippocastanum* in Fig. 11.15A.

Anticlinal divisions in mother cells, particularly in the xylem, have now been reported by a large number of workers in a wide variety of species. Invariably, those who analyzed cell sequences found that these cell files were not perpetuated in either the xylem or phloem. That is, unlike anticlinal divisions originating from fusiform initials, radial files arising from mother cells terminate when the mother cells cease dividing and mature as tissue cells.

6.1.1.1 Analysis

The terms anticlinal and pseudotransverse division are often used interchangeably, but there are some anticlinal divisions that are not pseudotransverse. Anticlinal refers to the division pattern in which partition walls are oriented perpendicular to, or at right angles to, the surface of the organ. Orientation of the partition wall might be either radial-longitudinal, transverse, or oblique. Pseudotransverse is a special type of anticlinal division in which the partition wall varies from slightly oblique to oblique (Fig. 6.11). However, in practice even those divisions in which partition walls are oriented strongly oblique are often considered to be pseudotransverse. On the contrary, some authors use the term pseudotransverse for anticlinal divisions in which wall orientation is essentially transverse.

The preferred technique for identifying the occurrence of anticlinal divisions and analyzing their consequences is to follow radial cell files in serial tangential sections. As stated so well by Chattaway (1951b), "... cell shape in the fully differentiated wood mirrors the conditions that existed at the periphery of the stem when those cells were differentiated from the cambial initials, so that the whole sequence from pith to bark gives a picture of the cambial changes that have taken place during growth from the seedling to the mature tree."

Klinken (1914) is often credited with first using this technique, although Müller (1875-76), Raatz (1892), Jost (1901), Zijlstra 1908), and others employed it as a standard technique many years earlier, as did Neeff (1914), a contemporary of Klinken. The method has been criticized (Priestley 1930b, Meeuse 1942) because it registers changes in cell shape that occur during both cambial and extracambial elongation rather than the true orientations of partition walls and the shapes of newly divided cells. Bannan (1964a, 1965a) recognized this dilemma and attempted to compensate for these changes by calculating correction factors which he applied to his data. Other workers (Klinken 1914, Evert 1963b, Srivastava 1963a, Cheadle and Esau 1964) confined their analyses to the phloem where changes in cell shape were minimized.

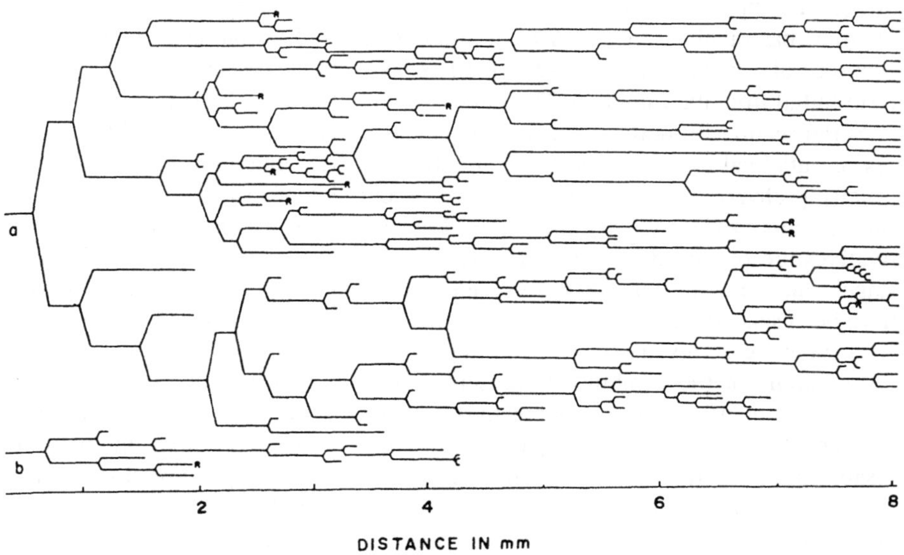

Fig. 6.4. Diagrammatic representation of two lineal files of tracheids in a *Chamaecyparis nootkatensis* stem showing the relative times and frequencies of anticlinal divisions of the fusiform initials. Interrupted sequences indicate disappearance of a sister initial cell after anticlinal division. The letter *R* denotes formation of a ray. File *b* was unsuccessful and was eliminated from the cambium. (Bannan 1950a)

Admittedly, the evidence for anticlinal cambial division patterns is circumstantial; but as pointed out by Cheadle and Esau (1964), the evidence "seem compelling or even unmistakable" if the trees are correctly sampled and the sections critically analyzed. Because it is often difficult to actually observe pseudotransverse divisions in sections of the cambium, many authors have assumed that the division patterns they observed in recent derivatives reflected those in the cambium. Hejnowicz (1961) tested this assumption in root tissues of *Larix decidua* and found that the obliquely oriented walls were indeed of primary origin and they did result from anticlinal divisions that became strongly oblique solely by intrusive growth.

The procedure of analyzing files of cambial cell derivatives has been used with great effectiveness by many investigators over the years. The following assumptions, as outlined by Wardrop and Dadswell (1952) for conifers, provide the basis for most of the conclusions that appear in the following sections: (1) the arrangement and length of the tracheids approximately recapitulate the arrangement and length of the fusiform initials of the cambium at the time the tracheids were cut off; (2) the appearance of two tracheids instead of one, as seen in serial tangential sections, has involved the occurrence of an anticlinal division; and (3) the disappearance of a tracheid from a radial file has involved the disappearance of a fusiform initial from the cambium. The same assumptions apply to the phloem.

Serial sections of both xylem and phloem derivatives have been used for analyses of anticlinal divisions and there are characteristics unique to each. Xylem is usually preferred because derivatives are produced more rapidly and the radial files are more extensive (Whalley 1950, Bannan 1964a). Also, because fewer derivatives are produced, phloem files occasionally have gaps between successive stages, and sometimes the files are severely distorted. These difficulties, however, have been successfully overcome by workers who carefully analyzed the phloem.

A typical case history of the succession of pseudotransverse divisions of fusiform initials in the cambium is presented diagrammatically in Fig. 6.4 representing 8 mm of secondary xylem in a *Chamaecyparis nootkatensis* stem. The diagram shows the time and frequency of pseudotransverse divisions, the point of disappearance of fusiform initials from the cambium, and the conversion of fusiform initials to ray cell initials. During the growth period, extending over a number of years in Fig. 6.4, the fusiform initials and their derivatives responsible for the production of lineal series a divided anticlinally a total of 192 times. Nine of the fusiform initials were transformed to ray cell initials, 157 were eliminated from the cambium, and only 27 remained as functional initials at the time of final examination. Fusiform initial b was less active. It underwent several successive anticlinal divisions, but its daughter initials did not survive and the file soon disappeared.

6.1.2 Cell Length at Division

Because early workers believed that cambial cells divided anticlinally by radial walls, they tended to emphasize tangential cell diameter as a criterion for time of division. Such a belief prompted Mischke (1890) to state: "When diameter growth of the cambial ring results in peripheral expansion and its cells have reached a (certain) tangential diameter, a radial division becomes necessary...." That cambial cell length was also considered a criterion, however, is evident in the writings of Nägeli (1858), Sanio (1873), and others.

Neeff (1920) claimed that no previous investigator had examined the variation in cambial cell size relative to anticlinal division. In normal-growing dicotyledonous trees, he found many cambial cells only one-half the length of others and all lengths in between. He concluded that the short cambial cells were daughter cells that had not yet completed elongation growth following anticlinal division. Apparently, Neeff overlooked Hartig's (1895c) work on tracheid length in *Pinus sylve-*

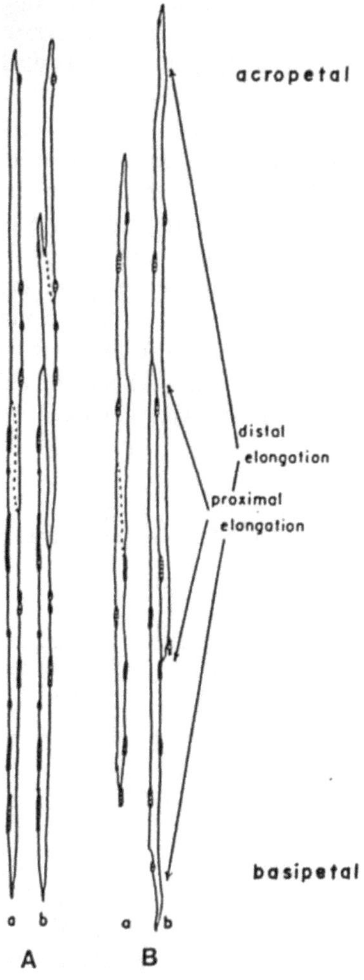

Fig. 6.5A,B. Drawings from tangential sections of *Chamaecyparis thyoides* xylem. **A** Aestival pseudotransverse division of an initial cell (*a*) soon followed by a similar division of the upper sister cell (*b*). **B** Proximal and distal elongation of sister fusiform initials after pseudotransverse division. (Bannan 1951b)

stris, where he stated that when initial cells of different lengths lay side by side, the shorter ones had probably just divided. Kleinmann (1923) conducted a similar study on *Phaseolus multiflorus*. The average length of a cambial cell was 90 μm, but those in the process of anticlinal division averaged 130 μm.

It is generally agreed that cell length is a factor, but not the only factor, in determining when a cambial initial cell will divide anticlinally. Bannan (1951b) found the mean cell length at division for *Chamaecyparis thyoides, C. lawso-*

niana, *Thuja occidentalis*, and *T. plicata* to be approximately 3.52 mm, 3.01 mm, 2.54 mm, and 2.20 mm, respectively. It was not unusual for two pseudotransverse divisions to occur consecutively, in which case one of the daughter cells divided before much elongation had accrued (Fig. 6.5A).

In another study of *T. occidentalis* (Bannan 1956) based on 2300 measurements, cell lengths immediately before pseudotransverse division ranged from 1.57 to 4.58 mm with an average length of 2.58 mm. To attain these lengths, the interval between divisions ranged from 1 to 8 years with an average of 3.7 years for all trees sampled. Mean elongation of the newly formed initials was approximately 550 μm during the first year after division, 290 μm in the second year, and 200 μm in the third year. A typical growth curve illustrating this trend is shown in Fig. 6.6. In *Chamaecyparis* (Bannan 1950b), average length of fusiform initials at the time of division increased from 1.35 mm near the pith to 3.22 mm in old stems. Table 6.2 summarizes the data for 15 coniferous species compiled by Bannan (1967a).

Table 6.2. Interspecific comparison of rate of anticlinal division and cell length at division in conifers[a] (After Bannan 1967a)

Species	Number of trees	Rate of division per cm xylem	Cell length mm
Abies balsamea	9	1.4	3.47
Cupressus sargentii	16	3.4	2.85
Juniperus virginiana	9	4.0	2.04
Larix laricina	16	2.2	3.28
Picea glauca	48	1.0	3.60
Pinus banksiana	13	1.6	3.15
Pinus contorta	12	1.1	2.71
Pinus lambertiana	11	1.3	3.85
Pinus resinosa	18	1.9	3.27
Pinus strobus	54	2.4	3.58
Pseudotsuga menziesii			
Coastal	17	1.4	4.02
Interior	18	1.7	3.39
Sequoia sempervirens	7	2.6	4.59
Thuja occidentalis	47	2.2	2.59
Tsuga canadensis	12	1.7	3.76

[a] Data based on outer growth rings 2-5 mm in width, at breast height, in trees 2.5-5.0 dm in diameter.

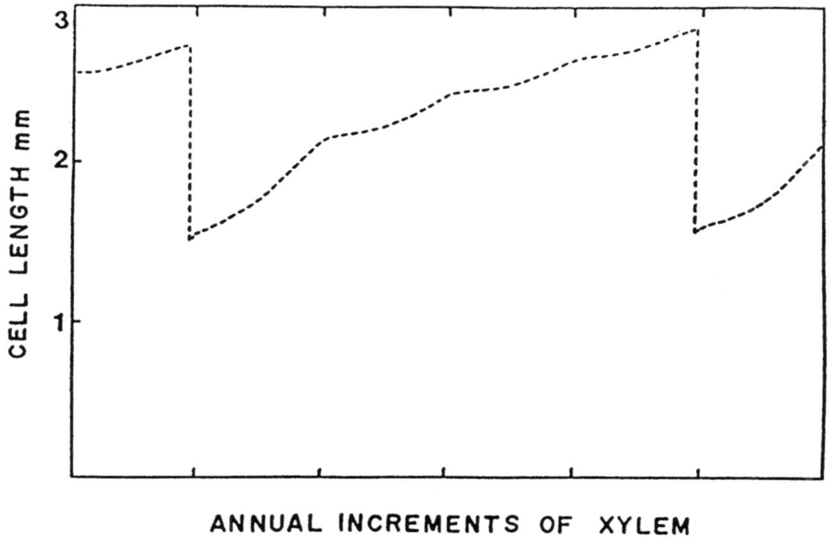

Fig. 6.6. Graph depicting the cycle of anticlinal divisions and elongation of the fusiform initials in recent growth rings of a mature *Thuja occidentalis* tree. The *trough* represents the decrease in length of a fusiform initial after anticlinal division. The *distance* between the trough and peak represents 4 years. The *minor peaks* indicate greater cell elongation in the last-formed part of each annual increment. (Bannan 1956)

Cell length at the time of pseudotransverse division also varied with growth rate in *T. occidentalis*. In the peripheral rings of over 100 trees 13 to 30 cm in diameter, cell length at division ranged from a mean of 2.98 mm in rings 0.60 mm and narrower to 2.64 mm in rings 2.00 mm and wider. Similar results were obtained from a later study of the same species (Fig. 6.7) (Bannan 1960a). Because the data in Fig. 6.7 were obtained by sampling different trees, additional observations were made on trees showing either a pronounced deceleration or acceleration in width of their peripheral rings in an attempt to reduce genetic influences. In 24 trees belonging to the first category, ring width was reduced from a mean of 1.48 mm to 0.66 mm in the outermost rings. The frequency of pseudo-transverse divisions dropped from 2.3/cm of radial increment to 1.8/cm, while mean cell length at division increased from 2.85 mm to 3.00 mm. In 16 trees belonging to the second category, ring width increased from a mean of 1.03 mm to 2.11 mm in the outermost rings. Data on the frequency of pseudotransverse divisions were not presented, but mean cell lengths at division decreased slightly from 2.77 mm to 2.68 mm.

The scatter of points around the fitted line in Fig. 6.7 is indicative of the considerable variability in cell length at the time of pseudotransverse division. Another figure published by Bannan (1954) also showed this inverse relation continuing to a ring width of about 0.60 mm, after which cell length at division dropped preci-

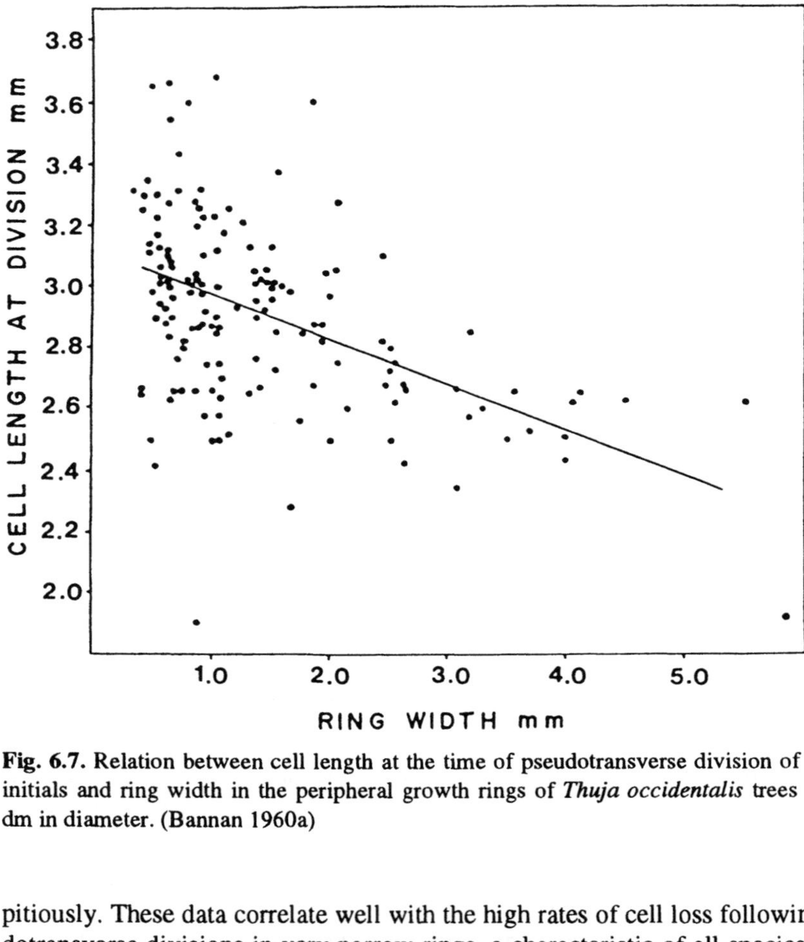

Fig. 6.7. Relation between cell length at the time of pseudotransverse division of fusiform initials and ring width in the peripheral growth rings of *Thuja occidentalis* trees 2.5 - 3.0 dm in diameter. (Bannan 1960a)

pitiously. These data correlate well with the high rates of cell loss following pseudotransverse divisions in very narrow rings, a characteristic of all species examined. Similar data were found for other conifers (Bannan 1963a, 1964b, 1970, Bannan and Bindra 1970b) (Sect. 6.52). Cell length at the time of division was also slightly greater on the narrow-ringed side of eccentric stems of several conifers subjected to prevailing westerly winds (Bannan and Bindra 1970a) (Fig. 6.48).

Just as in conifers, cell length also appears to be a factor in the timing of fusiform initial cell divisions in dicotyledons. The length of initials immediately before division is shown in Fig. 6.57 for *Pyrus communis* (Evert 1961). As a general rule, only the relatively long cells divided. The average length of dividing initials was 540 μm, and over 83% of the dividing initials were 454 μm and longer. A similar relation held for *Apocynum sibiricum*, a herbaceous dicotyledon (Cumbie 1969a). Although fusiform cell length averaged 300 μm in this species, approximately 75% of the dividing initials were in the 350 to 450 μm range. This relation did not hold for *Hibiscus lasiocarpus*, however (Cumbie 1963). In this species with short fusiform initials, length had little influence on whether an initial would divide. Anticlinal divisions were observed in cells varying greatly in length.

In some dicotyledonous species, age of growth ring apparently has little influence on cambial cell length at division. In a stem of *Leitneria floridana*, the mean length of dividing initials varied little from the 1st to the 26th growth ring (Cumbie 1967b). Interestingly, however, during this time period length of the anticlinal partition wall decreased from about 50 to 24% of the length of the mother cell (Table 6.6). This relation suggested to Cumbie that the length of the dividing initial cell exerted little or no influence on the relative length of the partition formed in oblique anticlinal divisions.

6.1.3 Cell Wall Orientation

The writings of early workers were imprecise regarding orientation of the anticlinal dividing walls of cambial cells. Most workers at this time simply stated that tangential expansion of the cambium took place by radial walls or radial divisions which one might assume meant radial longitudinal divisions. Thus, Sanio (1873) repeatedly mentioned "radiale Theilungen," or radial divisions, as did Nägeli (1858), Frank (1864), Sachs (1879), von Höhnel (1884c), Krabbe (1884), Hartig (1895b), Nordhausen (1898), Sieber (1912), and many others. This belief persisted for years, and both Neeff (1920) and Kleinmann (1923) cited several contemporary textbooks whose authors still wrote of anticlinal cambial divisions by means of radial walls, implying radial longitudinal. As an example, Rothert (1913) claimed that when a cambial cell divided by a radial wall, the resulting daughter cells were only one-half the width of the mother cell. Two decades later, Kohl (1933) published a drawing which he claimed showed a cambial cell dividing by "a radial cell plate."

Such a division process, although widely adopted, was not universally accepted. Some workers believed that cambial expansion occurred by transverse cell division (Cohn 1857, De Bary 1877, Mischke 1890). By transverse they meant that division of a cambial mother cell took place by a more or less horizontal wall through the middle of the cell. The two daughter cells then elongated and grew past one another, and in so doing the common horizontal wall was transformed to a sloping or oblique wall.

Questions sometimes arose as to whether orientation of the radial wall was transverse or oblique in the original cambial cell. Braun (1854) debated this question in his study of spiral grain, noting that the problem could only be solved by an investigation of the cambial layer which was difficult to do at that time. He tentatively concluded, however, that cambial cells at first divided by horizontal walls and these walls later became oblique. Raatz (1892) also questioned reports of "radiale Längstheilung," or radial longitudinal divisions. In enlarging cells of

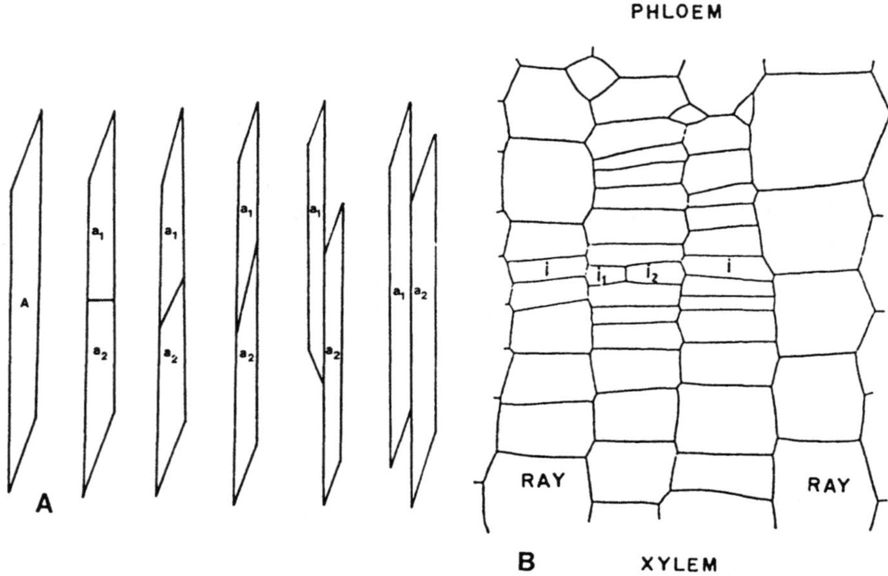

Fig. 6.8A,B. Orientation of the anticlinal dividing wall in *Tilia tomentosa* root cambia. **A** Tangential view of an anticlinal division by a horizontal wall and the subsequent shift to oblique orientation by elongation growth of daughter cells. 52x. **B** Transverse section showing the apparent inequality in the sizes of daughter cells (i_1 and i_2) after post-division elongation growth. 221x. (Redrawn from Neeff 1920)

Pinus austriaca, P sylvestris, and *P. strobus* he always observed oblique divisions, i.e., walls parallel to the radius but oblique to the stem axis. However, he was unable to obtain perfectly aligned tangential sections through the cambial zone. He finally concluded that the dividing wall was at first horizontal. However, as these new walls "pushed" against one another, their ends gradually became more pointed and obliquely oriented by sliding growth. Hartig's (1895b) investigation of the cambium in *Pinus sylvestris* confirmed Raatz' conclusion that anticlinal divisions did indeed occur by horizontal walls followed by extensive elongation of the daughter initials.

A similar conclusion was reached by Klinken (1914). In summarizing his results, he stated that the radial walls visible on transverse sections of conifer stems were formed when the horizontal wall that divided two superimposed initial cells became obliquely oriented by sliding growth of these cells. Although he claimed to have observed horizontal walls in cambia of *Taxus baccata,* he still believed that cambial cells of dicotyledons divided by radial walls. However, he was forced to admit that many dicotyledonous species exhibited a nonstoried cell structure. To account for this structure, he suggested that the effects of the radial divisions were distorted by subsequent sliding growth. The fact that Klinken envisioned two quite different patterns of anticlinal divisions led him to postulate a conifer type and a dicotyledon type of cambium. Neeff (1920) proved this distinction to be false by claiming that dicotyledons as well as conifers divided by horizontal walls (Fig. 6.8A). He offered several suggestions as proof, one of which was the fact that the new wall seldom divided a cambial cell equally, as would be expected if the division were by a true radial longitudinal wall (Fig. 6.8B).

Discrepancies between the results of Klinken (1914) and Neeff (1920) prompted Kleinmann (1923) to re-examine the division planes in dicotyledons. Serial tangential sectioning had by this time become a standard technique for studying anticlinal division patterns in cambial cell files, but he added a further improvement by sampling over a 24-h period. For example, in *Phaseolus multiflorus*, he found that the early morning between 03:30 and 07:30 h was best for observing nuclear divisions. He confirmed this finding in first-year shoots of *Aesculus hippocastanum* and *Sambucus nigra* and the roots of *Raphanus sativuus* with their enormous thickening growth. However, because the shoots of both *Aesculus* and *Sambucus* were still actively growing in length, he erroneously concluded that horizontal divisions contributed to cambial length growth and radial divisions to tangential increase. Kleinmann's results were therefore more confusing than clarifying. In general, he agreed with Neeff (1920). That is, cambial cells of both conifers and dicotyledons divided anticlinally by horizontal walls and he could find no differences between them. Yet, based on his own investigation of *Raphanus sativuus* and Klinken's (1914) observations on *Tilia tomentosa*, he concluded that cambial cells of dicotyledons could divide periclinally by tangential walls and anticlinally by either radial or horizontal walls. He de-emphasized the role of sliding growth in modifying these division patterns, and he unfortunately made no distinction between storied and nonstoried cambia.

Since Kleinmann's results did little to clarify the anticlinal division patterns in dicotyledonous woods, Beijer (1927) undertook a thorough study of how the dividing wall formed in species exhibiting either storied or nonstoried structures. He confirmed the fact that cambial cells in storied cambia divided by radial walls (Sect. 6.2). For his examination of a nonstoried cambium, he chose young rootwood of Alstonia spp., a rather unfortunate selection for this purpose. Despite the fact that the cambial cells displayed an irregular, nonstoried arrangement, he concluded that anticlinal divisions occurred by radial walls. He attributed the irregular arrangement to "intracambial length growth of the initials." In summarizing his results, he stated that a typical storied cambium occurred when the initial cells divided by radial walls, but they expanded only in width. On the other hand, a typical nonstoried cambium occurred when the initial cells divided by a horizontal wall but then elongated by sliding growth as in *Taxus baccata* (Klinken 1914). The cambia of most species lay between these two extremes. That is, most species divided by radial walls but the storied arrangement was soon distorted by length growth of the daughter initials.

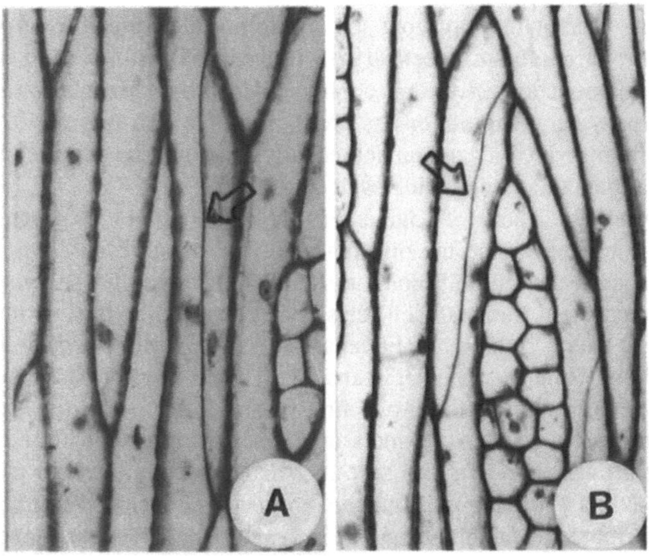

Fig. 6.9A,B. Anticlinal division patterns in fusiform initials of *Tectona grandis*. **A** Lateral anticlinal wall cutting off a small daughter initial from the side of the parent cell. **B** Oblique anticlinal wall at cell end resulting in sister initials of unequal lengths. Tangential sections, 176x. (Rao and Dave 1985a)

In keeping with this pattern of division, Beijer staunchly believed that all cambia originally consisted of unlayered, right-angled cells with horizontal end walls. Such a division pattern was widely accepted at this time (Jaccard 1910b, Haberlandt 1914, Neeff 1914, 1922, Büsgen and Münch 1929, Priestley 1930a, Sprengel 1936), although some workers were vague as to whether the dividing wall was at first strictly transverse or slightly oblique. For example, even Bailey (1920d, 1923) occasionally referred to anticlinal divisions as being transverse or oblique, although he most certainly meant pseudotransverse. An interesting illustration was presented by Kohl (1933), purportedly drawn from a photomicrograph, that showed radial and oblique cambial cell divisions in the same tangential section. The species was not identified.

The first report that I have been able to locate in which the term pseudotransverse division was used in reference to cambial initial cells was by Bailey (1920b,d). Unfortunately, he simply stated the fact without elaboration. The term was later adopted by others to describe the anticlinal division pattern in nonstratified cambia (Bannan 1950b, 1951a,b, Srivastava 1963a,b). This division pattern was sometimes referred to as oblique transverse (Lewis 1935), oblique anticlinal (Dodd 1948), and semitransverse (Bannan 1957a). As noted by Bailey (1920b),

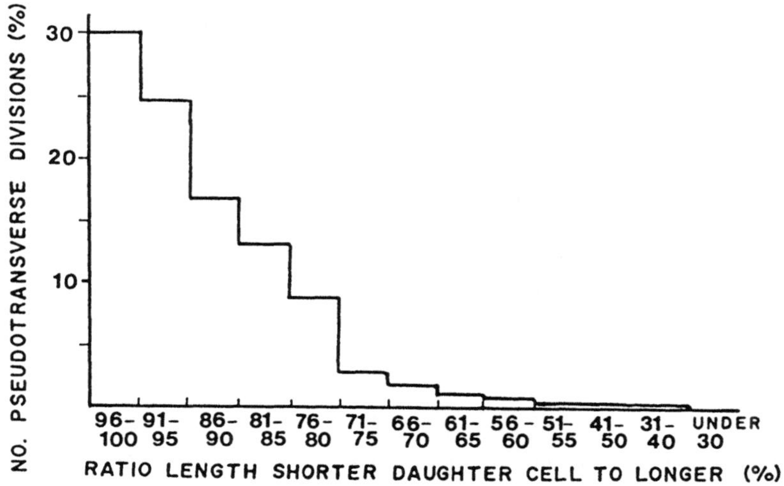

Fig. 6.10. Histogram comparing lengths of the two daughter fusiform initials immediately after pseudotransverse division. Data based on 950 divisions in xylem of *Chamaecyparis* and *Thuja*. (After Bannan 1951a)

the partition membrane usually intersected the side walls at acute angles rather than at right angles, as predicted by Sachs (1878) and later by Errera's (1888) laws.

Most pseudotransverse divisions occur near the cell's middle, but the pitch and length of the anticlinal walls fluctuate widely. The fact that the anticlinal wall was located in the median plane of the cell with few exceptions was noted by many workers (Hartig 1895b, Bannan and Whalley 1950, Bannan 1951b, Srivastava 1963a, Cumbie 1967b). Occasionally, the dividing wall was situated nearer one end, producing daughter cells of unequal lengths (Bannan 1950b). For example, Rao and Dave (1985a) found that asymmetric anticlinal divisons were rather common in *Tectona grandis* (Fig. 6.9B). However, the fact that such divisions were infrequent in temperate zone species is shown in Fig. 6.10. Bannan (1951a) found that the dividing wall was located in the median plane of the cell even in second and higher-order divisions that subdivided daughter cells into short ray cell initials.

In his extensive investigations of *Thuja occidentalis*, Bannan (1956) found that length of the anticlinal wall varied from almost transverse to strongly oblique (Fig. 6.11). In extreme cases, the wall extended one-third or more the length of the mother cell. When the wall was short it was more or less S-shaped, and when it was long it tended to undulate.

Anticlinal wall orientation in *Pyrus communis* was found by Evert (1961) to be most often strongly oblique, but it varied from almost transverse to nearly longitudinal. Wall orientation was considered strongly oblique in four dicotyledonous trees examined by Hejnowicz and Krawczyszyn (1969). Analyses of 197 anticli-

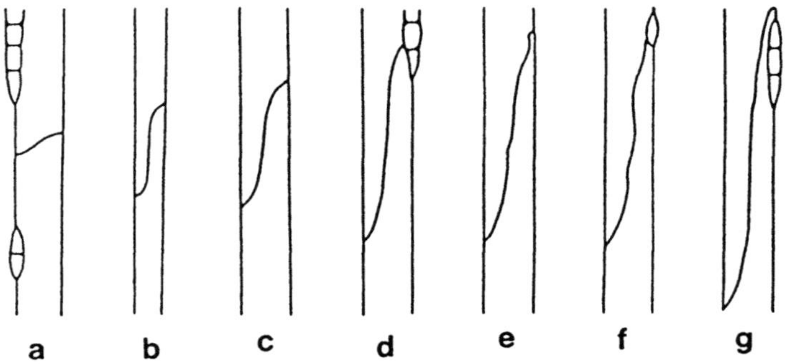

Fig. 6.11a-g. Range of variation in orientation and length of the pseudotransverse dividing wall in *Thuja occidentalis* cambial cells. (Bannan 1956)

nal cell divisions on the phloem side of *Liriodendron tulipifera* by Cheadle and Esau (1964) yielded 15.3% transverse, 35.5% slightly oblique, 31.5% oblique, 8.1% very oblique, and 9.6% lateral divisions (Table 6.3). Transverse to slightly oblique partition walls were more frequent in the shorter fusiform initials 100 to 300 µm in length. Conversely, oblique to very oblique partition walls were more frequent in the longer fusiform initials 400+ µm in length. There was no variation in partition wall inclination during the season independent of initial cell length. Initials with transverse or slightly oblique walls were twice as likely to be lost from the cambium as were those with oblique or strongly oblique walls.

Table 6.3. Orientations of anticlinal wall in length classes of fusiform initials in *Liriodendron tulipifera* (Cheadle and Esau 1964)

Dividing Length class µm	initials Number	Percentage of anticlinal dividing walls				
		Transverse	Slightly oblique	Oblique	Very oblique	Lateral
100-199	11	54.5	27.3	18.2	0.0	0.0
200-299	21	14.3	52.4	23.8	4.8	4.8
300-399	44	15.9	52.3	20.4	4.5	6.8
400-499	63	14.3	33.3	34.9	7.9	9.5
500+	58	8.6	20.7	41.4	13.8	15.5
Total and ___ percent	197	15.3	35.5	31.5	8.1	9.6

Not all radial longitudinal divisions are confined to species with storied cambia. Species with nonstoried cambia, such as *Hibiscus lasiocarpus* (Cumbie 1963), *Apocynum sibiricum* (Cumbie 1969a), *Gmelina arborea, Mangifera indica,* and *Tectona grandis* (Rao and Dave 1985a) were found to possess a high proportion of radial divisions. Of the oblique divisions, approximately 37% were almost transverse in *H. lasiocarpus* (Cumbie 1963). These divisions were most common at the beginning of secondary growth. The division patterns varied not only at different levels in the stem, representing different states of secondary development, but also along a stem radius. A relationship was also observed between fusiform cell length and orientation of the division wall, with radial divisions occurring most often in short initials and oblique divisions in the longest ones (Table 6.4). Walker and Cumbie (1968) attributed the patterns to the herbaceous habit and the relatively short lengths of the fusiform initials. They suggested that plants with a reduced growth habit might maintain a stable fusiform initial length similar to that of species with storied cambia but regulated by a different anticlinal division pattern.

Table 6.4. Type of anticlinal division in the cambium and mean length of the partition wall in four *Hibiscus* species (After Walker and Cumbie 1968)

	Radial		Lateral		Oblique	
Species	Percent[a]	Mean[b]	Percent[a]	Mean[b]	Percent[a]	Mean[b]
H. coccineus	22.2	85	29.6	58	48.2	45
H. grandiflorus	47.7	82	22.5	60	29.8	58
H. militaris	31.8	82	26.0	60	42.2	45
H. moscheutos	20.2	78	31.3	60	48.5	41

[a] Expressed as a percentage of total observed divisions.
[b] Expressed as a percentage of partition length to cell length.

Herbaceous dicotyledons differ greatly from arborescent species in the types of anticlinal divisions that prevail. Radial longitudinal divisions are essentially nonexistent in conifers, rare in nonstoried, arborescent dicotyledons, but common in herbaceous dicotyledons such as *Hibiscus* and *Apocynum.* Lateral divisions, in which a small segment divides from the side of an initial, are also rare in conifers (Bannan 1950b, 1957b), except in the case of injury (Smith 1967), somewhat common in dicotyledonous trees (Evert 1961, Cheadle and Esau 1964) and shrubs (Cumbie 1963), and especially common in herbaceous dicotyledons (Cumbie 1967b, 1969a,b). Interestingly, in *Polygonum lapathifolium,* radial divisions were absent but lateral divisions accounted for over one-half of all anticlinal divisions in the secondary xylem (Cumbie 1969b). Cumbie suggested that the high proportion of radial divisions in *Hibiscus* and lateral divisions in *Polygonum* might be accounted for by differences in cell elongation. Whereas the lengths of fusiform initials in *Hibiscus* remained relatively stable, those of *Polygonum* underwent considerable elongation after anticlinal division. Only a few woody shrubs and herbaceous dicotyledons have been examined, and additional analyses will undoubtedly reveal widely varied anticlinal division patterns among species.

As noted in Chapter 5.2, length of the tapering end wall of fusiform initials provides an estimate of the length and orientation of the anticlinal division wall. Some workers found that lengths of the tapering ends of daughter initials varied during the year, being greatest during the growing season when anticlinal divisions were more frequent (Ajmal and Iqbal 1987a). The seasonal variation can be significant, as in *Ficus rumphii* where length of the tapering end wall varied from 18 μm to 63 μm (Ajmal and Iqbal 1987b).

6.1.4 Length of Cell Plate

Orientation of the anticlinal dividing wall determines the length of the cell plate. The extent of overlap as shown in Fig. 6.11 varied from scarcely none to as much as 1570 μm, with a mean value of 490 μm in *Thuja occidentalis*. Consequently, length of the dividing wall averaged about 13% of the initial cell length (Bannan 1957b). Based on these data, Bannan (1956) calculated the mean length of the newly derived tracheids subsequent to pseudotransverse division to be 1550 μm. In *Chamaecyparis* spp., length of the dividing wall was considerably more variable, ranging from almost transverse to one-third the length of the fusiform initial (Bannan and Whalley 1950, Whalley 1950). The dividing wall on the xylem side averaged about 27% longer than that on the phloem side. In rationalizing this difference, the authors suggested that the first periclinal division after pseudotransverse division was probably toward the phloem. During the interval before the next periclinal division, toward the xylem, considerable elongation of the daughter initials presumably took place.

Like the extreme ranges, mean lengths of the dividing walls also differed between the two genera; the mean length for both *Chamaecyparis lawsoniana* and *C. thyoides* was 0.52 mm, and that for *Thuja occidentalis* and *T. plicata* 0.42 mm and 0.39 mm, respectively. These mean values were approximately one-fifth to one-seventh the lengths of the cells undergoing anticlinal division (Bannan 1951b).

In some species, a trend outward from the pith was noted with respect to the ratio of anticlinal partition wall length and cell length at the time of division (Bannan 1966b). For example, in *Picea glauca* the ratio of partition wall length to cell length was largest in young rings and it decreased with ring age from the pith (Table 6.5). With advancing age both partition wall and total cell length increased, but the ratio decreased. Average length of the partition wall also tended to vary with sampling position in the tree. In *Chamaecyparis* it averaged 19.0% of the

Table 6.5. Ratios of lengths of the anticlinal division partitions of cambial cells and lengths of cells undergoing division as determined from the xylem derivatives. (After Bannan 1964a, 1966b)

Species	Number of trees	Stem diameter dm	Partition length mm	Dividing cell length mm	Ratio[a] %
Chamaecyparis nootkatensis	22	2.0-6.2	0.41	2.56	16.0
Cupressus arizonica	19	2.0-6.0	0.38	2.17	17.3
Cupressus macnabiana	14	2.0-3.8	0.44	2.21	20.0
Cupressus sargentii	23	2.0-5.0	0.51	2.93	17.4
Juniperus virginiana	18	2.5-5.0	0.32	2.09	15.3
Libocedrus decurrens	16	2.5-5.0	0.50	3.26	15.3
Picea engelmannii	46	2.0-5.0	0.46	3.05	14.9
Picea glauca	50	2.0-5.0	0.55	3.51	15.7
Pinus contorta	12	2.0-4.0	0.53	3.41	15.6
Pinus edulis	10	2.0-5.8	0.38	2.46	15.5
Pinus resinosa	23	2.2-4.2	0.55	3.38	16.2
Pinus strobus	50	2.0-5.0	0.63	3.97	15.9
Pseudotsuga menziesii	72	2.5-5.0	0.57	3.60	15.8
Pseudotsuga menziesii	27	9.0-21.0	0.61	4.25	14.5
Sequoia sempervirens	16	9.0-18.0	1.05	6.48	16.2
Thuja occidentalis	100	2.5-5.0	0.46	2.86	16.0
Tsuga canadensis	16	2.5-5.0	0.61	3.89	15.7
Picea glauca[b]	3	1.0-2.0	0.522	3.00	1.74
" " [b]	2	-	0.260	1.33	19.6
" " [b]	3	-	0.281	1.29	21.7
" " [b]	3	-	0.216	0.92	23.5

[a] Ratio, partition length to cell length in percent.
[b] Based on growth ring numbers 10-19, 3, 2, and 1, respectively, outward from the pith.

length of the mother cell in the inner growth rings of seedlings, 19.4% in the inner rings of branches, 16.7% in the outer rings of branches, and 23.7% in the inner rings of roots (Bannan 1950b). These differences reflected, in part, differences in the lengths of the respective fusiform initial cells.

Data on the mean lengths of anticlinal partitions in 15 species, based on 20 000 measurements, are presented in Table 6.5. In compiling these data, Bannan (1964a) noted that certain corrections were necessary because measurements of partition lengths were made when evidence of a pseudotransverse division first appeared in the xylem. Consequently, some elongation growth had already taken place. Nonetheless, the data do provide interesting comparisons such as the remarkably similar ratio of partition wall length to cell length in all species. Also, the fact that species characterized by short cambial cells had short anticlinal partitions (*Cupressus arizonica* and *Juniperus virginiana*), whereas those characterized by long cambial cells had long anticlinal partitions (*Sequoia sempervirens*). In general, this relation was maintained despite variations in cambial cell length due to rate of growth, age of growth ring, and position within the tree. According to Bannan, the positive correlation between length of the anticlinal partition and

Table 6.6. Relative lengths of the partitions in oblique anticlinal divisions in *Leitneria floridana* (Cumbie 1967b)

Growth ring number	Number of divisions	Percentage of oblique partitions[a]			
		0-19	20-39	40-59	60+
1	187	7.5	17.1	49.7	25.7
2-10	211	9.0	39.8	39.3	11.9
22-26	104	34.6	56.7	8.7	0.0

[a] Relative lengths expressed as a percentage ratio of partition length to cell length.

cambial cell length has considerable significance. The greater the overlap of the proximal tips of the newly formed daughter initials, the earlier they will attain their full lengths and contribute to the cycle of periclinal division and radial expansion of the cambium.

Similar results were obtained by Srivastava (1963a) in his study of *Pinus pinea*. The partition varied from almost transverse and short in length to strongly oblique, extending over 25% the length of the initial cell. In old stems of *Picea abies* and *Pinus sylvestris*, the partition averaged about 15% of the initial cell length (Hejnowicz 1968). Calculations by Wilson (1963) in *Abies concolor* showed that 88% of the new surface area could be attributed to formation of the cell plate and 12% to radial increase of the original radial walls of fusiform cells.

Lengths of the partition walls in dicotyledonous trees averaged somewhat longer than those observed by Bannan (1964a) in conifers. In four species (*Alnus glutinosa, Aesculus hippocastanum, Acer pseudoplatanus* and *Betula verrucosa*) examined by Hejnowicz and Krawczyszyn (1969), the dividing walls averaged 25% the lengths of the mother cells.

Lengths of the partition walls in some dicotyledonous species also varied with age, or with growth ring number from the pith (Cumbie 1967b). As shown in Table 6.6 for *Leitneria floridana*, the partitions ranged from very oblique in the first

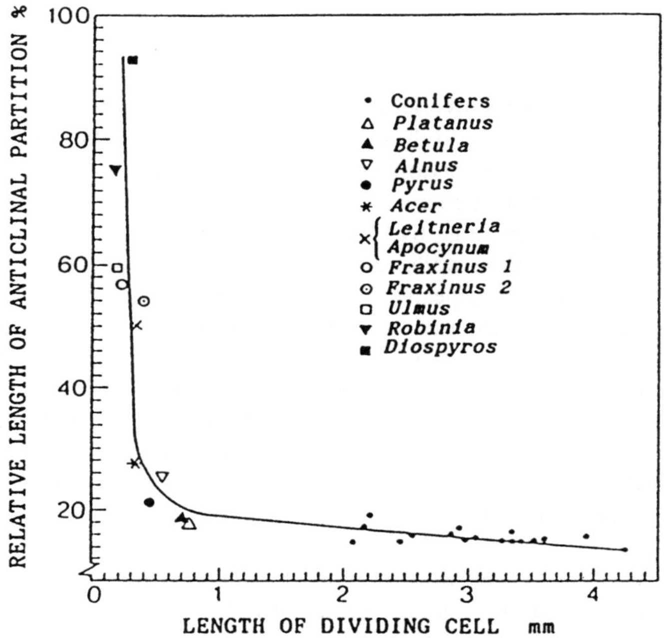

Fig. 6.12. Relationship between length of fusiform initial cells and relative length of the anticlinal partition in various woody species having either storied or nonstoried cambia. (Krawczyszyn 1977b)

growth ring of the secondary xylem to oblique to moderately oblique in the older growth rings. Lengths of the oblique partitions averaged 50, 38, and 24% of the lengths of the dividing cells in rings 1, 2-10, and 22-26, respectively. In the older growth rings, many dividing walls were almost transverse. Mean length of the partition formed by oblique anticlinal divisions averaged 50% of the length of the dividing cell in *Hibiscus* (Walker and Cumbie 1968), *Apocynum* (Cumbie 1969a), and *Polygonum* (Cumbie 1969b), all herbaceous dicotyledons. In *Hibiscus* (Walker and Cumbie 1968), lengths of the partitions varied slightly among species, and within a species they varied with the type of anticlinal division. For all four species, mean length of the dividing wall was 200 μm.

The relation between length of the fusiform initial and relative length of the anticlinally dividing wall in various woody genera is shown in Fig. 6.12 from Krawczyszyn (1977a). Because fusiform initials in storied cambia divide by more or less true radial walls, relative lengths of their dividing walls approach 100%. Conifers, on the contrary, have relatively short anticlinal partitions, while most nonstoried dicotyledons lie somewhere in between.

6.1.5 Cell Elongation

It has long been known that cambial cells dividing by anticlinal walls must somehow become reoriented by a process involving cell elongation. The need for such a process became particularly evident to those workers who believed that radial walls were at first horizontal in newly dividing cambial cells but then became obliquely oriented during subsequent periclinal divisions. Nägeli (1858), for example, noted that in addition to dividing by longitudinal walls (periclinal), cambial cells also divided by transverse walls (anticlinal). Dividing walls that appeared horizontal in cambial cells appeared sloping in sapwood cells, and he therefore assumed that these cells increased in length as their ends became pointed and "pushed forward." Both of these processes, division by transverse walls and length growth of the sapwood cells, were thought to be in balance so that within certain boundaries the lengths of the elements remained the same through all growth rings in the wood. Similarly, Radlkofer (1858) observed on tangential sections of *Cocculus laurifolius* that cells divided in "two thin halves," each of which "strives" to expand. Elongation of these prosenchymatous cells past one another was confined to the cell ends.

Analyses of radial cell files by later workers showed that elongation of the two sister initials following anticlinal division of a parent fusiform initial was not necessarily uniform. Not only might elongation of the sister initials differ from one another, but elongation of their proximal and distal ends might also differ.

6.1.5.1 Proximal End Growth

Newly formed fusiform initials resulting from an anticlinal division elongate both in the region of overlap and at their opposite ends, which Bannan (1951b) referred to as proximal and distal elongation, respectively (Fig. 6.5B). It was assumed that the proximal and distal ends behaved independently. For example, immediately following their origin, the sister initials in both *Chamaecyparis* (Bannan and Whalley 1950) and *Thuja* (Bannan 1951b) elongated more rapidly at their proximal region of overlap than at their distal ends; the reverse occurred rarely. During later growth, the combined elongation of the opposite ends soon equalled that in the region of overlap although the ultimate amount of growth varied widely. In *Chamaecyparis*, total proximal exceeded distal extension in 51% of the cases studied, whereas in 42% of the cases the reverse was true. In 63% of the cases, differences between the proximal and distal ends exceeded 30%.

Relative elongation of the proximal and distal ends of sister initials also varied with time of year, with proximal elongation being greater after aestival than vernal pseudotransverse divisions (Bannan 1951b). For example, Bannan found that proximal elongation of sister fusiform initials tended to increase progressively throughout the season following aestival pseudotransverse divisions (Table 6.7). When the annual increment was divided into four quarters, average elongation of the proximal cell ends during the fourth quarter exceeded that in the first quarter by 92 µm in *Chamaecyparis thyoides*, 216 µm in *C. lawsoniana*, but by only 3 µm in *C. plicata*. Total elongation in Table 6.7 represents the sum of proximal and distal elongation of sister initials, not their derivatives, during the course of one growing season. As stressed by Bannan, these data were derived from trees growing at different rates and simply indicated trends.

Table 6.7. Distal and proximal elongation of sister fusiform cambial cells during the year following their origin in aestival pseudotransverse division in *Chamaecyparis* and *Thuja* (Bannan 1951b)

Species	Quarter of the annual increment[a]								
	First		Second		Third		Fourth		Total
	Dist.	Prox.	Dist.	Prox.	Dist.	Prox.	Dist.	Prox.	Elong.
C. lawsoniana	95	105	125	78	238	227	369	321	1559
C. thyoides	180	130	183	193	234	185	318	222	1643
T. occidentalis	67	64	104	115	175	161	233	171	1092
T. plicata	72	73	117	113	180	121	163	76	915

[a] All data expressed in μm.

As noted earlier, a fair proportion of the anticlinal divisions in both conifers and dicotyledons have been classified as transverse although oblique or pseudotransverse divisions predominate. In stems of *Larix decidua* that had been either incised with a sharp knife or pricked with a needle, the wounded cambium responded by producing an exceptionally large number of anticlinal divisions with transverse partitions (Hejnowicz 1963a). The proximal ends of the daughter cells elongated by intrusive growth that was confined to a very narrow region of the wall bordering the radial edge of the daughter cells (Fig. 6.13). According to Hejnowicz, this type of cell elongation growth had not been reported elsewhere and it was not known to what extent it might be a wound response. Of those cells that divided by anticlinal walls, proximal elongation was identical to that in unwounded cambia. Smith (1967) also recorded an unusual number of transverse divisions (9.4%) in cambia of *Abies grandis* stems and branches that had been attacked by the balsam wooly aphid, *Adelges piceae* (Chap. 10.5.2.2), but he made no mention of how the daughter cells elongated.

Some authors have measured the lengths of overlap in cambial initials without reference to proximal or distal ends. For example, Ghouse and Yunus (1974a) found that mean length of the tapering ends of fusiform initials in four Ranalian species to be: *Annona squamosa* 62 μm, *Artabotyrys odoratissimus* 112 μm,

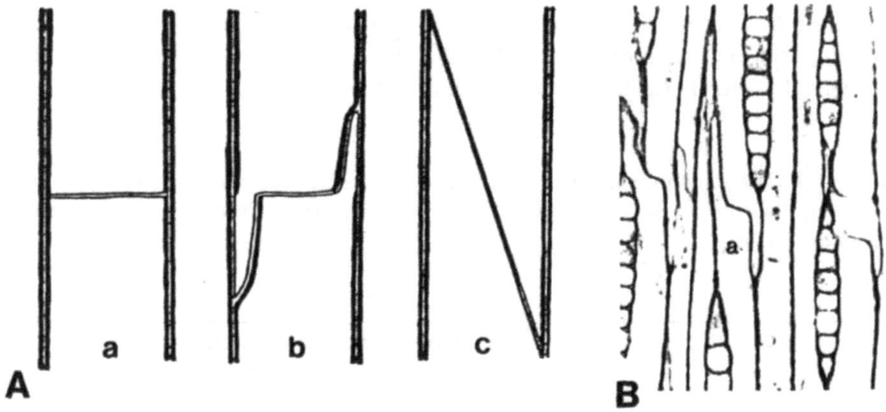

Fig. 6.13A,B. Unusual anticlinal divisions of fusiform initial cells in *Larix decidua* as seen in tangential sections. **A** Transverse division (*a*) followed by reorientation of dividing wall due to intrusive growth (*b*). Typical pseudotransveres dividing wall (c). **B** Light micrograph of (*b*) in **A**. (Hejnowicz 1963a)

Dillenia indica 260 µm, and *Polyalthia longifolia* 157 µm. Also, in six *Acacia* species length of the tapering end walls varied from 30 to 95 µm and in *Prosopis spicigera* it averaged 77 µm (Ghouse and Iqbal 1975). For data on lengths of the tapering ends of fusiform initials for many species, see Table 5.1.

6.1.5.2 Distal End Growth

Distal end growth of sister cells following anticlinal divisions (Fig. 6.5B), should not be confused with that following periclinal divisions (Chap. 7.1), although the mechanisms are essentially the same. One of the first workers to specifically describe the end or tip growth of fusiform initials following anticlinal divisions was Neeff (1914) in his study of cambial reorientation following decapitation in *Tilia americana* and *Picea abies.* Neeff demonstrated how the elongating cell tips could displace neighboring cells either radially or tangentially. By focusing in-depth on serial tangential sections, he demonstrated how the elongating tip of a new cambial cell penetrated between two tangentially adjacent cells (Fig. 6.14). The new initial cells eventually attained lengths comparable to those of the original mother cells. However, some daughter cells were either lost from the cambium or transformed to rays. In a similar way, Bannan (1956) demonstrated how sister initials resulting from a recent anticlinal division could elongate extensively and sometimes become separated by the intruding tip of a neighbor initial (Fig. 6.15).

Distal elongation of sister fusiform cells was also quantified by Bannan (1951b). Like proximal elongation, distal elongation varied with time of year. Invariably, distal elongation was greater after aestival than vernal pseudotransverse

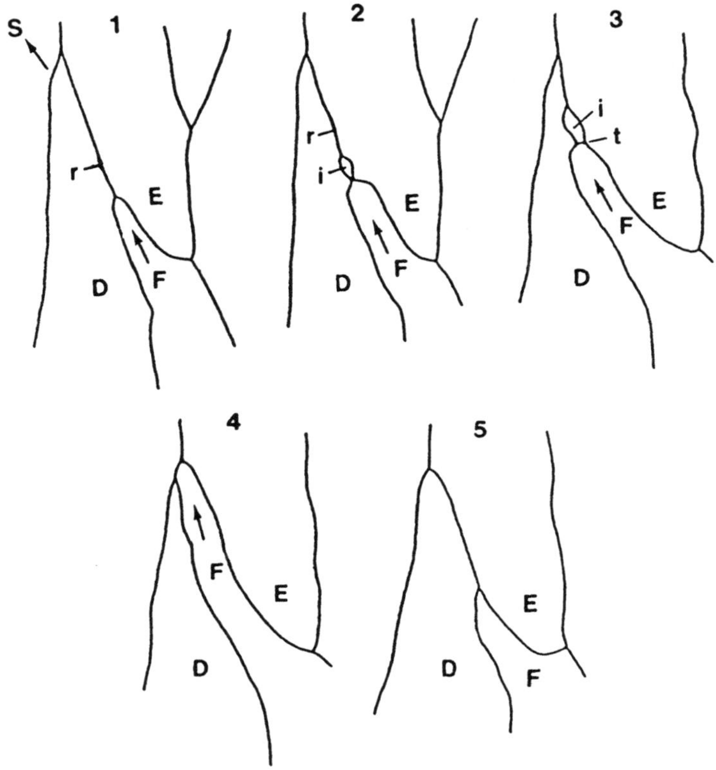

Fig. 6.14. Serial tangential sections of *Tilia americana* cambial zone illustrating the interpenetration of the elongating tip of cell *F* between walls (*r*) of adjacent cells *D* and *E*. Cell tip advance in direction *S* is preceded by formation of an intercellular space (*i*) and separation of the cell wall (*t*). (Neeff 1914)

divisions. Again, as shown in Table 6.7, distal elongation of sister fusiform initials tended to increase progressively throughout the season following aestival pseudotransverse divisions. For example, distal elongation during the fourth quarter exceeded that in the first quarter by 138 µm in *Chamaecyparis thyoides*, 274 µm in *C. lawsoniana*, 166 µm in *Thuja occidentalis*, and 91 µm in *T. plicata*. Despite the considerable variation due to tree size, growth rate, and other factors, Bannan concluded that both proximal and distal elongation tended to increase in the third and fourth quarters of the annual increment. Also, total elongation was greater, but in general less stable, in *Chamaecyparis* than in *Thuja*.

The tendency for distal elongation to be greater during the latter part of the growth ring might conceivably be related to the fact that both anticlinal divisions and cell losses tend to be greater at that time of year. In addition, as both Bannan (1951b) and Evert (1961) demonstrated, distal tip elongation tends to be greater when a new initial can occupy the space created by a failing initial. Conversely,

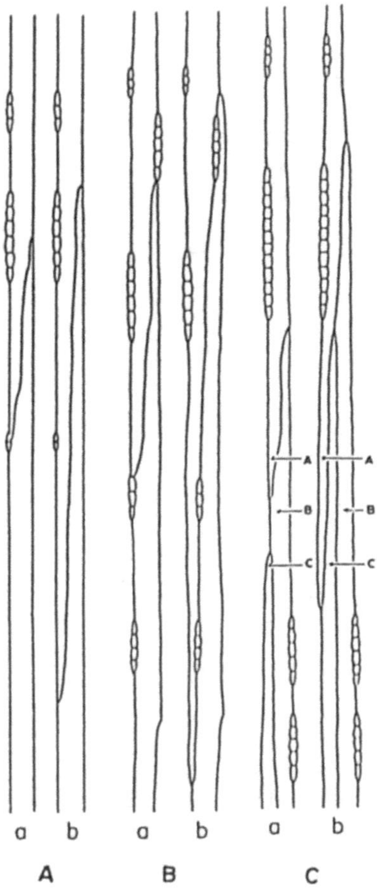

Fig. 6.15A-C. Drawings showing fusiform initial elongation following pseudotransverse division in *Thuja occidentalis*. **A** Sister cell elongation (*a* to *b*) during one growing season. Extent of overthrust was 425 μm. **B** Proximal tips of newly formed sister initials (*a*) elongated to lie on outer sides of neighboring rays (*b*). **C** Elongating distal tip of cell *C* (*a*) penetrated between the overlapping proximal ends of sister cells *A* and *B* (*b*). (Bannan 1956)

elongation is slow at the ends of cells lacking a "freed" space. Sometimes the space is occupied quite suddenly by tip elongation of a neighboring cell, as described by Cheadle and Esau (1964) in *Liriodendron tulipifera*. As shown in Fig. 6.16, a cell (A) divided anticlinally, but not equally (B). The upper cell (hatched) divided transversely forming a strand of ray cells which was quickly discontinued (C,D). The lower cell then elongated upward and completely reoccupied the space of the defunct cells (E). Several examples of the differential growth of sister cells

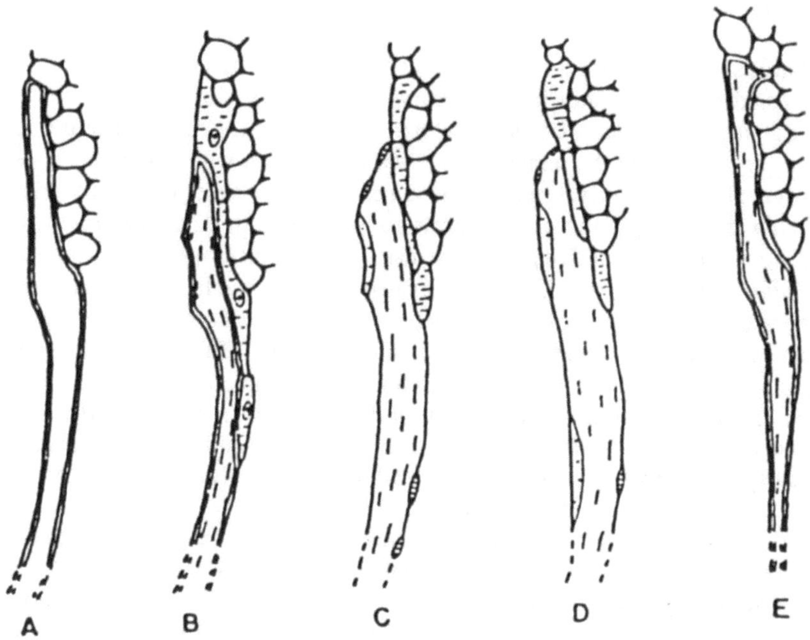

Fig. 6.16A-E. Drawings of serial tangential sections of *Liriodendron tulipifera* extending from the cambium (**A**) outward in the phloem (**E**) showing loss of an initial cell and invasion of its site by another cell. Details in text. 125x. (Cheadle and Esau 1964)

following a pseudotransverse division of a fusiform initial can be found in the case histories depicted in Fig. 6.57.

In *Chamaecyparis* (Bannan 1951b), rate of elongation at the opposite ends also varied widely, especially in young versus old stems. The average combined elongation at the two opposite ends of sister initials ranged from 0.70 to 1.25 mm in young stems and 0.36 to 1.19 mm in old stems during the production of 1 mm of xylem; 1 mm of xylem encompassed from 0.58 to 3.85 growth rings in young stems and from 0.3 to 11.1 growth rings in old stems.

6.1.5.3 Sister Cell Polarity

Elongation of the opposite ends of newly divided sister cell initials, commonly referred to as acropetal and basipetal elongation (Fig. 6.5B), also varies in both rate and extent. Sister initials of *Thuja occidentalis*, for example, displayed pronounced polarity. Measurements by Bannan (1956) showed that in 61% of the cases examined the major growth direction was downward. The growth differential was often considerable. In 72% of the cells examined elongation at one tip exceeded that at the other by 100 μm. Actual basipetal elongation exceeded acropetal by 20 to 30%. On the contrary, acropetal elongation prevailed in *Pinus strobus* with acropetal surpassing basipetal elongation by about 27% (Bannan 1968a).

An analysis of 20 gymnosperms involving about 28 000 pseudotransverse divisions revealed that both sister initials survived in only 33% of the cases and in 16% of the cases both initials failed (Bannan 1968a). An interesting feature appeared during analysis of the remaining 51% of the cases in which only one sister initial survived. Equal survival rates of the upper and lower sister initials occurred in only a few genera. The upper sister persisted more often in *Pinus, Pseudotsuga, Picea, Tsuga, Sequoia,* and *Juniperus* whereas the lower sister persisted more often in *Thuja, Libocedrus,* and *Cupressus* (Table 6.8). A correlation was also observed between the direction of cell elongation and sister cell survival. For example, when survival of the upper sister initial surpassed that of the lower, as in *Pinus strobus,* acropetal cell elongation predominated. On the contrary, when survival of the lower sister initial surpassed that of the upper, as in *Thuja occidentalis,* basipetal cell elongation predominated. Dormer (1980) re-evaluated these data and concluded that death of one sister cell contributed positively to survival of the other.

Branski (1970) observed a similar tendency for apical elongation to exceed basal elongation in newly derived sister initials of *Pinus sylvestris.* The anticlinal division was often unequal, in which case the lower initial was invariably shorter and was subject to a high rate of elimination. Chalk and Ortiz (1961) reported that length growth often occurred at one end only in fusiform initials of *Pinus radiata.* Hejnowicz (1961), on the contrary, could detect no evidence of polarity of fusiform tip growth in *Larix europea,* nor could he (Hejnowicz 1968) in *Picea abies* and *Pinus sylvestris.* However, his conclusions were based on estimated observations.

In the dicotyledon, *Liriodendron tulipifera,* Cheadle and Esau (1964) found evidence that fusiform initials grew more at one end than the other but their sampling method did not permit them to determine whether it was acropetal or basipetal. Nonetheless, they concluded that polarity was not a controlling factor in sister cell elongation. Evert (1961) had reached a similar conclusion in an earlier study of *Pyrus communis* in which he could detect no evidence of polarity in the elongation of fusiform initials.

Fusiform initials in storied cambia do not undergo extensive intrusive growth, but they do have pointed ends that elongate intrusively to a limited extent after radial anticlinal division. Zagórska-Marek (1975) found that growth at the opposite ends of fusiform initials in *Entandrophragma cylindrica* was not equal. In fact, in a population of 1000 cells, the ratio of initials elongating predominantly at either the upper or lower ends was 14:1. It was suggested that this preference might have contributed to the upward moving domain pattern in this species (Chap. 6.6.2). Forking of fusiform initial cells in *Tilia cordata* also exhibited a pronounced polarity (Wloch and Bilczewska 1987), with apical forking exceeding basal forking in a 1.5:1 ratio (Wloch and Zagórska-Marek 1982).

As to causality, Pfeiffer had stated in the 1904 edition of his *Pflanzenphysiologie* that cambial cells were apolar, possessing neither a shoot nor a root pole. Vöchting (1906) disagreed, maintaining that the polarity of cambial cells was inherent. For example, he believed that if the apical end of a cambial cell was severed, a new apical end would develop. Likewise, a new basal end would form if severed. Thus, the apical end would always remain apical and the basal end would always remain basal.

Table 6.8. Relative frequencies of failure and lineal continuation of sister fusiform initials after their origin in pseudotransverse divisions (Bannan 1968a)

		Percentage of pseudotransverse divisions			
Species	Number of divisions	Failure of both derived sister series	Continuation of lower sister series	Continuation of upper sister series	Continuation of both sister series
Abies balsamea	347	12	26	23	39
Cupressus sargentii	481	14	27	19	40
Juniperus virginiana	352	18	21	27	34
Larix laricina	719	17	28	29	26
Libocedrus decurrens	265	15	25	20	40
Picea engelmannii	951	13	23	31	33
Picea glauca	1871	15	24	27	34
Picea mariana	824	16	23	27	34
Pinus banksiana	1023	14	22	30	34
Pinus contorta	1363	14	23	31	32
Pinus edulis	271	16	21	27	36
Pinus flexilis	522	17	20	32	31
Pinus lambertiana	661	13	18	32	37
Pinus ponderosa	1658	15	24	31	30
Pinus resinosa	1730	18	24	30	28
Pinus strobus	7706	17	24	30	29
Pseudotsuga menziesii	1544	18	22	28	32
Sequoia sempervirens	556	18	21	27	34
Thuja occidentalis	4284	20	25	20	35
Tsuga canadensis	760	20	18	26	36

In summarizing his work on the polarity of cambial cells, Janse (1921) attributed no polarity to cambial cells in the tangential direction, unipolarity in the longitudinal direction, and bipolarity in the radial direction. Radially, he considered bipolarity to be indispensable for the cambium when it consisted of only one layer of meristematic cells giving rise more or less alternately to xylem and phloem. However, he recognized another possibility, probably in reference to anomalous cambia, where the cambium might consist of two layers of contiguous meristematic cells, one for the xylem and one for the phloem. In this case, the cambial cells would be equally unipolar in the radial sense. Such a concept is reminiscent of Hartig's (1853) biseriate cambium (Chap. 4.1.1).

Schüepp (1926) expressed the view that cells exhibiting strong polar growth, such as cambial cells, divide in two daughter cells whose form repeats as closely as possible that of the mother cell. Sister cells of an anticlinal division, although smaller than the mother cell, nonetheless assume the mother cell's form. Gravity appeared to be an obvious factor in their subsequent elongation. Bannan (1956), however, ruled out gravity by demonstrating that basipetal elongation persisted even in recurved branches. A study of 540 cell series in the upright, distal parts of eight recurved branches disclosed a difference of 38% in favor of basipetal elongation. In a similar study of 1200 cell series in the proximal portions of these branches, oriented inverse to gravity, the difference was still 21%. To account for this obvious effect of polarity, Bannan suggested a hormonal relationship and discussed the possible role of auxin in the regulation of cambial activity.

In a more recent paper, Edelstein (1982) concluded: "How polarity is induced in cells of the cambium is still a mystery." She suggested a molecular scheme capable of acting as a biochemical switch. The proposed model envisioned two structural protein monomers whose synthesis was in some way controlled autogeneously by feedback repression.

6.2 Radial Anticlinal Divisions

6.2.1 Storied Cambia

Storied, or stratified, cambia result from radial anticlinal divisions as opposed to oblique or pseudotransverse anticlinal divisions in dicotyledons; storied cambia are unknown in conifers (Carlquist 1988a). As a consequence, both the cambial initials and certain derivatives appear horizontally aligned, or stratified, when viewed on tangential sections (Fig. 6.17). Any one or all of the axially oriented cells of the xylem including the rays and axial parenchyma might become storied, although in many species the storied arrangement is confined to certain elements (Huber and Schmidt 1938, Metcalfe and Chalk 1983). As Carlquist (1988a) pointed out, storying is often overlooked because in some species only the vessel elements and axial parenchyma are storied.

Beijer (1927) attempted to clarify the terminology regarding storied cambia. Von Höhnel (1884a,b,c) had referred to these storied or tiered structures by several terms such as "Horizontalschichten," "Stockwerke," and "Etagenbau;" Hartig (1848), years earlier, had proposed the term "Stockwerke" to describe the superposed cell series in rays. Beijer selected von Höhnel's term "Etagenbau," modified it to "Etagencambium," and considered the new term as being the German equivalent of Bailey's (1923) term stratified or storied cambium. Schoute

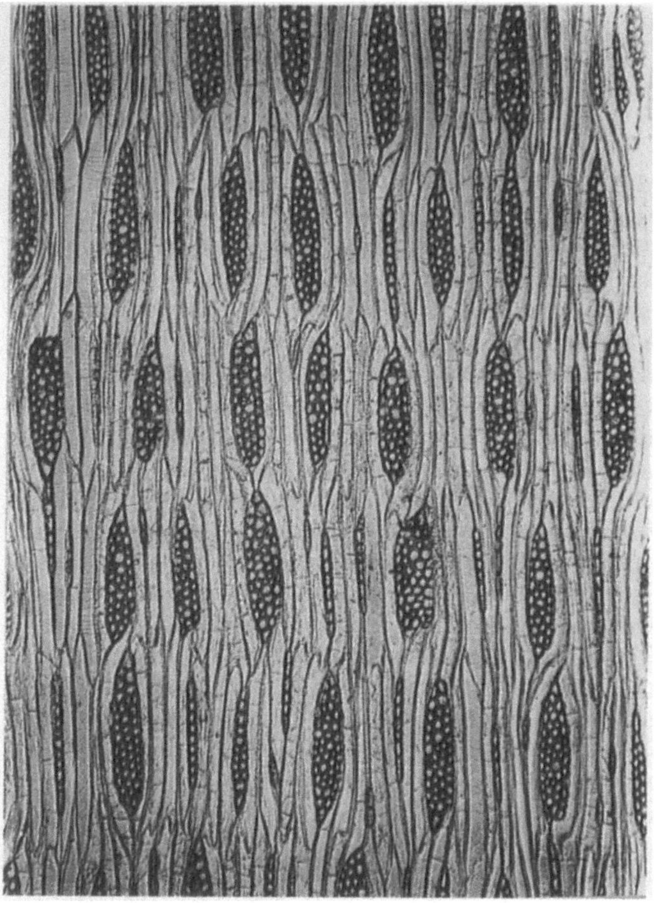

Fig.6.17. Tangential view of xylem from a specimen of *Tilia cordata* showing storied tracheary elements and rays. (Wloch and Szendera 1989)

(1902) had used the same term (Etagencambium) with reference to plants in which several primary cell layers contributed to cambial structure (Chap. 4.5.5). Beijer considered Schoute's choice of words to be unfortunate and suggested that priority be given to von Höhnel.

From time to time workers have attempted to account for storied structure on the basis of cell arrangement in the procambium. Von Höhnel (1884b), for example, observed in several tropical species that the storied structure was present in tissues near the pith and he was of the opinion that the structure was inherent in the primary tissues. Meeuse (1942) expressed a similar view by declaring that cell shape was derived from the original mother cells, i.e., the primary meristem or procambium. Klinken (1914), on the contrary, believed that it was not necessary for the primary tissue to exhibit this special structure. He maintained that storied structures resulted from the absence of sliding growth of cambial cells. To answer this question, Beijer (1927) prepared serial sections from pith to cambium, his "cambiopetal series," from a small root of *Herminiera elaphroxylon*. He found that the storied structure was not present in the procambium, but it did appear gradually in the cambium.

According to Record (1919), it is usually necessary for a stem to attain a considerable size or thickness before the storied structure occurs with regularity. However, in some stems of Leguminosae and Zygophyllaceae he found that the tiers were arranged with a fair degree of regularity in what appeared to be the second annual ring, located less than one-eighth of an inch from the pith. Cumbie (1984) also observed that the storied structure was initiated very close to the pith in *Aeschynomene virginica*, as did Butterfield (1973) in *Hoheria angustifolia*. Others have found that the presence of a storied procambium bears no consistent relation to a storied or stratified cambium. Many species with storied cambia, such as *Cochlearia armoracia* and *Herminiera elaphoroxylon* (Butterfield 1976) and *Glycine max* (Kang and Soh 1988), all exhibit nonstoried procambia. On the contrary, both *Ricinus communis* (Fahn et al. 1972) and *Aristolochia clematitis* (Pulawska 1982) possess storied procambia. However, neither species produces a typical storied cambium. According to Carlquist (1988a), a species that characteristically has storied wood usually begins secondary growth in a nonstoried condition.

Philipson et al. (1971), in their chapter *The Storeyed Cambium*, suggested that the appearance of storied structures might be related to an interrelation between radial and extension growth. That is, in many plants radial growth begins while primary extension growth is still underway. In this case, the nonstoried structure near the center of a stem that later becomes storied might be due, in part, to transverse and pseudotransverse divisions in a meristem that is still undergoing primary extension.

One of the first workers to specifically study storied cambia in detail was von Höhnel (1884 a,b,c), who examined over 80 species, mostly unidentified trees of tropical origins. He observed in woods with storied structures that the cambial cells not only were of almost equal lengths, but they were arranged in horizontal series both radially and tangentially. In many of these woods the rays were storied and each ray arose from a single cambial cell that had divided transversely. In some woods all elements were horizontally arranged, in which case positions of the vessel members and fibers not only corresponded exactly with the cambial cells but pits between the elements were also in perfect horizontal alignment. Von Höhnel relied on physical explanations to account for these storied structures. For example, he attributed the limited tip growth of cambial cells to their "sharply chisel-shaped ends" which permitted them to physically intrude between the radial walls of cell tiers both above and below because these walls were weakest. He believed that the sizes of wood elements as well as their arrangements were in some unexplained way regulated by bark pressure.

Many of the species studied by von Höhnel were re-examined by Klinken (1914), who believed that storied woods formed in the absence of sliding length growth. He questioned von Höhnel's conclusion that the pattern was established exclusively in the cambium, noting that the pattern might also occur in the region of differentiation, but only if sliding growth was absent. He postulated that a complete parallelism must exist between the sizes of successive elements in each radial series and the cambial initial. To test this assumption, Klinken compared the lengths of successive cambial derivatives in a radial series by examining serial tangential sections (Sect. 6.1.1.1). His search was not only to test for the existence of sliding growth but also the way in which it occurred. Klinken finally concluded that cambial cells in all dicotyledons divided by anticlinal radial walls, and the main difference between nonstoried and storied structures was the absence of extracambial sliding growth in the latter type.

Neeff (1920) questioned this conclusion and set out to investigate whether storied structure and cell rearrangement were indeed antagonistic phenomena, as implied by Klinken. He examined serial tangential sections of root wood of *Tilia tomentosa* and found evidence for considerable cellular rearrangement despite the storied structure. Cell reorientation in this species was facilitated by transverse anticlinal divisions of fusiform initials followed by intrusive tip growth, or by sliding growth according to Neeff. Analyses of stem wood revealed processes identical to those in root wood. Kleinmann (1923), in reviewing this work, was of the opinion that Neeff had placed far too much emphasis on the role of sliding growth. Results of his own investigations were more in line with those of von Höhnel in showing that storied structures were formed by radial-longitudinal divisions and limited sliding growth. Later workers have found that some species of *Tilia* exhibit both storied and nonstoried cambia (Sect. 6.6.2.2).

The first investigator to clearly distinguish storied from nonstoried woods on the basis of anticlinal division patterns was Bailey (1923). In nonstoried woods, he demonstrated that anticlinal divisions were of the pseudotransverse type and that daughter initials elongated intrusively, often to a considerable extent. Consequently, the fusiform initials varied in length and appeared irregularly arranged on tangential sections. In storied woods, anticlinal divisions were predominantly radial-longitudinal and the daughter initials underwent little or no intrusive growth. The fusiform initials were relatively short, of nearly uniform lengths, and more or less symmetrically grouped in parallel, horizontal series. Thus, tangential expansion of a nonstratified cambium occurred mainly by intrusive growth of the daughter fusiform initials whereas that of a stratified cambium occurred by radial division and lateral expansion of these initials.

Beijer (1927) accepted the idea that storied structures formed when cambial cells divided by radial-longitudinal walls and the daughter cells elongated little, if at all. His objective was to determine how storied structures were maintained and what role sliding growth, i.e., cell tip elongation, played in maintaining these structures. He employed the method of Klinken (1914) and examined root wood of ***Herminiera elaphroxylon*** by cutting a 7.8 mm radial segment in 104 tangential sections each about 75 µm in thickness. The fates of three tangentially aligned cells were followed. In section 1 of Figure 6.18, the innermost section, the dotted lines in cell 3 approximate the radial wall that produced daughter cells 3 and 3^a shown in section 3. Cell 4, shown dividing transversely, produced a ray. In section 36, about 2.8 mm farther out toward the cambium, cell 1 has also divided to produce cells 1 and 1^a and cell 2 is in the process of dividing by radial walls. Ray 4 now consists of five cells. The final section, number 104, shows a tangential array of 14 cells derived from the original 3 by 11 radial divisions. During this time, new rays l and n above and o, m, and k below the tier have appeared; o and m are only partially subdivided. This latter process is apparently typical of storied species in which the apical half of the fusiform initial subdivides to form ray cell initials followed somewhat later by the basal half. Tangential expansion of the cambium during the production of 7.8 mm of radial xylem was approximately 4x. The division patterns of stem wood of *H. elaphroxylon* were found to be essentially the same as those of root wood. Several other species exhibiting storied structure, such as *Cochlearia armoracia* and *Glycyrrhiza glabra*, were also found to divide by radial walls, but their division patterns were often less regular and in some species the cells underwent extracambial tip growth.

Beijer was also intrigued by the fact that the cells in horizontal tiers were not vertically aligned but appeared to be superimposed in zig-zag lines (Fig. 6.18). For example, as shown in section 1 of Fig. 6.18, the new radial wall in cell 3 did not form directly between points b and d but between end walls ab and cd. This division produced two daughter cells one of which was approximately normal in form and the other more or less rectangular. Beijer proposed five division schemes, two of which are shown in Fig. 6.19. Note that radial walls never intersected cell end points but always one of the two end walls. Each radial wall resulted in either a pointed or a trapezoidal end wall in an adjacent initial and a horizontal end wall in either one or both of the daughter initials. Beijer next described and illustrated how these end walls were presumably reshaped by sliding growth to conform with the storied structure. According to his proposed scheme, sliding growth in *H. elaphroxylon* was minimal, it was confined essentially to the cell tips, and it was believed to be intracambial.

In a recent paper on cell wall networks, Barlow's (1991) portrayal of wall orientation in epidermal cells (Fig. 6.20) is reminiscent of that described by Beijer (Figs. 6.18, 6.19). Barlow referred to this process as "off-setting" of adjacent cell walls at a three-way junction. He suggested that the phragmosome, in seeking the shortest route across the cell, avoided the three-way junction by intersecting the wall to either side.

Beijer's (1927) description of how the cambial initials divided appeared logical and it has been accepted, in general, by several later investigators (Butterfield 1973, Pulawska 1973, Cumbie 1984). One such pattern was clearly illustrated in Fig. 12 published by Derr and Evert (1967). However, in attempting to account for these intracambial adjustments, Beijer's (1927) conclusions became more speculative. He invoked the theoretical concepts of minimum area growth, surface for-

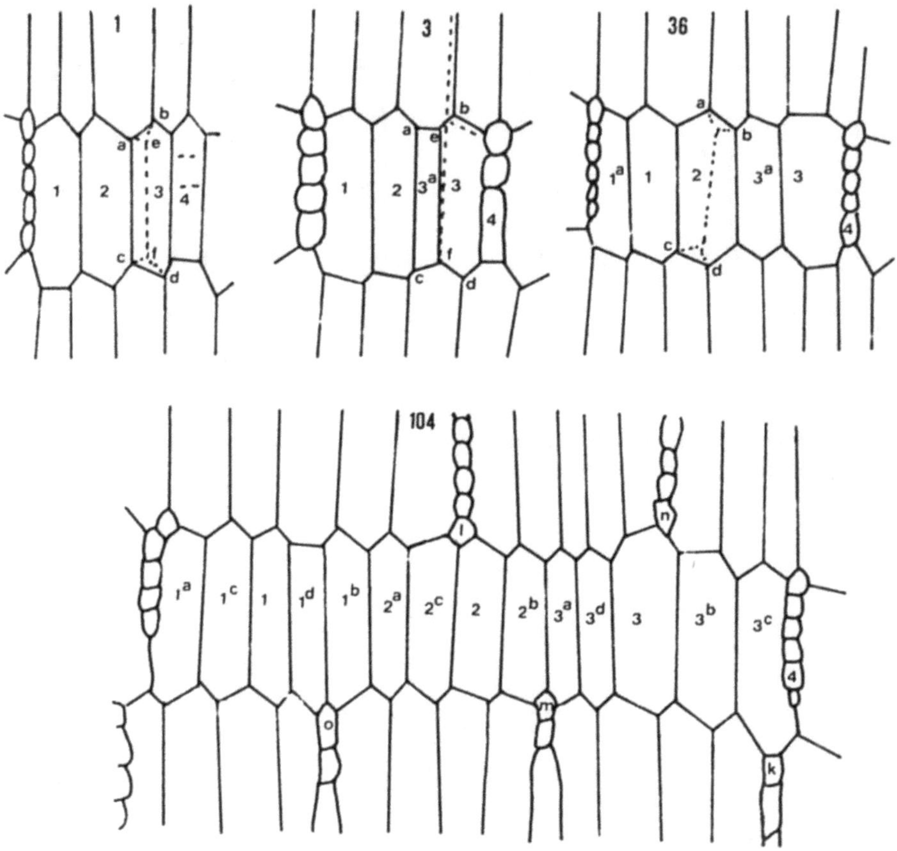

Fig. 6.18. Drawings from tangential serial sections of *Herminiera elaphroxylon* rootwood showing orientation of radial anticlinal dividing walls in a storied cambium. Note uniformity of ray spacing in Section *104*. See text for details. 130x. (After Beijer 1927)

ces, and cell division patterns postulated by Giesenhagen (1909) and others. Meeuse (1942), who did not believe in either sliding or intrusive growth, rejected Beijer's conclusions by simply stating that all the cellular adjustments could be accounted for by growth of the cambial initials in the radial direction.

Much of the early research on storied structures was based on fairly rigid interpretations of radial-longitudinal divisions and either the presence or absence of sliding tip growth. Later research showed that storied structures not only could be maintained under less rigid requirements but many gradations could be found between storied and nonstoried woods. For example, Janssonius (1931) found that the storied structures of tropical woods were not due to simple horizontal layering

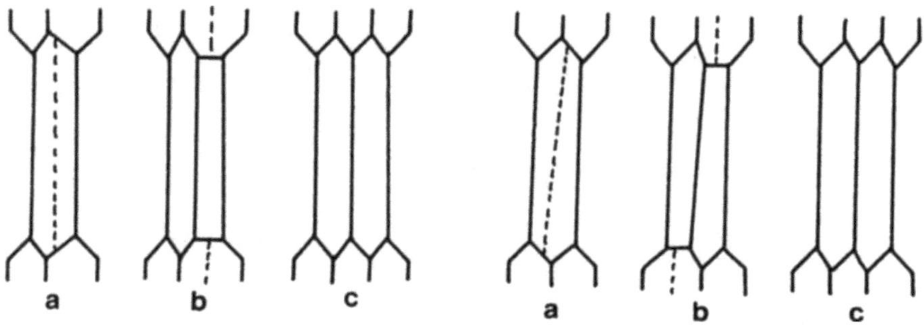

Fig. 6.19. Two possible schemes for radial anticlinal division in the storied cambium of *Herminiera elaphroxylon*. Note that radial walls always intersected a trapezoidal end wall, never a cell corner. (After Beijer 1927)

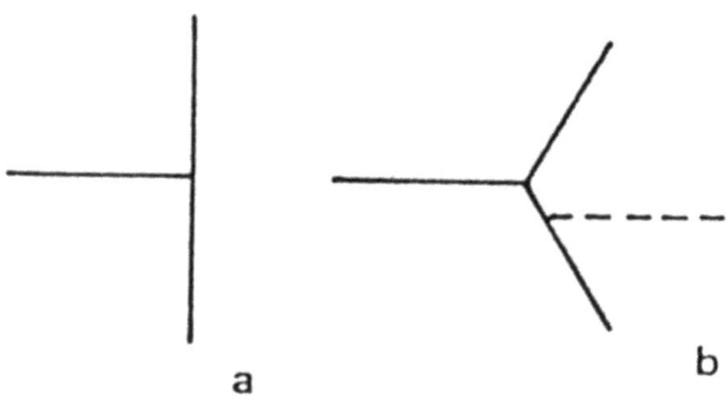

Fig. 6.20a,b. Off-setting of a new cell wall at a three-way junction in epidermal cells. **a** A wall makes a perpendicular attachment to a pre-existing vertical wall. **b** Subsequent expansion of the walls changes this attachment angle to 120°. The phragmosome (*broken line*), in searching for the minimum distance across the cell diameter, avoids the three-way junction. (Barlow 1991)

of cambial cells but to a complex of characteristics. This complex, being variable, contributed to many types and degrees of storied structure. Foremost among these characteristics was sliding growth, which occurred to varying degrees during extracambial fusiform development. Such a case was reported by Lawton (1972) for *Bombax buonopozense*, the phloem of which was storied except during very active growth when storying was disrupted. Another case was reported by Rao et al.

(1989) for *Holoptelea integrifolia*, a species with a clearly storied-type cambium that at times exhibited an intermediate condition because of relatively greater intrusive growth.

Some species have been considered semistoried because anticlinal divisions were both pseudotransverse and radial as in *Polyalthia longifolia* (Ghouse and Hashmi 1980) and *Ficus rumphii* (Ajmal and Iqbal 1987b, Sagheer and Iqbal 1987). In some species radial divisions produced isolated "pockets" of stratified cells. Other nonstoried tropical species also display a distinct tendency toward storied structure (Ghouse and Yunus 1974a, Iqbal and Ghouse 1985a), whereas still others, such as *Dalbergia sissoo*, are at first semistoried and only attain the storied condition in older growth rings. Carlquist (1985b) reported on members of the subgenus Misodendron in which the fusiform cambial initials were clearly storied but the fibers derived from them appeared more or less irregular. He attributed this change to the fact that the stems were small and true storying increased over time.

Chalk and Chattaway (1935) described storied woods in which the horizontal seriation was not perfect because some transverse divisions occurred in the cambial initials in addition to the typical radial divisions. One distinctly storied species, *Aesculus turbidata*, possessed an unusually wide range of vessel lengths. Upon further examination, they found that certain areas in the wood of this species, hence the cambium, were not storied. These were areas where the cambial initials presumably divided transversely, underwent intrusive tip growth, and contributed longer vessel members. In a somewhat similar way, Zagórska-Marek (1984) described how some stems of *Tilia cordata* appeared capable of maintaining a storied cambium despite a relatively high proportion of oblique anticlinal divisions.

As another example, Krawczyszyn (1977b) found that the cambium of *Fraxinus excelsior*, a typical nonstoried species, often appeared as a mosaic of storied and non-storied regions. The storied regions were due primarily to the arrangements of septate parenchyma. Cambial regions designated as storied had a preponderance of radial anticlinal divisions, although these divisions were often difficult to classify because many were intermediate between radial and oblique. Many short fusiform initials were also found in the areas of storied structure, and the short cells tended to divide radially whereas the long cells divided obliquely. Migrating domain patterns in trees with storied cambia are discussed in Section 6.6.2.2.

As a general rule, species with short fusiform initials tend to divide by radial anticlinal walls, whereas those with long initials divide obliquely. In conformance with this rule, Metcalfe and Chalk (1950) placed the transition at a cambial cell length of 0.4 mm. That is, in woody species with cambial cells of this length or less, the tendency would be for them to divide in the radial rather than in the transverse or oblique planes. This distinction appears to have phylogenetic significance (Bailey 1923, Chalk and Chattaway 1935, Record 1943, Bailey and Srivastava 1962).

In some species, such as *Aeschynomene virginica*, with well-developed storied structure, fusiform initials are short and radial divisions predominate (Cumbie 1984). Cumbie followed the cell division process in one stem during the production of 2.0 mm of xylem. In 50 radial files examined, 275 radial anticlinal divisions resulted in the production of 325 fusiform initials, 40 of which were transformed to rays. Interestingly, no fusiform initials were lost from the cambium, as is commonly found in the case of oblique divisions in nonstoried cambia. The number of radial divisions ranged from 0 to 12 in the 50 files, but most averaged 4 to 6 per file. For example, in Fig. 6.21, the fusiform initial that produced cell series a divided five times resulting in the formation of a horizontal tier of six fusiform initials. In a slightly different pattern, the initial that produced cell series f divided 12 times, resulting in the formation of 13 fusiform initials, one of which was transformed to a ray. Like Beijer (1927), Cumbie found that rarely did two cells located one above the other divide anticlinally at the same time. Yet, divisions in the vertically adjacent stories somehow appeared to be synchronized.

Aeschynomene hispida, a related species, was examined by Butterfield (1972). As in *A. virginica*, anticlinal divisions were almost exclusively radial. Mean lengths of the fusiform initials did not vary with age of growth ring (Fig. 6.22) nor did any of the xylem derivatives undergo appreciable intrusive growth during their differentiation. Butterfield (1972) analyzed tangential microsections of the xylem in one stem segment extending over a distance of 26 mm from the stem center. Within the tangential area examined, a total of 378 fusiform initials produced 105 new fusiform initials and 7 new rays. No fusiform initials were lost following radial division. Rays were formed by transverse divisions of fusiform initials. Although the majority of the rays, both uniseriate and multiseriate, usually remained within the stories, occasional rays developed at half-story heights while a rare few exceeded the story heights. Occasional transverse divisions also added initials in the vertical plane thereby contributing to a slight decrease in mean element length (Fig. 6.23).

In some storied woods all elements - the rays, vessel members, tracheids, wood fibers, and axial parenchyma - occur in horizontal seriation. Moreover, those woods with distinctly storied xylem usually show storied phloem elements (Record 1919, den Outer 1967, Anand et al. 1978). However, it is evident that other storied woods can accommodate considerable variability among their constituent elements while still maintaining a semblance of storied structure. Despite the fact that radial anticlinal divisions and the absence of intrusive growth are characteristics of species with storied cambia, some species apparently adapt to a variable proportion of oblique divisions (Chalk and Chattaway 1935, Zagórska-Marek 1984) and/or intrusive growth (Hejnowicz and Hejnowicz 1959, Krawczyszyn and Romberger 1979). When present, intrusive growth is most often confined to extracambial tip growth of the libriform fibers and fiber tracheids. The vessel segments and axial parenchyma usually show storying, as does the cambium (Carlquist 1975, 1988a, Philipson et al. 1971).

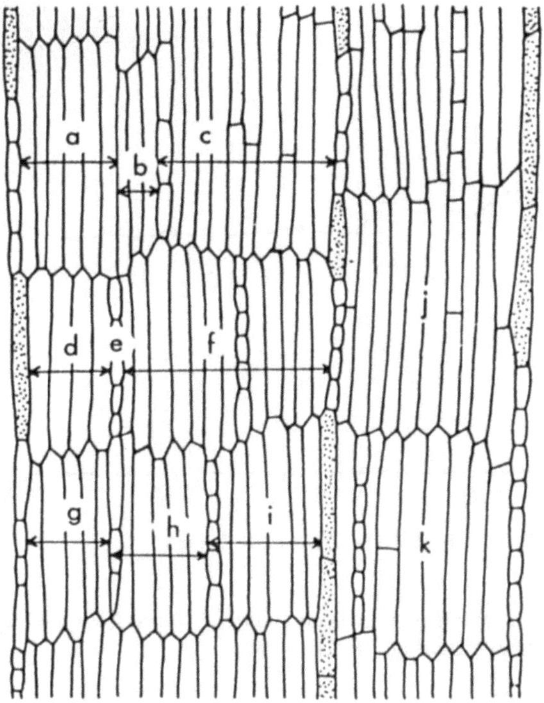

Fig. 6.21. Drawing of a tangential section of secondary xylem 2.0 mm from the primary xylem in *Aeschynomene virginica*. Letters indicate series of cells derived from individual fusiform initial cells by anticlinal divisions. Some primordial ray initials (*stippled*) have not yet transformed to rays. 125x. (Cumbie 1984)

As Beijer (1927) observed, not all radial divisions are symmetrical. In a similar way, many radial divisions in *Dalbergia sissoo* were found to result in daughter cells of unequal size and shape (Anand et al. 1978). This division pattern was presumably related to the high proportion of parenchyma cells having different structures in this species. Notwithstanding these unequal division patterns, the vessels, sieve elements, and most parenchymatous cells retained the storied structure. These variable patterns in the arrangement of fibers, rays, and other xylem elements contribute to the fact that a gradation exists between storied and nonstoried species, with many species exhibiting features characteristic of both types.

Many authors have compiled lists of woods with storied structures (Record 1919, Bailey 1923, Chalk and Chattaway 1935, Metcalfe and Chalk 1950, 1983, Carlquist 1988a), whereas others have classified them either taxonomically (Record 1936, 1943) or phylogenetically (Metcalfe and Chalk 1983). Record (1943) devised a key to woods with storied structures consisting of 149 genera from 29 families, mostly tropical. Takamatsu (1928) constructed a phylogenetic

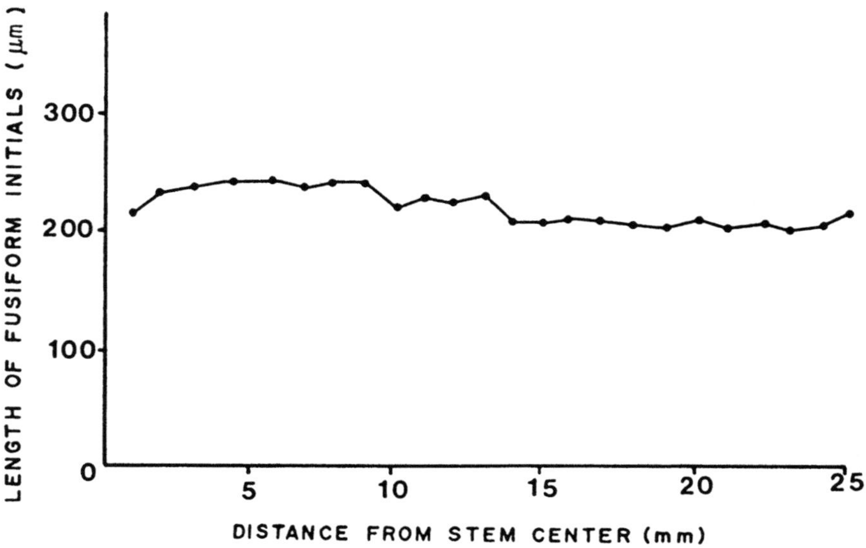

Fig. 6.22. Mean lengths of fusiform initials plotted against radial distance from the stem center in *Aeschynomene hispida*. (After Butterfield 1972)

tree by classifying families according to cambial cell arrangements. He examined 150 species from 50 families, mostly from Japan, and recognized four cambial types including regularly stratified and slightly stratified cambia. Den Outer (1986) prepared a list of five anatomical features, presumably characteristic of the cambium and xylem of storied woods. He pointed out that although storied structure is considered to be a highly specialized, advanced feature, the component elements of secondary phloem might be either specialized, intermediate, or even primitive. In a like manner, the component elements of the secondary xylem present a similar graded series. Thus, it is evident that across the range from well-developed storied to nonstoried structures, one might encounter a wide variety of intermediate types.

6.2.1.1 Storied Rays

Like the wood fibers, the rays of storied woods might be either storied or non-storied (von Höhnel 1884a,b,c, Record 1919, Philipson et al. 1971, Carlquist 1988a) (Fig. 6.17). In some species, two types of rays occur, small storied rays and larger nonstoried ones (Record 1919). Species with small, narrow rays usually exhibit the most conspicuously arranged storied rays, as for example, in *Pterocarpus santalinus* and *Picrasma excelsa* (von Höhnel 1884c). These rays retain their storied structures because the stratified fusiform initials do not undergo intrusive growth. Ray heights in this case can only be reduced by loss of ray cell initials (Barghoorn 1941a).

In species with large multiseriate rays, on the other hand, the rays are seldom storied and they show no distinct relation to the horizontal tiers of fusiform cambial initials (Barghoorn 1941a). This nonstoried structure results from the greater tendency for large rays to be dissected and irregularly dispersed (Fig. 6.24). Exceptions do occur and multiseriate, storied rays are maintained in some species, as, for example, in *Acacia catechu* (Ghouse and Yunus 1974c, Ghouse and Iqbal 1975), *Dalbergia sissoo* (Ghouse and Yunus 1972, 1973) and several other species of *Dalbergia* (Ghouse and Yunus 1974b), *Salvadora persica* (Sajwan and Paliwal 1976), and *Laurelia novae-zelandiae* (Metcalfe and Chalk 1983). In contrast to

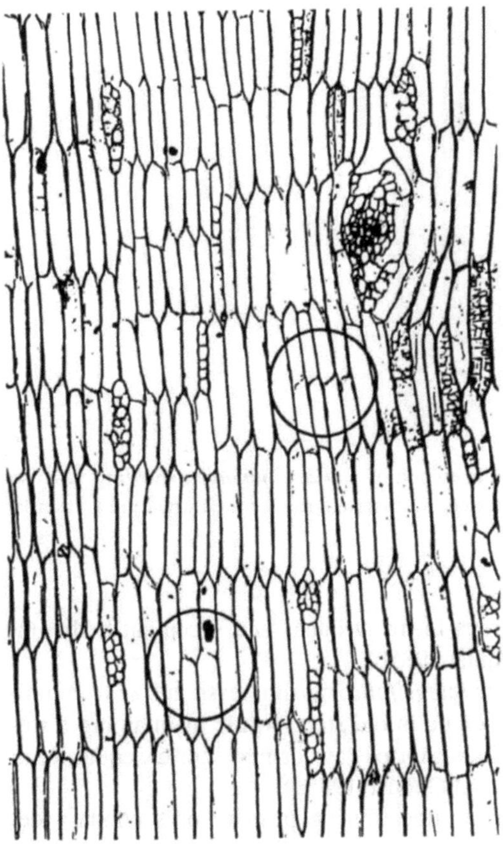

Fig. 6.23. Storied cambium in *Aeschynomene hispida* showing short stories produced by radial longitudinal divisions following transverse division of a fusiform initial (*circles*). 85x. (Butterfield 1972)

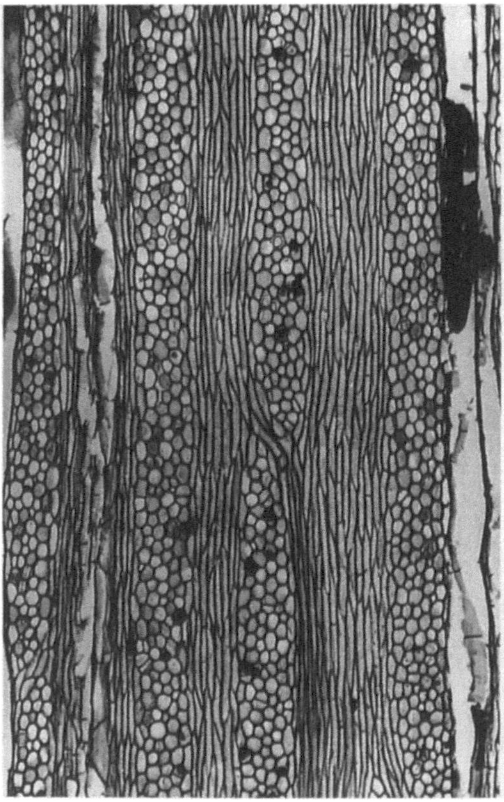

Fig. 6.24. Tangential section through the latewood of *Coriaria japonica* showing vascular elements and dissection of a large, nonstoried, multiseriate ray by the intrusion of vascular tracheids. (Carlquist 1985b)

most dicotyledons, the multiseriate rays in many lianas and succulent herbs undergo a low rate of subdivision (Carlquist 1985b).

In some species, such as *Holoptelea integrifolia*, rays increase in size from uni- and biseriate in young stems to tetraseriate in old stems while maintaining a storied structure (Ajaml et al. 1986a). Thus, storied rays can be shorter, equal to, or taller than the fusiform initials (Iqbal and Ghouse 1985a).

Storied ray structures can originate either during primary or secondary development. Cumbie (1984), for example, observed "primordial ray initials" that could be recognized in the procambium of *Aeschynomene virginica*. Primary rays, which are actually secondary tissues (Chap. 8.2.1), arose from these primordial ray initials and the storied structure therefore developed very close to the pith. In many other species the large primary rays are nonstoried and they are dissected into smaller, often uniseriate rays which then become storied. Dissection might take place by the reversion of ray cell initials to fusiform initials, as Chattaway (1933a) described in species of the Sterculiaceae (Fig. 8.27), or by the intrusive tip growth of fusiform initials as Wloch and Szendera (1989) described in Tilia cordata (Fig. 6.65). The previous discussion of Beijer's (1927) work included a description of the most common way in which new ray cell initials are added to rays in storied cambia, i.e., by subdivision of a fusiform cambial initial. These variable patterns of rays and other xylem elements contribute to the fact that a gradation exists between storied and nonstoried woods.

6.3 Fates of Initial Cells

After dividing anticlinally, the newly formed sister initials can follow any one of several different pathways or succumb to different fates. It is assumed here that the division occurred in the initial layer of the cambium and not in a mother cell, in which case the sister initials would be considered temporary residents of the cambium. The most logical fate of the new initials would be to survive as initial cells and to divide periclinally for an indefinite period of time. Eventually, however, every initial cell will either redivide anticlinally or succumb to another fate.

6.3.1 Survival of Initial Cells

The length of time a fusiform initial cell might reside in the cambium before dividing anticlinally will be influenced by a host of factors, as discussed in Section 6.5. However, two criteria of prime importance in determining survival rates are fusiform cell length and ray contacts.

6.3.1.1 Fusiform Cell Length

The relation between fusiform cell length and survival has been commented on by many investigators but examined most thoroughly by Bannan and Bayly (1956). Data on different survival scenarios are presented for 12 coniferous species in Table 6.9. Perusual of these data show that the longer the fusiform initial at the time of pseudotransverse division, the greater the probability that both daughter initials will survive. Conversely, the shorter the fusiform initial at the time of pseudotransverse division, the greater the probability that both daughter initials will fail. In the case where only one sister cell survived, the surviving initial usually elongated to a greater extent following loss of the other than when both sister cells survived (Bannan 1951b). The foregoing relationship held for all species examined and for both slow and fast-growing trees.

Further confirmation of this relationship was derived from an in-depth study of *Thuja occidentalis* (Bannan and Bayly 1956). When newly formed daughter initi-

Table 6.9. Relation between lengths of fusiform initials at pseudotransverse division and behavior of the derived daughter initials (After Bannan and Bayly 1956)

		Mean length of initials at division (mm)		
Species	Ring width mm	Both initials survive	One initial fails	Both initials fail
Abies concolor	1.0	7.15	5.73	4.20
Cedrus deodara	0.7	4.40	3.60	2.90
Chamaecyparis nootkatensis	0.14	3.42	3.03	2.60
Cupressus arizonica	0.8	2.73	2.38	2.30
Heyderia decurrens	0.6	5.33	4.56	3.78
Pinus resinosa	4.1	2.01	1.89	1.63
Pinus resinosa	3.0	3.76	3.34	2.55
Pseudotsuga menziesii	1.7	6.50	5.47	5.12
Sequoia sempervirens	0.3-0.9	7.10	6.47	5.06
Sequoiadendron giganteum	1.2	5.14	4.02	3.50
Thuja occidentalis	0.5-1.0	3.25	2.93	2.63
Thuja occidentalis	1.0-3.0	3.06	2.83	2.48
Thuja occidentalis	3.1	2.62	2.30	2.24
Thuja occidentalis	3.0-4.5	2.72	2.46	2.25
Thuja plicata	0.8-2.0	2.97	2.63	2.19
Tsuga canadensis	1.9	4.12	3.80	3.07

als were grouped in four size classes, survival ranged from a high of 90% in the quartile represented by the longest initials to a low of 12% in the quartile represented by the shortest initials; survival of initials in the middle one-half representing cells of intermediate length ranged from 38 to 50%.

The tendency for the longest daughter initials to survive following anticlinal division has been observed not only in conifers (Bannan and Whalley 1950, Hejnowicz 1961) but also in dicotyledons (Evert 1961). A particularly convincing example was described by Cheadle and Esau (1964) in Liriodendron tulipifera. As shown in Table 6.10, the percentage of fusiform initials surviving increased from 11.6% for those in the shortest length class to 77.8% for those in the longest length class. It should be mentioned that this analysis was based on the last-formed phloem increment in branch tissue. Consequently, the last column represented those initials persisting throughout the growing season without dividing. However, some initials were lost from the cambium as a result of anticlinal divisions. Here again, the longer the initials when they were lost, the more likely that they divided anticlinally; in this case, the original initial was lost in the sense that it was replaced by two new initials. Conversely, the shorter the initials when they were lost, the more likely it was that they were lost either outright or converted to ray cell initials. Thus, the lengths of fusiform initials have an important bearing on how they are likely to be lost. Hejnowicz (1961) chose to phrase his conclusion differently, based on analyses of *Larix decidua* cambia. He placed emphasis on elimination of the shortest initials, which favored survival of the longest initials by allowing stronger intrusive growth.

In summarizing the results of their analyses, Bannan and Bayly (1956) observed that increasing girth of the conifer cambium by anticlinal divisions was accompanied by continued selection of the largest initials. Maintenance of an adequate cell length was brought about by survival of the longest and elimination of the shortest initials, and this selection process contributed to the increase in mean cell length that occurred on a stem cross-section or with age of the tree.

6.3.1.2 Ray Contacts

The intracambial competition mentioned earlier favors not only long cells but also those with the most ray contacts (Fig. 6.25). Studies have shown that the number and type of ray contacts is more critical to the survival of the daughter initials than cell length. Of course, the two are usually related with the longest initials possessing more ray contacts than shorter ones. For example, in a study of *Chamaecyparis thyoides* by Bannan (1951b) the surviving sister cells were approximately 2090 µm long and had contact with an average of 6.1 rays having a combined total of 21.2 cells. Failing sister cells, on the other hand, were approximately 1940 µm long and had contact with an average of 4.9 rays having a combined total of 16.1 cells.

Table 6.10. Fates of fusiform initials in different length classes in *Liriodendron tulipifera* stems (After Cheadle and Esau 1964)

Dividing initials Length class µm	Number	Percentage of initials lost[a] Outright	To rays	By divisions	Persistent initials (%)
1-99	26	53.8	34.6	0.0	11.6
100-199	109	38.5	30.3	10.1	21.1
200-299	143	15.4	11.2	21.0	52.4
300-399	87	1.1	1.1	18.4	79.4
400-499	17	11.8	0.0	17.6	70.6
500+	9	0.0	0.0	22.2	77.8

[a] Based on lengths of first derivatives formed after initial divided by an anticlinal wall.

Table 6.11. Length and ray contacts of surviving and failing daughter fusiform initials shortly after their origin in pseudotransverse division (Bannan and Bayly 1956)

	Surviving fusiform initials			Failing fusiform initials		
Species	Length mm	Number of contact ray cells[a]	Number of ray cells/mm of length	Length mm	Number of contact ray cells[a]	Number of ray cells/mm of length
Abies concolor	3.75	97.5	26.0	2.66	53.0	19.9
Cedrus deodara	2.38	44.5	18.7	1.79	21.3	1.9
Chamaecyparis lawsoniana	1.90	22.0	11.6	1.56	14.6	9.4
Chamaecyparis nootkatensis	1.96	23.9	12.2	1.47	13.3	9.1
Cupressus arizonica	1.60	23.3	14.6	1.32	15.0	11.4
Heyderia decurrens	3.00	44.5	14.8	2.27	26.7	11.8
Pseudotsuga menziesii	3.65	65.6	18.0	2.91	44.1	15.2
Sequoia sempervirens	4.14	42.9	10.4	3.26	26.9	8.3
Sequoiadendron giganteum	2.88	45.1	15.7	2.02	25.7	12.7
Thuja occidentalis	1.89	16.8	8.9	1.60	10.3	6.4
Thuja occidentalis	1.85	19.7	10.6	1.54	12.6	8.2
Thuja occidentalis	1.60	19.8	12.4	1.33	12.2	9.2
Thuja plicata	1.81	27.2	15.0	1.43	15.4	10.8
Tsuga canadensis	2.45	41.4	16.9	1.97	25.2	12.8

[a] Mean number of ray cell initials bordering a newly formed daughter fusiform initial.

Similar data for 12 coniferous species were compiled by Bannan and Bayly (1956) (Table 6.11). In summarizing these data, the authors noted that surviving initials had contacts with 70% more ray cells than failing ones, or 30% more ray contacts per millimeter of length. Although the number of ray contacts might either increase or decrease during subsequent elongation of a fusiform initial, the ray contacts at the time of origin appeared most important to its survival.

One of the earliest studies of ray contacts was made by Janse (1887). His interest was not with cell survival but with water transport between rays and tracheids. Janse examined a large number of gymnosperms and found that less than 1% of the tracheids were without ray contacts. On the average each tracheid was in contact with from one to six rays, and each ray was from one to four cells high although some ranged up to 20 cells high. Dodd (1948), in his investigation of cambial cell shape in *Pinus sylvestris*, also found some initials without ray contacts.

The average number of ray contacts per cell was 1.6 for the 50 initial cells examined (Fig. 5.4).

Usually, the longest cells survive and the shortest fail. Bannan and Bayly (1956) found that short cells seldom survived even when they were provided with good ray contacts. Cells of intermediate lengths, on the other hand, varied widely in their rates of survival. In these cells the area of ray contact was of critical importance, and the survivors generally had more extensive ray contacts than the failures, both in an absolute sense and in proportion to cell length.

Fusiform initial survival depends not only on cell length and ray contact but also on the type of ray contact as shown in Table 6.12 for 12 coniferous species (Bannan 1965b). The fact that initials in contact with fusiform rays divided more frequently than those in contact with uniseriate rays was attributed to the larger size of the former type ray. Fusiform rays as a rule also had more parenchyma cells in contact with the cambial initials than uniseriate rays did. In *Pinus* this difference ranged from 60 to 100%, but in *Picea* and *Pseudotsuga* the disparity was only about 15 to 20%.

Both cell length and extent of ray contact were also found to be related to survival and loss of initials from the cambium after anticlinal divisions in *Pyrus communis* (Evert 1961). On the average, failing initials were 34% shorter, with 37% fewer ray contacts than surviving initials. Likewise in *Liriodendron tulipifera*, Cheadle and Esau (1964) found that 40% of all initials lost outright had no ray contacts, 40% had only lateral, and 16% only terminal contact with small rays. In *Leitneria floridana*, Cumbie (1967b) observed that 75% of the failing initials were without ray contacts and surviving initials had 42% more ray contacts than failing ones.

6.3.2 Failure of Initial Cells

Failure of a fusiform initial can occur outright by cell elimination, by various kinds of segmentation, and by conversion to ray cell initials. These are all considered failures in the sense that the fusiform initial is itself eliminated from the cambium. Elimination or disappearance of a fusiform initial means that the initial differentiated and matured as a permanent xylem or phloem element. Before being eliminated, an initial usually divides periclinally from one to several, sometimes many times, and it often undergoes size reduction.

An interesting feature of woods with storied structures is the extremely low level of cell failure. Fusiform initials in storied woods divide predominantly by radial anticlinal divisions. Because the initials undergo little or no intrusive growth, there would be no advantage to cell elimination. In fact, cell elimination followed by intrusive growth would soon destroy the storied structure. Several workers who followed xylem development through extensive radial series reported that no cells were lost from the cambia of storied woods after radial anticlinal divisions (Butterfield 1972, Pulawska 1973, Cumbie 1984). However, others observed some cell loss (Rao et al. 1989).

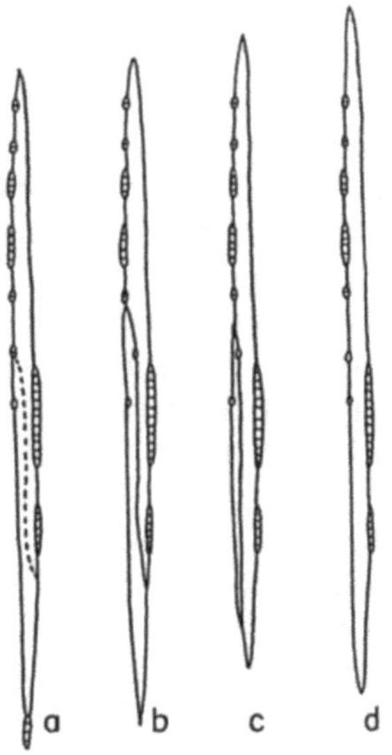

Fig. 6.25a-d. Diagrams illustrating effect of ray contacts on sister initials after their origin in pseudotransverse division. The upper initial with extensive ray contacts survived whereas the lower initial with few ray contacts failed. (Bannan 1957c)

6.3.2.1 Cell Elimination

Cell loss was one characteristic of cambial activity rather widely recognized by early workers. Although they did not fully understand the processes that brought it about, they did appreciate the consequences. Radlkofer's (1858) studies with *Cocculus laurifolius* provide good examples. He observed that wood cells often elongated eight or more times after their origin in the cambium, and they did so at the expense of neighboring cells which often became smaller and died out. When he examined two adjacent cell series in a radial wood bundle, he found that the length of one cell increased three times over five cell derivatives, whereas the adjacent cell became shorter and disappeared at the end of the fourth derivative. These trends were repeated in the opposite direction in the bast, indicating that the divisions arose in the cambial initials. Nägeli (1858) agreed completely with Radlkofer's conclusions that cell shortening and death of one radial cell file explained the increase in length of an adjacent cambial file. Moreover, he believed that this process contributed in some way to the fact that cambial cells, after lengthening for a period of time, "henceforth average the same length."

Elimination of one cell file during the elongation growth of an adjacent file was illustrated as well as described by Neeff (1914) in *Tilia* and *Picea*. In the transverse section of *Picea abies* shown in Fig. 6.26, cell files 5 and 7 were eliminated by being virtually squeezed from the cambium by elongation and expansion of adjacent cell files. Such a process of cell elimination has been described by other investigators (Raatz 1892, Nordhausen 1898, Beijer 1927).

Table 6.12. Comparison of rate of pseudotransverse division in fusiform initials touching fusiform rays and initials touching uniseriate rays (Bannan 1965b)

Species	Number of trees	Rate of division/cm xylem		Difference in division rate (%)
		Initials touching fusiform rays	Initials touching uniseriate rays	
Picea engelmannii	46	1.54	1.42	8.4
Picea glauca	64	1.47	1.39	6.6
Pinus aristata	10	2.86	2.34	22.2
Pinus banksiana	29	2.12	1.81	16.9
Pinus contorta	45	2.26	1.95	15.9
Pinus edulis	12	4.70	4.02	16.6
Pinus flexilis	11	3.14	2.76	13.5
Pinus lambertiana	15	1.57	1.36	15.7
Pinus ponderosa	43	2.94	2.48	18.3
Pinus resinosa	50	3.02	2.54	19.1
Pinus strobus	59	3.06	2.64	15.9
Pseudotsuga menziesii				
Coastal	40	1.56	1.52	3.0
Interior	41	1.95	1.78	9.3

The nature and extent of cell failure following anticlinal divisions were prominent processes in all of Bannan's extensive investigations. His studies of Chamaecyparis (Bannan 1950b) revealed that during the first year after anticlinal division of a fusiform initial, either one or both of the two daughter cells vanished from the cambium in 43.3 and 26.4% of the cases, respectively. Thus, of the 1100 anticlinal divisions examined, both daughter initials survived the first year in only 30.3% of the cases. A summation analysis for all positions in one tree showed a net gain of only 162 fusiform initials resulting from 1123 anticlinal divisions. Of these divisions, 1035 fusiform initials either disappeared entirely from the cambium or were converted to ray cell initials.

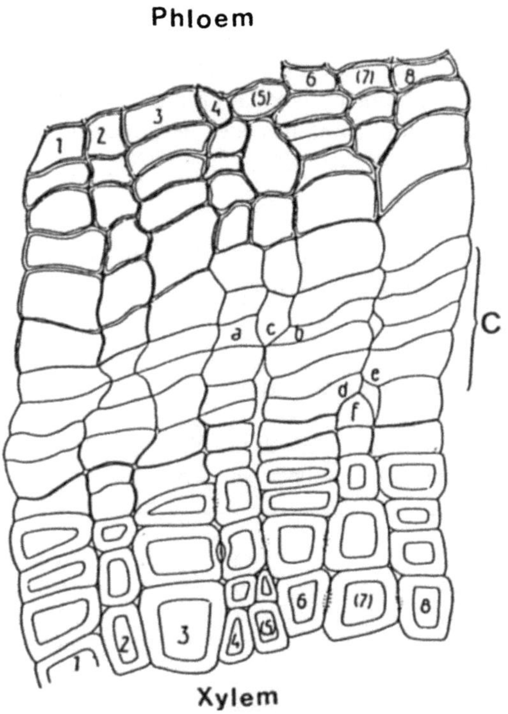

Fig. 6.26. Schematic representation of a transverse section of *Picea abies* stem. Cell files *5* and *7* are in the process of being eliminated in both the phloem and xylem following failure of their respective initial cells *c* and *f*. 264x. (Neeff 1914)

Other studies (Bannan 1951b) showed that the disappearance of initials varied seasonally, with the highest loss occurring in the final one-quarter of the growth ring (Table 6.13). Since this was also the time of year when both the frequency of anticlinal divisions and elongation of the fusiform initials were greatest, Bannan suggested that these events might be correlated with the disappearance of initials. Some daughter initials abruptly stopped dividing periclinally, whereas others underwent a period of reduction before disappearing from the cambium (Bannan 1951a). If the anticlinal division occurred early in the season when radial growth was vigorous, a daughter initial might continue dividing periclinally for an extended period before being eliminated. On the contrary, if the division occurred late in the growing season when radial growth had subsided, the daughter initial would more likely be subjected to abrupt elimination (Bannan 1953). The reduction in cell size preceding disappearance from the cambium occurred in all dimensions, as shown in Fig. 6.27 for *Thuja occidentalis*. These reductions in cell dimensions were simply reflections of the changes that took place in the cambial initials. However, it should not be assumed that cell loss occurs only during the season in which the fusiform initial was formed. An initial might persist and divide periclinally for many years before being eliminated. In fact, Bannan (1956) suggested that in some extremely slow-growing trees this process might continue for several decades, during which time the initial would undergo a gradual decrease in length.

Table 6.13. Time of loss of fusiform cambial cells in relation to the annual increment in *Chamaecyparis* and *Thuja* (Bannan 1951b)

		Percentage loss			
Species	Number of observations	First quarter	Second quarter	Third quarter	Fourth quarter
C. lawsoniana	122	17.0	12.0	13.0	58.0
C. thyoides	276	14.0	27.0	29.0	30.0
T. occidentalis	295	9.5	17.0	30.5	43.0
T. plicata	172	10.0	18.0	33.0	39.0

Elimination of initials has also been found to occur more frequently in very old stems (Hejnowicz 1961) where loss of initials tends to be compensated for by a higher frequency of anticlinal divisions and a greater rate of intrusive elongation (Hejnowicz 1968). That is, intrusive growth is strongest when neighboring cells are undergoing elimination. In another study, Hejnowicz (1967) attempted to account for the loss of fusiform initials mathematically by analyzing the interrelationship among initial length, frequency of anticlinal divisions, and intrusive growth.

Srivastava (1963a,b) examined the loss of fusiform initials, which he called declining initials, from the cambium of many coniferous species. Although his analyses were confined to the phloem, his conclusions with regard to cell loss were essentially the same as those of Bannan. This was also true of several dicotyledonous species. In both *Pyrus communis* (Evert 1961) and *Pyrus malus* (Evert 1963b), as well as *Leitneria floridana* (Cumbie 1967b), the loss of a fusiform initial was generally preceded by a progressive shortening of the initial before it eventually lapsed into maturity. In *Pyrus communis* (Evert 1961), cell loss was exceptionally high with approximately 50% of all new initials being eliminated outright from the cambium over a 7-year period. Failing initials not only were reduced in length but they often produced asymmetric periclinal divisions before expiring. Progressive shortening and eventual loss is also of common occurrence in tropical species. This loss can be significant during periods of high cambial activity (Rao and Dave 1985a).

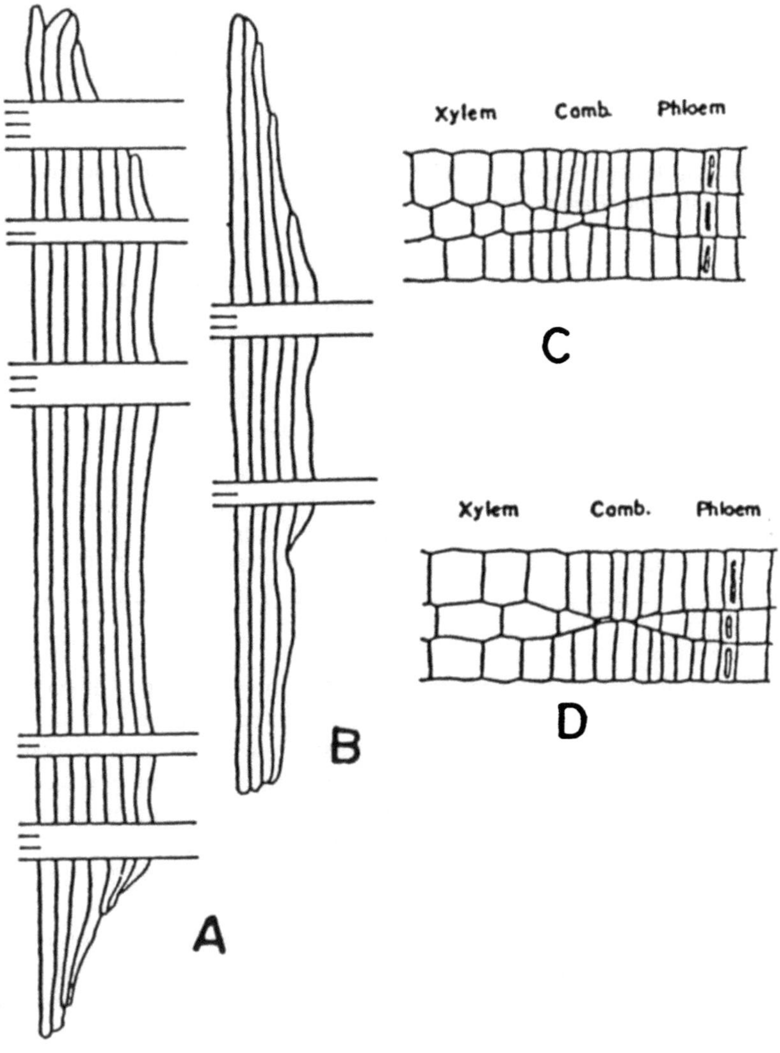

Fig. 6.27A-D. Loss of a fusiform initial from the cambium is often preceded by a reduction in size. Drawings of radial (**A,B**) and transverse (**C,D**) sections of *Thuja occidentalis* cambial zone showing reduction in length and tangential diameter, respectively, in files of failing initial cells. (Bannan 1953)

In *Liriodendron tulipifera* (Cheadle and Esau 1964), 20.7% of the new initials were lost outright in the last-formed phloem increment. As shown in Table 6.10, cell loss was highest in the shortest initials. In addition, however, initials arising from divisions with transverse or slightly oblique partitions were twice as likely to be lost outright or to rays as those with more oblique partitions. Further analysis of the fates of 311 original fusiform initials showed a net gain of only 9 initials during the course of a year. Of the 394 anticlinal divisions that occurred during that time, 105, or 26.6%, were lost outright from the cambium.

A somewhat different pattern occurs in woody shrubs and herbaceous dicotyledons. During the formation of the inner secondary xylem of the woody shrub *Leitneria floridana*, Cumbie (1967b) observed that almost all new fusiform initials survived. The production of new initials proceeded at a rate just about adequate to maintain circumferential expansion of the cambium. In older cambia, up to 26 years of age, the failure rate had increased to only about 20% of all anticlinal divisions. Cumbie attributed this difference, in part, to the relatively short length of the fusiform initials in this species; that is, with the tendency for dicotyledons with short fusiform initials to divide less frequently. In two herbaceous dicotyledons, *Apocynum sibiricum* (Cumbie 1969a) and *Polygonum lapathifolium* (Cumbie 1969b), no fusiform initials were eliminated during the production of seconday xylem, except by conversion to ray cell initials.

6.3.2.2 Segmentation to Parenchyma

Some fusiform initials give rise to longitudinal files of parenchyma cells by a series of transverse divisions. These parenchymatous cells were the "Zellfasern," or cell fibers, of Hartig (1848, 1859a), or simply the "Holzparenchymzellen," or wood parenchyma cells, of Radlkofer (1858), Schacht (1860), Sanio (1863a), and Frank (1864) which originated by transverse division of "cambial fibers" (Fig. 6.28A). The vertically oriented parenchymatous elements were also called "Strangparenchym," or strand parenchyma (Fig. 6.28B), in contrast to the horizontally oriented "Strahlenparenchym," or ray parenchyma (Hartig 1891). In discussing formation of the parenchymatous tissue in dicotyledonous trees, Küster (1899) suggested that the cambial cells must first be transformed to parenchyma before segmentation could begin. He believed that the stimulus for this transformation was in some way related to bark pressure. Neeff (1920) also considered the so-called "Fächerung," or compartmentation, of these cells to be extracambial, i.e., to occur within the cambial zone but not in the initial cells (Chap. 7.1.1). Thus, most longitudinal parenchyma cells were thought to arise from extracambial divisions, in which case they would not be perpetuated in a radial series.

It is now known that strand parenchyma are formed by transverse divisions of fusiform cambial derivatives. However, fusiform parenchyma are derived directly from a fusiform cambial initial without transverse subdivision (Panshin et al. 1964). These latter cells are the so-called "Ersatzfasern," or substitute fibers in the older literature (Sanio 1863a). In both cases, the entire strand or cell is approximately the same shape and size as the original mother cell.

Some cell structures that appear to be axial parenchyma have different origins. In *Ribes*, Stern et al. (1970) observed isolated strands of parenchyma two to four cells high that were shorter than the associated vessel elements. They suggested that these isolated parenchyma strands were not derivatives of fusiform initials but of ray cell initials. Kuroda (1986) reported similar septate cells derived from ray

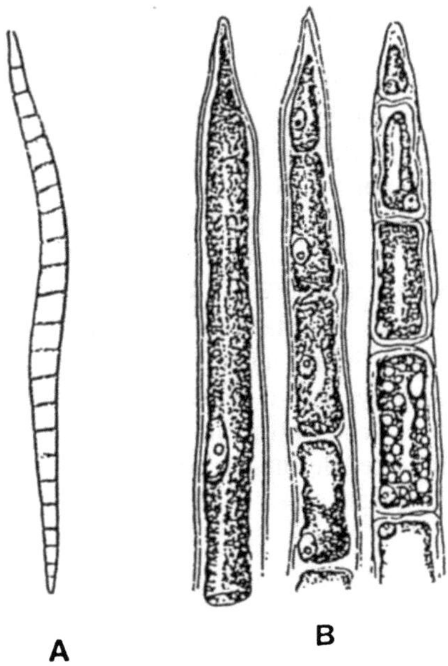

Fig. 6.28. A Sanio's (1873) conception of how a bast cell of *Xanthoxylon* redivided to form a file of crystal-containing cells. The bast cell conformed in size and shape to the cambial mother cell. **B** Origin of strand parenchyma from a cambial mother cell by horizontal wall formation. (Hartig 1891)

cell initials in wounded tissue of *Populus euramericana*. These cells were easily confused with strand parenchyma derived from fusiform cells.

Segmentation to parenchymatous cells often precedes or accompanies failure of a fusiform initial. The initial is reduced in size during successive periclinal divisions and the final derivatives produced undergo segmentation (Whalley 1950, Bannan 1953). A typical case history depicting this process in *Ginkgo biloba* is shown in Fig. 6.29A. The declining fusiform initial was subjected to simultaneous size reduction and segmentation to parenchyma before maturing and being eliminated from the cambium.

Srivastava (1963b) found that the radial files of axial parenchyma in the xylem of *Ginkgo biloba* were simply declining files of fusiform initials. These files often appeared as various sized plates of axial parenchyma cells depending on the time period through which the fusiform initials declined and continued dividing periclinally toward the xylem. Similar parenchymatous plates were observed and described by Chrysler (1913) in the roots of many conifers. As might be expected, segmentation of fusiform initials to parenchymatous cells results in a rich variety of cell structures and combinations, depending on the species and on whether the transformations occurred on the xylem or phloem side.

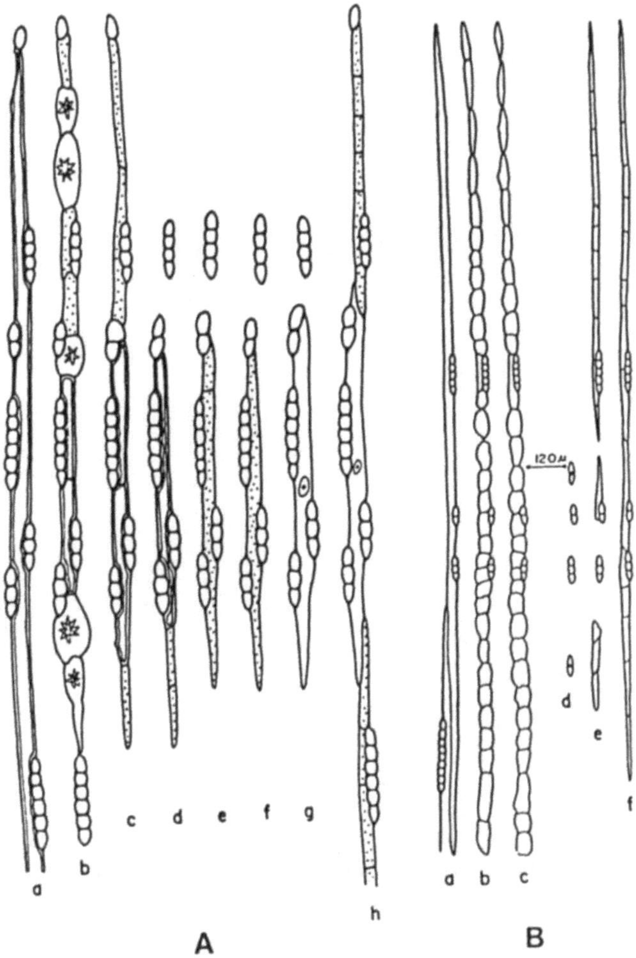

Fig. 6.29A,B. Tangential sections showing derivatives in declining files in an old *Ginkgo biloba* tree. **A** Successive xylem (*a-f*) and phloem (*h*) derivatives; cambial initial (*g*). Tracheary cells with double walls, crystal cells with druses, starch-containing cells *stippled*, albuminous cells without nuclei, ray cells *in outline*. 80x. **B** Successive xylem (*a-c*) and phloem (*f-e*) of a fusiform initial that declined and left two new ray initials in the cambium (*d* with nuclei). Additional xylem and ray cells (not shown) were produced in the 120 µm interval between *c* and *d*. 50x. (Srivastava 1963b)

In many conifers, strands of axial or longitudinal parenchyma arise by transverse divisions of daughter cells resulting from periclinal divisions of cambial fusiform initials (Record 1934, Panshin et al. 1964). The strands of terminal parenchyma in *Pseudotsuga menziesii* were shown by Grillos and Smith (1959) to be derivatives of the last-formed xylem mother cells that divided transversely in the

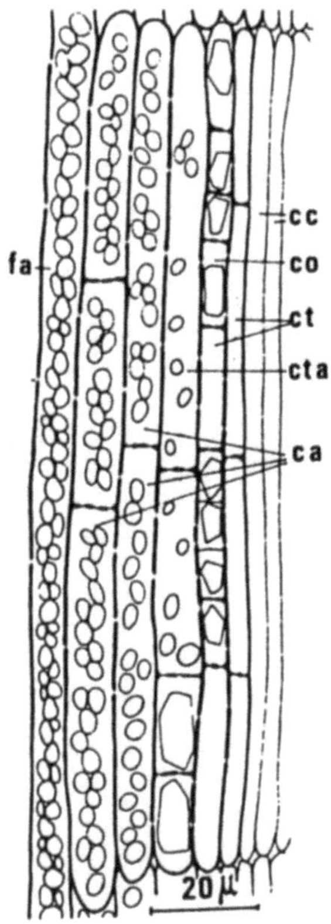

Fig. 6.30. Terminal parenchyma cells of *Robinia pseudoacacia*, shown in radial section, formed by segmentation after derivation from the cambium (*cc*). *ca* Starch-containing cells; *co* crystals of cambium oxlate; *ct* cells with tannin; *cta* cells with tannin and starch; *fa* starch-containing fibers. (Czaninski 1968)

autumn. The parenchyma strands matured the following spring. The characteristic strands of terminal parenchyma in *Larix* were also produced by post-cambial transverse divisions of xylem mother cells (Meier 1973).

One type of axial parenchyma cell in conifers that is often overlooked is the epithelial cell associated with resin canals. As described in Chapter 9.2.2.2, the epithelial initials are derived from cambial derivatives that divide first longitudinally and then transversely (Mayr 1893, Bannan 1936b, Werker and Fahn 1969).

In dicotyledons, axial parenchyma can be classified in two main groups, apotracheal parenchyma, which is typically independent of the vessels, and paratracheal parenchyma, which is associated with vessels or vascular strands (IAWA 1964). Kribs (1937) recognized nine different types of wood parenchyma and Hess (1950) attempted a classification of the many different patterns.

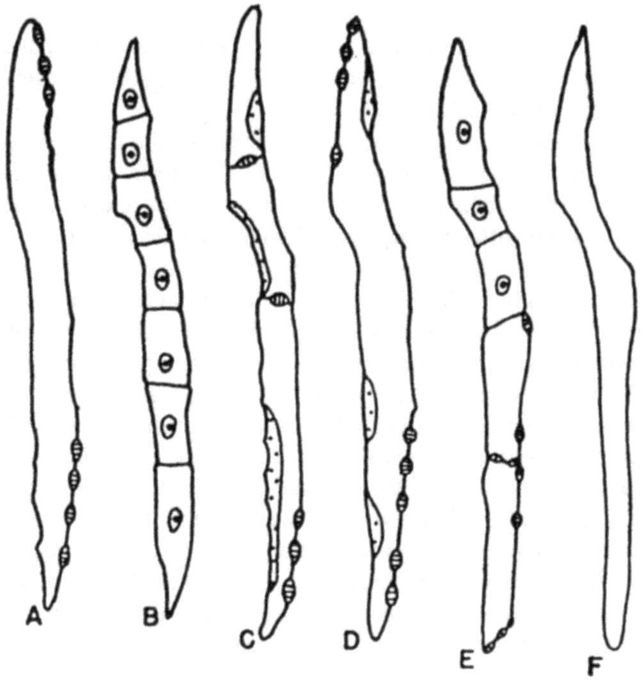

Fig. 6.31A-F. An assemblage of phloem cells derived from a single fusiform initial cell in *Liriodendron tulipifera* shown in tangential view. **A,D** Single sieve-tube members. **B** Parenchyma strand. **C** Sieve-tube members. **E** Combination of two sieve-tube members and three parenchyma cells (with nuclei) resulting from divisions within phloem initials. **F** Cambial cell. 100x (Cheadle and Esau 1964)

Exceptionally detailed classification schemes for wood parenchyma have been proposed by Hess (1950) and Wagenführ (1961).

Strands of vertical parenchyma in dicotyledonous species are not only arranged in many different patterns but they perform different functions. To cite but one example, the xylem of *Robinia pseudoacacia* possesses both terminal (Fig. 6.30) and paratracheal parenchyma. The latter cells, also known as "vessel associated cells" (Czaninski 1968), are formed in the spring in association with the vessels by transverse division of cambial derivatives (Czaninski 1964). The two types of parenchyma differ in time of formation, position in the xylem, and also in function. Many species, such as *Cedrela odorata* (Jane 1934), *Terminalia tomentosa* (Chowdhury 1936), *Gmelina* spp. (Chowdhury 1953), and others (Fahn 1974) produce initial parenchyma that appear at the beginning of a growth ring. Chowdhury (1953) believed that the initial parenchyma in many other species might have been misinterpreted as terminal parenchyma. In *Citrus sinensis*, Schneider (1952) found that terminal parenchyma cells were formed by maturation in the spring of xylem mother cells produced the previous fall. Initial and terminal parenchyma are difficult to differentiate from one another once they have been deposited within a growth ring. To avoid confusion, Hess (1950) suggested that both types be referred to as marginal parenchyma.

On the phloem side, segmentation is a natural consequence of the differentiation process, and phloem parenchyma formed by septation of fusiform derivatives are commonly found (Abbe and Crafts 1939). Evert (1960, 1963a) and Cheadle and Esau (1964) referred to the different types of elements all derived from a single fusiform initial or phloem mother cell as an "assemblage." If the assemblage was derived from a mother cell, the file would be considered temporary (Chap. 4.5.1.2) and the mother cell initial would, of course, be eliminated from the cambium. These assemblages were extremely varied and many possible combinations of phloem cells might be encountered (Figs. 4.25, 6.31). In *Pyrus communis*, Evert (1960) found that phloem parenchyma usually occurred in strands with the number of cells in a strand varying with the length of the initial cell, the degree of obliquity of the partition wall, and the ultimate fate of cells in the strand (Fig. 6.32). The parenchyma strands of various files were often adjacent to one another so that in transections they appeared in tangential bands, a common occurrence in many species. In *Liriodendron tulipifera*, it was not unusual for entire assemblages to enter the dormant period in a partially developed state and to become fully functional the following season (Cheadle and Esau 1964).

6.3.2.3 Segmentation to Rays

A common way in which a fusiform initial can be lost or eliminated from the cambium is by segmentation to ray cell initials. These initials can either persist as an independent ray, unite with other ray cell initials in a wide variety of ways, or function temporarily through a relatively few periclinal divisions before being eliminated from the cambium. Because of fluctuations in the composition of rays over time, it has been suggested that the term ray initials be confined to the individual cells, the ray cell initials, that divide periclinally rather than to the ray as a whole (Cheadle and Esau 1964, Schmid 1976, Metcalfe and Chalk 1983, Cumbie 1984) (Chap. 8.3.1).

Rays can arise by segmentation of a fusiform initial in three different ways: either by complete segmentation of the initial, or by partial segmentation involving the cutting off of a cell or cells from either the tip (tip origin) (Fig. 6.36), or the side (lateral origin) of the initial (Fig. 6.9B). Complete segmentation results in either the immediate elimination of the initial from the cambium by partitioning and transformation to ray cell initials, or in a gradual size reduction of the fusiform initial before partitioning and transformation of the segments to ray cell initials (Fig. 6.29B). Partial segmentation results in the cutting off of a small daughter initial which might or might not undergo further reduction in size and be eliminated. Rays can arise by all three methods in many species of both gymnosperms and dicotyledons. As an example, analyses of cell lineages in branches of *Pyrus communis* (Evert 1961) revealed that new rays arose by complete segmen-

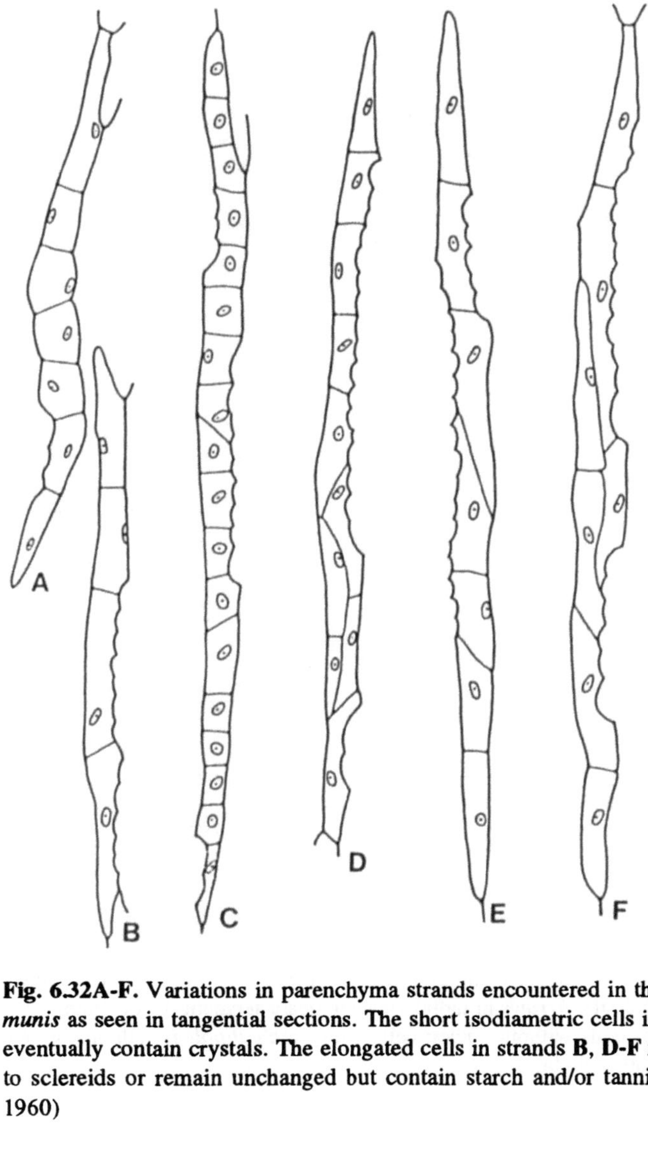

Fig. 6.32A-F. Variations in parenchyma strands encountered in the phloem of *Pyrus communis* as seen in tangential sections. The short isodiametric cells in strands **A** and **C** might eventually contain crystals. The elongated cells in strands **B**, **D-F** might either differentiate to sclereids or remain unchanged but contain starch and/or tannin deposits. 180x. (Evert 1960)

tation and reduction of fusiform initials in 48% of the cases, by segmentation from the tip in 39% of the cases, and by segmentation from the side in 13% of the cases. Segmentation in the first case usually resulted in a vertical file of three to nine ray initial cells (Fig. 8.17), whereas in the latter two cases only one or two new ray initials were formed.

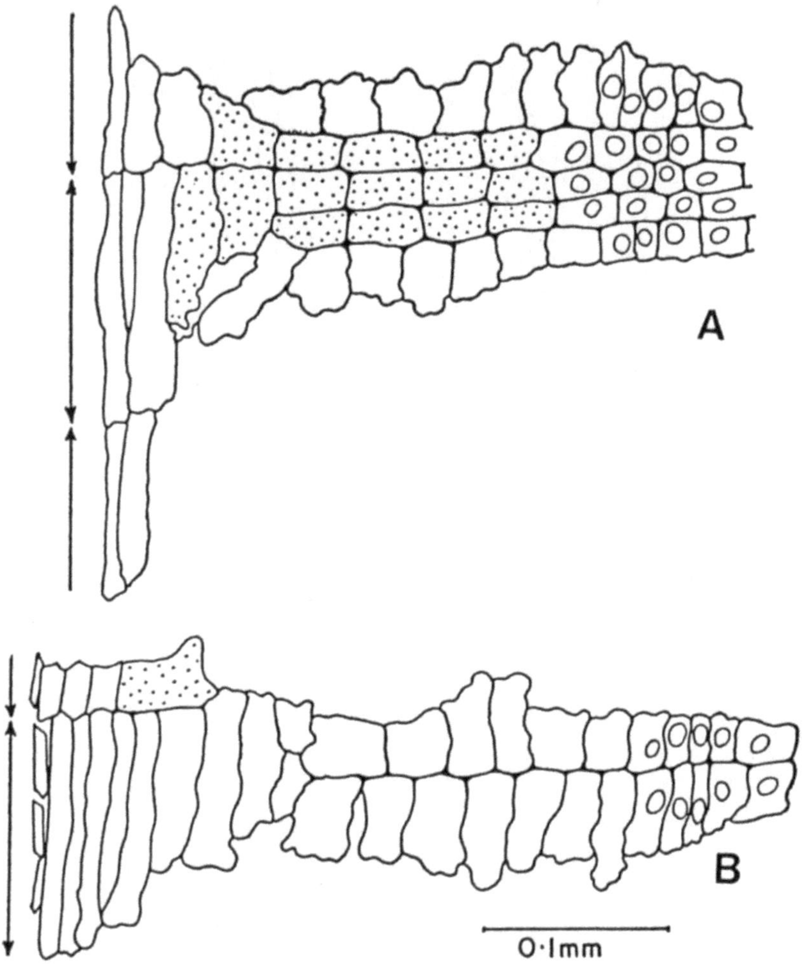

Fig. 6.33A,B. Decline of two fusiform initials (**A,B**) of *Pinus pinea* and their complete segmentation to ray cell initials. *Arrows* in radial views indicate segments of declining fusiform initials. Albuminous cells without nuclei; cells with starch *stippled*; cells in cambial zone with nuclei. (Srivastava 1963a)

6.3.2.4 Complete Segmentation

The general process by which cambial cells are segmented to form rays has been known for many years (Hartig 1855a, Müller 1875-76, Velten 1875, De Bary 1877). For example, Hartig (1855a) observed that some rays did not extend inward to the pith, but originated in the cambium. That is, they were formed by segmentation of "cambial fibers." Schacht (1860), in describing the process, assumed that one to several vertical cambial cells divided horizontally and the new mother cells formed a ray. Consequently, new rays could arise only in the cambium. Identical statements were made by Rothert (1913) and Neeff (1914). In his study of *Quercus fordii*, Abromeit (1884) described how one cambial initial gave rise to five ray cells by transverse divisions. Von Höhnel (1884b,c) observed numerous small rays, all more or less the same size, on the tangential surfaces of many woods, particularly in members of the Caesalpinaceae. Each of these rays was derived from segmentation of a single cambial initial cell, and the apparent horizontal arrangement in many of these species was traceable to the stratified structure of the cambium. Beijer (1927) described a similar origin of rays in the stratified cambium of *Herminiera elaphroxylon* in which the "gestreckte Initiale," or elongated initials, divided by transverse divisions.

A variation of complete segmentation is shown in Fig. 6.29B for *Ginkgo biloba*. Here the fusiform initial was partitioned into numerous small segments, especially on the xylem side, but only two survived to be transformed to ray cell initials. It is highly problematic how long these new initials would continue to survive in the cambium.

Examples of reduction and segmentation of fusiform initials prior to transformation to ray cell initials were described by Srivastava (1963a) in the phloem of *Pinus pinea* (Fig. 6.33). Segmentation was almost always preceded by pseudotransverse division of a fusiform initial followed by the decline of one daughter initial. As a rule, only one or two of the segments were transformed to ray cell initials; the remaining segments were further reduced and eliminated from the cambium. Modifications of the above process occurred in a seemingly infinite array with the segments contributing not only to the creation of new rays but also to the increase in size of existing rays. One example is shown in Fig. 6.34, where a declining fusiform initial underwent segmentation and reduction in size. Two of the segments were transformed to a single, new ray cell initial while an existing ray was increased in size when two other segments were transformed to ray cell initials and added to its margins. In another example, Srivastava described how ray size was increased when two rays became fused after the intervening fusiform initial was eliminated from the cambium.

According to the observations of Barghoorn (1940a), complete segmentation of a fusiform initial to form rays was exceedingly rare in the older, secondary body of conifers. It occurred more commonly in rapidly growing young stems and particularly in roots. Whether an entire fusiform initial in the stem was converted to ray cell initials was difficult for the author to determine. A more likely outcome was a reduction in size during or after septation because the new ray was always shorter than the fusiform initial. Segmentation in the root was different, and it was not unusual for a fusiform initial to be completely transformed to a ray. Irrespective of the number of cells in the new ray, the ray initial cells usually separated into one-celled rays as the cambium enlarged tangentially. Vertical fusion of these one-celled rays might reoccur during subsequent growth. A similar process was observed in *Pinus strobus* by Bannan (1934a), where rays derived by segmentation of fusiform initials soon separated into a number of small rays one cell in height.

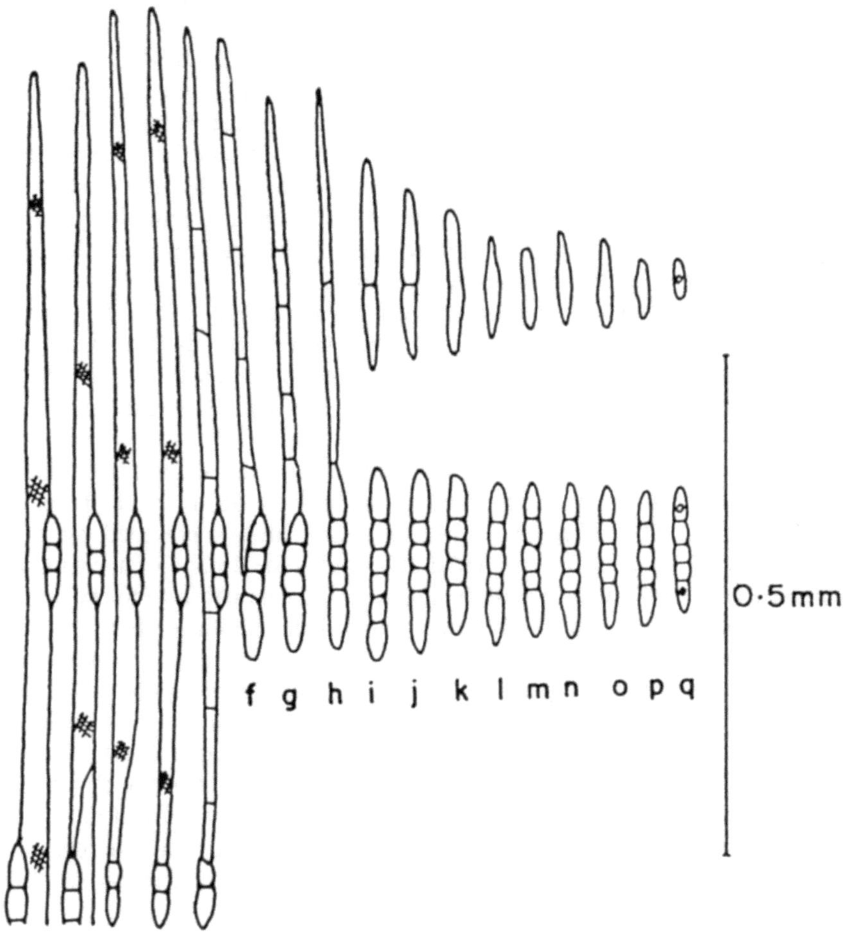

Fig. 6.34. Successive phloic derivatives in a radial series of a declining fusiform initial of *Pinus pinea* that eventually gave rise to three ray cell initials (shown with nuclei at q). Sieve cells with *cross-hatched* areas; derivatives of the declining fusiform initial and new ray cell initials without nuclei; cells in established rays with nuclei. (Srivastava 1963a)

The accumulated evidence obtained from a study of many hundreds of ray origins in *Chamaecyparis, Thuja,* and other conifers indicated that the majority of new rays in secondary tissues arose by subdivision of fusiform initials (Bannan 1934a, 1950b, 1951a, 1953, 1957c). After dividing anticlinally into more or less equal halves, the daughter cells might develop in one of several directions. Either one or both of the daughter cells might undergo further segmentation by second and third order divisions. Cell loss of these new ray cell initials was usually high

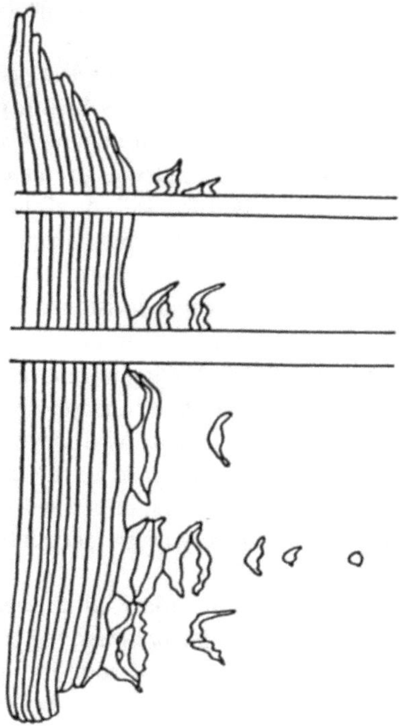

Fig. 6.35. Diagram in radial view showing sequence of cell changes associated with complete disappearance of a fusiform initial from the cambium of *Chamaecyparis thyoides*. (Bannan 1950b)

during subsequent periclinal divisions with only one or two surviving even under the most favorable conditions (Fig. 6.35). Cambial cells that were about to cease production of fusiform elements often underwent segmentation, after which the segments either were lost by maturation or converted to ray cell initials (Bannan 1953, Savidge and Farrar 1984).

The process was not much different in dicotyledons (Barghoorn (1940b). Uniseriate rays often arose by the partitioning of a fusiform initial either in whole or in part, with short initials being the most likely to be completely segmented. As in conifers, this type of ray origin occurred more frequently in young, rapidly growing trees than in older, slow-growing ones. Many examples were described and illustrated by Evert (1961) in *Pyrus communis*, Cheadle and Esau (1964) in *Liriodendron tulipifera*, and Rao and Dave (1985a) in three tropical species. The process of segmentation and reduction in dicotyledons proceeded much as it did in conifers (Fig. 8.17). Also, as in conifers, a substantial number of the new ray cell initials arising from segmentation of a fusiform initial were lost outright from the cambium by elimination of initials either from the ends or middle parts of the rays. Such losses occurred even when the new ray was in contact with older rays, thus indicating that contact with other rays was not necessarily a factor in determining the number of ray cell initials retained or lost.

Cumbie (1967b) documented the process of ray formation in *Leitneria floridana*, a woody shrub. Relatively short fusiform initials, averaging 143 µm in length, were transformed to ray cell initials without any appreciable changes in length. Subsequent transverse divisions, occurring in rapid succession, produced rays with 2 to 12 initials. This method of ray origin accounted for 33% of all new rays formed in the secondary xylem. Fusiform initials of intermediate lengths, averaging 206 µm long, usually were reduced in length to 30 to 135 µm before being transformed to rays consisting of 2 to 7 initials by transverse divisions. This method accounted for 47% of all new rays. Relatively long fusiform initials, averaging 279 µm in length, were transformed in two ways: (1) following a transverse division, one daughter cell either persisted as a fusiform initial or disappeared. The other daughter cell was transformed to a ray with 2 to 6 initials by transverse divisons; (2) occasionally, the entire fusiform initial was transformed to ray cell initials by transverse divisions followed by loss of initials from the upper and lower ends to yield a 3- to 8-celled ray. These two methods of ray origin accounted for 14 and 6% of all new rays, respectively.

Not all rays formed by the complete segmentation of a fusiform initial were subjected to reduction and loss of ray cell initials. In *Hibiscus lasiocarpus*, ray cell initials thus formed rarely decreased in length, and outright loss of such initials was never observed (Cumbie 1963). Occasionally, however, one of the shortest initials decreased in length and formed a single ray cell initial. On rare occasions, these were lost from the cambium. As a general rule, initials that gave rise to rays were shorter than related initials in a series. Subsequent transverse divisions formed vertical series of 2 to 7 ray cell initials. A similar process was observed in *Apocynum sibiricum*, a herbaceous dicotyledon (Cumbie 1969a). Ray formation occurred by simple division of fusiform initials. Although complete loss of fusiform initials following anticlinal division was never observed, almost one-half of the daughter cells were converted to ray cell initials in the secondary xylem of the outer stem. In another herbaceous dicotyledon, *Polygonum lapathifolium*, rays were completely absent from the inner secondary xylem (Cumbie 1969b). During subsequent cambial activity, some fusiform initials were converted to ray cell initials; phylogenetically, this is considered a highly specialized condition (Barghoorn 1941a). As in *A. sibiricum*, ray formation occurred by simple transverse division, usually of the shortest fusiform initials, and with little or no change in length of the dividing initial by the loss of ray cell initials.

Ray formation in species with storied cambia proceeds in much the same way as it does in species with nonstoried cambia (Chattaway 1933a, Butterfield 1972, Cumbie 1983). In fact, the origin of new rays by segmentation of fusiform initials was commonly found to occur in members of the Sterculiaceae (Chattaway 1933a). Primary rays of the herbaceous dicotyledon, *Aeschynomene virginica*, arose by subdivision of one to three adjacent "primordial ray cell initials," resulting in the formation of 4- to 12-celled rays (Cumbie 1984). Secondary rays were

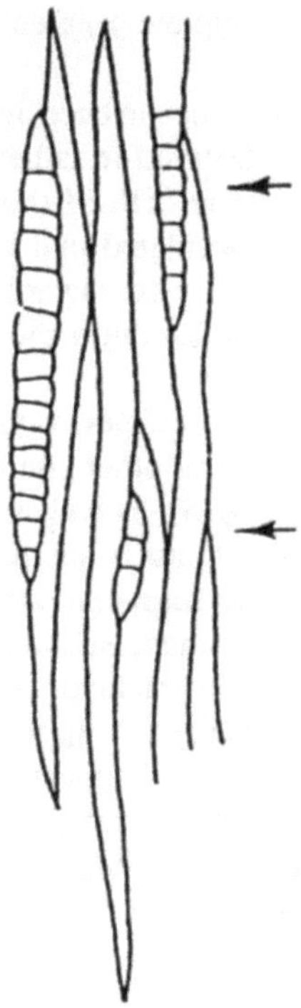

Fig. 6.36. Origin of new rays in *Populus alba* in tangential view. Tip origin *upper arrow*; lateral origin *lower arrow*. 300x. (Müller 1875-76)

derived from transverse divisions of fusiform initials. The first division occurred in the middle of the cell, followed by one or more divisions in each of the daughter cells. In a few instances, transverse divisions took place in only one of the daughter cells, the other continuing to function as a short fusiform initial. An identical process was described by Beijer (1927) for *Herminiera elaphroxylon.*

6.3.2.5 Partial Segmentation

Tip Origin: During tip origin of a ray, a small segment of a cambial fusiform initial is cut off by an anticlinal division. The segment might then redivide by horizontal walls from one to several times and persist as a small ray for varying lengths of time (De Bary 1877, Mischke 1890, Hartig 1895b, Rothert 1913, Elliott 1935). Müller (1875-76) described and illustrated the result of this process in *Gleditsia horrida* and *Populus alba.* He noted that each ray possessed its own "mother cell" in the cambial layer. He claimed that one could determine the age of a ray in certain deciduous trees by the number of vertical cell series comprising the ray. Conversely, by following the ray through tangential sections one could often determine which wood cell, i.e., fusiform initial, gave rise to the ray. As shown in Fig. 6.36, one new ray was derived from a segment cut off from the tip, and another from the side of the fusiform initial rather than by complete segmentation of the initial.

Velten (1875) considered tip origin of rays to be of common occurrence in woody species. He described another type of ray origin he had observed in radial sections of *Taxus baccata* rootwood. The "Kopfzelle," or beginning cell, of this ray was almost identical in shape to the usual ray cell. Velten speculated that it might have formed when a cambial cell "evaginated" either upward or downward, with the result that the daughter cell so produced began to function as a new mother cell.

Schmidt (1889), noting that previous workers were unsure as to how new rays arose, set out to investigate the process in *Pinus sylvestris* and several other conifers. According to Schmidt, both Müller (1875-76) and Velten (1875), as well as De Bary (1877), who relied on their observations, had been unable to ascertain the true origin of secondary rays because they could not follow the rays in radial sections through the tender cambial region. In all the species he examined, Schmidt observed that the first ray cell appeared to be the upper end of a tracheid. As shown in Fig. 6.37A, total length of tracheid b plus the superposed ray cell approximated that of tracheid a. Moreover, the upper radial wall of the ray cell appeared coterminus with that of tracheid a. He referred to these rays as "Markstrahleinsatz," or inserted rays, a term suggested to him by C. Müller; an "Einsatzzelle" would be the equivalent of Velten's (1875) "Kopfzelle."

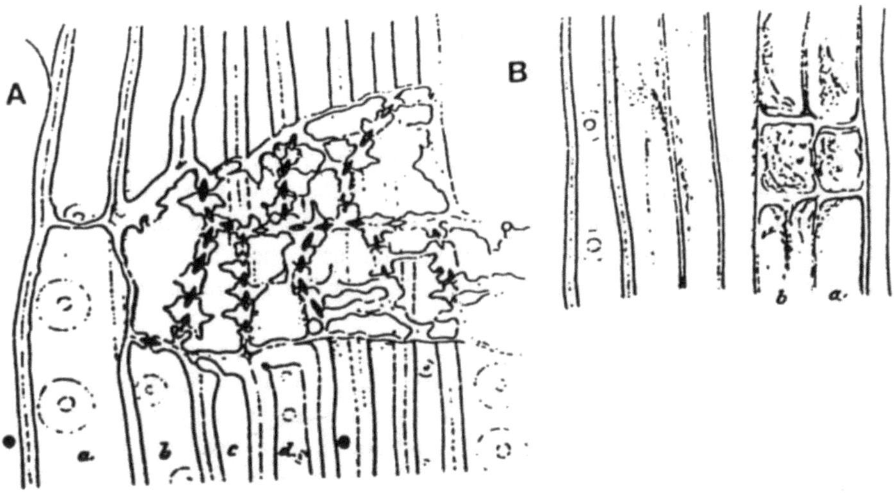

Fig. 6.37A,B. Origin of a new ray in *Pinus sylvestris* in radial view. **A** The new ray cell in tracheid *b* appeared to be derived from the tip of tracheid *a*, reflecting an earlier anticlinal division in the cambial initial. Cambium to the right. **B** Once formed, the new ray cell and the fusiform initial from which it was derived divide more or less independently. (Schmidt 1889)

As to origin, Schmidt subscribed to the theorem "Nature makes no leaps," implying that ray origin was not a sudden change but simply part of the constant "struggle" among neighboring elements. Thus, a cambial cell would exert its energy primarily in the production of tracheids, and they would grow in the direction of least resistance by expanding toward the cambium. A new ray arose when a cambial cell divided transversely near one end (Fig. 6.37B). The first ray cell would be distorted by growth of the tracheids as in Fig. 6.37A. However, later ray cells would be more successful in the "struggle" and they would expand horizontally. Once a ray cell formed, both it and the associated fusiform initials divided more or less independently (Fig.6.37B). Schmidt concluded by stating that the "Einsatzzelle" of a ray did not arise by the immediate form change of a cambial cell, but as the end result of a series of introduced forms.

Although Schmidt succeeded in demonstrating how a new ray might arise by partial segmentation from the tip of a fusiform initial, the process he described was quite unrealistic. Nonetheless, Klinken (1914) accepted Schmidt's concept of ray origin as being correct. In his study of *Taxus baccata*, Klinken concluded that some small rays in the secondary body arose by separation from either the upper or lower ends of fiber initial cells by means of transverse divisions. Another possibility, the shrinkage or disappearance of an initial cell leaving only a ray initial as a residue, he considered to be of infrequent occurrence.

Whereas Schmidt and Klinken attempted to describe the process of tip origin of a new ray, other workers simply recognized it as one method by which a new ray might be initiated (Neeff 1920, Beijer 1927, Bannan 1934a). Barghoorn (1940a), on the other hand, not only described the cambial changes but also the cytological details involved in ray origin. He found that the division giving rise to a new ray cell initial occurred shortly after an anticlinal division of the fusiform initial. These divisions might involve the entire apex of the initial or they might occur at varying distances from the end (Fig. 6.38). In the latter case, Barghoorn included within the category of tip origin the small rays that other workers would classify as being of lateral origin. However, they differed from typical lateral divisions in being situated near the cell tips rather than centrally in the dividing initials. Formation of a ray cell initial in this manner involved a complex series of nuclear movements as follows: (1) an original central position of the nucleus in a free fusiform initial; (2) an anticlinal cell division, followed by migration of the daughter nuclei to the centers of the resulting cells; (3) a periclinal division of the two fusiform initials, with both nuclei remaining in a central position; (4) a migration of the nucleus of one fusiform initial to the apical region; (5) the formation by cell division of a ray cell initial near the region of the preceding anticlinal division; (6) a remigration of the nucleus of the fusiform initial to the center of the cell; (7) the normal periclinal division of both ray cell and fusiform initials.

Barghoorn (1940a) never observed a nucleus near the cell end preceding a periclinal division. Consequently, migration of the nucleus to the cell end prior to an anticlinal tip division must have occurred with extreme rapidity. He firmly believed that a high proportion of new rays in conifers arose by tip origin, an opinion

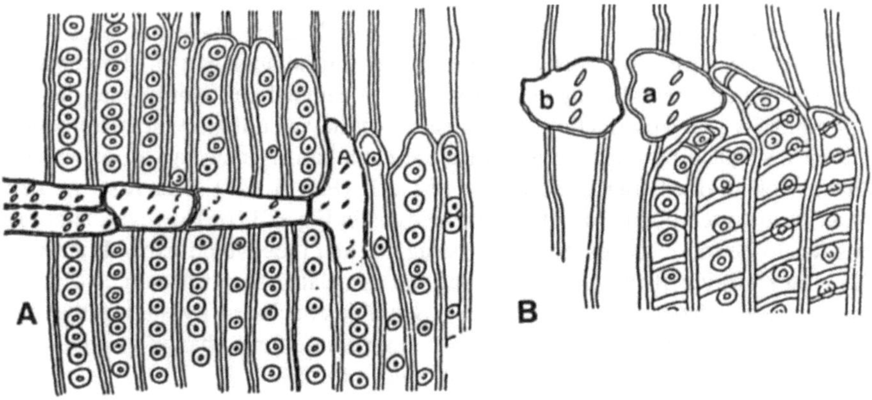

Fig. 6.38A,B. Tip origin of new rays in *Ginkgo biloba* (**A**) and *Amentotaxus argotaenia* (**B**) in radial view. **A** Note the sudden transition in shape from the first-formed (**A**) to the succeeding ray cells. Length of tracheids increased following new ray origin. Cambium to the left. **B** Tracheary element length was reduced following origin of new ray (*a,b*). 210x. (Barghoorn 1940a)

that was questioned by later workers, who found rays arising in this way to be either rare or infrequent (Bannan 1950b, 1951a, Whalley 1950).

In *Liriodendron tulipifera*, Cheadle and Esau (1964) could find no evidence of new ray cell initials arising from the cutting off, or "pinching off," of terminal portions of fusiform initials. Evert (1961) did, however, observe this process in *Pyrus communis*. Although approximately 39% of the anticlinal divisions occurred in this manner, it was not an important source of new ray cell initials because of the extremely high rates of cell loss. On the contrary, tip origin was an important source and lateral origin an insignificant source of new rays in *Polyalthia longifolia* (Ghouse and Hashmi 1980).

Lateral Origin: During lateral origin of a ray, a small cell of varying size is cut from the side of a fusiform initial by an anticlinal division (Figs. 6.9B, 6.36, 6.39). As in tip origins, the source of the new ray was recognized for many years although the process by which the ray originated was not well understood (Velten 1875, De Bary 1877, Rothert 1913). Müller (1875-76) described the part cut off as being "muschelförmig," because of its characteristic shell-shaped contour in *Gleditsia horrida* and *Populus alba.* Later workers simply called the shorter cell a "segment" (Barghoorn 1940a, Whalley 1950). During anticlinal division, the curving partition intercepts only one lateral wall in two places resulting in one daughter cell as long as the original dividing initial and the other daughter cell somewhat shorter. The shorter cell might either persist as a single ray cell initial, subdivide in two or more ray cell initials, or undergo several periclinal divisions and then disappear. Bailey (1920d) illustrated the division process in his paper on

Table 6.14. Types of anticlinal divisions in the cambium of *Leitneria floridana* (Cumbie 1967b)

Growth ring number	Number observed	
	Oblique	Lateral
1	187	107
2-10	211	78
22-26	104	5

Esau (1938) described the process by which new ray cell initials were cut out of a fusiform initial as being comparable to that of a companion cell cut out of a sieve-tube mother cell in *Nicotiana tabacum.* Barghoorn (1940a) considered lateral origin of new rays in conifers to be simply a modification of the process that occurred at cell tips, although of less frequent occurrence. Whalley (1950) found lateral ray origin to be rare in *Thuja occidentalis,* as did Srivastava (1963a) in *Pinus* pinea and Bannan and Bindra (1970b) in *Picea glauca,* where the proportion ranged from about 1% in stems to 11% in horizontal roots.

In *Chamaecyparis* spp., only 2.6% of all anticlinal divisions were of this type (Bannan 1950b, 1951a). Length of the lateral cell ranged from very short to one-third the length of the fusiform initial. In all cases, it lay opposite the central part of its larger sister initial, indicating the approximate position of the dividing nucleus. Short initials derived from lateral divisions were usually lost from the cambium and of all such divisions only about 33% survived as ray cell initials (Bannan 1950b). Because only a small portion of the fusiform initial was involved, Bannan considered it an "uncommon mode of ray origin which spared the destruction of the fusiform initials."

Lateral ray origins apparently are uncommon during normal growth in most conifers, especially in older stems (Bannan 1950a,b, Whalley 1950, Hejnowicz 1961). Meier (1973) found no evidence of either tip or lateral ray origins in the three conifers he examined growing in alpine regions. In the 14 species investigated by Bannan (1957b), divisions off the side of fusiform initials accounted for only about 2% of all anticlinal divisions. Exceptions did occur, however. In one specimen of *Thuja occidentalis,* 9% of all anticlinal divisions were lateral, and in a localized area of one growth ring the frequency increased to 29%. The fact that derivatives of these lateral divisions occurred in both xylem and phloem verified their origin from cambial initials. Failure rate was also high in all conifers examined. Approximately 74% of the lateral divisions failed after a few periclinal divisions, 21% were reduced to one or more ray cell initials, and 5% survived to become fusiform initials. The length of the lateral partition wall for each of the foregoing cases was 23, 33, and 43%, respectively, of the length of the mother cell.

Again, the relationship between cell length at origin and survival rate was demonstrated.

In contrast to the foregoing, Braun (1955) observed both tip and lateral ray origins in *Pinus sylvestris*. An analysis of 100 secondary ray origins revealed that 45 arose from the tip and 55 from the lateral wall of the fusiform initial. Smith (1967) also observed an unusual number of rays originating from both the ends and sides of fusiform initials in *Abies grandis* branches infested with *Adelges piceae*. Of the six new rays originating in a segment of the growth ring formed before aphid infestation, five arose by lateral divisions, and of the eleven new rays originating in a segment of the growth ring formed after aphid infestation, six arose by lateral divisions and four from the ends of fusiform initials. Although the data base was small, the number of rays originating by either tip or lateral divisions was extraordinarily large. These divisions were the primary sources of new rays and essentially no new rays were formed by segmentation of declining fusiform initials.

Lateral divisions apparently occur with greater frequency in dicotyledons than in conifers. In *Pyrus communis* (Evert 1961) and *Liriodendron tulipifera* (Cheadle

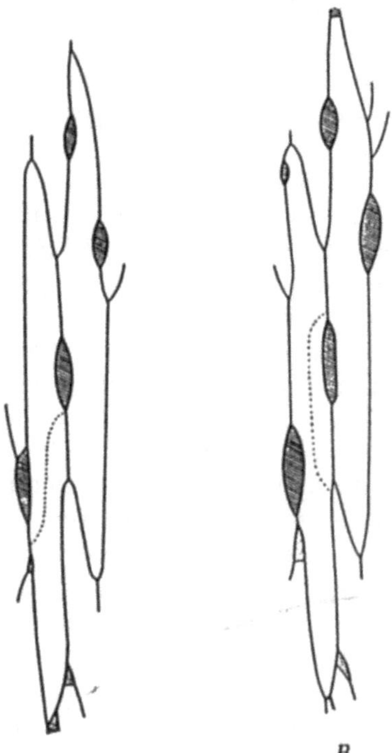

Fig. 6.39A,B. Drawings in tangential view showing typical pseudotransverse division (**A**) and lateral division (**B**) in *Pinus strobus*. Transverse diameter 170x; vertical ca. 38x. (Lewis 1935)

Table 6.15. Analysis of length of fusiform cambial initials dividing laterally as daughter cell segments in *Leitneria floridana* (Cumbie 1967b)

	Dividing initials length (μm)		Daughter cells length (μm)	
Growth ring number	Range	Mean	Range	Mean
1	252-432	340	72-306	185
2-10	270-468	362	90-279	190
22-26	315-450	353	144-279	192

and Esau 1964), approximately 13 and 9.6%, respectively, of all anticlinal divisions were lateral (Table 6.3). In the latter species, the percentage of lateral divisions tended to increase with length of the fusiform initial. Yet, despite the greater number of lateral divisions, new rays were seldom derived from lateral divisions irrespective of the length of the daughter initial. Lateral divisions were also common in *Hibiscus lasiocarpus*, where they made up 15.5% of the divisions in the stem segment near the plant base (Cumbie 1963).

In *Leitneria floridana*, a woody shrub, Cumbie (1967b) found that lateral divisions were particularly abundant in the first growth ring of the secondary xylem, where they comprised 37% of all anticlinal divisions (Table 6.14). In some areas of the stem the frequency increased to 50%, but it rapidly decreased in older growth rings. The length of the lateral partition wall averaged about 55% of the length of the dividing initial irrespective of growth ring age. However, variation around this average value was considerable, as shown by ranges in the lengths of daughter cell segments resulting from lateral divisions in Table 6.15. As a general rule, the shortest segments cut off by lateral divisions were transformed to rays. These transformations were more numerous in the youngest growth rings, which is understandable considering the scarcity of lateral divisions in older growth rings (Table 6.14).

6.4 Tangential Increase of Cambium

As the cambium continues adding new cells radially to the xylem during its seasonal course of growth, it must also expand tangentially. A vertical or longitudinal component is absent in most woody species because once a cambium is initiated all internodal elongation ceases. It was assumed by many early investigators that tangential growth of the cambium was due primarily to radial anticlinal divisions. However, as noted in the previous sections, most divisions are not radial and by no means do all the daughter cells of anticlinal divisions survive and contribute to increase in girth of the cambium. Cell loss can be exceedingly high, and only those cells that survive either as fusiform or ray cell initials can contribute to this increase. Bailey (1923), while investigating this problem, listed the following five factors that either alone or in combination might contribute to the increase in girth: (1) an increase in the number of fusiform initials; (2) an increase in the length of these cells; (3) an increase in the tangential diameter of these cells; (4) an increase in the number of ray cell initials; (5) an increase in the diameter of these cells.

To this list might be added expansion of the pith. Although pith expansion is only effective during the seedling stage of most arborescent trees, it might continue for several years in many woody herbs (Braun 1970). In most arborescent plants growing under normal conditions, pith size remains relatively constant, as Ladell (1963) found in *Pinus*.

6.4.1 Nonstoried Cambia

6.4.1.1 Fusiform Initials

One of the first to investigate the question of tangential expansion of the cambium was Nägeli (1864) in his study of the Sapindaceae. He attempted to calculate exactly how many radial divisions were required to increase the "cylindrical cambial mantle" by a given amount. To do so, he assumed uniform radial and tangential cell dimensions and a tangential increase in cell number by symmetric radial divisions with no loss of cambial cells. By simple calculations and derivation of a formula, Nägeli showed that if the cambial cell in each radial row divided in two daughter cells, the number of new xylem cells added in the radial direction would equal the number already present on the radius. That is, "... when the radius of the cambial ring has a length of 50 or 100 or 1000 wood cells, then the length of the radial rows must increase by 50 or 100 or 1000 cells in order to double once." As an example, he assumed a 2-m thick, pithless stem with wood cells averaging 30 µm in diameter. Table 6.16 shows, for successive radial increments, the number of cells that must be added to the stem radius if the cambial cell in each radial file divided once in the tangential plane. These were theoretical calculations and Nägeli, of course, recognized them as simple approximations. The important point to be derived from these calculations, however, was the clear demonstration that each cambial cell on the circumference of a large stem need divide only infrequently to keep pace with radial growth.

Many later investigators accepted this explanation. Sanio (1873), for example, stated that in *Pinus sylvestris* either new cambial cells must increase to extraordinary tangential widths, or new cambial cells must be inserted between the older ones. His observations revealed that the farther the cambium receded from the pith, the greater the tendency for new cambial cells to become inserted between older ones both above and below. De Bary (1884) simply restated Nägeli's conclusions.

Table 6.16. Number of cells that must be added to the cambium during tangential expansion in stems of different radii (After Nägeli 1864)

	Radial increase (μm)						
	30	60	90	150	300	600	
Tangential cell increase[a]	1	1	1	1	1	1	
Radial cell increase	1	2	3	5	10	20	
	Radial increase (mm)						
	1	5	10	50	100	500	1000
Tangential cell increase[a]	1	1	1	1	1	1	1
Radial cell increase	33	166	333	1666	3333	16 666	33 333

[a] Assuming that the cambial initial in each radial row divided once in the tangential plane.

Detlefsen (1878) attempted to explain the process mechanistically by suggesting that tangential cell divisions and form changes in the cambial cells were somehow brought about by stresses resulting from radial expansion. This explanation was a consequence of the widespread belief at the time in the role of bark pressure as a regulator of cambial activity.

Notwithstanding the fact that most investigators recognized intrusive growth, which they interpreted as sliding growth (Sect. 6.7.1), it was nonetheless implied in their writings that tangential increase of the cambium was due primarily to an increase in the number of cells resulting from anticlinal divisions (De Bary 1884, Rothert 1913). Several workers, such as Hartig (1895b) and Klinken (1914), recognized the additional contribution of sliding growth. However, they were not directly concerned with the question, nor was Neeff (1920). Yet Neeff demonstrated that after cambial cells had divided anticlinally, the daughter cells elongated by sliding growth. These processes permitted the cambial surface to expand tangentially at the same time the cambium was adding cells radially. Kleinmann (1923), in contrast, while admitting to the contribution of sliding growth, expressed the opinion that tangential expansion of the cambium was mainly the result of "radial-anticlinal divisions." Thus, whereas Neeff emphasized cell elongation, Kleinmann stressed cell number.

It remained for Bailey (1923) to make the proper distinction between an increase in cell number and cell size, i.e., cell elongation by intrusive growth. He

Table 6.17. Increases in cambial cell sizes with tree age in *Pinus strobus* (After Bailey 1923)

	One-year (μm)	60-years (μm)
Radius of woody cylinder	2000	200 000
Circumference of cambium	12 566	1 256 640
Av. length of fusiform initials	870	4000
Av. tangential diameter of fusiform initials	16	42
Av. tangential diameter of ray cell initials	14	17

pointed out the errors in Nägeli's (1864) formulas in which the sizes of cambial initials were treated as constants. On the contrary, Bailey showed that the average length of fusiform initials in nonstratified cambia, such as *Pinus strobus* (Table 6.17), increased substantially over a 60-year period. He concluded that neither fusiform cell number nor size by itself could account for increase in girth of the cambium, although cell number appeared more important. Priestley (1930a) recalculated Bailey's data, noting that the original 724 fusiform initials increased in number to somewhere between 23 100 and 32 000 during the 60 years of growth. On the basis of these new calculations, he concluded: "This works out at a division once in about 15 years in any individual initial"! The results of all the aforementioned investigations were reviewed by Meeuse (1942). In keeping with his strong belief in symplastic growth, he stressed the role of fusiform cell shape and changing cell size in tangential expansion of the cambium.

Expansion of the cambium in dicotyledonous species differs little from that in conifers, with the exception of rays, which in some species might contribute one-half or more to the cambial surface (Bailey 1923). In *Leitneria floridana*, Cumbie (1967b) attributed the increase in cambial circumference to anticlinal division of fusiform initials, intrusive growth of daughter initials together with elimination of some initials, and the formation of ray cell initials from fusiform initials. A discussion of this aspect of cambial circumferential increase can be found in Carlquist (1988a).

It is clear that both an increase in the number of fusiform initials and intrusive growth of new initials following anticlinal divisions contribute to tangential expansion of the cambium. However, as pointed out by Bannan (1957a,c, 1962a), there is no direct relation between the frequency of anticlinal divisions and cambial expansion because of the large numbers of initial cells lost from the cambium. This conclusion was confirmed by Hejnowicz and Branski (1966) in a quantitative analysis of cambial expansion. In this study, growth rings at ages 4,

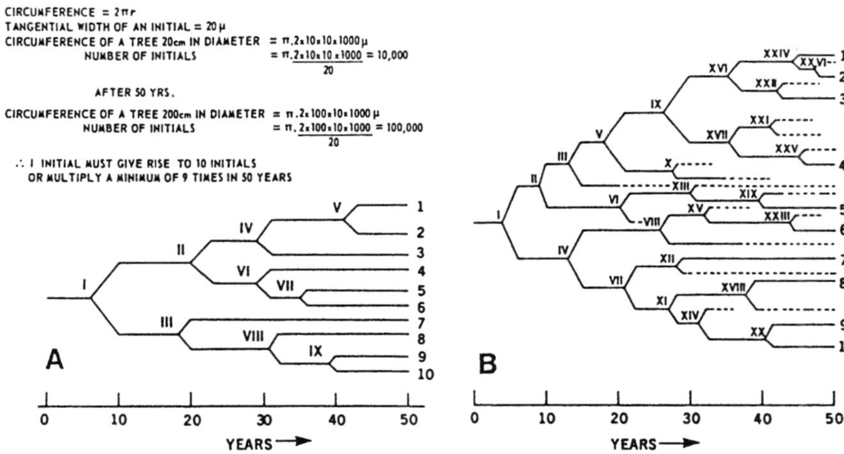

Fig. 6.40A,B. The expected increase in number of fusiform initials over a 50-year growth period. **A** *Roman numerals* preceding bifurcations denote anticlinal divisions with all daughter cells surviving. **B** Similar to **A**, but showing more frequent anticlinal divisions followed by either loss of daughter initials (*lines that terminate before reaching the right end*) or conversion to ray cell initials (*broken lines*). (Srivastava 1973)

11-12, and 15-17 were examined in a 29-year-old stem of *Thuja occidentalis*. The frequency of pseudotransverse divisions was high, but so were the rates of cell loss and intrusive growth in all growth rings. Intrusive growth exceeded cell loss in the youngest growth ring, whereas cell loss exceeded intrusive growth in the older growth rings. However, in all cases, the sum of pseudotransverse divisions and intrusive growth surpassed cell loss, consequently the cambium expanded tangentially. According to the author's calculations, the increase in cambial surface during the production of 4 mm of radial xylem was 65, 11, and 9% for each of the sampled growth regions. The increase in total length of cambial initials by intrusive growth after pseudotransverse division was 59, 8, and 3% for the same growth regions. Agreement between the two sets of values was considered to be remarkably good.

Additional calculations were made by Srivastava (1973). He assumed a tree that grew from 20 cm to 200 cm in diameter during a 50-year period. He further assumed a cambium restricted to fusiform initials that maintained a constant tangential width of 20 µm. As shown in Fig. 6.40A, a young tree with 10 000 initial cells in the cambial ring would require 100 000 initial cells when it reached a diameter of 200 cm. This value could be attained if each initial cell divided anticlinally nine times over a 50-year period. However, as pointed out here and else-

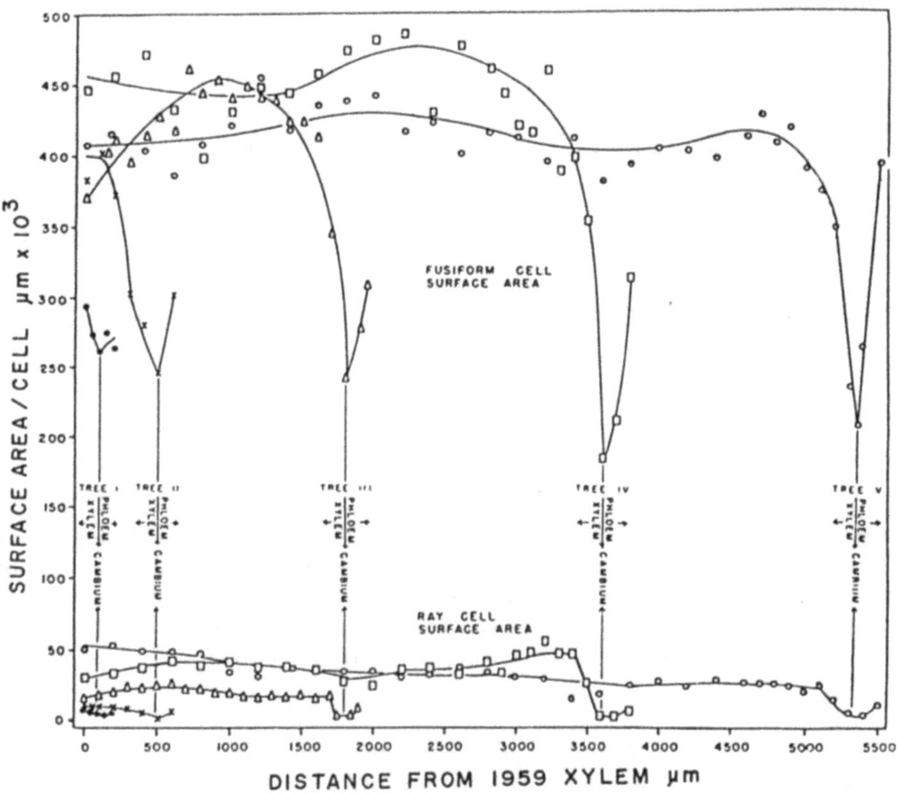

Fig. 6.41. Cell wall surface areas of cambial cells and their derivatives in five *Abies concolor* trees sampled at monthly intervals during the 1960 growing season. (Wilson 1963)

where (Sect. 6.5), the frequency of anticlinal divisions is much greater than nine, and the excess cells are lost from the cambium as shown in Fig. 6.40B.

An analysis of cell wall surface in cambial derivatives of *Abies concolor* was made by Wilson (1963). A seasonal growth curve was constructed for the 1960 increment by calculating values for cambial zone cells and derivatives from five sample trees collected at monthly intervals (Fig. 6.41). Cell wall surface area of xylem derivatives exhibited the typical pattern, increasing to mid-season and then decreasing precipitously at season's end. However, this increase was due almost entirely to radial cell expansion, since tracheid radial diameter increased almost 400% and elongation only 13%; tangential enlargement was negligible. Wilson concluded that even though tracheid elongation over the long term might be significant, as Bailey (1923) demonstrated, its short-term contribution to increase in girth of the cambium appeared to be minimal. Other attempts to measure cambial surface have been made by Lescaffette (1951) and Anucin (1960).

The third point mentioned by Bailey (1923), tangential diameter of the fusiform initials, contributes to peripheral increase of the cambium in a small, but not necessarily an insignificant way long-term. Sanio (1873) was aware of these changes in tracheid dimensions even though he did not consider them in his calculations. For example, his measurements of tangential tracheid diameters at the base of the current-year shoot of a 2-year-old *Pinus sylvestris* averaged 0.012 mm, whereas those of a 100-year-old tree averaged 0.026 mm. This twofold difference was considered to be a relatively insignificant contributing factor. Meeuse (1942) also considered the effect of tangential cell expansion to be "hardly noticeable," and Amos et al. (1950) found that it remained almost constant in *Eucalyptus gigantea*. Bailey's (1923) calculations (Table 6.17) showed that tangential diameter of the tracheids of *Pinus strobus* increased between two and three times during the 60-year period, an increase he considered "by no means a negligible factor" in expansion of the cambium. In *Pseudotsuga menziesii*, tangential diameter of fusiform initials increased from 14 μm in a 1-year-old stem to 40 μm in a 200-year-old stem (Grillos and Smith (1959). The fact that tangential tracheid diameters increase with tree age has been verified by many workers (Frank 1864, Sanio 1872, Essner 1882, Bertog 1895, Bannan 1942, 1944, Cown 1975, Wheeler et al. 1966, Terrazas 1991) (Fig.6.42).

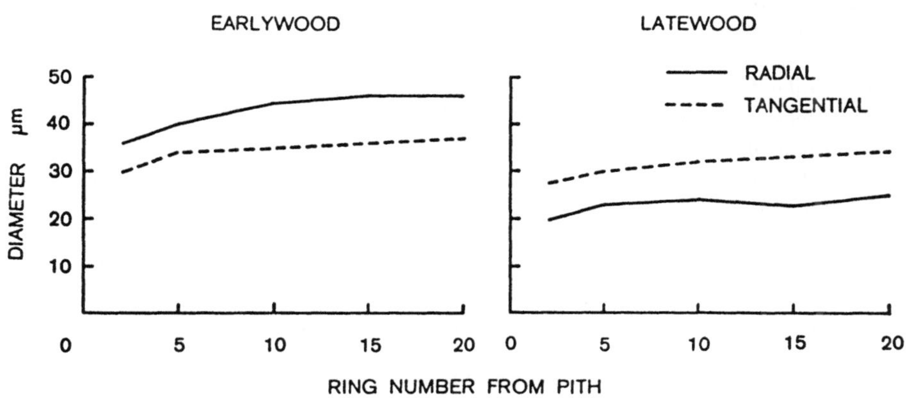

Fig.6.42. Variation with age from the pith in radial and tangential diameters of earlywood and latewood tracheids in a 26-year-old *Pinus radiata* stem. (Cown 1975)

Table 6.18. Lengths and tangential diameters of earlywood tracheids in the peripheral wood at breast height in coniferous trees 20-50 dm in diameter (After Bannan 1965c)

Species	Number of trees	Length mm	Diameter µm	Length/width ratio
Abies balsamea	12	3.36	35.4	95:1
Abies concolor	19	3.51	40.8	86:1
Chamaecyparis nootkatensis	6	2.32	28.2	82:1
Cupressus arizonica	16	1.70	23.2	73:1
Cupressus glabra	16	1.81	22.4	81:1
Cupressus sargentii	9	2.91	31.8	91:1
Juniperus osteosperma[a]	5	1.59	23.5	68:1
Juniperus virginiana	8	1.98	25.1	79:1
Larix laricina	7	3.54	42.1	84:1
Larix occidentalis	6	3.37	44.4	76:1
Libocedrus decurrens	7	2.97	34.2	87:1
Picea engelmannii	15	2.89	36.6	79:1
Picea glauca	15	3.34	36.8	91:1
Picea mariana	5	3.11	32.5	96:1
Pinus aristata[a]	11	1.80	26.6	68:1
Pinus banksiana	8	3.14	34.9	90:1
Pinus contorta	17	2.90	35.4	82:1
Pinus edulis[a]	9	2.27	32.3	70:1
Pinus flexilis[a]	8	2.37	32.6	73:1
Pinus lambertiana	5	3.98	44.5	90:1
Pinus ponderosa[a]	9	3.37	35.6	95:1
Pinus ponderosa	16	3.33	38.6	86:1
Pinus resinosa[a]	15	3.21	36.7	88:1
Pinus resinosa	16	3.36	37.4	90:1
Pinus strobus[a]	10	3.86	39.0	99:1
Pinus strobus	12	3.73	41.8	89:1
Pseudotsuga menziesii				
(Coastal B.C.)	10	3.62	41.5	87:1
(Interior B.C.)	23	3.14	38:1	82:1
Thuja occidentalis	25	2.59	28.6	90:1
Tsuga canadensis	12	3.63	38.8	94:1
Sequoia sempervirens	11	4.12	43.7	95:1

[a] Ring widths 0.1-0.5 mm; all others ring widths 1-2 mm.

Although the increase in tangential cell diameter is generally small, a cumulative effect can be detected when measured over a period of years. Bannan (1965c) examined the relationship between tracheid lengths and widths in many species (Table 6.18). Both values increased with distance from the pith. However, the tracheid length increase exceeded that of width, so that over time the length/width ratio also increased.

From the foregoing discussion it is obvious that tangential expansion of the non-stratified cambial surface in older stems is due primarily to an increase in the number of cells comprising the cambium. This increase is brought about by anticlinal division followed by elongation of the daughter fusiform initials. The increase in tangential diameter of the fusiform initials with increasing distance from the pith is a minor contributing factor.

Tangential increase of the cambium in nonstoried species was summarized by Srivastava (1963b) in his study of *Ginkgo biloba*. In young gymnosperm stems, all the factors cited by Bailey (1923) appeared to be operative. However, in old stems, an increase in the number of fusiform initials and ray cell initials played a dominant role and all other factors were more or less insignificant.

6.4.1.2 Ray Cell Initials

The contribution of rays to tangential expansion of the cambium varies from minimal (Beijer 1927, Barghoorn 1940b) to extensive, depending on the species examined. Essner (1882) believed that the tangential width of conifer tracheids increased while the number of rays decreased with growth ring age outward from the pith. He therefore suggested that tracheid width, together with tracheid length, had to increase to compensate for the decrease in the number of rays. Mischke (1890) arrived at a somewhat different conclusion. He believed that rays had to contract tangentially by compression to accommodate the tangential division of cambial cells and subsequent expansion of tracheids.

Beijer (1927) listed four characteristics of rays that might contribute to tangential increase in the cambium: (1) Increase in ray width; (2) transition of a ray cell initial to a fusiform initial; (3) increase in the number of ray cells by division or splitting of larger rays; (4) origin of new rays from fusiform initials.

The characteristics cited by Beijer are discussed more thoroughly in Chapter 8. With regard to the first characteristic, it should be pointed out that the widths and heights of large multiseriate and aggregate rays can increase considerably in many dicotyledonous trees. The ray volumes of some tropical species exceed 70% of the cambial surface (Fig. 8.36) (Chap. 8.3.11).

Measurements by Bailey (1920d, 1923) indicated that tangential diameters of ray cells increased from 14 µm in the cambium of a 1-year-old *Pinus strobus* stem to 17.0 µm in that of a 60-year-old stem, a relatively insignificant increase considering the much larger increase in fusiform cell length (Table 6.17). Bailey's (1923) calculations showed that although ray cell initials did not expand to any considerable extent in *Pinus strobus*, their contribution in terms of cell number could not be ignored. During the 59-year interval, the number of ray cell initials increased from 70 to 8796. Priestley (1930a), in recalculating Bailey's data, showed the total peripheral increase in ray surface area to be about 150-fold. The actual contribution of rays to cambial expansion is often difficult to determine because new rays are formed by the transformation of fusiform initials. It is perhaps for this reason that some investigators have reported that a relatively constant ratio is maintained between rays and fusiform initials (Braun 1955, Evert 1961). Moreover, as Wilson (1963) demonstrated in *Abies concolor*, ray cells continued to elongate in the zone of differentiation after tracheid radial expansion had ceased.

6.4.2 Storied Cambia

Tangential expansion in species with storied cambia can be accounted for primarily by the increase in cell number following radial-longitudinal divisions. Thus, Nägeli's (1864) calculations would conform more closely to storied than to nonstoried cambia. Beijer (1927), in fact, arrived at a similar conclusion in his investigation of species with storied cambia. He attempted to demonstrate in several species how a doubling of the stem radius must be accompanied by a doubling of the number of radial cell files, i.e., every initial must theoretically divide once. Thus, the older the plant, the longer the time required for the number of radial cell files to double.

An identical pattern of circumferential increase was observed in *Aeschynomene hispida* by Butterfield (1972). In this species, tangential cambial expansion was due exclusively to radial-longitudinal divisions that divided the parent fusiform initial in two equal halves. No fusiform initials were lost from the cambium and the occasional transverse division did not add cells in the tangential direction. Thus, in a young stem 1 mm in diameter, a 1 mm increment of radial growth would require that every fusiform initial must divide at least once for the cambial circumference to double. As radial growth proceeded, however, the percentage of cells that need double decreased. For 1 mm of radial growth between 20 and 21 mm from the stem center, only one fusiform initial in every 20 must divide. The theoretical frequency of anticlinal divisions in a storied cambium, together with the recorded results obtained from *A. hispida*, are shown in Figure 6.43. Cumbie (1984) confirmed this pattern in *A. virginica*.

Not all species with storied cambia conform to this perfect pattern. For example, in *Hoheria angustifolia* Butterfield (1973) found that in addition to the increase in cambial cell number by radial-longitudinal divisions, the tangential diameter of the initials also increased as the cambium expanded. Both processes were also operative in *Robinia pseudoacacia* (Derr and Evert 1967). One might also anticipate other contributing factors to tangential cambial expansion of storied woods, such as changes in ray size and distribution as observed by Cumbie and Mertz (1962).

In species with storied cambia, the xylem elements usually conform in length to those of the cambial initials. This relationship is maintained throughout the life of

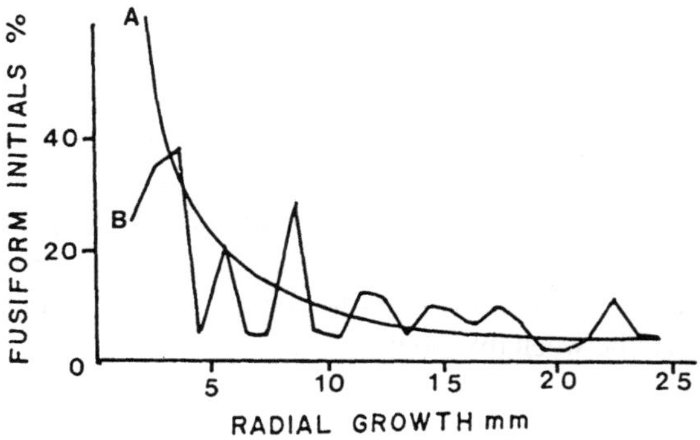

Fig. 6.43. *Line A* The theoretical percentage of fusiform cambial initials that need to divide anticlinally per millimeter of radial growth in order to supply the needs of the expanding cambial circumference. *Line B* The measured percentage of fusiform initials of *Aeschynomene hispida* that divided radial longitudinally per millimeter of radial growth. (After Butterfield 1972)

the tree. As pointed out by Bailey (1920a) and others (Anand et al. 1978), xylem derivatives cannot undergo extensive post-cambial elongation without destroying the storied structure. Moreover, fusiform initials of storied cambia rarely elongate as the stem enlarges (Bailey 1944a).

6.5 Frequency of Anticlinal Divisions

Tangential increase of the cambium involves both an increase in the number of cells and in their sizes, with number of cells being of greatest importance. The rate at which initial cells divide, or the frequency of division, is therefore a critical factor in tangential expansion of the cambium. Attempts to calculate the frequency were often no more than crude estimates. Using Bailey's (1923) data, for example, Priestley (1930a) came to the unlikely conclusion that each cambial initial need divide anticlinally no more than once every 15 years. Not only was this report questioned by subsequent investigators, but more realistic values were obtained by actually observing both the occurrences and the types of anticlinal divisions on serial sections. Such analyses showed that anticlinal divisions occurred throughout the growing season but the frequency of division varied widely in response to a range of influencing factors, most of which were strongly interrelated. Several of the more important influencing factors were age of tree, rate of growth, location within the growth ring, and position in the tree, as well as differences among species.

6.5.1 Age of Tree

The interrelationships among influencing factors becomes clearly evident when the effect of tree age is examined. As Hartig (1895b) concluded from his data, tracheids in the inner growth rings of *Pinus sylvestris* trees were short because they were derived from young cambial initials which had recently divided anticlinally. On the contrary, tracheids in the outer growth rings were longer because they were derived from cambial initials that divided less frequently, thus allowing more time for the initials to elongate between divisions.

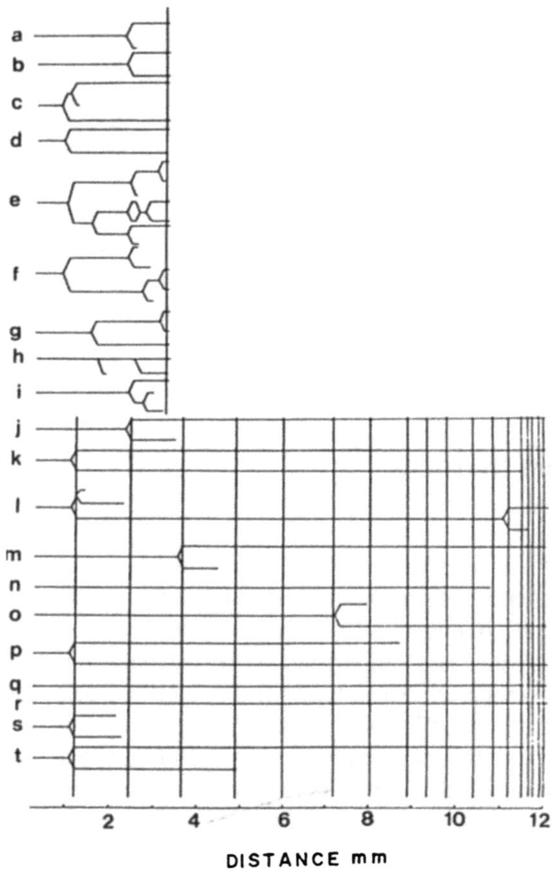

Fig. 6.44. Case history diagrams showing relative times and frequencies of pseudotransverse divisions of fusiform initials in a stem of *Thuja occidentalis*. Series *a-i* are from the inner growth rings 1 cm from the pith, and series *j-t* from the outer growth rings between 12 and 23 cm from the pith; note the tendency for divisions to occur in the last-formed parts of the growth rings. *Incomplete arms* indicate elimination of a sister initial, and the enclosed box (*e*) indicates the temporary division of a mother cell. (Bannan 1960b)

Table 6.19. Frequency of pseudotransverse division and cell length at division in a stem of *Thuja occidentalis* 23 cm in diameter (Bannan 1960b)

Distance from pith (cm)	Ring width mm	Frequency of divisions[a]	Cell length at division (mm)
1	3.0	5.1	1.64
4	3.1	3.4	2.42
6	4.3	1.3	2.50
9	1.2	2.3	2.53
11	0.8	2.8	2.40
	0.2	4.0	2.52

[a] Frequency of division/cm of xylem.

The age effect was examined in greater detail in stems of *Chamaecyparis* by Bannan (1950b). In young stems, the frequency of anticlinal divisions was greater than in old stems. Whereas three or four anticlinal divisions per unit area per year occurred in vigorous young stems, only one per year occurred in old stems with comparable rates of growth. Losses of cambial initials due to cell elimination were also smaller in the young stems so that new initials were added to the cambium at a faster rate. An analysis of the net increase in fusiform initials revealed that during the development of 4 to 6 mm of secondary xylem near the stem center, i.e., while the tree was young, an average of 4.5 anticlinal divisions was required to add one fusiform initial to the cambium. In the older stem, when it was 20 to 60 cm in diameter, an average of 43 anticlinal divisions was required to add one fusiform initial to the cambium. Therefore, to produce an equivalent radial increment, almost ten times as many anticlinal divisions were required in the old stem compared to the young stem because of the greater cell loss in the former.

A similar comparison of the frequency of anticlinal divisions in the inner and outer wood of a *Thuja occidentalis* stem is shown in Fig. 6.44. In Series a-i of Fig. 6.44, representing the inner growth rings 1 cm from the pith, anticlinal divisions occurred in each cell file, although considerable variability was evident among files. In Series j-t, representing the 18 outer growth rings between 12 and 23 cm from the pith, anticlinal divisions were sparse and many years often elapsed between divisions. On one radius of this tree the frequency of anticlinal divisions decreased from 5.1 per cm of xylem increment at 1 cm from the pith to 0.2 to 2.8 per cm of xylem increment in the peripheral growth rings. In the young rings, cell survival exceeded cell loss resulting in a net gain of 30 to 50% of the potential, but in the older rings cell survival barely maintained a status quo.

According to Bannan (1960b), these data indicated that during the first few years of growth, when the stem perimeter was increasing rapidly, anticlinal cambial divisions occurred in a rapid sequence, the survival ratio of newly formed initials was high, and a continuous increase in mean cell length followed. All these

Table 6.20. Frequency of pseudotransverse division and cell length in the peripheral growth of *Thuja occidentalis* trees of similar growth rates but differing in size (Bannan 1960a)

Diameter dm	Number of trees	Ring width mm	Frequency of division[a]	Cell length at division
1.5-2.0	16	1.44	2.3	2.68
2.5	54	1.42	2.1	2.87
3.0	35	1.42	2.3	2.90
4.0-5.0	23	1.43	2.5	2.84

[a] Frequency of division/cm of xylem.

phenomena facilitated accommodation of the cambium to circumferential expansion. With increasing age and a diminishing rate of perimeter growth, the frequency of anticlinal divisions declined, the survival ratio decreased, and the increase in mean cell length slackened.

Trends similar to these were observed in stems of *Cedrus*, *Picea*, and *Pseudotsuga* by Bannan (1960b), in *Ginkgo biloba* by Srivastava (1963b), in *Pinus sylvestris* by Branski (1970), and in branches of *Thuja occidentalis* by Whalley (1950). Similar trends were also observed in the dicotyledons *Leitneria floridana* (Cumbie 1967b) and *Apocynum sibiricum* (Cumbie 1969a). However, as noted elsewhere, other factors such as rate of growth, position in the tree, or inherent differences often compensate for the effect of age. In such cases, older growth rings sometimes take on characteristics usually attributed to younger rings. Some old trees yielded astonishingly contrary results. For example, in the outermost growth rings of *Sequoia sempervirens* and *Sequoiadendron giganteum*, ranging from 2 to 6 m in diameter, Bannan (1960a) observed up to five pseudotransverse divisions per cm of radial increment.

Comparison of the data in Tables 6.19 and 6.20 tend to verify the interrelation of tree age and growth rate for trees of *Thuja occidentalis*. In Table 6.19, the frequency of anticlinal divisions decreased outward from the pith as growth ring age increased, but this decrease was accompanied by a corresponding decrease in ring width. In the extremely narrow rings outward from the pith, the frequency again increased in a typical manner in association with declining growth rate. However, when rate of growth was held constant (Table 6.20), the frequency of anticlinal divisions remained relatively stable irrespective of tree size, hence tree age.

6.5.2 Rate of Growth

Rate of growth or width of growth ring is perhaps the most important factor contributing to variability in the frequency of anticlinal divisions. A high frequency of anticlinal divisions results in shorter fusiform initials, consequently shorter xylem

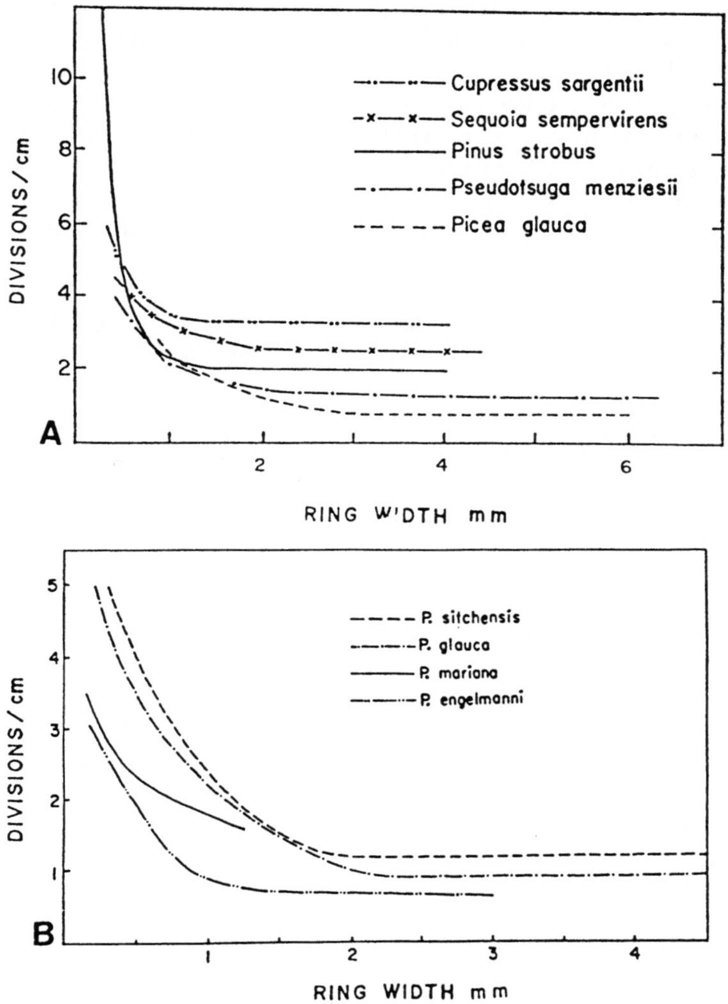

Fig. 6.45A,B. Smoothed curves showing the relation between rate of anticlinal division and ring width in the outer growth rings of mature trees of five different conifers (**A**) (Bannan and Bindra 1970b) and four species of *Picea* (**B**) (Bannan 1963a).

elements. Because of the influence of xylem cell length on wood properties, the effect of rate of growth on xylem cell length via the frequency of anticlinal divisions has engendered considerable interest over the years. Many years ago, Hartig(1895b) demonstrated that the greater the increase in circumferential growth of the tree, the more vigorous the increase in cell number by transverse (anticlinal) divisions, and the shorter the average length of the cambial initial cells, and vice versa.

The most comprehensive and detailed studies of this relationship have been conducted by Bannan on many north-temperate zone conifers (Bannan 1962a, 1963a,b, 1964a,b, 1965a, 1966a, 1967b, 1970). As a general rule, the frequency of anticlinal divisions per cm of xylem increment remained relatively stable over a wide range of ring widths down to a ring width of about 1 mm, at which time the frequency increased precipitiously in narrower rings. Smoothed curves for five species are shown in Fig. 6.45A from Bannan and Bindra (1970b). The ring width at which the frequency of anticlinal divisions began to abruptly rise varied among

Table 6.21. Base rates of anticlinal division of fusiform initials when annual rings exceed 2 mm, and cell lengths at division when ring width is optimal for cell length (Bannan 1970)

Species	Number of trees	Anticlinal division rate/cm xylem	Cell length at division (mm)
Abies balsamea	21	1.6	3.72
Abies concolor	30	2.0	3.91
Chamaecyparis nootkatensis	13	2.1	2.68
Cupressus arizonica	17	4.6	2.23
Cupressus macnabiana	13	5.6	2.33
Cupressus sargentii	48	3.3	3.19
Juniperus virginiana	21	4.1	2.21
Larix laricina	22	2.2	3.83
Larix occidentalis	9	1.7	3.84
Libocedrus decurrens	16	1.9	3.38
Picea engelmannii	34	0.7	3.15
Picea glauca	79	1.0	3.66
Pinus aristata[a]	9	2.1	2.04
Pinus banksiana	36	1.3	3.55
Pinus contorta	37	1.0	3.29
Pinus edulisa	22	3.8	2.60
Pinus lambertiana	19	1.3	4.47
Pinus ponderosa	33	2.0	3.67
Pinus resinosa	44	2.0	3.77
Pinus strobus	182	2.3	4.12
Pseudotsuga menziesii			
Coastal	30	1.3	4.12
Interior	56	1.8	3.39
Thuja occidentalis	148	2.1	2.95
Tsuga canadensis	29	1.8	3.95
Sequoia sempervirens	19	2.5	5.00

[a] Rings less than 1 mm wide.

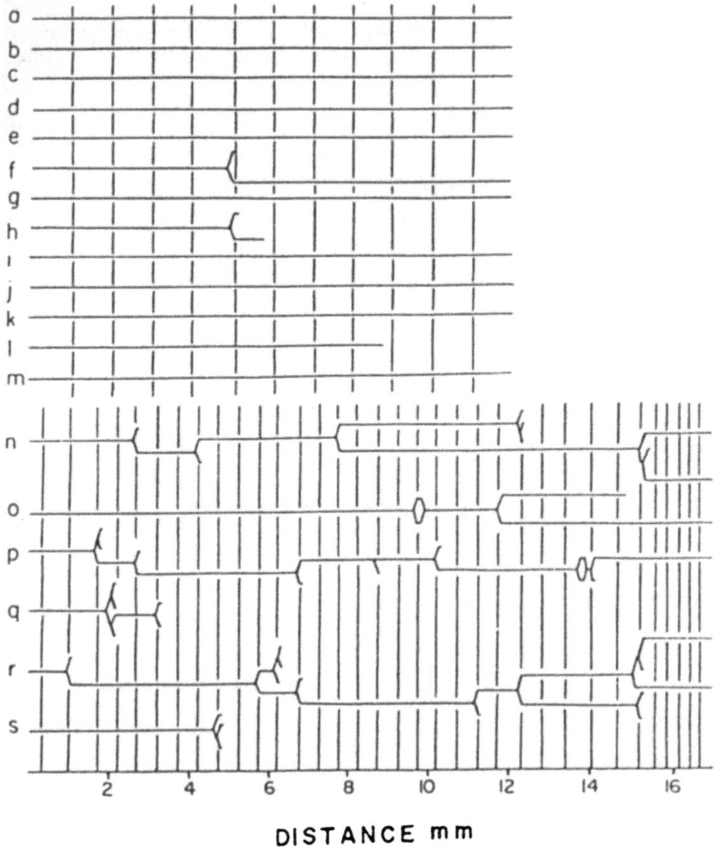

Fig. 6.46. Case history diagrams of individual cambial initials in the recent growth rings of two *Pseudotsuga menziesii* trees, each 2.1 m in diameter. The rate of division was 0.1 and 2.3/cm of xylem for series *a-m* and *n-s*, respectively. Details as in Fig. 6.44. (Bannan 1964a)

genera, occurring at an extremely narrow 0.3-mm ring in *Thuja occidentalis* (Bannan 1962a) and at a relatively wide 2.0-mm ring in *Picea glauca* (Bannan 1963a). The frequency of anticlinal divisions over the stable range of values also varied among genera, ranging from a low of about one division per cm of xylem increment in *Picea glauca* to almost four per cm in *Cupressus sargentii*. Differences within genera were almost as great as those among genera, as shown in Fig. 6.45B for four species of *Picea*. The stable values, indicated by the level portions of the curves, were referred to as "base rates of pseudotransverse divisions" (Bannan 1964b, 1970). The base rates for 24 species are shown in Table 6.21. Despite the remarkable similarity in the general trends within and among genera, the smoothed curves conceal considerable variability. Other data published by Bannan (1964b) showed a broad scatter of points around the smoothed curve.

A large number of factors contribute to variability within a species. At times the variability is apparently inherent and no meaningful cause can be found to account for it. For instance, Bannan (1964a) analyzed cell files in the outer growth rings of two *Pseudotsuga menziesii* trees, both of which were about 2.1 m in diameter. In one (Fig. 6.46, Files a-m), the rate of anticlinal division was only 0.1 per cm of xylem increment whereas in the other (Files n-s) the rate was 2.5 per cm. Differences in growth rates between the two trees were insufficient to account for the disparity in the rates of division. Stem size itself, other than that attributable to

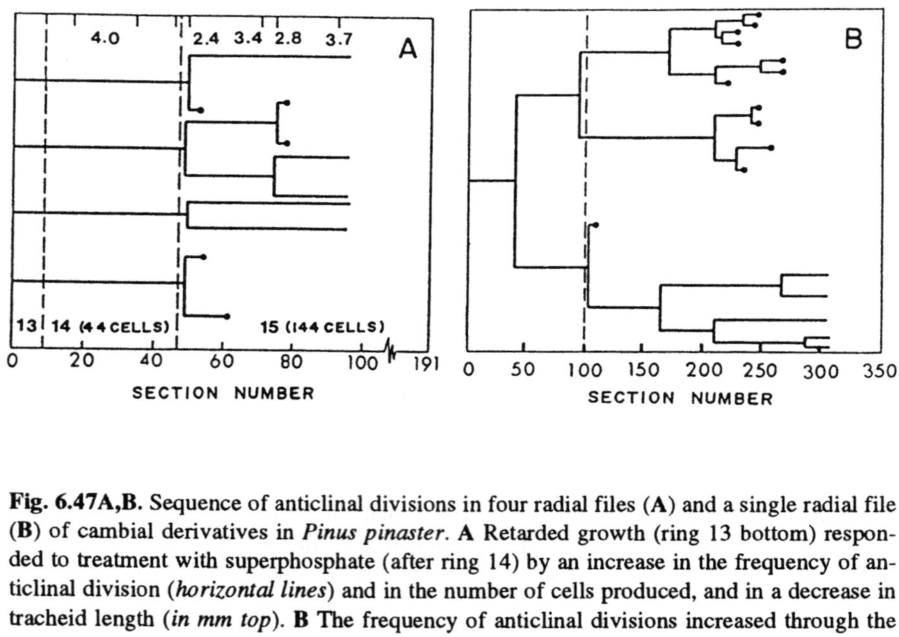

Fig. 6.47A,B. Sequence of anticlinal divisions in four radial files (**A**) and a single radial file (**B**) of cambial derivatives in *Pinus pinaster*. **A** Retarded growth (ring 13 bottom) responded to treatment with superphosphate (after ring 14) by an increase in the frequency of anticlinal division (*horizontal lines*) and in the number of cells produced, and in a decrease in tracheid length (*in mm top*). **B** The frequency of anticlinal divisions increased through the zone of compression wood that began at the *vertical broken line*. *Black dots* represent elimination of a tracheid file. (After Wardrop and Dadswell 1952)

age, seemed to contribute only slightly to the variability. Comparison of two size categories of *Picea sitchensis* stems, 25 to 30 cm and 92 to 214 cm in diameter, again revealed only slight differences in the rates of anticlinal division (Bannan 1963a). Similar results were obtained when four size categories of *Pinus strobus* stems were examined (Bannan 1967b). Bannan (1963a) concluded that the maintenance of anticlinal divisions at a relatively high level in the peripheral growth of some large trees was an indication that factors associated with radial increment exerted a greater effect on the rates of multiplicative division than circumferential expansion did.

Evidence to support this conclusion can be found in situations where rapid changes in rates of growth have markedly altered the rates of anticlinal division, as demonstrated by Wardrop and Dadswell (1952) in *Pinus pinaster*. One tree, which had been severely retarded, responded to treatment with superphosphate by a considerable increase in radial growth. This growth increase was accompanied by a rapid rise in the frequency of anticlinal divisions, from zero in the 14th to six in the first 50 sections of the 15th growth ring (Fig. 6.47A).

Locality or site can also be a source of variation. The frequency of anticlinal divisions for 12 *Pseudotsuga menziesii* trees of similar sizes and rates of growth but growing on different sites varied from 0.4 to 4.0 per cm of radial increment (Bannan 1964a); but when the data for *P. menziesii* trees from two widely separated locales were plotted, the scatter of data points made comparison almost meaningless.

That a reduction in growth rate can also affect the frequency of anticlinal divisions was demonstrated by Bannan (1964b) in *Pinus contorta*. In the successive ring sequences, referred to as A, B, and C in Table 6.22, mean ring width decreased from 0.67 mm to 0.29 mm and then recovered to 0.56 mm. The abrupt narrowing of the growth rings in sequence B was paralleled by an increase in the frequency of anticlinal divisions from 1.6 to 3.1 per cm of xylem increment, whereas the subsequent recovery was accompanied by a decrease in the frequency from 3.1 to 1.3 per cm.

Table 6.22. Correlations between frequencies of pseudotransverse divisions and fluctuations in widths of annual growth rings in five *Pinus contorta* trees (After Bannan 1964b)

Successive ring sequences	Number of rings/sequence	Mean ring width (mm)[a]	Mean frequency of division[b]
A	4-11	0.67	1.6
B	6-11	0.29	3.1
C	6-9	0.56	1.3

[a] Measured in peripheral growth rings in trees 25-30 cm in diameter.
[b] Frequency of division/cm of xylem increment.

6.5.2.1 Eccentric Stems and Branches

Considering the fact that the frequency of anticlinal divisions is influenced by rate of growth, it is not unexpected to find a frequency differential on opposite sides of

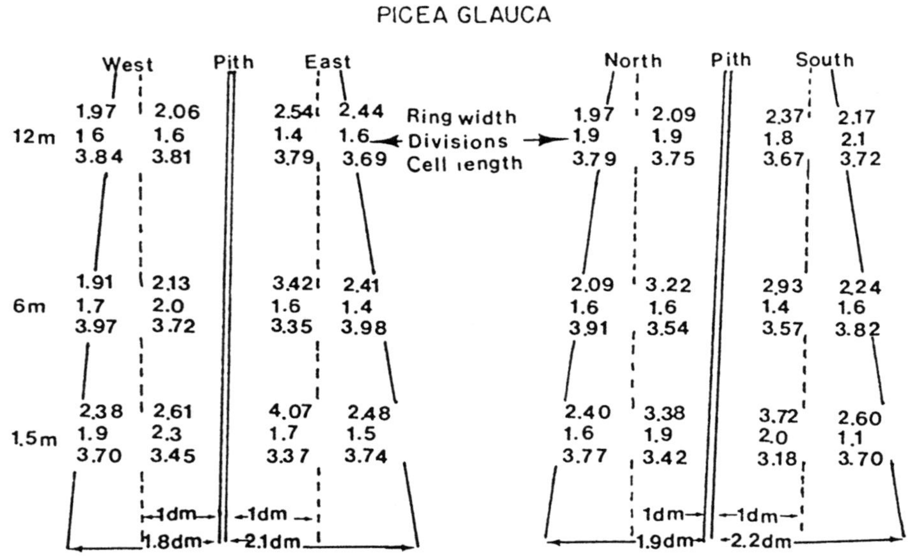

Fig. 6.48. Mean data for ring width (mm), rate of anticlinal division of fusiform initials (per cm of xylem increment), and cell length at division (mm) at 1 dm from the pith, at the stem periphery, and at three height levels in three *Picea glauca* trees. Prevailing westerly winds induced slightly eccentric stems. (After Bannan and Bindra 1970a).

eccentric stems and branches. Hartig (1895b) reported on this effect in eccentric stems of *Pinus sylvestris*. On the lower stem side, growth rings were wider, the frequency of anticlinal divisions was greater, and the cells were shorter than on the upper side. These differences became more accentuated as both eccentricity and ring width increased with age of the tree. Based on these observations, Hartig concluded that the slower the growth and less frequent the anticlinal divisions of initial cells, the longer the derivative tracheids in the growth ring.

The obvious relation between tracheid length and ring width in eccentric stems of conifers was noted by Wardrop and Dadswell (1950). To account for the shorter tracheids on the wide-ringed, compression wood sides of inclined stems, they relied on the statement of Priestley (1930a) that the frequency of anticlinal cambial divisions increased with rate of growth. This work was explored in greater detail in stems of *Pinus pinaster* (Wardrop and Dadswell 1952). A young tree, which had been suddenly bent, exhibited a pronounced increase in radial growth associated with the development of eccentric growth rings and the formation of severe compression wood (Fig. 6.47B). The frequency of anticlinal divisions increased dramatically from two in the 100 sections before compression wood formation, to three in the next 100 sections, and to 11 in the remaining 106 sections. Only five of the 17 daughter cells survived. Although the frequency of anticlinal divisions increased concomitant with the increase in rate of growth, the increased frequency was associated with a decrease in the lengths of fusiform initials at the time of division. Similar results were obtained by Kennedy and Farrar (1965) and Yoshizawa et al. (1985).

A more thorough investigation of this relationship was made by Bannan and Bindra (1970a) in three coniferous species, *Picea glauca, Pinus contorta,* and *P. strobus*. The trees examined had slightly eccentric stems, presumably due to prevailing westerly winds. The data for three *Picea glauca* stems are shown in Fig. 6.48. Ring widths were narrow on the westerly, windward side at all levels in the stem. Concomitantly, cell length at the time of division and the frequency of anticlinal divisions were both slightly greater on the windward side. Differences between the north and south sides were in some cases almost as great as those between the east and west sides. However, as pointed out by the authors, the trees were originally selected because of their apparent normality and the slight eccentricity was not noted until after felling.

A voluminous literature exists on the eccentric growth of both gymnosperm and angiosperm trees. Many authors have made reference to the fact that lengths of reaction wood cells are related to the frequency of anticlinal divisions of cambial fusiform initials just as the lengths of normal wood cells are (Chap. 7). Yet relatively few have actually analyzed the process. The factors determining tracheid length variation in compression wood of conifers were thoroughly reviewed by Timell (1986). He concluded: "From the combined evidence now available it seems inescapable ..." that the reduction in tracheid length in compression wood is a consequence of the increased number of anticlinal divisions.

6.5.3 Location Within the Growth Ring

The frequency of anticlinal divisions also varies with position in the growth ring, or what might be considered a seasonality factor. Not only do anticlinal divisions occur more frequently in the inner rings of stems and branches, but they are more likely to be scattered throughout the growth ring. Although it is not uncommon to find similar patterns during periods of exceptionally rapid growth at all ages, it is more common to find that anticlinal divisions tend to occur later in the season in older rings. During normal growth of both *Chamaecyparis* and *Thuja*, Bannan (1951b) found that only rarely did anticlinal divisions occur during the first-formed part of the growth rings. Such differences are well illustrated by comparison of cell series a-i with j-t in Fig. 6.44 representing the inner and outer growth rings, respectively, of a *Thuja occidentalis* tree. In the young, inner rings, the anticlinal divisions were scattered throughout the rings, whereas in the older, outer rings, they were concentrated in the last-formed part of the rings. The foregoing was not an isolated incidence, as similar patterns were found in trees of other species (Table 6.23).

Table 6.23. Distribution of pseudotransverse divisions in annual growth rings of three species of *Picea* and *Pseudotsuga menziesii* (After Bannan 1963a; 1964a)

Species	Stem diameter dm	Ring width mm	Location of divisions in ring[a] Second quarter	Third quarter	Final quarter
Picea glauca	2.5-5.0	7.9	20	43	37
Picea glauca	2.5-5.0	2-3	8	37	55
Picea glauca	2.5-5.0	0.4-1	4	30	66
Picea engelmannii	2.5-5.0	2-3	1	9	90
Picea sitchensis	2.5-5.0	2-3	2	42	56
Picea sitchensis	9-20	1-3	5	15	80
Pseudotsuga	2.5-5.0	0.5-1	3	8	88
menziesii	2.5-5.0	1-2	2	16	81
	2.5-5.0	2-4	4	36	59
	2.5-5.0	4-7	11	36	52
	9-21	0.5-1	4	6	88
	9-21	1-2	3	12	81

[a] Expressed as a percentage of total divisions.

This general pattern, in which anticlinal cambial cell divisions tend to occur in the latter part of the growth ring, has been verified in both gymnosperms (Bannan 1950b, 1963a, 1964a, Chalk and Ortiz 1961, Smith 1967, Meier 1973) and dicotyledons (Evert 1960, 1961, 1963, Catesson 1962a, 1964, Cumbie 1967b, Derr and Evert 1967, Alfieri and Evert 1968, Iqbal and Ghouse 1985b, 1987, Rao and Dave 1985a) (Fig. 6.57). A similar late-season decrease in the surface area of fusiform cambial initials was observed by Wilson (1963) in *Abies concolor*, which he attributed to anticlinal divisions. Moreover, as Braun (1963) demonstrated in *Pinus sylvestris*, new ray origins tended to be more numerous in the latter one-half of the growth ring (Fig. 8.41A).

The foregoing pattern can occur in young rings as well as old ones if growth is slow, as Bannan (1950b) found in seedlings of *Chamaecyparis*. For example, when seedling growth rings were narrow, averaging only 0.2 to 0.4 mm in width, 75% of the anticlinal divisions occurred in the final one-third of the ring. In growth rings of older trees of some species, there was a tendency for anticlinal divisions to be concentrated even more strongly in the final part of the ring (Table 6.23).

Newman (1956) questioned Bannan's reports by pointing out that he had observed many anticlinal divisions occurring during the early stages of growth ring development. However, Newman's observations were made on material exhibiting "vigorous cambial activity," and under these conditions such a pattern would not be unusual. Bannan (1961), in fact, had reported such trends in several fast-growing trees. For example, in a stem of *Cedrus deodara* with rings 1 cm wide, 30% of the anticlinal divisions occurred in the first one-half, 30% in the third quarter, and 40% in the fourth quarter of the growth rings. Smith (1967) noted a similar response in his studies of *Abies grandis*. Under normal growth conditions, anticlinal divisions tended to occur in the last one-third of the growth rings. However, they were distributed more uniformly throughout the growth rings when cambial activity was stimulated following infestation by *Adelges piceae* (Chap. 10.5.2.2). Rapid growth due to applications of fertilizer might also have been responsible for the lack of pattern in *Pinus radiata* reported by Hartley (1960). In this case, anticlinal divisions occurred throughout the growth rings and there was no decrease in tracheid length at season's end.

The final anticlinal divisions of the season before cambial dormancy sets in might terminate in several different ways (Bannan 1955). If the last anticlinal division occurs while both xylem and phloem production are in progress, a double radial file might extend for a short distance on both sides of the cambium (Fig. 6.49A). If the final division in a radial file occurs late in the season, the cell doubling might be confined to the initial cell and one to several mother cells (B,C), and if the anticlinal division occurs after radial growth has essentially ceased, the partition wall might appear only in the initial cell (D,E).

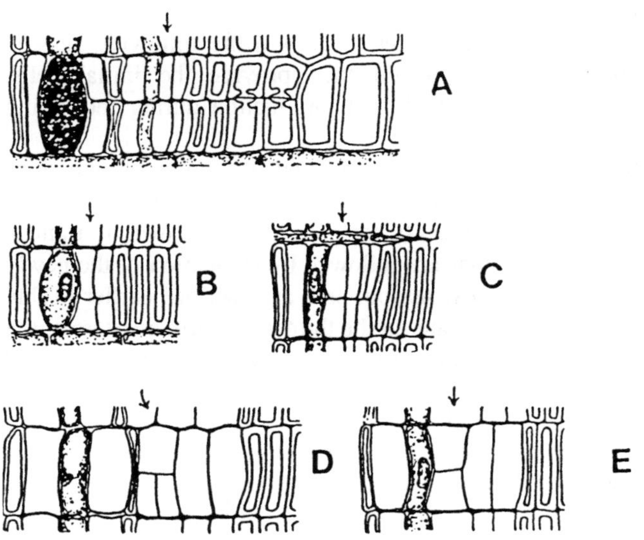

Fig. 6.49A-E. Drawings of transverse sections of the cambial zone of *Thuja occidentalis* trees illustrating the appearance of anticlinal divisions of fusiform cambial initials (*arrows*) at different times at the end of the growing season. See text for details. (Bannan 1955)

Table 6.24. Rate of pseudotransverse division and cell length at division at various heights in a *Pseudotsuga menziesii* tree 13 dm in diameter. (Bannan 1964a)

Height m	Ring width mm	Division rate per cm xylem	Cell length mm
1.8	1.50	17.8	2.28
6.1	0.89	3.0	4.40
15.2	0.78	3.5	4.58
24.4	0.72	2.4	4.70

The times at which anticlinal divisions occur during the year are by no means constant. Based on studies of the sequence of periclinal divisions, Bannan (1951b, 1955, 1964a) concluded that the anticlinal partition walls which appeared in the first-formed earlywood were due in part to anticlinal divisions that actually took place at the conclusion of growth the previous season. Thus, these anticlinal divisions were aestival. This fact becomes apparent by examination of the data of Table 6.23 for *Pseudotsuga menziesii*. Localization of anticlinal divisions occurred in the final quarter of all growth rings but it was most pronounced in the rings less than 1 mm wide of slow growing trees. Similar distributions of anticlinal divisions were recorded for several species of *Chamaecyparis* (Bannan 1950b) and *Picea* (Bannan 1963a). According to Bannan, daughter cells derived from these aestival anticlinal divisions would immediately begin producing new radial files by periclinal division at the onset of growth the following season. These late-season anticlinal divisions would then provide the initial cells, consequently the mother cells, for the rapid cell production during the spring surge of growth.

Concentration of anticlinal divisions in the latewood of conifer growth rings explains, in part, the characteristic increase in tracheid length during the season as observed and recorded by so many workers (Chap. 7.1.2.1). As shown by Bannan (1954), these aestival anticlinal divisions are followed by elongation of the newly formed sister initials throughout the next season; and, as described earlier (Table 6.7), elongation of fusiform initials is greater in the third and fourth quarters of the year. This elongation pattern usually continues for several years, or until the initial again divides anticlinally.

6.5.4 Position in the Tree

6.5.4.1 Height in the Stem

Variations in the frequency of anticlinal divisions attributed to sampling position within a growth ring are repeated at different height levels. However, the differences appear to be quantitative and related to the relative rates of circumferential and radial stem expansion at various height levels in the stem. Thus, Hartig (1895c) concluded that in a 190-year-old *Pinus sylvestris* tree the frequency of transverse (anticlinal) divisions at the 6.3 m level was less and the average cell length was greater than at the 1.3 m level because the circumferential increase in stem diameter was less at the higher level.

The variation in anticlinal divisions with height in the stem was examined in considerable detail by Bannan (1964a) in *Pseudotsuga menziesii*. In general, most differences could be accounted for by other factors such as age of ring and rate of growth. Yet, in one sample of trees 25 to 50 cm in diameter, the frequency of anticlinal divisions tended to decrease from 3.2 per cm of xylem increment at the 1.5 m level to 2.0 per cm at the 6.6 m level. The reduction in frequency of anticlinal divisions favored the development of longer tracheids.

Data for one tree, 13 dm in diameter at stump height, are presented in Table 6.24. A pronounced decrease in the frequency of anticlinal divisions occurred at the 24.4 m level although the increase in cell length was slight. The aberrant data for the 1.8 m level were probably due to the gnarled grain present at this level in the stem. Somewhat similar data were presented for various heights in the stems of Sequoia spp. (Bannan 1966a), *Picea glauca*, *Pinus contorta*, and *P. strobus* (Bannan and Bindra 1970a) (Fig. 6.48).

6.5.4.2 Branches

Branches of trees have often been the source of material for investigations of anticlinal divisions of cambial cells because sampling was both easy and less injurious to the tree. The patterns and consequences of anticlinal divisions in branch material differ little from those in stem material when consideration is given to rates of radial and circumferential branch expansion, age, eccentricity, and other contributing factors (Whalley 1950, Evert 1960, 1961, 1963a, Cheadle and Esau 1964). Many investigations have been undertaken to compare the lengths and widths of xylem elements in branches and roots with those of stems. These differences involve primarily extracambial and not intracambial changes in cell dimensions. As a general rule, xylem elements in both branches and roots are smaller than those of stems (Dinwoodie 1961).

Some indication of the variability in branch material was provided by the data of Bannan and Bindra (1970b). As shown in Fig. 6.50A, the mean length of fusiform initials at the time of division increased to about the 6th to 10th years, and then gradually decreased as ring widths narrowed. The rate of anticlinal division of fusiform initials was high during development of the first annual ring, as in stems, but then declined coincident with the peak in fusiform cell length. This decrease was followed by a rise in the rate of division on both the upper and lower branch sides. The authors attributed the higher rate of anticlinal division on the upper branch side to the extreme narrowing of the growth rings.

The data of Bannan and Bindra suggest that consideration must be given to the possibility of reaction stimuli and reaction wood in branch material, as discussed

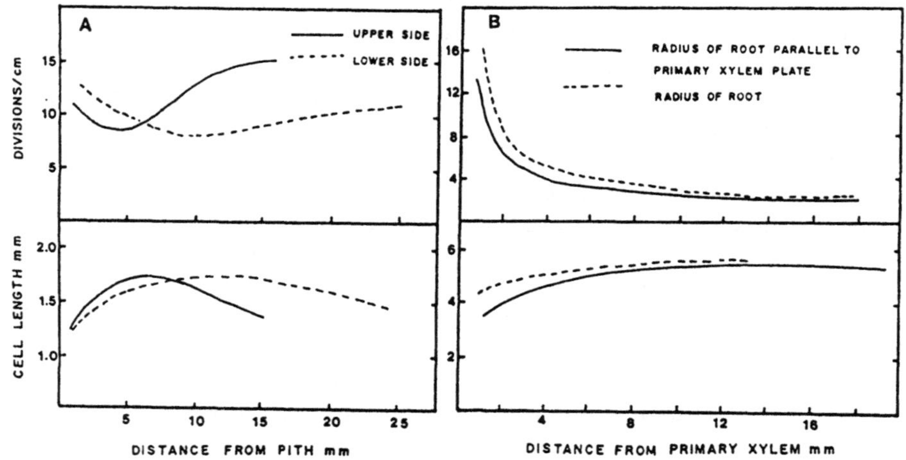

Fig. 6.50A,B. Sequential changes in the frequency of anticlinal division and cell length at time of division of fusiform initials in the upper and lower sides of branches (**A**) and horizontal roots (**B**) of *Picea glauca*. **A** Extension of the curves on the lower branch side is due to wider growth rings. (After Bannan and Bindra 1970b)

in greater detail by Timell (1986). In view of the fact that branch growth is often eccentric, one might assume that the differential growth on the upper and lower sides would influence both the frequency of anticlinal divisions and the sizes of the derivatives. Therefore, caution should be exercised when assessments of cambial division patterns for a species are based on branch material.

6.5.4.3 Branch Crotches

Branch crotches are regions of erratic cell development (Kny 1877a). Jost (1901) was intrigued with the question of how the cambium developed at a branch insertion, and he could find nothing in the literature directly relating to it. In posing the question, he assumed a branch inserted at right angles to the stem as shown in Fig. 6.51A. Over time, the cambium would progress from C_1 to C_2. Since ab advanced to a'b' and de to d'e', a point would be reached where the cambium at bcd collided at a'd'e'. He then asked, where is this point and how does it occur?

In the above example, cambial contraction would occur equally on the upper and under sides of the branch insertion. However, as branch angle increased, cambial contraction would become more accentuated on the upper side while lessen-

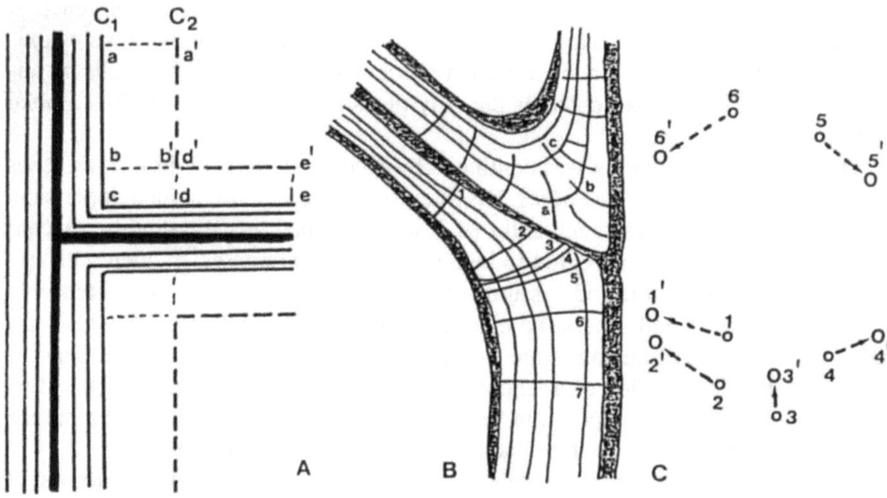

Fig. 6.51A-C. Theoretical readjustment of tissues in branch crotches of *Pinus sylvestris* as branch angle increased from the horizontal. See text for details. (After Jost 1901)

ing only slightly on the under side (Fig. 6.51B). Based on the earlier findings of Sanio (1873) on *Cytisus laburnum*, Jost assumed that tracheids of *Pinus sylvestris* were essentially the same length as the cambial initials from which they were derived. This assumption allowed him to estimate the lengths of cambial cells at different positions within the branch crotches.

To account for cambial contraction on the branch under side, Jost examined variations in tracheid length, tracheid inclination angles, and ray orientations. In the latter case, the extent of cambial contraction was estimated by the radial convergence of rays, as shown in Fig. 6.51B (rays 2 to 5). The only reasonable explanation he could offer was reorientation of the cambial cells brought about by sliding growth of the radial walls. Despite his misgivings of the concept of sliding growth, he concluded that such a process must exist. The repositioning of rays, as observed on serial tangential sections, was used to demonstrate the effects of this process. For example, during the deposition of 1 cm of new xylem in Fig. 6.51C, rays 1 and 6 were reoriented to positions 1' and 6', respectively, corresponding to the radial convergence of rays 2 to 5 in Fig. 6.51B. This reorientation resulted in a "cambial contraction" of about 25%, which Jost attributed to "sliding growth of all cambial cells." Inexplicably, however, as the cambium contracted both tracheid length and ray height increased in the radial direction.

Analysis of the branch upper side was more complicated because cells in this region were often swirled, like the cambium in the center of "Knäuelbildungen," or gnarl formations. Jost first considered the mechanical resistance of the rind to radial expansion of the xylem and concluded that the rind must yield to the expanding xylem. However, as the xylem expanded radially, it contracted just as it did on the under side, but much more drastically, as indicated by the convergence of rays a and b in Fig. 6.51B. Nonetheless, the general patterns of growth were similar on both the upper and under sides and reorientation of the cambial cells by sliding growth was again accepted as the only plausible explanation for the cambial contraction.

This line of investigation was continued by Jaccard (1910b) on branch crotches of ***Pinus sylvestris*** and ***Picea abies***. His interest centered on the branch upper side, the internal angle of the branch axil that Jost had studied less thoroughly. Jaccard found the tracheids in this region to be distorted and to undergo frequent changes in orientation. Many tracheids assumed the appearances of those associated with compression wood and others the characteristics of sclerenchyma. The resin canals also resembled traumatic canals rather than those of normal wood. Jaccard attributed the reorientation of tracheids to sliding growth and the abnormal xylem cell characteristics to compression brought about by rapid growth in the branch axil situated between the strongly inclined branch and the main stem.

In a study of branch crotch formation in ***Malus***, MacDaniels (1923) arrived at conclusions similar to those of Jost (1901), although no references were cited. MacDaniels also observed a pronounced shortening and distortion of wood elements in the crotch region and related these anomalies to the cambial initials from which they were derived. He, too, surmised that a distinct shortening of the cambial cells was probably due to contraction and "crowding" of the cambium as more layers of wood were deposited toward the outside. However, unlike Jaccard (1910b), MacDaniels discounted sliding growth, believing that such a process would split the middle lamellae and plasmodesmata. Because cambial cells in this region were "in a plastic, unlignified condition," he suggested that adjustments might take place by "the change in form of the cells."

In addition to the shortening and distortion of xylem elements, branch crotches are also sites of complex cellular reorientations such as the distorted and whirled arrangement of tracheids observed by Jaccard (1910b) and Böhlmann (1970a,b,c,d) and the "circular vessels" observed by others (Sect. 6.6.2.5). These "whirlpools" of xylem elements have been observed in regions where two or more branches diverge from the stem in both conifers and dicotyledons (Lev-Yadun and Aloni 1990). In *Quercus calliprinos* and *Q. ithaburensis*, branch crotches were also found to contain localized regions in which migrating domains of grain reorientation appeared (Fig. 6.63) and aggregate rays were dissected to smaller rays (Lev-Yadun and Aloni 1991b).

Böhlmann (1970a,b,c,d) conducted a series of histological studies on branch junctions in both coniferous and dicotyledonous trees. He described the vascular pathways in general terms, noting not only the distorted, whirled arrangement of elements but also the presence of upward-directed elements in the region of the branch crotch. Two fundamental types of vascular union were recognized. In the first, typical of ***Juglans***, ***Fraxinus***, and the Cupressaceae, a zone of parenchymatous tissues separated the branch and the main axis (Fig. 6.52A). The second type, represented by ***Betula***, ***Fagus***, ***Quercus***, ***Populus***, and the Pinaceae, initially pos-

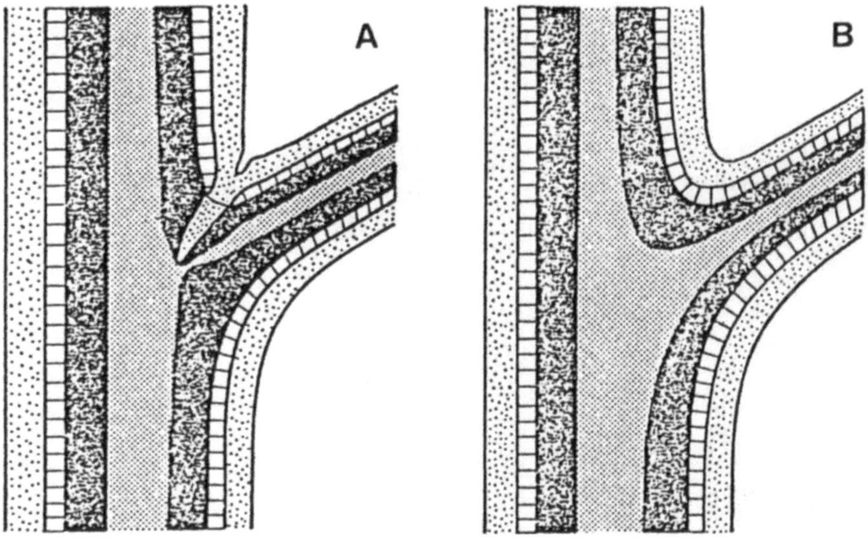

Fig. 6.52A,B. Two functional types of vascular branch union. **A** A zone of parenchymatous tissue separates the branch and main axis. **B** Vascular tissue is in complete continuity between the branch and main axis. (Böhlmann 1970d)

sessed a direct connection between the branch and the main axis (Fig. 6.52B) that became separated by a disturbance zone during the second year.

Similar work on the nodal vasculature of *Populus deltoides* was undertaken by Larson and Richards (1981), Richards and Larson (1981), and Larson and Fisher (1983). These authors demonstrated that the primary and secondary xylem in lateral buds and branches, respectively, was independent of that in the main axis. That is, xylem did not differentiate into the bud or branch from the main axis. Rather, vascular tissues from the branch lower side extended downward to both overlie and intermingle with elements of the main axis, whereas those from the branch upper side traversed the branch crotch and extended upward where they united with elements of the main axis. Meanwhile, the downward-directed xylem vasculature of the main stem essentially bypassed the branch junction.

This work was extended by Shigo (1985) on branches of mature trees from a large number of species. Removal of bark from a *Quercus rubra* branch junction in June clearly revealed the vessel arrangement (Fig. 6.53). Xylem tissues developing mainly from the lateral and under sides of the branch formed a basal collar. While the collar was forming, it was concomitantly being enveloped by the basipetally developing xylem of the stem as shown in Fig. 6.54. Like other vascular patterns, the arrangement and reorientation of these vascular tissues in the region of branch junctions must be viewed simply as reflections of the patterns displayed by the cambial initials.

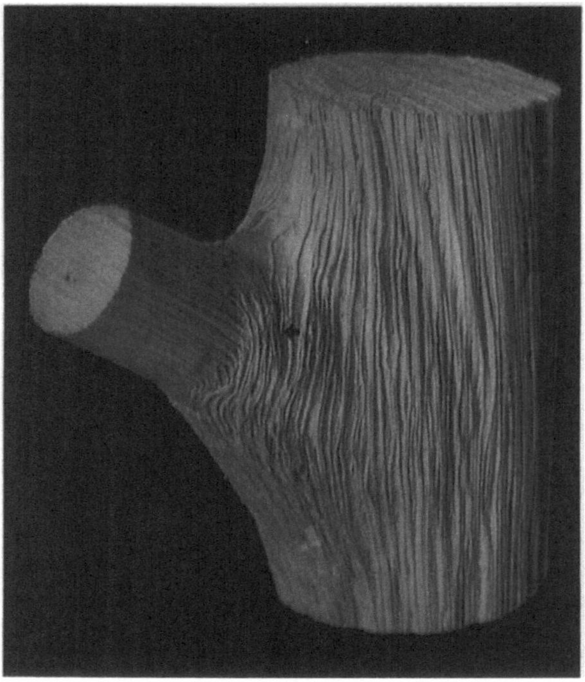

Fig. 6.53. A small trunk section of *Quercus rubra*, from which the bark was peeled, showing arrangement of the large earlywood vessels at the branch collar (*arrow*). (Shigo 1985)

6.5.4.4 Roots

The patterns and consequences of anticlinal divisions in root material differ little from those in stem material when contributing factors are taken into consideration. For example, in the root cambium of *Herminiera elaphroxylon*, Beijer (1927) showed that when an anticlinal division occurred in a xylem mother cell it was not perpetuated. However, when the anticlinal division occurred in a cambial initial, derivatives of the two daughter cells were present in both the xylem and phloem. The response of roots to external stimuli and growth conditions is also similar to that of stems, as shown by studies of root grafting (Chap. 10.3.6.2) and root decapitation (Chap. 10.3.8).

Bannan (1950b) and Whalley (1950) both noted a tendency for fusiform initials in roots to be slightly longer at the time of anticlinal division than those in stems. Also, Bannan (1950b) observed that the frequency of anticlinal divisions in roots tended to be more erratic, although he thought it was of little significance. This work was extended by Bannan and Bindra (1970b), who found that cell length at the time of anticlinal division varied greatly in fusiform initials of vertical and horizontal roots of *Picea glauca*. In vertical roots, cell length at division seldom exceeded 2 mm, a value similar to that in branches (Fig. 6.50A). In horizontal roots,

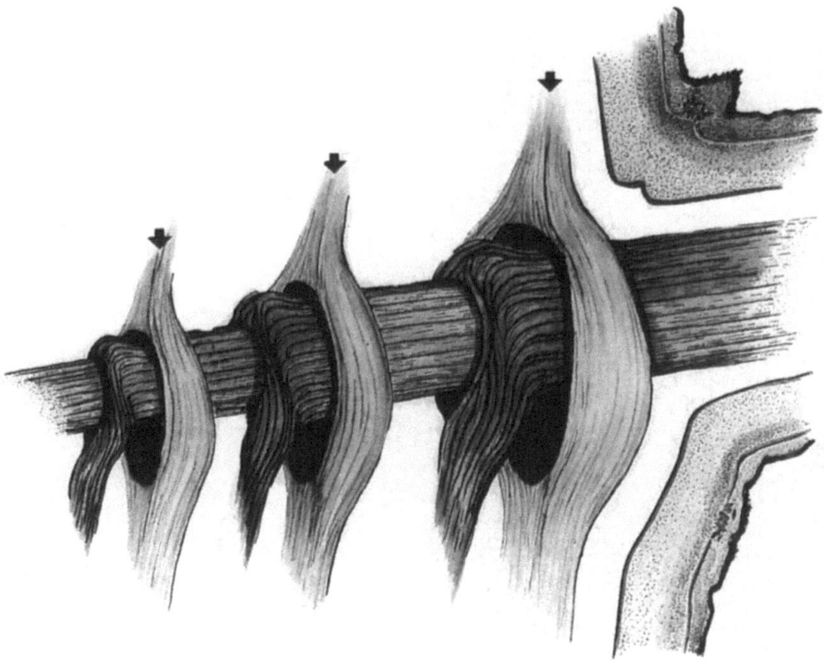

Fig. 6.54. Schematic diagram showing how xylem tissues of the trunk (*arrows*) overlie and envelop those of the branch. The collars for each growth ring are depicted as if they had been pulled apart. (Shigo 1985)

in contrast, cell length at division averaged about 5 mm after the first year of growth (Fig. 6.50B). The rate of anticlinal division was also dissimilar in vertical and horizontal roots. Whereas the rate was more or less maintained at about 20 divisions per cm of xylem increment in vertical roots, it decreased from that level in early growth to slightly less than 3 divisions per cm in later years in horizontal roots (Fig. 6.50B). The longer fusiform initials and slower rates of anticlinal division in horizontal roots was attributed to slower growth and narrower growth rings.

Hejnowicz (1961) included an intensive analysis of root wood in his study of *Larix decidua*. The results of this study showed that all the changes Bannan had described in the cambia of stems could be confirmed in the cambia of roots. These changes included anticlinal division, ray cell initiation, intrusive growth, and loss of initials. The only difference between root and stem cambia in the trees examined was the tendency for the anticlinal divisions of the latter to conform to a left-hand spiral.

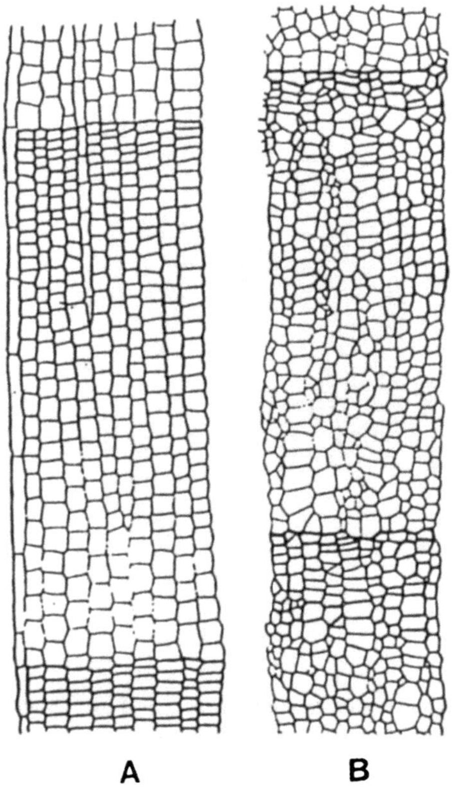

Fig. 6.55A,B. Xylem from normal stem wood (**A**) and stumpwood (**B**) of *Abies alba* showing distortion of radial cell alignment in the latter. Transverse sections 54x. (Neeff 1922)

6.5.4.5 Living Stumps

Living stumps, which should not be confused with stumps that produce sprouts, are commonly encountered in forest stands (Graham and Bormann 1966). Their persistence, despite the lack of a vegetative crown, has been attributed to root grafts with neighboring living trees (Chap. 10.3.6.2). Many examples of long-lived stumps can be found in the literature. Lanner (1961), for example, found one *Pseudotsuga menziesii* stump that had persisted for 87 years, and he cited Newins, who in 1916 reported a 200-year-old living stump of the same species.

Because of their connections with neighboring trees via root grafts, living stumps produce annual growth rings in the wood below the cut surface. The rings are often abnormal in structure (Fig. 6.55), narrow in width, and they might be discontinuous or missing on stump sides opposite that of the host tree (Liphschitz et al. 1987).

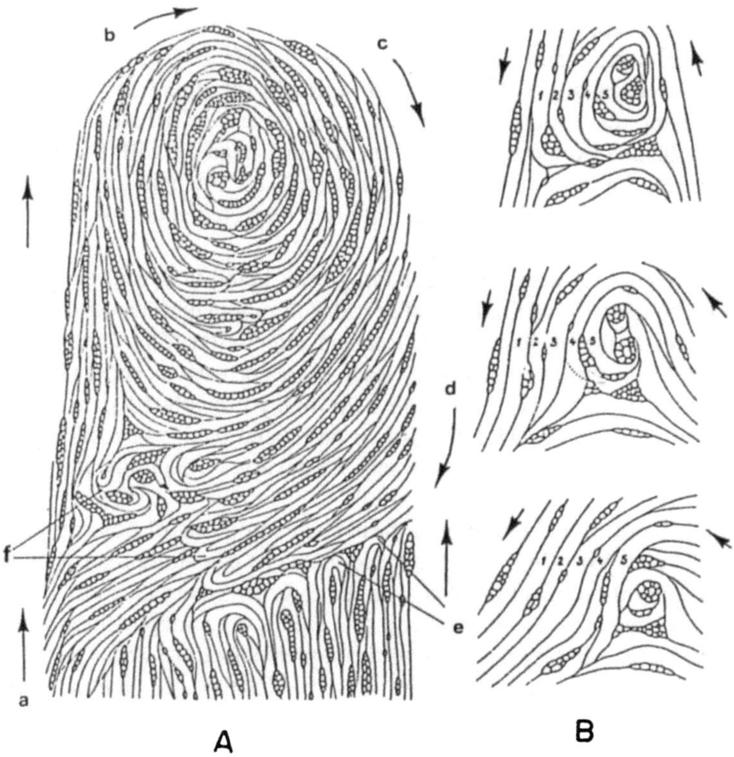

Fig. 6.56A,B. Drawings of tangential sections through stumpwood of *Abies alba*. **A** Whirled arrangement of elements in xylem 10 cm below the stump callus. 29x. (Neeff 1914). **B** Serial sections showing dissolution of a cell vortex outward in the xylem. 27x. (Neeff 1922)

As in other tree wounds, callus sometimes grows over the cut surface of the living stump. Cambium from the stump grows into the callus pad but it rarely produces normal secondary tissues. This is also true of buds that sometimes arise within the stump callus. Such buds often show many anomalies which are indicative of extreme growth disorders (Dostál 1967).

Jacquiot (1966) examined living stumps of *Quercus* and found that the wound callus consisted of unlignified parenchymatous tissue more or less irregular in structure. Included within this tissue were islets of lignified cells containing vessels that differed significantly from normal ones. Regenerated stumpwood therefore differed from typical wound callus in which the newly formed wood eventually includes all cell types. Jacquiot attributed these differences to the absence of inductive influences emanating from a living crown.

A living stump of *Abies alba* that had persisted after cutting a 127-year-old tree 31 years earlier was investigated by Neeff (1914). In addition to the 1.5 cm wood layer produced on the stump, an extensive callus developed over the cut surface. Xylem elements in the stump as well as in the callus were severely distorted, as shown in a tangential section taken 10 cm below the cut surface (Fig. 6.56A). The tracheids appeared to circulate in the form of a vortex in the direction of the arrows. Neeff interpreted this pattern on the basis of cell polarity. That is, where the apical poles of tracheids developing in one direction met head-on with the apical poles of those developing in the opposite direction (e,f), the tracheid tips became bent and distorted. These tracheids resemble the circular elements described in Section 6.6.2.5.

Neeff (1922) investigated these phenomena in greater detail in living stumps of *Abies alba* and *Picea abies*. Serial tangential sections through stumpwood showed progressive, but ever-changing, patterns of tracheid orientation. The vortex illustrated in Fig. 6.56A was simply a temporary stage in this process as shown by the dissolution of a similar vortex in Fig. 6.56B. The many convoluted patterns of tracheid orientation described by Neeff could all be related back to events in the cambium. As cambial cells divided "transversely" they varied directions, not arbitrarily, but uniformly.

As in his earlier work, Neeff interpreted these events in terms of polarity phenomena. He agreed with Vöchting, who proposed that cells can attain their normal lengths and forms only if the plant body is attached to growing organs both above and below, i.e., to a shoot and a root. According to Neeff, the stump callus served as a substitute for the lost shoot; but it was an incomplete substitute and, lacking a true apex, the callus was unable to maintain polarity. That is, the cambial cells were not in polar unity with those of the callus. Loss of the vegetative apex also resulted in suppression of directional growth of the cambial cells. The cambial cells compensated for this loss by expressing their own polarity, which gave rise to the vortices and convoluted patterns of growth. Neeff, of course, was unfamiliar with the role of growth hormones (Sect. 6.6.4).

6.5.5 Variations Among Species

The frequency or rate of anticlinal division apparently differs among species. However, it is obvious from the data and observations presented in the foregoing sections that meaningful comparisons are difficult to make because of the many uncontrollable variables. Some evidence of species differences can be found in Tables 6.5 and 6.21. Inter- and intrageneric trends relating frequency of anticlinal division to rate of growth are shown in Fig. 6.45A and B, respectively; similar data plots were presented by Bannan for all species investigated in his numerous publications. Although the trend lines for all species were similar in shape, they differed in detail. Both the base rates of anticlinal divisions and the ring widths at which the base rates increased precipitiously varied quite widely among species examined. Species differences in the frequency of anticlinal division undoubtedly also exist due to other factors such as age of growth ring, position in tree, etc.

However, comparisons of this kind should be accepted with caution. In summarizing his data, Bannan (1961) stated that, on the whole, xeric species had higher rates of anticlinal divisions and shorter cells than mesic species.

The frequencies with which anticlinal divisions occur within the cambium exert a decisive influence on the length of the resulting daughter initials. Consequently, differences in the frequency of division among species will ultimately contribute to differences in lengths of both the fusiform initials and the derivative cells.

6.5.6 Biennial Frequency

An interesting source of variation in the frequency of anticlinal division was reported by Evert (1961) in *Pyrus communis*. As shown in Fig. 6.57, anticlinal divisions tended to concentrate in alternate years. Branch samples from four other trees revealed similar alternate patterns, and in some years the frequency of division exceeded that in either the preceding or following year by two to three times. Although no conclusive cause for this phenomenon was established, Evert suggested that it might be related to the alternate bearing of fruit. That is, similar factors might be responsible either directly or indirectly for both phenomena.

6.6 Aberrant Anticlinal Divisions

Normal anticlinal divisions might be viewed as those in which two sister initials elongate in such a way that a relatively straight-grained cambium is produced. Aberrant anticlinal divisions, on the contrary, would be those in which the two sister initials elongate in such a way that some form of distorted or disturbed grain is produced. The term aberrant, as used here, does not imply that the division patterns are abnormal, but simply that they differ from those generally regarded as normal.

6.6.1 Localized Domains

Cambial domains are characterized by localized regions of varying size where cells behave in a like manner and often in marked contrast to cells in adjacent regions. This, of course, excludes regions where differences are due to fissures, grooves, wounds, eccentricity or other anomalous causes. Under a very broad interpretation, although not truly a domain, one might include the differential frequency of anticlinal divisions of adjacent files of fusiform initials. Such differences have been recorded in many species and depicted diagrammatically in case histories. Good examples can be seen in Fig. 6.4, where file a produced a large

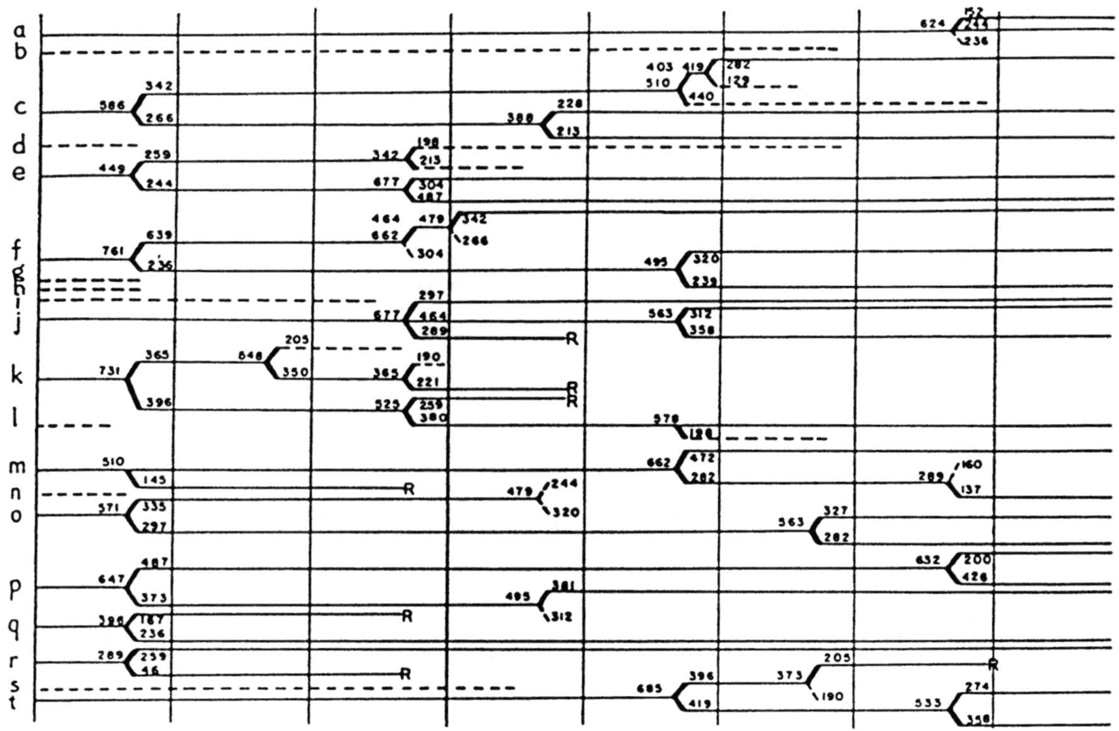

Fig. 6.57. Case history diagram of anticlinal cell divisions of fusiform initials in *Pyrus communis* illustrating the tendencies for divisions to occur not only in the latter part of the annual growth rings but also in alternate years. *Forking of a horizontal line* represents division of an initial. *Broken lines* indicate failing initials and the positions at which they disappeared from the cambium. *Numerals* indicate the lengths of fusiform initials at the time of division and the approximate lengths of the daughter initials after division (in µm). The letter *R* signifies transformation of a fusiform initial to a ray cell initial. *Vertical lines* identify seven annual growth increments; not drawn to scale. (Evert 1961)

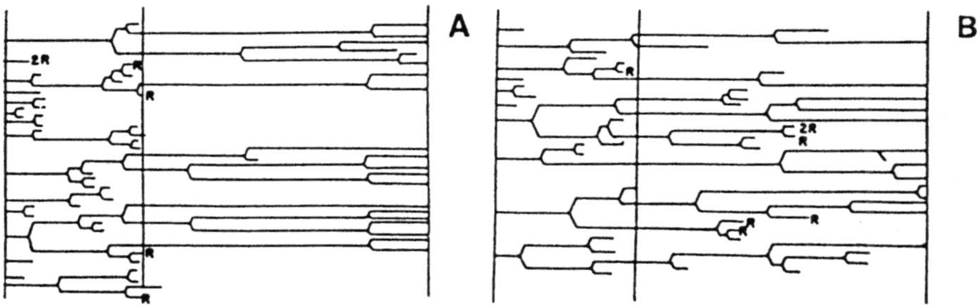

Fig. 6.58A,B. Case history diagrams of anticlinal divisions of fusiform initials in two sectors of the cambium, tangentially 2 mm apart, of a *Chamaecyparis thyoides* stem. *Vertical lines* separate the fourth and fifth annual growth rings. Details as in Fig. 6.4. (Bannan 1950b)

number of anticlinal divisions during the 8 mm of radial growth, whereas file b produced relatively few. Similar examples can be found by examination of Figure 6.57.

A more common interpretation of a cambial domain would refer to an area or sector of the cambium. Bannan (1950b) described and illustrated two such sectors, just 2 mm apart, encompassing the fourth and fifth annual rings of *Chamaecyparis thyoides*. In one sector (Fig. 6.58A), frequent anticlinal divisions in the fourth growth ring were followed by a high rate of cell failure. After 30 anticlinal divisions, the original 15 fusiform initials were reduced to eight. A more favorable survival ratio during the following years allowed recovery to 18 initials. In the other sector (Fig. 6.58B), cambial behavior was more typical of the species with the survival/failure ratio resulting in a slight increase in the number of fusiform initials over the 2-year span. Bannan (1956) also described how localized growth conditions often permitted cambial cells to elongate predominantly upward in one sector and predominantly downward in an adjacent sector. Although such sectors were small and of short duration, it was evident that some form of growth coordination allowed certain cells to behave in a like manner.

Sectorial domains commonly consist of regions of cambium in which the fusiform elements are predominantly inclined in the same direction, either to right or left of the vertical. Such domains are delimited not only tangentially within the cambium but also radially, in which case persistent inclination in one direction contributes to spiral grain development (Sect. 6.6.2.3). Sectorial variations in fusiform element inclination have been described in *Chamaecyparis* spp. (Bannan 1950b), *Thuja occidentalis* (Bannan 1954, 1956), *Larix decidua* (Hejnowicz 1961), *Picea* spp. (Bannan 1963a), *Pseudotsuga menziesii* (Jones 1963, Bannan

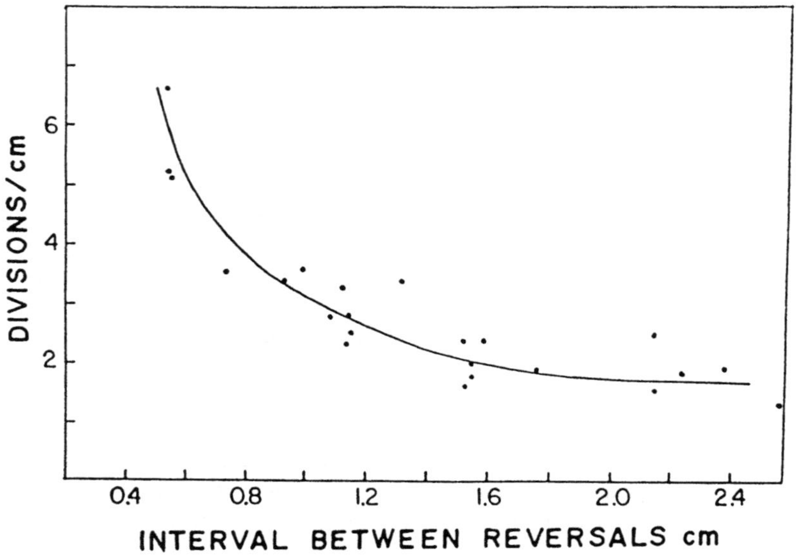

Fig. 6.59. Interval between reversals in orientation of anticlinal divisions and frequencies of divisions as determined at breast height in various coniferous species in trees 2-6 dm in diameter. (Bannan 1966b)

1964a), *Pinus contorta* (Bannan 1964b), and *Abies grandis* (Smith 1967). Meier (1973) found that 85 to 100% of the anticlinal divisions were inclined in the same direction in the three conifers he investigated from alpine sites.

The size of a cambial sector with similar element inclination varies widely. As found by Bannan (1966b), the area was sometimes so small that it consisted of only a few neighboring cells and at other times so large that it was difficult to determine. As a crude estimate, he calculated that the mean vertical extent of cambial sectors with similarly inclined partition walls was approximately 2.9 dm in *Pinus*, 2.4 dm in Pinaceae other than *Pinus*, and 1.5 dm in Cupressaceae.

The trees examined by Bannan in his extensive investigations were for the most part straight-grained. Nonetheless, he did observe sectors of the cambium in which directions of the anticlinal dividing walls deviated from the preferred orientation (Table 6.25). The proportion of nonconforming divisions was usually low, amounting to 10% or less in most species examined. Sometimes the sectors were small and the nonconforming divisions were of short duration. At other times, these divisions occurred over extensive areas and persisted through several annual rings. From time to time, the direction of cell orientation, or tilt, reversed from right to left and vice versa. Although the interval between reversals was gradual, examination of extensive data from many trees and species revealed an overall relationship with frequency of anticlinal division (Fig. 6.59). In general, the greater the frequency of anticlinal division, the shorter the interval between reversals in cell orientation.

Table 6.25. Extent of deviation from the preferred orientation of pseudotransverse divisions in various conifers[a] (After Bannan 1966b)

Species	Number of trees	Percentage of trees[b] 0-9	10-19	20-40	Indet.	Overall %[c]
Abies balsamea	15	73	20	7	0	6.5
Abies concolor	22	68	23	9	0	7.3
Cupressus macnabiana	13	38	38	24	0	12.4
Cupressus sargentii	19	37	37	16	10	11.9
Juniperus virginiana	12	59	33	8	0	10.3
Larix laricina	20	80	15	0	5	5.0
Picea engelmannii	35	51	29	20	0	10.8
Picea glauca	45	64	27	9	0	5.7
Pinus aristata	6	0	68	16	16	23.0
Pinus banksiana	27	59	26	11	4	11.1
Pinus contorta	46	22	30	44	4	20.8
Pinus edulis	9	56	44	0	0	9.3
Pinus flexilis	11	73	18	9	0	7.6
Pinus lambertiana	13	61	15	24	0	12.2
Pinus ponderosa	32	44	31	19	6	12.0
Pinus resinosa	29	55	24	21	0	10.0
Pinus strobus	69	70	25	5	0	7.6
Pseudotsuga menziesii						
Coastal	32	59	18	9	14	8.2
Interior	34	47	29	18	6	10.0
Thuja occidentalis	49	59	39	2	0	8.3
Tsuga canadensis	25	72	24	4	0	5.8
Sequoia sempervirens	12	58	37	8	0	7.6

[a] Measured in peripheral growth rings of trees 2-6 dm in diameter.

[b] Percentage of trees with pseudotransverse divisions deviating from the preferred orientation to the extent indicated.

[c] Overall percentage of nonconforming divisions.

6.6.2 Migrating Domains

The most detailed investigations of domain patterns have been conducted by scientists in Poland. In fact, Hejnowicz (1964) introduced the term domain to refer to the patterns observed in cambial sectors of several conifers brought about by changes in anticlinal divisions. A domain referred to a cambial region in which morphogenetic events were more or less uniform. The domain patterns were initiated in the cambium, they differed in size, and the patterns changed with time.

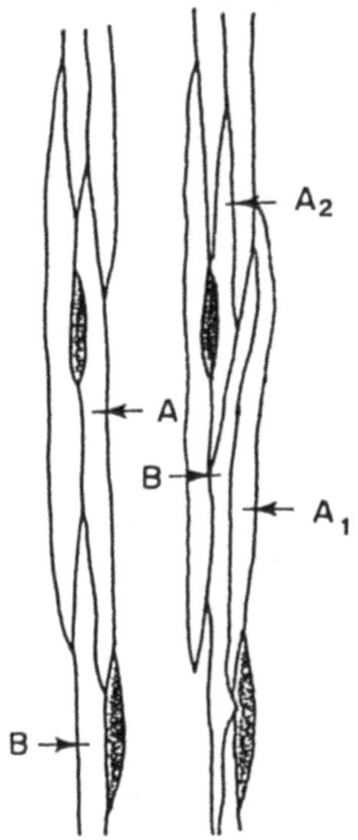

Fig. 6.60. Coordinated pseudotransverse division and intrusive elongation result in a change in the angle of inclination of a *Pinus sylvestris* tracheid. (Hejnowicz 1968)

The most conspicuous domain patterns were associated with changes in grain angle, i.e., the inclination of xylem elements relative to the axis of the stem. Grain inclination was conveniently designated as either S- or Z-type, with the middle stroke of each letter indicating the direction of inclination as observed on the outer surface of the wood or cambium; S- and Z-types corresponded to the usual left- and right-handed (chiral) designations, respectively. Thus, a domain was defined as a region of the cambium within which the anticlinal divisions of fusiform initials and related morphogenetic events were mostly oriented in one direction, either S or Z (Hejnowicz and Romberger 1973).

Hejnowicz (1968) investigated events taking place in the cambium that were related to changes in grain angle in stems of *Picea abies* and *Pinus sylvestris*. The rate at which alterations in grain inclination occurred varied widely in both species, with slope changing in some cases from 0° to 25° during the radial production of only 1 mm of xylem. Reversals in slope from S- to Z-type and vice versa were also common, although in a given sector either one or the other type dominated. Orientation of the anticlinal dividing wall in initial cells was closely associated with grain inclination. That is, partitions were Z-directed in sectors with Z-

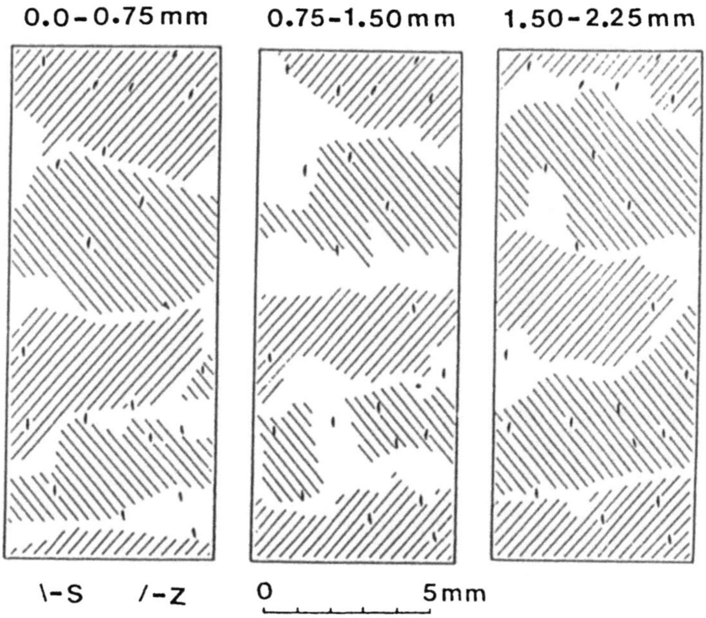

Fig. 6.61. Maps showing the domain patterns and areas of reorientation at successive time intervals in tangential sections of tumorous *Picea abies* xylem. Note the tendency for the small rays to resist reorientation. (Hejnowicz 1971)

type grain angles and reversals in orientation of the dividing wall were associated with changes in grain angle. Moreover, the higher the frequency of anticlinal divisions, the higher the rate of grain alteration. The rate of intrusive growth was also associated with both the frequency of anticlinal divisions and the rate of change in grain angle. As shown in Fig. 6.60, cell A divided by a Z-directed partition. Cell B then elongated intrusively along the oblique partition thereby splitting daughter cells A_1 and A_2 as well as preserving the Z-type orientation by its tip growth. Intrusive growth of cell A_1 was consequently forced to conform to this orientation. Cell loss of daughter cells following anticlinal division was also high, a factor favoring rapid intrusive elongation of surviving initials. Although frequent anticlinal divisions led to shorter cambial initials, the lower mean cell length was largely offset by a higher rate of cell loss and greater rate of intrusive growth. Reversals in the inclination of fusiform initials were preceded by changes in orientation of the anticlinal dividing walls, but no explanation could be offered for the latter change.

When cambia from specimens of tumorous *Picea abies* containing patches of wavy-grained xylem were examined (Hejnowicz 1971), they were found to differ from normal cambia by having higher frequencies of oblique anticlinal divisions and higher rates of intrusive growth of the sister fusiform initials. A particular orientation, whether S- or Z-type, migrated consistently upward and rather erratically

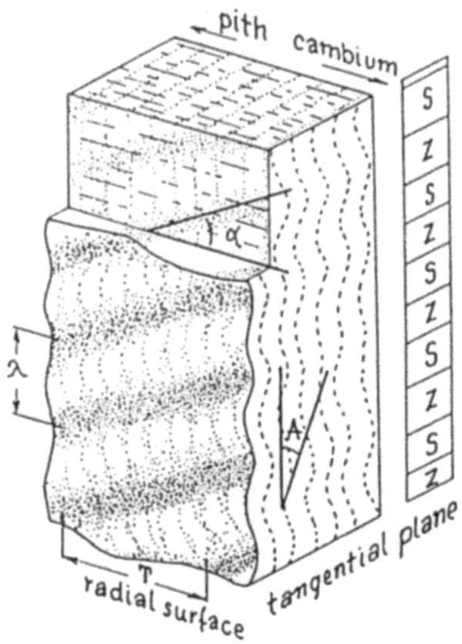

Fig. 6.62. Schematic of block of wavy-grained xylem with radial surface split and partially cut. *A* Amplitude of grain undulation or deviation from the longitudinal axis of the block; *lambda* length of the grain wave; *alpha* angle between undulations in the crest or trough and ray direction; *T* period of grain undulation. *S* and *Z* designations to the *right* indicate the domain pattern that existed in the cambium when the wood on the tangential surface was formed. (Hejnowicz 1990)

either to right or left (Fig. 6.61). A consequence of migration of the domain pattern was the occurrence of periodic reversals of the S- and Z-type orientations, producing what appeared to be a wave pattern of the grain. Slow migration of the domain pattern caused the wave pattern of the grain to be systematically displaced upward in successive layers of the xylem.

Similar domains were found to occur in dicotyledonous trees. Hejnowicz and Krawczyszyn (1969) found that the terminal layer of the late xylem (TLX) of the outer annual ring provided a good estimate of the morphogenetic events in the cambium. This approach enabled them to determine the patterns resulting from anticlinal divisions of fusiform initials and subsequent events through a large number of annual rings in eight different species. Anticlinal partitions were always oblique and a conformity existed between the orientation of the partition and the direction of intrusive growth. Moreover, the direction of inclination was not random. Rather, it occurred over relatively large surface areas of cambium, indicating a pattern of domains characterized by directional uniformity. The xylem fibers of the camphor tree (*Cinnamomum camphora*) was also found by Lo (1989) to conform to either the S- or Z-configuration, contributing to development of interlocked grain.

Gamma irradiation of *Acer rubrum* trees during a complete growing season temporarily suppressed, but did not destroy, the domain pattern (Romberger et al. 1980). Suppresssion was brought about by a several-fold increase in events associated with intrusive growth of fusiform initials. The local domain pattern was restored after radiation treatment ceased.

As evidence accumulated regarding the nature of domains together with the fact that domains migrated, it was soon realized that migrating domains might explain the grain patterns in trees. Wavy, interlocked, and spiral grain patterns were presumed to be manifestations of the same phenomena; they were similar qualitatively but differed quantitatively. A block of wavy-grained xylem is shown schematically in Fig. 6.62. Interlocked grain differed from wavy-grain by exhibiting an infinitely long grain wave (λ) and an angle (α) of about 90° (Hejnowicz 1974a, 1990).

6.6.2.1 Wavy Grain

Wavy grain is characterized by changes in grain direction approximately at right angles to the longitudinal axis of the tree. It frequently occurs at the root-stock, where large lateral roots depart from the main axis (Hartig 1901, Jaccard 1910a, 1915a), in the vicinity of large branches (Kienitz 1878, Lev-Yadun and Aloni 1991b) (Fig. 6.63), and also in the main stem of many species (Hartig 1901, Hejnowicz and Romberger 1973, Pyszynski 1972, 1977a). A review of the literature on wavy grain can be found in Schultze-Dewitz and Götze (1986).

"Wimmerwuchs," or wavy grain, was examined by Hartig (1901) where it occurred mainly in the vicinity of the rootstock in *Alnus*, *Fagus*, *Populus*, and *Quercus* but also throughout the main stem in old trees of *Acer* and *Castanea*. Because he could not relate wavy grain to external causes, Hartig tentatively attributed it to locally confined regions of bark folding due either to lateral roots departing at the stem base or to stem curvature. The bark folds, in turn, presumably exerted pressure on the cambium, thus disturbing cell division and distorting orientation of the derivatives.

A similar explanation was offered by Jaccard (1910a) for the "plisse" and by Jaccard (1915a) for the "Wellenholz," or wavy-grain, that occasionally was found in the rootstocks of *Picea abies* stems. Such growth is rare in conifers and Jaccard cautioned that it should not be confused with either compression wood, undulating grain, or "Haselwuchs" (Chap. 8.4.1.4). Like Hartig (1901), he found that it did not occur in young stems, but only near the bases of older trees. He attributed the wavy grain to passive, longitudinal pressure exerted on the cambium by weight of the stem and crown. This compression of the cambial cells presumably restricted their elongation growth with the result that the differentiating tracheids were forced to overlap by the process of sliding growth.

A quite different type of wavy grain was described by Kalberlah (1898) in *Tetrastigma scariosum*, a tropical liana. Here, the wavy grain was confined to the phloem fibers and it occurred in the radial plane. Kalberlah referred to this unique structure as "Verkoppelung," literally coupling or linkage of the fibers. He attri-

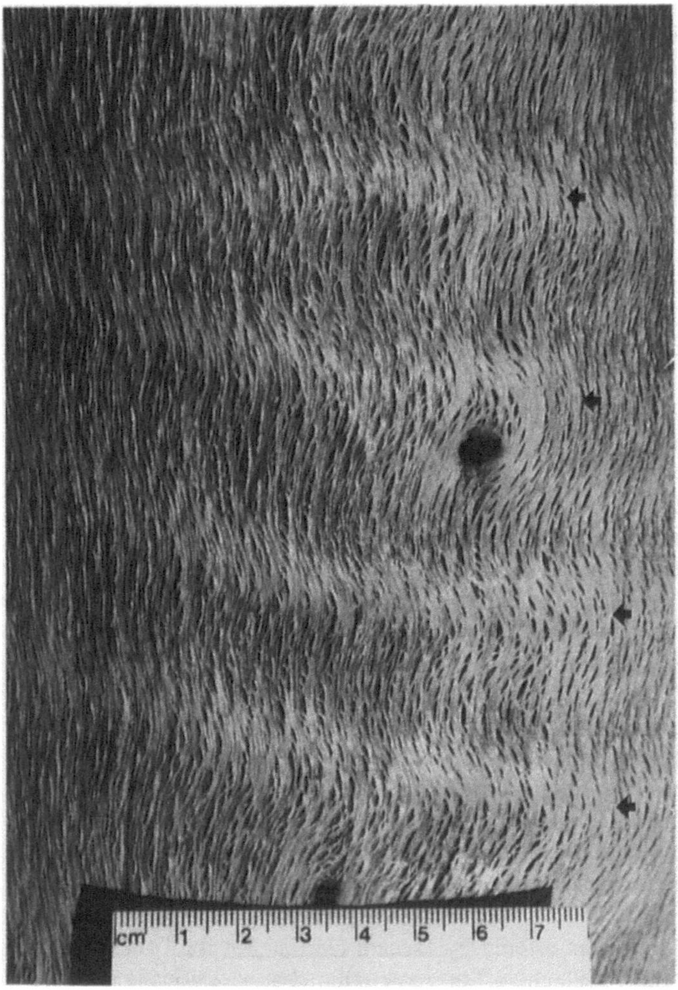

Fig. 6.63. Wavy grain in a branch junction of *Quercus ithaburensis* after bark removal. Note that xylem ridges (*arrows*) have a low density of aggregate rays where cambial activity was stronger. (Lev-Yadun and Aloni 1991b)

buted this structure to the presence of fibers of two different lengths, the longest of which became distorted radially because of space restrictions during their elongation.

Success in analyzing the wavy grain pattern in *Picea abies* (Hejnowicz 1971), prompted Hejnowicz and Romberger (1973) to examine wavy, or curly, grain in several dicotyledonous species. Species examined were *Acer platanoides*, *A. pseudoplatanus*, *Betula verrucosa*, and *Fraxinus excelsior*. Growth rings near the pith, the first 10 to 20 annual rings, were invariably straight-grained. Wavy grain developed gradually, but, once established, it persisted through all subsequently formed wood. As in previous studies, the terminal layer of late xylem in successive annual rings was selected as being representative of the patterns of cell orientation in the cambium. Such studies in *F. excelsior* showed that the orientation of fusiform initials within specified areas of cambium changed with time. These changes, contributing to either S- or Z-oriented events, included oblique anticlinal divisions of fusiform initials, overlapping of intrusively growing tips, oblique splitting of rays by intruding fusiform initials, and uniting of adjacent rays by elimination of the oblique fusiform initials between them. Again, these events were non-random and in any one particular area, or domain, strong preference for either an S- or Z-orientation of all phenomena prevailed.

Based on these data, the authors proposed that the cambia of trees generally consisted of a mosaic of S- and Z-oriented domains. The domains varied in size and shape and they migrated with time. Migration rates and directions could be measured with reference to rays (Fig. 6.61). The rays were reoriented more or less passively during grain reversals by rotating around their centers in response to the changing inclination of the fusiform initials, but they maintained their radial positions in the stem. The rays therefore served as reference points for observing orientational changes in the fusiform initials (Fig. 6.64), and in this way migration patterns of the domains could be plotted and analyzed. During migration the domain patterns moved; the cells remained in place. Grain patterns were determined jointly by the size and arrangement of domains, the frequency of morphogenetic events occurring within them, and the rate of domain migration. When the frequency of morphogenetic events was low, the nonrandom orientational tendencies of the domains would have little effect and the grain would remain straight. If, however, the frequency of events was high relative to the migration rate of the band-like domains, a wave pattern would develop. Thus, to produce wavy grain the domains must migrate upward relative to the rays. The upward-directed migration pattern was present in all dicotyledons examined, with one exception. In *Aesculus hippocastanum*, Pyszynski (1972, 1977a) found that the domain pattern progressed downward.

A different type of wavy grain was described by Bosshard and Stahel (1969) in *Populus robusta*. The waves appeared as undulations in the annual growth rings resembling the dips and undulations related to the occurrence of broad rays in other dicotyledons (Chap. 8.3.4.2). However, in this case, peaks and valleys of waves in the xylem corresponded exactly with similar structures in the bark. As to a causal mechanism, the authors entertained the possibility of differences in bark pressure and also differential rates of cambial activity. Cursory measurements showed no differences and they therefore concluded that the cause remained unknown. A theoretical consideration of wave propagation in the cambia of trees is included in the text by Dormer (1980).

6.6.2.2 Interlocked Grain

Nonstoried Cambia: Interlocked grain, or cross-grain (Schultz 1831), results from an accentuated domain pattern and it represents a modified form of wavy grain as well as a special type of spiral grain. Interlocked grain has usually been thought of as a condition in which the grain spirals in one direction for several years and then reverses direction to spiral oppositely (Haygreen and Bowyer 1982). This view,

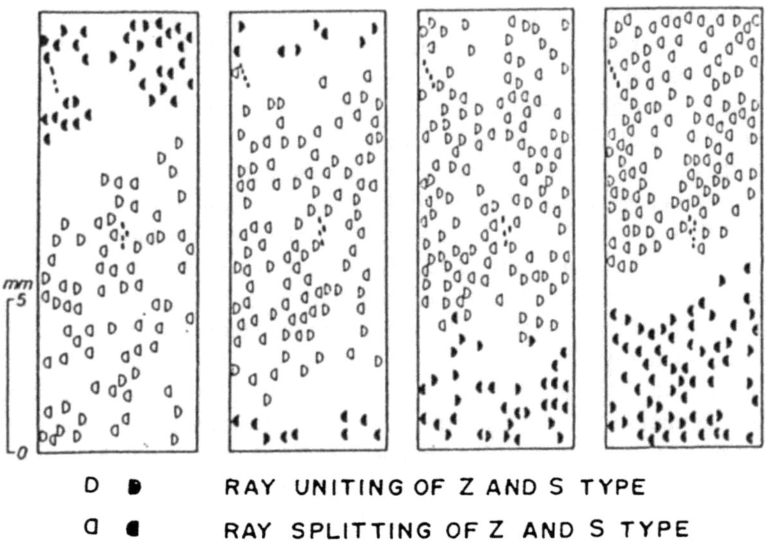

Fig. 6.64. Maps showing the splitting and uniting of rays during 4 successive years in the same tangential area of *Betula verrucosa* cambium. *Solid symbols* indicate S events, *open symbols* Z events. The domains migrated upward during new xylem formation. (Hejnowicz 1973)

implying that the entire growth sheath conforms uniformly, has been questioned (Krawczyszyn and Romberger 1980). Webb (1967), for example, found that interlocked grain in *Liquidambar styraciflua* resulted from "... repeated changes in both direction and angle of spiral among growth rings brought about by changes *within* a single ring with increasing height above the ground" (author's italics). Webb used orientation of the sclariform perforation plates of vessels to estimate orientation of the anticlinal divisions of fusiform initials. In trees with severe interlocked grain, the alternation between left- and right-handed perforation plates agreed, in general, with similar alternations in spiral grain. However, large deviations in perforation-plate orientation in straight-grained trees suggested that other unexplained factors were also involved.

Interlocked grain can also be interpreted in terms of migrating domains due to the relation between rate of movement of the domain pattern and domain height

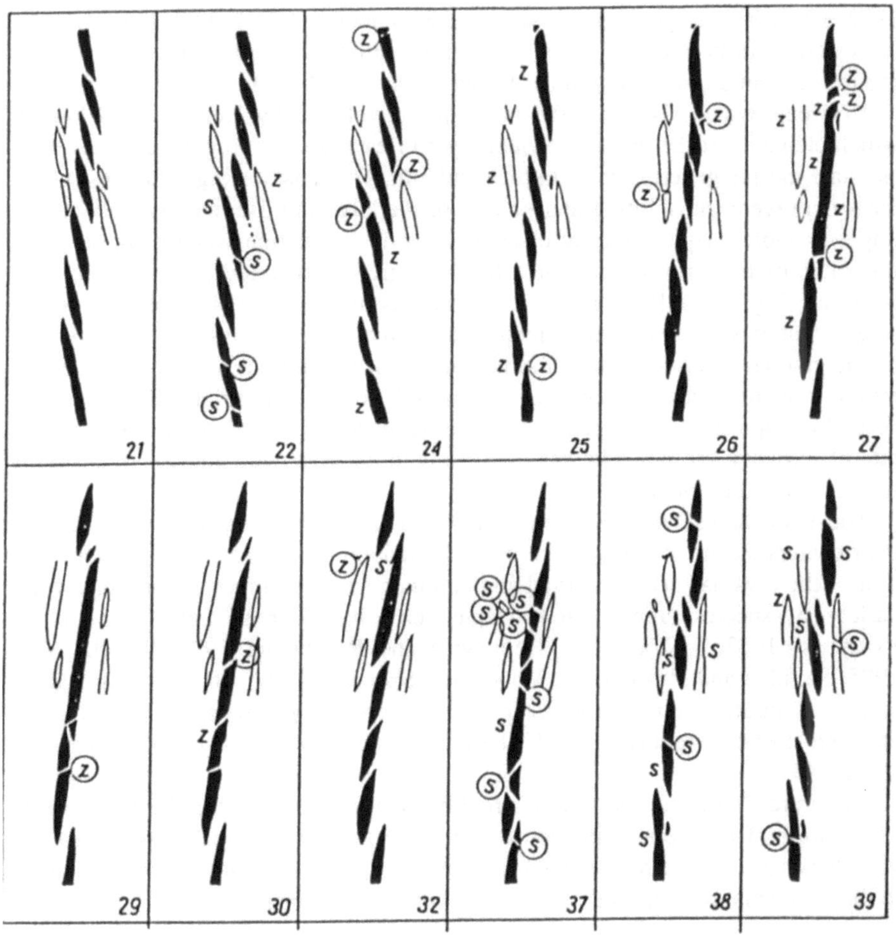

Fig. 6.65. Ray rotation during changes in grain inclination in *Platanus acerifolia. Circled letters Z* and *S* denote ray splitting to the right and left, respectively. *Uncircled letters Z* and *S* denote ray uniting to right and left, respectively. Note that the centers of rotating rays remain in the same position relative to the vertical axis. Each section (35 μm in thickness) is based on the terminal xylem layer of a successive annual growth ring. Total ray height is approximately 10 mm. (Krawczyzyn 1971)

(Wloch and Wawrzyniak 1990). The longer the domain pattern, or grain wave, the faster it moves. In interlocked grain, the wave length is long, the wave peaks are arranged almost vertically parallel to the stem axis, and grain inclination periodically changes direction in successive annual layers.

Grain patterns in *Platanus acerifolia* were investigated by Krawczyszyn (1971, 1972, 1973, 1977a) and Krawczyszyn and Romberger (1979). As in *Liquidambar*, the wood grain in Platanus tends to spiral, but it changes directions in successive groups of annual rings many times during radial growth of the tree.

Krawczyszyn (1971) used the technique developed by Hejnowicz (1964) in which elements of the terminal xylem layer of each growth ring were analyzed to estimate morphogenetic events in the cambium. The most distinguishing events in the xylem were the splitting and uniting of the large multiseriate rays. Ray splitting was more frequent during changes in grain inclination when daughter fusiform initials from recent anticlinal divisions grew intrusively between cells of a large ray (Fig. 6.65). Rays were split by strands of elongated elements and not by single elements as commonly encountered (Chap. 8.3.6). Rays were united both by the gradual withdrawal and elimination of fusiform initials and, less frequently, by the transformation of some fusiform initials to ray cell initials. Thus, the process of ray splitting was dependent on the frequency of anticlinal division and the rate of intrusive growth, whereas that of ray union was related to the intensity of fusiform initial elimination.

The rays themselves appeared to respond passively, simply rotating around their centers during the changes in grain inclination (Fig. 6.65). This passive rotation of the rays has been observed and commented on by investigators for a wide variety of species (Jones 1963, Hejnowicz 1968, Harris 1969, 1973, 1981, Pyszynski 1972, 1977a, 1990, Hejnowicz and Romberger 1973, Wloch 1976, 1985, Savidge and Farrar 1984, Wloch and Szendera 1989). In some cases, such as spirally girdled stems of *Pinus radiata*, the intrusively growing fusiform initials appeared to be "wrapped around" the ends of the more slowly reorienting rays (Harris 1973).

An interesting condition was reported by Bhat and Bhat (1983) in *Anacardium occidentale*. The fibers responded either more rapidly or more slowly than the vessels because, in the region of spirality reversal, the fibers were oriented parallel to the stem axis, whereas the vessels were oriented diagonally.

Storied Cambia: The morphogenetic events associated with interlocked grain in species with non-storied cambia occurred in domains in which intrusive growth of fusiform initials in one direction was correlated with the withdrawal and elimination of initials in the opposite direction (Krawczyszyn 1972). Similar domains are present in species with storied cambia. For example, the stratified arrangement of axial parenchyma in *Entadrophragma cylindrica* permitted Hejnowicz and Zagórska-Marek (1974) and Zagórska-Marek (1975) to follow cambial events during the formation of interlocked grain. The most significant event was the change in contact between the ends of fusiform initials in superposed stories. Cell ends of fusiform initials appeared to "creep" past the ends of cells in the adjoining story. The direction of creep was the same for all cells within a domain for a period of time and it then reversed (Fig. 6.66). Intrusive growth involved the formation of a new pointed tip and the disappearance of the previous tip. In many cases, a forked tip was produced when a new tip formed before the old one had completely withdrawn. Cells with forked tips comprised about 18% of the population. However, in 16.8% of the cells, forking occurred only at the upper end suggesting a strong polarity of intrusive growth.

The authors concluded that the general mechanisms for reorientation of initial cells were similar in both storied and nonstoried cambia. Although intrusive growth played an important role in both types of cambia, it was manifested in different ways. In nonstoried cambia, intrusive growth was believed to be long-lasting. Fusiform initials might continue elongating for many years, and they often did so by growing into spaces vacated by failing initials. In storied cambia, on the other hand, intrusive growth was confined to one edge of the cell tip for a certain period before reversing and forming a point on the opposite edge. In this case, length of the fusiform initial did not increase over time and for this reason there was no special need for the elimination of cells.

As demonstrated by Hejnowicz and Zagórska-Marek (1974), in species with storied cambia neither the xylem fibers, vessel members, nor the strand parenchyma showed evidence of elongation during their differentiation. Moreover, there was no evidence of long-term changes in element length radially in the stem. These facts were used by Krawczyszyn and Romberger (1979) to examine the intrusive growth process more closely in storied cambia. Their analyses revealed that the fusiform fibers of *Entandrophragma cylindrica* and *E. utile* were actually about 5% longer when they were oriented at maximal S- and Z-orientations than when axially oriented. The differences became apparent because intrusive growth in these species did not obliterate the effects of small differences in cambial cell length. Despite intrusive growth of the cell tips during grain reversals, overall fusiform initial length did not change because tip growth on one side was balanced by withdrawal on the other side.

Intermediate-Type Cambia: Some species exhibit cell patterns that can be interpreted as either storied or non-storied (Sect. 6.2.1). One such species with interlocked grain is *Tilia cordata* (Fig. 6.17). Although normally classified as a storied wood (Metcalfe and Chalk 1983), cambia from different trees of this species varied widely in the presence or absence of storied structure (Wloch and Bilczewska 1987). Individual trees might vary in the proportion of radial vs. oblique anticlinal divisions, the degree to which fusiform initials exhibited intrusive growth, and the size and arrangement of the rays, all of which influenced wood structure.

As demonstrated by Zagórska-Marek (1984), the cambia of *Tilia* trees can accommodate a certain proportion of pseudotransverse divisions and intrusive growth without significant loss of the storied pattern. During the last 17 years of growth in a 50-year-old tree, for example, about 80% of the anticlinal divisions were found to be pseudotransverse. Grain inclination cycled between the S- and Z-types. In the years when a change in grain occurred, either an S- or Z-type of division dominated. Intrusive growth of the daughter cells occurred in several different ways, and movement of cells between stories was not uncommon. Thus,

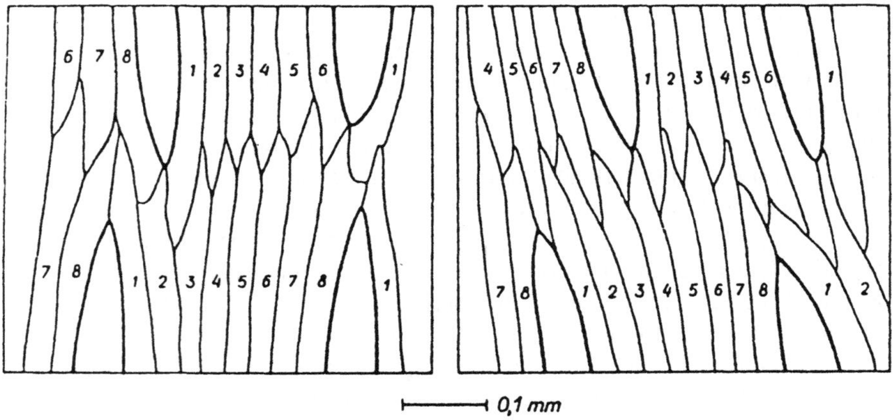

Fig. 6.66. The same tangential area of cambium at two stages of leftward cell reorientation in *Entandrophragma cylindrica*. The ***groups of initials*** are numbered to aid in identification. The oppositely directed ends of initials in the storys changed positions in relation to one another. Radial separation of sections, 1.2 mm. (Zagórska-Marek 1975)

Tilia possessed anticlinal division characteristics of both storied and nonstoried cambia. The rate of change in cell inclination was slow, averaging only 1°/mm of radial growth, and the storied pattern was maintained because the number of radial divisions exceeded pseudotransverse. However, in girdled stems of *Tilia* (Hejnowicz and Zagórska-Marek 1974), the rate of change was faster, averaging about 4°/mm of radial growth. In this case, the proportion of pseudotransverse divisions exceeded radial divisions and the storied pattern was lost.

Changes in cell inclination were brought about in *Tilia* in much the same way as previously described in *Entandrophragma*, i.e., by localized intrusive growth on the edge of one cell tip, and then reversing to the opposite edge. However, in *Tilia* intrusive growth was more variable and often extensive, leading to a high proportion of forked cells. For example, in Fig. 6.67 from Wloch and Zagórska-Marek (1982), the fusiform initial in radial row a grew intrusively on the left edge of its basal end while the initial in row b did the same on the right edge of its apical end. The net result was forking of both cell ends followed by a change in the direction of cell inclination. In one cambial area examined, 31% of the initials had at least one cell end forked. Forking of apical cell ends was more prevalent, exceeding that of the basal ends in a 1.5:1 ratio. Forking, however, was transient and it was eliminated in a relatively few periclinal divisions (Fig. 6.68). Thus, intrusive growth leading to forking was believed to occur either in the initial cell of a given row or in the mother cells arising from division of the initial cell.

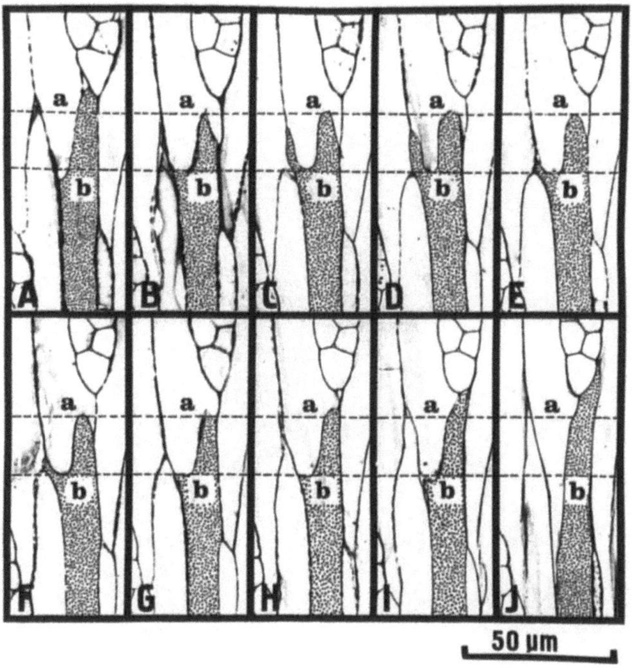

Fig. 6.67A-J. Series of tangential sections showing reorientation of two forked cell ends in two radial rows of cambium (*a,b*) in *Tilia cordata.* Sections 3 µm thick, every second section in series is shown. Lower cell row *shaded* for clarity. (Wloch and Zagórska-Marek 1982)

When forking and intrusive growth predominated in a given direction, either S or Z for a period of time, a change in grain alignment would follow. Further research on this problem by Wloch (1987), Wloch and Bilczewski (1987), and Wloch and Wawrzyniak (1990) led to the conclusion that intrusive growth was the most important factor contributing to cell inclination in *Tilia.* Cell ends that forked and underwent intrusive growth on a story border were called "active" ends (Fig. 6.69). During any one time period only one end of the fusiform initials, either the apical or basal, would be active within a small cambial region called a "microdomain." As the ends of initial cells in one microdomain ceased activity, those of inactive initials in adjacent microdomains were activated. This frequent change of domain pattern was referred to as "fibrillation," and the appearance and disappearance of the active cell ends in a cross-wise direction contributed to the migration of wavy and interlocked grain patterns. Mean wave length of the grain was 1 m. Cycles of grain orientation in different *Tilia* stems was variable; in some stems it was 6 years (Wloch and Zagórska-Marek 1982), in others 2 to 4 years (Wloch 1987, Wloch and Bilczewska 1987, Wloch and Wawrzyniak 1990), and in still others 2 to 14 years (Wloch 1985).

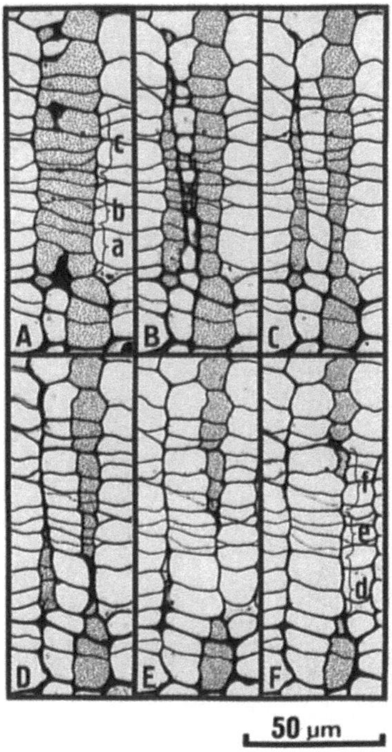

Fig. 6.68A-F. Series of transverse sections of *Tilia cordata* showing a row of cambial cells (*shaded*) the lower ends of which are forked. In **A** three groups of sister cells (*a,b,c*) with Z-type reconstruction of forked ends are indicated. In **F** the three groups of sister cells (*d,e,f*) are without forked ends. Sections 3 µm thick, every second section in series is shown. (Wloch and Zagórska-Marek 1982)

The size and arrangement of rays also differed in storied and nonstoried cambia of *T. cordata* and these features could in turn be related to changes in grain alignment. A nonstoried arrangement of rays was always found in the first ten or so annual rings from the pith (Wloch and Szendera 1989). These rays tended to develop in a nonstoried arrangment of large, multiseriate rays that exceeded the storied heights of fusiform initials. Repeated splitting and uniting often reduced the heights of these large rays to that of the stories, resulting in a storied ray arrangement (Fig.6.17). Prolonged splitting reduced some rays to a uniseriate condition, in which case it was unusual for both uniseriate and multiseriate rays to be present in the same tree or in the same region of the cambium. Some trees formed storied rays early, others not at all.

Ray splitting was due to intrusive growth of the fusiform initials inclined in either the S- or Z-configuration. Ray uniting occurred when the inclination reversed direction and passed through the vertical in much the same way as in *Platanus acerifolia* (Fig. 6.65). During ray splitting, the fusiform initials appeared to rotate

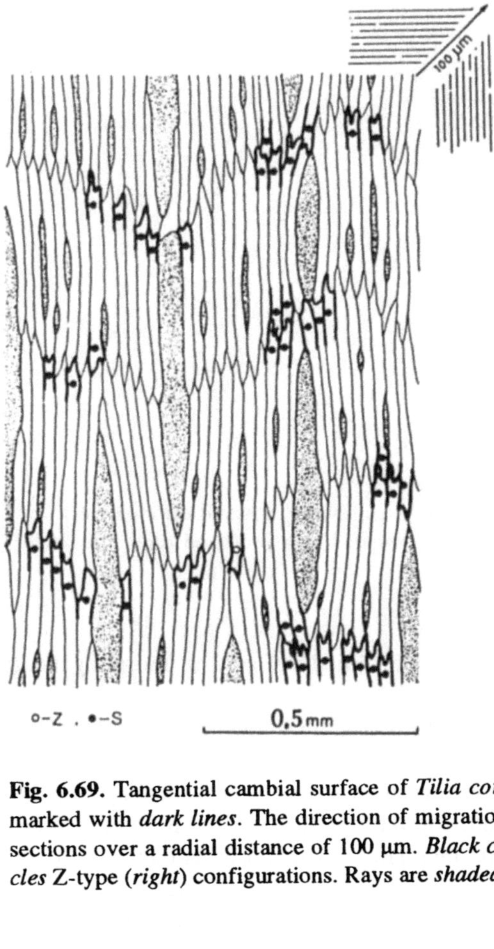

Fig. 6.69. Tangential cambial surface of *Tilia cordata.* Cell ends exhibiting migration are marked with *dark lines*. The direction of migration was reconstructed from 21 consecutive sections over a radial distance of 100 µm. *Black circles* indicate S-type (left) and *open circles* Z-type (*right*) configurations. Rays are *shaded*. (Wloch and Bilczewska 1987)

slightly around the ray centers (Wloch 1985). Repeated splitting and uniting of the large rays combined with a rapid restructuring of the cambium contributed to a high proportion of short rays (Wloch and Szendera 1989). The higher the proportion of short rays with small seriation, i.e., number of cells, the more pronounced the storied structure and the greater the inclination of the grain. Whereas large rays tended to resist a change in grain inclination, short rays simply rotated passively around their central axes.

6.6.2.3 Spiral Grain

Spiral grain occurs when the inclination of xylem elements persists in a preferred direction over the entire surface of a stem, or a large part thereof, for a prolonged period of time. In a sense, it represents the accentuated formation or exaggerated development of a domain pattern. The changes in xylem element inclination are

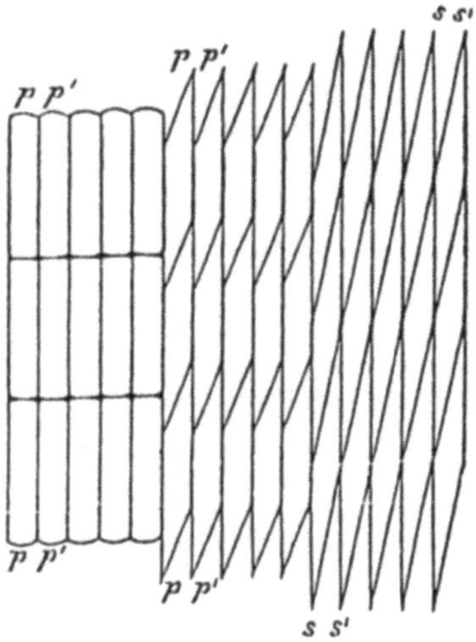

Fig. 6.70. Three stages in development of spiral grain. Neatly stacked cambial cells (*left*) first assumed pointed ends (*middle*) and then elongated in a preferred direction (*right*). (Braun 1854)

similar in both phenomena, and they can be directly related to morphogenetic events in the cambium.

Braun (1854) was perhaps first to relate spiral grain to events in the cambium. Citations in his review of the literature suggest that previous explanations were either teleological or at best suppositions based on simple observations. Braun examined a large number of species, both conifer and dicotyledon, primarily to determine the preferred angle of grain inclination. His observations led to the conclusion that "Drehung der Bäume," or the twisting of trees, was not a phenomenon of the stem itself but of the individual growth layers. He knew that wood cells, produced by the cambium, elongated during their development; but to do so, they must somehow avoid one another and they could only do this by acquiring pointed ends and pushing or sliding past one another. For example, in Fig. 6.70 cambial cells were assumed to be neatly stacked vertically (pp') with horizontal dividing walls. During the next stage of development, the cells acquired pointed ends but they were still vertically aligned. During the final stage, the elongated cells not only retained their vertical alignment but they also attained an oblique orientation (ss'). Spiral grain would therefore result if this deviation continued uniformly either to the right or left so that the cells eventually became inclined.

Although Braun did not specify that the first stage in Fig. 6.70 represented the cambium, one might assume that it did. He made measurements comparing the lengths of cambial and wood cells in many species. Moreover, he attempted to demonstrate that the shorter the cambial cell relative to its width, the greater the degree of inclination of its products. He was only partially successful in this attempt because of the extreme difficulty he experienced in obtaining suitable sections of the cambium. Braun's insistence that the "Drehungswinkel," or angle of fiber inclination relative to the vertical, increased with decreasing cell length was in conflict with a later conclusion of Sanio (1873). Sanio was of the opinion that sloping grain was the result of the increase in cambial cell length that occurred with age, or in the direction of radial growth.

Raatz (1892), without reference to the work of Braun (1854), assumed that the dividing wall of a cambial cell was at first horizontal but it became oblique as the daughter cell underwent sliding growth. These sloping walls were sometimes oriented preferentially to either right or left, and when this occurred during a period of radial growth the inclined fibers could impart a "torsion" to the stem. Raatz' results applied to *Pinus nigra* var. *nigra, P. sylvestris*, and *P. strobus*. Like Raatz, Hartig (1895c, 1901) also found that daughter cells of an anticlinal division elongated at the position of the dividing wall and "glided past one another." The ends of these new cells were literally shoved between neighboring cells, often in a preferential direction. Thus, Hartig confirmed the conclusions of Raatz. However, unlike Raatz, who believed that the dividing wall was at first horizontal, Hartig showed that it was oblique. This dividing wall could also be inclined to the right or left and it too could contribute to spiral growth. Büsgen and Münch (1929) referred to this condition as a "torsion of fibers."

At a later date, Klinken (1914) made the observation in *Taxus baccata* that 90% of the observed anticlinal divisions were inclined in the same direction. Although he speculated that this phenomenon might be due to some form of internal regulation, he did not relate it in any way to spiral growth. Neeff (1920) confined his remarks to a footnote in which he stated that the anatomical phenomenon of spiral growth depended in the narrowest sense on the arrangement of cambial cells. He suggested that the casual explanation might be found in the "extraordinary increase in horizontal divisions of cambial cells." Earlier Neeff (1914) had shown that tip growth of cambial cells was not a simple stretching but a polar characteristic of the cell involving sliding growth. He arrived at this conclusion by observing that new cells formed after decapitation of a stem apex were reoriented with their long axes directed toward an outgrowing lateral bud. Kohl (1933), apparently unaware of the previous work by Raatz (1892), also attempted to relate spiral grain to orientation of the anticlinal dividing walls. He expressed the opinion that the oblique dividing wall determined the pitch of the path that daughter cells would take during their elongation. Once established, this path would in some obscure way be perpetuated by subsequent cambial cells.

The most convenient and efficient method for examining grain inclination and tracheid realignment is the analysis of serial tangential sections. As described in Chapter 7.1.1, when certain conditions are met, tracheid lengths and orientations in the xylem closely approximate those of fusiform initials in the cambium of conifers. Using this approach, Jones (1963) examined 800 serial tangential sections cut at 20 μm to analyze a stem of *Pseudotsuga menziesii* in which grain inclination changed 90° in the outer 1.6 cm of xylem. Where the grain was

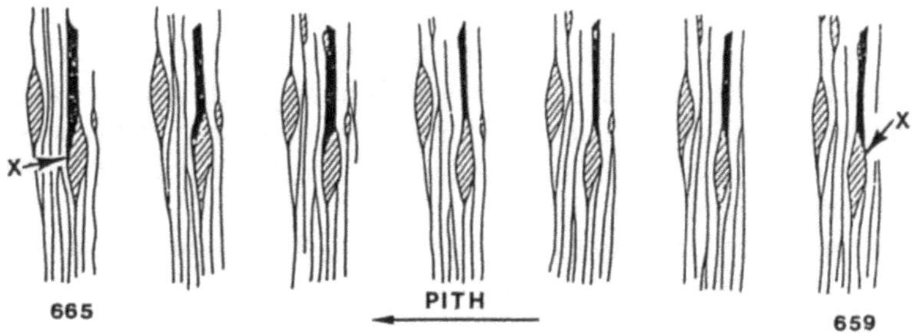

Fig. 6.71. Tangential sections of *Pseudotsuga menziesii* showing changes in orientation of a tracheid tip from the left-hand to the right-hand side of a ray. Serial sections *659-665* cut at 20 µm. 108x. (Jones 1963)

straight, left- and right-handed anticlinal divisions were about equal when based on orientation of the dividing wall. However, in the radial region of grain change, 64% of the anticlinal divisions were right-handed. Jones also observed a tendency for the tips of tracheids to shift from left- to right-handed sides of rays concurrent with the preponderance for right-handed divisions (Fig. 6.71). The redirected tracheid tips presumably accentuated the change in grain angle. During the change in grain angle, the rays gradually accommodated to the new orientation, but their response appeared to be passive and they simply rotated around their centers. A similar slow response in ray orientation was reported by Harris (1973) and Savidge and Farrar (1984).

The passivity of rays and their eventual realignment during spiral grain formation was examined in *Picea abies* and *Pinus sylvestris* by Pyszynski (1990). He hypothesized that wind action on tree crowns caused them to rotate and to induce torsional stresses in the stems. These stresses acting asymmetrically on the cambium at different times of the year and positions within the stem resulted in regions of either S- or Z-oriented types of cells. According to this hypothesis, the rays were "pulled" along by torsion of the rotating wood column. Although the grain spiral reflected the torsional effects, the direction of grain orientation was ultimately determined by events occurring in the cambium.

Referring back to Jones (1963), his data suggested that events leading to reorientation of the fusiform initials following anticlinal divisions might be responsible for initiating changes in grain angle. If these events continued in the same direction over the entire bole of the tree, or an extensive area thereof, they could contribute to the formation of spiral grain. Other workers arrived at similar conclusions (Trendelenburg and Mayer-Wegelin 1955, Hejnowicz 1961, 1964, 1968, Bannan 1963a, 1964a,b, 1966a,b, Hejnowicz and Romberger 1979, Harris 1981).

An alternative interpretation was offered by Savidge and Farrar (1984) based on the reorientation of the xylem in diagonal bridges formed by partial girdles in stems of *Pinus contorta*. They found that repeated periclinal divisions resulted in shortening of all fusiform initials in regions they called microdomains. Cell populations within these microdomains, which they attributed to "imperfect periclinal divisions," presumably changed orientations abruptly whereas those within larger regions occupied by other fusiform initials changed orientations less rapidly. The authors did not believe that shortening of fusiform initial length was, in itself, sufficient cause for reorientation but neither was subsequent tip elongtion. They could find no strong evidence for invoking preferential cell tip orientation in spiral grain development. Other workers, however, have found evidence for the redirection of xylem elements by intrusive growth in girdled stems (Harris 1969, 1973, Kirschner et al. 1971, Zagórska-Marek and Little 1986).

Other explanations have been offered to account for the initiation and perpetuation of spiral grain in trees (Kubler 1991), but the foregoing offered by Jones and others appears most plausible at this time. The literature on spiral grain is extensive and most of it is well beyond the limits of this treatise. For a review of the terminology relating to spiral grain the paper by Meyer (1949) is suggested; for a well-documented review of the entire subject, the interested reader should consult Harris (1989).

Some plants, such as tropical woody lianas, exhibit a form of spiral grain as a consequence of normal growth. Particularly vivid examples appeared in Wigand (1856). One unidentified specimen of Malphighiaceae from Brazil was characterized by an extremely irregular upward spiraling xylem that appeared deeply furrowed in transverse section. In this stem the fibers were sharply inclined upward and for short distances they actually reversed directions. In another nonclimbing woody species from Brazil, simply labeled "Caffeestamm," regions of strongly inclined fibers overlay the older, straight-fibered stem. No information was provided as to the origin of this stem. Hartig (1862) also described how a "spiral body" such as a *Lonicera* vine caused the new growth to assume the direction of the spiral. Similar reorientations of the xylem fibers along spiral constrictions caused by *Celastrus scandens* and *Vitis* spp. were described by Lutz (1943). Spiral ligations have also induced sloping grain, such as those applied to *Ulmus campestris* by de Vries (1876) and *Pinus radiata* by Harris (1969).

6.6.2.4 Whirled Grain

The wave patterns that are found in stumps (Fig. 6.56) and in tumorous woods (Chap. 10.5.1) occur in complicated arrangements with the cells twisting and turning in often unpredictable ways. The first xylem derivatives to differentiate in wound-induced callus often occur in a whirled arrangement (Bangerter 1984). They presumably assume this shape because of rapid intrusive growth of the first-formed fusiform initials that develop from callus mother cells.

Wloch (1976), when investigating tumorous wood of *Picea abies*, referred to these areas as "whirling." As a rule, whirling involved continuous changes in tracheid arrangement in successive wood layers, indicating that analogous changes took place in the cambium. Unlike wavy grain, where changes in grain angle were more or less gradual, changes in cell orientation in whirled areas occurred with great rapidity. For this reason, it was difficult to establish domain patterns.

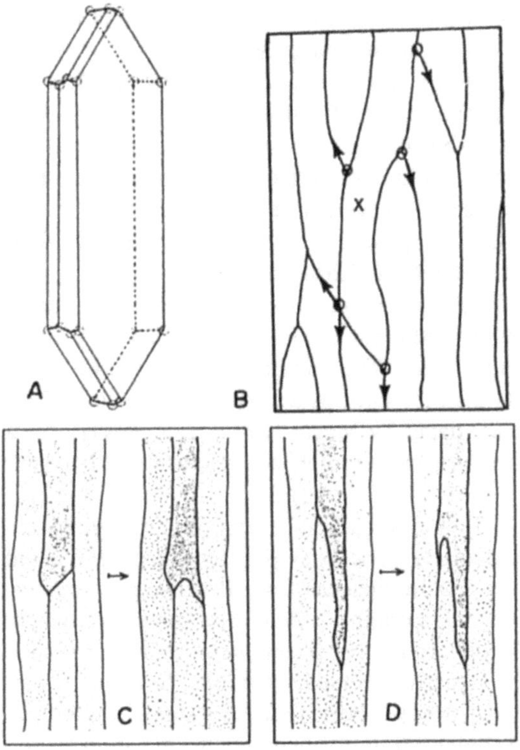

Fig. 6.72A-D. Forking of cambial initials in *Picea abies*. **A** Diagram of cambial cell illustrating various edges and corners (*circles*). **B-D** Tangential sections. **B** Sites of eventual intrusive growth indicated by *circles*, and directions of growth of radial edges by *arrows*. **C-D** Diagrams of cell forking due to intrusive growth at a lateral edge close to a apical edge (**C**) and at a greater distance from a apical edge (**D**). (Wloch 1976)

Intrusive growth was considered the basic mechanism for cell reorientation in tumorous wood. Forking, which occurred in the cell tips of fusiform initials, was common when the initial cells undergoing excessive intrusive growth encountered an obstacle such as another cell tip or a ray. Intrusive growth might occur at any edge of the initial cell (Fig. 6.72A), but usually close to a cell corner (Fig. 6.72B). If forking occurred at a radial edge, the growing end would intrude between the radial walls of cells belonging to two adjacent stories. If, however, intrusive growth occurred at a lateral edge (Fig. 6.72C-D), then the growing end might intrude between the tangential walls of other cells, creating localized "whirls." Cells at the middle of the whirled region were severely distorted with some cell tips bending 180° (Fig. 6.73). Many of these forked and bent cells were visible only in the tangential plane as they were short-lived and rapidly eliminated from the cambium by periclinal divisions. Interestingly, the slope of the rays remained unchan-

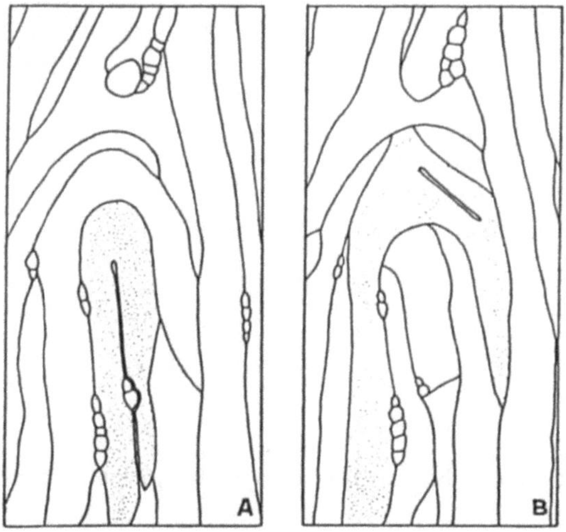

Fig. 6.73A,B. Tracheids from a whirled area of *Picea abies* xylem in tangential section. **A** Both cell ends growing in same direction with bent ends in contact. **B** A closed cavity within a distorted cell end. (Wloch 1976)

ged even in regions of severely disturbed cell arrangement. Apparently the rays, which respond passively to alterations in cell orientation, were unable to follow the rapidly changing patterns in the tumorous wood.

6.6.2.5 Circular Vessels and Tracheids

Circular vessels and tracheids represent special types of whirled grain patterns. Sachs and Cohen (1982) induced circular vessel formation in *Pisum sativum* and *Phaseolus vulgaris* by wounding. These circular patterns were induced in stem cambia above transverse wounds where the normal transport channels to the roots had been disrupted. The authors suggested that when disruption occurred, the original polar auxin transport underwent reversals and the resulting inductive signals therefore created unique patterns in the cambial cells.

Although it had been assumed that circular vessels rarely occurred in uninjured, intact plants, Hejnowicz and Kurczynska (1987) discovered them above axillary buds in eight different dicotyledonous species (*Acer platanus, A. platanoides, Betula verrucosa, Fraxinus excelsior, Populus nigra, Sorbus aucuparia,* and *Tilia cordata*) (Fig. 6.74). The circular vessel members were qualitatively the same as those of normal vessels but shorter due to the shorter cambial initials from which they were derived. The authors suggested that these vessels occurred in vortices resulting from circular polarity in the cambial region above the axillary bud.

This line of research was continued by Lev-Yadun and Aloni (1990) with a number of coniferous and dicotyledonous species. In all species examined, they observed pronounced changes in vascular organization at the zone of bifurcation

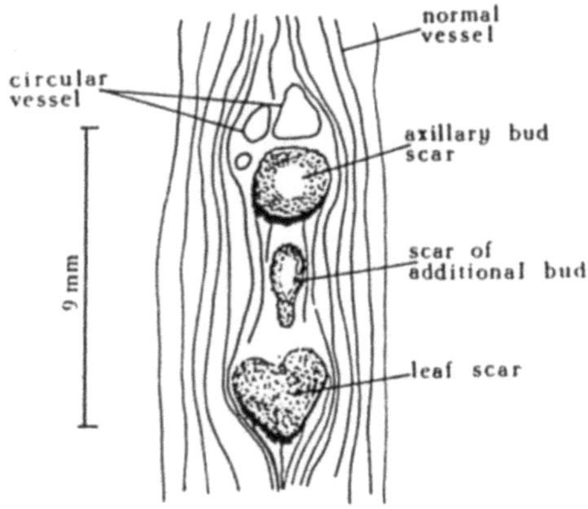

Fig. 6.74. Typical location of circular vessels above axillary buds in dicotyledonous trees. (Hejnowicz and Kurczynska 1987)

of two or more branches (Fig. 6.75A). Both tracheids and vessels differentiated in spiral patterns that continuously increased with enlarging stem girth. Circular cell orientations arose in the contact zone between two opposing vascular patterns (Fig. 6.75B-C). To account for these circular patterns in branch junctions of woody species, the authors adopted a modified version of the polar auxin transport theory proposed by Sachs and Cohen (1982).

Kurczynska and Hejnowicz (1991) also relied on the hypothesis of Sachs and Cohen (1982) for their interpretation of circular vessels. Extensive investigations by Sachs (1984) had shown that vascular differentiation involved the polar flow of a signal from the leaves towards the roots. The influence of leaves on vessel differentiation, in turn, could be replaced by a localized source of auxin. Blockage of polar auxin transport by incisions in the stem caused diversion and sometimes a complete reversal of signal polarity and consequently of vessel differentiation.

Kurczynska and Hejnowicz (1991) induced circular vessels in dormant terminal segments, without buds, of *Fraxinus excelsior* by applications of auxin in vitro. The auxin-treated segments resumed cambial activity and produced new xylem near the basal ends of the internodes. Circular vessels differentiated near the cambial initial layer but always in isolation from the contemporaneously formed longitudinal vessels. The circular vessels were usually short, indicating a higher frequency of anticlinal divisions in the specific region of the cambium in which they arose. The prevailing form assumed by the circular vessels was a pair of cells connected by two perforations. This circular configuration, according to the authors, could only be explained on the basis of a circular flux of a signal for vessel differention, presumably auxin, acting on the cambial derivatives.

Circular xylem elements have also been found in many other situations such as branch junctions (Lev-Yadun and Aloni 1991b), root grafts (Neeff 1922), stem grafts (Küster 1899), living stumps (Neeff 1914), mechanically induced wounds

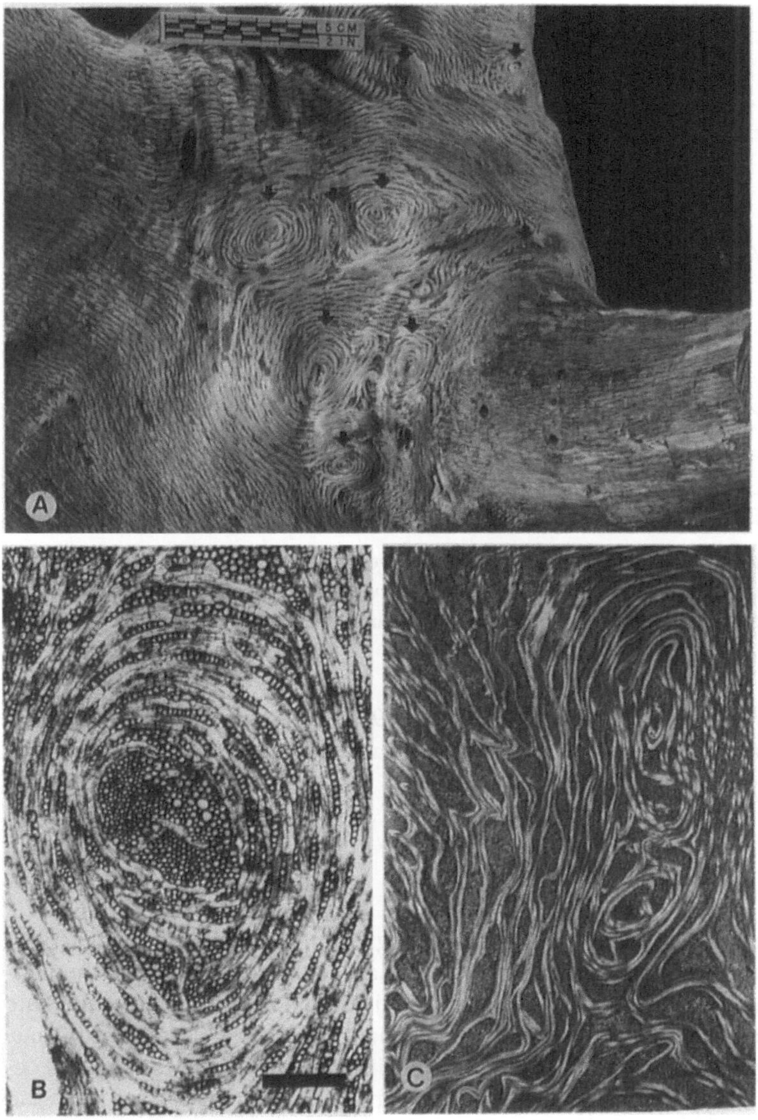

Fig. 6.75A-C. Circular vessels at branch junctions. **A** Tissue orientation of late-formed xylem at junction between the trunk and two lateral branches in *Quercus ithaburensis* exposed by bark removal. Circular vessels *arrows*. **B** Photomicrograph in tangential view of circular xylem elements in a branch junction of the same species showing small uniseriate rays surrounding a large round ray. *Bar* 250 μm. **C** Curved patterns of narrow vessels in branch junction of *Platanus acerifolia*. 8.8x. (**A,C** Lev-Yadun and Aloni 1990; **B** 1991b)

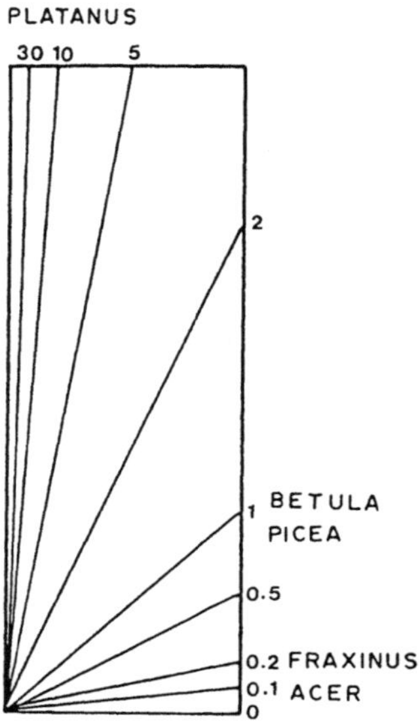

Fig. 6.76. Relation between velocity of domain movement in the cambium and the inclination of stripes on the radial surface of species with wavy (*Acer* to *Betula*) and interlocked (*Platanus*) grains. The *numbers* correspond to domain movement in mm/mm of xylem produced. (After Krawczyszyn 1972)

(Aloni and Wolf 1984, Lev-Yadun and Aloni 1991c), and insect-produced wounds (Liphschitz and Mendel 1987). Many other occurrences were cited by Lev-Yadun and Aloni (1990).

6.6.3 Wave Pattern of Domains

As mentioned earlier, domains are localized regions of the cambium characterized by the non-random orientation of fusiform initials and the occurrence of morphogenetic events. Within a domain the cambial initials and their derivatives oscillate between the S- and Z-type orientations. Domain boundaries change with time, consequently the sizes and shapes of the domains also change. These changes occur in such a way that the domain pattern appears to migrate upward, or occasionally downward, even though the cells themselves remain in place. It is this upward or downward movement of the domain pattern that contributes to grain patterns in wood. These patterns have been interpreted in terms of complex wave phenomena, and they have been described both mathematically and theoretically (Hejnowicz 1973, 1974a,b, 1975, Pyszynski 1977b, 1980, 1990, Hejnowicz and Romberger 1979, Krawczyszyn and Romberger 1980, Zagórska-Marek and Hejnowicz 1980). For a more detailed explanation of these wave phenomena, the interested reader should consult the excellent review of Harris (1989) and the more

Table 6.26. Domain patterns in trees with different grain types (Krawczyszyn 1972)

Grain type in xylem	Genera	Domain height mm	Velocity of domain movement mm/mm of xylem produced
Curly	*Acer* *Fraxinus*	2-8	ca. 0.1 (Hejnowicz and Romberger 1972)
Wavy	*Betula* *Picea*	ca. 10	ca. 1.0 (Hejnowicz 1971)
Interlocked	*Platanus*	80-280	6.5-12 (Krawczyszyn 1972)

recent interpretation of Hejnowicz (1990). Our concern here is not with the theory but how these events relate to the cambium.

The three principal grain patterns - wavy, interlocked, and spiral - appear to be modifications of a common pattern and they differ from each other quantitatively rather than qualitatively. When the frequency of morphogenetic events is low and nonrandom, straight or normal grain is produced, and when the frequency of morphogenetic events is high, that is, many events occur in the cambium during one season, the domain pattern is manifested as a wavy or serpentine grain pattern in the wood (Hejnowicz 1975). Thus the frequency with which these morphogenetic events occur determines the velocity of domain migration and the height of the domain pattern. Domain height, in turn, determines certain attributes of the cambial morphogenetic wave perceived as grain changes in the wood (Table 6.26). A low domain, or a short wavelength, contributes to wavy grain whereas a high domain, or a long wavelength, contributes to interlocked grain. In addition, slow migration between domains results in an almost horizontal arrangement, whereas rapid movement results in an almost vertical one. The relation between slope and velocity of domain movement for species with wavy grain (*Acer* to *Betula*) compared to species with interlocked grain (*Platanus*) is shown in Fig. 6.76. Since the differences between the two grain types is believed to be quantitative, analyses of many species would most likely show one type grading into the other (Hejnowicz and Romberger 1979).

In some species, it has been possible to relate grain migration patterns to domain height, or wave length, and to length of the period between changes. In general, the higher the domain, the faster the domain migration pattern. The shortest recorded domain heights in cambia producing wavy grain, in *Fraxinus* and *Acer*, were only a few millimeters (Hejnowicz and Romberger 1973), and the tallest in *Entandrophragma*, exceeded 1 m (Zagórska-Marek and Hejnowicz 1980). Domain heights in *Platanus* averaged 8 to 28 cm (Krawczyszyn 1972) and in *Aesculus* about 25 cm (Pyszynski 1977a). In *Tilia*, the domain patterns were more variable (Wloch and Szendera 1989). Although it was possible to derive migration patterns for individual trees, it was unlikely that these patterns would apply to other trees of the species.

6.6.4 Theoretical Interpretations

Numerous investigations have provided both circumstantial and direct evidence of hormonal involvement in the alignment of cambial fusiform initials and consequent grain patterns (Sachs 1986, Aloni 1987a, Harris 1989, Lev-Yadun and Aloni 1990, 1991b). Zajaczkowski et al. (1983, 1984) and Wodzicki et al. (1987) proposed a model of hormonal regulation of cambial activity. They reviewed evidence from many experiments suggesting that auxin transport was predominantly polar and that the basipetal transport channels in tree stems were mainly in the cambium and phloem. On the basis of these experiments, they proposed a concept of wave phenomena in which the propagation of auxin wave fronts induced patterns of both cambial cell division and differentiation. Thus, a two-dimensional wave front that hypothetically influenced the cambial initial layer would be modified in complex ways to form a three-dimensional wave front that influenced development of the differentiating elements. However, because of the many factors affecting the postulated wave fronts that were not yet understood, the authors were unable to relate the auxin waves to domains of cell orientation. In a short abstract, Zajaczkowski and Dodd (1987) attempted to explain many of these morphogenetic processes by viewing the cambium as a supra-cellular system regulated by relative values of potential energy. Obviously, these theoretical models require more rigorous experimental data.

A study by Harris (1969) suggested that the flow of metabolites, including hormones, might be preferentially directed to induce changes in grain angle. For example, he suggested that in young stems of *Pinus radiata* the spiral flow of metabolites might conceivably be initiated by the helical phyllotactic arrangement of needle traces serving the short shoots. In later work, Harris (1973) attempted to demonstrate that preferential orientation of the anticlinal dividing wall was determined during mitosis. He envisaged a skewing of the mitotic spindle from the cell axis brought about by stresses in the cytoplasm. This effect in turn was believed to determine the direction of chromosomal separation at anaphase and subsequent orientation of the new cell wall.

Another study with direct pertinence to the present topic was conducted by Zagórska-Marek and Little (1986) on *Abies balsamea*. They investigated fusiform initial orientation in the helical bridge that remained after stems were girdled. The first responses to girdling were an increase in the number of S- and Z-oriented anticlinal divisions without any change in their ratio and a decrease in fusiform initial length. As these events increased in frequency, the anticlinal dividing walls and intrusively growing ends of the daughter initials both became reoriented in the direction that would eventually result in the fusiform initials being aligned parallel to the bridge angle.

Applications of indole-3-acetic acid (IAA) to the upper edge of a helical bridge inhibited fusiform initial reorientation, mainly by decreasing the frequency of oriented intrusive growth events. Exogenous IAA also decreased the percentage of fusiform initials reorienting more than 17° as well as the total amount of reorientation (Table 6.27). These responses were, in turn, associated with an increase in the frequency of anticlinal divisions and a decrease in fusiform initial lengths. Thus, exogenous applications of IAA tended to maintain the fusiform initials in a pre-girdling condition, i.e., in the direction of flow of endogenous IAA before girdling. The authors concluded that their data supported the hypothesis that fusiform initials become oriented parallel to the direction of IAA transport. However, in a related study, Kurczynska (1986) was unable to detect any intrusive growth of fusiform initials in *Fraxinus excelsior* internodes treated with IAA.

In one of his more recent papers, Hejnowicz (1990) differentiated between cyclic and noncyclic events in the cambium. Cyclic events included oblique anticlinal divisions, intrusive growth of fusiform initials and their consequences, and fusion of rays. A non-cyclic event would be the twisting of periclinal partitions, i.e., when the partition is not flat but distorted at its ends in the manner of a screw propeller. Cyclic events conformed to the domain type whereas noncyclic events did not.

To explain phase shifts, Hejnowicz proposed that the cambium functions as a system of coupled, chiral oscillators. Coupling would affect the phase and period of the oscillations. High amplitude, manifested in a high frequency of events, would favor coupling and bring about a phase gradient. Steepness of the gradient, with reference to a traveling wave, would determine the length of the chirality wave but not its period. Strong coupling would increase duration of the oscillation cycle, helping to explain why traveling chirality waves have long periods. On this

Table 6.27. Effect of exogenous IAA and position across the bridge on fusiform initial reorientation in *Abies balsamea* trees[a] (After Zagórska-Marek and Little 1986)

	Fusiform initial reorientation in the S direction			Fusiform initial length (μm)	
Treatment	Bridge position	Percent tracheids[b] oriented > 17°	Amount degrees	Before girdling	After girdling
Control	Top	78	45	1087	483
	Middle	86	45	1064	440
	Bottom	78	45	1000	432
IAA	Top	18	20	951	451
	Middle	43	38	1029	591
	Bottom	0	<17	978	525

[a] IAA (indole-3-acetic acid) applied to 45° S-oriented bridges on trees grown for 15 weeks in favorable controlled-environmental conditions.

[b] Radial width of xylem in which the tracheid long axis was oriented more than 17° from the stem long axis x 100/total radial width of xylem.

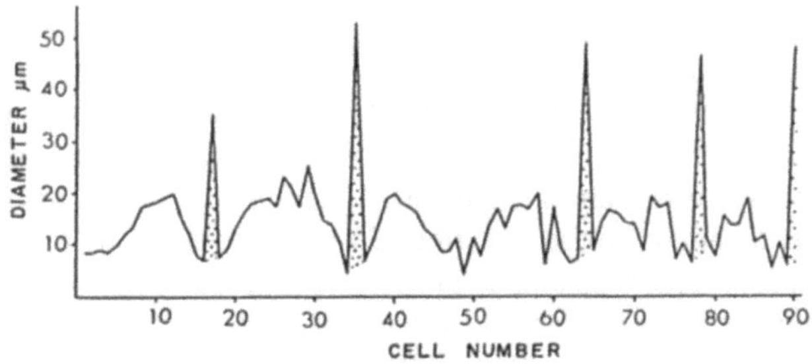

Fig. 6.77. Radial diameter changes of the derivative cells of a cambial mother cell of *Acer pseudoplatanus* over time. The *stippled peaks* indicate individual vessels, *area in between* are primarily fibers. (Stieber 1985)

basis, the normal cambium producing straight-grained wood and characterized by a low frequency of events would be interpreted as being in a state of weak coupling between chiral oscillations. On the contrary, a cambium producing wavy grain would be in a state of strong coupling.

Stieber (1985) proposed a wave theory of cambial activity directed more to cell organization than to cell orientation, although parallels do exist. He was concerned with the fact that cells in adjacent radial files often differed in size and/or structure in regular wave-like patterns. For example, in a sample of *Acer pseudoplatanus*, the vessels occurred in a more or less zig-zag formation. When cells derived from a single cambial mother cell were followed along a radial file, the vessel groups appeared at fairly regular intervals dispersed among fiber cells (Fig. 6.77). Adjacent radial files displayed a somewhat similar pattern but the cells were displaced radially to produce the zig-zag vessel pattern. Thus, adjacent mother cells in the cambium might be performing different functions, one producing a vessel, another a fiber, or fibers of different sizes. Stieber referred to these regions of similar tissue production as "function areas" designated spatially either as f or F. He concluded that a definite relationship existed between vessel and fiber-forming factors, and that the F-movements were only possible if stimulators which determine the nature of cambial production moved tangentially across the cambium. According to this "rhythmic profactor theory," the characteristics of a cell derivative were determined by "factors" which in turn required that two profactors of rhymical nature must occur coincidentally in appropriate phase. The profactors were assumed to move past one another with different wave frequencies and speeds. By projecting these profactors in the form of concentric rings, Stieber constructed models that presumably could account for all possible patterns of wood structure.

In a more recent series of papers, Aloni (1987a,b 1991) proposed several hypotheses to account for the role of auxin both in xylogenesis and xylem differentiation. These hypotheses, while not specifically addressing the wave phenomena in trees, clearly demonstrate the important function of auxin and other hormones in the regulation of cambial activity.

6.7 Mechanism of Cell Elongation

In previous sections it has been demonstrated that elongation of the sister fusiform initials following anticlinal division contributes significantly to both tangential increase of the cambium and to various patterns of organization in the cambium and its derivatives. Much research, as well as speculation, has been undertaken with the objective of elucidating the mechanism of cambial cell elongation. Three main hypotheses have evolved over time; namely, sliding growth, symplastic growth, and interpositional or intrusive growth.

6.7.1 Sliding Growth

Krabbe (1886) is often credited with introducing the concept of "gleitende Wachsthum," translanted into English as gliding or more frequently sliding growth (Scott 1888) and into French as "croissance transgressive" (Jaccard 1910a,b). Although Krabbe was by no means the first to describe the process, he was most certainly the first to examine it in depth and thereby develop it as a hypothesis for cell elongation. Previous workers had also observed that cambial cells elongated following division, and they often described this process as either the pushing, or sliding, of one cell past another in elements of both the developing wood (Trécul 1853a, Braun 1854, Nägeli 1858, Radlkofer 1858, Sanio 1863b, 1872, 1873, Frank 1864, Winkler 1872, Velten 1875, Vöchting 1875, de Vries 1876, De Bary 1877, Russow 1883b, von Höhnel 1884a) and bast (Braun 1854, Schacht 1856, Radlkofer 1858, Sanio 1863b, De Bary 1877, von Höhnel 1884a,b). Thus, Trécul (1853a) concluded that the "utricles," his cambial cells, formed a radial series that were at first intimately united by their walls. Later, these cells elongated at their extremities in such a way that they became interspersed between neighboring cells situated above and below them. Similarly, Braun (1854) described how the new wood and bast cells "avoided" one another during elongation by their "more or less pointed ends pushing between one another," and von Höhnel (1884b) described how the chisel-shaped ends of cells in storied cambia sought the radial spaces between cells because these walls were weakest. Klinken (1914), at a later date, criticized Krabbe (1886) for more or less ignoring earlier work on the subject, particularly that of Sanio (1872).

An interesting account of the cell elongation process, directly related to the present topic, was that of Radlkofer (1858) in his studies of *Cocculus* and *Cissampelos*. He observed that cells of the rind expanded more or less uniformly in three dimensions and those of the wood elongated vertically eight or more times after their origin from "rind parenchyma." The new wood cells were produced after internodal elongation had ceased; thus, cell stretching was no longer a factor. He then asked how do these long cambial cells arise from the short rind parenchyma cells? He suggested that: "A way appears possible for the formation of a new cambium; the direct transformation of rind cells, or in the transformation of newly formed cells to cambial cells by the predominant growth in length together with simultaneous growth of their pointed ends and suppression of their sides - perhaps even complete suppression and resorption of cells lying in their way." The significance of Radlkofer's description resided in his insistence that elongation of cambial cells was confined to their ends when growing past neighboring cells and that this growth was at the expense of neighboring cells, which often became smaller. Nägeli (1858) agreed completely with Radlkofer's observations but disagreed as to the process. Whereas Radlkofer viewed cell elongation as a more or less passive process, Nägeli considered it to be an active one. That is, in describing how cambial cells attained the "spindelform" of sapwood cells, he visualized that the cell ends actively "vorbeischieben," or push past one another; but unlike Radlkofer, Nägeli did not believe that the cell tips occupied a void created by failing cells.

Years later, Sachs (1882), in a somewhat related way, differentiated between apical and intercalary growth. Apical growth occurred when surface growth attained a maximum at any one spot so that this part of the cell wall projected as a point. Intercalary growth occurred when deposition of new substance was localized in a zone of the cell wall. As this zone extended, new cell wall material was intercalated by degrees within the existing wall by a process of "intussusception."

Krabbe (1886) was of the opinion that differentiation of wood and bast elements resulted in volume increases requiring sliding growth. That is, growth processes were accompanied by mutual displacement of certain cells or groups of cells. As an example, he assumed a stem with a radius of 50 mm and a circumference of 314 mm. A 1 mm increase in radius would require a 6 mm increase in circumference. Assuming further that 1 mm of circumference was equivalent to 30 cell files, then each cell file must increase tangentially 0.0006 mm during 1 mm of radial growth. According to Krabbe, this increase was too large to ignore and the change could only be explained by sliding growth. Thus, sliding growth together with cell division was necessary for growth of a tissue and its change in form. He first considered vessel development, noting that in trees such as *Quercus* and *Fraxinus* the large vessels were often surrounded by 30 or more cells. After tentatively ruling out turgor pressure, resorption of cell wall parts, and several other potential processes, he concluded that the diameter increase of such large vessels could only be accommodated in this milieu of neighboring cells by some form of sliding growth. He observed that many of the cells bordering a large vessel were inserted between neighboring cells by growth of their tips, although some new cells arose by periclinal divisions.

Krabbe next addressed the question of what caused the sapwood cells in one radial file to appear between those in adjacent files as observed on transverse sections of stems. He concluded that these cells were nothing more than the severed ends of cambial cells, i.e., enlarging cells in the cambial zone that were inserted between other cells by tip growth (Fig. 6.78A). Consequently, these rows of small cells did not arise from new cell divisions but were the result of sliding growth. He next asked how these cells found room for elongation growth. In response, he noted that elongation of cambial cells, just like the expansion of vessels, required the accommodative shifting and form variation of neighboring cells. This elongation was facilitated by the pointed ends of cambial cells that served as wedges (Fig. 6.78B). As soon as a cambial cell began to elongate, its tips became more pointed, forcing form changes in both the cambial cell and its immediate neighbors.

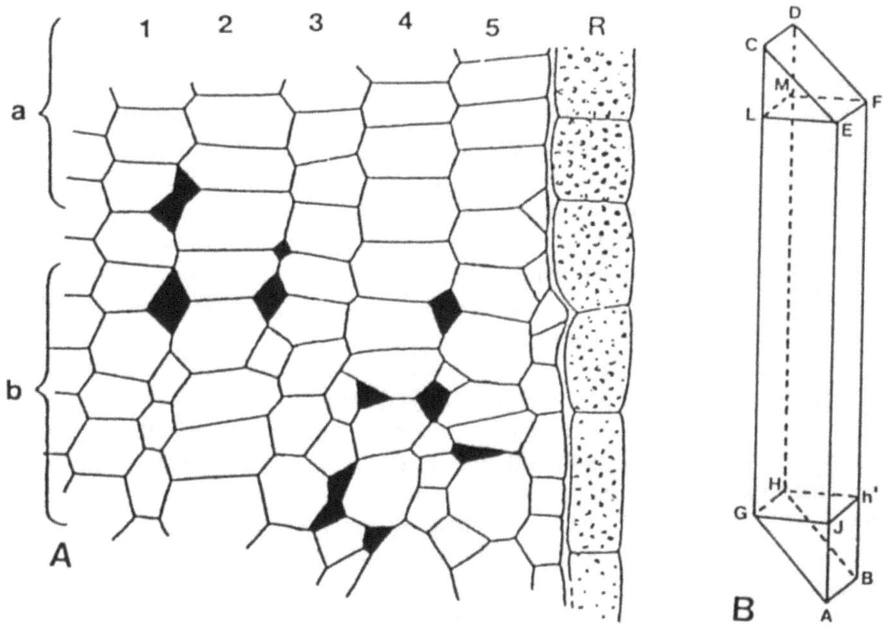

Fig. 6.78. A Transverse section of *Fraxinus excelsior* extending through the cambium (*a*) and developing xylem (*b*). *Shaded* cells are the severed tips of fibers that have elongated by sliding growth and appear to be inserted between cells in neighboring files. **B** Shape of a cambial cell presumably showing how the upper cell end elongated from the plane *LMEF* to form a wedge-shaped tip *CD* and a similar tip *AB* at the lower end (After Krabbe 1886)

According to Krabbe, a vessel wall expanded by sliding over the walls of adjacent libriform fibers which were no longer growing. He then extrapolated this sliding growth process not only to elongating libriform and bast fibers but to all elements of the thickening ring in general. Nordhausen (1898), working under the guidance of Schwendener, questioned both the process of and the necessity for sliding growth. Using Krabbe's illustrations, he attempted to show that neighboring cells accommodated to vessel expansion by supplemental divisions, as Krabbe himself had shown (Fig. 4.2B). However, Nordhausen claimed there was no need for sliding growth, and in his own investigations of vessel expansion in

Aesculus and *Populus* he could find no evidence for it. Ray cells were also believed to expand without the intervention of sliding growth. Ray cells of *Populus* and other woody species were judged to be rectangular in the cambium but to assume the form of a parallelogram during expansion. This change in form was not due to sliding growth but to division and expansion of neighboring libriform fibers that literally dragged the ray cells along. In so doing, the tangential walls of ray cells were transformed from a vertical to an oblique orientation. As another example, Nordhausen referred to the "Langstäbe," or trabeculae, frequently observed in many woods (Chap. 4.5.2). He claimed that if sliding growth had occurred, these radial series of trabeculae would be broken, but he had never observed such disruptions.

Raatz (1892), while not addressing the problem of sliding growth directly, alluded to it in defense of his concept of a "Wendepunkt" (Chap. 4.1.3). He observed that adjacent radial files did not function in perfect harmony. That is, in order to avoid shifting of daughter cell walls, one radial file must remain relatively inactive while the neighboring file was active. This requirement resulted from the fact that cells in each radial file were stretched following tangential division, first to the wood and then to the bast side. Therefore, according to Raatz, each radial file must have its own "Wendepunkt," or turning point.

The concept of sliding growth, as visualized by Krabbe (1886), assumed that sliding growth occurred over much of the cell wall. In fact, Krabbe had emphatically stated that protoplasmic continuity could not exist among cells involved in the process of sliding growth, although he assumed that continuity could be reestablished after completion of cell elongation. It was perhaps for this reason that Jost (1901) expressed "a certain distrust" of the "indirect reasoning for sliding growth" of cambial cells, even though he accepted it as the only plausible explanation. Haberlandt (1914) also pointed out this weakness, adding that the concept might be clarified if it could be shown that the protoplasmic connections at the cell tip were of secondary origin. Such a possibility had been suggested earlier by Kienitz-Gerloff (1891).

Krabbe's insistence that the entire wall surface was subjected to sliding growth was surprising in view of the reports of both earlier workers and his contemporaries. Radlkofer (1858), Nathanshon (1898), and others had demonstrated that elongation growth of cambial cells and their derivatives was confined to the cell tips. Moreover, von Höhnel (1884b,c) had described how the small ends of cambial cells arose by "tip growth" and how the chisel-like ends of these cells "seek" the spaces between neighboring cells. Localization of elongation growth at the ends of cambial cells was later demonstrated by Mischke (1890), Nathanshon (1898), Strasburger (1901), Grossenbacher (1914), Klinken (1914), Neeff (1914), Bailey (1920d, 1923), Beijer (1927), Kerr and Bailey (1934) and Sinnott and Bloch (1939). Neeff (1914) concluded that during the process of elongation the middle lamella between adjacent cell walls was digested, thus permitting the cell tips to intrude by physical separation of the walls. Ziegenspeck (1931) attempted to define the process histologically. When he applied certain histological stains to microsections, such as phloroglucinol-HCl, neither the cambium nor the "zone of sliding growth" produced a color reaction. The differentiating xylem, however, produced a weak red color.

The general concept whereby cambial cells and their immediate derivatives elongate by sliding growth of their cell tips was accepted by most investigators for many years (Mischke 1890, Kienitz-Gerloff 1891, Raatz 1892, Hartig 1895b, Jost 1901, Jaccard 1910b, 1915a, Grossenbacher 1914, Haberlandt 1914, Neeff 1914, 1920, 1922, Bailey 1920b,d, 1923, Klinken 1914, Kleinmann 1923, Schüepp 1926, Beijer 1927, Büsgen and Münch 1929, Kerr and Bailey 1934, Record 1934, Lewis 1935, Sprengel 1936, Huber and Schmidt 1938). As evidence in support of sliding growth, Huber and Schmidt (1938) demonstrated how the elongating fibers of *Swietenia mahagoni* were bent when they encountered a ray (Fig. 6.87D). Schüepp (1966) maintained that cambial cells were especially suited for detecting sliding growth because the derivatives occurred in a fixed radial series. Moreover, sliding growth allowed a cell to express its "individuality" and to strive for its "own form" within a given tissue. Yet Dodd (1948) had earlier criticized the concept because all previous workers had failed to adequately take into account the changes in cell shape.

6.7.2 Symplastic Growth

Until about 1930, the concept of sliding growth had never been critically examined, primarily because the process appeared obvious to most investigators and no other plausible alternative had been proposed. As pointed out by Priestley (1930a,b), sliding growth in a complex of living tissue had never been actually observed, but only inferred from the successive stages of cell arrangement on microsections. One of Priestley's main objections was the fact that he could not reconcile the distribution of plasmodesmata and pits with a process of elongation whereby cells or parts of cells slipped past one another. He assumed that once plasmodesmata were ruptured, the connections were unlikely to be reestablished. Similar objections had been raised earlier by MacDaniels (1923) and were later repeated by Majumdar (1942).

Priestley (1930b) offered an alternative hypothesis that he called symplastic growth in which the adjustments in cell shape and position presumably occurred without the necessity for sliding growth. He first attempted to show how "cambial elements" derived from the apex elongated during internodal extension by assuming that these cells were simply "... pulled out and elongated by the continued elongation of the parenchyma in which they were embedded." He then described how such a process might also take place during the elongation growth of cambial cells and their derivatives. During this process, the two adjacent cell walls and the intervening middle lamella presumably behaved as a single unit, or a common framework. Priestley defined symplastic growth as the "... mutual adjustment of cell position (that) takes place between semifluid protoplasts, separated from one another by plastic walls, which move as a common framework under tension and without any slip between the wall surfaces of neighboring cells." He suggested that this mode of growth was characteristic of all meristematic tissues, whether they formed an apical growing point or an intercalary cambium. In a later paper (Priestley and Scott 1939), he again denied the existence of sliding growth in cambial cells.

The concept of symplastic growth was greatly expanded by Meeuse (1942) in an extremely detailed survey and analysis of the growth processes of meristematic cells and their derivatives. He critically evaluated Krabbe's (1886) concept of sliding growth and also Sinnott and Bloch's (1939) concept of intrusive growth. Meeuse's conception of sliding growth is illustrated diagrammatically in Fig. 6.79A. During the differentiation of cambial cells (1) to wood fibers (2), he assumed that the radial walls would either lengthen or shorten as indicated by the ar-

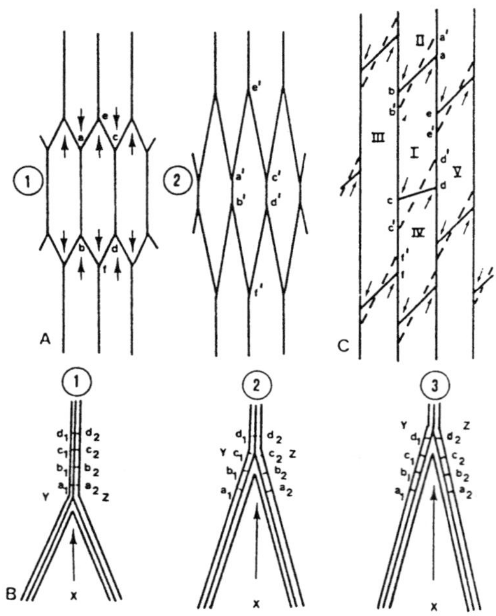

Fig. 6.79A-C. Meeuse's interpretation of Krabbe's (1886) theory of longitudinal sliding growth (**A**), Sinnott and Bloch's (1939) theory of intrusive growth (**B**), and Priestley's (1930b) theory of symplastic growth (**C**). **A** Cambial cells in longitudinal section (*1*) and the same cells after differentiation to xylem fibers and elongation in direction of arrows (*2*). **B** Three successive stages in which cell *X* intrudes between cells *Y* and *Z* by splitting the common middle lamella. **C** Diagram of fusiform initials in tangential view. *Solid lines* indicate walls at an early stage of development. *Dashed lines* show elongation of the initials in direction of the *arrows*. (After Meeuse 1942)

rows. Thus, radial walls ab and cd (1) would shorten to a'b' and c'd', respectively (2), while walls ae, ce, bf and df would elongate. According to Krabbe's descriptions, these changes should involve sliding growth.

Meeuse's interpretation of intrusive growth is presented diagrammatically in Fig. 6.79B. In this growth process, the intruding tip of elongating cell X presumably separated the walls of cells Y and Z, thus destroying the protoplasmic connection at points a_1-a_2 through d_1-d_2 between these two cells. Both of the aforementioned concepts were rejected by Meeuse because he assumed that the required growth process would either disrupt or destroy primary pit fields, plasmodesmata, crassulae, and trabeculae. He claimed that no evidence for the splitting of primary pit fields in two corresponding halves had ever been demonstrated.

Meeuse concluded his evaluation by stating that: "All changes in intercellular relationships among the cells of the cambial zone must therefore be explained without assuming sliding or intrusive growth. These changes come about very gradually." As an alternative, he invoked Priestley's (1930b) concept of symplastic growth. His interpretation of this process is illustrated in Fig. 6.79C in which cell abcd is a new cambial initial derived from an anticlinal division along partition wall cd. Subsequent elongation growth occurred in the direction of the arrows with elongation progressing either upward or downward from the nonprime to the prime letters.

It is not possible to present or even to briefly summarize the numerous detailed analyses offered by Meeuse. Most of his arguments were negatively directed toward the concepts of sliding and intrusive growth, and he offered little in the way of positive evidence as to why the concept of symplastic growth should be accepted.

The concept of symplastic growth gained only limited acceptance following Priestley's (1930b) publication. Majumdar (1941) considered it a possible explanation for the width growth of differentiating vessels and elongation growth of the middle part of fibers that underwent cell adjustments but limited elongation. Wardrop (1949) simply stated that after anticlinal division, sister-cell elongation might possibly involve symplastic growth. Whitmore (1962) listed symplastic growth as one factor that relieved tangential stresses in the expanding inner bark. However, Clowes and Juniper (1968) stated that dicotyledons with storied cambia required only symplastic growth for wall adjustments following multiplicative divisions by radial walls.

6.7.3 Intrusive and Interpositional Growth

All previous attempts to describe elongation growth of cambial initial cells and their derivatives were based on indirect methods of measurement because no means had been found to observe the process directly. Sinnott and Bloch (1939) devised a technique whereby the growth of young, meristematic root cells of grasses (*Phleum pratense*) could be directly observed under high magnification. This material was ideal for the purpose because cells in the surface layer of the roots divided by anticlinal walls resulting in rectangular cells arranged in regular longitudinal rows. To determine the nature of the growth process, it was necessary to identify specific points on the cell wall and to measure the distance between them at successive periods of growth. All evidence obtained by this method of analysis indicated that cells did not slide along one another but were held so firmly together that points in adjacent cell walls which were originally opposite each other were still opposite after the cells had greatly elongated. This type of growth was referred to as "differential wall growth," and the authors suggested that it was the method by which intercellular adjustments, due to unequal cell enlargement, were brought about in root and similar meristematic tissues.

Other types of differential wall growth were also recognized. For example, previous workers had shown that elongation of a cambial cell was sharply localized at its ends with essentially no elongation elsewhere in the wall. For this type of localized growth Sinnott and Bloch (1941) proposed the term "intrusive growth," a type of differential growth that occurred in specialized meristematic cells such as the cambium and its derivatives (Fig. 6.80). They considered Priestley's (1930b) symplastic growth to be somewhat analogous to intrusive growth. However, according to their interpretation, Priestley confined adjustments in the cell wall to early and late stages of cell growth and not to the entire period of cell development.

Following a review of the three foregoing cell elongation processes, sliding, symplastic, and intrusive, Majumdar (1941) claimed that none of the authors made a distinction between elongation of cambial initials and mother cells. This claim was, of course, in error because each of the authors had considered growth of cells derived from mother cells. In dicotyledons with stratified cambia, where mother cell elongation was assumed to be limited, Majumdar suggested that cell adjustment was undoubtedly due to symplastic growth as visualized by Priestley. On the contrary, in dicotyledons displaying nonstratified cambia, where cell elongation might be many times that of the cambial initial cell, he suggested that intrusive growth took place at both ends of the fiber mother cell. He found no evidence to support the concept of sliding growth.

The denial of active cell tip elongation according to the concept of symplastic growth was questioned by Schoch-Bodmer (1945), based on her observations of the growth of pollen tubes. To confirm the presence of tip growth, she examined developing bast fibers in *Linum perenne*. The fiber tips were immature, thin-walled, and rich in plasma, while the fiber middle had developed relatively thick walls. These fiber characteristics were applicable to both the sliding and symplastic growth theories. In keeping with the former theory, the elongating cell tips would be simply inserted between the walls of neighboring cells, whereas in the latter theory the cell tips would elongate by mutual accommodation in the absence of slip between cell walls. In contrast, Schoch-Bodmer assumed that the tip dissolved the middle lamella of neighboring cells, split their walls, and without slid-

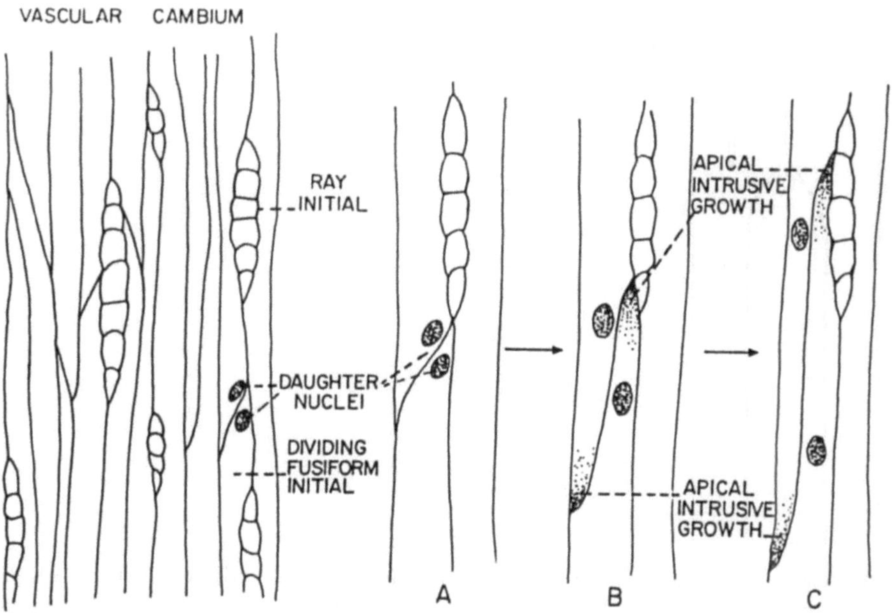

Fig. 6.80A-C. Tangential view of cambium showing pseudotransverse division of a fusiform initial (**A**) followed by subsequent elongation of daughter initials by apical intrusive growth (**B**,**C**). (Brown 1970)

ing, grew into the interstices. New wall material was then deposited by the neighboring cells by intussusception (Fig. 6.81). Thus, "interpositional growth," as she called this process, was considered synonymous with intrusive growth (Schoch-Bodmer and Huber 1949a). However, the authors (Schoch-Bodmer and Huber 1951) were careful to point out that tip growth was not necessarily synonymous with either interpositional growth or intrusive growth. Whereas the expression tip growth usually referred to growth of the outermost end of a cell, the latter two expressions also inferred the splitting of middle lamellae as well as tip growth of neighboring cells in unison. Roelofsen (1959) agreed completely with the view of Schoch-Bodmer and Huber by stating that "present knowledge of cell-wall structure and growth agrees very well" with the concept of intrusive growth.

As further evidence in favor of interpositional growth, Schoch-Bodmer and Huber (1948, 1949a,b) considered the tip growth of forked cells (Sect. 6.7.5), which could not be adequately explained by either sliding or symplastic growth. In fact, they concluded that neither sliding nor symplastic growth could lead to

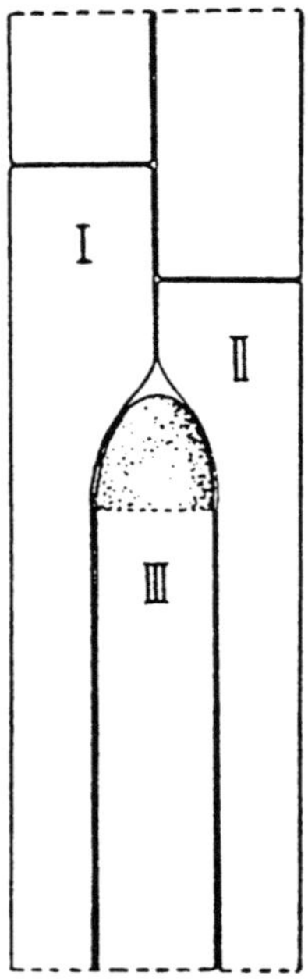

Fig. 6.81. Schematic diagram of interpositional (intrusive) growth in which cell *III* elongates by localized apical growth (*shaded*), splits the middle lamella of neighboring cells *I* and *II*, thus changing the shapes of these cells. The membrane of *III*, formed by local apical growth, coalesces with walls of cells *I* and *II*. (Schoch-Bodmer and Huber 1952)

actual forking, but simply to the formation of funnel-like structures (Schoch-Bodmer and Huber 1952). They showed that during elongation of a normal cambial cell after anticlinal division, additional tip growth occurred by interpositional growth. However, if an elongating cell encountered an obstacle, the direction of growth was either changed or suppressed. When tip growth was suppressed by an obstacle, interpositional growth again occurred, but the cell tip was either deflected around the intruding cell, molded to conform to that cell, or bifurcated to produce a forked cell. When growth was suppressed, the cells did not alternate tangentially as in normal cambia, but were actually opposed by being directly over one another. Consequently, it was necessary that the forked cell split the middle lamellae of two neighboring cells to complete its interpositional growth. In such cases, Schnepf (1986) postulated that the plasma membrane might function as a sensor that recognized variations in pressure and tension of the cell apex.

Interpositional growth has been equated with intrusive growth as a process involved in the elongation of cambial cells and their derivatives. It should be pointed out, however, that some of the early studies were conducted on the bast fibers of *Boehmeria* (Aldaba 1927) and *Linum* (Esau 1942, 1943, Schoch-Bodmer and Huber 1951) and these fibers are derivatives of protophloem. Nonetheless, their elongation growth is accomplished in essentially the same way as that of a cambial derivative as Schoch-Bodmer and Huber (1948, 1949a.b) demonstrated in *Sparmannia africana*. Schoch-Bodmer and Huber (1951) admitted the possibility of limited symplastic growth in the cell middle and sliding growth at the cell ends during the extensive elongation process of bast fibers, but they maintained that only interpositional growth could account for the majority of the elongation. They listed eight characteristics favoring localized interpositional growth that occurred at both ends of these fibers.

In contrast to protophloem fibers derived from procambium that often elongate to great lengths, the phloem elements derived from the cambia of many species elongate slightly if at all (Bannan 1955, Evert 1960, Cheadle and Esau 1964); Radlkofer (1858) commented on this fact in his studies of *Cocculus laurifolius*. Other authors, however, arrived at different conclusions (Ghouse and Yunus 1975, Khan and Khan 1983). Ghouse and Sabir (1974), for example, claimed that phloem fibers of *Erythrina indica* and *Pongamia glabra* elongated eight to nine times that of the fusiform initials. According to the authors, these phloem fibers elongated by intrusive growth after completing their symplastic growth phase. A similar conclusion was reached with species of *Acacia* and *Prosopsis* (Ghouse and Iqbal 1977b, 1979). The extensive literature on this subject has been reviewed by Esau (1969), who interpreted all cell elongation in terms of intrusive growth.

6.7.4 Objections to Intrusive Growth

Critics of intrusive growth have based their objections on three main points: (1) lack of evidence for active intrusive tip growth; (2) the splitting of bordered pits; and (3) the presumed rupture of protoplasmic connections. Each of these objections has been overcome by subsequent workers.

6.7.4.1 Intrusive Tip Growth

References to tip growth of cambial initials and their derivatives abound in the literature, but in many cases such growth either is simply assumed to exist or is based on indirect evidence. One of the most common sources of indirect evidence is the interpenetration of the tips of fusiform initials and their immediate derivatives between cells of adjacent files (Fig. 6.78). Direct evidence indicating that elongation growth of fusiform cambial initials and their immediate derivatives is confined to the cell tips comes from a variety of sources.

Whether the middle lamellae and parent cell walls of cambial cells are ruptured, digested, or both during intrusive tip growth has not been decisively determined. For example, differentiating xylem precursor cells of *Pinus radiata* and *Aesculus hippocastanum* were found by Barnett and Harris (1975) and Barnett (1981), respectively, to possess swollen, vesicle-containing regions in the middle lamellae. Barnett (1981) suggested that this material might be involved in loosening the walls to permit growth of the cell tips between adjacent cells. In a quite different example, Mio and Matsumoto (1982) observed in conifer tracheids that the fragile parent cell walls were often ruptured during radial enlargement and these fragments appeared in the intercellular spaces of compression wood. Although not specifically stated, it was implied by the authors that rupture of parent cell walls might also occur at the tracheid tips, but fragments were seldom found there due to post-cambial elongation of the tracheids.

One clear example of intrusive growth was presented by Wenham and Cusick (1975) in their study of several dicotyledons exhibiting nonstoried structures. Like other workers, they found that intrusive growth began in the cambial zone and reached its peak in the cell expansion zone. When they followed a radial file of fibers all derived from a common cambial initial, they observed that intruding tips not only elongated to different lengths but they did so independently of one another. Each fiber followed its own "intrusive pathway" along the tangential walls of adjacent cells, and the amount of elongation seemed to depend both on intrusive pathways available to the tip and the extent to which these potential pathways could be exploited. As shown in Fig. 6.82A, a fiber tip with two tangential walls to exploit during intrusive growth had a greater growth potential than a fiber with only one tangential wall available; a fiber with no tangential walls in which to intrude was usually the shortest.

The authors concluded that the course and extent of intrusive growth of any one fiber was determined principally by the spatial relationships existing between it and the surrounding fibers. Cells which were displaced by intrusive growth at one level were themselves intruding at another level. After the initial phase of intrusion, the subapical part of the tip enlarged by "transverse intrusion" of the surrounding cells. In this way, the fiber tip acquired and maintained its tapered form. As noted by others, the intruding tip was at first chisel- or wedge-shaped (Fig. 6.82B), but its form changed during later radial growth. During the final stages of elongation, the fiber tip became more rounded, presumably as secondary wall development and maturation processes extended to the tip. Many other workers have also commented on the presence and arrangement of interpenetrating cell tips during intrusive growth (Bannan and Whalley 1950, Srivastava 1963b, Srivastava and O'Brien 1966a, Wloch 1976, Fujita et al. 1984). The tips of cambial zone cells presumably intrude in the same way.

Additional evidence confirming the existence of intrusive tip growth is the extreme thinness of primary walls at the cell tips (Schoch-Bodmer 1945, Wardrop 1952, Bannan 1956, Evert 1960) and, the progressive thickening of cell walls towards the tips during fiber maturation (Schoch-Bodmer and Huber 1951, Wardrop and Dadswell 1952, Yoshizawa et al. 1987).

The absence of parent walls near the tips of differentiating tracheids of *Pinus pinaster* as found in both normal wood by Wardrop (1952) and in compression wood by Wardrop and Dadswell (1952) also suggested to these authors that growth was confined to the cell tips. In maturing tracheids of this species, the lumen remained in contact with the primary wall at the extreme cell tip by means of

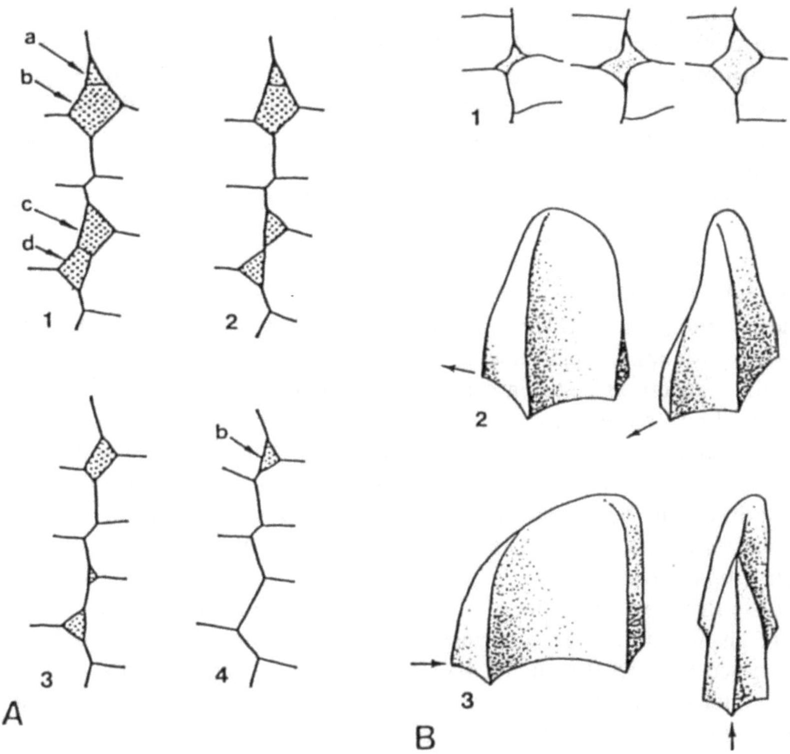

Fig. 6.82A,B. Drawings of *Salix viminalis* fibers. **A** Transverse sections through the ends of mature fibers. Fiber *b*, which intruded between two tangential walls, is longest at maturity. Fiber *a*, with no tangential walls, is shortest. *c* and *d*, each with one tangential wall abutting, are intermediate in length. **B** Shapes of fiber tips during intrusion drawn from photomicrographs. *1* Successive levels in transverse section at 14 µm thickness. 2000x. *2,3* Drawings from two perspectives of two fiber tips. *Arrows* indicate direction of stem radius toward the cambium. Note the strongly concave faces of tips during intrusive growth. 3750x. (Wenham and Cusick 1975)

a minute terminal canal, again suggesting that the final stages of growth were confined to the cell tip. These observations and others, all of which were supplemented with photomicrographic evidence, supported the concept of intrusive rather than symplastic growth.

Perhaps the most convincing evidence in support of intrusive tip growth is the concentration of protoplasm in the elongating tip. Krabbe (1886) commented on the fact that "plasma" collected at both ends of elongating cells. Aldaba (1927) and Schoch-Bodmer (1945) both observed protoplasmic continuity extending to the tips of elongating bast fibers in *Linum*, as did Schoch-Bodmer and Huber (1951) in *Sparmannia africana*. In *Thuja occidentalis*, Bannan (1956) described dense accumulations of cytoplasm in the tips of fusiform initials and the outermost xylem mother cells. This phenomenon is clearly evident in a photomicrograph of a fusiform initial of *Pyrus communis* prepared by Evert (1960). As pointed out by Evert (1960), the rounded shapes of fusiform initial tips and their dense cytoplasm are features usually associated with intrusive growth. Another characteristic of elongating cells, such as those in the cambial zone, is the forward migration of the nucleus, which often assumes a position near the cell tip (Schnepf 1986).

6.7.4.2 Primary Pit Fields

One of the favorite arguments leveled against both sliding and intrusive tip growth has been the assumption that such growth processes would inevitably disrupt protoplasmic connections between cells. It was assumed that the pit connections would rupture first, followed by severing of the plasmodesmata. Russow (1883a), for instance, explained at great length his perception of how pits were disrupted and protoplasmic connections severed and re-established during cambial cell division. It has long been known that cambial cells show some evidence of incipient pit formation, and critics claimed that these pits would be ruptured during active tip elongation.

One of the strongest arguments for precocious pit formation has been the appearance of pit-like structures on dormant cambial cell walls. For example, Elliott (1935) observed that radial walls of *Acer pseudoplatanus* cambial cells were pitted in the winter condition, and Huber (1939) described "fine pitting" on cambial cell walls of *Betula verrucosa*. The radial walls of all cambial zone cells were thickened and showed numerous primary pits fields in *Pseudotsuga menziesii* during dormancy (Grillos and Smith 1959). The latter authors made no distinction between the initiating layer and xylem and phloem mother cells. This distinction is important, however, because one difficulty that all investigators have experienced is identifying the dormant cambial initials on tangential sections.

Dormant cambial cells often display a characteristic "beaded" appearance which has been interpreted as being due to incipient primary pit fields (Fig. 6.83A). In *Pyrus communis*, the radial walls appeared beaded due to thickened wall areas alternating with deeply depressed primary pit fields (Evert 1960). The beaded appearance in cambial cells of *Aesculus hippocastanum* was found by Rao and Catesson (1987) to be due to deposition of cell wall material between primary pit fields, especially on the radial walls during dormancy induction. Cockerham (1930) had made a similar observation years earlier in *Acer pseudoplatanus*, as did Catesson (1961) in the same species. Similar appearing beaded walls were observed in *Robinia pseudoacacia* (Derr and Evert 1967), *Populus tremuloides* (Evert and Kozlowski 1967, Davis and Evert 1968), *Acer negundo* (Tucker and Evert 1969), and *Picea abies* (Timell 1980). The tangential walls of dormant *Fraxinus excelsior* cambial cells were described by Kurczynska (1986) as being indented due to the presence of primary pit fields.

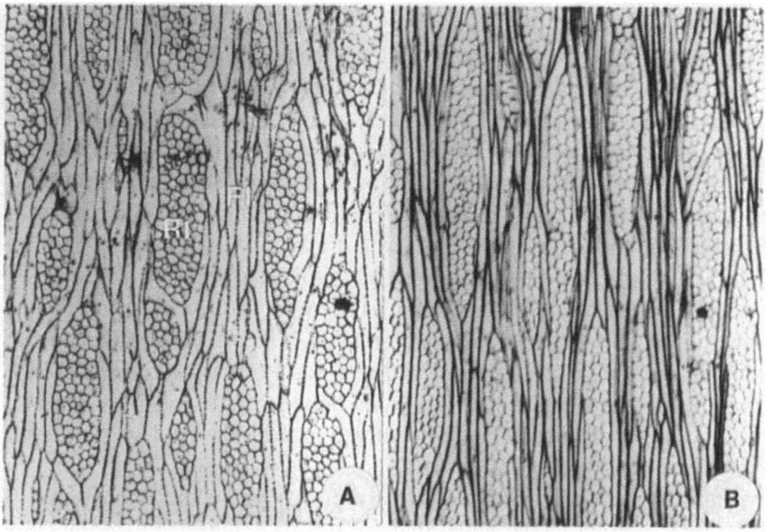

Fig. 6.83A,B. Tangential sections of cambium in *Tectona grandis*. **A** Dormant cambium showing short fusiform initials (*FI*) with "beaded walls" due to incipient primary pit fields. *RI* Ray cell initials. **B** Active cambium showing disappearance of "beaded walls" attributed to elongation and tangential stretching of the fusiform initial cell walls. (Rao and Dave 1981)

The beaded appearance of cambial cell walls during dormancy is not confined to temperate zone species. Many tropical species exhibit dormant periods of varying length. The beaded appearance of cambial cell walls attributed to incipient pitting has not only been described but clearly illustrated for many species by investigators in India as, for example, in *Dalbergia sissoo* (Paliwal and Prasad 1970, Venugopal and Krishnamurthy 1989), *Polyalthia longifolia* (Paliwal et al. 1975), *Salvadora persica* (Paliwal et al. 1976, Sajwan and Paliwal 1976), *Azadirachta indica* (Sharma et al. 1979b), *Tectona grandis* (Rao and Dave 1981), *Gmelina arborea* (Dave and Rao 1982a), *Mangifera indica* (Dave and Rao 1982b), *Mimusops elengi* (Ghouse and Hashmi 1983), *Prosopis spicigera* (Iqbal and Ghouse 1985b), *Holoptelea integrifolia* (Rao and Dave 1985b, Rao et al. 1989), *Ficus rumphii* (Ajmal and Iqbal 1987b, Sagheer and Iqbal 1987), *Streblus asper* (Ajmal and Iqbal 1987a), *Acacia nilotica* var. *telia* (Iqbal and Ghouse 1987) and *Ficus religiosa* (Siddiqi 1991). Most of the foregoing reports of beaded cambial cell walls did not distinguish between either radial and tangential walls or the main cell body and cell tip. However, it might be assumed that the reports referred to radial walls because most observations were made on tangential sections. The apparent disappearance of the beaded walls during active growth has been attributed to tangential stretching of the radial walls (Fig. 6.83B) (Rao and Dave 1981).

Contrary to the foregoing observations on dormant cambial cells, Murmanis and Sachs (1969) reported that bordered pits did not appear on radial walls until the differentiating tracheids began enlarging in *Pinus strobus*. However, once formed, the pit membranes were traversed at an early age by plasmodesmata. Their observations agreed with those of Wardrop and Dadswell (1952) in that the initial pit border formed before there was any evidence of secondary wall deposition. Thus, the pit border was considered to be a primary structure. However, Wardrop and Dadswell (1957) and Fengel (1966) found that the initial pit membrane was deposited on the primary wall, making it a secondary structure. Barnett and Harris (1975) also observed that incipient pit formation occurred just outside the cambial zone in *Pinus radiata*. According to Barnett (1979, 1981), these sites were determined during the phase of cell enlargement. The pit fields divided with the rest of the cell and they were enlarged and restored to their original size with de novo formation of additional plasmosdesmata between cell divisions. Barnett believed these findings in *P. radiata* to be contrary to the long-held view that pit sites are predetermined in the cambial initials.

Irrespective of whether or not incipient pits are present on the main body of cambial cells, several workers have noted that the tips of elongating cambial cells were generally free of pits (Klinken 1914, Janssonius 1931, Schoch-Bodmer and Huber 1952, Wardrop 1954, Barnett and Harris 1975). Moreover, it has been reported that new pits formed readily at the completion of tip growth (Grossenbacher 1914, Foster 1947, Bannan 1950b) and that the arrangement of new pits completely coincided in adjacent elements (Bannan and Whalley 1950, Bannan 1956). The absence of crassulae at the elongating cell tip was also noted by Bannan and Whalley (1950).

6.7.4.3 Plasmodesmata

Plasmodesmata are highly complex, intercellular passages serving communication and transport functions between cells (Beebe and Turgeon 1991). Although plasmodesmata were believed to be present in cambial cells, critics claimed that these structures would be severed during active tip elongation. Several workers, cited by Schenck (1885), believed that in certain cases an intercellular cytoplasmic communication might exist. Schenck questioned these reports by suggesting that there was only an "approaching" of the plasma strands and not a continuous connection between neighboring cells. Krabbe (1886) admitted that the "Plasmastränge," or plasma strands, would be broken according to his concept of sliding growth, but he also assumed that they could be readily re-established. Kienitz-Gerloff (1891), Krüger (1892), and Haberlandt (1914) all concurred, maintaining that the new protoplasmic connections were of secondary origin. Gardiner (1907) accepted sliding growth but only by assuming that the disrupted connecting threads would be reunited later on the "hit and miss" principle. Strasburger (1901), who proposed the term "Plasmodesmen" for these plasma strands, was at first doubtful, but later relented and admitted that supplementary strands might form. In lending support to this idea, Grossenbacher (1914) attempted to demonstrate in *Pyrus malus* how pits and protoplasmic connections at the tips of cambial cell derivatives formed at the completion of sliding growth.

One of the few investigators to actually examine the cambium in detail was Kuhla (1900) in his comprehensive study of mistletoe (*Viscum album*). He claimed that "Plasmaverbindungen," or plasma connections, were numerous in all cambial cells where they occurred only in the pit areas. They were particularly prevalent on tangential cell walls and characteristically less so on radial walls. Average data from many samples showed 11.5 μm^2 of pit area with 15.1 plasma connections for each 100 μm^2 tangential wall area, and 5.2 μm^2 of pit area with 6.5 plasma connections for each 100 μm^2 radial wall area. On the basis of these data, Kuhla rejected Krabbe's (1886) assumption that plasma connections could not occur between cambial cells because of sliding growth. Kuhla extended his observations by comparing the number of pit areas and plasma connections in the walls of cambial cells with those in the walls of "Ersatzfasern," or substitute fibers, one of the first xylem derivatives produced by the cambium of *V. album*. No differences were found on the tangential walls of the two cell types, but the radial walls of the Ersatzfasern, when compared to the cambial cells, possessed many more pit areas and plasma connections. These observations added further confirmation to the idea that plasma connections were of secondary origin.

Hill's (1901, 1908) research was concerned mainly with development of the sieve plate. He experienced great difficulties in trying to demonstrate connecting protoplasmic threads in cambial cells of both conifers and dicotyledons. These threads could not be detected with certainty until the pit-closing membranes began to differentiate, again suggesting secondary formation. However, in *Picea abies*, he claimed that these connecting threads could be "... clearly seen in the tangential walls of medullary-ray cells right through the cambium."

Observations on the cambium were included from time to time in the studies of other investigators. Livingston (1935), for example, was unable to satisfactorily demonstrate plasmodesmata in cambial cells of *Nicotiana tabacum*. In one preparation, examination of the cambial tissue strongly indicated the presence of numerous threads but the evidence was inconclusive. Kerr and Bailey (1934), on the contrary, reported numerous plasmodesmata uniformily distributed or aggregated in the primary pit fields of cambial cell walls. According to Meeuse (1941), a number of workers had reported plasmodesmata as being present in all meristematic tissues of higher plants, including the cambium. However, in keeping with his own strong belief in symplastic growth, Meeuse rejected the idea that plasmodesmata were ruptured by sliding growth, as well as the contention that these protoplasmic connections, once ruptured, might reform, as suggested by Sinnott and Bloch (1939).

Although a variety of techniques had been used over the years in attempts to visualize plasmodesmata, all had employed swelling agents of one type or another. Livingston (1964) examined undehydrated, thus essentially unaltered, cambial tissues of *Pinus strobus*. He observed plasmodesmata to be present in all primary pit fields in the walls of both fusiform and ray cell initials. Sections cut from living cambia and immediately examined in water in an unstained condition showed characteristic groups of paired beads adhering to opposite sides of the paired pit fields. These beads were presumed to be masses of coagulated protoplasm that adhered to the ends of plasmodesmata when the main protoplasmic mass was torn away during sectioning. Because Livingston examined the cambium in the winter condition, he could not address the question of how the plasmodesmata were affected by intrusive growth.

The advent of electron microscopy was at first believed to be a decisive means for settling the foregoing controversy. Wardrop (1954), employing this technique, demonstrated in *Pinus* radiata that new plasmodesmata formed at the cell tips when new cellular contacts were made following cell division. Moreover, the distribution of pits at the tips of both tracheids and fibers was consistent with the assumption that primary pit fields were regions of growth, consequently areas of cellulose synthesis. Later reviews strongly favored secondary formation of plasmodesmata in cambial cells (Frey-Wyssling 1959, Clowes and Juniper 1968, Carr 1976, Jones 1976, Robards and Lucas 1990). However, as Carr (1976) pointed out, just how the distributional pattern of the primary pit fields in cambial cells might arise was still not known, and as Jones (1976) cautioned, some situations where secondary plasmodesmata might be expected to occur, such as in the radial walls of cambial cells, still awaited elucidation. However, recent evidence has shown conclusively that graft unions of stock and scion are interconnected by true plasmodesmata that are secondarily formed (Kollmann and Glockmann 1991). One might therefore assume secondary formation of plasmodesmata in the elongating tips of cambial cells.

Many early photomicrographs of conifer cambia published by various authors were re-examined by Barnett and Harris (1975) and in no case could they find evidence for plasmodesmata on radial walls. In addition, "many thousands of electron microscope sections of *Pinus radiata* cambium" examined by Barnett (1972, 1973) showed that neither pits nor plasmodesmata were present in the radial walls of fusiform cambial cells destined to become tracheids (Fig. 6.84). Barnett and Harris (1975) expressed the opinion that if plasmodesmata were present in the radial walls of conifer cambial cells, they would present a problem in maintaining continuity during sliding growth. Thus, they concluded that plasmodesmata are not the means by which perfect coordination is achieved between adjacent tracheids in their production of bordered pits. Indeed, plasmodesmata were absent from radial walls of *P. radiata* cambial cells, and many other reports claiming the presence of plasmodesmata in other conifers were believed to be artifacts (Barnett 1979). Apparently, this rejection also applied to the work of Murmanis (1971), who claimed to have found that plasmodesmata were numerous on radial walls, but rare on tangential walls, in dormant *Pinus strobus* cambia.

Reports have demonstrated conclusively that plasmodesmata are present in cambial zone cells of dicotyledonous species. Robards and Kidwai (1969) obser-

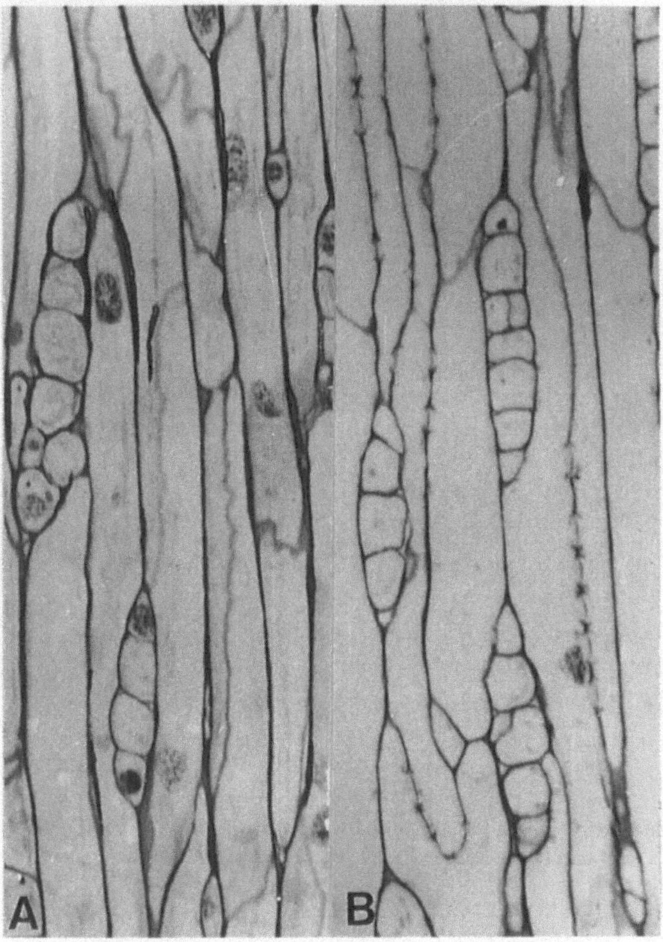

Fig. 6.84A,B. Tangential sections through the cambial zone of *Pinus radiata*. **A** Note the near uniformity in thickness and absence of "beading" in the radial walls of fusiform initial cells. **B** Pit borders developing at periphery of the pit fields at an early stage of differentiation into tracheids. (Barnett and Harris 1975)

ved plasmodesmata on tangential walls of both fusiform and ray cell initials in *Salix fragilis*, and Kidwai and Robards (1969) reported seeing them on both radial and tangential walls of both types of initials in *Fagus sylvatica*. Rao and Dave (1983c) noted that plasmodesmata traversed the cambial cell walls and occurred on radial walls between fusiform and ray cell initials in *Tectona grandis*. In dormant cambia, they were restricted to primary pit-fields on the radial walls. Plasmodesmata in the radial wall between two fusiform cambial cells in *Aesculus hippocastanum* can be seen in Fig. 6.85 from Barnett (1981).

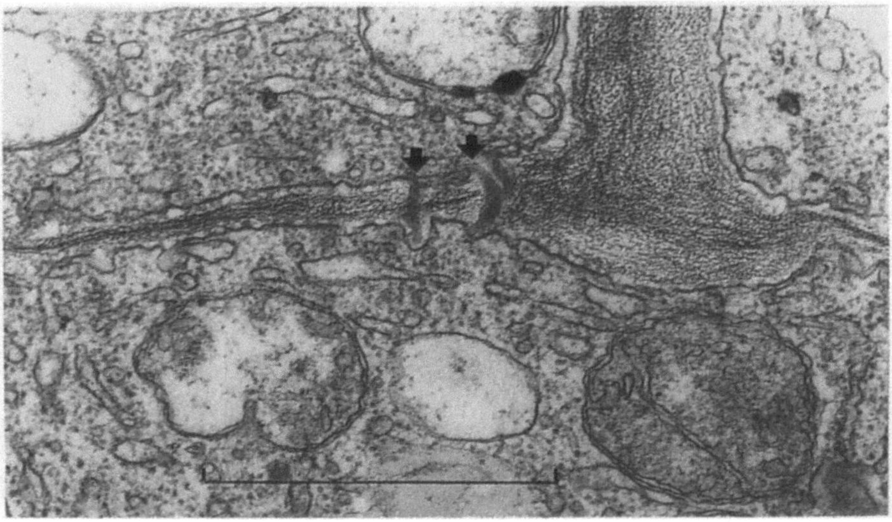

Fig. 6.85. Plasmodesmata (*arrows*) in a radial wall between two fusiform cambial cells of *Aesculus hippocastanum*. EM transverse section. Bar 1 μm. (Barnett 1981)

Evert and Deshpande (1970) observed plasmodesmata traversing both the tangential and radial walls of nondividing cambial cells in *Tilia americana* and *Ulmus americana*. The plasmodesmata were more or less scattered throughout the tangential walls and confined to the primary pit-fields on the radial walls. In her study of *Acer pseudoplatanus*, Catesson (1974) stated that the thin fusiform cell walls were "punctuated by plasmodesmatal fields." Sennerby-Forsse et al. (1984) reported plasmodesmata occurring in tangential and radial walls between initial cells of dormant cambia in several species of *Salix*.

Plasmodesmata have also been observed in ray cell initials, where they are often particularly well developed (Catesson 1990). Numerous plasmodesmata were reported on tangential walls of ray cell initials in *Abies balsamea* (Timell 1979) and *Pinus radiata* (Barnett and Harris 1975). They were observed on tangential walls of ray cell initials (Rao 1985) and also between ray cell and fusiform initials (Barnett 1981) in *Aesculus hippocastanum*.

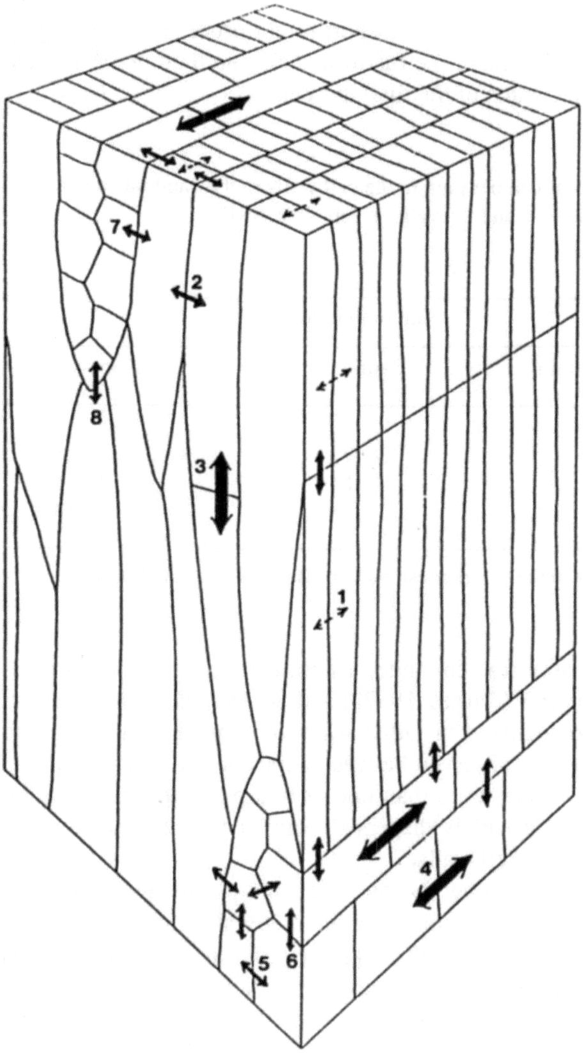

Fig. 6.86. Diagram showing the plasmodesmatal contacts between fusiform and ray cells in the cambial zone of *Fraxinus excelsior*. Plasmodesmatal contacts were identified between fusiform cells in the tangential (*1*), radial (*2*), and transverse (*3*) walls; between ray cells in the tangential (*4*), radial (*5*), and transverse (*6*) walls; and between fusiform and ray cells in the radial (*7*) and transverse (*8*) walls. *Arrow widths* are proportional to the plasmodesmatal counts in Table 6.28. (Goosen-de Roo and Creyghton-Schouten 1981)

Goosen-de Roo and Creyghton-Schouten (1981) recognized eight different categories of plasmodesmatal connections between initial cells in the cambial zone of *Fraxinus excelsior* (Fig. 6.86, Table 6.28). The authors concluded that transport in the radial direction through plasmodesmata was only possible in the rays and not between fusiform cells. From the rays, transport might proceed in the tangential direction through radial cell walls of the fusiform cells and then in the vertical direction through the transverse walls of the fusiform cells.

Table 6.28. Number of plasmodesmata per μm^2 in the cell walls in the cambial zone of Fraxinus excelsior (After Goosen-de Roo and Creyghton-Schouten 1981)

	Wall orientation		
Cell wall contact	Tangential	Radial	Transverse
Fusiform/fusiform	0.05 (1)[a]	2.8 (3)	3.8 (3)
Ray/ray	9.1 (4)	3.3 (5)	2.3 (6)
Fusiform/ray	-	3.4 (7)	3.1 (8)

[a] Numbers in parentheses refer to the plasmodesmatal contacts shown in Fig. 6.86.

The foregoing line of research was continued by Barnett on a number of dicotyledons. In both *Pyrus communis* (Barnett 1987a) and *Sorbus aucuparia* (Barnett 1987b), pit fields could be recognized in radial walls of cambial zone cells and plasmodesmata were occasionally observed between adjacent fusiform initials. In *S. aucuparia* it was unusual to find more than one or two plasmodesmata per pit field, possibly because they were widely dispersed. Similar pit fields were never encountered in tangential cambial cell walls.

Barnett (1987b) speculated as to how the pit fields and plasmodesmata might be maintained during cambial cell division. The symmetrical arrangement of pit fields between radially adjacent cambial cells suggested to him that new pit fields might arise during periclinal division when a new tangential wall formed. Nonetheless, all evidence favored secondary formation of plasmodesmata, i.e., de novo origin during phases of cell enlargement. Thus, the question as to how protoplasmic continuity is maintained during cambial cell division, although clarified to some extent, remains unresolved.

6.7.5 Forked Cells

Cells with forked or gabled tips have been frequently reported in the literature in both xylem (Sanio 1863b, Müller 1890) and phloem (Meeuse 1942, Schoch-Bodmer and Huber 1949a,b, 1951) and their formation has been associated with elongation of the cell in question. Schultz (1831), for example, observed wood cells in *Vitis vinifera* with long, sharp points which occasionally became charac-

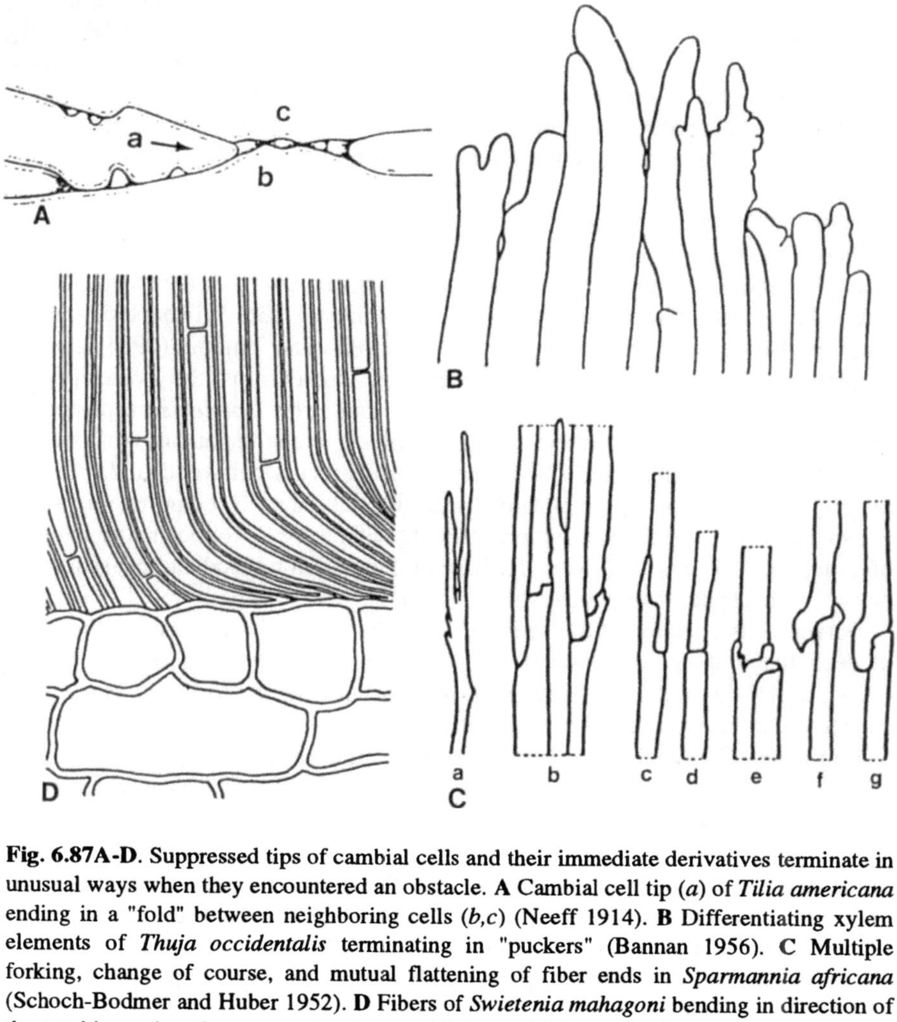

Fig. 6.87A-D. Suppressed tips of cambial cells and their immediate derivatives terminate in unusual ways when they encountered an obstacle. **A** Cambial cell tip (*a*) of *Tilia americana* ending in a "fold" between neighboring cells (*b,c*) (Neeff 1914). **B** Differentiating xylem elements of *Thuja occidentalis* terminating in "puckers" (Bannan 1956). **C** Multiple forking, change of course, and mutual flattening of fiber ends in *Sparmannia africana* (Schoch-Bodmer and Huber 1952). **D** Fibers of *Swietenia mahagoni* bending in direction of the cambium when they encountered a ray. (Huber and Schmidt 1938)

teristically forked during "stretching growth" of the cambial cells. Later investigators used the appearance of forked cells, which might occur in the fusiform initials but more commonly in the differentiating elements, to justify either sliding (Jaccard 1910b, Neeff 1914), symplastic (Priestley 1930b, Meeuse 1942), or intrusive (Schoch-Bodmer and Huber 1949b, 1951, Bosshard 1951, 1952, Wardrop and Dadswell 1952, Liese and Ammer 1958, Wloch 1976, Rao and Dave 1985a, Iqbal and Ghouse 1987, Lo 1989) growth. Thus, Raatz (1892) and Neeff (1914) concluded that if tip growth of a cambial cell was hindered by another cell, usually a ray, then it might either stop elongating, bend into a fold (Fig. 6.87A), or more often grow around the obstacle by forking. These obstacles to tip elongation might result either in actual forking, in "puckers" (Fig. 6.87B), or in mutually flattened fiber ends in addition to accentuated forms of bifurcations (Fig. 6.87C). Abnormal cell tips of the latter type were frequently encountered by Smith (1967) in *Abies grandis* branches infected with *Adelges piceae*. In some cases, the fiber tips, rather than forking, simply elongated passively in the direction of ray expansion (Fig. 6.87D).

Priestley (1930b) expressed the opinion that forked cells were commonly found in dicotyledons but were of rare occurrence in conifers. While perhaps less prevalent in conifers, they are by no means rare, as many reports to the contrary indicate (Müller 1890, Groom and Rushton 1913, Bannan 1956, Wilson 1963). Wardrop and Dadswell (1952), Yoshizawa et al. (1985), and Yumoto (1984) all presented excellent photomicrographs of bifurcated tracheid tips, the former two in compression wood tracheids of *Pinus pinaster* and *Larix leptolepis*, respectively, and the latter in atypical tracheids of *Picea glauca*, which the author attributed to frost injury. Approximately 20% of the tracheids in normal wood of *Larix leptolepis* had abnormal tips compared with 50% of those in compression wood (Yoshizawa et al. 1985). The authors attributed this increase to the greater number of anticlinal divisions in the cambial zone of compression wood and consequently on the limited space for tip elongation.

Forked cells were found by Neeff (1914) to be especially abundant in the stubs of decapitated shoots and wound wood in *Tilia americana* and also by Wloch (1976) in woody tumors of *Picea abies*. In the latter case, the forked cells were associated with an increased frequency of anticlinal division and accentuated intrusive growth of cambial initial cells. The fork might occur near the cell tip (Fig. 6.72C) or at some distance behind the tip (Fig. 6.72D), depending at which point the intruding cell encountered its neighbor. This latter type of forking was also observed by Rao and Dave (1985a) in *Mangifera indica* caused by intrusive growth of the lateral radial walls of fusiform initials. These authors suggested that growth, once arrested at the apical ends of fusiform initials, continued at the lateral walls. Many forked cells were also encountered during formation of interlocked grain in *Entandrophragma cylindrica*, a species with storied structure (Hejnowicz and Zagórska-Marek 1974).

6.7.6 Reconciliation

As envisioned by the authors, the hypotheses of sliding growth and intrusive-interpostional growth are conceptually quite different. Yet, when based on the interpretations of later investigators, the distinctions often tend to become blurred. Krabbe (1886), for example, described how neighboring cells separated as elongating cell tips, thin-walled, and rich in protoplasm, grew between them. Although Krabbe was unsure as to how cell separation occurred, Neeff (1914) later suggested that it might be due to digestion of the middle lamella. As the concept of sliding growth matured, workers began to accept the idea that elongation was localized more or less at the cell tip and that the middle lamella between adjacent cells in some way separated in advance of elongation. This conception of sliding growth was described by Eames and MacDaniels (1925, 1947). As pointed out by Majumdar (1941), if this were the case, then there would be little difference between the concepts of sliding and intrusive growth. However, most proponents of sliding growth envisioned elongation of the cell tip to be essentially a physical process, whereas those favoring intrusive growth viewed it as a physiological process. Arguments for the latter view were presented by Hejnowicz (1980). By demonstrating that the cell tip was not rigid and therefore a locus of minimum stress, he argued that it could not act as a mechanical wedge. Rather, he suggested that the middle lamella was under tension and intrusive growth might therefore occur simply by dissolving the cementing substances. The latter idea had been discussed earlier by Mühlethaler (1950). A theoretical treatment of certain aspects of intrusive growth can be found in Dormer (1980).

The foregoing processes occur in all cells that elongate, and interpretations often vary according to the cell type, as well as the extent and form of elongation during differentiation. For example, to account for the differentiation of sclereids with their extensive arm growth, Arzee (1953) adopted symplastic growth, whereas Foster (1947) rejected symplastic growth in favor of interpositional growth. Amer (1977) believed intrusive growth was responsible for elongation growth of fibers derived from fusiform initials in the thorns of *Gleditsia triacanthos*. Kaussmann (1963) suggested that while symplastic growth played a major role during primary tissue differentiation, interpositional growth was responsible for the length growth of cambial initials, fibers, and tracheids and the diameter growth of vessels. Similarly, Rodin (1967) suggested that bundle cap fibers in the leaves of *Gnetum* elongated by symplastic growth early in development and then continued their elongation by intrusive growth. Other authors believed that two growth processes might be involved, depending on the cell part undergoing elongation. As an example, when growth occurred in localized regions below the cell tip, Roelofsen (1959) suggested that intrusive growth might be supplemented to some extent by sliding growth, a view also expressed by Bannan and Whalley (1950), Bosshard (1952), Srivastava and O'Brien (1966a), and Philipson et al. (1971); Wenham and Cusick (1975) were of the opinion that intrusive growth might be supplemented by symplastic growth. With regard to fusiform initials and their derivatives, either intrusive or interpositional growth has been the process of choice by most recent investigators. Yet sliding growth, or less often symplastic growth, is often considered a possibility to explain either elongation or expansion of wall membranes below the cell tip.

Frey-Wyssling (1950), in discussing these concepts, expressed the opinion that according to a true theory of sliding growth, a tissue would lose its entity and disintegrate into individual cells. In the extreme case, these cells would no longer form a tissue but a cell colony. Pure symplastic growth, on the other hand, would be impossible in view of the enormous changes in cell shape during differentiation. Consequently, the truth must lie somewhere between these two extremes. That is, one should recognize that whereas the main cell body must remain united during differentiation to assure entity of the tissue and allow correlated histological growth, the cell tips could behave independently resulting in a cell shape quite different from that of either the mother cell or neighboring cells. He deplored the "tedious" terminology such as extension, tip, and width growth for the observed phenomena and sliding, symplastic, intrusive, and interpostional growth for explanation of the observed facts. According to Frey-Wyssling, all these terms could be "subsumed under the concept of growth in surface," consisting of "plasticizing of the cell wall and actual growth by intussusception." When both these phenomena occurred diffusely, extension growth would occur, and when they continued locally for some time tip or width growth would follow. Symplastic growth would then simply mean that two adjacent cell walls had the same rates of surface growth. Intrusive or interpositional growth, on the other hand, would imply large differences in the rates and amounts of surface growth due to the pronounced differences in intussusception in neighboring cell walls.

The physico-chemical aspects alluded to by Frey-Wyssling lead to considerations of cambial cell elongation and wall formation that go well beyond the confines of this treatise. Despite the plea of Frey-Wyssling, the terms sliding, symplastic, intrusive, and interpositional growth persist in the literature. Although useful in conveying conceptual images of cell growth processes, the terms are nonspecific and are easily misinterpreted. When used, the term symplastic growth should refer to wall growth of the cell body as described above by Frey-Wyssling. The term sliding growth should be dropped from usage as it not only implies a physical growth process, but as generally used, it applies to growth of the entire cell. Intrusive and interpositional growth are considered synonyms. They should be used with reference to growth of the cell tip. As defined by Esau (1977), intrusive growth is: "A type of growth in which a growing cell intrudes between other cells that separate from each other along the middle lamella in front of the tip of the growing cell." Although just how the middle lamella separates is not stated, present evidence suggests a physiological process.

Chapter 7

Periclinal Cambial Divisions

7.1 Nonstoried Cambia

A periclinal division occurs by a tangentially oriented wall that usually, but not always, divides the cambial initial in two equal halves. Periclinal divisions are also referred to as additive divisions because they add new cells radially to the secondary body of the plant (Berlyn 1982, Metcalfe and Chalk 1983). The fact that cambial cells divide in this manner was recognized long before the true nature of the cambium itself was fully appreciated. Radlkofer (1858), for example, described the process in *Cocculus laurifolius* as active division in the radial direction that formed dividing walls parallel to the periphery of the stem.

Periclinal divisions occur in cambial mother cells as well as initial cells. The xylem mother cells, which are more numerous than the initial cells, contribute the greatest number of new cells to the radial increment. Phloem mother cells also divide periclinally, but less often than those of the xylem (Esau and Cheadle 1955). Gregory (1971) estimated that periclinal divisions in *Picea glauca* exceeded anticlinal divisions by a ratio of 278:1.

The active cambial zone is subjected to constant change in the number, size, and shape of its constituent cells. These changes in cell dimensions have been subdivided into intracambial (Klinken 1914) and extracambial (Krabbe 1886). Although interpreted in slightly different ways by different workers (Haberlandt 1914), intracambial changes might be thought of as those changes that affect the cambial initials directly. Thus, intracambial changes would persist in the cambium to be passed on through all subsequent periclinal divisions. Extracambial changes are those that affect cambial derivatives. Such changes do not persist in the cambium. Cambial mother cells often produce a variable number of derivatives, some of which might undergo substantial changes, but neither the mother cells nor their progeny persist in the cambium.

7.1.1 Intracambial vs. Extracambial

Most intracambial changes in cell dimensions are the consequences of anticlinal divisions described in previous sections. Thus, all changes in length and breadth of fusiform initials that are passed on through periclinal divisions are the results of the readjustments that occur in the cambial initial layer following each anticlinal division. Because of the longitudinal orientation of the periclinal dividing wall, all daughter cells of fusiform initials are more or less identical in size and shape. However, during each periclinal division the readjustments in the initial layer that accompanied the previous periclinal division are passed on to the daughter cells. In this way a fusiform initial that elongates by intrusive growth through a series of successive periclinal divisions will pass on this added increment to each of its daughter cells. Or, looked at another way, the initial that persists as the fusiform initial will retain the new dimensions. Thus, the length that a tracheary element in the secondary xylem will attain depends on (1) the length of the cambial mother cell, and (2) the amount of elongation that the tracheary element undergoes during its differentiation (Bailey and Tupper 1918).

Exceptions and anomalies do occur in the periclinal division process, in which cases the daughter cells might be smaller in size and/or different in shape than those of the parent initial. One example, cited by Evert (1961) and Cheadle and Esau (1964), would be the asymmetric periclinal divisions that sometimes occur in failing initials. Another example would be the transformation of fusiform initials to ray cell initials (Chap. 8.3.4.2).

The present section concerns the changes that occur in length of the fusiform cambial initials over time. For a discussion of fusiform initial size variations among species see Chapter 5.2.

Intracambial changes that occur in the lengths of the fusiform initials can be classified as either within-ring or age-related. Within-ring, or seasonal, are the growth patterns that occur during a growth period. Age-related, or perennial, are the year-to-year patterns that occur radially on a stem section over time. Within-ring changes are usually attributed to the frequency and timing of anticlinal divisions among fusiform initials during the growing season. In the absence of anticlinal divisions, fusiform initial lengths would increase incrementally by intrusive tip growth to keep pace with tangential expansion of the cambium. However, as described elsewhere (Chap. 6.5.3), anticlinal divisions occur throughout the growing season. Many investigators have noted a tendency for such divisions to occur with greater frequency at season's end when periclinal division rates are declining. This tendency results in a pattern of increasing fusiform initial lengths followed by a gradual or abrupt decline at the approach of cambial dormancy. The pattern is not invariable. It can be modified by fluctuations in the rate of growth that might either alter the frequency of anticlinal divisions or contribute to the complete absence of pattern. An example from Bannan (1965a) appears in Fig. 7.1. It shows the amount of fusiform initial elongation that might be anticipated during a growing season in trees with different rates of growth. Carlquist (1988a)

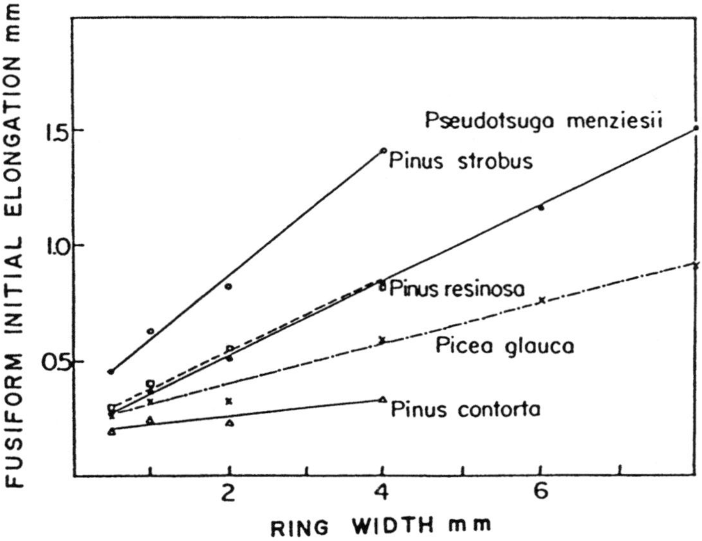

Fig. 7.1. Relationship between annual elongation of fusiform cambial initials and growth ring width in the outer increment of trees 25-50 cm in diameter. (Bannan 1965a)

questioned such patterns by suggesting that fusiform initials do not elongate during formation of the annual increment.

Data on the seasonal patterns of cambial initial lengths most often have been obtained from estimates of cambial derivatives, as described later. Relatively few investigators have made such measurements directly on cambial initials. Investigators in India, however, have been particularly attentive to seasonal patterns of fusiform initial lengths in tropical species. Unlike temperate zone species that often exhibit the aforementioned seasonal pattern, tropical species are more likely to exhibit a pattern that is more closely attuned to the vicissitudes of weather that occur during the exceptionally long growing seasons. Thus, in some species fusiform initial lengths have been found to attain a maximum during the winter dormant season and decrease to a minimum at some time during the periods of active cambial activity (Paliwal et al. 1976, Iqbal and Ghouse 1985b, Ajmal and Iqbal 1987a,b, Rao et al. 1989). Dave and Rao (1982b) even reported a summer minimum in *Mangifera indica*, a species exhibiting continuous cambial activity. In other species a contrary pattern has been observed, with the maximum occurring during the period of active cambial activity and the minimum during the winter rest (Paliwal and Prasad 1971, Paliwal et al. 1975, Sharma et al. 1979a, Dave and Rao 1981, Rao and Dave 1981). For example, in *Chorisia speciosa* even though the fusiform initials experienced their greatest elongation near the season's end in September, the minimum length occurred during the winter months of January-February when the trees were completely without foliage (Sharma et al. 1980).

Table 7.1. Size variation of cambial initials of *Strebulus asper* during different months of the year (After Ajmal and Iqbal 1987a)

Month	Fusiform initials (μm)			Ray cell initials (μm)	
	Length	Width	Tapering end length	Vertical diameter	Horizontal diameter
January	342	18	29	20	26
February	338	17	34	24	29
March	332	19	30	21	27
April	337	20	38	22	28
May	338	20	35	25	30
June	328	21	39	21	31
July	319	21	38	20	27
August	312	23	34	22	26
September	317	21	32	24	31
October	324	20	36	22	29
November	328	18	32	19	27
December	332	17	35	18	26

The patterns described by these authors often displayed exceptionally large fluctuations at all times of the year as, for example, in *Strebulus asper* (Table 7.1). Such fluctuations were apparently associated with the ratio of anticlinal to periclinal divisions that occurred during periods of growth acceleration and retardation. Because of these fluctuations, Dave and Rao (1982a) and Rao and Dave (1985a) concluded that no strong correlation existed between fusiform initial lengths and the season of cambial activity in *Gmelina arborea*.

Many of the aforementioned authors found that tangential widths of fusiform initials also varied seasonally. In general, there appeared to be an inverse relation with fusiform initial length, although the maxima and minima seldom coincided. Data on *Chorisia speciosa* led Sharma et al. (1980) to speculate that an increase in temperature and rainfall contributed to an increase in fusiform initial length and a decrease in tangential width. Such a relationship might imply that when the frequency of anticlinal divisions abated, the fusiform initials would become more slender as they continued to elongate by intrusive tip growth.

Age-related changes, or those due to aging of the cambium, represent the annual changes accumulated over time. Despite the tendency for fusiform initial lengths to decrease at season's end, there is nonetheless a residual increase that is carried over from year to year. Thus, the age-related pattern is one of increasing fusiform initial length with the fluctuating within-ring pattern superimposed upon it. Like the within-ring patterns, data on the age-related patterns have been derived primarily from measurements of cambial derivatives.

Measurements of fusiform initial lengths and tangential widths in trees of different sizes have shown an increasing trend outward from the stem center. For example, Bailey (1923) found that fusiform initials increased in length from 870 to 4000 µm and in tangential diameter from 16 to 42 µm in 1-year-old vs. 60-year-old *Pinus strobus* trees (Table 6.17). Similarly, Grillos and Smith (1959) found that fusiform initials increased in length from 800 to 4500 µm and in tangential diameter from 14 to 40 µm in 1-year-old vs. 200-year-old *Pseudotsuga menziesii* trees. In *Citrus sinensis*, fusiform initials increased in length from 180 to 212 µm, while stem diameter increased from 20 to 198 µm (Khan et al. 1983). However, in *Bauhinia parviflora*, Khan et al. (1981) reported that fusiform initial lengths attained a maximum in young stems and then decreased to a more or less stable value in mature stems (Fig. 7.2).

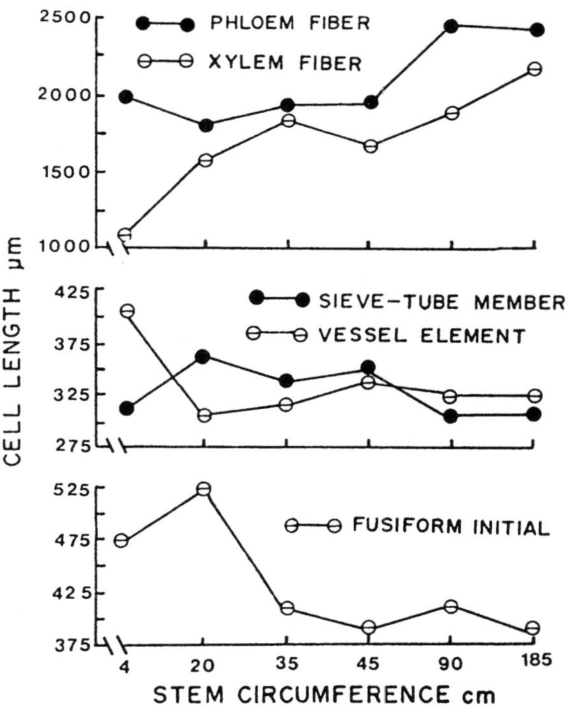

Fig. 7.2. Variation in length of the fusiform initials and their xylem and phloem derivatives in relation to stem circumference in *Bauhinia parviflora*. (After Khan et al. 1981)

Trends similar to the above have been reported for other tropical species growing in India. For example, in *Delonix regia* (Ghouse and Hashmi 1981) and *Polyalthia longifolia* (Ghouse and Hashmi 1980), fusiform initial lengths increased from 322 to 462 µm and from 325 to 436 µm, respectively, during an increase in stem circumference from 30 to 300 cm. Fusiform initial lengths also increased from 160 to 203 μm in *Dalbergia sissoo*, a species with storied cambium, during an increase in stem circumference from 12 to 301 cm. However, like the seasonal patterns, the patterns related to cambial aging undergo fluctuations due either to inherent physiological or to external growth conditions. Thus, Ghouse and Iqbal (1977a) found that the fusiform initials of *Prosopsis spicigera* increased from 220 to 306 μm and then decreased to 260 μm during an increase in stem radius from 3 to 286 mm. These authors attributed the decline in fusiform initial length with age to the "exhausted" condition of the meristem rather than to the frequency of anticlinal divisions. In all the investigated species, tangential widths of the fusiform initials increased slightly as stem sizes increased.

Extracambial changes occur during the differentiation and maturation processes. In some cases, extracambial changes are absent or minimal. Thus, a newly formed xylem derivative should faithfully reproduce the general contour and dimensions of the parent cambial initial and, barring further changes during its differentiation, it should retain these dimensions to maturity. Most derivatives, however, undergo varying degrees of dimensional change during differentiation. Knowledge as to how and to what extent these changes occur has permitted investigators to reconstruct associated changes in the cambial layer. This approach is not without error and uncertainty. Unless preceded by thorough testing, the results should be considered either as approximations or indicative of trends. Cautions and limitations regarding the procedures have been issued by many investigators, such as Rendle and Clarke (1934a,b), Philipson et al. (1971) and Metcalfe and Chalk (1983).

The literature on the patterns of variation among xylem and phloem derivatives is voluminous and an exhaustive survey could easily become a book-sized undertaking. Consequently, the following sections will consist of brief surveys of the patterns in lengths of cambial derivatives that might reflect similar patterns in lengths of the fusiform initials. A more thorough coverage of the general patterns of tracheid and fiber length variation in forest trees can be found in Zobel and van Buijtenen (1989).

7.1.2. Gymnosperm Xylem

7.1.2.1 Tracheids

It has long been assumed that conifer tracheids are essentially the same length as the cambial cells from which they are derived (Frank 1864), and this relationship has been used in turn to estimate cambial cell length (Hartig 1895b). Mischke (1890) tested this assumption in *Pinus sylvestris* and found that during differentiation early and latewood tracheids elongated approximately 18 and 17%, respectively. Klinken (1914) repeated this test in *Taxus baccata* and, in general, concurred with Mischke's conclusions. However, whereas Mischke was simply trying to demonstrate the existence of sliding tip growth, Klinken wanted to test whether this growth was intracambial or extracambial. He concluded that it was extracambial, but he also considered it to be insignificant. This conclusion permitted him to measure the lengths of tracheids and to use these values to estimate the lengths of cambial cells.

Several other workers, such as Strasburger and Winkler, were also of the opinion that all differences in tracheid length were attributable to changes that took place during differentiation. Bailey (1920c) conducted an intensive examination comparing the lengths of cambial initials and tracheids in the adjacent xylem of a number of gymnosperms. His measurements were made on serial, tangential sections of cambium and xylem and cross-checked by measurements from macerations. As shown in Table 7.2, lengths of cambial initials in the species examined closely resembled, but were usually slightly less than, those of tracheids in the last-formed growth layer of the xylem. Bailey concluded that by allowing for a 5 to 10% error, depending on the species, it would be possible to use the tracheids of gymnosperms as indices of the approximate lengths of the cambial initials.

Table 7.2. Comparative lengths of cambial fusiform initials and tracheids in selected gymnosperms (After Bailey 1920a)

Species	Cambial initials Mean length (mm)	Tracheids Mean length (mm)
Abies nordmanniana[a]	1.1	1.2
Cedrus libani[a]	2.0	2.1
Ginkgo biloba	2.2	2.2
Juniperus virginiana	2.2	2.3
Larix decidua	4.0	4.2
Picea abies	3.3	3.6
Pinus strobus	3.2	3.4
Pseudotsuga menziesii[a]	1.2	1.2
Taxus cuspidata[a]	1.1	1.3
Thuja occidentalis[a]	1.5	1.7
Tsuga canadensis[a]	1.4	1.5
Sciadopitys verticillata[a]	1.2	1.3
Sequoiadendron giganteum[a]	3.7	3.8

[a] Material obtained from small branches or small stems.

Lewis (1935) also found extracambial elongation of tracheids in *Pinus strobus* to be about 5%, as did Bannan (1957c) in a number of other conifers. Extracambial growth was confined to the tips of the xylem mother cells and differentiating

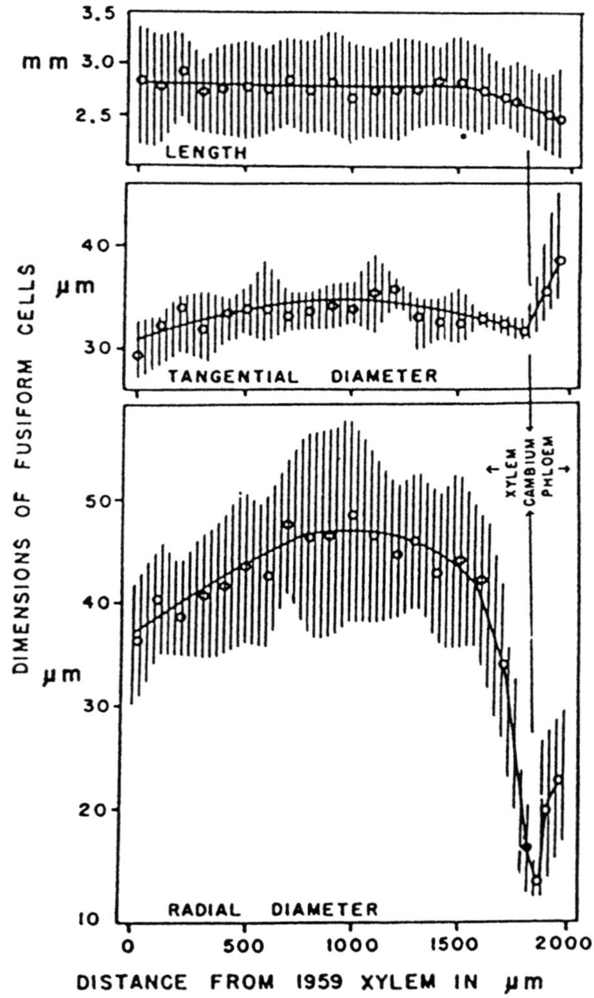

Fig. 7.3. Lengths, tangential diameters, and radial diameters of fusiform cells in the 1960 xylem and phloem increments of a *Abies concolor* tree. Each *circle* represents the mean value for 40+ cells. *Vertical lines* indicate the ranges of variation. (Wilson 1963)

tracheids (Fig. 4.8B-E). In a more recent study of *Larix laricina*, Yang et al. (1986) estimated cambial initial lengths on the basis of measured tracheid lengths and found the latter to be 10 to 15% longer than the fusiform initials. Based on these and similar data, many later workers attempted to estimate the patterns of fusiform initial length variation within trees from tracheid length measurements. The main patterns are the age-related, or sometimes called pith-to-bark, pattern that extends radially on a transverse stem section, and the within-ring pattern that extends radially across a growth ring. Although these basic patterns are repeated at different positions within stems, branches, and roots, they sometimes exhibit rather large variations, as demonstrated by Wilson (1963) in *Abies concolor* (Fig. 7.3).

Age-Related Patterns: The wood of trees was subjected to intensive study by early forest botanists and it was a well-known fact that dimensions of the xylem elements varied greatly within individual trees (Schacht 1856, von Mohl 1862, Sanio 1863b, Frank 1864). However, this knowledge was not systematized until Sanio (1872) formulated a series of "laws" in which he described the patterns of tracheid size variation within trees of *Pinus sylvestris*. Of Sanio's five laws, the one that concerns us here has been translated by Bailey and Shepard (1915) as follows: "In the stem and branches the tracheids everywhere increase in size from within outward, throughout a number of annual rings, until they have attained a definite size, which then remains constant for the following annual rings." The remaining four laws concern the patterns of tracheid size variation at different positions within the stem and also in branches and roots.

Sanio (1872) had investigated a 110-year-old *Pinus sylvestris* tree and found that tracheid length increased from less than 1 mm to about 2.65 mm during the first 30 years and then remained more or less constant. From this statement one might conclude that a cambial initial cell attained a constant value. Sanio addressed this issue by stating, "Finally, the question remains to be answered, whether the size increase of wood cells on a given cross-section depends on the size increase of the cambial cell or whether the size of the latter remains unvaried and the young wood cells attain a significant length only by additional elongation growth." Based on his measurements of radial cell series, the length difference between cambial and wood cells was only slight. Therefore, on a priori grounds, Sanio concluded that in the wood of *Pinus* the longer the wood cell, the longer the cambial cell from which it arose. Klinken (1914) took issue with Sanio's conclusion that tracheid length increased to a constant value. What Sanio meant to say, according to Klinken, was that during the first 30 years the initial cells attained successively greater lengths before dividing.

Despite the significance of Sanio's laws, they are neither invariable nor universal. Subsequent authors have found it necessary to modify these laws to conform more closely with actual observations and measurements due to variability not only among species but also with tree age, rate of growth, position within the tree, and inheritance as well as a host of growth conditions. Although Sanio's laws were based on data from *Pinus sylvestris*, later investigators tested them not only on other conifers but also on many dicotyledons. A tracheid length pattern conforming to Sanio's first law, derived by Bissett and Dadswell (1949) in *Pinus radiata*, is shown in fig. 7.4.

Shepard and Bailey (1914) and Bailey and Shepard (1915) tested Sanio's laws in five conifers and could find no confirmation of the first law. Although tracheid lengths increased rapidly during the first years of growth, they tended to fluctuate

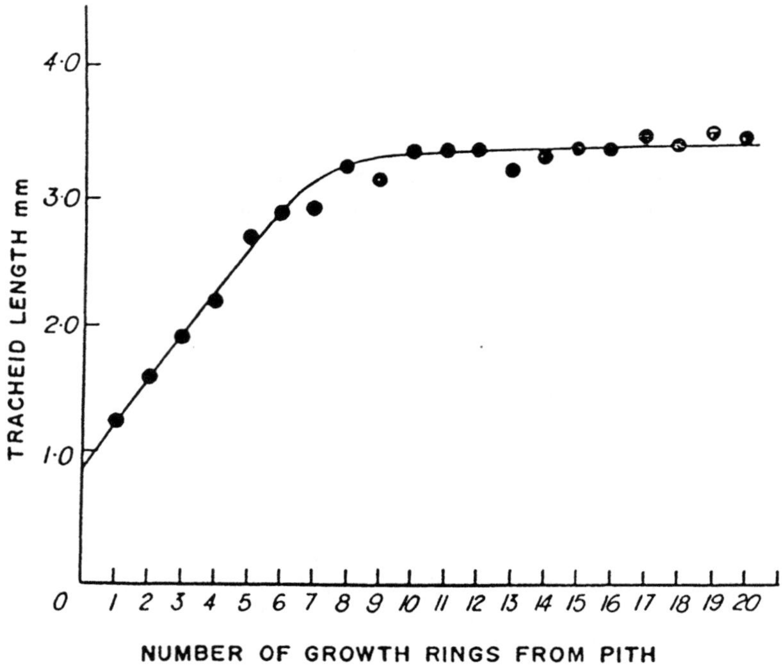

Fig. 7.4. An example of Sanio's (1872) first law in *Pinus radiata* showing the rapid increase in tracheid length from the pith outward to a maximum value after which it stabilized. (Bissett and Dadswell 1949)

thereafter and no constant values were achieved. More extensive tests with many additional species by Bailey and Tupper (1918) showed general agreement with Sanio's first law but data for individual trees often exhibited wide deviations from the norm due to environmental factors. Since that time, Sanio's laws have been repeatedly tested for either verification or modification in a host of species and under widely varying growth conditions. Space permits discussion of just a few of the many papers that relate to patterns of tracheid length variation.

Several workers, such as Gerry (1915), Lee and Smith (1916), Chalk (1930), and Bethel (1941), confirmed Sanio's first law as modified by Bailey and Shepard (1915). The same relationship was found by Liang (1948) in his study of *Larix*. However, to account for the age-related pattern, Liang adopted Bailey's (1920c) proposal that the variation in tracheid length from the pith outwards was similar to the variation in fusiform initial length from year to year. He related fusiform initial length to the rate at which new initials were added by anticlinal divisions, and this in turn to tangential increase of the cambium. That is, when the percentage increment of the cambial perimeter decreased, the rate of anticlinal divisions also decreased, and the average length of the fusiform initials correspondingly increased.

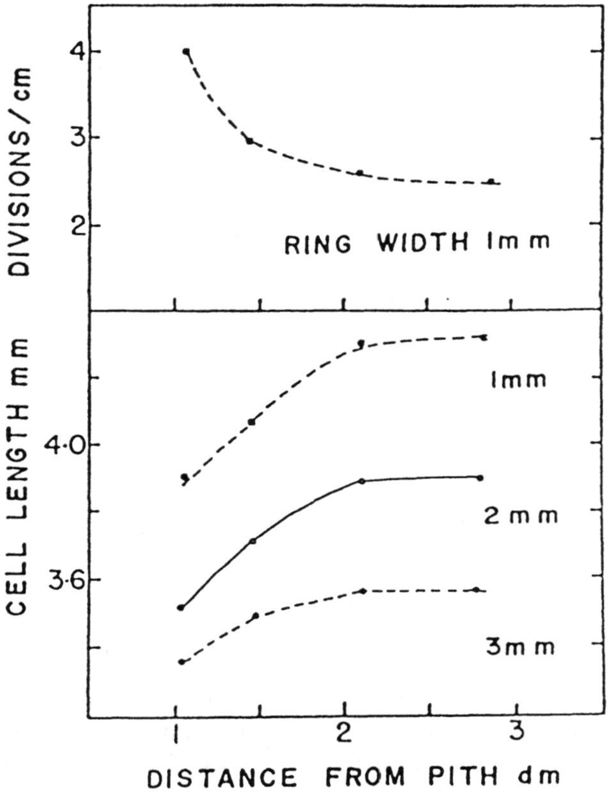

Fig. 7.5. Relation between rate of anticlinal division, cell length, and distance from the pith in *Pinus strobus* trees with uniform growth rates. (Bannan 1967b)

Age-related patterns of tracheid length variation in gymnosperms have been thoroughly reviewed by such authors as Trendelenburg (1939), Spurr and Hyvärinen (1954), Dinwoodie (1961), Larson (1963b), and Timell (1986). Relatively few authors have discussed the relation between tracheid length and fusiform initial length. The most extensive research in this area has been conducted by Bannan (1965c, 1967a,b, 1970), who compared the age-related pattern in tracheid length to that of the fusiform initials as previously described. That is, as a tree increases in age and circumference, the frequency with which fusiform initials divide anticlinally decreases. Concomitantly, as the frequency of anticlinal divisions decreases, the time between divisions increases permitting the initials to elongate intrusively.

The commonly observed pattern described by Bailey and Shepard (1915) is further modified by rate of growth. For example, as Bannan (1967b) demonstrated for *Pinus strobus* (Fig. 7.5), the age-related increase in tracheid length began to stabilize as the frequency of anticlinal divisions decreased in association with de-

Table 7.3. Tracheid length variation in the first-formed earlywood and last-formed latewood within a growth ring of selected species of gymnosperms (After Bissett and Dadswell 1950)

Species	Average tracheid length Earlywood mm	Latewood mm	Difference[a]
Abies grandis	3.4	3.8	11
Athrotaxis selaginoides	3.1	3.2	3
Callitris glauca	3.8	3.8	0
Larix laricina	2.7	2.8	4
Picea sitchensis	3.9	4.1	5
Pinus radiata	3.2	3.6	12
Pinus sylvestris	3.0	3.2	7
Pseudotsuga menziesii	4.5	5.0	11

[a] Difference expressed as a percentage of earlywood.

creasing ring widths. In reviewing these trends, Philipson and Butterfield (1967) suggested that the age-related pattern was due to the interaction of three factors: (1) the frequency of anticlinal divisions; (2) the relative sizes of the parent and daughter cells; and (3) the preferential loss of short initial cells. The age-related pattern can be further modified by varying rates of growth that occur in successive growth rings over a period of years, as described in the next section.

Within-Ring Patterns: It has long been known that the lengths of tracheids in conifers vary not only with age on a stem radius but also seasonally within individual growth rings. For example, Bertog (1895) observed that the earlywood tracheids of *Abies alba* were slightly shorter than those of latewood, but he was not completely convinced of the pattern because many tracheids were broken during maceration. Gerry (1915) claimed that tracheid lengths were longest in the earlywood and decreased gradually toward the growth ring boundary in both *Pinus palustris* and *Pseudotsuga menziesii*. Lee and Smith (1916), on the contrary, found that earlywood tracheids of *P. menziesii* were shorter than those of latewood, although the shortest tracheids occurred at the growth ring boundary. Tracheid length measurements by Kribs (1928) for *Pinus banksiana* and Chalk (1930) for *Picea sitchensis* confirmed the finding of Lee and Smith that latewood tracheids were longer than those of earlywood.

The patterns of within-ring variation in tracheid length were re-examined by Bissett and Dadswell (1950) in eight gymnosperms. Tracheid measurements were made on macerations of tangential microsections cut serially at 80 μm thickness

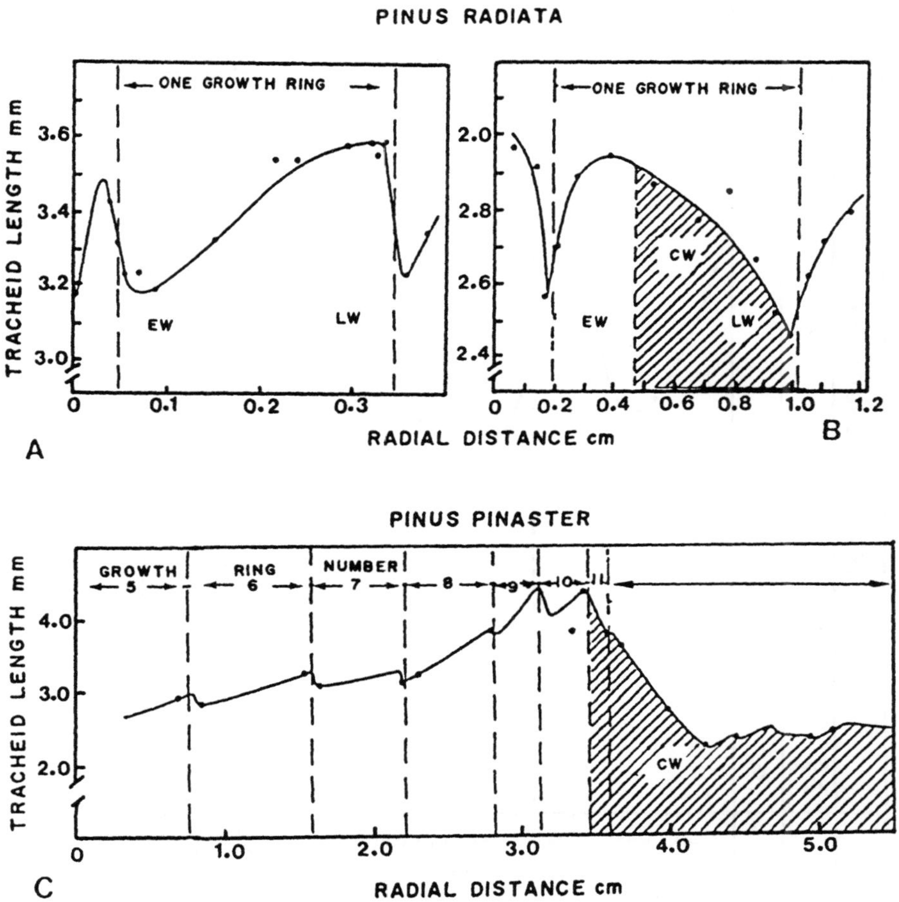

Fig. 7.6A-C. Variation in tracheid length within growth rings of conifers. **A,B** Single growth rings of *Pinus radiata* showing the typical increase in tracheid length from early-wood (*EW*) to latewood (*LW*) followed by the precipitious decrease at the growth ring boundry in normal wood (**A**), and the sharp decline in tracheid length at the onset of compression wood (*CW*) formation (**B**). **C** The same patterns shown in **A** and **B** extending over a series of growth rings in *P. pinaster*. Note the tendency for tracheid length to increase with growth ring age despite the year-end decline. Growth rings were not distinguishable in the compression wood zone. (After Bissett and Dadswell 1950)

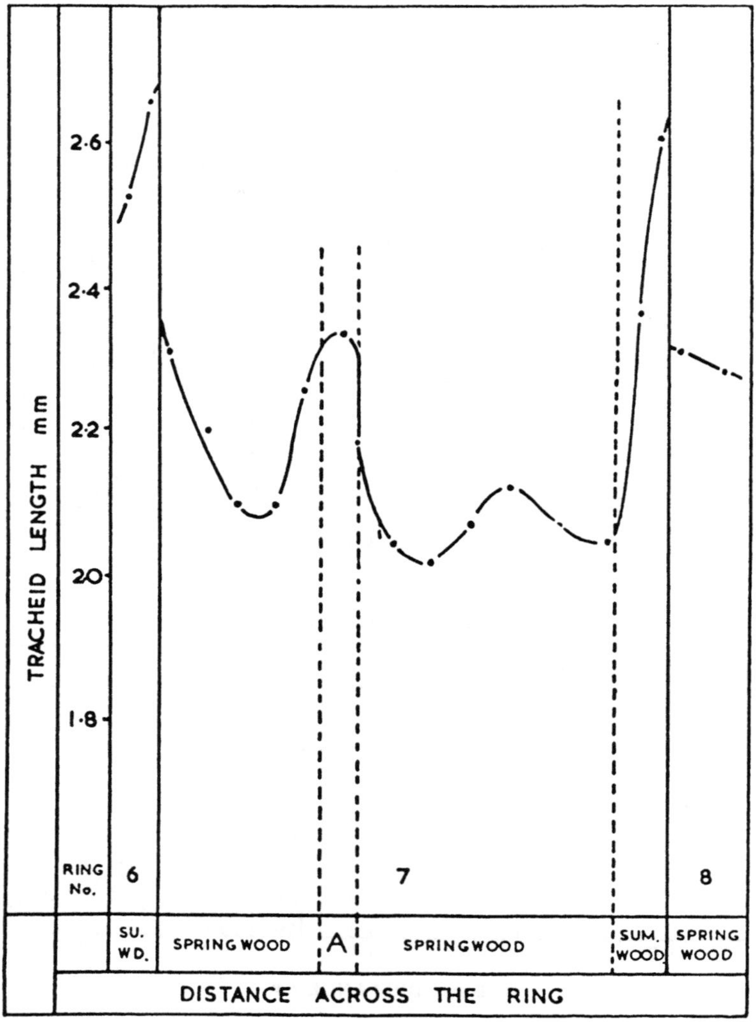

Fig. 7.7. Variation in tracheid length across a false growth ring in *Picea sitchensis*. Tracheid length increased during the formation of false latewood (summerwood) at *A* in growth ring number 7, after which it reverted to the typical earlywood (*springwood*) pattern. (Dinwoodie 1963) (By permission of the Controller of Her Britannic Majesty's Stationery Office: British Crown copyright)

through each selected growth ring. They, too, confirmed that latewood tracheids were longer (up to 12%, Table 7.3) than those of earlywood (Fig. 7.6A). The authors considered this difference to be caused by a "time factor" related to radial growth. That is, during rapid radial growth early in the growing season cambial cell divisions (type unspecified) were frequent, but the daughter cells had insufficient time to elongate before undergoing further division. Later in the growing season, when the cell division rate diminished, the daughter cells had a longer time available for intrusive tip growth. These results were confirmed in additional studies by Bissett et al. (1951).

Macerations are nonspecific with regard to the radial files of elements and they can also result in many broken tracheids. To overcome these shortcomings, Bannan (1955) made serial tangential microsections through growth rings of *Thuja occidentalis* and measured tracheid lengths directly on different radial files of elements. With this method he could determine the seasonal change in tracheid length along a single radial file. He, too, found latewood tracheids to be longer than those of earlywood, but the degree of change was related to rate of growth. Lengths of latewood tracheids exceeded those of earlywood by 2 to 3% in trees with rings 0.6 to 0.7 mm wide and by 4 to 27% in trees with rings 2.0 to 3.0 mm wide. Unlike Bissett and Dadswell (1950), who attributed the within-ring pattern simply to the frequency of cambial divisions, Bannan attributed it specifically to the frequency of anticlinal divisions of these initials. He (Bannan 1951b) had demonstrated previously that anticlinal divisions occurred most frequently toward the end of the growing season, thus reducing mean fusiform initial lengths. The following season, the initial cells elongated with the amount of elongation tending to be greater during the second half of the growing season.

Sometimes the anticlinal divisions occurred at the very end of the growing season after periclinal divisions had ceased (Bannan 1965c). When this happened, the double radial file, consisting of sister fusiform initials, first appeared some distance outward in the earlywood. The explanation offered for this pattern was based on the observation that the first periclinal divisions in the spring were usually in the xylem mother cells rather than in the initials proper (Bannan 1955). Other fluctuations in the within-ring pattern of tracheid length due to growth conditions could be similarly accounted for by the frequency and timing of anticlinal divisions and subsequent elongation of the sister initials (Bannan 1968c).

A few examples will suffice to illustrate the variations due to growth conditions. Chalk and Ortiz (1961) examined a fast-growing *Pinus radiata* tree from Kenya by a method similar to that of Bannan (1954). As in conifers growing in more temperate regions, tracheid length increased from earlywood to latewood but the transition at the ring boundary was more gradual. This difference was most likely due to the occurrence of anticlinal divisions not only at season's end but also at the beginning of the next season's growth ring. Superimposed on the typical earlywood to latewood pattern were fluctuations due to the presence of false growth rings. False rings were characterized by a decrease in mean tracheid length of the false earlywood, which in turn was associated either with an increased frequency of anticlinal divisions in the cambial mother cells, or to an increase in tracheid length in the case of false latewood, as found by Dinwoodie (1963) (Fig. 7.7). Thus, Chalk and Ortiz agreed with Bannan that anticlinal division of mother cells was probably the most important factor causing the typical reduction in tracheid length of the earlywood in *P. radiata*. They added, however, that during latewood formation in this species, mother cells elongated rapidly after an anticlinal division, and the intrusive growth of these initials accentuated the earlywood-latewood differential. A similar pattern was reported by Seth et al. (1988) in *Cedrus deodara* and Seth et al. (1989) in *Pinus wallichiana*, in which case tracheid length increased in each false latewood band and decreased in each false earlywood band.

Like false earlywood, the presence of compression wood within individual growth rings often causes abrupt decreases in tracheid length (Lee and Smith 1916, Bissett and Dadswell 1950, Timell 1986, Seth et al. 1989) (Fig. 7.6B-C). This, too, can be related to an increased frequency of anticlinal divisions attributed to the more rapid rate of growth.

The assumption of a correction factor in relating tracheid lengths to fusiform initial lengths, as suggested by Bailey (1920c), implies uniform extracambial elongation for both earlywood and latewood tracheids. Chalk and Ortiz (1961) suggested, however, that elongation of latewood tracheids might exceed that of earlywood in *Pinus radiata*. Dinwoodie (1963) also emphasized the role of intrusive tip growth in contributing to the increased length of latewood tracheids in *Picea sitchensis*. Although extracambial tip growth can be a source of tracheid length variation, it is far less important than the frequency of anticlinal divisions.

Over the years, investigators have conducted numerous studies of tracheid lengths in conifer stems and arrived at contradictory conclusions. Latewood tracheids have been reported to be either equal to, shorter, or longer than those of earlywood. Taylor and Moore (1981) once again analyzed this relationship in a carefully conducted study of *Pinus taeda*. Although some patterns were revealed, analysis of a large number of samples led the authors to conclude that no consistent pattern of variation existed within mature growth rings of this species. The number of cells produced by the cambium, however, was negatively correlated with tracheid length in both earlywood and latewood. The significance of this correlation suggested that cells differentiating from a rapidly dividing cambium did not elongate as much as those produced by a more slowly dividing cambium.

Many other investigators have reported a decrease in tracheid length with increasing ring width or rate of cambial cell division (see Dinwoodie 1961, Taylor and Moore 1981). Since it has also been demonstrated that the frequency of anticlinal divisions increases with increasing rate of growth, one might logically assume that the many variable patterns in tracheid length reported in the literature can probably be related to seasonal fluctuations in the rates of cambial cell division. Megraw (1985) interpreted the tendency for tracheid length to fluctuate in mature growth rings of *Pinus taeda* in this way. For example, the high proportion

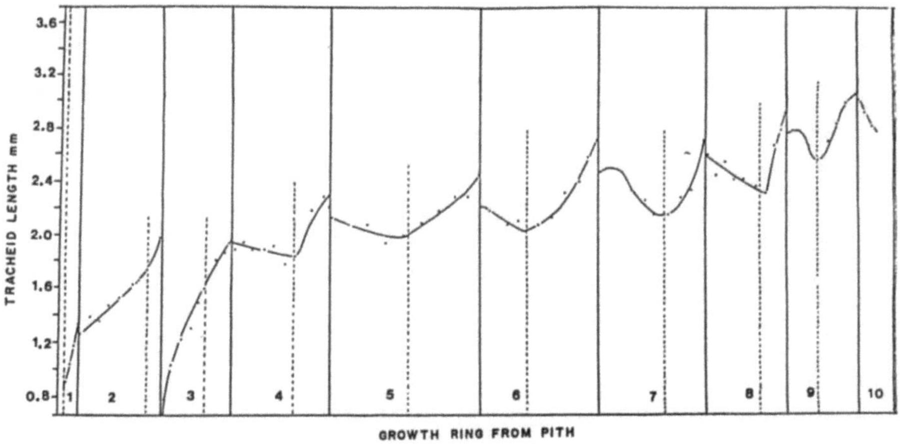

Fig. 7.8. Variation in tracheid length across ten complete growth rings in *Picea sitchensis* illustrating the seasonal patterns of tracheid length variation superimposed on the longer term pattern. The *vertical dashed lines* indicate the approximate earlywood-latewood transitions. (After Dinwoodie 1963)

of earlywood in the wide inner growth rings requires a larger number of anticlinal divisions, which reduces the average fusiform initial length. In growth rings situated farther outward from the pith, the proportion of earlywood to latewood tends toward equality, resulting in a reduced frequency of anticlinal divisions in the earlywood zone. This interpretation agrees with Bannan's (1960b) finding that the frequency of anticlinal divisions not only decreased rapidly with distance from the pith but also as growth ring widths declined.

The seasonal, or within-ring pattern of tracheid length is superimposed upon the long-term, or age-related pattern first recognized by Sanio (1872). Despite the decline in tracheid length usually observed at the end of the growth ring, the age-related pattern prevails. Thus, a record for the first years of growth in a typical conifer might show a persistent increase in tracheid length with a somewhat erratic seasonal pattern superimposed upon it, as observed in *Pinus pinaster* (Fig. 7.6C) and even more clearly in *Picea sitchensis* (Fig. 7.8).

7.1.3 Dicotyledon Xylem

7.1.3.1 Vessels

Like the tracheids of conifers, the vessel members of most dicotyledons are approximately the same length as the cambial initials from which they are derived (Fig 7.9A).

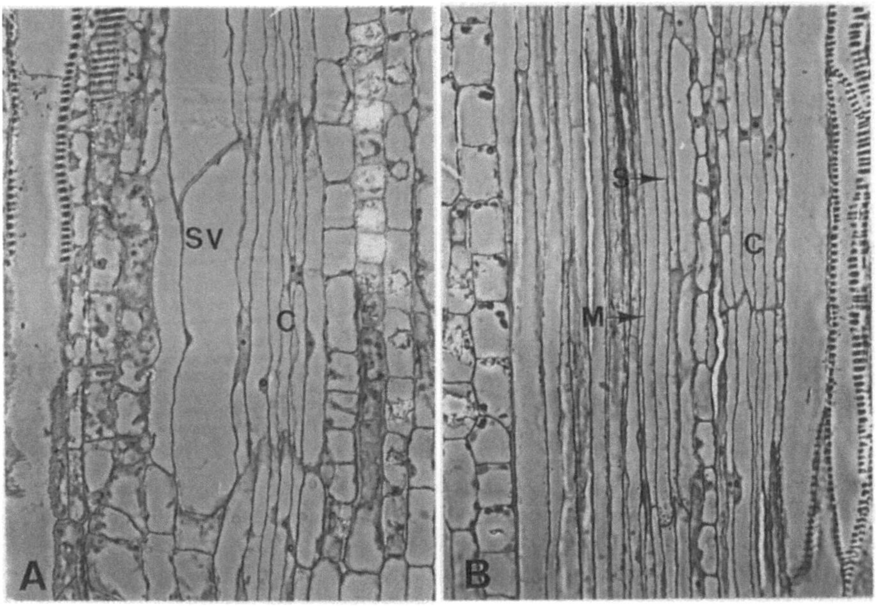

Fig. 7.9. Radial longitudinal sections through the cambial zone of *Populus deltoides*. The newly derived secondary vessel element (SV) in A and the sieve element (S) in B are approximately equal in length to the cambial cells (C) from which they were derived. In B, the metaphloem elements (M) are considerably longer than the secondary sieve element (S). Bright field A, 174X; B, 276X (Isebrands and Larson 1963)

Although this fact has been known for years, it was difficult for early workers not only to make these determinations but also to interpret them correctly. For example, Hartig (1859a) claimed that the "Holzröhren," or wood vessels, originated from the "growing together" of derivatives of "cambial fibers" that did not grow anymore. Although this statement might imply that the vessel members comprising the vessel did not elongate, Sanio (1863a) pointed out that Hartig's reference to growing together might also have meant the lateral union of two adjacent cambial cells.

Cohn (1857), in his investigations of *Vitus vinifera*, was of the opinion that vessels formed when certain cambial cells elongated to form a long tube with pointed ends that "pushed" past one another. In each of these tubes, a series of short cells arose by transverse cross walls. The short cells then elongated, the cross walls were "resorbed," and the vessels matured. However, the vessel segments were often shorter than the cambial cells. Sanio (1863a) also found vessel segments to be shorter than the cambial cells in *Cytisus laburnum*, measuring about 0.14 and

0.16 mm, respectively. He attributed this difference to maturation of the vessel segment. That is, whereas the newly derived vessel segment had a sharply oblique end wall just like the cambial cell, the mature vessel segment appeared shorter because the cross wall became horizontal during subsequent expansion. Nägeli (1858), on the contrary, found vessel elements to be approximately the same length as the cambial cells. He explained this relation by assuming that cambial cells simply divided by longitudinal walls; consequently the daughter cells retained their lengths. Sanio (1873) also arrived at this same conclusion in his later work, as did de Vries (1876c) and Kny (1886), and by the time of Hartig (1891) it was an accepted fact.

Bailey (1920a) examined 54 angiosperm species in 14 families and compared lengths of vessel segments with those of cambial initials. Selected data from Bailey's paper are shown in Table 7.4. In most species, the cambial initials were approximately the same lengths as the vessel segments. They tended to be slightly shorter than vessel segments in *Alnus, Euptelea, Myristica, Liquidambar, Rhizophora,* and *Nyssa,* species with primitive-type vessels, and slightly longer than these elements in species with highly specialized conducting systems. Again, by allowing for a 5 to 10% error, Bailey suggested that the vessel segments of arborescent and fruticose dicotyledons might be used to estimate the approximate lengths of the cambial initials.

Table 7.4. Mean length of cambial initials, vessel segments, and fiber tracheids in selected dicotyledons (After Bailey 1920a)

Species	Cambial initials Mean length (mm)	Vessel segments Mean length (mm)	Fiber tracheids Mean length (mm)
Acer rubrum	0.49	0.49	0.84
Alnus incana	0.60	0.66	1.89
Betula populifolia	0.94	0.89	1.31
Carya glabra	0.56	0.43	1.13
Fraxinus americana	0.29	0.31	0.96
Liquidambar styraciflua	0.70	0.76	0.96
Nyssa sylvatica	0.83	1.25	1.76
Populus sp.	0.49	0.50	0.90
Prunus serotina	0.46	0.45	0.99
Quercus alba	0.53	0.46	1.00
Robinia pseudoacacia	0.17	0.18	0.87
Sassafras officinale	0.39	0.39	0.61
Sterculia foetida	0.37	0.35	2.11
Ulmus americana	0.35	0.33	1.53

Chalk and Chattaway (1934) continued this line of investigation and found that it held true only for diffuse-porous woods. That is, the total length of a vessel member was normally equivalent to that of the cambial initial from which it was derived in these species. Total length included the vessel tails, since they found no evidence for extracambial growth of the vessel extremities. Essentially the same relationship prevailed for the latewood vessels of ring-porous species. However, earlywood vessels were usually shorter than the cambial initials, ranging from 13 to 47% shorter in the 12 species examined (Chalk and Chattaway 1935). This difference was not due to shorter cambial initials in the earlywood, but to some peculiar readjustment of the end walls that accompanied the tremendous lateral expansion of these vessels. Chalk (1970) later reaffirmed this observation in the earlywood vessels of *Fraxinus*. A different interpretation was offered by Swamy et al. (1960). They suggested that the shorter vessel segments in ring porous woods might be due to more frequent anticlinal divisions of the fusiform initials.

The findings of Bailey (1920a) and Chalk and Chattaway (1934, 1935) provided justification for many investigators to use vessel member lengths to estimatedo the lengths of fusiform initials in both diffuse- and ring-porous species (Hejnowicz and Hejnowicz 1958, Kaeiser 1964, van der Graaff and Baas 1974, Carlquist and Hoekman 1985, Mahmooduzzafar and Iqbal 1986) as well as in those species with storied cambia (Bailey 1944a, Esau and Cheadle 1955, Hejnowicz and Hejnowicz 1959, Butterfield 1972, Pulawska 1973, Krawczyszyn and Romberger 1979, Metcalfe and Chalk 1983, Cumbie 1984).

Table 7.5. Relation between fusiform initial lengths and vessel element lengths in several representative species in India

Species	Fusiform initial length (μm)	Vessel element length (μm)	Reference
Polyalthia longifolia	405	366	Sharma et al. 1979b
Jacaranda mimosifolia	288	191	Anand et al. 1981
Putranjiva roxburghii	1048	960	Anand et al. 1981
Eucalyptus gracilis	310	265	Ajmal et al. 1985
Eucalyptus tereticornis	273	149	Ajmal et al. 1985

Relative to the above, Butterfield (1973) called attention to the fact that no one had critically evaluated this relationship. To do so, he compared fusiform initial lengths measured on microsections to vessel member lengths measured on macerations in each of 25 samples of *Hoheria angustifolia*, a species with storied

Table 7.6. Lengths of vessel segments and parenchyma strands in selected diffuse- and ring-porous species (After Süss 1967)

Group[a]	Species		Vessel segment	Parenchyma strand	Vessel segment[b]
			Mean length (μm)		Relative length %
I	*Quercus robur*	LW[c]	555	550	100.9
	Betula pendula		915	920	99.5
	Acer platanoides		290	290	100.0
II	*Fagus sylvatica*	EW	550	520	105.8
		LW	615	520	118.3
		Mean	583	520	112.1
	Platanus acerifolia		670	640	104.7
	Sambucus racemosa		335	300	111.7
III	*Quercus robur*	EW	440	470	93.6
	Ulmus glabra	EW	205	250	82.8
		LW	245	250	98.0
		Mean	225	250	90.0
	Fraxinus excelsior	EW	240	295	81.4
		LW	290	295	98.7
		Mean	265	295	89.8
	F. pennsylvanica	EW	255	300	85.0
		LW	290	300	96.7
		Mean	273	300	91.0

[a] Group I, vessel segments (VS) and parenchyma strands (PS) about the same lengths; Group II, VS longer than PS; Group III, VS shorter than PS.
[b] Length of vessel segment/length of parenchyma strand x 100.
[c] LW, latewood; EW, earlywood.

cambium. Results of a statistical analysis revealed a strong relationship with a highly significant correlation coefficient at the 99% confidence level. This evaluation confirmed previous conclusions and provided strong support for using vessel member lengths to predict fusiform initial lengths.

Several investigators in India have greatly expanded the list of species in which the lengths of derivative elements were compared to lengths of fusiform initial cells. Although vessel element lengths approximated those of the fusiform initials in some cases, the former were more often shorter; a few representative examples are presented in Table 7.5. The authors attributed these differences to the fact that the end walls of fusiform initials had tapering tips, whereas those of vessel elements ranged from slightly oblique to almost transverse (Sharma et al. 1979b,

Anand et al. 1981), or to more frequent anticlinal divisions (Ajmal et al. 1985). In species without distinct growth rings, no specific pattern of vessel segment length could be detected (Swamy et al. 1960).

As noted, in some tropical species vessel elements are shorter than the fusiform initials. Similar differences exist in many temperate zone species. For example, many authors have observed variations in vessel element lengths across growth rings, all of which cannot be accounted for simply by differences in fusiform initial lengths (Bosshard 1951, Hejnowicz and Hejnowicz 1958, Meyer-Uhlenreid 1958, Swamy et al. 1960, Süss 1967, Chalk 1970). Süss (1967) used the lengths of parenchyma strands as an indicator of the lengths of fusiform initials. His data showed that vessel element lengths might be either about the same length, shorter, or longer than the associated parenchyma strands (Table 7.6). Moreover, vessel elements in latewood were invariably longer than those in earlywood in ring-porous species. Therefore, Bailey's (1920a) assumption of a 5 to 10% difference in the lengths of vessel elements and cambial initials should not be accepted without question. Like Chalk and Chattaway (1935), Süss attributed changes in lengths of vessel members during their differentiation primarily to transformations that occurred in formation of the end walls. However, because these changes alone could not account for the extreme shortening of earlywood vessel members of ring porous woods some other mechanism must be involved.

In a later study of *Fagus sylvatica*, Süss and Müller-Stoll (1984) found that vessel element lengths not only increased from earlywood to latewood but to varying extents at different height levels in the stem (Table 7.8). These comparative values were based on an assumed cambial initial length of 500 μm derived from the means of parenchymatous strands in the first-formed earlywood of the 1955 growth ring. The authors concluded that vessel element length was not necessarily a reliable indicator of fusiform initial length.

Transformations in the vessel end walls apparently account for much of the variation between lengths of vessel segments and fusiform initials. However, the generally observed increase in vessel element length from earlywood to latewood in nonstoried woods is probably due to differences in the frequency of anticlinal divisions of the fusiform initials, as pointed out by Swamy et al. (1960) and Metcalfe and Chalk (1983). Thus, while it is very likely that lengths of vessel elements approximate those of fusiform initials, the relationship should not be accepted without adequate testing. Considerable variation can result simply from the method employed in measuring vessel lengths and vessel tails, as discussed by Chalk and Chattaway (1934). Another source of variation inevitably arises from the comparison of vessel element lengths measured in macerations with those measured on microsections. However, the greatest source of error arises from the assumption of a direct relationship.

Polyploid trees, as a rule, have longer vessel elements and therefore longer fusiform cambial initials than diploid trees. This proved to be the case in *Fraxinus americana*, in which vessel elements from hexaploid and tetraploid trees were on average 43 and 27% longer, respectively, than those from diploid trees (Armstrong 1982). Armstrong attributed these differences to the longer cambial initials in the polyploid trees.

Vessel element lengths, hence cambial initial lengths, also exhibit ecological trends. As one example, Baas (1986) and Baas et al. (1983) demonstrated that vessel element length decreased with latitude and altitude of provenance, i.e., with cooler environments. Moreover, species growing in xeric habitats had shorter vessel elements than those growing in mesic habitats. Vessel elements in shrubs, as a rule, were shorter than those in trees, and those in dwarf specimens were substantially reduced (Baas et al. 1984). Carlquist (1962, 1989) attributed many of these aberrant patterns in dicotyledonous woods to paedomorphosis, or the retention of juvenile characteristics during mature growth. In such species, the vessel elements become shorter rather than longer as growth progresses, due to shortening of the fusiform cambial initials.

A number of authors have found that a correlation exists between vessel element length and ray height (Ferreirinha 1965, van der Graaff and Baas 1974, Giraud 1980, Carlquist and Hoekman 1985) (Fig. 7.10). Such a correlation might be expected in those species in which rays are derived directly by the segmentation of fusiform initials. As shown in Fig. 7.10, a correlation also exists between vessel element length and fiber length. This, too, is a commonly observed relationship (Bergman 1949, Ferreirinha 1965, Giraud 1980, Rao and Dave 1981,

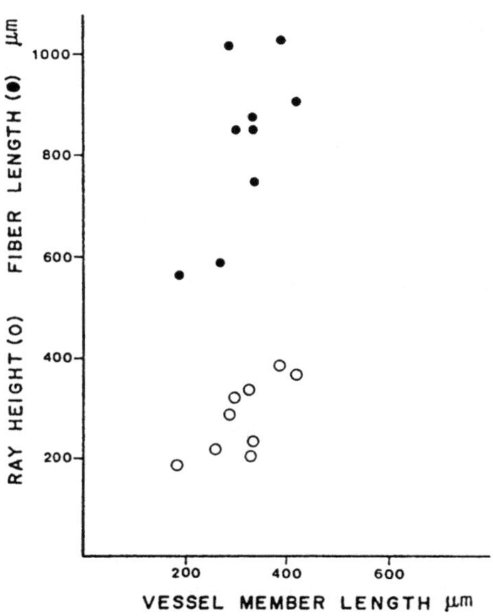

Fig. 7.10. Correlations between mean fiber length, mean ray height, and mean vessel member length in *Melia azedarach*. (van der Graaff and Baas 1974)

Carlquist and Hoekman 1985, Carlquist 1988a). Even though fibers undergo extracambial intrusive growth, a mutual relationship is apparently maintained since both elements are originally derived from the same cambial initials.

7.1.3.2 Vesselless Dicotyledons

The absence of vessels in dicotyledons is assumed to be a primitive characteristic. Caspary (1862) confirmed earlier reports dating back to 1742 claiming that members of the genus *Drimys* were vesselless. He also provided a list of other species that he considered to be without vessels.

According to Bailey (1944b), fusiform initials in the nonstratified cambia of the vesselless Winteraceae periodically divided anticlinally, they elongated extensively, and commonly attained lengths of 5000 μm in the outer parts of old stems. Tracheids derived from these initials elongated only slightly during differentiation. Consequently, they resembled the cambial initials in both length and width.

Tracheids in the Winteraceae are the longest tracheary elements found among the dicotyledons (Bailey and Tupper 1918). A similar primitive type of cambium in *Amborella trichopoda* was described by Bailey (1957) as having long fusiform initials with extensively overlapping end walls.

The average length of tracheids in the vesselless dicotyledons exhibits a very wide range of values. According to Carlquist (1975), tracheid lengths appeared to be closely related to plant size. Thus, the longest tracheids were found in *Trochodendron araloides*, the tallest tree in the vesselless group. Additional data can be found in Patel (1974) and Carlquist (1988a).

The wood of two species of *Ephedra* from high alpine localities was found by Carlquist (1988b) to be near-vesselless. This structure was interpreted as being an adaptive condition to a severe climatic habitat, and it was thought to be partially reversible when grown under more moderate conditions.

7.1.3.3 Libriform Fibers

Xylem fibers of dicotyledons, unlike the vessels, undergo varying amounts of extracambial intrusive elongation. As a general rule, fiber tracheids and libriform fibers are considerably longer than the cambial initials from which they are derived (Kny 1886, Dinwoodie 1961, Khan et al. 1979b, Sharma et al. 1979b, Fujita et al. 1984, Ajmal et al. 1985).

Some rather crude, but revealing, measurements were made by Braun (1854). In *Aesculus hippocastanum* and *Sambucus nigra* the length of cambial cells ranged from 1/5 to 3/10 mm and 1/5 to 1/3 mm, respectively, and that of wood cells ranged from 1/3 to 1/2 mm and 1/2 to 2/3 mm, respectively. In both species, the wood cells increased more than 50% in length and their ends became more pointed. Cohn (1857) made similar measurements in *Vitis vinifera*, *Ampelopsis hederifolia*, and several other vines. Frank (1864) also demonstrated how cambial cell lengths, his cambial fibers, increased from the first to second growth rings in *Quercus pedunculata* and *Taxus baccata*. A concomitant increase occurred in the wood fibers and tracheids, respectively. As shown in Fig. 7.11 from Süss (1967), extracambial elongation might occur at either one end or both ends of a xylem fi-

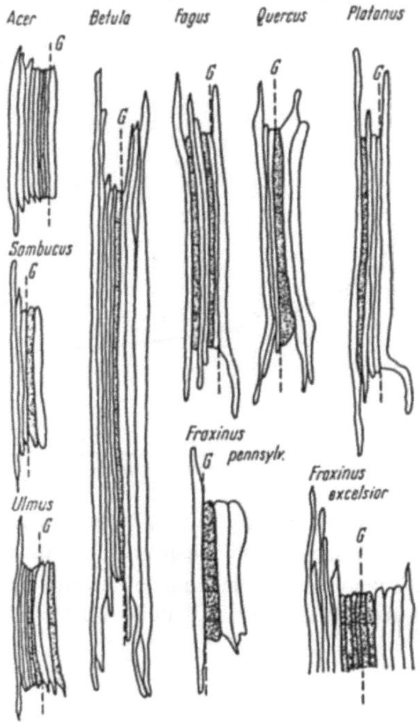

Fig. 7.11. Drawings of macerations in radial section illustrating elongation of xylem fibers relative to those of parenchyma strands (*shaded*) in different species. Last latewood cells to the *left* and first earlywood cells to the *right* at the growth ring boundary (*G*). Parenchyma strands were assumed to approximate the lengths of cambial initials. (Süss 1967)

wood fibers and tracheids, respectively. As shown in Fig. 7.11 from Süss (1967), extracambial elongation might occur at either one end or both ends of a xylem fiber.

Chattaway (1936) compared mean fiber lengths with mean cambial initial lengths in 276 genera representing 111 different families. An interesting relationship was revealed by this exercise. In genera with short cambial initials the fibers were approximately 2.5 to 3.5 times their initial lengths, whereas in genera with longer cambial initials they were only about 1.2 times their initial lengths (Fig. 7.12B). When mean fiber length was plotted against mean cambial initial length, a straight line relation was obtained (Fig. 7.12A). Chattaway could offer no explanation for this relationship other than pointing out that more specialized species tended to have shorter cambial initials, consequently shorter fiber lengths.

In his extensive analyses of vessel member lengths in angiosperms, Bailey (1920a) also measured the lengths of fiber tracheids. A few representative examples comparing fiber length to cambial initial length made by Bailey appear in Table 7.4, and a similar comparison made by Süss (1967) in Table 7.7. In the latter

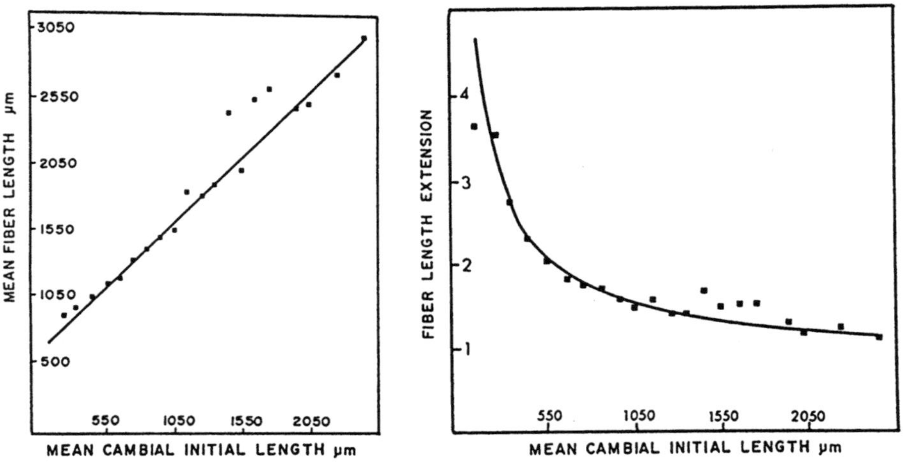

Fig. 7.12A,B. Comparisons of xylem fiber and cambial initial lengths in 276 genera representing 11 families. **A** Relation between mean fiber length and mean cambial initial length. **B** Relation between fiber length extension (expressed as the ratio of mean fiber length to mean cambial initial length) and mean cambial initial length. (Chattaway 1936)

Table 7.7. Comparisons of the mean maximal lengths of xylem fibers and the mean lengths of parenchyma strands (Süss 1967)

Species	Xylem fibers Length (µm)	Parenchyma strand Length (µm)	Xylem fibers Relative increase (%)
Acer platanoides	675	290	133
Betula pendula	1440	920	57
B. populifolia	1310	940	40
Carya ovata	1300	520	150
Fagus sylvatica	1375	520	164
Fraxinus americana	960	290	231
F. excelsior	1070	295	263
F. pennsylvanica	975	300	225
Liquidambar styraciflua	960	700	37
Platanus acerifolia	1640	640	156
Quercus alba	1000	530	89
Q. robur	1205	470	156
Robinia pseudoacacia	870	170	412
Ulmus americana	1530	350	337
U. glabra	1260	250	404

case, the cambial initials were assumed to be comparable in length to the parenchyma strands. The variation in fiber length at different height levels in a *Fagus sylvatica* stem is shown in Table 7.8.

Not all xylem fibers are subjected to extensive elongation growth. Bosshard (1952) described short fibers associated with the large earlywood vessels in the wood of *Fraxinus excelsior*. The middle parts of individual fibers were essentially the same lengths as the cambial initials from which they were derived, but the intrusive ends were abruptly constricted. Chalk (1970) found similar short fibers to be present in all eight species of *Fraxinus* he examined. Some of these fibers had rounded ends and they evidently had experienced no intrusive growth. Mean lengths of the short pore-zone fibers averaged 450 μm compared to normal fibers of earlywood and latewood that averaged 1240 and 1280 μm, respectively. Chalk referred to these short fibers as "fusiform fibers."

The assumption that vessel members do not undergo extracambial elongation has been used by a number of workers to evaluate the patterns of variation in xylem fibers. For example, in a study of 49 dicotyledonous trees, Bergman (1949) found a highly significant correlation between fiber length and vessel element length. Other workers have drawn conclusions regarding fusiform cambial initials by comparing the changes in lengths of fibers with those of vessels. Hejnowicz and Hejnowicz (1958) used this approach to calculate what they called the "coefficient of intrusive growth" for the xylem fibers of *Populus tremula*. Based on data from their study, they concluded that "... the length of fusiform initials is regulated by the relative age of cambium without regard for absolute age or the part of the trunk, provided that the part closest to the ground is disregarded."

Age-Related Patterns: Like conifer tracheids, xylem fibers of dicotyledons exhibit the same age-related pattern described by Bailey and Shepard (1915). Also like conifer tracheids, this age-related pattern can be accounted for by changes in the year-to-year pattern of anticlinal divisions associated with age, or distance from the pith, and modified by rate of growth. However, in general, the fibers of dicotyledons undergo more extensive extracambial elongation than conifer tracheids do. Sanio's first law was re-examined by Shimaji (1950) on fibers of a number of dicotyledons. Although the law was confirmed, fluctuations occurred in the pattern, which Shimaji attributed to growth tensions on the cambium.

Within-Ring Patterns: Early research had shown that fiber length in dicotyledons varied across growth rings from earlywood to latewood just as tracheid length did in conifers. This within-ring pattern was re-examined in 28 angiosperm species by Bissett and Dadswell (1950). In all species showing distinct growth rings, both diffuse and ring-porous, a definite increase in fiber length was observed from the first-formed earlywood to the last-formed latewood within an individual growth ring. At the growth ring boundary, where the latewood of one ring gave way to earlywood of the next, an abrupt decrease in fiber length occurred.

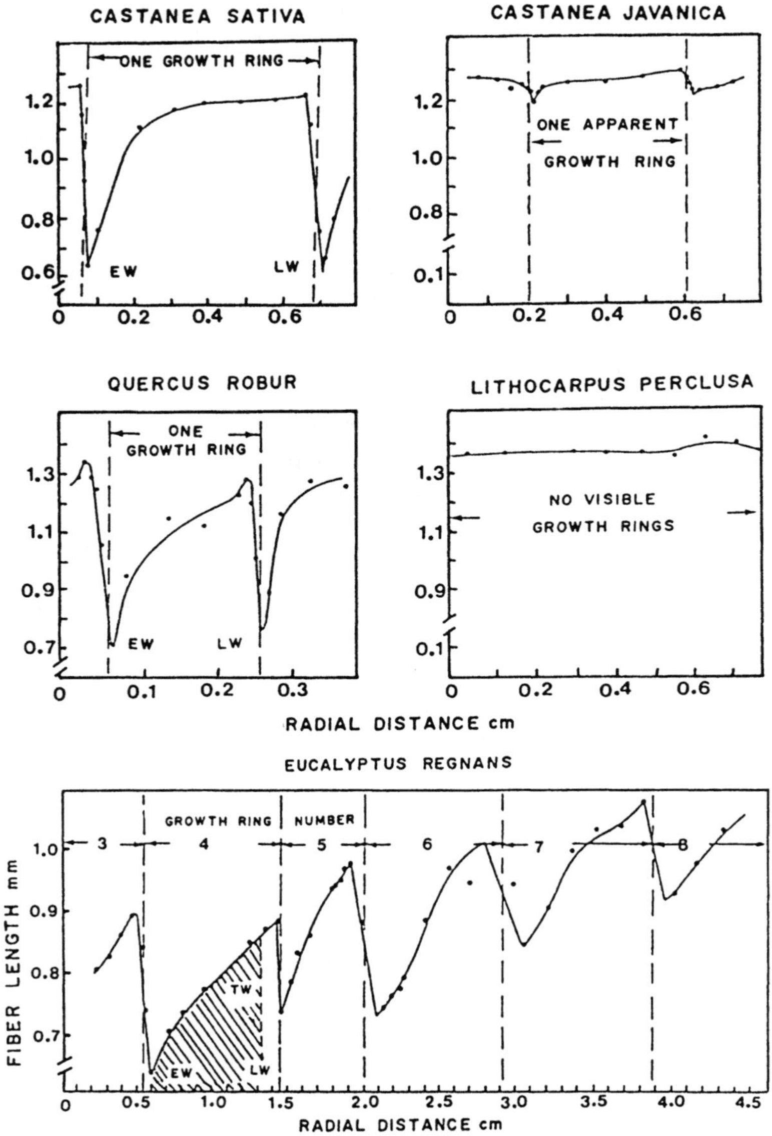

Fig. 7.13. Variations in fiber length within growth rings of dicotyledons. *Top Castanea sativa* and *Quercus robur*, temperate-zone, ring-porous species with well-defined growth rings. *C. javanica* and *Lithocarpus perclusa*, tropical-zone, diffuse-porous species without well-defined growth rings. *Bottom Eucalyptus regnans*. The within-ring pattern of fiber length in the growth ring containing tension wood (*TW*) is the same as in normal wood, and the upward trend with increasing growth ring age is maintained. *EW* earlywood; *LW* latewood. (After Bissett and Dadswell 1950)

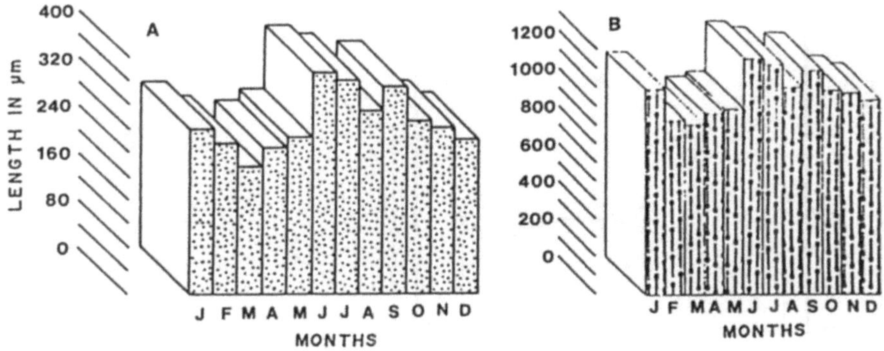

Fig. 7.14A,B. Histograms showing seasonal variation in mean lengths of fusiform initials (**A**) and xylem fibers (**B**) of *Tectona grandis*. (Rao and Dave 1981)

Certain tropical species without distinct growth zones, such as *Castanea javanica* and *Lithocarpus perclusa*, displayed little, if any, variation in fiber length either across a single growth ring or a radial series of growth rings. However, their temperate-zone counterparts with distinct growth rings, such as *Castanea sativa* and *Quercus robur*, showed the characteristic angiosperm pattern (Fig. 7.13). As they had postulated for the tracheids of conifers, the authors believed that the increase in fiber length from earlywood to latewood was due to a "time factor." That is, because cambial divisions were more numerous early in the growing season, less time was available for elongation between divisions than later in the season. However, whereas in conifers the difference between earlywood and latewood elements seldom exceeded 15%, in angiosperms it often exceeded 200%. Unlike the abrupt decrease in tracheid length during compression wood formation in conifers, fiber lengths continued to increase through the tension wood zone of *Eucalyptus regnans* (Fig. 7.13).

Similar within-ring patterns were observed by other investigators (Amos et al. 1950, Dinwoodie 1961, Süss 1967). For example, Bosshard (1952) confirmed the pattern in the ring-porous species *Fraxinus excelsior* and Hejnowicz and Hejnowicz (1958) in the diffuse-porous species *Populus tremula*. Both authors attributed the differences partly to differences in fusiform initial lengths but primarily to extracambial intrusive elongation of the fiber derivatives.

In the tropical species *Tectona grandis*, Rao and Dave (1981) observed a remarkable relationship between fusiform initial lengths and fiber lengths, both of which varied seasonally. During the dormant period, xylem fibers were 3.7 to 4.1

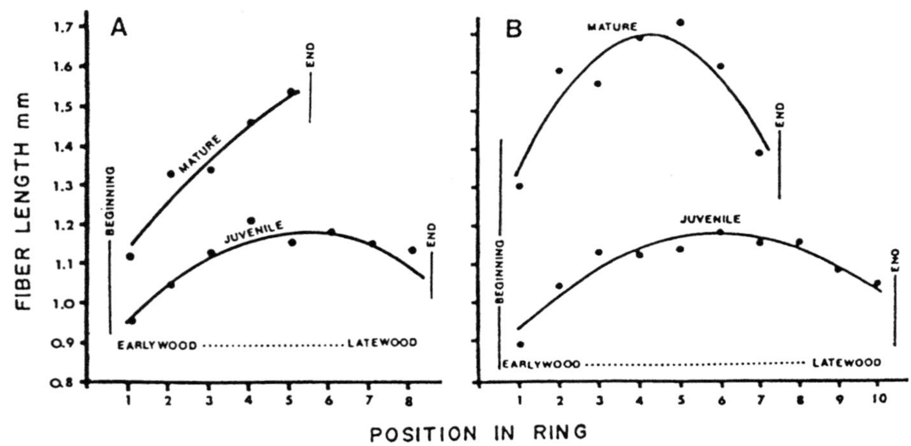

Fig. 7.15A,B. Relationship of fiber length to position within growth rings of juvenile and mature wood of *Quercus stellata* (**A**) and *Carya tomentosa* (**B**). Each *point* on the abscissa represents 250 tangential sections 250 μm in thickness. (Taylor 1977)

times longer and during the active period 3.3 to 3.5 times longer than the fusiform initials (Fig, 7.14). A similar, but less clear, relationship was found in *Gmelina arborea* (Dave and Rao 1982a) and *Mangifera indica* (Dave and Rao 1982b). The latter species has evergreen foliage and cambial activity is continuous throughout the year. Nonetheless, the trees exhibited growth pauses during which time the lengths of both the fusiform initials and derivative xylem fibers showed slight declines. A rather close relationship between fusiform initial length and fiber length has also been reported by other investigators in tropical species (Anand et al. 1978, 1981, Sharma et al. 1979b, Rao and Dave 1985b).

The within-ring pattern is neither perfect nor consistent. As shown by Taylor (1977) in seven southern hardwoods, variable patterns can be found both within individual trees and among species. In general, a more or less typical pattern prevailed in juvenile growth rings near the stem center but variability increased in the mature growth rings of older trees. In some species, such as *Carya tomentosa*, maximum fiber lengths occurred in the middle of the growth ring and in others, such as *Quercus stellata* (Fig. 7.15), it occurred at the end of the growth ring.

As in conifer tracheids, the within-ring pattern of fiber length variation is superimposed on the age-related pattern. Thus, fiber length increases from year to year, at least during juvenile growth, despite the intra-seasonal fluctuations (Fig. 7.13). Obviously, lengths of xylem fibers cannot be used to directly estimate the lengths of fusifom cambial initials.

7.1.3.4 Living Wood Fibers

Living wood fibers in dicotyledons are simply variations of the dead fiber types (Wolkinger 1969, 1970, 1971). Living wood fibers are of two types; namely, nonseptate and septate.

Nonseptate: Sanio (1863a) recognized "Ersatzfasern", or substitute fibers, in the ground tissue of many woody species. These were thick-walled prosenchymatous cells derived from parenchyma and they were easily confused with true xylem fibers. In many species, the substitute fibers retain their parenchymatous characteristics, and they are known as fusiform parenchyma cells (Metcalfe and Chalk 1983). Nonseptate wood fibers are libriform fibers and in some cases fiber tracheids that retain their living protoplasts. Such cells are commonly found in woody shrubs growing in unfavorable habitats (Fahn and Lesham 1963). They also occur abundantly in wide tangential bands surrounding the vessels in roots of *Acer saccharum* (Gregory 1978). In *Tamarix aphylla*, these wood fibers retained their living protoplasts and nuclei for 16 to 21 years, or until the initiation of heartwood formation (Fahn and Arnon 1963).

Septate: Septate wood fibers are formed by transverse division of the protoplast either during or after completion of the secondary wall (Record 1934, IAWA 1964). These fibers are the "gefächerte Holzzellen," (chambered wood cells) or the "Fächerprosenchym" (segmented prosenchyma) described by Sanio (1857b). He demonstrated how these cells arose by transverse division of a fiber cell, but not an initial cell. For example, he stated, "The undivided fiber cell is primary, however not the divided." Thus, transverse division of the thin-walled wood parenchyma cells occurred early whereas division of the thick-walled, segmented wood cells occurred later.

Because of the late development of septae in living fibers, all junctions are made with the parent's secondary wall. Consequently, the septae remain primary and they consist almost entirely of pectin (Wolkinger 1969). Each septate cell within the fiber might retain its protoplasm and nucleus for many years. The most characteristic feature of septate fibers is their richness in starch, and both starch deposition and dissolution parallel that of the ray cells. Thus, the primary functions of living wood fibers are storage and mechanical strength with no participation in the hydrosystem of the tree (Wolkinger 1970). As a general rule, Wolkinger (1969) found that living wood fibers were shorter than living nonseptate fibers and the latter were shorter than dead fibers.

The development of septate fibers in *Lagerstroemia speciosa* and *Protium serratum*, ring- and diffuse-porous species, respectively, was described by Purkayastha (1958). After being cut off from the cambium, septa formed by nuclear div-

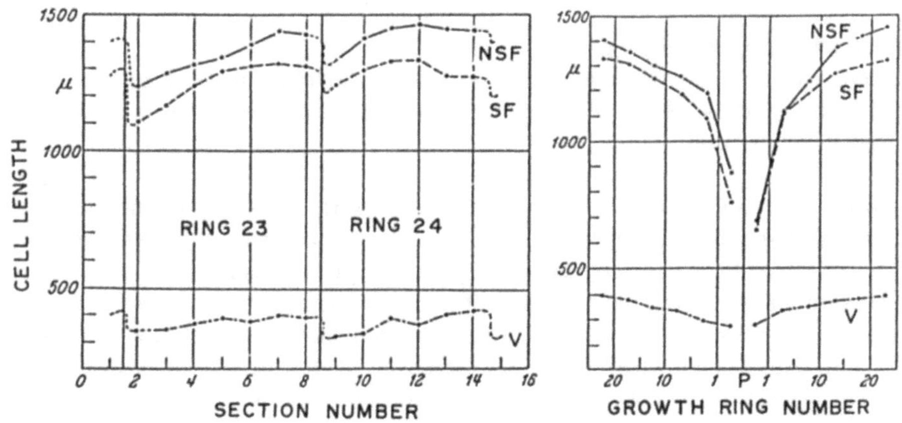

Fig. 7.16. Lengths of vessels (*V*), septate (*SF*), and nonseptate (*NSF*) fibers in two growth rings (*left*), and their variation with growth ring number from the pith (*P*) (*right*) in *Tectona grandis*. (Parameswaran 1964)

ision followed by cytokinesis at right angles to the fiber walls. Because the septa formed after secondary walls were laid down and fibers had attained maturity, further extracambial elongation was impossible. Septa formation in *Ribes sanguineum* proceeded in much the same way (Parameswaran and Liese 1969).

In a 25-year-old stem of *Tectona grandis*, Parameswaran (1964) found that the lengths of both the septate and nonseptate fibers increased about 24% and vessel element lengths about 15% across the growth ring from earlywood to latewood (Fig. 7.16A). All three xylem elements increased in length with increasing growth ring age, with the septate fibers always remaining shorter than the nonseptate fibers (Fig. 7.16B). In *Pithecolobium dulce*, in contrast, the septate fibers were always longer than the non-septate ones (Govindarajalu and Swamy 1955). The variability in length of septate fibers due to extracambial elongation makes them poor candidates for estimating fusiform initial lengths.

Septate fibers are more commonly found in tropical than in temperate-zone species. Lists of families in which these fibers occur can be found in Metcalfe and Chalk (1950), Chowdhury et al. (1956), and Carlquist (1988a). Wolkinger (1971) found living wood fibers in 71 families, 224 genera, and 413 species of dicotyledons.

7.1.3.5 Axial Parenchyma

Septation in axial parenchyma occurs during primary wall development. Consequently, secondary wall formation of the septae takes place concomitant with formation of the parent cell wall. The axial parenchyma strands in many dicotyledons have been assumed by some workers to be the same length as the fusiform initials from which they are derived (Chalk and Chattaway 1934, Butterfield 1972, Sharma et al. 1979b, Metcalfe and Chalk 1983). Krawczyszyn and Romberger (1979) compared parenchyma strand lengths in *Entandrophragma* spp. with vessel lengths. By assuming that vessel elements did not elongate, they found that parenchyma strand length overestimated fusiform initial length by as much as 7.7%. Yoshimura (1979) and Yoshimura et al. (1984) examined this relationship in *Fraxinus mandshurica*. Although parenchyma strand length increased slightly from earlywood to latewood, the change was considered by the authors to be insignificant. Fiber length changes within the growth ring, however, were substantial. The authors concluded that these intra-annual variations were due to extracambial elongation of the xylem fiber derivatives and not to elongation of the fusiform initials.

Table 7.8. Lengths of wood elements, and their absolute and relative lengths of tip growth in the 1955 growth ring at different stem heights in a *Fagus sylvatica* tree (Süss and Müller-Stoll 1984)

Stem height (m)	Ring width (mm)	Xylem element length (μm)			Tip growth (μm)			Tip growth, percent of cambial initial[a]		
		PS[b]	VM[c]	XF[d]	PS	VM	XF	PS	VM	XF
0.0	6.5	530	562	1272	30	62	722	6.0	12.4	154
1.3	5.1	549	596	1293	49	96	793	9.8	19.2	159
6.0	4.6	558	580	1240	58	80	740	11.6	16.0	148
12.0	3.9	531	566	1182	31	66	682	6.2	13.2	136
15.0	4.1	529	563	1154	29	63	654	5.8	12.6	131
18.0	2.8	543	577	1183	43	77	683	8.6	15.4	137

[a] Based on assumed lengths of cambial initials, 500 μm.
[b] Parenchyma strand
[c] Vessel member
[d] Xylem fiber

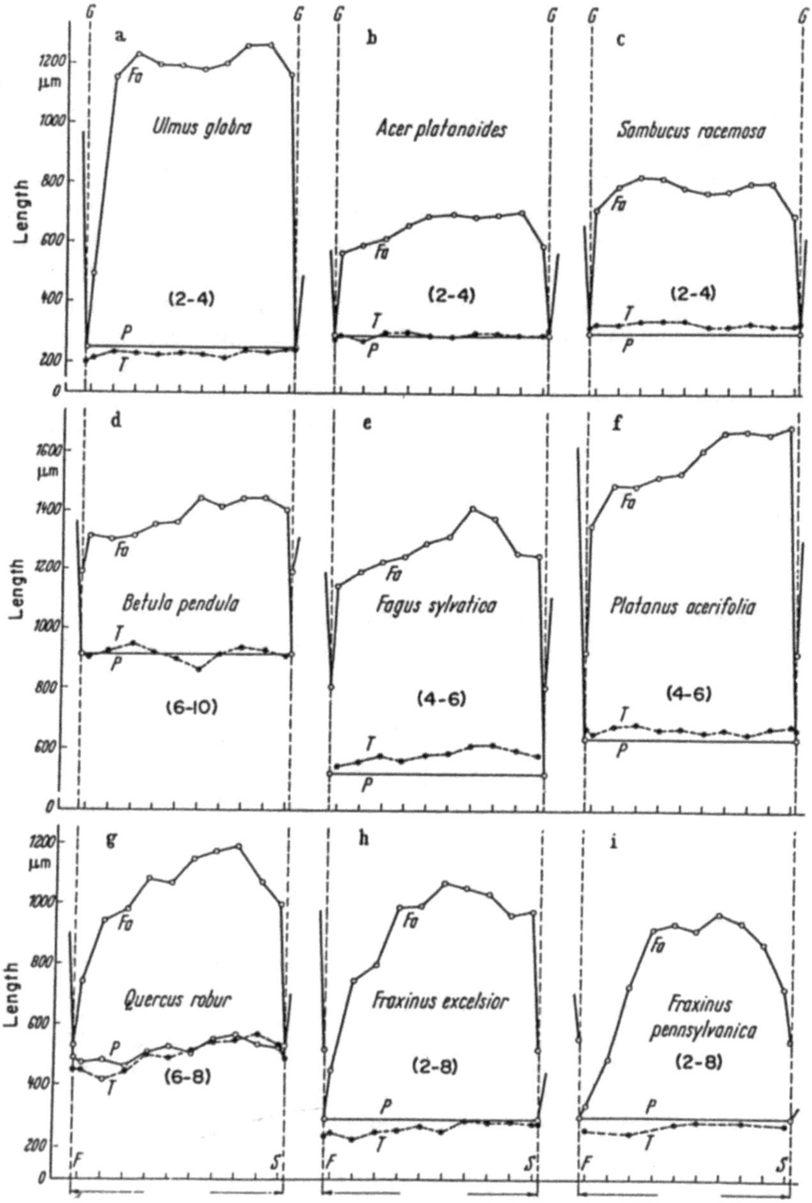

Fig. 7.17a-i. Variation in mean lengths of xylem fibers (*Fa*) within a growth ring compared to the mean lengths of parenchyma strands (*P*) and vessel members (*T*) for nine species. Growth ring boundaries at *G*; earlywood (*F*): latewood (*S*). Values in *parantheses* represent the mean number of cells in a parenchyma strand. (After Süss 1967)

One of the most detailed investigations of parenchyma strand lengths was conducted by Süss (1967) on five diffuse- and four ring-porous species. As shown in Fig. 7.17, practically no variation in parenchyma strand length existed within growth rings with the exception of *Quercus robur*. Additional measurements on this species suggested that the elongation, which averaged 18.7% from earlywood to latewood, occurred before subdivision of the fusiform derivatives by transverse walls.

Lengths of the parenchyma strands varied greatly among species, as did the number of cells per strand (Fig. 7.17). Highest values for both variables occurred in diffuse-porous and lowest in ring-porous species. These trends supported the hypotheses of Bailey (1920a) and Kribs (1937) that the lengths of fusiform cambial initials decrease with increasing specialization of the xylem; see Carlquist (1988a) for more on this relationship.

A later study by Süss and Müller-Stoll (1984) on *Fagus sylvatica* confirmed the minimal change in parenchyma strand length across a growth ring with the exception of upper stem levels within the crown, where an increase in length from earlywood to latewood occurred. The lengths of parenchyma strands were found to undergo variable amounts of extracambial tip elongation at different heights in the stem (Table 7.8). Consequently, the authors cautioned that one should not generalize in assuming that parenchyma strands accurately reflect the lengths of fusiform initials. The data in Table 7.8 should also be accepted with caution, for the authors assumed a standard fusiform initial length of 500 μm in making their calculations. This standard was not based on measurements of cambial initials, but on the mean length of parenchyma strands associated with the first earlywood of the 1955 growth ring at different tree heights.

A further precaution to be considered in using lengths of parenchyma strands in estimating lengths of fusiform initials is the fact that many parenchyma strands are derived from failing initials (Chap. 6.3.2.2). Thus, parenchyma strand lengths might not be representative of the fusiform initials giving rise to other xylem and phloem derivatives.

7.1.4 Dicotyledon Phloem

7.1.4.1 Sieve Elements

The lack of apical elongation is presumed to be a characteristic of sieve elements in both dicotyledons and gymnosperms (Esau 1969). Thus, sieve element lengths have been found to agree, in general, with those of fusiform initials in both dicotyledons (Esau and Cheadle 1955, Evert 1961, Cheadle and Esau 1964, Isebrands and Larson 1973, Khan et al. 1981, Fujita et al. 1984) (Figs. 7.2; 7.9B) and gymnosperms (Mischke 1890, Brown 1912, Klinken 1914, Abbe and Crafts 1939, Meeuse 1942, Bannan 1955, Grillos and Smith 1959, Bailey and Srivastava 1962, Srivastava 1963a,b, Wilson 1963, den Outer 1967, Gregory 1971). In fact, Klinken (1914) concluded that none of the phloem cells of *Taxus baccata*, neither the sieve elements, parenchyma, nor bast fibers, showed any evidence of extracambial elongation. However, in a radial series from the outer phloem toward the cambium, phloem cell length increased. This increase was related to increasing length of the cambial initial cells and not to extracambial elongation of the derivatives. A similar pattern of increasing sieve element length from the cambium outward was observed in *Prosopis spicigera* by Iqbal and Ghouse (1977). These authors attributed the increase to the combined effects of cambial initial length and intrusive growth.

It is well to keep in mind, however, that sieve elements are far less reliable than vessel members for estimating fusiform initial cell lengths. Esau and Cheadle (1955) examined this relationship in 91 species representing 44 families of dicotyledons and found that in many species anticlinal divisions of the phloem mother cells resulted in sieve elements considerably shorter than the fusiform initials. That is, the phloem mother cells were the same length as the fusiform initials, but subsequent anticlinal divisions of some phloem mother cells produced a population of short sieve elements that lowered the mean element length as shown in Table 7.9. As noted by den Outer (1967), the sieve cells of conifers can also display considerable variation.

Table 7.9. Length comparisons of cambial initials, sieve elements, phloem fibers, and vessel members in *Liriodendron tulipifera* (After Cheadle and Esau 1964)

Type of cell	Length (μm) Range	Average
Cambial initial	205-678	482
Sieve element	204-589	386
Phloem fiber	244-725	468
Vessel member	262-659	464

The data of Zahur (1959) indicated that sieve-element mother cells divided anticlinally in a large proportion of the woody dicotyledons he examined (Table 7.10). Evert (1960, 1961, 1963a) invoked the same explanation to account for the relatively short sieve elements found in some phloem regions of *Pyrus communis* and *P. malus* (Table 7.11), as did Sharma et al. (1979b) in *Polyalthia longifolia*. Unreliability was due to the occasional, sometimes frequent, anticlinal divisions of the phloem mother cells and *not* to intrusive growth of the sieve elements.

The sieve elements of many tropical species present somewhat different patterns than temperate-zone species. Lawton (1972) examined phloem development in seven species native to Nigeria. In four species, *Albizia adianthifolia*, *Antiaris*

Table 7.10. Occurrence of secondary septation of sieve-tube mother cells before differentiation in the Polypetalae and Gamopetalae. (Zahur 1959)

Group	Number of species studied	Sieve-tube mother cells Nonseptate	Septate
Polypetalae	304	200 (66%)	104 (34%)
Gamopetalae	120	24 (20%)	96 (80%)

Polypetalae: Corolla with petals separate.
Gamopetalae: Corolla with petals united.

africana, Holarhena floribunda, and *Ricinodendron heudelotii,* lengths of the sieve elements not only exceeded those of the fusiform initials but they also varied seasonally. The author suggested that the increase was probably due to post-cambial adjustments of the end walls by symplastic growth, a process that presumably preserved protoplasmic connections. Ghouse and Yunus (1975) also found sieve tube members in certain *Dalbergia* species to be longer than the cambial initials from which they were derived. Although the main bodies of the sieve elements did not elongate, the cell ends elongated by intrusive growth that produced peculiar apical structures and "tails." Similar increases in sieve element lengths by the formation of tails were reported in other tropical species by Ghouse and Iqbal (1977b, 1979). On the contrary, Sharma et al. (1979b) found the sieve elements of *Polyalthia longifolia* to be slightly shorter than the fusiform initials, due to a shifting of the end walls to a more transverse position. Sieve elements were also found to be shorter than the fusiform initials in *Jacaranda mimosifolia* and *Putranjiva roxburghii* (Anand et al. 1981).

As shown in Fig. 7.18, the lengths of sieve elements in two tropical species with nonstoried cambia, *Mangifera altissima* and *Shorea squamata,* decreased from the cambium to the periderm (Parameswaran and Liese 1974). A similar pattern was reported for *Antiaris africana* by Lawton (1972) and in several species by Ghouse and Iqbal (1977b). If one accepts the aforementioned correlation between fusiform initial and sieve element lengths, then such increases can be attributed to the well-known age effect. That is, lengths of the sieve elements reflect lengths of the fusiform initials which increase with increasing age of the phloem bands.

7.1.4.2 Phloem Fibers

Phloem fibers are derived directly from either fusiform initials or phloem mother cells. They undergo immediate intrusive growth and mature in functional phloem.

Fiber sclereids, which are often indistinguishable from phloem fibers at maturity, arise from parenchymatous cells in the nonfunctional phloem. They, too, undergo intrusive growth as well as cell wall thickening (Zahur 1959, Evert 1963a, den Outer 1967, Nanko et al. 1977).

Phloem fibers of conifers do not as a rule undergo any appreciable elongation (Klinken 1914, Bannan 1955, Srivastava 1963a, Esau 1969), although den Outer (1967) claimed otherwise. Phloem fibers of dicotyledons, on the other hand, show variable behavior, with some undergoing considerable elongation and others failing to elongate (Esau 1969). Schacht (1860), for example, found that bast fibers of some dicotyledons were five to ten times longer than the cambial cells, whereas

Table 7.11. Lengths of fusiform cambial initials and sieve elements of *Pyrus communis* and *P. malus*[a] (After Evert 1960, 1963a)

Group	*Pyrus communis* Cambial initial	*Pyrus communis* Sieve element	*Pyrus malus* Cambial initial	*Pyrus malus* Sieve element
	Length (μm)		Length (μm)	
Longest	711	693	717	677
Shortest	252	252	268	214
Average	551	460	475	437

[a] 200 cambial initials and 200 sieve elements chosen at random for each species.

Sanio (1863a) found that bast fibers of *Magnolia tripetala* were only slightly longer than the "cambial fibers," measuring 0.44 and 0.30 mm, respectively. In *Liriodendron tulipifera*, phloem fibers varied in length but typically were comparable in length to the cambial initials (Cheadle and Esau 1964). In *Populus euramericana*, the phloem fibers elongated from 210 to 240% after their origin from cambial initials (Nanko et al. 1977). In *Erythrina indica*, they were on average nine times as long as the fusiform initials from which they were derived (Ghouse and Sabir 1974).

An examination of phloem fiber lengths in five species in the family Myrtaceae by Khan and Khan (1983) showed that mature fibers averaged from 1.4 to 2.5 times longer than the fusiform initials. Species with the longest fusiform initials produced the longest fibers and vice versa. This rule did not, however, prevail in the genus *Dalbergia* (Ghouse and Yunus 1975). Although fusiform initial lengths were comparable (190-196 μm) among four species examined, phloem fiber lengths varied from a mean of 496 μm in *D. lanceolaria* to 1000 μm in *D. sissoo*. Similarly, no correlation was found in several species of *Acacia* and *Prospis* (Ghouse and Iqbal 1977b).

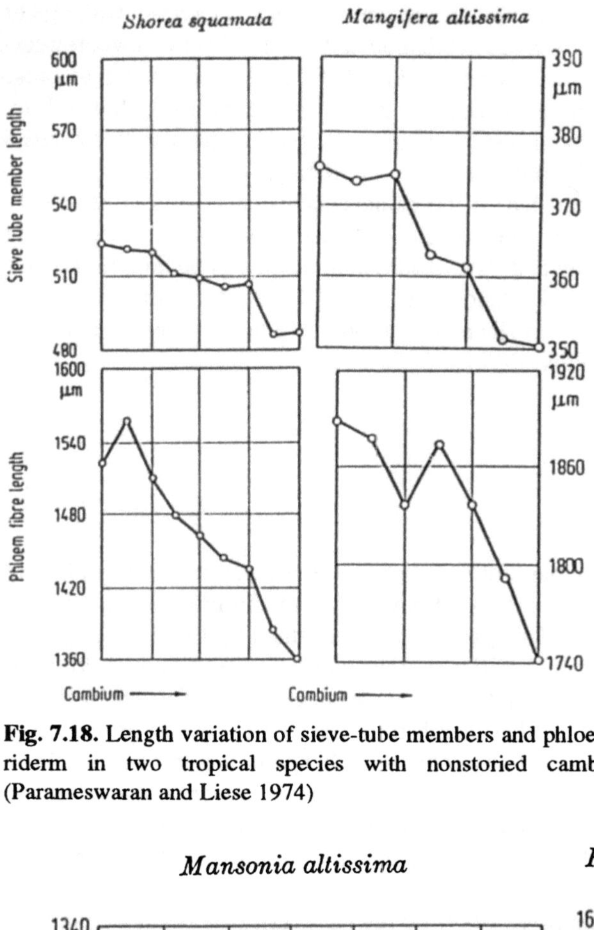

Fig. 7.18. Length variation of sieve-tube members and phloem fibers from cambium to periderm in two tropical species with nonstoried cambia. Periderm to the *right*. (Parameswaran and Liese 1974)

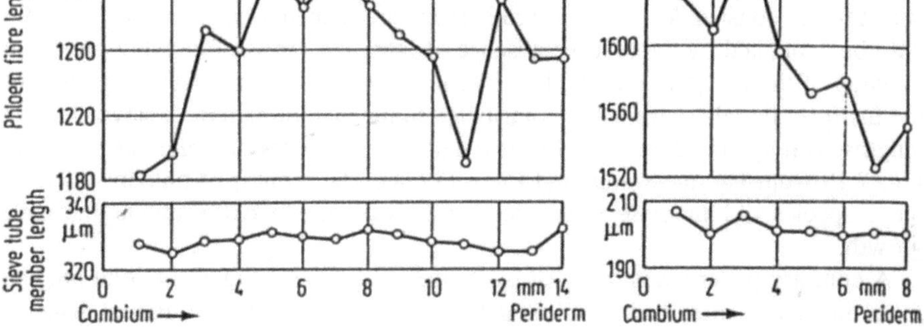

Fig. 7.19. Length variation of phloem fibers and sieve-tube members from cambium to periderm in two tropical species with storied cambia. (Parameswaran and Liese 1974)

Like many vascular elements, phloem fiber lengths tend to increase with age of the tree. Parameswaran and Liese (1974) demonstrated this effect in several tropical species, both nonstoried and storied (Fig. 7.18, 7.19). According to these authors, extracambial elongation by intrusive growth was confined to the cell tip; length of the cell body, which corresponded to the original cambial cell, remained unchanged. Measurements of phloem fibers in ***Robinia pseudoacacia***, another species with storied cambium, led to the same conclusion (Fujita et al. 1984). Elongation of phloem fibers in other species has also been judged to be bipolar (Ghouse and Sabir 1974, Khan and Khan 1983) and in some to be both mono- and bipolar (Ghouse and Yunus 1975, Ghouse and Iqbal 1979). As in the case of xylem fibers, the lengths of phloem fibers are too variable to serve as reliable indicators of fusiform initial lengths.

7.2 Storied Cambia

Age-Related Patterns: On the basis of his extensive investigations, Bailey (1944b) generalized that the fusiform initials of species with storied cambia did not elongate as the stem enlarged. This pattern was confirmed by Chalk et al. (1955) in *Pterocarpus angolensis*, which differed from nonstoried woods in lacking the characteristic increase in fiber length with increasing age or distance from the pith. Other than a sharp rise in fiber length from the first to second growth rings near the pith, there was no sign of any consistent change in length of either fibers or parenchyma strands along a stem radius. Similar patterns were obtained in two other storied species, *Aeschynomene elaphroxylon* and *Nesogordonia paparifera* by Chalk et al. (1955) and in *Entandrophragma cylindrica* by Hejnowicz and Zagórska-Marek (1974).

Venugopal and Krishnamurthy (1989) reviewed a series of studies conducted on *Dalbergia sissoo*. Variation in the lengths of fusiform initials was relatively insignificant, ranging from 165 μm in young trees to 220 μm in older ones. The year-to-year consistency in the lengths of xylem elements in these species could only be explained by the relative absence of intrusive growth following anticlinal division of the fusiform initials. In a like manner, absence of intrusive growth of the fusiform initials presumably contributed to the lack of pattern in sieve element lengths with age in *Mansonia altissima* and *Pterocarpus marsupium* as shown in Fig. 7.19 from Parameswaran and Liese (1974). The lengths of phloem fibers in these species did increase with age, but it was due to extracambial intrusive growth.

A contrary pattern was reported by Ajmal et al. (1986b) in *Holoptelia integrifolia*. In this species, fusiform initial length increased with tree age or distance from the pith as in nonstoried woods, but there was no significant change in either fusiform cell width or length of the gabled end walls.

Within-Ring Patterns: Chalk et al. (1955) examined the within-ring variation in fiber length in a storied hardwood, *Pterocarpus angolensis*. They found that fiber lengths varied significantly within growth rings of all widths. However, unlike most species in which maximum fiber length occurred near the growth ring boundary, it culminated at a mid-ring position. Although fiber length exhibited an abrupt decline at the growth ring boundary, this decline was insufficient to account for the pronounced reduction in length at the beginning of the next ring (Fig. 7.20).

As an additional test of this pattern, Chalk et al. (1955) measured the within-ring lengths of parenchyma strands, which presumably conform to lengths of fusiform initials in storied woods. No consistent variation in length of these strands

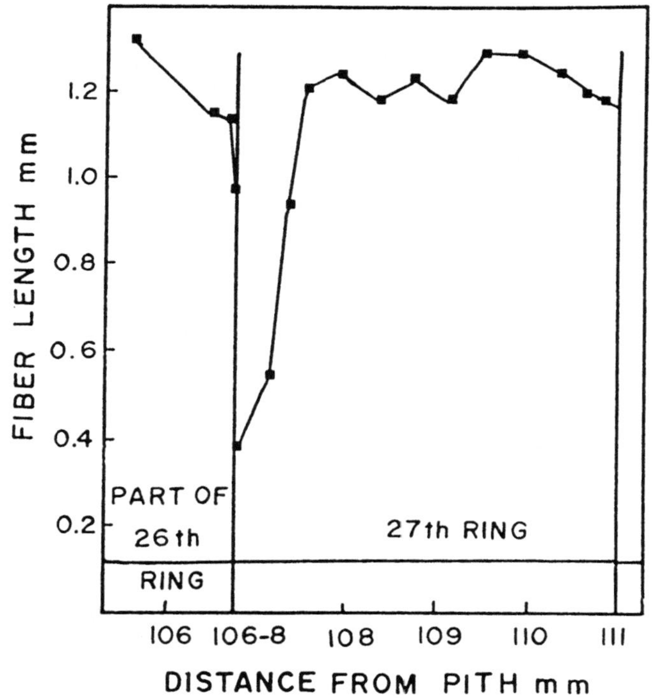

Fig. 7.20. Variation in fiber length within a growth ring of *Pterocarpus angolensis*, a species with storied cambium. (After Chalk et al. 1955)

was found, although the change in fiber length was considerable (Table 7.12). The authors therefore concluded that the changes in fiber length within growth rings were due to post-cambial intrusive growth of the fibers and not to the lengths of the fusiform initials from which they were derived.

A slightly different pattern was found in growth rings from a 36-year-old stem of *Robinia pseudoacacia* by Hejnowicz and Hejnowicz (1959). A distinct seasonal change in fiber length was observed, with a minimum in the first-formed earlywood as in nonstoried woods. Although lengths of earlywood fibers did not increase from the pith outward, lengths of fibers in other parts of the growth ring did increase, but at a much slower rate than in nonstoried species. Like Chalk et al. (1955), the authors concluded that the seasonal within-ring increase in fiber length of this species was due to extracambial intrusive growth and not to an increase in length of the fusiform initials as in the nonstoried species *Populus tremula* (Hejnowicz and Hejnowicz 1958).

Seasonal changes were also found in fusiform initials of *Salvadora persica* by Paliwal et al. (1976). During the active growth period, the fusiform initals first elongated to about 156 μm and later their lengths declined to about 120 μm as a result of radial divisions. However, these fluctuations are far less than those reported for nonstoried species. In another tropical species, *Holoptelia integrifolia*, Rao and Dave (1985b) found that both fusiform initial lengths and widths varied seasonally, attaining maximum values during the dormant season. Swamy et al. (1960) also reported that fiber lengths in *Cipadessa baccifera* varied seasonally within the growth ring as in nonstoried woods. Vessel lengths, however, remained constant.

Table 7.12. Lengths of fibers and parenchyma strands within one growth ring of *Pterocarpus angolensis*, a species with storied cambium (After Chalk et al. 1955)

Distance from ring origin (μm)	Fiber length μm	Fiber elongation[a] X	Parenchyma[b] strand length (μm)
22			276.5
380	1170	4.24	
737			280.5
1095	1283	4.60	
1452			273.0
1810	1272	4.67	
2167			273.5
2525	1156	4.35	
2882			261.0
3240	1157	4.13	
3597			280.0

[a] Fiber length/parenchyma strand length.

[b] Parenchyma strands assumed to be the same length as fusiform cambial initials.

7.3 Anomalous Cambia

Anomalous cambium and anomalous secondary growth are the terms "... under which have been grouped cambial conformations, cambial products, and cambial numbers which differ from the most common 'normal' condition, namely, a single cylindrical cambium that produces phloem externally and xylem internally" (Carlquist 1988a).

The literature on anomalous cambia has been adequately covered by Carlquist (1988a) and Philipson (1990) and no attempt will be made to repeat it here. However, the question might be posed as to whether the general trend of increasing xylem element length with age would be repeated in each of the vascular rings

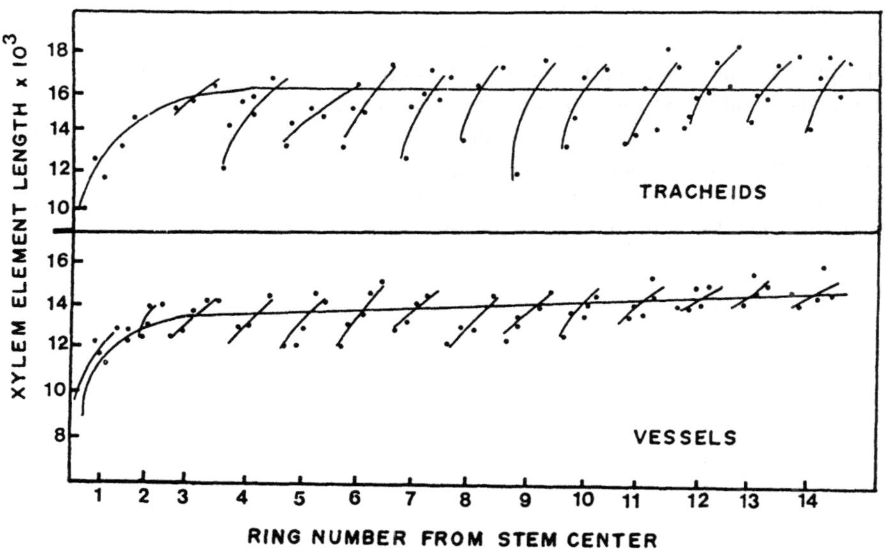

Fig. 7.21. Variation in length of xylem elements with ring number from the stem center in *Gnetum ula*, a species with anomalous cambia. Each *vertical line* represents a new vascular ring derived from an independent cambium. (After Krishnamurthy and Venugopal 1984)

produced by successive cambia in species with anomalous growth. This possibility was examined by Krishnamurthy and Venugopal (1984) in two gymnosperms, *Cycas circinalis* and *Gnetum ula*, and the dicotyledon, *Dalbergia lanceolaria*. In these species, the first cambium arose in the normal position and many successive cambia arose external to it. Xylem element lengths increased rapidly in rings near the pith and then either continued to increase gradually as in *D. lanceolaria* or more or less stabilized as in in *G. ula* (Fig. 7.21). However, each successive cambium exhibited an internal pattern similar to the seasonal trends in species with normal cambia. That is, the last-formed elements of one ring were longer than the first-formed elements of the succeeding ring. The authors attributed these patterns to the fact that the first fusiform initial cells were shorter because the cambium for each new growth ring arose in parenchymatous tissues rather than being a continuation of the previous cambium.

Tracheids were also found to increase in length in the successive cambia of *Cycas* examined by Terrazas (1991). This trend toward increasing tracheid length was attributed to the fact that successive cambia were long-lived. That is, rather than each cambium ceasing activity when a new one formed, several cambia remained active simultaneously.

The same pattern was observed in stems of *Phytolacca dioica* (Wheat 1977). In this species, a new cambium formed in the two to four cell layers of parenchymatous cells situated between the outermost discernible secondary phloem and the ring of fiber cells. The first-formed sieve elements in each new ring were of the same lengths as the adjacent parenchymatous cells that gave rise to the new cambium. However, the lengths of new sieve elements increased progressively as the cambium aged. Similar trends were observed in the xylem.

Chapter 8

Rays

8.1 Medullary Rays - Historical

The term medullary ray is derived from the fact that the pith was for many years called the medulla. Consequently, the parenchymatous tissue extending from pith to cortex between primary vascular bundles became known as a medullary ray or pith ray. According to Hartig (1855a), the term "Markstrahl," or pith ray, was applied not only to the parenchymatous tissue occurring between vascular bundles in woody plants, but also to the cellular tissue that occupied the same position in the growing shoot. Because of this relationship, many early workers assumed that the rays of woody plants arose by compression of the originally separated fiber bundles. Conversely, new fiber bundles were thought to arise by the intervention of new ray cells. That is, when new ray cells were initiated in the "cambial layer," they split a vascular bundle, thus forming two bundles in the place of one (Göppert 1842, Hartig 1854a).

Some workers referred to the procumbent cells of medullary rays as "Spiegelfaser," or mirror-fibers, because in many cases the oppositely arranged ray cells resembled horizontal fibers rather than parenchymatous cells (Hartig 1837a, 1855a). The term "Spiegel" was used by Sachs (1879) and Abromeit (1884) in a different sense with reference to the reflectivity of the broad rays of *Quercus* when viewed on a smooth transverse wood surface. The term is still commonly used to describe broad rays in the German furniture industry (Kisser 1967). Clarification of the different cell types, as well as the pitting between cells, was made by Schulz (1883). A fairly detailed review of the early literature on rays can be found in Herbst (1894), although his primary emphasis was on diagnostics.

Early workers were by no means clear in their writings as to the nature of the tissues occurring between vascular bundles during the early stages of development. Nägeli (1858) referred to the tissue situated between primary vascular

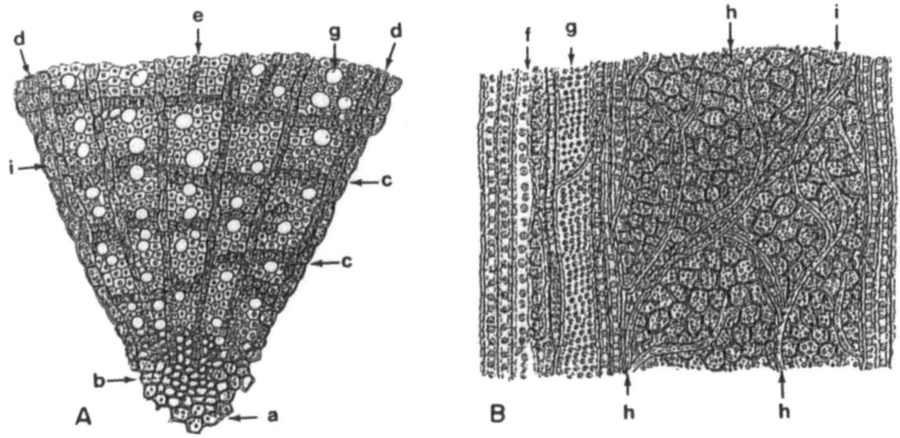

Fig. 8.1A,B. *Casuarina equisetifolia.* **A** Transverse section showing "concentric medullary rays" (*c*), or apotracheal parenchyma. **B** Radial section showing ramifying xylem elements (*h*) splitting *a* large "centrifugal medullary ray." *a* Pith; *b* medullary sheath; *d* large medullary rays extending to pith; *e* small centrifugal medullary rays; *f* pitted prosenchymatous element; *g* pitted vessel; *i* starch grains. (Göppert 1842)

bundles as "Markverbindung," or pith-connecting tissue. Russow (1872) adopted the older term Markverbindung for the ground tissue occurring between the first-year conducting bundles of all gymnosperms and many dicotyledonous plants. Vöchting (1873-74), however, objected to Russow's use of this term as opposed to primary Markstrahlen (primary rays) by pointing out that the between-bundle part of the thickening ring, which in many cases became a primary ray, might in other cases be transformed directly to wood or bast, while in still other cases the primary ray itself might be transformed to cambium. Thus, according to the views of these workers, tissue originally designated as a primary or medullary ray might later differentiate to one of several forms of tissue. For these reasons, the early literature is confusing and often contradictory with respect to the primary between-bundle tissue that might later either remain as some form of interfascicular tissue or differentiate as ray tissue or even as interfascicular cambium. To add to the confusion, authors were not always consistent in their terminology of ray tissues. Göppert (1842), for example, referred to the concentric bands of xylem parenchyma in *Casuarina equisetifolia* as "rayons médullaires concentriques," or concentric medullary rays, (Fig. 8.1A). These tissues are in fact bands of apotracheal parenchyma (Metcalfe and Chalk 1950).

To investigate the primary rays of *Pinus nigra* var. nigra, Hartig (1855a) collected winter buds and allowed them to dry out before hand sectioning. He observed that the pith and cortex of the upper bud region were separated by radially arranged cells, which he called "cambial parenchyma," presumably because he believed these cells were derived from the original "cambial" layer of the apical meristem along with the fiber bundles. The cambial parenchyma differed from both the brick-like ray cells and the typical pith and rind parenchyma cells. It was within this layer of cambial parenchyma that new fiber cells were thought to arise, not by displacement or removal of cambial cells, but by division and metamorphosis of these cells.

When Hartig examined sections from beneath the bud, he noted how the wood ring had closed resulting in one-celled rays in *Pinus* and several-celled rays in *Fagus* and *Quercus*. Farther down in the axis he observed new rays forming in previously differentiated fiber tissue. These new rays penetrated radially through all layers of the fiber bundle, even the thick-walled cells of the rind and spiral fibers of the pith. Meanwhile, the number of fiber bundles also increased. The new bundles did not originate between previously formed fibers, but arose by transformation of these fibers in the same way as the fibers originally arose by transformation of cambial parenchyma. Hartig attributed the origin of these new bundles to a process of "retrograde metamorphosis." He (Hartig 1859a) could not accept the view that new fiber bundles originated in ray tissue. This negation was presumably in response to von Mohl's (1858) assertion that medullary rays had the "power" of forming new wood bundles. However, it was not clear whether von Mohl referred specifically to medullary rays or to the between-bundle parenchymatous tissues.

In Nägeli's (1858) classification scheme, the cambial ring of most dicotyledonous plants was viewed as having the ability to expand without limit. Cambial strands derived from the apical meristem were supplemented by additional strands which, by lateral expansion, eventually closed the cambial cylinder. If the strands were closely adjacent, they often merged, but most often they were separated by medullary rays. Seldom were all cambial bundles united in an unbroken ring. As the cambial ring continued its unrestricted radial expansion by cambial division, the medullary rays kept pace by means of a "Strahlenmeristem," or ray meristem. Nägeli referred to primary rays that extended to the pith as parenchyma rays and secondary rays that did not reach the pith as "Holzstrahlen," or wood rays. However, Nägeli erred in considering medullary rays as being simply extensions of the pith, which he regarded as Markverbindung, or pith-connecting tissue, a viewpoint contested by later workers.

Considerable confusion existed in the early literature regarding not only the origin and nature of ray tissue but also its terminology (Nicoloff 1911). Hartig (1855a), in considering different ray origins, suggested the term primary rays for those originating from cambial parenchyma, secondary rays for those metamorphosing from mature fibers, and tertiary rays for those originating in older wood from the cambial layer. The search for an acceptable terminology continued with Sanio (1863b). He noted that the medullary ray tissue extending from pith to bark in dicotyledonous plants had been referred to as large medullary rays by Kisser, Meyen, Schleiden, and Unger, primitive rays by Schacht (1856), primary rays by Hartig (1855a), and parenchyma rays by Nägeli (1858). The tissue that did not extend to either pith or bark but began in the secondary body was referred to as small rays by Kisser, Meyan, Schleiden, and Unger, secondary rays by Hartig (1859a) and Schacht (1860), and either incomplete parenchymatous rays, or wood rays by Nägeli (1858).

Sanio (1863b) rejected the concept of either a primitive or a primary ray because the rays were already present when the bundles became separated by the "Scheidegewebe", or sheathing tissue. He therefore decided to call this tissue, consisting entirely of parenchymatous cells, an interfascicular ray. In some species, such as *Quercus*, *Fagus*, and *Carpinus*, he recognized primary fascicular rays that extended inward to the "Markkrone," or pith boundary. In other species, such as *Clematis* and *Berberis*, he recognized secondary fascicular rays that were confined more or less to the outer fascicular tissues and did not extend to the pith. *Fagus sylvatica* possessed both primary and secondary fascicular rays in addition to interfascicular rays. *Corylus* and *Carpinus* possessed not only primary and secondary fascicular rays but also special adventitious rays that subdivided the interfascicular wood of the secondary body.

De Bary (1877) adopted the older terms large and small medullary rays for rays that either did or did not extend inward to the pith, respectively. However, like Nägeli (1858) and Sachs (1882), De Bary made no distinction between primary rays and ground tissue. That is, he considered the primary rays, along with primary cortex and pith, to be a homogeneous ground or fundamental tissue. Similarly, d'Arbaumont (1881) considered the "rayons médullaires," or medullary rays, to form within the bands of "arrested parenchyma," or ground tissue, separating the primary vascular bundles. He believed that these rays provided a direct means of communication between pith and cortex.

It is clear from the foregoing discussion that medullary ray and associated terms were applied rather loosely by early workers to designate the parenchymatous tissue extensions from pith to cortex. These terms often were applied irrespective of either their origins in the primary body or their subsequent differentiation in the secondary body. That is, there was no clear understanding as to how the so-called medullary ray tissue was transformed at the onset of secondary development, although there was a perceptive recognition on the part of some workers that such a transformation had taken place.

8.2 Classification of Rays

8.2.1 Primary Rays

The term medullary ray was rejected by Bailey (1910a, 1911). He also objected to the concept of a "so-called interfascicular cambium" within which "the imagined primary ray originated." He expressed the opinion that all rays were of secondary origin and they were only indirectly, if at all, related to the fundamental or ground tissues emanating from the pith. This is a valid argument if one assumes that the primary ray, like the interfascicular cambium, is initiated at the onset of secondary vascular development and that its products are secondary elements and tissues. That is, although incipient ray tissue might be present in the primary plant body, it assumes the true characteristics of ray tissue in the secondary body.

Based on the foregoing objections, Bailey, and later Jeffrey (1917), suggested replacing the term medullary or primary ray with wood ray when referring to the parenchymatous tissue of the xylem that originated with secondary growth. By similar reasoning, Alexandrov and Alexandrova (1929) suggested the equivalent German term "Holzstrahlen," a term proposed years earlier by Nägeli (1858). Despite these objections, the term medullary ray is still used by some workers with reference to the parenchymatous ground tissue separating procambial bundles in the primary vascular cylinder and its extension in the secondary body. Such rays occur not only in many herbaceous plants, but also in certain families of woody dicotyledons (Metcalfe and Chalk 1950).

A common current practice is to refer to rays as being of either primary or secondary origin (Thompson 1910, Chattaway 1933a, Barghoorn 1940a,b, Philipson and Ward 1965, Cumbie 1967a, 1969a, 1983). When used in this way, primary refers to a ray originating at the beginning of secondary growth in a position opposite the so-called medullary ray in the primary body. As defined by the IAWA (1964) in the Multilingual Glossary, a primary ray in wood anatomy is "... a ray originating in the primary tissues and extended by cambial growth. Commonly used for any ray that can be traced inwards to the pith." A more accurate definition could be achieved by substituting in the first sentence the phrase "... a ray whose *precursor cells* originate in the primary tissues," and in the second sentence the phrase "... that can be traced inwards to the beginning of the first growth ring."

8.2.2 Secondary Rays

As defined by the IAWA (1964), a secondary ray in wood anatomy is "... a ray derived from the cambium (i.e., originating after the development of secondary xylem), and not extending inwards as far as the pith." Again, a more accurate definition could be achieved by substituting the phrase "... and not extending inwards as far as the beginning of the first growth ring."

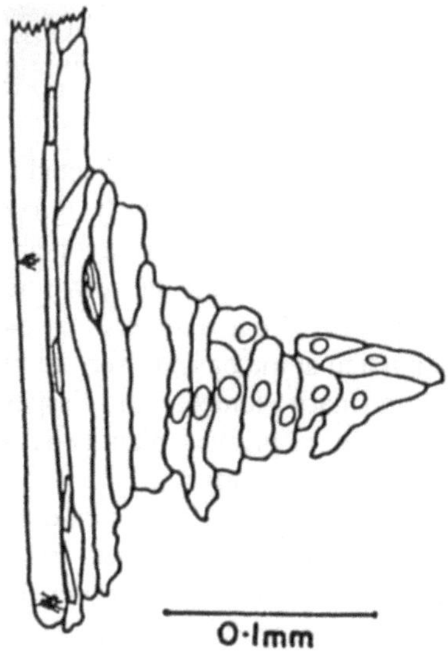

Fig. 8.2. Radial section of *Pinus pinea* showing the ultimate derivatives of a declining fusiform initial and the early, erect derivatives of a newly established ray. (Srivastava 1963a)

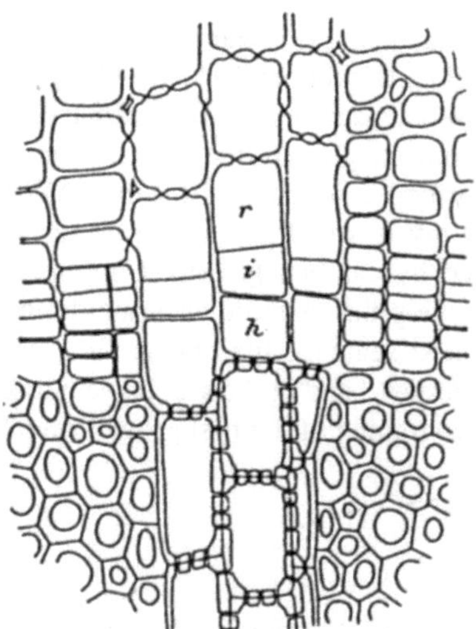

Fig. 8.3. Ray cell division in *Cytisus laburnum* in transverse view as visualized by early botanists. Ray cell initial (*i*) gave rise to xylem (*h*) and then to phloem (*r*) derivatives without the intervention of a mother cell. 315x. (Haberlandt 1886)

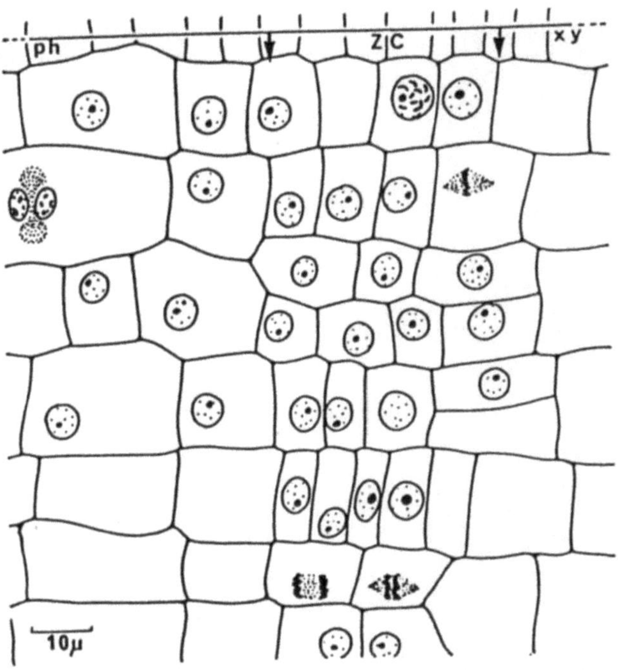

Fig. 8.4. Part of a ray in *Acer pseudoplatanus* in radial section. Note that divisions occur in more than one cell within a tier of the cambial zone (*ZC*) delimited by the *arrows*. (Catesson 1964)

Secondary rays are initiated by the segmentation of fusiform cambial initials. Segmentation at times involves the entire fusiform initial and at other times a partial segmentation of the initial giving rise to either tip origin or lateral origin (Chap. 6.3.2.1). Secondary rays can increase in size by the outright merger of small rays, by the acquisition of small by large rays, and by cell divisions within the rays (Barghoorn 1941a). Conversely, large rays can be reduced in size by various types of ray splitting. In most species, the rays are irregularly dispersed throughout the xylem. However, in some dicotyledons they are storied, in which case they appear horizontally aligned on tangential microsections.

An acceptable terminology has always presented a problem in studies of ray anatomy. In one attempt at understanding, Hartig (1848) suggested the term "Stockwerk," or tier, for ray cells superimposed in a vertical series and "Lager," or layer, for those situated next to one another horizontally. Currently, different ray types are defined by appropriate prefixes and adjectives. The number of ray cells in the tangential direction is indicated by a prefix, either uniseriate, biseriate, triseriate, or multiseriate. Some authors consider all rays biseriate or larger to be multiseriate. Ray cells aligned in the vertical direction are said to be arranged in either rows, tiers, or series and numbered accordingly.

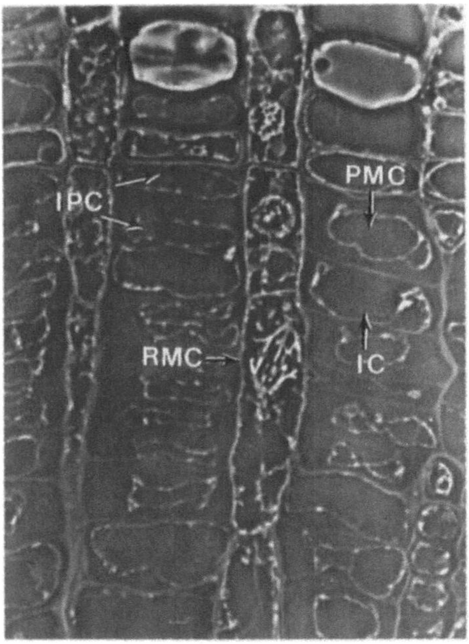

Fig. 8.5. Ray cell division in *Cryptomeria japonica* as seen in transverse section in polarized light. The dividing cell was judged to be a ray mother cell (*RMC*) rather than a ray cell initial because of its greater length. *IC* Initial cell; *IPC* immature phloem cell; *PMC* phloem mother cell. 352x. (Itoh et al. 1968)

Kribs (1935) distinguished between homocellular and heterocellular rays and Carlquist (1988a) recently enlarged on this scheme. As redefined by Esau (1977) and Metcalfe and Chalk (1983), a homocellular ray is composed of cells of one form only; in dicotyledons of procumbent, square, or upright cells and in conifers of parenchyma cells only. A heterocellular ray is composed of more than one form; in dicotyledons of either procumbent, square, or upright cells and in conifers of parenchyma and ray tracheids. A homogeneous ray system then becomes one in which all ray tissues are homocellular, and a heterogeneous ray system either one in which all ray tissues are heterocellular or one consisting of homocellular and heterocellular rays combined. Heterocellular rays usually are derived from heterocellular ray cell initials (Wodzicki and Brown 1973).

The foregoing terms have not been applied consistently by all workers, consequently individual definitions and usages differ (Huber 1949, Reinders-Gouwentak 1949). Some workers have also categorized rays according to other criteria (Metcalfe and Chalk 1983). An exceptionally detailed classification scheme for ray cells appeared in Wagenführ (1961). Excellent illustrations and detailed descriptions of the Kribs' (1935) ray types can be found in Carlquist (1988a). A better understanding of these different ray types, as well as the relationship between primary and secondary ray origins, can be achieved by examining the processes of ray development in selected species and species groups. As will be seen in later sections, all the variations in ray development can be related directly to events arising in the cambium.

8.3 Ray Dynamics

Rays are by no means static. Chattaway (1951b) stressed the fact that changes observed in mature rays actually occur at the cambium, either as a result of changes in the ray cell initials themselves or in the daughter cells very soon after they are cut off from the cambial initials. Rays constitute highly complex tissue systems, and even though the individual elements often appear to differ little from one another, they nonetheless function in a variety of ways, serving as storage, conductive, and secretory systems. Their main function is to facilitate lateral transport between xylem and phloem. Ray cell initials usually produce tissues on both sides of the cambial layer, thus preserving the continuity of radial conduction between xylem and phloem.

During ontogeny, rays obviously undergo profound physiological and functional changes in addition to anatomical and structural changes. Those of primary concern here involve increases and decreases in ray size that can be directly related to events initiated within the cambium.

8.3.1 Secondary Ray Origin

New ray tissue must be constantly produced to keep pace with the increasing tangential expansion of the stem. With few exceptions, new secondary rays are uniseriate irrespective of whether they are initiated by partial or total segmentation of fusiform initials (Chattaway 1951b, Braun 1955, 1963). Carlquist (1988a) reported that a large number of dicotyledonous species have uniseriate rays exclusively at the onset of secondary growth. Only rarely and under unusual circumstances are biseriate rays initiated. One exception might be the first-formed rays at the very beginning of secondary development (Fig. 8.38). Rays arising by tip or lateral origins usually consist of single cells, although two-tiered rays are not uncommon. Rays arising by complete segmentation of a fusiform initial might consist of several to many cell tiers (Chap. 6.3.2.1).

With the exception of cells in the primary rays of some gymnosperms which are procumbent or horizontally oriented (Schmidt 1889, Bannan 1934a, Barghoorn 1940a), the new ray cells in most woods are at first erect, or vertically oriented (Chattaway 1951b, Braun 1970) (Fig. 8.2). During subsequent development of the ray, new derivatives can be either erect, procumbent, or a combination of the two forms. Cells comprising new rays are usually parenchymatous in dicotyledons but can be either parenchymatous or tracheidal in gymnosperms (Braun 1970).

Parenchymatous ray cells that are in functional contact with the adjacent xylem cells by large, numerous pits are referred to as "contact" cells. Similar cells with no pit contacts are referred to as "isolation cells." In two exceptionally detailed and superbly illustrated publications, Braun and den Outer (1964) and Braun (1970) have categorized many woods on the basis of the "contact-isolation-differentiation" status of their rays.

8.3.1.1 Ray Spacing

Attempts have been made to account for the spacing of rays as they appear on tangential sections with some authors proposing inhibiting influences (Bünning 1952), and others either some type of stimulus arising in the phloem (Carmi et al. 1972) or hormonal gradients (Aloni 1987b, Lev-Yadun and Aloni 1990, 1991b, 1992b).

Bünning (1965) observed that ray spacing conformed to a regular pattern. He therefore proposed that each ray formed a "Meristemoide", or center of activity, that inhibited the initiation of other meristemoids in its immediate vicinity. Thus, new rays could only form when tangential cambial expansion permitted sufficient space between rays to nullify the inhibition. Data for this hypothesis relating to ray initiation were based, in part, on Beijer's (1927) research on the storied cambium of *Herminiera elaphroxylon* (Chap. 6.2.1).

Carmi et al. (1972) disputed this hypothesis. Although they admitted that ray cell initials might indeed interact, this effect played only a part in the spacing of rays. They suggested that rays differentiate along channels of movement of a stimulus from phloem to the differentiating xylem or vice versa. Thus, according to their hypothesis, the size and number of rays would be a function of the volume of the differentiating vascular system. The control of spacing, rather than being situated in the cambium, would reside in either the xylem or phloem, where excess stimulus that was not drained by existing rays would flow through the cambium. This excess stimulus would then contribute either to an increase in the size of existing rays or to the initiation of new rays. As partial verification, the authors pointed out that fusiform initials poorly served by rays experienced the highest frequencies of failure.

Lev-Yadun and Aloni (1992b) expanded on the foregoing hypothesis by suggesting that the hormone ethylene somehow controlled differentiation of rays in the cambial zone. They proposed that ethylene influenced ray initiation and size by its negative effect on polar auxin transport. The disturbance to axial auxin flow, in turn, enabled the radial signal flow that determines ray formation to become relatively stronger thereby inducing new rays and enlarging existing ones.

8.3.1.2 Ray Cell Initials

Rays are perpetuated by either ray initials or ray cell initials (Cheadle and Esau 1964, Schmid 1976, Metcalfe and Chalk 1979, 1983, Cumbie 1984). The former term implies that all initials comprising the ray divide simultaneously, whereas the latter term refers to the division of a single initial cell within the ray. Iqbal and Ghouse (1990) suggested that the two terms be used as synonyms to avoid confusion. However, I prefer the term ray cell initials with reference to the individual cells within a cambial ray.

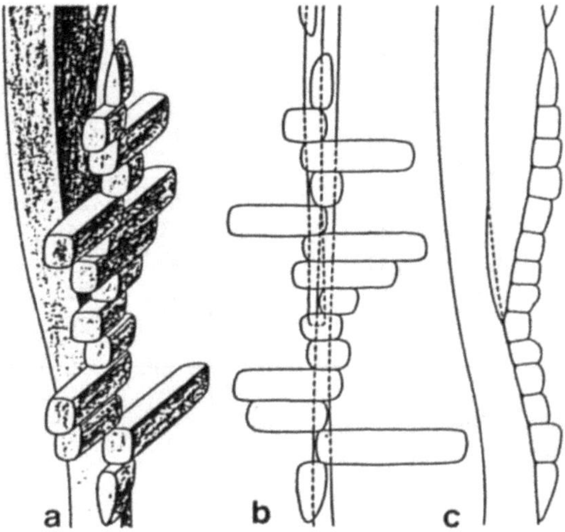

Fig. 8.6a-c. Model depicting the structure of a ray in the cambial zone of a typical species in the Pinaceae. Note that the ray consists of a mixed population of cells in various states of division and elongation. **a** Sectional; **b** radial; **c** tangential view. (Wodzicki and Brown 1973)

The older literature suggested that the immediate derivatives of ray cell initials did not as a rule redivide (De Bary 1877, Krabbe 1884, Haberlandt 1886, 1914, Strasburger et al. 1908) (Fig. 8.3). That is, there were no so-called ray mother cells. This condition often prevails in slow-growing trees (Meier 1973). However, as Bannan (1953) observed, periclinal divisions were not restricted to the ray cell initials of rapidly grown *Thuja occidentalis* trees. Consequently, the dividing zone producing ray cells was as wide as that producing fusiform elements, although cell numbers were not necessarily the same. Catesson (1964) also demonstrated that periclinal divisions in *Acer pseudoplatanus* were not confined to the ray cell initials (Fig. 8.4) as did Itoh et al. (1968) in *Cryptomeria japonica* (Fig. 8.5). Imagawa (1985a) concluded from his study of ray development in *Larix leptolepis* that most divisions occurred in the elongated ray mother cells and not in the iso-

Table 8.1. Numbers and sizes of rays at different heights in stems of *Picea omorica* (After Jaccard 1915b)

Height in stem mm	Number of rays[a] Average	Minimum	Maximum	Number of cells[b]
0.1	44	40	49	26
0.5	42	39	45	-
1.0	37	34	44	-
2.0	38	32	42	17
3.0	38	35	47	27
4.0	38	35	44	22
5.0	36	33	45	18
6.0	35	33	40	25
7.0	39	37	43	25
8.0	36	30	41	23
9.0	38	34	42	24
10.0	38	35	42	26

[a] Number of rays in microscopic field of 0.75 mm^2.
[b] Number of cells in longest ray measured.

diametric ray cell initials. Likewise, Chattaway (1933b) observed in *Guazuma tomentosa* and several other species that the erect tile cells kept pace with elongation of the procumbent cells by more frequent divisions of the ray mother cells, not the ray cell initials (Fig. 8.55). In species of *Argyrodendron*, the enlargement of multiseriate rays involved tangential divisions of the marginal and sheath cells which Chattaway (1951b) suggested might be extracambial. Such extracambial divisions were infrequent and most divisions were confined to the ray cell initials.

Attempts have been made to identify the ray cell initials. One of the most commonly used criteria is the presumed narrowness of the initial cells relative to the derivatives (Kuroda 1986). However, as Wodzicki and Brown (1973) pointed out, ray cell initials are rarely isodiametric. A sectional view of the active cambium might reveal a mixed population of cells including nondividing and dividing cells in various states of radial elongation (Fig. 8.6). That is, different parts of a ray cell initial might be subjected to different rates of radial elongation because they are in radial contact with fusiform initials in various states of growth. The observations of Wodzicki and Brown on division characteristics of ray cell initials supported the concept of a uniseriate cambium.

8.3.2 Ray Numbers

The largest numbers of new rays are formed during the early juvenile years when peripheral expansion of the cambium is greatest. Thus, many workers have found that the number of rays per unit tangential area is greatest either in the first year (Essner 1882, Fischer 1885, Kucera and Kucera 1967) or first several years of growth (Bertog 1895, Zijlstra 1908, De Smit 1922, Linnemann 1953, Fujiwara and Nakayama 1978, Ghouse and Hashmi 1980, 1981). Thereafter, the number of rays tends to either decline or fluctuate around some mean value. These fluctuations often can be related either to sampling position within the tree (Essner 1882, Hartig 1894, Bertog 1895, Jaccard 1915b, De Smit 1922, Bannan 1937, Holdheide 1955, Kärkkäinen 1973, Fujiwara and Nakayama 1978) (Tables 8.1, 8.2), or to external conditions that influence growth rate of the tree (Essner 1882, Fischer 1885, Forsaith 1920, De Smit 1922, Bannan 1937, Carmi et al. 1972, Gregory 1977, Fujiwara and Nakayama 1978) (Table 8.3). Süss (1989) reported that the number of rays, both uniseriate and fusiform, were about twice as numerous in branchwood than in stemwood of *Pinus aristata*.

Table 8.2. Distribution of xylem ray tissue in roots, branches, and adult stems of seven coniferous species (Bannan 1937)

Species	Number of rays/mm^2			Number of ray cells/mm^2		
	Roots	Branches	Stems	Roots	Branches	Stems
Abies balsamea	24	64	36	220	285	330
Larix laricina	26	70	36	205	305	315
Picea sp.	37	76	42	255	290	295
Pinus strobus	28	54	33	105	145	180
Taxodium distichum	50	75	36	90	225	210
Thuja occidentalis	63	67	56	135	140	185
Tsuga canadensis	48	77	42	215	250	295

Reports on the effects of rate of growth on ray numbers by different authors are quite variable. For example, Bannan (1937) found that rays occurred in the largest numbers in narrow rings of conifers where ray height was lowest. In contrast, he (Bannan 1954) reported a positive relation between ray number and ring width in *Thuja occidentalis*. In the three ring-porous species (*Castanea sativa, Fraxinus excelsior*, and *Sassafras officinale*) examined by White and Robards (1966), the number of rays was approximately 50% greater in fast- as opposed to slow-growing trees (Table 8.1).

In a contrary report, Gregory and Romberger (1975) could detect no differences in ray numbers with increasing growth rate in the conifers they examined. Although the number of ray cell initials increased in wider growth rings, this relationship was not due to an increase in ray numbers but to ray heights (Fig. 8.12). A similar relation was found by Gregory (1977) in *Acer saccharum*. In this species, changes in the number of cells per unit area of cambium conformed rather closely to changes in growth ring width, whereas the number of rays per unit area either remained fairly stable or declined slightly (Fig. 8.7).

The increase in growth rate above a stem girdle was shown by Grill and Gigler (1985) to influence the number of rays in *Thuja occidentalis*. Ray cell number in control stems averaged 170 to 210/mm^2, whereas above the girdle it averaged 310 to 370/mm^2. Although most of these additional ray cells simply increased the height of existing rays, about 26% contributed to new ray formation. Girdling stems of *Pinus halepensis* resulted in a 44% increase in ray number, whereas wounding the cambium with a razor blade resulted in a 139% increase in ray number (Lev-Yadun and Aloni 1992b). Both procedures produced a few extraordinarily tall rays.

Table 8.3. Analyses of ray sizes and dimensions in narrow and wide growth rings of *Castanea sativa*, *Fraxinus excelsior*, and *Sassafras officinale* (After White and Robards 1966)

	Castanea		*Fraxinus*		*Sassafras*	
	Narrow	Wide	Narrow	Wide	Narrow	Wide
Mean ring width (mm)	2.8	12.2	3.6	8.0	1.1	7.2
Ray width (mean cell number)	1.1	1.5	2.7	3.4	2.3	3.3
Ray width (μm)	14.9	17.7	30.3	45.4	28.8	54.8
Ray height(mean cell number)	9.3	9.7	11.9	12.2	13.2	11.6
Ray height (μm)	170.0	171.0	195.0	212.0	305.0	266.0
Rays/mm^2	75.0	116.0	44.0	65.0	41.0	65.0
Mean cell width (μm)	13.5	11.9	11.2	13.5	12.4	16.4
Mean cell height (μm)	18.3	17.6	16.2	18.4	23.0	23.1

Partial girdles in *Ailanthus altissima* that left a bridge of living tissue were also found by Carmi et al. (1972) to result in an increase in ray number. The number of ray cells in a microscopic field of view (1.8 mm^2) increased from 23 in control plants to 36 in the lateral edges of unconfined bridges, and to 46 in confined bridges in which radial growth was accentuated by confining tangential growth by a metal barrier. Similar girdles made in stems of the same species by Lev-Yadun and Aloni (1992b) resulted in a 174.5% increase in ray number above the girdle.

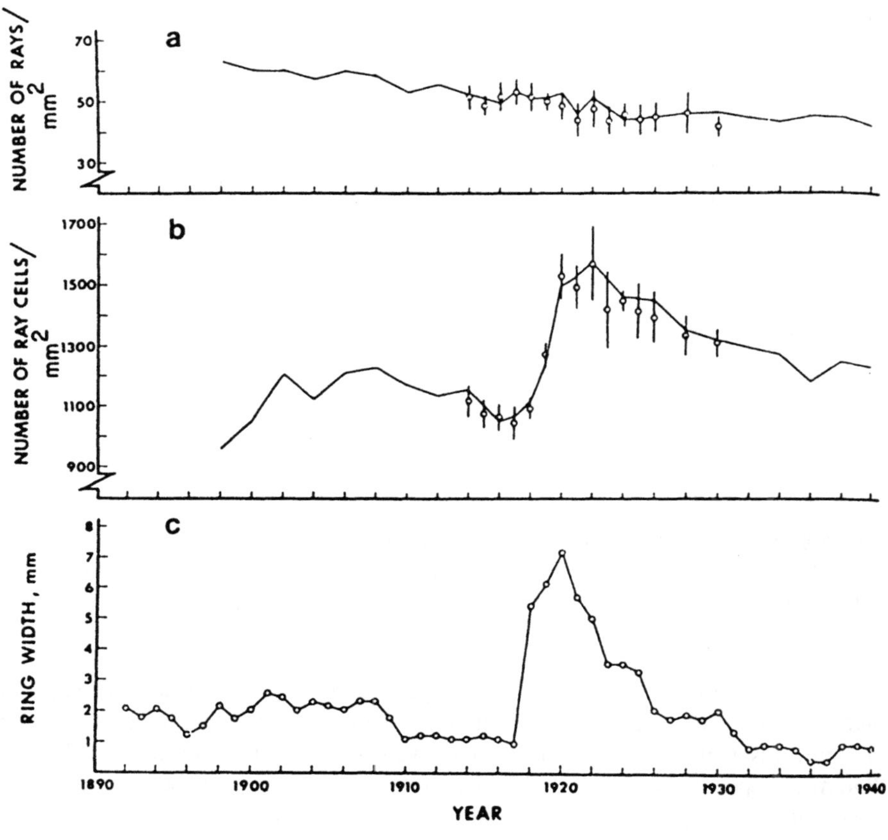

Fig. 8.7a-c. Comparison of the numbers of rays (**a**) and ray cells (**b**) per mm^2 of tangential xylem surface with ring width (**c**) over a 50-year period in a *Acer saccharum* tree. *Points connected by the vertical lines* in **a** and **b** represent values based on the same ray within a given tangential area in successive growth rings. (Gregory 1977)

The number of rays per mm^2 more than doubled and existing rays became larger. In both *Pinus halepensis* and *A. altissima*, the increase in ray number was a wound response and the effect diminished with increasing distance from the wound site.

Scattered reports in the literature suggest that the number of rays might also vary seasonally in some species. Although Fischer (1885) could find no differences in the numbers of rays between earlywood and latewood in *Pinus abies*, Fen-

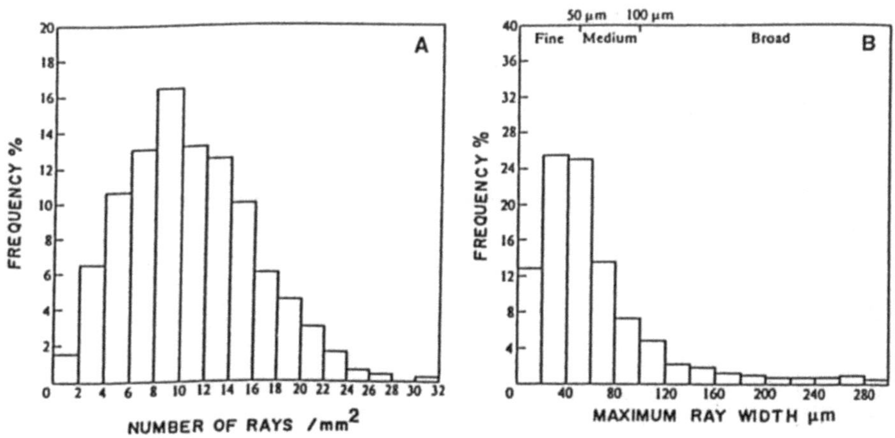

Fig. 8.8. **A** Frequency distribution of the number of rays/mm^2 of tangential surface for 1800 dicotyledonous species. **B** Frequency distribution of maximum ray widths from 650 species of dicotyledons. (Metcalfe and Chalk 1983) (Courtesy Oxford University Press)

gel (1966) reported a definite within-ring pattern in *Fagus sylvatica*. In *Mangifera indica* and *Tectona grandis*, ray numbers increased from minima of 98 and 57/cm^2, respectively, when the cambia were dormant to maxima of 125 and 85/cm^2, respectively, when the cambia were active (Rao 1988a). A frequency distribution of the number of rays/mm^2 of tangential surface, based on 1800 dicotyledonous species, is shown in Fig. 8.8A.

8.3.3 Division of Ray Cell Initials

One of the simplest ways that rays increase in height is by the vertical elongation of the marginal ray cell initials (Chattaway 1933a, Barghoorn 1940b, Huber 1949, Evert 1961, Meier 1973) (Fig. 8.9) In conifers, such elongation occurs more commonly in the phloem parts of rays (Meier 1973).

An unusual type of ray-size increase by cell division was described by Lev-Yadun and Aloni (1991a) in *Suaeda monoica* in the family Chenopodiaceae. The first-formed xylem of this species is temporarily rayless, but large, multiseriate rays arise during later stages of ontogeny. Rays arose by the gradual transition of axial parenchyma to ray cells, and a continuation of this process added to both ray height and width. Further division of ray cell initials within the multiseriate rays

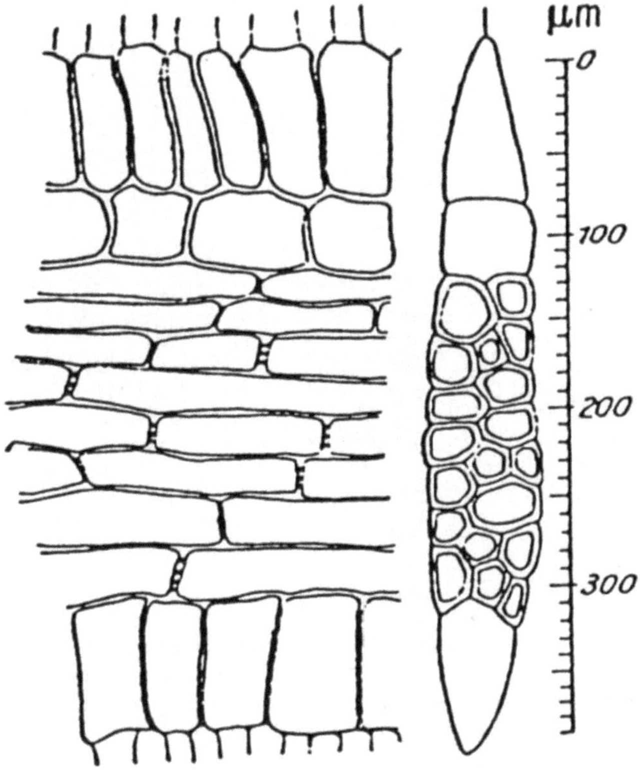

Fig. 8.9. Radial (*left*) and tangential (*right*) sections of a heterogeneous ray of *Daniellia thurifera* illustrating the increase in ray height by elongation of the marginal cells. (Huber 1949)

created "ray centres" consisting of small procumbent or square cells embedded in larger, more or less axially elongated cells. Origin of several such ray centers within a large multiseriate ray gave rise to a "polycentric ray," a specialized type of aggregate ray.

Most secondary rays increase in height either by transverse division of ray cell initials, by the merger of adjacent rays, or by the addition of segments from declining fusiform initials (Fig.8.10). Divisions of ray cell initials increase ray heights by adding cells either to the borders or interiors of existing rays. Uniseriate rays commonly increase in height by simple, transverse anticlinal divisions of the terminal cells in both gymnosperms (Bannan 1934a, 1953, 1965b, Barghoorn 1940a, Braun 1955, 1970, Srivastava 1963a,b, Smith 1967, Gregory and Romberger 1975), and dicotyledons (Beijer 1927, Braun 1955, Evert 1961, Ghouse and Hashmi 1980, Cumbie 1983, Ajmal and Iqbal 1987a,b, Rao 1988a). In *Abies alba*, the anticlinal divisions usually resulted in two cells unequal in size (Kucera and Kucera 1967). Since new cells most often are added to the ends of

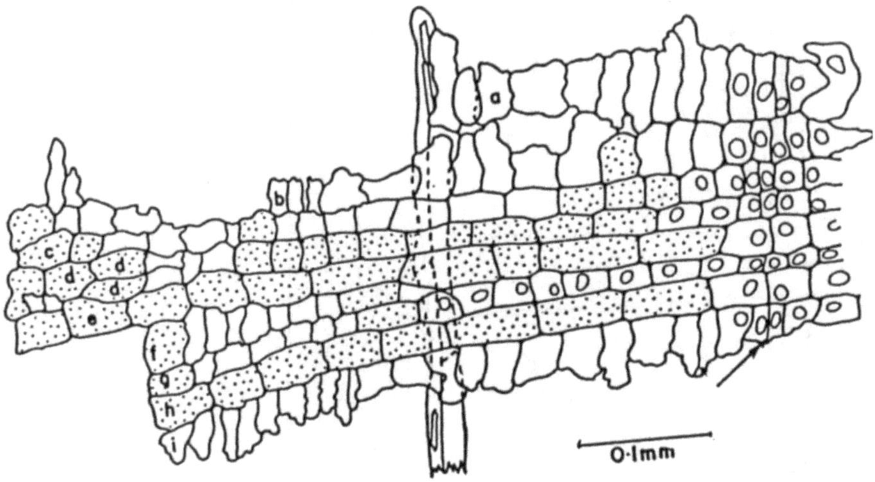

Fig. 8.10. Increase in ray height in *Pinus pinea* by the addition of a new tier of ray cells (*a*) transformed from the upper segment of a declining fusiform initial. *Arrow* indicates cambium, phloem to *left*. Albuminous cells without nuclei; cambial cells with nuclei; starch-containing cells *stippled*. (Srivastava 1963a)

conifer rays as border cells, they are usually tracheids (Barghoorn 1940a, Braun 1970). Barghoorn (1940a,b) maintained that the interior ray cells never divided in conifers but they did in dicotyledons. Although it is unusual for an interior cell of a low, uniseriate ray to divide by transverse anticlinal walls, such cells in tall, uniseriate rays occasionally do so in both gymnosperms and dicotyledons (Braun 1955).

During his investigation of *Taxus baccata*, Klinken (1914) had observed that individual rays tended to converge towards a fairly constant height. That is, low rays increased in cell number whereas high rays decreased, although the two processes operated independently. Barghoorn (1940a) confirmed this tendency in other conifers. Yet other workers observed that ray heights increased with increasing age. As a general rule, ray heights increase with age of the tree in both gymnosperms (Essner 1882, Fischer 1885, Jost 1901, Weinstein 1926, Bannan 1937, Schultze-Dewitz 1959, Kucera 1968, Kärkkäinen 1973) and dicotyledons (Zijlstra 1908, De Smit 1922, Braun 1955, 1963, 1970, Sebastine 1955, Noskowiak 1978). Typical age patterns are shown in Fig. 8.11. Rays in the first-year growth ring are high because of the predominating influence of prior primary growth. In the second-year ring, these high rays are split and the age-related changes become established.

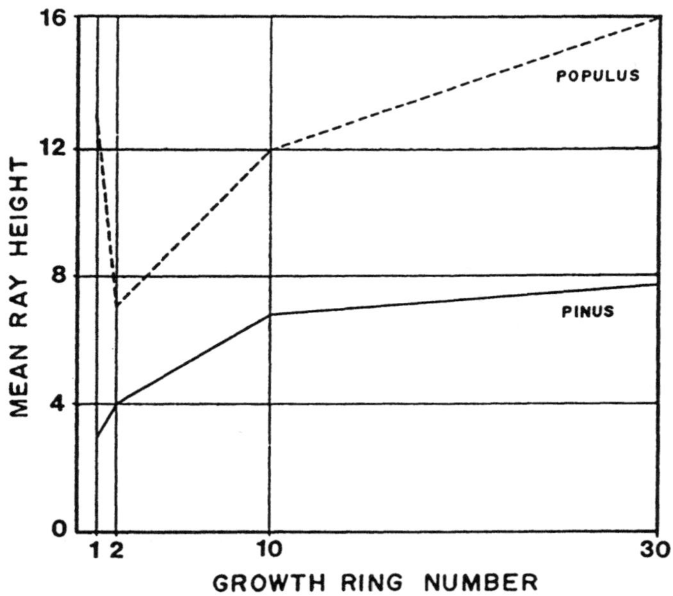

Fig. 8.11. Variation in mean ray height with increasing growth ring age in *Populus alba* and *Pinus sylvestris*. Ray height is expressed as the number of vertical cell tiers in uniseriate rays. (Braun 1970)

Like all general rules, considerable variation is introduced in the age pattern due to species differences and growth conditions. Thus, ray height can vary seasonally within a growth ring (Essner 1882, Huber and Prütz 1938, Braun 1955, 1970, Fengel 1966, Ajmal and Iqbal 1987a,b, Rao 1988a) as well as among growth rings. In the latter case, ray heights appear to be positively related to rate of growth (Fischer 1885, Bertog 1895, Bannan 1937, 1954, 1965b, White and Robards 1966, Carmi et al. 1972, Gregory and Romberger 1975). That is, in successively older growth rings, rays tend to increase in height slowly when radial growth rate is low and rapidly when growth rate is high. Declines in ray height, on the other hand, are usually due to growth conditions that seriously reduce growth ring widths (Bannan 1934a, 1937, 1954). Schultze-Dewitz (1961) found that both ray height and the number of ray cells were greater in dominant than in codominant *Picea abies* trees (Table 8.4). However, no significant relationship could be found between the two variables (Schultze-Dewitz 1992).

Ray heights also vary with vertical position in the tree stem (Hartig 1848, De Smit 1922, Iqbal and Ghouse 1987), although some authors considered the differences to be slight (Essner 1882, Fischer 1885). In *Pinus sylvestris*, Kärkkäinen (1973) recorded greater ray heights at stump level than higher in the tree. In some trees, a secondary increase in ray height occurred in the upper stem. Schultze-Dewitz (1992) summarized the data of several investigators in a table showing

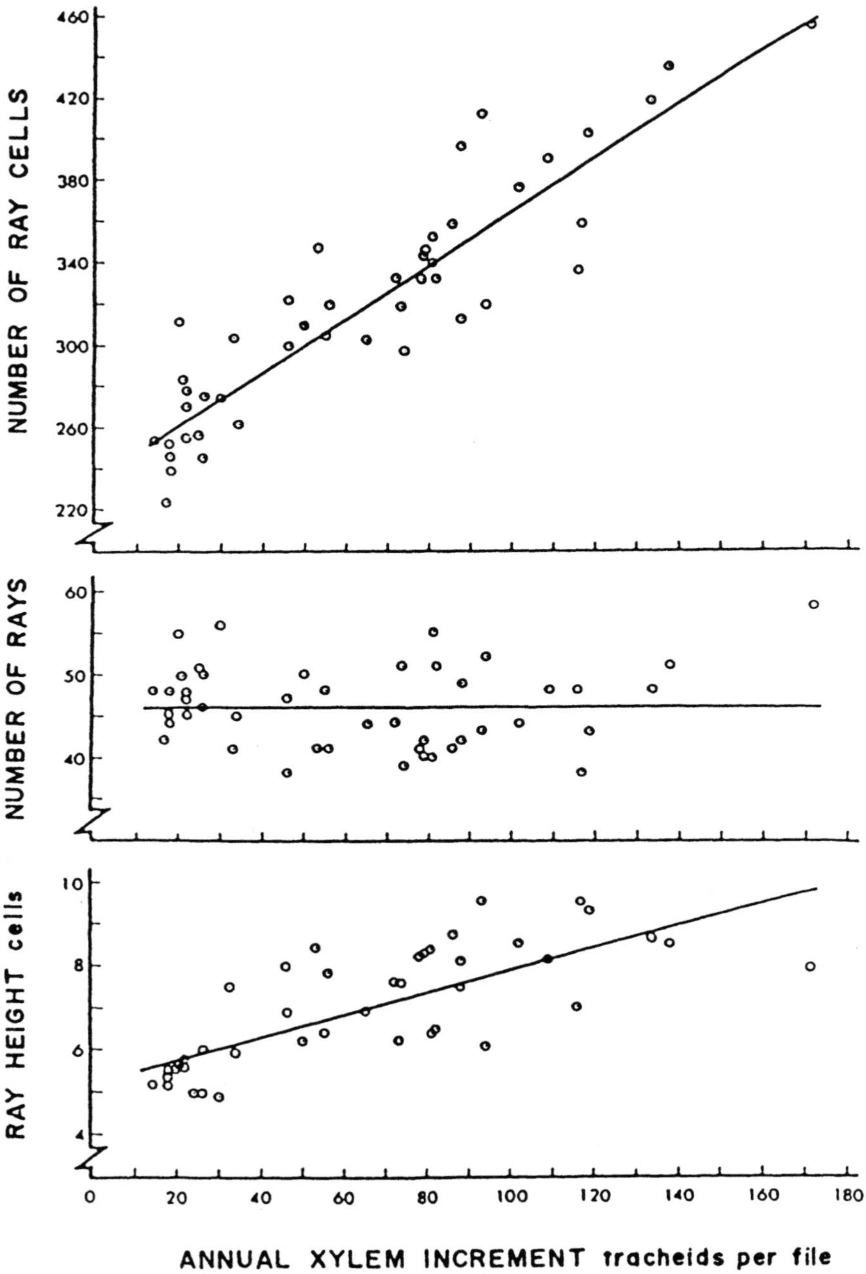

Fig. 8.12. Number of ray cells (*upper*), number of rays/mm^2 tangential area (*middle*), and the number of cell tiers per ray (*lower*) in the conducting phloem of *Picea glauca* as a function of xylem radial growth rate (number of tracheids per radial file produced annually). (Gregory and Romberger 1975)

how the heights and numbers of rays per unit area varied with both stem position and crown class in conifers.

Because some reports of increased ray numbers and ray heights were based on studies of either aphid infestations or other injuries that induced localized regions of radial growth, Gregory and Romberger (1975) decided to analyze ray cell production in uninjured, normal trees. From a 50-year-old stand of *Picea glauca*, the last-formed growth ring from 45 trees was examined. The data showed that the number of ray cells per unit tangential area was related to growth rate, based on the number of tracheids per file in the annual increment (Fig. 8.12). The number of rays was not related to growth rate, whereas ray height was. Consequently, the number of cells per ray increased with rate of growth.

Table 8.4. Influence of crown class on the height and number of rays in *Picea abies* from the Harz Mountains[a]. (After Schultze-Dewitz 1961)

Age	Ray height (μm)			Number of ray cells[b]		
years	Dominant	Codominant	Suppressed	Dominant	Codominant	Suppressed
52	183	134	99	90	78	47
72	212	151	119	139	82	61
94	232	184	157	109	89	74
112	220	175	140	125	110	76

[a] Based on average values for the years 1939-1958 from stem samples collected at breast height.
[b] Number of ray cells/0.442 mm^2.

Similar results were obtained with samples from a 73-year-old *Abies balsamea* tree. The latter tree had been released by an adjacent clear-cutting in 1964, and growth rate accelerated in subsequent years. Ray heights due to increased ray cell numbers increased dramatically with growth rate (Fig. 8.13). New rays one and two cells in height were added in abundance at the end of each growing season, as indicated by the peaks in ray cell numbers in Fig. 8.13, but many of these new rays did not survive. New ray cell initials were added to the ends of uniseriate rays by transverse, anticlinal divisions. No ray mergers were observed in this study. In a similar manner, Braun (1963) found that the increase in ray height usually occurred at season's end by anticlinal divisions of ray cell initials (Fig. 8.41).

Ray height increases do not continue indefinitely. As trees approach maturity ray heights tend to stabilize, as suggested by Klinken (1914), and during periods of appreciable growth decline ray heights are often reduced. Ray heights diminish as cell production in the marginal rows becomes more sporadic, and cell production finally ends as the ray cell initials involved mature (Bannan 1954).

Despite the many sources of variation, ray heights of different species fall within rather broad ranges. As Essner (1882) demonstrated, ray height by itself is of little diagnostic value, although it is sometimes useful when combined with other criteria (Metcalfe and Chalk 1983). Heights of the uniseriate rays of conifers

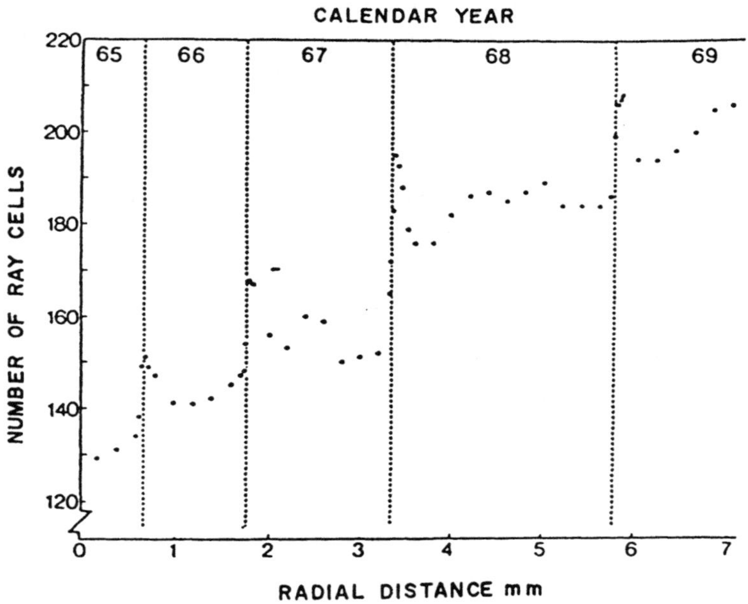

Fig. 8.13. The total number of ray cells in 25 xylem rays as a function of radial distance and calendar year in a 73-year-old *Abies balsamea* tree. Note the dramatic increase in ray height associated with increased ring width following release from competition in 1964. Note also the tendency for new ray cells to be initiated at the end of each growing season. (Gregory and Romberger 1975)

as a rule tend to remain low in comparison to the extremely tall rays in some dicotyledons. The mean ray height in *Pinus sylvestris*, for example, was four to six cells, and in *Abies alba* four to six cells in uniseriate and eight to eleven cells in biseriate rays (Fig. 8.14). However, in *Taxodium distichum* and *Sequoia sempervirens*, rays 40 to 60 cells high have been reported (Panshin et al. 1964).

In contrast to the conifers, the uni- and biseriate rays in *Corylus avellana* ranged from 2 to 60 (Nicoloff 1911), the multiseriate rays in *Ulmus fulva* from 4 to over 100 (De Smit 1922), and those of *Prosopis spicifera* from 1 to 110 cells in height (Ghouse and Iqbal 1977a). Rays in tropical dicotyledons are often exceedingly tall, with ray heights tending to increase with tree age. These vertically elongated rays of varying heights have been referred to as "ray initial units" (Iqbal and Ghouse 1985a).

Ray heights for different species of Pinaceae are shown in Table 8.5. Similar data on ray heights have appeared in many publications for both gymnosperms (Essner 1882, Bannan 1937, 1954, 1965b, Huber 1949, Braun 1955, Kucera 1968) and dicotyledons (Trendelenburg and Mayer-Wegelin 1955, Ghouse and Yunus 1974a, Ghouse and Iqbal 1975, Paliwal et al. 1984).

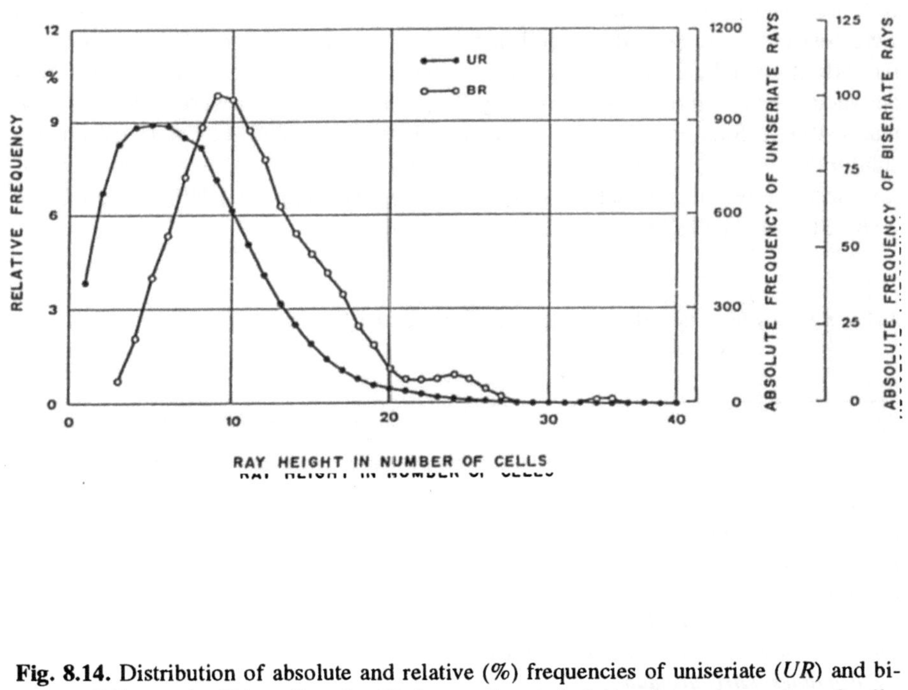

Fig. 8.14. Distribution of absolute and relative (%) frequencies of uniseriate (*UR*) and biseriate (*BR*) rays in *Abies alba*, classified according to heights of rays in number of cells. (Kucera and Bosshard 1975)

8.3.4 Ray Mergers

Ray mergers, or the uniting of rays, most generally occur through either elimination or conversion of the intervening fusiform initial cells separating two closely adjacent rays. Both types occur commonly in dicotyledons, whereas conversion of fusiform initials occurs rarely and usually only in cases of injury in conifers.

8.3.4.1 Gymnosperms

The general process whereby rays merge by the elimination of fusiform initials might be modified in a variety of ways in gymnosperms. Several of these modifications were exemplified in the work of Klinken (1914) on *Taxus baccata*. He demonstrated how adjacent uniseriate rays could merge in such a way that the newly merged ray became biseriate. Such mergers were temporary and the rays soon reverted to the uniseriate condition. The new ray, however, was now much taller. The simplest type of merger involved two superposed rays. As shown in Fig. 8.15A, a six- and a seven-tiered ray merged. They became partially biseriate and, after the loss of one cell, a 12-tiered, uniseriate ray was formed. In a somewhat more complex modification, two closely adjacent rays, both six-tiered, merged laterally to form a temporary biseriate ray (Fig. 8.15B). In this case, five cells were lost for a net gain of one cell in forming a new seven-tiered, uniseriate ray. An extremely tall, 23-tiered ray was formed when a 13-tiered ray was inserted between a seven- and a three-tiered ray (Fig. 8.15C). No cells were lost during the readjustment. Such tall rays seldom survived very long, and they were eventually subdivided either by a process of cell elimination or ray splitting. Similar types of ray merger have been observed in other conifers (Barghoorn 1940a, Braun 1955, 1970, Srivastava 1963a, Smith 1967).

Table 8.5. Ray heights in different species of Pinaceae (Srivastava 1963a)

Species	Usual maximal height[a]	Unusual height[a]
Abies concolor	10-12	
Cedrus deodara, Abies magnifica	16-20	25+
Pinus aristata	8-12	20+
Pinus ponderosa, P. radiata	8-12	
Pseudotsuga menziesii, Picea pungens	6-10	14+
Pinus murrayana, P. lambertiana, P. pinea, Picea abies, Larix decidua,	5-8	10+
Pinus sabiniana, P. muricata, P. jeffreyi, P. attenuata, P. flexilis, P. monticola	3-6	7+

[a] Ray height is expressed in terms of number of cells.

The aforementioned ray mergers often result in border tracheids becoming internal cells (Barghoorn 1940a). The ray cell initials that formerly produced border tracheids now produce internal tracheids. Such rays have been called "polymere Markstrahlen," or polymeric rays (Huber 1949, Braun 1955). They are not confi-

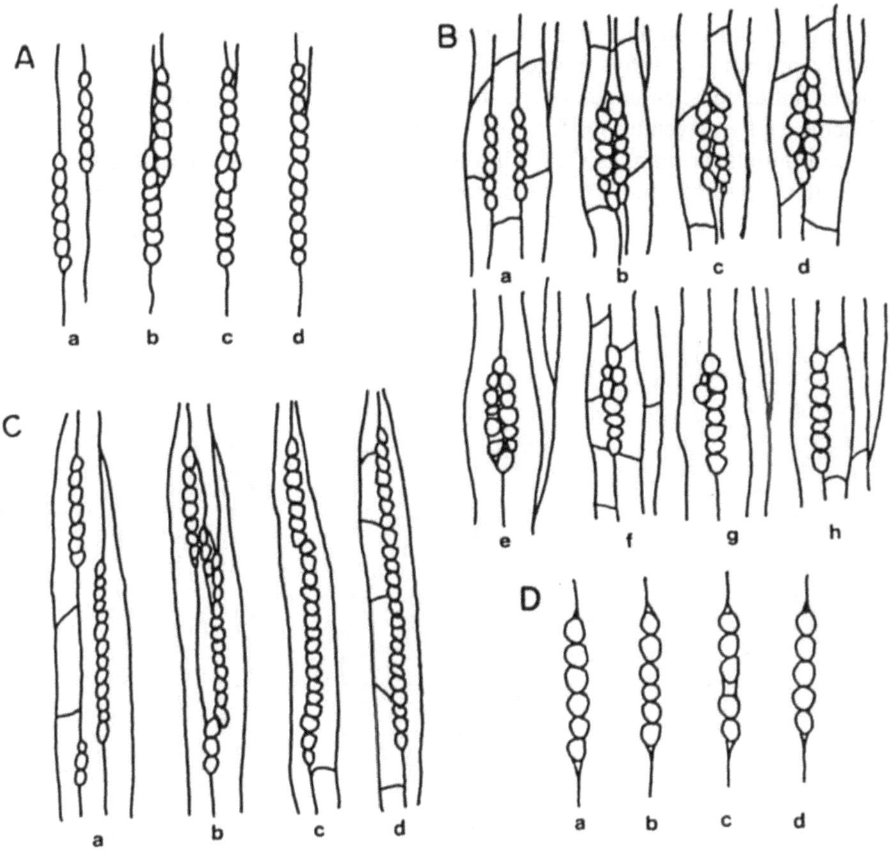

Fig. 8.15A-D. Examples of increases (**A-C**) and decreases (**D**) in ray height of *Taxus baccata*. See text for details. (After Klinken 1914)

ned to conifers but also occur in dicotyledons where a subtype has been referred to as an "articulated ray" (Huber 1949). The transposed ray cell initials, i.e., the former border ray cell initials, usually are either converted to the production of parenchymatous cells or eliminated. Thompson (1910), in his study of *Pinus resinosa*, described these former border tracheids as being "pinched out" when they were eliminated in this way.

External conditions can result in the formation of abnormal or aberrant multiseriate rays in conifers. For example, Bannan (1950a) described how broad rays sometimes formed by the merger of uniseriate rays in *Chamaecyparis*. Although Bannan could not account for their formation, these rays appear similar in form and structure to the multiseriate rays that are produced in several conifers by dwarf mistletoes (von Tubeuf 1930a, Srivastava and Esau 1961a,b) (Chap. 10.5.6). In the latter case, the broad rays are also initiated by the merger of uniseriate rays and continue to enlarge by further mergers of the newly formed multiseriate rays. In the simplest type, described by Srivastava and Esau (1961a,b), two closely superimposed rays merged by the intrusive growth of one or both of the border initial cells. In another type, merger was brought about by either the displacement of intervening fusiform initials or subdivision of these initials to form ray cell initials, or both, as shown in Fig. 8.16. These aberrant merging processes in conifers resemble those that occur normally in dicotyledons.

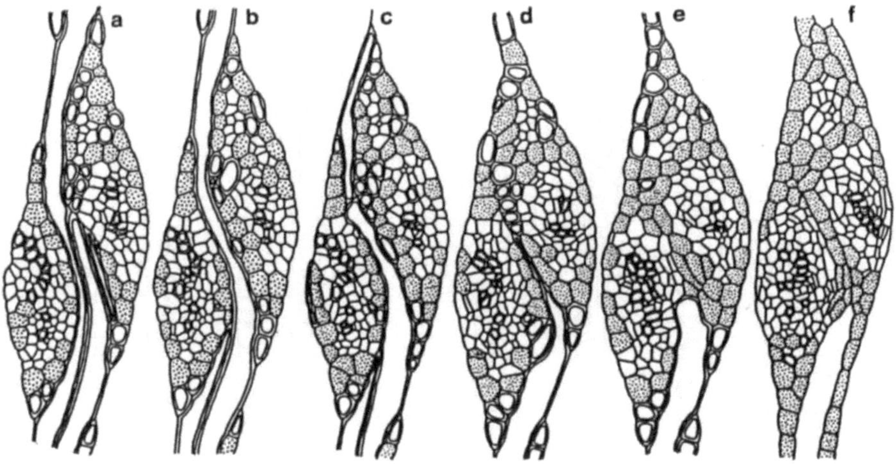

Fig. 8.16a-f. Mistletoe infected rays, drawn from serial tangential sections, from the xylem of *Pinus contorta* var. *murrayana* illustrating lateral fusion of multiseriate rays by the displacement, decline, and division of fusiform initial cells. (Srivastava and Esau 1961b)

8.3.4.2 Dicotyledons

The merger of rays in dicotyledonous species with predominantly uniseriate rays is brought about in much the same way as in gymnosperms (Braun 1955, Cheadle and Esau 1964). However, it can be far more complex in species with multiseriate rays. In the simplest type of merger, the border cell initials of two superposed rays expand vertically and close the gap between them (Fig. 8.17A). A variant of this type involves the fusion of two superposed rays by the simultaneous increase in their respective heights together with the rapid elongation of their border cell initials (Fig. 8.18). Similarly merged rays were called "articulated rays" in *Homa-*

lium laurentii and *Lonicera tatarica* by Huber and Schmidt (1938) and Braun (1955), respectively. By far the most common types of ray mergers in dicotyledons involve either the squeezing out and elimination of intervening fusiform initials or the conversion of fusiform initials to ray cell initials.

Elimination of Fusiform Initials: Merging of rays by the elimination or withdrawal of intervening fusiform initials has been documented by many authors (Eames 1910, Elliott 1935, Chattaway 1948, Braun 1955, 1970, Evert 1961,

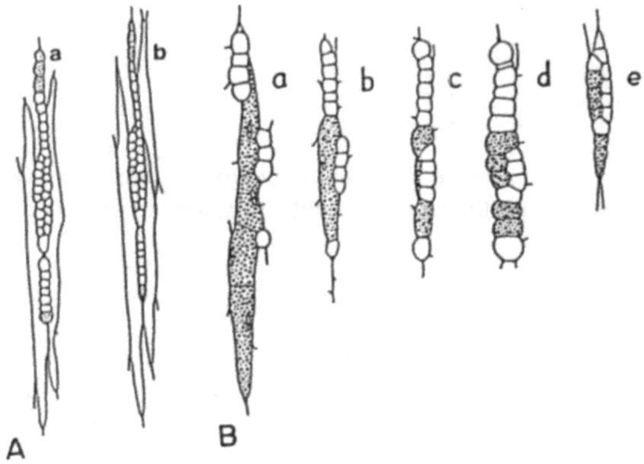

Fig. 8.17A,B. Ray mergers in *Pyrus communis* in tangential view. **A** Merger by the marginal cells of two superposed rays. 320x. **B** Division of a fusiform initial to form ray cell initials (*shaded*) and the merger of these cells with those of existing rays. Section *a* represents ray cell derivatives and section *e* ray cell cambial initials. 125x. (Evert 1961)

Cheadle and Esau 1964, Krawczyszyn 1971, 1972, Hejnowicz and Romberger 1973, Rao 1988a, Wloch and Szendera 1989). One way in which such a merger might occur was described and illustrated by Barghoorn (1940b). In this particular case, the uniseriate ray was first split by the intruding tip of a fusiform initial (Fig. 8.19). As the intruding tip later either withdrew or was eliminated, one ray cell initial enlarged many times its original size and then divided anticlinally to form three new ray cell initials.

Another way in which such a merger might occur was illustrated by Evert (1961) in *Pyrus communis*. As shown in Fig. 8.17B, a declining fusiform initial divided transversely. Three of these segments were eliminated, leaving a single segment in contact with three uniseriate rays. Further division of this segment to form ray cell initials, together with rearrangement of these initials, finally gave rise to a biseriate ray.

Conversion of Fusiform Initials: Ray mergers resulting from the conversion of fusiform to ray cell initials is also of rather common occurrence in dicotyledons

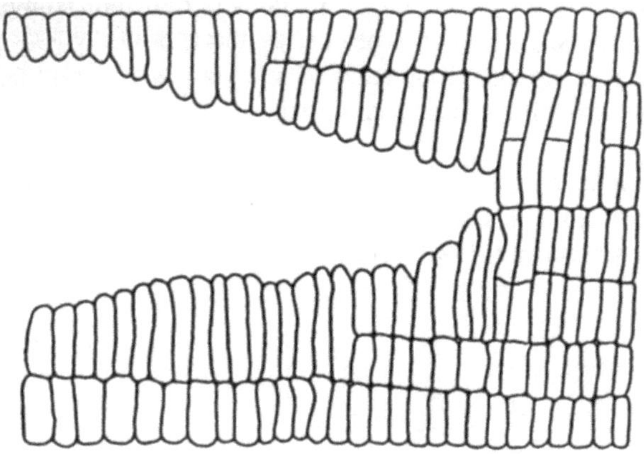

Fig. 8.18. Fusion of two rays in *Casearia nitida* by the vertical elongation of marginal ray cell initials and their derivatives. Radial longitudinal section with cambium to the *right*. 53x. (Barghoorn 1940b)

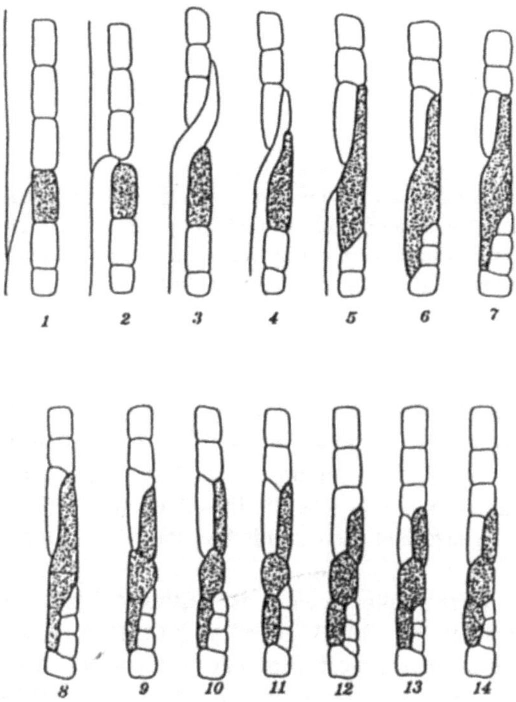

Fig. 8.19, *1-14*. *Viburnum odoratissimus*. A intruding fusiform initial first split the ray (*1-4*) but then withdrew (*5*). Concomitantly, a ray cell initial (*shaded*) enlarged (*3-7*), and then underwent a series of anticlinal divisions (*8-10*) to form three new ray cell initials (*11-14*). Serial tangential sections of xylem approximately 20 μm thick. 43x. (Barghoorn 1940b)

(Eames 1910, Chattaway 1933a, Braun 1955, 1970, Cheadle and Esau 1964, Ghouse and Iqbal 1977a, Ajmal and Iqbal 1987a,b), and the two processes often occur concomitantly (Krawczyszyn 1972, Ghouse and Yunus 1973). For example, merger of the broad rays in *Platanus acerifolia* involved both the withdrawal of fusiform initials and the conversion of fusiform to ray cell initials (Krawczyszyn 1971). As in most mergers, first evidence of the impending union was a gradual shift or approach of one ray to another (Fig.8.20A). Withdrawal of some fusiform

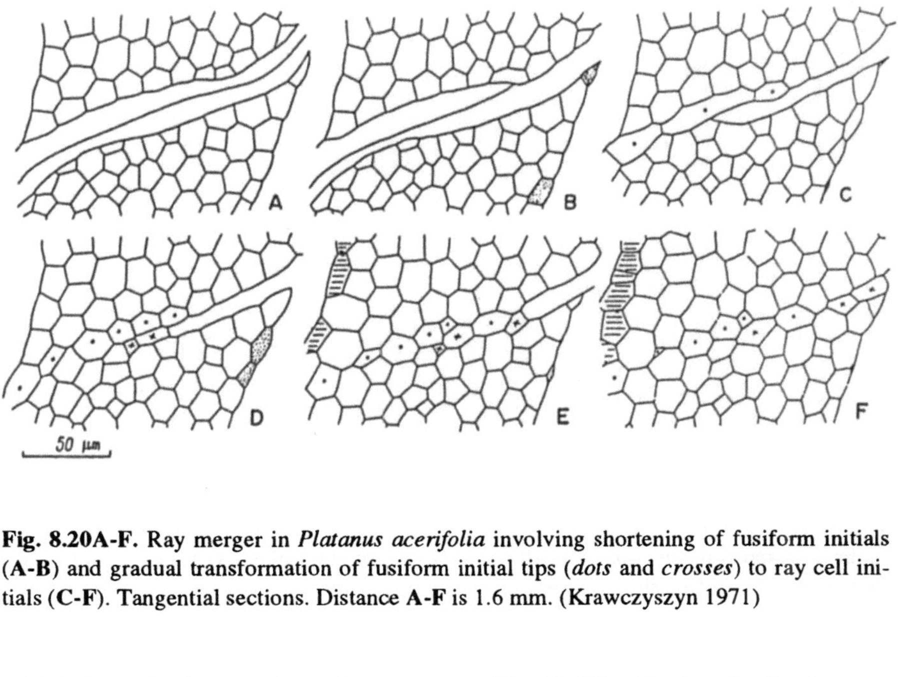

Fig. 8.20A-F. Ray merger in *Platanus acerifolia* involving shortening of fusiform initials (**A-B**) and gradual transformation of fusiform initial tips (*dots* and *crosses*) to ray cell initials (**C-F**). Tangential sections. Distance **A-F** is 1.6 mm. (Krawczyszyn 1971)

initials brought the rays into closer contact (Fig. 8.20B). During the final stage, the remaining fusiform initials were converted to ray cell initials thus completing the merger (Fig. 8.20C-F).

8.3.5 Loss of Ray Cell Initials

Just as ray heights can increase by the addition of new ray cell initials, they can decrease by the loss of initials. Rays of all sizes can be reduced in height by the intrusion of fusiform initials and other types of ray splitting. A decrease in height of uniseriate rays usually occurs by the loss of initial cells on the ray borders (Bannan 1934a, 1937, 1954, 1965b, Srivastava 1963a, Cheadle and Esau 1964, Smith 1967, Nabuchi et al. 1979), although it might also involve interior cells (Klinken 1914, Barghoorn 1940a, Cheadle and Esau 1964, Kucera and Kucera 1967) (Figs. 8.15D, 8.21).

Ray decline generally occurs in narrow growth rings of trees under stress. For example, in old stems of *Thuja occidentalis*, Bannan (1954) observed that mean ray height declined in the extremely narrow growth rings. He attributed this decline to the sporadic elimination of ray cells due to the loss of generative capacity and eventual maturation of the ray cell initials involved. In another case, he (Bannan 1934a) described the condition of an old *Pinus resinosa* tree that had been injured by fire. The growth rings preceding the fire were wide, but afterwards very narrow. The rays traversing these narrow rings suffered considerable cell loss which was gradual but consistent. As the initial cells producing border

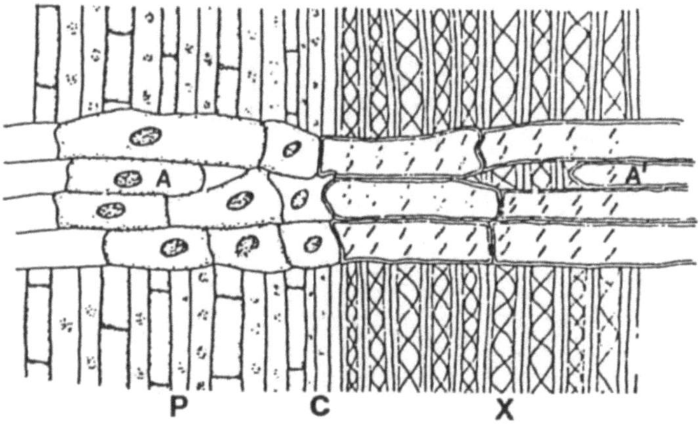

Fig. 8.21. Loss of a interior ray cell initial from the cambium (*C*) of *Taxus baccata* resulted in the corresponding deletion of a cell file (*A-A'*) in both the phloem (*P*) and xylem (*X*). Radial section. 164x. (Barghoorn 1940a)

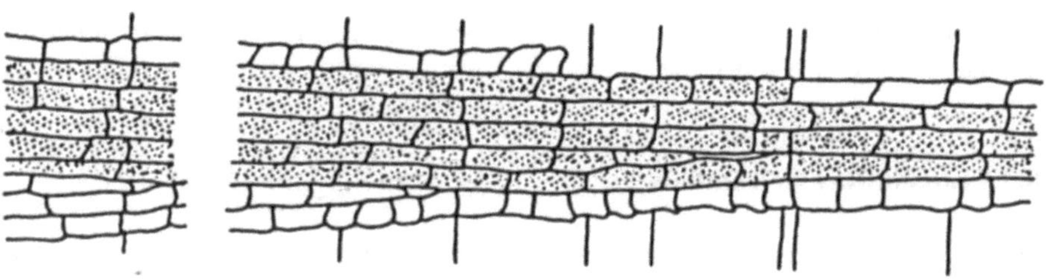

Fig. 8.22. Ray in old stem of *Pinus resinosa* 15 cm from pith showing dying of marginal tracheary rows and development of ray tracheids (*unshaded*) in newly exposed parenchymatous rows. Radial section. (Bannan 1934a)

tracheids were lost, the submarginal initials that formerly produced parenchyma cells began producing tracheids (Fig.8.22). In one small area of cambium, 42 rays were eliminated in the 5 years following the fire.

Similar patterns were observed in narrow growth rings of many other conifers (Bannan 1937). For example, average ray height in a large *Pseudotsuga menziesii* tree decreased from 10.6 cells 70 cm from the pith to 7.7 cells in the older, narrower rings 105 cm from the pith. The decrease in ray heights was brought about mainly by the maturation of tracheary rows along the ray margins which were replaced by conversion of formerly submarginal parenchymatous cells to tracheids.

8.3.6 Ray Splitting

Large primary rays seldom persist during secondary growth of most woody species. The most common way for large rays to be broken into smaller units is ray splitting, either by the intrusion of fusiform initials into the rays or by the reversion of ray cell initials to fusiform initials. A less common method is simply the elimination of a ray cell initial in the middle of a tall ray resulting in two shorter rays (Klinken 1914, Kucera and Kucera 1967).

8.3.6.1 Intrusion of Fusiform Initials

Splitting of rays by the intrusion of fusiform initials into rays has been a matter of record for many years (Sanio 1863a). One pattern, as visualized by Göppert (1842) for *Casuarina equisetifolia*, is shown in Fig. 8.1B. Jost (1901) questioned several of these early reports concerning the splitting of rays in his study of *Fagus sylvatica*. He agreed that the isolated, tangential microsections commonly used gave the impression that vessels and fibers invaded the rays, but serial tangential sections often showed that these disruptions were simply "bridges." That is, if the invasion was due to the elongating tip of a cambial mother cell, the disruption would be temporary. Segmentation of a ray would only persist if the invading cell was a fusiform cambial initial. Thus, not only did Jost establish the fact that subdivision of large rays occurred in the cambium, but he also claimed that the process by which the ray elements were displaced involved sliding growth of the fusiform initials. Klinken (1914) and Neeff (1914) also attempted to verify sliding growth in this way.

Ray splitting by the intrusive growth of fusiform initial cells has been widely reported in both gymnosperms (Schmidt 1889, Klinken 1914, La Rivière 1916, Barghoorn 1940a, Bannan 1941, 1950a, 1953, Braun 1955, Kucera and Kucera 1967, Smith 1967, Harris 1969, Savidge and Farrar 1984) and dicotyledons (Jost 1901, Zijlstra 1908, Eames 1910, Bailey 1911, 1912, Thompson 1911, Holden 1912, Neeff 1914, 1922, Beijer 1927, Chattaway 1933a, 1951b, Sprengel 1936, Barghoorn 1940b, 1941a, Schoch-Bodmer and Huber 1952, Braun 1955, 1970, Holdheide 1955, Evert 1961, Cheadle and Esau 1964, Hejnowicz and

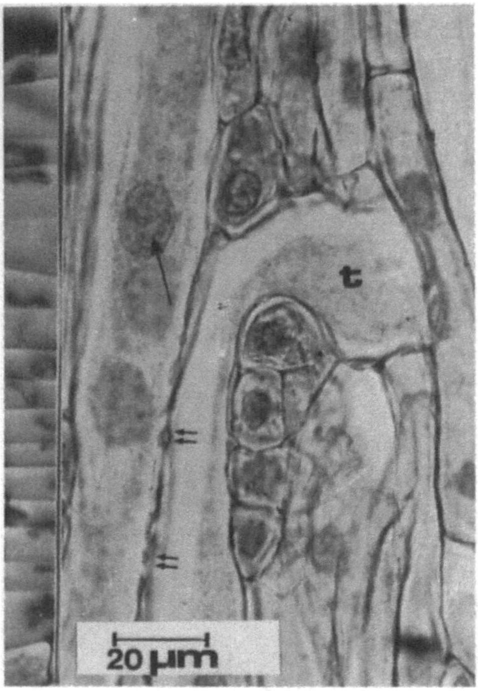

Fig. 8.23. Tangential section of *Pinus contorta* in the vicinity of a stem girdle. The intrusive tip of fusiform cell (*t*) is shown splitting a biseriate ray in the zone of xylem differentiation. The *large arrow* identifies the spherical nucleus and the *double arrows* the developing bordered pits in an adjoining fusiform cell. (Savidge and Farrar 1984)

Krawczyszyn 1969, Krawczyszyn 1971, Hejnowicz and Romberger 1973, Hejnowicz and Zagórska-Marek 1974, Ghouse and Hashmi 1977b, 1980, 1981, Khan et al. 1983, Iqbal and Ghouse 1985b, 1987, Ajmal and Iqbal 1987a,b, Lo 1989, Wloch and Szendera 1989). Thus, uniseriate as well as multiseriate rays of various widths are dissected in this manner. In some species, particularly those in which the stem structure is typically herbaceous, ray breakup might be limited and the broad primary rays tend to persist (Carlquist 1988a).

The rays of most conifers, being primarily uniseriate, are readily split by intruding fusiform initials (Fig. 8.23). Rays less than four cells high are rarely split. At times, the total cell number remains unchanged, whereas at other times, a ray cell initial might be eliminated. Bannan (1941, 1950a) encountered what he considered unusual rays in several conifers. These rays normally originated very close to the pith. At first they were uniseriate and resembled ordinary rays but later they expanded both vertically and laterally. After expansion, some rays were dissected into a number of small rays by the intrusion of adjoining fusiform initials. As elongation of the fusiform initials proceeded, they underwent periodic anticlinal divisions to produce daughter initials, and these cells continued the process of growing between and separating the ray cell initials.

In dicotyledons, the splitting of rays ranges from simple (Fig. 8.24) to complex. Zijlstra (1908) described how the large rays of *Fagus sylvatica*, *Quercus robur*, and *Aristolochia ornithocephala* were repeatedly dissected by intruding fibers as the rays progressed outward in the stem. Dissection of a broad ray, as shown in Fig. 8.25 from Zijlstra (1908), usually began by the penetration of fusiform initials in the broad, middle region of the ray as also observed by others (Barghoorn 1941a, Braun 1955, Krawczyszyn 1971). The smaller ray segments split off in this way seldom retain either their form or their individuality. Ray development is a dynamic process. Small segments are occasionally lost from the cambium, while

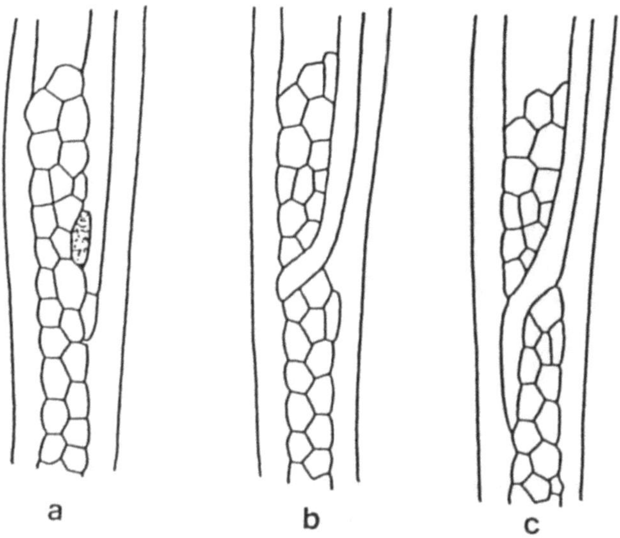

Fig. 8.24a-c. Serial tangential sections of xylem from *Trochodendron aralioides* showing ray splitting by the apical elongation of a fusiform cambial initial. The ray cell initial (*shaded*) was lost from the cambium, facilitating the intrusion of the fusiform initial. The interval between sections **a** and **c** is approximately 75 μm. 100x. (Barghoorn 1940b)

others can increase in size by internal cell divisions and/or mergers with other rays. Ajmal and Iqbal (1987a,b) found that the abundance of ray types varied seasonally in *Streblus asper* and *Ficus rumphii*. Large rays resulting from ray mergers were more numerous in late fall and winter, whereas small rays resulting from ray splitting were more numerous during the active growing season.

Additional, more detailed, examples of the dynamic development of rays can be found in Section 8.4 and Chapter 6.6.2. In the latter case, it was pointed out that ray splitting was far more common in stems undergoing frequent changes in grain orientation. Splitting of rays is also a common consequence of the reorientation of cambial cells following decapitation of the shoot or wounding of the cambium. Particularly clear examples of this effect were described and illustrated by Neeff (1914) in *Tilia americana*, *Salix babylonica* and several other species (Chap. 10.3.8). After decapitation, the cambial initials rather quickly became reoriented in the direction of the newly dominant shoot. Reorientation of the rays, however, was much slower and they appeared passive in their response. Small rays yielded to the reorientation stimulus but large rays resisted. As a consequence the reoriented tips of fusiform initials penetrated the large rays splitting them into smaller fragments (Fig. 10.25). Many of these smaller ray segments were later lost from the cambium. Similar results were observed in decapitated roots of *Tilia tomentosa* and *T. euchlora* (Neeff 1922).

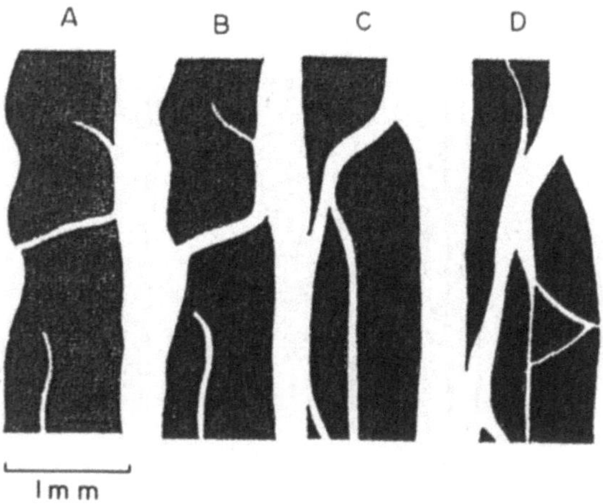

Fig. 8.25A-D. Tangential sections at different radial positions in a broad ray in a 5-mm-thick stem of *Aristolochia ornithocephala*. Dissection began by intrusive growth of fusiform cells in the middle of the ray followed by intrusive growth of other fusiform cells throughout the ray. Section **A** nearest the pith. (After Zijlstra 1908)

Lev-Yadun and Aloni (1991b) mentioned that the aggregate rays of *Quercus calliprinos* and *Q. ithaburensis* were dispersed as smaller rays in regions of partial girdles. They claimed this to be the first study showing a decrease in ray size in woody plants in response to experimental treatment. However, they apparently overlooked Neeff's comments to this effect although, they cited one of his papers. Mäule (1896) had earlier demonstrated the same effect in girdled stems of *Rosa centifolia* (Fig. 10.13).

8.3.6.2 Reversion of Ray Cell Initials

Ray splitting by the reversion, or conversion, of ray cell initials to fusiform initials is considerably less common than the aforementioned method. Bannan and Bayly (1956) stated that ray initials "... do not, except under unusual circumstances, expand to fusiform initials in conifer cambium." No examples of this type of ray spitting in normal conifer cambia have been found in the literature, perhaps because the process is unsuited for splitting uniseriate rays. One example of an unusual circumstance was reported by Hejnowicz (1963b) in which ray cell initials of *Larix decidua* were transformed to fusiform initials following mechanical incision of the cambium.

Jost (1901) suggested that ray splitting might involve the reversion of ray cell initials to fusiform initials in the dicotyledons he examined, but he provided no confirmation of the process. Years later, during a study of cambial growth in rootwood of *Tilia euchlora*, Neeff (1922) described what he interpreted to be such a reversion. As shown in Fig. 8.26, the ray was first split by the intrusive growth

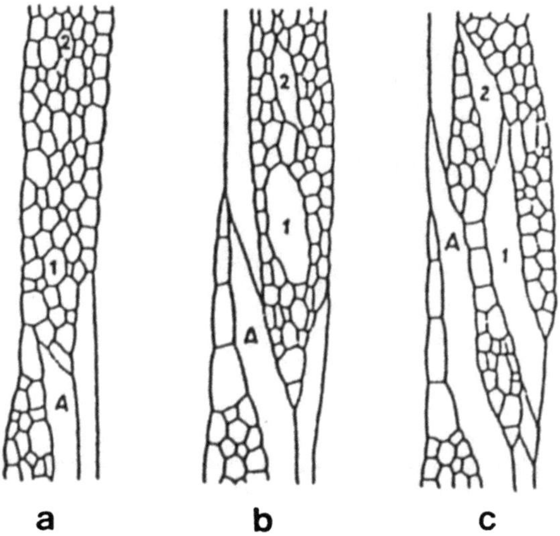

Fig. 8.26a-c. Ray splitting by transformation of ray cell initials to fusiform initials in *Tilia euchlora*. A fusiform initial (**A**) first split the ray followed by the reversion of ray cells *1* and 2 to new fusiform initials. Nonsequential tangential sections. 57x. (Neeff 1922)

of fusiform initial A. However, it was also split by the expansion and gradual transformation of ray cell initials 1 and 2 to fusiform initials.

A more definitive confirmation of the process was provided by Chattaway (1933a), who observed this type of ray splitting in several species and described the process in detail. In one species, *Cola togoensis*, not only did such reversions occur with extraordinary frequency, but they were considered the principal means

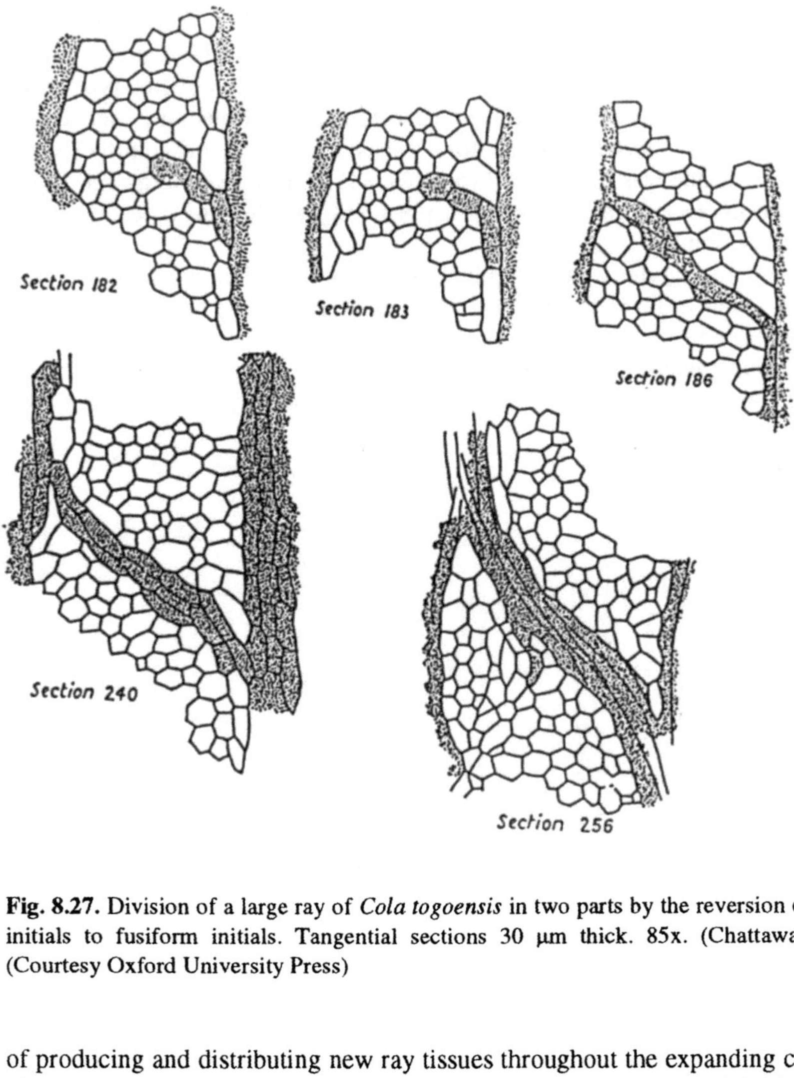

Fig. 8.27. Division of a large ray of *Cola togoensis* in two parts by the reversion of ray cell initials to fusiform initials. Tangential sections 30 μm thick. 85x. (Chattaway 1933a) (Courtesy Oxford University Press)

of producing and distributing new ray tissues throughout the expanding cambium. The reversion began with swelling of certain contiguous ray cell initials in the broad, middle portion of a multiseriate ray (Fig. 8.27, Sect. 182, 183). As the line of swelling cells spread across the central part of the ray, the involved ray cell initials gradually elongated (Sect. 186, 240). Eventually, the ray cell initials were transformed to a band of fully elongated fusiform initials completely separating the ray in two parts (Sect. 256). Chattaway (1948) observed similar ray splitting in species of *Banksia*.

Barghoorn (1940b, 1941a) cited and illustrated several instances in which ray cell initials reverted to fusiform initials in dicotyledonous species. The reversion

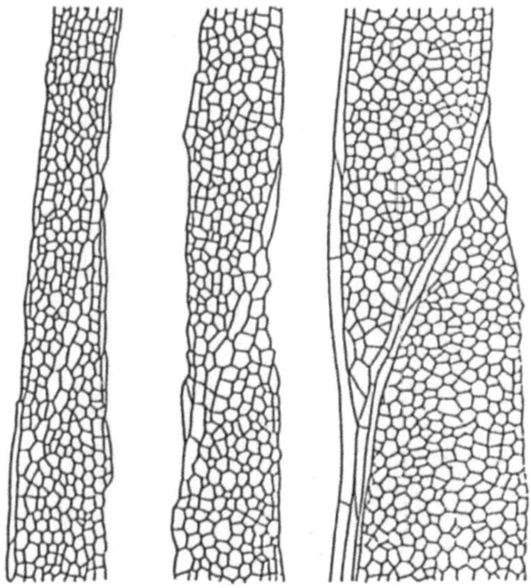

Fig. 8.28a-c. Tangential sections illustrating the splitting of a large ray in rootwood of *Platanus acerifolia* by the transformation of ray cell initials (**a,b**) to fusiform initials (**c**). (Süss and Müller-Stoll 1973)

usually began in the central part of a multiseriate ray when one to several contiguous ray cell initials elongated during successive periclinal divisions. These ray cell initials gradually assumed the characteristics of fusiform initials, and as they continued to elongate in a more or less vertical direction they eventually split the ray. Another example cited by Barghoorn (1940b) did not split the ray but reduced its height. In this case, a marginal ray cell initial reverted to a fusiform initial and, in so doing, caused the elimination of two additional ray cell initials from the cambium. Barghoorn considered the transformation of ray cell initials to fusiform initials to be the most significant factor in reducing the heights of very tall, primary, multiseriate rays, as well as in the dissection of large secondary rays.

According to Braun (1955), the broad rays of *Fagus* and *Quercus* were most commonly split by fusiform initials intruding from the ray margins. Occasionally, however, the process began within the ray, in which case certain ray cell initials were gradually transformed to fusiform initials. Sprengel (1936) attributed the breakup of broad rays in the tumorous growth on *Quercus* stems to the transformation of ray cell initials to fusiform initials (Chap. 10.5.1). Other species in which authors claimed to have observed reversion include *Acacia nilotica* var. *telia* (Iqbal and Ghouse 1987), *Citrus sinensis* (Khan et al. 1983), *Ficus rumphii* (Ajmal and Iqbal 1987b), *Gmelina arborea* (Rao 1988a), *Polyalthia longifolia*

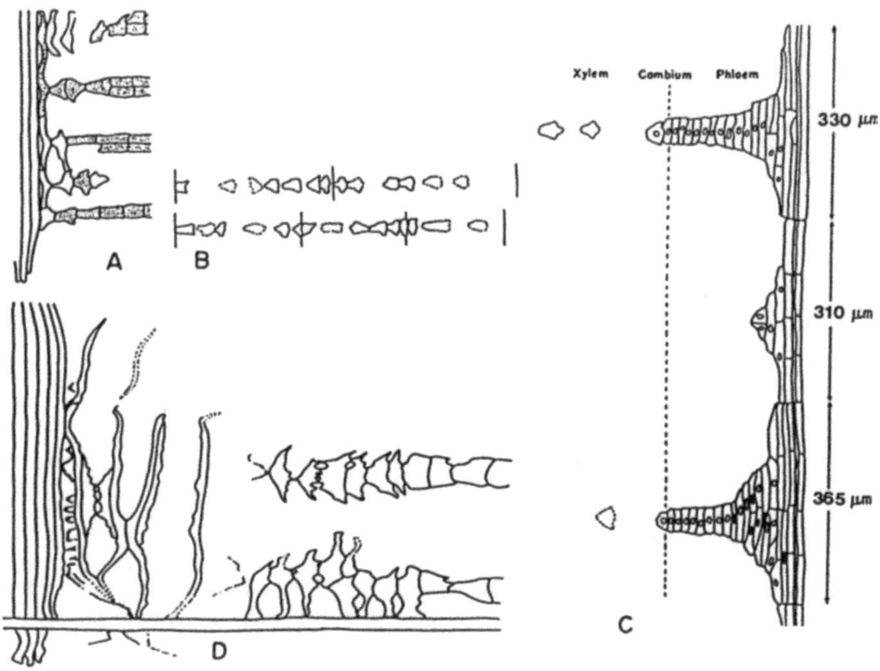

Fig. 8.29A-D. Discontinuous rays in conifers in radial view. **A,B** *Pinus strobus*. **A** Origin of a ray 0.5 mm from the pith consisting of contiguous cells. Parenchyma cells *shaded*; tracheids *unshaded*. **B** Origin of a ray 11.5 cm from the pith consisting of discontinuous and "ghost" tracheids (*dotted outlines*). **C** *Thuja occidentalis*. Conversion of declining fusiform initials to ray cell initials. The upper and lower rays are discontinuous in the xylem; the middle ray has aborted. **D** *Chamaecyparis nootkatensis*. Discontinuous xylem cell production in a new ray. Note also the transition from vertical to radial cell alignment. (**A-B, C,** and **D** from Bannan 1934a, 1953, and 1951a, respectively)

(Ghouse and Hashmi 1977b, 1980), and *Streblus asper* (Ajmal and Iqbal 1987a). In fact, Sagheer and Iqbal (1987) not only observed frequent transformations of ray cell initials to fusiform initials in *Ficus rumphii* but they claimed the initials to be "interchangeable."

Some workers have failed to find any evidence of the reversion of ray cell initials to fusiform initials. Evert (1961) never observed such a reversion in his study of *Pyrus communis*, nor did Cheadle and Esau (1964) in *Liriodendron tulipifera*; both studies were conducted on phloem rays. Srivastava and O'Brien (1966a) simply stated that the transition of fusiform to ray cell initials and vice versa was a well-documented phenomenon, whereas Esau (1965) dismissed it as a less-common type of ray splitting. Catesson (1964), although admitting the possibility of such a transformation, apparently never actually observed it in *Acer pseudoplatanus*.

One might expect the process to occur more frequently in trees with rapidly changing grain patterns (Chap. 6.6.2). However, Krawczyzyn (1971) did not observe this type of ray splitting in *Platanus acerifolia*, and Hejnowicz and Romberger (1973) considered it as playing an insignificant role, if any, in ray splitting of the species they examined. Süss and Müller-Stoll (1973), on the other hand, did observe reversion of ray cell to fusiform initials in *P. acerifolia* (Fig. 8.28). They suggested that the large rays in rootwood of this species might be due to less frequent ray splitting than in the stemwood.

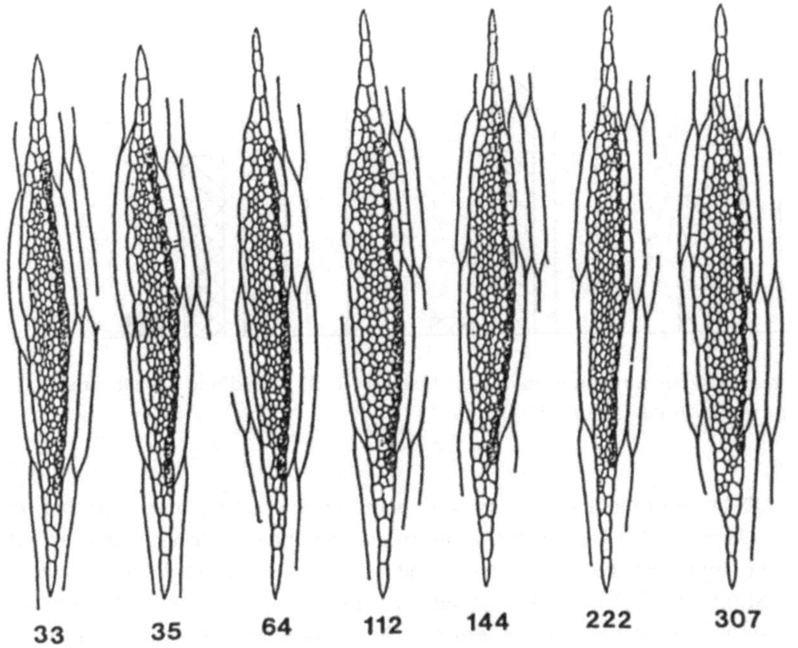

Fig. 8.30. Formation of sheath cells by the addition and subdivision of adjacent fusiform initials and their conversion to ray cell initials in *Pterocymbium javanicum*. *Numbers at bottom* identify the tangential sections 30 µm in thickness. 43x. (Chattaway 1933a) (Courtesy Oxford University Press)

8.3.7 Discontinuous Rays

Hartig (1859a) described what he called "aussetzende Markstrahlen," or discontinuous rays, that terminated within the xylem of *Fagus* and several exotic species.

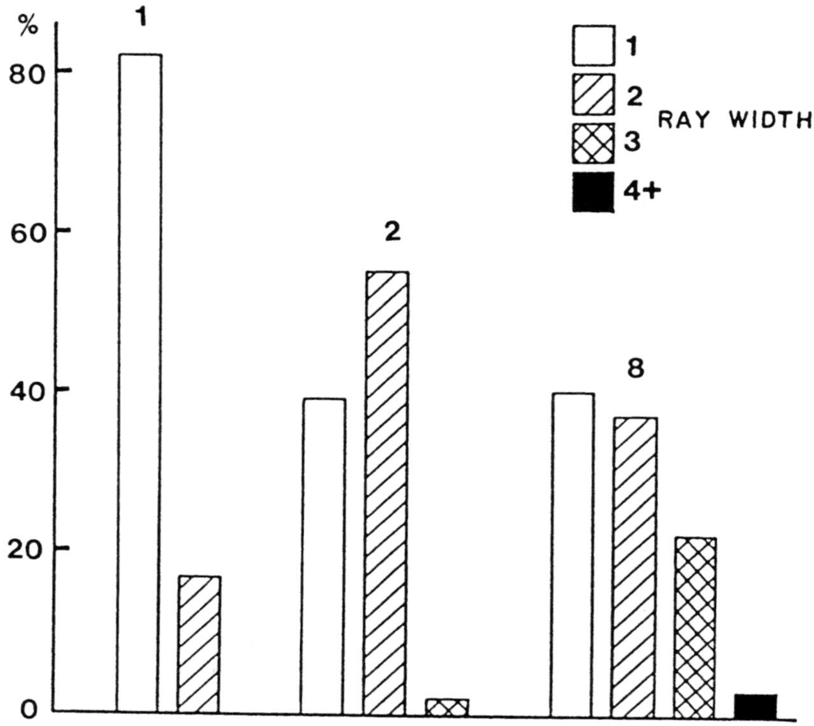

Fig. 8.31. Frequency distribution of rays *1*, *2*, *3*, and *4+* cells in width in growth rings *1*, *2*, and *8* in *Lonicera tatarica*. (Braun 1955)

Abromeit (1884) interpreted similar rays as being analogous to false rays that he observed in *Quercus*. De Bary (1877) questioned whether these rays did in fact terminate and suggested that, more likely, they simply curved out of the plane of sectioning. Similarly, Jost (1901) believed that Hartig's conclusion was based on incomplete sampling. Jost noted that while selected tangential sections might give the impression of the rays being invaded by vessels and fibers from the surrounding wood, serial tangential sections always showed the rays to be radially continuous.

Actual discontinuous rays were described by Bannan (1934a) in *Pinus strobus*. New rays arising in young growth rings next to the pith consisted of contiguous cells (Fig. 8.29A), but in older wood considerable gaps occurred between the last tracheids and the first cells of the succeeding rays (Fig. 8.29B). The new rays were usually one cell high and the cells were mostly tracheary interspersed with "ghosts." Ghost cells were ray tracheids, occasionally parenchymatous cells, that had lost their protoplasm shortly after differentiation. These cells were visible in radial sections solely by their wall outlines. Similar discontinuities in newly formed rays, along with occasional ray abortions (Fig. 8.29C), were observed by Bannan (1936a, 1941, 1951a, 1953) in other species (Fig. 8.29D). These discontinuities apparently resulted from the fact that the new ray cell initials, arising from declining fusiform initials, were unable to divide as rapidly as the surrounding fusiform initials. Consequently, ray cell production was at first sporadic. During the conversion of fusiform initials to ray cell initials, the initial cell for only one segment ordinarily survived to perpetuate the ray, while those for all remaining segments matured (Fig. 8.29C).

In addition to the discontinuities that often occurred at this time, the new ray cells underwent adjustments that changed their forms to a radial alignment (Fig. 8.29D). The gaps between ray cells were at first filled by tangential expansion of fusiform initials (Bannan 1953). Somewhat later, when the ray cell initials had shortened, the ray cells tended to adhere to one another by the median portions of their tangential walls. Eventually, after further shortening and form adjustments, periclinal division rates stabilized and a continuous, uninterrupted ray formed between xylem and phloem.

Similar discontinuities in newly formed rays were observed in other gymnosperms by Barghoorn (1940a) and Srivastava (1963a,b). In *Ginkgo biloba*, Srivastava (1963b) noted that the new ray cell initials often divided preferentially toward the phloem, resulting in discontinuities only on the xylem side, a condition reported earlier by Chrysler (1913) in *Abies* and *Pinus* and Chrysler (1915) in *Cedrus*. An analogous condition, known as "disjunctive ray cells," has been reported by several authors (Carlquist 1988a). In this case, the ray cells appeared to have been pulled apart prior to maturation, leaving the partially separated cells interconnected by finger-like projections. Strasburger referred to these structures as "gedehnte Markstrahlen" or stretched rays (Ziegler 1964).

8.3.8 Ray Failure

Not all rays are perpetuated indefinitely in the cambial zone. In addition to those lost by mergers with other rays, some rays are lost outright from the cambium (Neeff 1922, Beijer 1927). Braun (1955), in his study of *Pinus sylvestris*, considered loss of an entire ray to occur infrequently. Small rays are, however, sometimes lost. As described by Beijer (1927) in several dicotyledonous species that he examined, the rays were separated by the tip growth of adjacent cambial cells. Many of these smaller, separated rays were later lost from the cambium. He referred to this process as "Wandbrechungen," or wall-breaking, and he attributed it, along with intracambial cell elongation, to the irregular (non-stratified) cell arrangement in certain species such as *Alstonia*.

8.3.9 Ray Widths

Like ray heights, ray widths in dicotyledons can increase either by anticlinal divisions of initial cells within the ray (Chattaway 1933a, 1948, 1951b, Braun 1955, 1970, Holdheide 1955, Braun and den Outer 1964, Cumbie 1967b, Khan et al. 1981) or by ray mergers. Another method, reported by Neeff (1914) in *Tilia americana,* involved the division and conversion of adjacent fusiform initials to ray cell initials that were then added to the sides of the ray. This process apparently occurs in many dicotyledons. As described by Metcalfe and Chalk (1983), the resulting ray cell initials are at first taller than the adjacent, procumbent ray cell initials and their overall shape conforms to that of the original fusiform initial. Because of their position on the ray margins, these newly added cells are often called "sheath cells" (Chattaway 1933a, Barghoorn 1940b, Huber 1949) (Fig. 8.30). The obverse situation was also described by Barghoorn (1940b) in *Cochlospermum vitifolium,* where marginal ray cell initials were transformed to a fusiform initial.

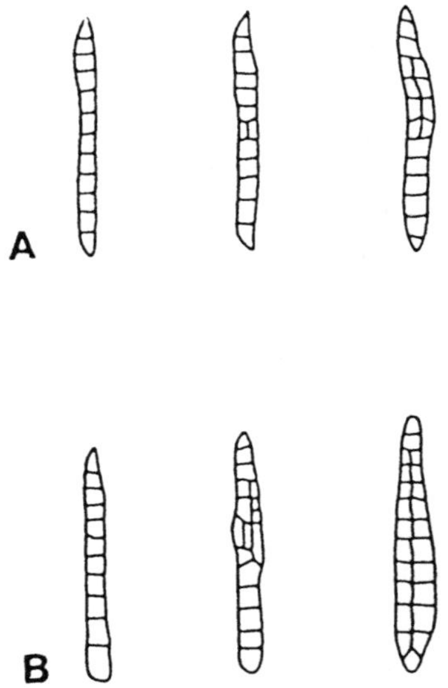

Fig. 8.32A,B. *Fagus sylvatica* showing in tangential view how width growth of two new secondary, uniseriate rays (**A,B**) began with cell divisions in the ray middle and then spread toward the ends. (Braun 1955)

Ray development, as described by Braun (1955) in *Lonicera tatarica,* typifies a pattern found in many dicotyledons. New secondary rays were primarily uniseriate at their origin and the multiseriate condition was gradually achieved over time (Fig. 8.31). As shown in *Fagus sylvatica* (Fig. 8.32), broadening of the rays began in the middle and extended toward the ends. The original homocellular ray

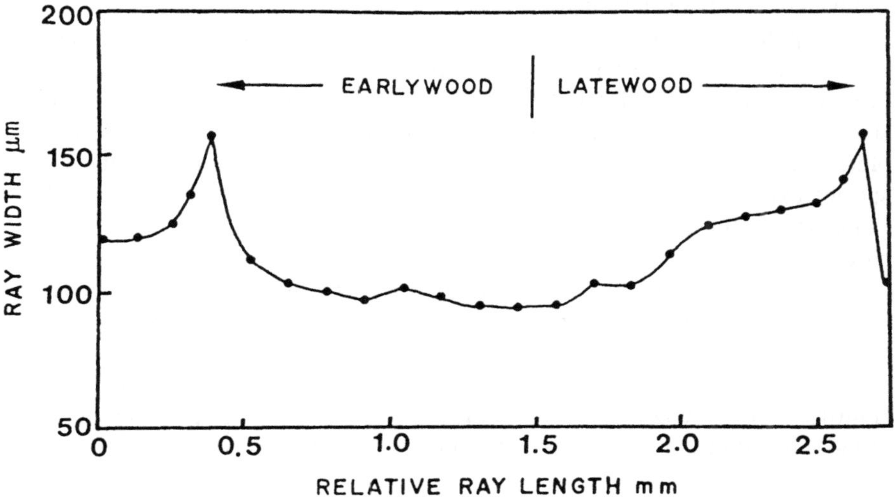

Fig. 8.33. Variation in ray width within a single growth ring of *Fagus sylvatica*. *Peaks* indicate growth ring boundaries. (After Fengel 1966)

thus became heterocellular with a mixture of vertically and horizontally oriented cells. Carlquist (1978) noted that the wider the multiseriate ray in any particular species, the greater the likelihood that procumbent cells will be present. In species with broad multiseriate rays, this process of internal anticlinal cell division is supplemented by ray mergers. Excellent illustrations of ontogenetic ray development in *Bursera simaruha*, from uniseriate near the pith to multiseriate near the cambium, can be found in Carlquist (1988a).

As a general rule, the widths of multiseriate rays increase with increasing age or distance from the pith during the juvenile years and then tend to fluctuate thereafter (Bertog 1895, De Smit 1922, Vasiljevic 1950, Linnemann 1953, Sebastine 1955, Ghouse and Iqbal 1977a, Khan et al. 1983). Zijlstra (1908), for example, observed that the large rays in both *Fagus* and *Quercus* increased in width with age of the cambium. He related this increase to greater growth activity associated with crown expansion. Similarly, De Smit (1922) found that the multiseriate rays in *Ulmus fulva* fluctuated appreciably in size throughout the tree, whereas the size of the uniseriate rays remained fairly stable. However, as pointed out in Section 8.3.6, large rays seldom remain static but undergo constant changes in both size and form. For example, Cumbie (1983) noted that as the rays of *Bocconia vulcania* increased in width they were occasionally broken into smaller fragments. Nonetheless, overall ray size continued to increase by internal cell divisions. Some of the changes in ray width are apparently due to rate of growth, with fast-grown trees producing significantly wider rays than slow-growing trees

(White and Robards 1966). As a general rule, rays are wider in rootwood than in stemwood (Metcalfe and Chalk 1983).

It is quite common to find ray widths increasing with tree age in tropical species (Iqbal and Ghouse 1985a). For example, in *Acacia nilotica* var. *telia*, uni- and biseriate rays comprised about 80 to 85% of the total ray population in young stems, whereas tri- and tetraseriate rays comprised about 63% of the population in older stems. In the Verbanaceae, ray widths also increased with age but more so in species exhibiting tree forms than in climbers and shrubs (Ghouse et al. 1980). Thus, species in the latter two growth forms possessed narrower rays than tree species.

It is also not unusual to find ray widths varying seasonally in tropical species. Iqbal and Ghouse (1985b), for example, found that the relative proportion of rays tetraseriate and larger in *Prosopis spicigera* tended to increase in the winter months and to decrease in summer. This difference was due to more frequent ray splitting during the active growing season. In several other tropical species, Rao (1988a) found no correlation between width of rays and cambial periodicity.

Table 8.6. Percentage of ray tissue in relation to ray size in the dicotyledons (Chalk 1955)

Maximum ray size	Number of samples	Ray tissue (%)
1-seriate	29	16.0
2-seriate	50	16.8
3-seriate	49	19.2
4-seriate	36	21.4
5-seriate	22	22.2
6-seriate	12	20.1
7-seriate	8	25.7
8- to 10-seriate	8	31.6
11-seriate and over (mean 16-seriate)	12	33.6

Ray width was found by Fengel (1966) to vary seasonally within a growth ring of *Fagus sylvatica*, with the widest parts of rays occurring at the growth ring boundaries (Fig. 8.33). Many other workers also observed that maximum ray widths occurred at the growth ring boundaries (De Smit 1922, Linnemann 1953, Schulz 1957, Süss and Müller-Stoll 1970, Kucera et al. 1980, Ajmal and Iqbal 1987a,b). Nordhausen (1898) attributed this effect in *Salix fragilis* to the fact that at season's end tip growth of the libriform fibers declined. Consequently, with fe-

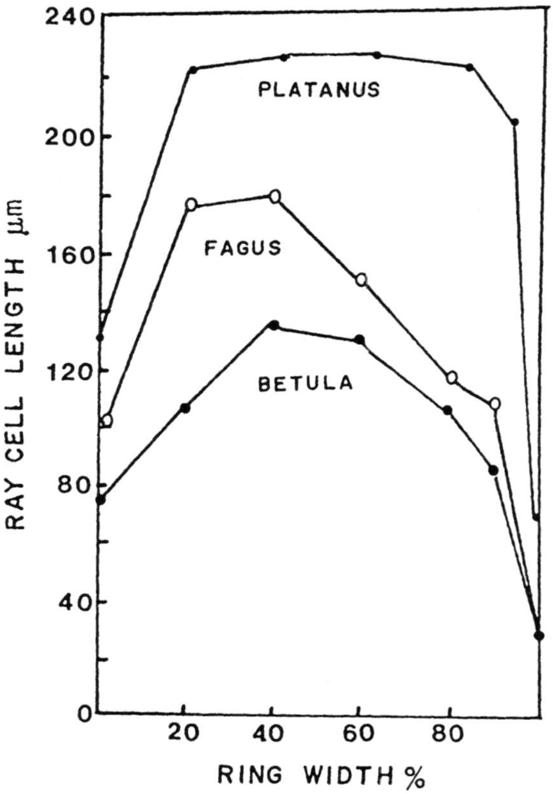

Fig. 8.34. Variation in the radial lengths of ray cells within a single growth ring of *Platanus acerifolia*, *Fagus sylvatica*, and *Betula pendula*. (After Süss and Müller-Stoll 1970)

wer interpenetrating fiber tips the still expanding ray cells were able to expand tangentially.

During the so-called "dilatation" of the rays, the parenchyma cells have been found to become larger, more rounded, and to possess a greater volume of intercellular spaces (Kucera 1985b). Dilatation of rays becomes particularly prominent in the cortex of some dicotyledonous trees where the rays broaden by both cell division and enlargement as the stem expands tangentially (Koeppen 1889, Esau 1969). Schneider (1955) attributed the dilatation of rays in *Citrus* to a "dilatation meristem." Whitmore (1962), who referred to the dilatation tissue as "expansion tissue," attempted to show that this tissue might be used as an indicator of the xylem/phloem ratio. The amount of expansion tissue and the degree of bark fissuring both increased as the xylem/phloem ratio decreased. In more recent work, Lev-Yadun and Aloni (1992a) experimentally demonstrated the possible involvement of ethylene in regulation of the dilatation meristem.

When large numbers of species or families are examined, ray widths are found to extend over a considerable range. In one analysis of 293 species (Table 8.6), Chalk (1955) found that 33.6% of the species possessed rays 11-seriate or larger. Another analysis of 650 species by Metcalfe and Chalk (1983) classified species according to absolute ray widths rather than ray seriation (Fig. 8.8B). Lists of dicotyledonous families categorized according to broad, narrow, and mixed ray widths appear in Metcalfe and Chalk (1950, 1979). Additional tabulations of ray widths can be found in Vasiljevic (1950), Ghouse and Yunus (1974b), Ghouse and Iqbal (1975), Paliwal et al. (1984), and Trendelenburg and Mayer-Wegelin (1955).

8.3.10 Ray Cell Size

According to Bailey (1923), ray cell initials, unlike fusiform initials, do not elongate appreciably in the vertical direction with either increasing age or distance from the pith. A somewhat similar view was expressed by De Smit (1922), who claimed that the increase in ray size with age was due more to an increase in ray cell number than to cell size in *Ulmus fulva*, although cell size also increased to some extent. Other workers have reported that heights, often called lengths, of ray cell initials either remained essentially stable (Ghouse and Yunus 1973, Ghouse and Iqbal 1977a, Ghouse and Hashmi 1980, 1981, Khan et al. 1983, Iqbal and Ghouse 1987), increased slightly (Kärkkäinen 1973, Kucera 1985b, Ajmal et al. 1986b), or even decreased slightly (Khan et al. 1981) with increasing age of the stem. Seasonal variations are usually slight or insignificant (Ajmal and Iqbal 1987a,b) (Table 7.1), although Paliwal et al. (1976) reported that initial cells in the storied rays of *Salvadora persica* increased in the vertical dimension from a minimum of 14 μm during periods of active cambial development to a maximum of 20 μm during cambial dormancy.

Ray cell size in the horizontal direction has been found to vary within a growth ring (Süss and Müller-Stoll (1970). In both *Betula pendula* and *Fagus sylvatica*, radial length of ray cells increased to a maximum at about 40% of the ring width and then rapidly decreased, whereas in *Platanus acerifolia* the maximum occurred as a plateau (Fig. 8.34). These growth patterns closely coincided with similar patterns in vessel diameter, and they were undoubtedly associated with cell expansion that occurred during differentiation rather than a change in dimensions of the ray cell initials.

Ray cell sizes have been found to vary with changes in rate of growth. The widths of rays in *Castanea sativa*, *Fraxinus excelsior*, and *Sassafras officinale* broadened as growth ring widths increased by increases in cell size as well as cell number (White and Robards 1966). The rays of *F. excelsior* also showed concomitant increases in height due to larger cell dimensions. In the family Misodendraceae, described by Carlquist (1985a), radial cell elongation appeared to be the chief source of increase in the radial dimension of rays, with the ray cell initials dividing tangentially only infrequently. Members of this family are parasitic flowering plants from South America.

Data on ray cell sizes in many different temperate zone species can be found in Trendelenburg and Meyer-Wegelin (1955) and Nyrén and Back (1960b). Data on ray cell sizes in tropical species are extensive and they have been reported for species with non-storied cambia (Ghouse and Yunus 1974a, Ghouse et al. 1976, 1980, Paliwal et al. 1976, 1984, Ghouse and Hashmi 1977b, 1980, 1981, Anand et al. 1978, 1981, Yunus et al. 1978, Sharma et al. 1979a, Jamal and Ghouse 1981, Dave and Rao 1982a,b, Bartwal et al. 1983, Ajmal et al. 1985, 1986a, Ajmal and Iqbal 1987a,b, Iqbal and Ghouse 1987), and also for species with storied cambia (Ghouse and Yunus 1974a, Yunus et al. 1978, Ajmal et al. 1986b, Rao and Dave 1985b).

Table 8.7. Distribution of ray proportions in the dicotyledons (Chalk 1955)

Ray tissue percentage of total	Chalk[a]		Myer[a]	Total
	Number of species	Percent of species	Percent of species	Percent of species
0-8	9	3.7	5.7	4.1
9-16	82	34.2	49.0	36.8
17-24	87	36.3	30.2	35.2
25-32	39	16.2	13.2	15.8
33-40	14	5.9	0.0	4.7
1-48	6	2.5	1.0	2.4
49-56	2	0.8		0.7
57-64	1	0.4		0.3

[a] Based on 240 species (Chalk 1955) and 53 species (Myer 1922).

8.3.11 Ray Volume

All rays in the secondary xylem are derived from periclinal divisions of ray cell initials. Consequently, by calculating the area occupied by rays on tangential sections of xylem as well as cambium, one can obtain a close approximation of the ratio of ray cell initials to fusiform initials in the cambium. This ratio is often referred to as ray volume, as opposed to fusiform initial volume (Chap. 5.2.1). As Hartig (1894) pointed out, a high ray volume can be due to either unusually broad rays or a large number of relatively small rays. However, species (Huber and Prütz 1938) and sometimes individual trees (Scukanec and Petric 1977) with exceptionally high ray volumes usually possess broad rays; as a general rule, rays are better developed and occupy larger proportions of the xylem in woody than in herbaceous dicotyledons (Nicoloff 1911). A rapid method of calculating ray volumes was proposed by Gregory (1981).

In an analysis of 38 gymnosperms by Myer (1922), ray volumes ranged from a minimum of 3.4% in *Thuja occidentalis* to a maximum of 11.7% in *Pinus heterophylla* with an overall average of 7.08%. For those species with fusiform rays, the average ray volume was 0.62% greater than for those without. Of the 53 dicotyledons examined, *Tilia heterophylla* was lowest with 5.30% and *Quercus densiflora* highest with 42.26% ray volume; average ray volume for the species examined was 17.04%. Chalk (1955) expanded on this work with a study of 240 dicotyledonous species, mostly tropical species not examined by Myer. A summary of the data from both studies is shown in Table 8.7. Chalk also observed a tendency for the proportion of ray tissue to increase with ray size as shown in Table 8.6. In the 240 species studied by Chalk, the proportion of ray tissue ranged from 5.3 to 59.3%, with an average value of 16%.

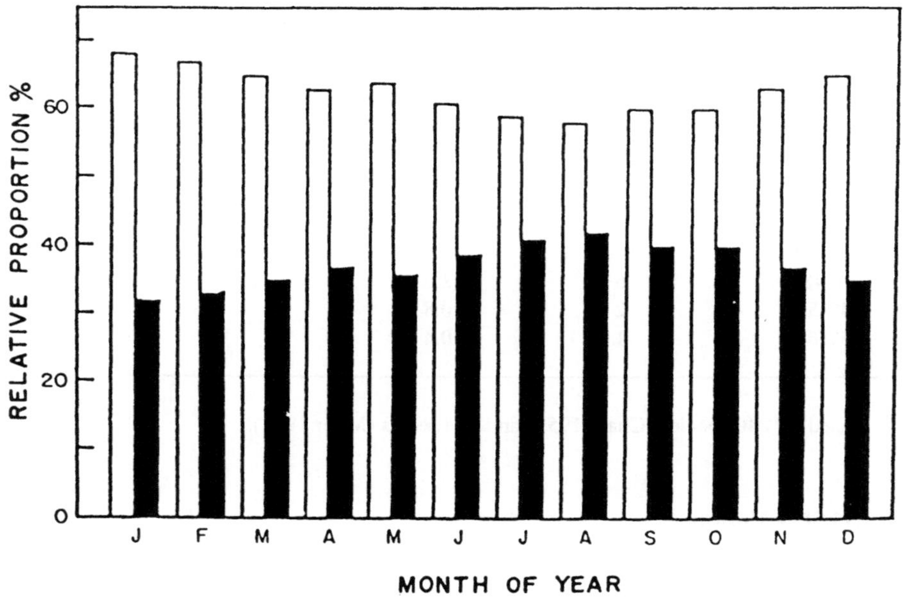

Fig. 8.35. Relative proportions of fusiform (*unshaded*) and ray cell (*shaded*) initials in different months in *Ficus rumphii*. Data are averages for 2 years. (Ajmal and Iqbal 1987b)

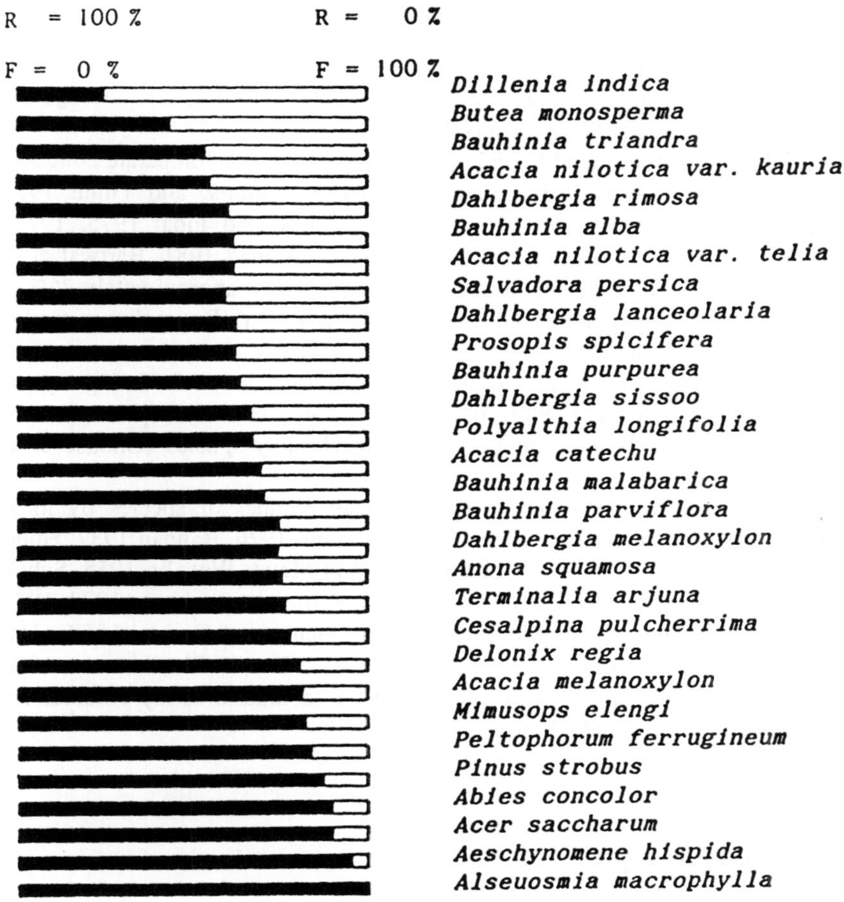

Fig. 8.36. Variations in ray volume among selected species based on the proportions of ray cell (*R*) and fusiform (*F*) initials measured on tangential sections. (Catesson 1980)

Myer (1922) claimed variation within a species to be "usually not over 4 or 5 per cent." This value should be questioned, because ray volumes vary with tree age (Hartig 1894, Bertog 1895, Eichhorn 1895, Braun 1955, Kucera and Kucera 1967, Kucera 1968, Furqan and Ahmad 1981, Ajmal et al. 1986b), position within the tree (Hartig 1894, 1895b, Mayr 1893, Bertog 1895, Eichhorn 1895, Kärkkäinen 1973, Iqbal and Ghouse 1990), growth conditions (Hartig 1894, 1901, Mayr 1884, Eichhorn 1895, Geiger 1915, Forsaith 1920, Myer 1922, Linnemann 1953, Schulz 1957, Schultze-Dewitz 1959, 1961, White and Robards 1966, Koltzenburg 1969, Gregory 1977, Margaris and Papadogianni 1977), and season of the year (Iqbal and Ghouse 1987) (Fig. 8.35). For example, in *Ulmus fulva* De Smit (1922) found that total ray volume attained a maximum at the stem base, it then decreased at mid-stem and reached a second optimum value within the crown. The maximum at the stem base was due to ray size and within the crown to ray number. Although specific data have not been compiled, the many papers I have read suggest that roots possess a much greater ray volume than stems. This difference is undoubtedly related to the fact that roots perform greater storage and lesser mechanical functions than stems do.

As a rule, ray volumes tend to be higher in tropical than in temperate zone species, and values exceeding 30% are not uncommon (Ghouse and Yunus 1973, 1974a,b, 1976, Ghouse et al. 1975a,b, 1976, Ghouse and Iqbal 1975, 1977a, Paliwal et al. 1976, Yunus et al. 1978, Furqan and Ahmad 1981, Bartwal et al. 1983, Ajmal and Iqbal 1987a,b). In fact, in *Dillenia indica* approximately 75% of the xylem consists of ray tissue (Ghouse and Yunus 1974a). Ray volumes tend to increase with age in these species, and it is often traceable to the constant conversion of fusiform to ray cell initials rather than to multiplication of existing ray cell initials (Ghouse and Hashmi 1980). In the Verbenaceae, tree-form species possessed higher ray volumes than either climbers or shrubby species (Ghouse et al. 1980).

Ray volumes have been measured in a large number of species by many workers such as Simon 1902, Bailey 1923, Weinstein 1926, Bannan 1937, Huber and Prütz 1938, Klauditz 1942, Trendelenburg and Mayer-Wegelin 1955, Schulz 1957, Nyrén and Back 1959, Panshin et al. 1964, Petric and Scukanec 1973, 1975, Ghouse et al. 1975a,b, Yunus and Yunus 1976, Ghouse and Hashmi 1977a, Hug 1979, Jamal and Ghouse 1979, 1981, Khan et al. 1979a, 1981, Catesson 1980, Ajmal et al. 1986b) (Fig. 8.36). A more extensive tabulation of ray volumes has been compiled by Iqbal and Ghouse (1990).

8.4 Selected Examples

As pointed out at the beginning of this chapter, the relations between primary and secondary ray origins as well as the formation of different ray types can most easily be described by examining the processes of ray development in selected species and species groups. The selected examples that follow are by no means exhaustive of the numerous ray types that exist in nature (Carlquist 1988a). Nonetheless, they do provide an overview of the role of the cambium in ray origin and development according to the processes described in previous sections.

As noted previously, primary refers to the first ray originating at the beginning of secondary growth, and secondary to a ray originating later during secondary growth.

8.4.1 Gymnosperms

8.4.1.1 Primary Rays

In most woody species, medullary ray tissues are ontogenetically, but not necessarily morphologically, continuous with those of the secondary body. In those species in which a complete cambial cylinder forms, the ray cells assume new characteristics at the onset of cambial activity. Because the way in which this change occurs was not known to early workers, Velten (1875) set out to determine how the first secondary ray cells, i.e., ray cell initials, were derived from the medullary ray tissue between primary vascular bundles. In his opinion, defining the first secondary ray cell was one of the most difficult tasks of microscopic technique. The first cell to begin a new ray was called a "Kopfzelle," literally a head or summit cell, which he identified as the first ray cell in a radial series enclosed on three sides by nonray cells. Radial sections from rootwood of *Taxus baccata* served as study material. In one case, "... the Kopfzelle was like a trumpet with a long neck, whose mouth was turned toward the pith." In only a few cases did he observe that the first ray cell in the wood corresponded with the typical horizontal form of a ray cell. He speculated whether, during the building process, a "Urcambiummarkstrahlenzelle," or primary ray cell, might have "evaginated" either upward or downward, thus producing a new, functional ray mother cell. He also thought it remarkable that both the beginning of a new ray and appearance of the first earlywood cells in the secondary xylem occurred concomitantly.

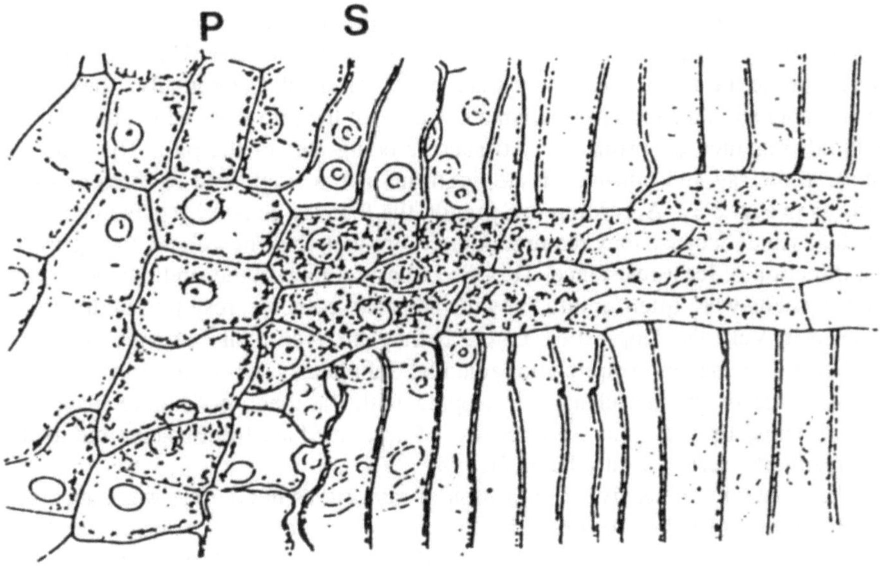

Fig. 8.37. Origin of a new ray in *Pinus sylvestris* consisting of three cells in a vertical series. Primary (*P*) and secondary (*S*) tissues in radial section. (Schmidt 1889)

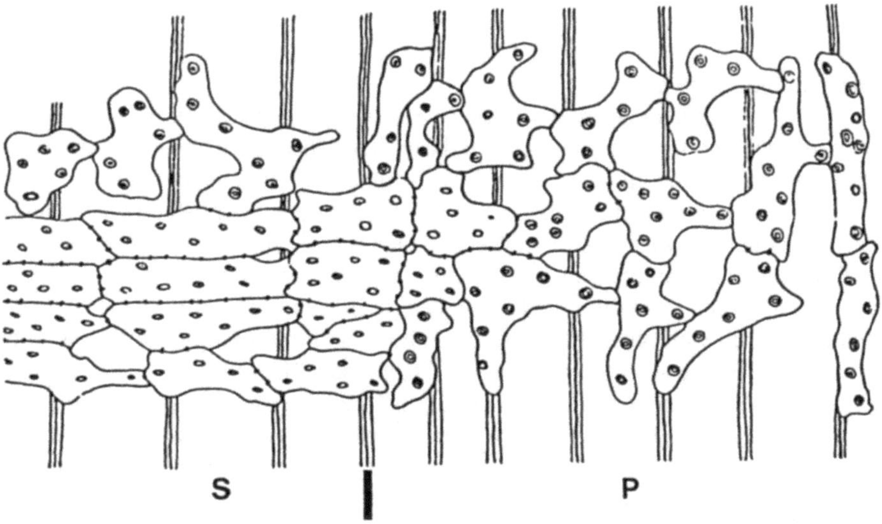

Fig. 8.38. Origin of a new ray in *Cedrus libanitica* illustrating the transition from vertically elongated cells in the primary (*P*) to radially elongated ray cells in the secondary (*S*) tissues. Note the "amoeboid" appearance of the "primordial ray cells." 150x. (Barghoorn 1940a)

Schmidt (1889) questioned Velten's conclusions, claiming that Velten had never established actual origin of primary rays in the first-year growth ring. Schmidt examined ray formation in first-year shoots of *Pinus sylvestris*. The first ray cells were simple in form and, other than being oriented horizontally, they were remarkably similar to the parenchymatous cells of the pith from which they arose; Schmidt regarded all primary ray tissues as being part of the pith. The juncture between the ray and pith-like tissue might involve a single cell or three or more cells as in Fig. 8.37. According to Schmidt, the first ray cells were presumed to be tracheids and, because they required more room than their neighbors, they grew in the direction of least resistance. This direction was toward the rind. However, the first ray cells were often "injured" by the adjacent tracheids, resulting in distortion as seen in Velten's "Kopfzelle." Later, as they survived their "struggle" with the tracheids, they began to stretch horizontally. Schmidt's view of secondary ray origin was one of gradual change in keeping with the theorem "Nature makes no leaps." Thus, a ray did not arise because of an immediate form change but as the end result of a series of introduced factors.

The most comprehensive examination of ray origins in conifers was conducted by Barghoorn (1940a). Although he found no foliar rays in conifers, he did observe distinct differences in cell size and shape between the fascicular and interfascicular regions. Most rays arose in the interfascicular regions from certain vertically elongated parenchymatous cells which he referred to as "primary ray cells." These interfascicular cells differed from the "primary ray cells" that origi-

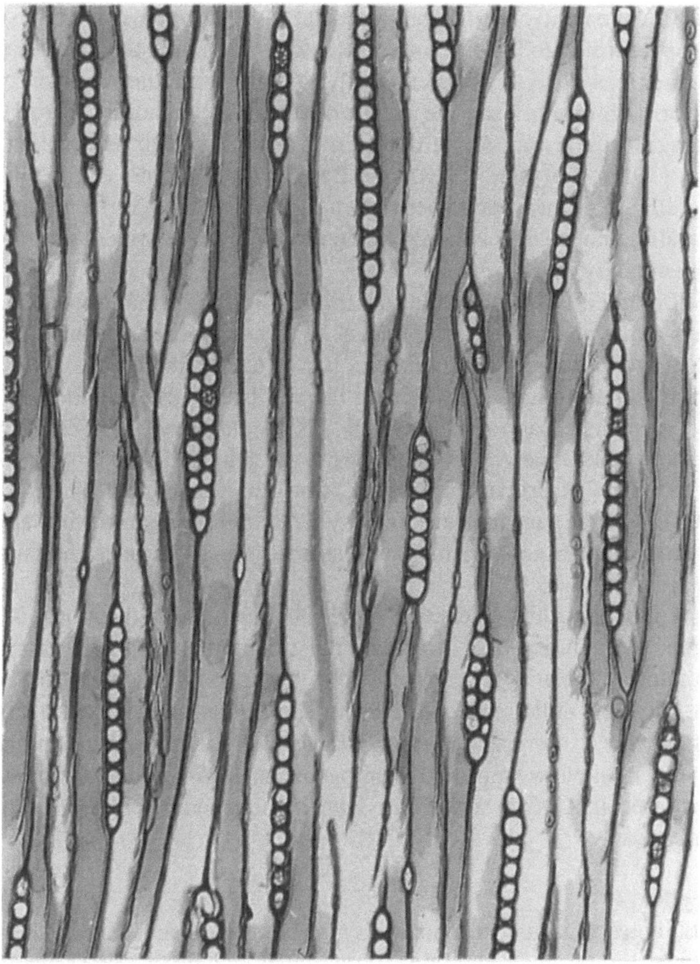

Fig. 8.39. Tangential section of *Abies alba* xylem showing the presence of biseriate rays. 56x. (Kucera and Bosshard 1975)

nated in the fascicular segments. In the latter, the incipient ray cell initials were more elongated vertically and they were arranged in extensive vertical strands closely appressed between protoxylem tracheids of the primary xylem (Fig. 8.38). Clearly, the two types of ray differed ontogenetically. Primary ray cells that arose in the interfascicular gaps presumably were derived from certain primordial cells of the growing point that did not undergo extensive vertical elongation. In contrast, those that arose in the fascicular segments apparently were derived from vertically elongated procambial cells by a series of anticlinal divisions.

Barghoorn (1940a) apparently recognized incipient ray cell initials in primary tissues of the stem. According to his descriptions, ray cells arising in the fascicular and interfascicular regions differed ontogenetically. Yet the transition to primary rays was similar for both origins, and the first-formed ray cells underwent rapid and extensive changes in size and shape during differentiation. Earlier workers, such as Thompson (1910), Chrysler (1913), and Bannan (1934a), had also reported that the first cells of primary rays were often highly irregular in shape and more or less vertically elongated in contrast to the more regular, horizontally oriented cells of the secondary rays.

Barghoorn (1940a) cited two factors that contributed to the formation of very low, uniseriate rays in the young stems of conifers: (1) the independent elongation and development of individual ray cell initials in the vertically contiguous series; and (2) the failure of certain primary ray cells in the vertical series to divide periclinally and to be perpetuated as ray cell initials. As a consequence, a series of rays one, two, or three cells in height originated from an elongated strand of vertically contiguous members of primary ray tissue. Another factor, cited by Strasburger et al. (1908), was the fact that the first ray cells, unlike fusiform initials, did not undergo further anticlinal divisions but assumed almost at once the characteristics of ray cell initials.

On the basis of more recent evidence, Soh (1990) distinguished among the types of ray origins in gymnosperms. In *Pinus*, ray cell initials originated from short cells in the interfascicular region but not from the fascicular region. In *Sequoia*, in contrast, ray cell initials originated from short cells interspersed alternately between long cells of the axial strands. The general structure of the initial procambium in tangential view appeared heterogeneous in *Sciadopitys*, homogeneous and nonstoried in *Ginkgo* and *Pinus*, and homogeneous and storied in *Podocarpus* and *Sequoia*.

8.4.1.2 Secondary Rays

Barghoorn (1940a) noted that new secondary rays generally arose as single cells in conifers, and it was rare to find ray origins of more than one initial. He assumed that most secondary rays arose by divisions from either the tip or side of a fusiform initial, a view contested by later workers (Bannan 1950b, 1951a, Whalley 1950). Secondary rays can also arise by complete segmentation of fusiform initials, but such divisions are usually confined to the youngest growth rings in conifers where cambial expansion is rapidly underway.

A less frequent type of secondary ray origin, characteristic of the phloem of Pinaceae, is the derivation of ray cell initials from radial plates (Sect. 8.5). The newly formed, single ray cell initials are generally elongated in the vertical plane, a shape presenting a striking contrast to subsequently formed ray cells.

Barghoorn (1940a) also claimed that anticlinal divisions giving rise to new biseriate rays had never been observed in the Coniferae except in abnormal cases; e.g., like those described by Bannan (1941, 1942, 1950a) and Srivastava and Esau (1961b) (Fig. 8.16). Similarly, Hartig (1848) had reported many years earlier that

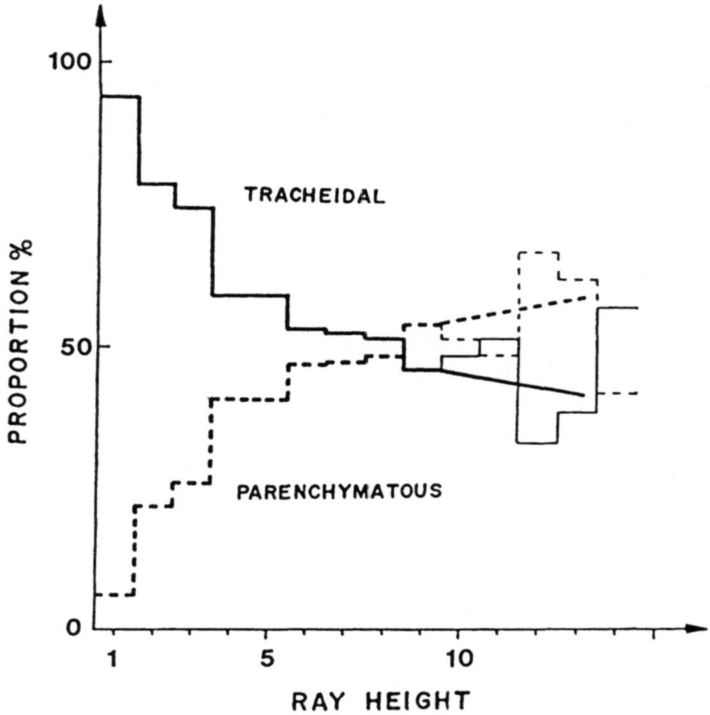

Fig. 8.40. Proportion of tracheidal and parenchymatous cell rows in uniseriate rays of different heights in *Pinus sylvestris* stems. Data for rows *9* and *higher* have been smoothed due to low frequency of occurrence. (Nyrén and Back 1959)

the rays in all conifers examined by him were uniseriate, with only an occasional ray being biseriate. On the contrary, Bannan (1965b) observed that some rays might be biseriate in species lacking fusiform rays such as *Cupressus* and *Sequoia*. Biseriate rays in older wood, derived from the cambia of large coniferous trees, frequently were encountered by Nyrén and Back (1959) and Gray (1973). In fact, nonfusiform, biseriate rays have been found to occur in all living gymnosperm families with the exception of Ginkgoaceae (Kucera and Bosshard 1975) (Fig.8.39). Aggregate rays, consisting mainly of multiseriate rays 2 to 15 cells wide and partly of uniseriate rays, have recently been reported in *Thujopsis dolabrata* var. *hondai* (Sugawa and Fujii 1992).

Unlike the first cells of a primary ray which are almost invariably parenchymatous (Bannan 1941, Huber 1949), the first derivative of a new secondary ray cell initial is usually a tracheid (Schmidt 1889, Mischke 1890, Huber 1949, Nyrén and Back 1959). Some workers concluded that, although tracheids predominate, the first cell might sometimes be parenchymatous (Thompson 1910, Bannan 1934a, 1941, 1942, 1950a, Barghoorn 1940a, Braun 1955, den Outer 1967). Huber

(1949) reported frequent transformations among initial cells in these new rays, particularly at the beginning of a new growth ring. That is, a ray cell initial that produced a ray tracheid might produce, at the next periclinal division, a parenchymatous cell and vice versa. Although he was unable to predict the fate of a ray cell initial in *Pinus radiata*, Barnett (1981) suggested that the presence of bands of microtubules might signify the incipient cell wall thickening of a potential ray tracheid.

In contrast to Barghoorn (1940a), who was concerned primarily with the origin of secondary rays in the Coniferae, Braun (1955) emphasized their subsequent development. He found that mean ray height in *Pinus sylvestris* increased rapidly from the second to about the tenth year and then very slowly in subsequent years (Fig. 8.11). Consequently, ray height increased gradually despite the tendency toward reduction of tall rays. Braun was deeply concerned with what he called "Umprägungen," the impression of a feature or characteristic on the cambial derivative. That is, a ray cell initial that produced a ray tracheid might again divide to produce either a tracheid plus a parenchyma cell, two parenchyma cells, or any

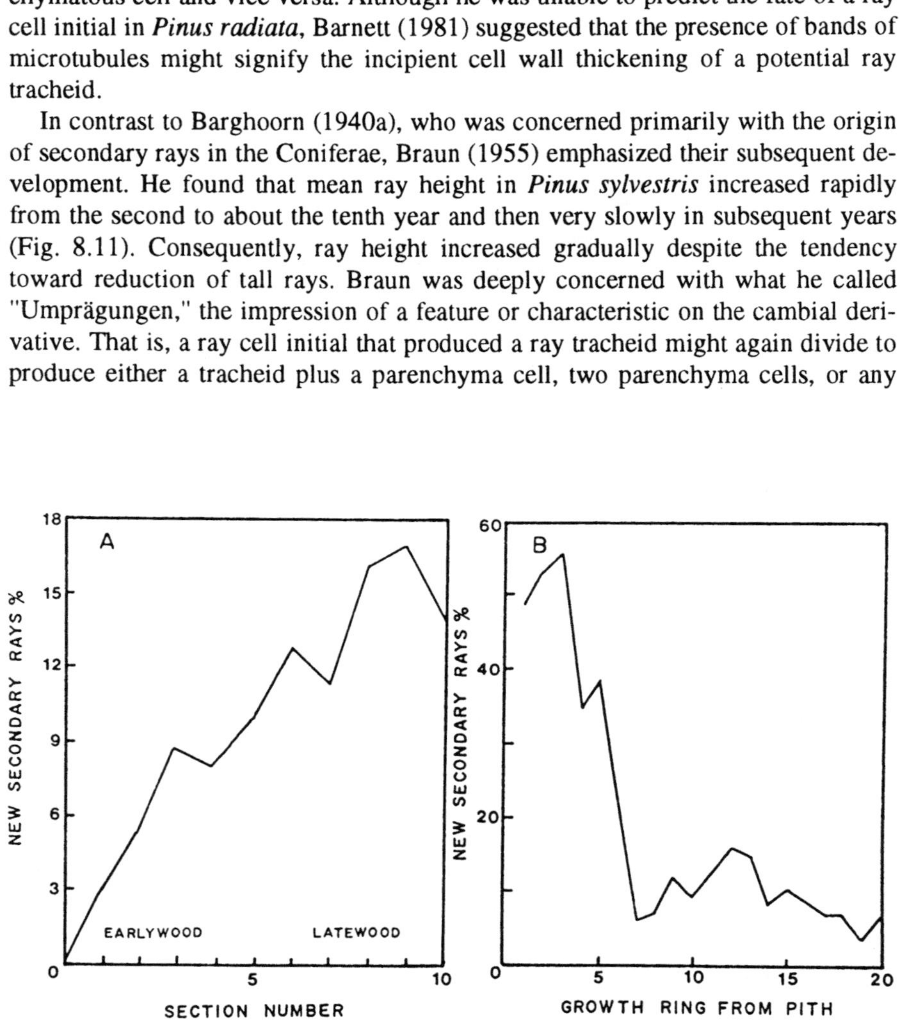

Fig. 8.41A,B. New ray origins in *Pinus sylvestris*. **A** Percentage of new secondary rays originating at different positions within one growth ring divided in ten zones. Note tendency for new rays to arise in the latter part of the growth ring. Data based on over 1000 rays. **B** New rays as a percentage of total rays on a given tangential area in each growth ring from the pith outward. (After Braun 1955)

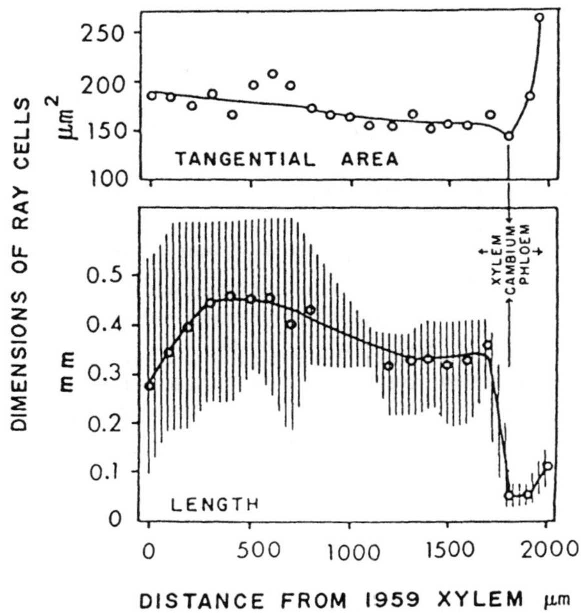

Fig. 8.42. Tangential cross-sectional areas and radial lengths of ray cells in the 1960 xylem and phloem increments of a *Abies concolor* tree. *Vertical lines* indicate the range of variation. (Wilson 1963)

number of combinations. Braun's analyses led to the formulation of three rules: (1) with age constant, tracheid number decreased with ray height, e.g., in the 20th growth ring the proportion of tracheids decreased from 83% in one-celled to 47.5% in 12-celled rays; (2) with ray height constant, tracheid number increased with age, e.g., in a 6-celled ray the proportion of tracheids increased from 41% at age 5 to 56% at age 50; and (3) cell divisions and "Umprägungen" were most frequent at the beginning and ends of a growth ring. However, as shown by Bannan (1965b) in several conifers, the proportion of tracheids in uniseriate rays decreased with increasing ring widths. Conversely, there was a tendency for the proportion of tracheids to increase in trees with declining growth rates (Bannan 1934a).

In randomized samples of *Pinus sylvestris*, Nyrén and Back (1959) found that tracheids occupied 45% and parenchyma cells 35% of the ray volume; the remaining 2% of the volume consisted of resin canals and intercellular channels. In uniseriate rays, 94% of those one cell high, 78% of those two cells high, and 40 to 45% of those nine cells high were tracheidal (Fig. 8.40). Parenchyma cells were added during later development of the ray, usually occupying the inner rows of the ray.

Braun (1955) found that the origin of new rays was highest during latewood formation (Fig. 8.41A) and particularly during the juvenile period of growth (Fig.

8.41B) in *Pinus sylvestris*; these trends were confirmed by Kucera and Kucera (1967) in *Abies alba*. Divisions of the initial cells in existing rays were also more numerous at the end of the vegetative season. These new initials were added to the borders, thus increasing the heights of the rays. Most secondary rays arose as single cells which often merged readily to form taller rays. At times, two several-celled rays merged resulting in former border tracheids becoming internal tracheids. Such rays, with border tracheids mixed internally with parenchymatous cells, were referred to as "polymere Markstrahlen", or polymeric rays (Huber 1949, Braun 1955). These rays would correspond to the heterocellular ray classification of Kribs (1935). Initial cells of the former border tracheids seldom persisted as such when they became internal cells, and they either began producing parenchymatous cells or were eliminated from the ray. Although primary rays were frequently reduced to two- and three-celled rays by splitting, Braun found that secondary rays were seldom split. Also, in contrast to fusiform initials, the outright loss of either ray cell initials or entire rays occurred infrequently in *Pinus sylvestris*.

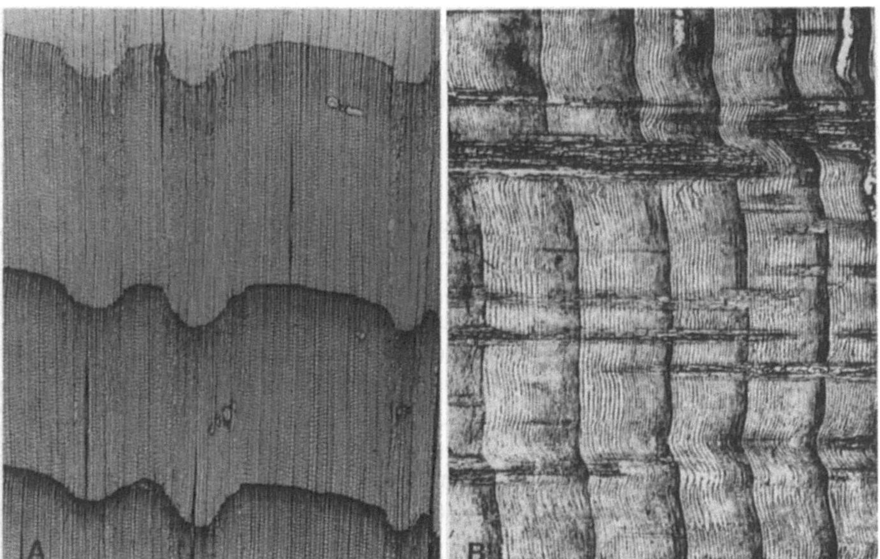

Fig. 8.43A,B. "Haselwuchs" in conifers. **A** Indentations in growth rings of *Picea abies* of unknown cause that dip towards the pith. Transverse section, 5.1x. **B** Radial section of *Pinus jeffreyi* showing distorted wavy grain pattern with dips toward the pith at *left*. 5.9x. (Ziegler and Merz 1961)

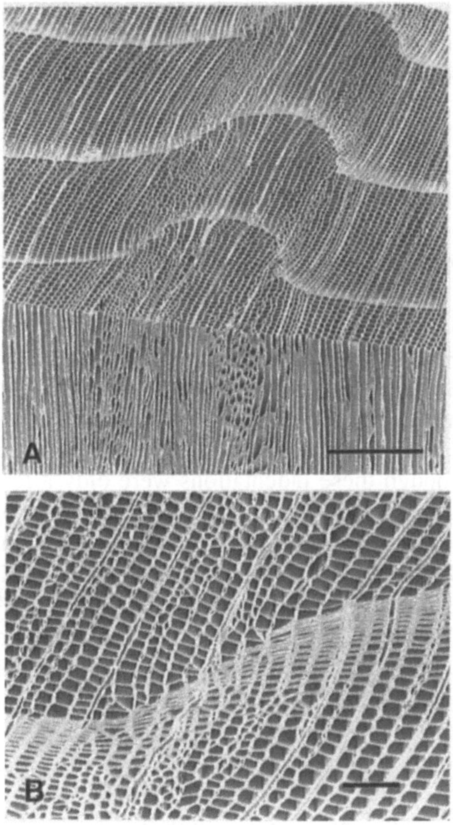

Fig. 8.44A,B. "Indented rings" in *Picea sitchensis*. **A** Transverse and tangential faces with indentations toward the cambium. *Bar* 500 μm. **B** Annual ring boundary showing distorted and irregularly arranged tracheids. *Bar* 100 μm. (Ohtani et al. 1987)

In a somewhat related study, Wilson (1963) demonstrated a seasonal pattern in the radial lengths of xylem ray cells in *Abies concolor* (Fig. 8.42). Wilson's measurements were made on all cells within the uniseriate rays and, as shown by the shaded area, the standard deviation was quite large. Nonetheless, concomitant with the decline in xylem ray cell length was a similar decline in tangential ray cell area. Although the author offered no explanation for these patterns, they could conceivably be correlated with late-season divisions of ray cell initials as described by Braun (1955).

8.4.1.3 Fusiform Rays

Rays that contain radial resin canals are called fusiform rays. As pointed out long ago by Hartig (1848), it was thought for many years that these were the only broad rays that normally occurred in conifers. It is now known that biseriate and occasionally broader rays do occur in some conifers, although they are not common. Fusiform rays are invariably present in species of *Picea, Pinus, Larix*, and *Pseudotsuga* (Brown and Panshin 1940) (Fig. 9.15). For a discussion of fusiform rays, see Chapter 9.2.2.3.

8.4.1.4 Haselwuchs

An abnormal type of growth called "Haselwuchs," or hazel growth, is found in certain conifers (Ziegler and Merz 1961). This growth type was so named because of its resemblance to the broad-rayed wood of hazel (*Corylus*). When it occurs in *Picea*, it is referred to more specifically as "Haselfichte," or hazel spruce.

The wood of "Haselwuchs", as described by Ziegler and Merz (1961) in *Picea abies* was characterized by indentations that occurred at irregular intervals in the growth rings (Fig. 8.43A). These indentations, which dipped towards the pith, did not arise in the first growth rings but during later life of the tree by unknown causes. Once formed, they persisted at the same radial position in successive growth rings, often through 10 to 30 rings. Although these indentations were only a few millimeters high, they nonetheless disrupted the normal course of development of the fusiform initials, resulting in inclined tracheids resembling those in wavy grain (Fig. 8.43B). Each indentation extended tangentially across 20 to 30 cell files and, within this disturbed region, the number of uni- and biseriate rays was 40 to 50% greater than in the undisturbed regions. However, because the rays in the disturbed region were shorter than normal, ray volume was only 15 to 20% greater. Ziegler and Merz attributed the distortion of tracheids in the disturbed region to an increase in the number of rays. These rays bear only a superficial resemblance to the false rays of *Corylus* and other dicotyledons and the term Haselwuchs is therefore an unfortunate misnomer.

Like some types of wound tissues, a significantly higher number of trabeculae were found associated with the indentations of hazel growth in *Picea abies* by Grosser (1986). In more recent work, Schultze-Dewitz and Götze (1986) distinguished "Haselwuchs" from a similar-appearing xylem structure, "Wimmerwuchs," or wavy grain (Chap. 6.6.2.1). They proposed that whereas "Haselwuchs" was induced by endogenous processes, "Wimmerwuchs" was due to external factors of a mechanical nature.

A similar growth type found in *Picea sitchensis* was described as "indented rings" by Fukazawa and Ohtani (1984) and Ohtani et al. (1987). These indentations varied from 1.5 to 3.0 mm in width and 20 to 80 mm in height (Fig. 8.44A). Like those of *P. abies*, the indentations never extended inward to the pith and, once formed, they continued radially through many growth rings. Tracheids on the margins of the indentations were inclined not only markedly in the radial direction, but also slightly in the tangential direction. The fact that the tracheids also varied in size and were irregularly arranged contributed to a disordered and complicated annual ring boundary (Fig. 8.44B). Rays traversing this disordered zone were temporarily diverted to conform with the orientation of the marginal tracheids. As noted by others, biseriate rays lacking resin canals and tracheids with trabeculae (Fig. 4.30) were commonly observed in the marginal zones of the indentations. Similar indented growth rings in *Thujopsis dolabrata* var. *hondai*, due to the presence of aggregate rays, were reported by Sugawa and Fujii (1992).

Bosshard (1976a) noted a relationship among the Haselwuchs in *Picea* described by Ziegler and Merz (1961), the asymmetric or fluted stems of *Carpinus* described by Holdheide (1955), and the dips in the cambial circumference observed by many workers in the Fagaceae. In each case, the cambium was locally retarded to varying degrees in the vicinity of the rays, suggesting the involvement of a common growth principle. Bosshard offered this relationship as an example of "functional tropism," a structure-function concept of tissue development.

More recent work by Lev-Yadun and Aloni (1991c) provided evidence that hazel growth might be due to a disturbance of the normal hormonal balance following wounding. These authors attempted to induce hazel growth experimentally in *Pinus halepensis* by making parallel longitudinal incisions approximately 200 mm long through the bark and into the active cambium with razor blades. Examination of the wound sites immediately after wounding showed the typical wound responses characterized by curved and distorted tracheids, more numerous and misaligned rays, and traumatic resin canals. Two years after wounding, indented growth rings similar to hazel growth occurred at the sites of wounding. At this time, the indented xylem consisted of curved tracheids, irregularly oriented rays, and normal-sized resin canals. The authors did not determine whether the indented rings would persist through many growth rings, as often found in typical hazel growth.

8.4.1.5 Pseudorays

Pathological infections of conifer cambia frequently result in hypertrophy characterized by excessive xylem growth of irregular form. It is of interest that in all conifers the hypertrophy caused by pathogens is accompanied by a significant increase in ray development. Von Tubeuf (1930a,b) described a tumor-like hypertrophy that occurred on stems and branches of *Pinus sylvestris* from certain forest regions in Germany (Chap. 10.5.1). Although the wood possessed many anatomical and structural characteristics of pathologically infected wood, von Tubeuf could find no visible evidence of infection. He surmised that the growth resulted from mutations.

Von Tubeuf noted that disturbance of the cambium often began in growth rings near the pith with vigorous cell divisions and proliferation of parenchymatous ray cells followed by formation of intercellular gaps and development of traumatic resin canals. In some trees the rays appeared to flare outward from the pith through many growth rings. Tracheids adjacent to these rays were twisted and irregularly formed, with some resembling compression wood cells. The ray-like tissue was further accentuated by the lateral spread of precociously formed heartwood. Because of the similarity of these broad radial bands of tissue to the false rays of *Carpinus*, *Alnus*, and certain other dicotyledons, von Tubeuf referred to them as "Scheinmarkstrahlen," or pseudorays. At the termini of these ray-like structures, cambial development was often hindered, resulting in pronounced indentations in the xylem similar to those described for Haselwuchs. Sprengel (1936) observed similar rays in a later study of tumorous growth on *Pinus* stems. In *Cryptomeria japonica*, ray structure in the regions of cambial surface projections and depressions were also found to be abnormal (Imamura 1978). Rather than typical uniseriate rays, many biseriate rays of irregular form and structure were produced.

An interesting case of patholgically induced multiseriate rays in *Pinus contorta* var. *murrayana* was described by Srivastava and Esau (1961a) and Srivastava (1973) (Chap. 10.5.6). Seeds of the dwarf mistletoe, *Arceuthobium americanum*, germinated on the host branch and from there sent out haustorial appendages that grew in the cortex and cambial layer. The cambial initials responded by dividing into smaller cells that surrounded the haustorial appendage, or sinker. When a sinker encountered a ray, the ray cell initials responded by producing a large, multiseriate ray with the sinker embedded within it. Once formed, these large rays were perpetuated and were accommodated to the prevailing growth conditions by the usual processes of ray mergers and splitting (Fig. 8.16).

The term "pseudoray" is not confined to conifers. Middleton (1987) used the term with reference to the large aggregate rays in *Nothofagus* consisting of groups of small, narrow rays separated by fibers.

8.4.2 Dicotyledons

8.4.2.1 Primary Rays

In their study of the origin of foliar rays in angiosperms, Sinnott and Bailey (1914, 1922) noted that radially disposed sheets of parenchyma, commonly known as medullary rays, varied greatly in number, width, height, and distribution among different species. In one group of trees, shrubs, vines, and herbs the rays were all narrow and more or less uniformly distributed. With the exception of foliar gaps, the secondary xylem formed a continuous unbroken cylinder. In another group of trees, shrubs, vines, and herbs with sharply defined primary bundles, the gaps between the primary xylem bundles were confronted in the secondary xylem by wide sheets of ray parenchyma. In the first annual ring, these wide interfascicular rays, which were actually multiseriate rays flanking the leaf-trace segments, often extended a considerable distance down the stem.

Jeffrey and Torrey (1921a,b) had maintained that the so-called foliar storage ray consisted of both the flanking interfascicular rays and the intervening nodal parenchyma. Sinnott and Bailey (1922), however, considered these to be two structurally distinct tissues which merged in various ways in different species. Dissection of the secondary vascular system depended to a great degree on the manner and extent to which these two tissue systems merged. Thus, they observed many species that lacked foliar rays but possessed instead narrow rays and essentially a continuous vascular cylinder. In contrast, many other species possessed high multiseriate, foliar rays and vascular cylinders dissected into a series of discrete, wood segments. Formation of these broad rays apparently is not confined to foliar organs. Similarly structured rays have been found in association with latent, adventitious root primordia in many species of both conifers and dicotyledons (Fink 1982).

As in gymnosperms, primary rays in dicotyledons are sometimes designated as being either fascicular or interfascicular in origin (Jeffrey and Torrey 1921b, Barghoorn 1940a, Cumbie 1960, Philipson and Ward 1965, Braun 1970, Sieber and Kucera 1980). Most primary rays are associated with the interfascicular regions situated between vascular bundles. Here, the rays usually arise as elongated masses of parenchymatous tissues that extend a considerable vertical distance within an internode (La Rivière 1916, 1921, Jeffrey and Torrey 1921b, Sinnott and Bailey 1922, Chattaway 1933a, Braun 1955), or sometimes the entire length of an internode (Kalberlah 1898, Zijlstra 1908, Bailey 1911, Eames 1911, La Rivière 1916, 1921, Philipson and Ward 1965, Kucera et al. 1980, Sieber and Kucera 1980, Dobbins 1990). Barghoorn (1940a) and Cumbie (1969a, 1984)

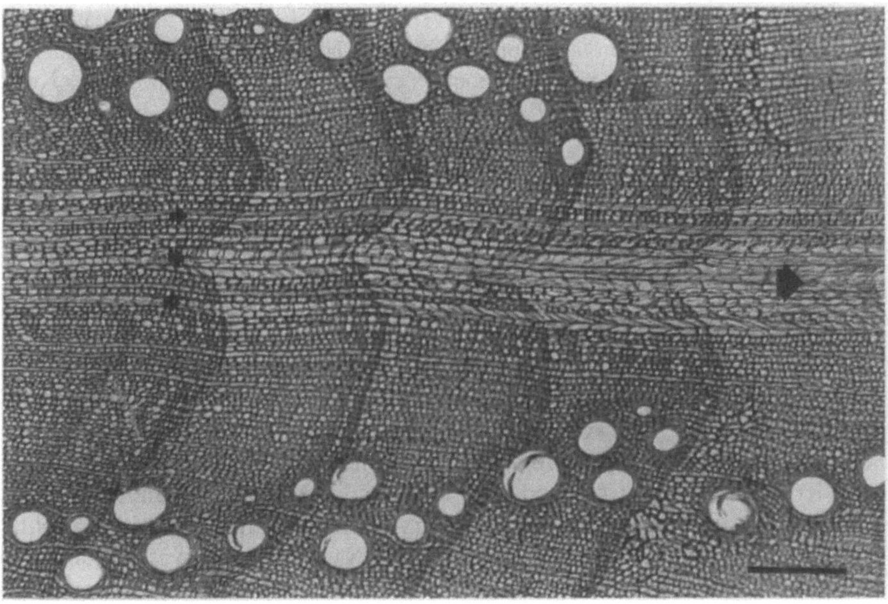

Fig. 8.45. Transverse section showing how small rays (*small arrows*) in the juvenile wood merge to form a large aggregate ray (*large arrow*) in more mature wood in a slow-growing branch of *Quercus ithaburensis*. *Bar* 250 µm. (Lev-Yadun and Aloni 1991b)

spoke of these rays as arising from primordial ray initials that could be identified in the procambium. Barghoorn (1940a) recognized three types of ray organization in dicotyledons based on the position of origin in the primary body: (1) multiseriate and uniseriate rays arising from the interfascicular regions; (2) multiseriate rays only, mainly from the interfascicular regions, with few rays extending from the fascicular regions; (3) uniseriate rays only, extending from both the fascicular and interfascicular regions.

However, as noted by Barghoorn (1941a), the condition at origin is not necessarily preserved during further secondary development. Uniseriate rays might be lost either by merging in various ways to form multiseriate rays or by merging with existing multiseriate rays. Contrarily, large multiseriate rays might be dissected to yield smaller rays, some of which are uniseriate.

8.4.2.2 Secondary Rays

A distinction is often made between aggregate and compound multiseriate rays in dicotyledons. An aggregate ray is: "A group of small, narrow xylem rays appearing to the unaided eye or at low magnifications as a single large ray" (IAWA 1964). At higher magnifications libriform fibers can be seen separating the narrow rays (Fig. 8.45). Because of the included files of fibers, such rays are also called false rays. Compound, or fusion rays, are less well defined, but they usually appear as aggregate rays in which the intervening files of libriform fibers have been eliminated. The distinction is seldom recognized and the terms are often used synonymously. Although the term compound ray is now considered obsolete, I will use it when appropriate to preserve the author's intended meaning.

Fagaceae: In most species of *Fagus* and *Quercus* the rays are of two distinct sizes, uniseriate and multiseriate, whereas in *Castanea* and *Castanopsis* the rays are exclusively uniseriate (Metcalfe and Chalk 1950). In one species of *Quercus*, *Q. incana*, the rays were found to be mostly uniseriate but with a few biseriate and larger rays in older wood ranging from 1 to 30 cells in height (Paliwal et al. 1984).

An interesting paper by Zijlstra (1908) attempted to make sense of the confusion regarding ray origin up to this time. He pointed out that most textbook illustrations were schematic and not very accurate. Because the authors depicted rays in young stems, they showed uniform rays with little information as to how they might vary in height along a stem radius. This shortcoming was perhaps excusable considering the scarcity of good data then available in the literature. Previous authors had usually stated that the medullary ray tissue between primary vascular bundles, the "primäre Markverbindungen" of Nägeli (1858), was divided into many smaller rays when new vascular bundles, the "Zwischenbündel" of Sanio (1863b), arose within this tissue. However, these observations were invariably made on isolated tangential sections. A case in point was Essner (1882), who measured ray heights in several conifers, but his values were not comparable because he did not measure the same rays in different growth rings within a species.

Zijlstra (1908) adopted De Bary's (1877) designations of "gross," or large, for rays that developed in continuity with the pith and "klein," or small, for those that first arose during diameter growth of the stem. He noted that in textbooks, even these terms were confused. He deplored use of the terms primary and secondary as used by previous workers, because primary rays that persisted into the second year would also possess secondary elements.

To overcome the shortcomings cited above, Zijlstra sampled old trees and followed specific rays through tangential sections. The sections were sawed at regular intervals from pie-shaped, radial sectors of the stem, smoothed with sandpaper, and stained for visibility. The radial sector from a 56-year-old stem of *Fagus sylvatica*, 36 cm in diameter, was divided into 25 tangential sections at 1-cm intervals; from a 360-year-old stem of *Quercus robur*, 77 cm in diameter, a 140-year ring sequence from the middle of the radial sector was divided into 18 tangential sections. Large ray segments derived from medullary rays could be followed backwards to their origin near the pith. Large rays were in turn split into many smaller rays in their outward traverse towards the cambium. Most often, this splitting was brought about by sloping layers of fibers penetrating a larger ray. In older growth rings the rays were of mixed sizes and the large rays could only be identified by following segments backwards to their origins.

When Zijlstra measured the heights of all rays in a ray complex, produced by the successive splitting of a large ray, he found that cumulative ray height increased with age. This meant that over time cumulative ray height was greater than the height of the original medullary ray; the latter was equal to internode length in the primary stem. Although he could account for part of this increase in ray height with age by the merger of small rays and the addition of new cells to existing rays, he was unaware of the processes by which new rays were initiated. Zijlstra also examined an 8-year-old stem section of *Aristolochia sipho*, a liana. The processes by which the large rays were split into smaller parts were the same as those in *Fagus*, with the exception that the splits not only occurred later and more sparingly but the resulting rays were larger.

Sprengel (1936) was also of the opinion that the broad rays in *Quercus* arose in the first growth ring. The pith of a *Quercus* stem often has five stelar points. Between each of these points, two broad rays were initiated that continued in parallel outward to the periphery. Uniseriate rays arose later and many of these merged to form broad rays, but never with the same uniformity as the five original rays.

Unlike Zijlstra (1908), who believed that the large primary rays in the Fagaceae and several other families were derived directly from medullary rays, Bailey (1910a, 1911) rejected the idea of medullary rays entirely. He examined a large number of woody species of the Cupuliferae and demonstrated that in most cases rays originated either in association with leaf traces or as uniseriate rays at the beginning of secondary growth. When he examined transverse and tangential serial microsections through nodal regions, he observed that the leaf trace produced a disturbing effect on the surrounding woody tissue as it passed outward towards the leaf. This effect was expressed as a decrease in the number of vessels and an increase in the proportion of storage tissue, particularly the enlargement and multiplication of rays.

The extent to which the parenchymatous tissue developed in relation to the leaf trace varied with different species. In *Castanea*, *Ostrya*, and certain species of *Alnus* and *Betula*, the tissue was confined to the immediate vicinity of the leaf trace. In other woods with well-developed aggregate rays, such as *Corylus* and *Quercus*, the parenchymatous tissue extended for some distance above and to a considerable distance below the level of leaf-trace insertion. This relationship with the leaf trace was often obscured and difficult to establish in species possessing well-developed aggregate rays early in ontogeny. Difficulties arose not only because of the precocious origin of the aggregate rays but also because new uniseriate rays were constantly forming in the secondary body. These large aggregate rays, formed in association with leaf traces, often produced a segmented appearance in the central vascular cylinder. Bailey contended that this segmentation had been erroneously interpreted in the past to indicate that the woody cylinder originated from a ring of so-called primary bundles. What Bailey was referring to here was the segmentation of these large sheets of parenchymatous tissue into numerous uniseriate rays in the first, or first several, growth ring(s) from the pith.

Rays that developed from these large sheets of parenchymatous tissue, originating in association with leaf traces, were called foliar rays (Bailey 1912). As a rule, rays that arose in association with lateral leaf traces developed more strongly, presumably because of some restrictive influence of the axillary bud on the median trace (Bailey 1911). In some species, such as the deciduous oaks, the "lateral leaf-trace rays" extended downward through several internodes. The sheets of storage tissue that formed were relayed from node to node in more or less continuous lines along the stem in conformance with the phyllotaxy of the plant. The rays thus formed resulted in regions of retarded growth in the cambial cylinder which appeared as dips, a phenomenon observed by many previous workers (De Bary 1877, Wieler 1886b, Kny 1890, Nordhausen 1898, Eames 1910) (Sect. 8.4.1.4). Because these dips were related to leaf traces at different nodes, they often extended through several internodes down the stem. Eames (1911) confirmed Bailey's observations and arrived at similar conclusions regarding the leaf trace in his work on the Rosaceae, as did Thompson (1911) in several other dicotyledons.

Eames (1910), in a follow-up study of ray formation in *Quercus*, differentiated between the black (Erythobalanus) and white (Lepidobalanus/Leucobalanus) oak groups as classified by Bailey (1910a). In seedlings of the black oak group (*Q. rubra*, *Q. velutina*, *Q. coccinea*, and *Q. ilicifolia*), approximately the first 15 annual rings exhibited a progressive compounding characterized by aggregation and fusion of the linear, uniseriate rays; these were typical false rays. As compounding continued, a broad homogeneous ray developed. The rows of fibers and wood parenchyma separating individual rays disappeared as fusiform initials were either lost from the cambium or transformed to ray cell initials. These changes occurred gradually, and eventually the ray cambium consisted entirely of ray cell initials.

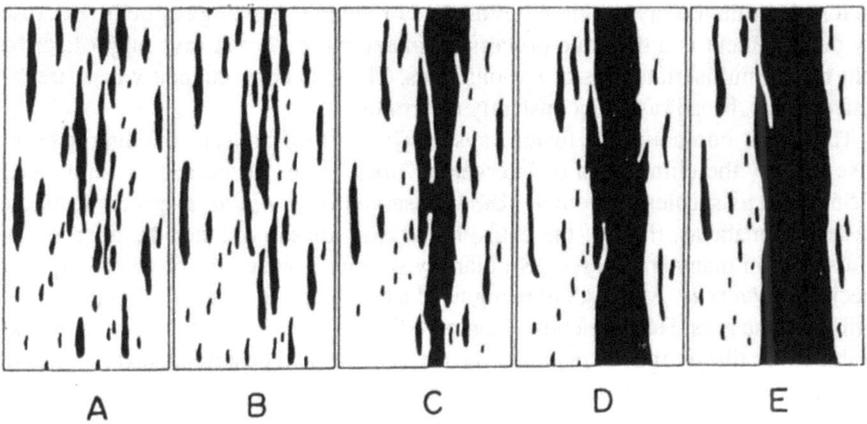

Fig. 8.46A-E. Origin of broad, fusion rays in *Quercus robur* by ray mergers, ray splittings, and ray insertions. Sections **B**, **C**, **D**, and **E** are situated 240, 360, 480, and 600 μm from **A**, respectively. 42x. (Braun 1970)

In seedlings of the white oak group (*Q. alba*, *Q. bicolor*, and *Q. prinoides*), only uniseriate rays were present in the xylem with no sign of either aggregation or fusion for the first 15 to 20 years. When large rays did appear they formed as in the black oaks. However, formation was abrupt rather than gradual and the compounding phase was confined to one or two growth rings. At times, a large ray seemed to appear suddenly at the beginning of an annual ring, in which case fusiform initials that had previously produced fibers were rather abruptly transformed to ray cell initials. This process could be reversed by wounding in *Quercus* (Bailey 1910b). Wounded cambia reverted to the formation of uniseriate rays, followed some time later by renewed aggregation and compounding by the same processes as in normal cambia.

According to Riedl (1937), the broad secondary rays in older wood of *Quercus* are formed by the merger of superimposed uniseriate rays with the so-called false rays being inserted as a transitional stage. Braun (1955) followed development of the broad rays by analyzing serial tangential sections of *Quercus robur* (Fig. 8.46). Formation of a false ray began with the merger of superimposed uniseriate rays and by cell division within the biseriate and larger rays. False rays are therefore formed when the fusiform cambial cells between the narrow rays are literally squeezed out to the point where only one or two rows of fusiform initials remain. Other processes involved in the formation of broad rays are ray mergers, ray splittings, and ray insertions. The sequential sections in Fig. 8.46 show how these processes contribute to the rapid formation of a broad ray that not only increases in breadth but also in height. Once formed, broad rays continue adding small rays to their borders. The ends of the broad rays gradually close, and when height growth of the ray is complete, the ends become pointed. Concomitantly, the entire ray becomes more compact as the cambial initials that formerly produced fibers enclosed within the ray are either eliminated or begin producing ray cells. Because ray development is a dynamic process, segments of the broad rays might later be split off as multiseriate rays of various sizes. Thus, as new secondary rays are initiated, older, broad rays are constantly reforming.

The broad compound or fusion rays of *Quercus* differ from the aggregate or false rays by the elimination of intervening fiber tissues in the former. In a study of 55 *Quercus* species, Abromeit (1884) observed in *Q. dilata*, a species from the western Himalayas, that "... the entire broad rays appeared under the microscope to dissolve in many small rays." Similar ray structures were also observed in other species of *Quercus*. Although unrecognized as such at this time, Abromeit was describing false rays. He erroneously compared these rays to the "aussetzende Markstrahlen," or discontinuous rays, that Hartig (1859a) had described years earlier in *Fagus*. In that paper, Hartig also differentiated between species like *Fagus* with broad "closed" rays and those such as *Carpinus* with broad rays "interspersed with" rows of fibers, i.e., false rays.

New secondary rays in *Quercus* are uniseriate, usually one or two cells high, and they arise either by tip or lateral origin (Braun 1955) (Chap. 6.3.2). Like *Pinus sylvestris* (Fig. 8.41A), most new rays in Quercus are initiated near completion of the vegetative period. Braun (1955) examined development of the uniseriate rays in *Quercus robur* by measuring the heights of 100 rays in each of several growth rings (Table 8.8). Despite cell divisions and ray mergers, uniseriate ray heights appeared to stabilize between 5 and 15 cells because of the high frequency of ray splitting by elongating fibers.

Table 8.8. Variation in the number of uniseriate rays with ray height in growth rings of different ages in *Quercus robur* (Braun 1955)

Growth ring number	1	2	8	20	48
Ray height			Number of rays[a]		
1-4 cells	10	27	24	18	17
5-15 cells	49	58	67	70	68
Over 15 cells	41	15	9	12	15

[a] Based on 100 rays for each growth ring.

The rays of *Quercus* roots are morphologically different from those of the stem (Riedl 1937). In *Quercus pedunculata*, xylem rays near the stem base differed little from those in the upper stem, although a gradual transition occurred with increasing distance from the stem base. Typical false rays, consisting of aggregations of uniseriate parenchymatous rays separated by fibers, were commonly

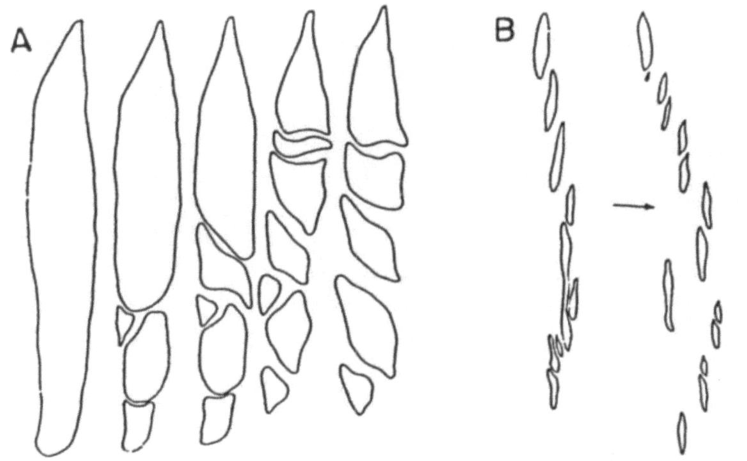

Fig. 8.47A,B. *Fagus sylvatica.* **A** Splitting of a broad primary ray into smaller multiseriate rays. **B** Vertically arranged rays, as in **A**, dispersed laterally by continued cambial expansion. Radial distance between tangential sections, 1.5 mm in **A**; 1 cm in **B**. (Braun 1955)

found in lateral roots. Because uniseriate rays were more numerous, however, the proportion of parenchymatous tissue was much greater than in the stem. Also, in contrast to the stem, vessels were uniformly distributed with no evidence of ring porosity. Deep roots had no false rays, only uniseriate rays whose height ranged up to 20-cell series. The nearer the root end, the greater the proportion of parenchymatous ray tissue in the root.

In contrast to the false rays in *Quercus robur*, Braun (1955) observed in *Fagus sylvatica* that multiseriate, or fusion, rays could arise by primary origin as well as by the secondary merger of uniseriate rays. The primary rays were extraordinarily high, sometimes measuring 22 mm in the first growth ring. In 2-year-old rings, 63% of the rays were uniseriate, 25% biseriate, 6% triseriate, and the remainder multiseriate. In 8-year-old rings, the proportion of uni- and biseriate rays decreased while that of multiseriate rays increased. As Jost (1901) showed, ray height increased concomitant with ray width on the stem cross-section. And, as shown by Kucera et al. (1980), this trend contributed to formation of two populations, one consisting of small uni- to triseriate and the other of 8- to 20-celled multiseriate rays.

Ray sizes in *F. sylvatica* also apparently vary seasonally. Values for both ray height and width, as well as the number of cells within a ray, declined precipitiously in the first-formed earlywood. According to Fengel (1966), both the cross-sectional areas and the widths of rays exhibit maxima in the latewood at the growth ring boundary (Fig. 8.33).

As in *Quercus*, Braun (1955) found that the high primary rays of *Fagus* were split although the products were still multiseriate rays (Fig. 8.47A). Ray splitting

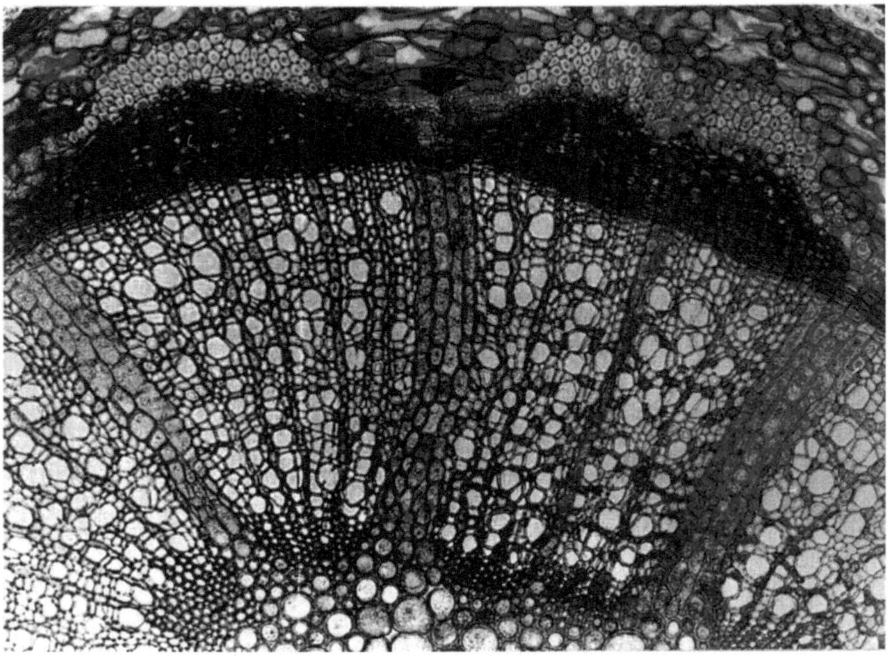

Fig. 8.48. Beginning stages of "Keilwuchs", or wedge growth (*arrow*), in the cambial region of the first growth ring of a *Fagus sylvatica* stem. Transverse section. 71x. (Kucera et al. 1980)

in both genera progressed either outward from the middle of the ray or inward from the margins by the transformation of ray cell initials to fusiform initials or vice versa. At first the split rays were vertically arranged, but with continued cambial expansion they dispersed tangentially (Fig. 8.47B). Secondary rays in *Fagus* usually arose adjacent to strand parenchyma in the form of uniseriate rays two cells high. These rays increased both in breadth and height by cell division. They also increased in height by the merger of superimposed rays, but, unlike *Quercus*, tangentially adjacent rays seldom merged. Also, unlike *Quercus*, where a sharp distinction could usually be made between the low, uniseriate and the broad, multiseriate rays, in *Fagus* the two ray types intergraded (Braun 1955).

Fagus sylvatica characteristically exhibits indentations at the growth ring boundary in the region of the broad rays (Fig. 8.48). Bosshard (1976a) called this formation "Keilwuchs," or wedge growth, and considered it to be analogous to the Haselwuchs in *Picea* (Sect. 8.4.1.4) and the "Spannrückigkeit" of *Carpinus*. Kucera et al. (1980), following the development of these indentations, found that they were evident in the first-season's growth ring. They originated in the cambial region and, whereas the normal cambium consisted of three to four, thin-walled

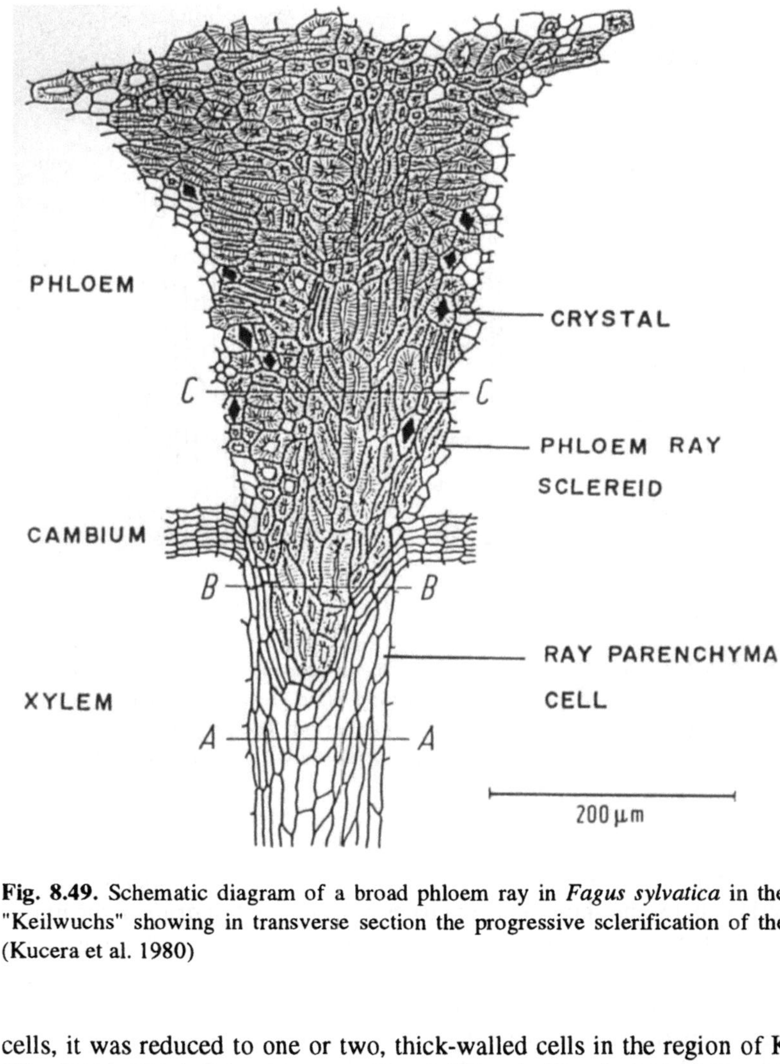

Fig. 8.49. Schematic diagram of a broad phloem ray in *Fagus sylvatica* in the region of "Keilwuchs" showing in transverse section the progressive sclerification of the ray cells. (Kucera et al. 1980)

cells, it was reduced to one or two, thick-walled cells in the region of Keilwuchs (Fig. 8.49). The indentations were formed by sclerification which began in the innermost cells of the broad phloem rays. As shown in Fig. 8.49, once initiated, sclerification progressed steadily inward until it formed a wedge of tissue that intruded to varying depths in the xylem. Similar indentations in the cambial region of *Nothofagus* have been associated with aggregate rays (Middleton 1987).

Betulaceae: An interesting sequence of ray development can be found in the genus *Alnus* which presents a range of ray structures (Bailey 1911). The woods of both *Alnus* and *Castanea* are characterized by possessing numerous small, uniseriate rays similar in form and structure to those of many conifers. This ray type is found in *A. acuminata*, a species in which the rays remain uniseriate during ontogeny

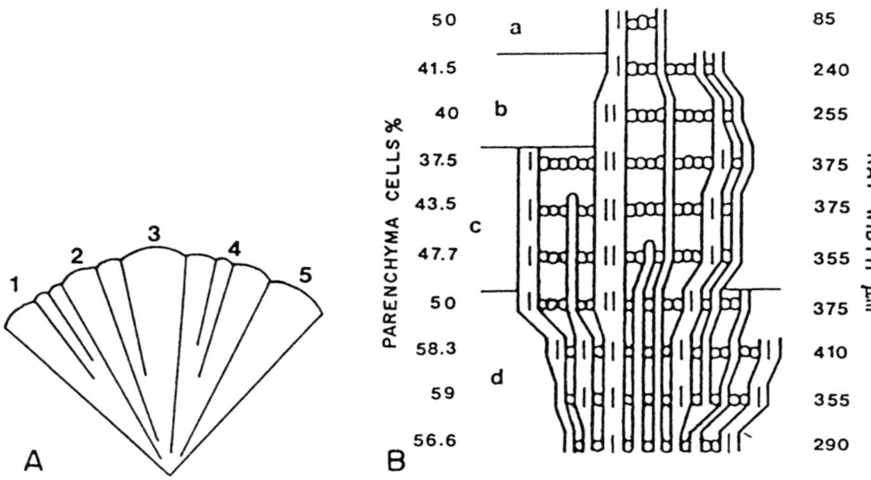

Fig. 8.50A,B. Fluted stem in *Carpinus betulus*. **A** Eleven-year-old stem in partial transverse section showing undulating stem periphery due to suppressed growth of false rays. **B** Origin of a false ray during four phases (*a-d*) over a 10-year period. The proportion of parenchymatous cells (*circles*) and the width of the false ray are shown for each year. (Holdheide 1955)

without any tendency toward aggregation. However, in *A. rubra*, many rays are bi-and triseriate and some show signs of coalescing, and in *A. maritima*, the uniseriate rays are either aggregated or fused creating a more or less homogeneous mass of ray tissue. Finally, the wood of *A. rhombifolia* possesses completely fused aggregate or compound rays comparable to those found in *Quercus*. In all these species Bailey reported that nonaggregated, uniseriate rays were present in the first-formed annual rings of seedling plants and the different degrees of aggregation occurred during ontogeny. Moreover, all these changes occurred within the cambial layer and the described patterns were simply reflected in derivatives in the xylem.

False rays are commonly found in members of the Cupuliferae, and especially so in the Betulaceae. They form by the aggregation of many smaller rays that appear uniform to the unaided eye but on close examination are found to be an agglomeration of closely packed small rays (Holdheide 1955, Braun 1963). They are also called aggregate rays (Jeffrey 1917, Barghoorn 1941a) or sometimes diffuse rays (Myer 1922). Compound rays, on the other hand, are usually considered to be more compact with few, if any, nonray inclusions (Metcalfe and Chalk 1983). The terms aggregate and compound rays have not been used consistently, and indeed often interchangeably in the Betulaceae. Holdheide (1955) did not observe false rays in *Betula* but he did find them in *Alnus*.

As Bailey (1910a, 1911, 1912) demonstrated, the parenchymatous leaf-trace tissue in all species of Betulaceae that he examined was dissected in such a way that uniseriate rays appeared in the first, or first several, growth rings from the pith. Mature ray types therefore depended on the rate and degree to which these simple rays subsequently became aggregated and fused. For example, the large so-called primary rays reported by many earlier workers in both the Betulaceae and Fagaceae were shown to be formed by the fusion of numerous small rays (Bailey 1910a). In the Betulaceae, the compounding process did not extend beyond the preliminary stages resulting in the false rays characteristic of these species. For this reason, the aggregate or false rays of *Alnus japonica*, as described by Jeffrey (1917), were formed not only by small rays separated by fibers but often included in their midst was the transverse section of a leaf trace. As in *Quercus*, wherever one of these large rays crossed an annual ring, it exerted a depressing effect on the cambium, resulting in a dip at the ring boundary. Compounding attained a higher level of perfection in the Fagaceae, but it also varied more widely among species.

False rays have been implicated in the formation of asymmetric or fluted stems, particularly in certain members of the Cupuliferae (Bailey 1910a). Stem asymmetry, which is especially common in the genus *Carpinus* (Fig. 11.25), has been described by workers such as T. Hartig, Nördlinger, von Tubeuf, and Wiesner (see Rubner 1910a). Wigand (1856) had observed longitudinal swellings in stems of *Carpinus* and attributed them to more favorable growth on one part of the stem as opposed to restricted growth on another.

Von Tubeuf had observed that growth depressions often formed beneath branch insertions in older trees of *Alnus*, *Betula*, *Fagus*, and *Populus*. He speculated that these depressions were sites of false ray development where xylem growth was suppressed in contrast to the adjacent "Stutzleisten," or lateral ridges, where growth was promoted. Rubner (1910a) continued this line of investigation under the direction of von Tubeuf by examining asymmetric *Carpinus* stems in which he related the occurrence of "Scheinstrahlen," or false rays, to depressions in the stem periphery. That is, xylem growth was inhibited and phloem growth was promoted in regions of the stem periphery where false rays were most crowded (Figs. 8.50; 11.25). Like Bailey (1910a), Rubner (1910a) observed the beginnings of these rays in 1-year-old wood. The primary emphasis of Rubner's research, however, was not with the development of false rays but with the occurrence of discontinuous and missing growth rings (Chap. 11.8). It should be pointed out at this time that the false rays in *Carpinus* and other Betulaceae differ in some respects from those in *Quercus*, although it is of interest that the dips in the annual rings of *Quercus* as well as the depressions in the stem periphery of *Carpinus* are both associated with rays.

"Spannrückigkeit," asymmetric or fluted stem growth, was also examined in *Carpinus betulus* from the standpoint of false ray formation by Holdheide (1955).

As noted by Holdheide, the stem circumference in this species enlarges irregularly when cambial initials on certain parts of the stem actively divide while on other parts they apparently rest. This differential cambial activity produces an undulating stem periphery consisting of ridges and depressions. Once initiated, the ridges continue to enlarge. The ridges are bounded by false rays which tend to suppress growth (Fig. 8.50A). Holdheide differentiated between false or apparent rays and true or parenchyma rays that extended to the pith. False rays were not uniformily distributed but tended to be concentrated in growth depressions in the xylem. False rays also differed from other wood rays in having few or no vessels in their immediate vicinity, a characteristic noted by Bailey (1911) and described in detail by Rubner (1910a). Thus, a false ray in *Carpinus* consists of parenchymatous rays separated by narrow, radial bands of fibers.

Incipient false ray formation was observed by Holdheide (1955) in first-year growth rings of *C. betulus*. These rays were initiated between the primary xylem bundles, a region free of vessels on the xylem side and sieve elements on the phloem side. False rays developed rapidly, in one case increasing from three in a first-year to 15 in a second-year shoot. Development of a typical false ray is depicted schematically in Fig. 8.50B. Phase a shows first-year origin of the false ray in the region where vessel formation was suppressed in the xylem and fiber formation promoted in the phloem. A step-wise formation of secondary parenchymatous P-rays increased the size of the false ray during the second and third years (Phase b). P-ray formation continued and then stabilized during the fourth to seventh years (Phase c), followed by a period of contraction or consolidation (Phase d). This latter phase was the most conspicuous with regard to abnormal diameter growth because convergence of the P-rays indicated loss of cambial initials in the intervening fiber bands. Thus, convergence of the P-rays occurred passively, and in deep stem depressions it was followed by diminished activity of the P-rays themselves.

According to Holdheide, the false rays of *Carpinus* are not necessarily stable regions although they can be under certain conditions such as the extreme cases depicted by Rubner (1910a). In very actively growing stems, ray convergence (Fig. 8.50B, Phase d) might be delayed for a prolonged period, in which case the proportion of P-rays would be reduced and width of the intervening fiber bands increased. False rays might also wander laterally or oscillate from broad to narrow and vice versa, a condition Holdheide attributed to the "determination region in the cambium." Although false rays occur in other members of the Betulaceae, those in *Carpinus* are expressed most clearly.

As to causality, most early workers attributed false rays to nutritional deficiencies because of their occurrence beneath branch insertions (Hartig 1891, Rubner 1910a), although Nördlinger (1871) invoked the additional factor of bark pressure. Holdheide (1955) viewed false rays as being more or less self-perpetuating due to the unique anatomical structure of both the xylem and phloem. This pattern was true of the phloem in particular, because of its high proportion of fiber elements which effectively blocked the cambium from contact with active

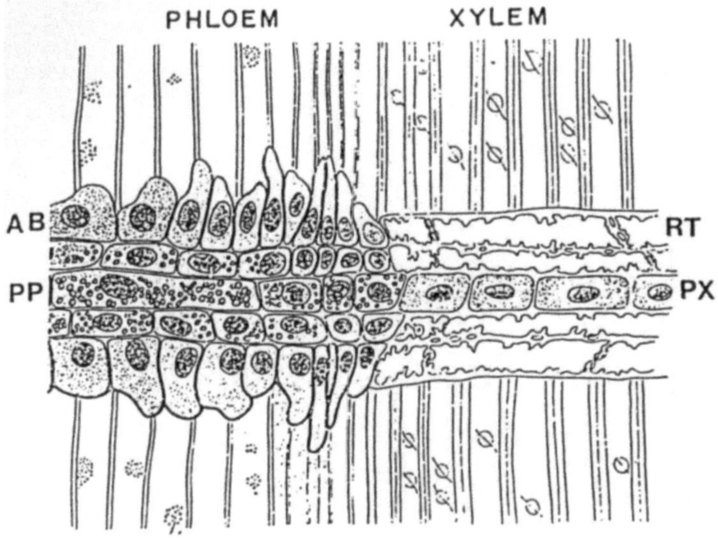

Fig. 8.51. Radial section from a *Pinus* stem at the junction between xylem and phloem showing erect, albuminous cells (*AB*) on the margins of a phloem ray. *RT* Ray tracheids; *PX* and *PP*, starch-filled ray parenchyma cells in xylem and phloem, respectively. 200x. (Strasburger et al. 1908)

sieve elements. Korovin and Zuikhina (1985) attributed the abnormally enlarged wood rays in *Alnus incana* and *Betula pubescens* to disturbances in the cambium which contributed to figured zones in the wood. The sporadic occurrence of aggregate rays in *Alnus rubra* was associated with frost injury and branch traces by Noskowiak (1978) and those in four species of *Betula* with pith flecks and branch traces by Bhat (1983). Cambial injury resulting in a disturbance known as stem pitting led to the formation of aggregate rays in *Betula alleghaniensis* (Thomas and Wall 1986).

Salicaceae: Dicotyledonous species and genera differ not only in the amount of leaf-trace parenchyma produced but also in the nature of the ray tissue derived from it. The major differences are in the rate and extent to which the leaf-trace tissue becomes dissected and whether and how it reforms as large rays. Many species, such as most members of the Salicaceae, possess uniseriate rays when mature. Of all the species of *Populus* examined by Holden (1912), only one was biseriate. Of the *Salix* species examined, there were almost as many biseriate as uniseriate. Multiseriate rays were found to occur in association with leaf traces in seedlings of all species, but they were quickly reduced to small rays. For example, broad bands of parenchymatous leaf-trace tissue in seedlings of *Populus deltoides* were dissected to rays three to four cells wide in the first annual ring. In older shoots, the leaf-trace tissue was reduced to biseriate rays in the first two growth

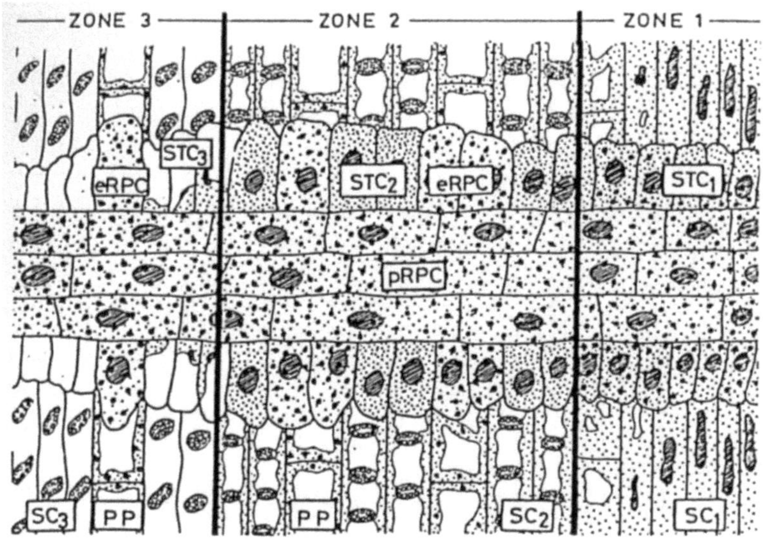

Fig. 8.52. Radial sectional drawing illustrating the functional stages of Strasburger cells (*StC₁*, *StC₂*, and *StC₃*) in a secondary phloem ray of *Larix decidua*. In *Zone 1*, the sieve cells are immature (*SC₁*), in *Zone 2* they are mature (*SC₂*), and in *Zone 3* they have died (*SC₃*). The *heavy black dots* in the erect (*eRPC*) and procumbent (*pRPC*) ray parenchyma cells indicate the presence of starch. *PP* Phloem parenchyma. (Sauter 1980)

rings and uniseriate thereafter. In other species of *Populus* and *Salix*, dissection to biseriate and sometimes to uniseriate rays was brought about by the intrusion of vessels during formation of the first annual ring. The dips in the cambial cylinder observed by Bailey (1911) in association with lateral leaf trace rays in *Quercus* did not occur in *Salix*. In this genera, ray development associated with the median leaf trace was extensive but it differed little in structure from that of the lateral traces.

In *Populus alba* (Braun (1955), another species with uniseriate rays, ray height increased dramatically for about the first 8 years and then continued to increase slowly with age, as in *Pinus sylvestris* (Fig. 8.11). The low ray heights during the first few years were due to the rapid formation of new rays whereas the increases in ray heights during later years were due to cell divisions within rays and to the mergers of superimposed rays. The rays of *Populus tremuloides*, while remaining uniseriate in the conducting phloem, were found to undergo dilatation in the non-conducting phloem (Davis and Evert 1968).

8.4.2.3 Rayless Woods

In some species a closed cambium forms without the usual formation of rays. Hartig (1859a) commented on this fact and provided a tentative list of families in which such species might be found. Regnault (1860) stated that certain members of the Crassulaceae were rayless. A more comprehensive list has recently been compiled by Carlquist (1988a).

Some species are completely rayless from the beginning of secondary development (Falkenberg 1876, Barghoorn 1941a, Aymard 1968, Esau and Cheadle 1969, Paliwal and Srivastava 1969, Pulawska 1973, Carlquist 1981, Metcalfe and Chalk 1983). Other species begin rayless but later develop low and narrow rays in the outer xylem (Barghoorn 1941a, Cumbie 1963, Carlquist 1970, Gibson 1978), or gradually lose their rays by a process of cell elimination (Cumbie 1983). The early formed xylem of *Suaeda monoica* is temporarily rayless but large, multicellular rays develop later (Lev-Yadun and Aloni 1991a). Raylessness was viewed by Carlquist (1988a) as a delay in the occurrence of transverse divisions in ray cell initials, or even the entire absence of such divisions. The fact that some woods are at first rayless but gradually produce rays during secondary growth was viewed by Carlquist (1962, 1988a) as an indication of paedomorphosis, the retention of juvenile characteristics in the adult plant.

Phylogenetically, the absence of rays in the inner and their presence in the outer secondary xylem of some species has been interpreted as the first stage in a trend toward the develoment of complete raylessness. That is, rays are lost first from the inner secondary body (Barghoorn 1941b). This loss can sometimes be observed in transverse sections by the vestiges or ghosts of rays in the process of elimination.

According to Barghoorn (1941b), the absence of rays is correlated with a reduction in cambial activity and often with a tendency toward the herbaceous habit of growth. Thus, the rayless condition can be found in many plants that have undergone dwarfing or modifications in relation to unfavorable environments as well as in some forms possessing anomalous secondary thickening. Carlquist (1975, 1988a), in general agreement with Barghoorn, has interpreted these modifications as a phyletic increase in herbaceous plant size or woodiness. That is, there is a tendency for the plants to sacrifice ray tissue in favor of mechanical strength. Raylessness is also associated with short fusiform initials which are often found in many herbaceous plants. In species with longer fusiform initials, there is a greater likelihood that fusiform initials in the interfascicular areas will subdivide to form rays (Carlquist 1970).

8.5 Radial Plates and Erect Cells

The so-called erect cells occur on the margins of phloem rays in many gymnosperms (Fig. 8.51). Erect cells have been equated with albuminous cells (Strasburger et al. 1908) and the terms have often been used interchangeably. However, this identity has been questioned because not all erect cells are albuminous cells (den Outer 1967). In the recent literature, albuminous cells have been called Strasburger cells on functional grounds (Sauter 1980) (Fig. 8.52). The Strasburger cells of gymnosperms, which lack companion cells, are believed to be the functional equivalents of companion cells in angiosperms. Ontogenetically, these cells differ. Whereas Strasburger cells are derivatives of ray cell initials, companion cells are derived from extracambial divisions of sieve elements (Esau 1969).

In contrast to normal ray cells, the long axes of erect cells are oriented vertically rather than horizontally. They are the products of erect ray cell initials, which are also oriented vertically on the margins of the cambial ray (Thompson 1910, 1912a,b, Chrysler 1913, 1915, Bannan 1936a, Abbe and Crafts 1939, Barghoorn 1940a). Although the erect ray cell initials give rise to erect cells on the phloem side, they give rise to either marginal tracheids or parenchyma on the xylem side (Fig. 8.51).

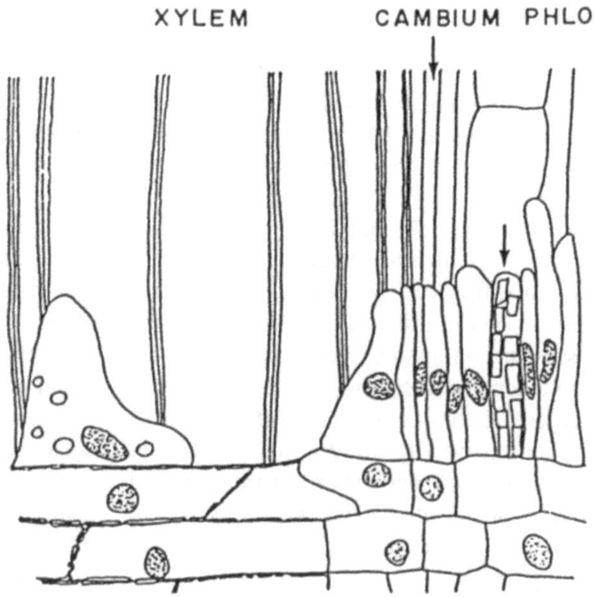

Fig. 8.53. Erect cells in a *Cedrus libani* root showing in radial section scattered production on the xylem side. One erect phloem cell contains crystals (*arrow*). (Chrysler 1915)

Erect cell initials are formed by the conversion of an adjoining fusiform initial to a ray cell initial which is then added to the cambial ray margin. One way in which fusiform initials are converted to ray cell initials is by their progressive reduction in size as a result of asymmetric periclinal divisions (Bannan 1951a, 1953, Grillos and Smith 1959) (Chap. 6.3.2.3). During the reduction process, many odd-shaped cells might be added to the margins of the rays on the xylem as well as on

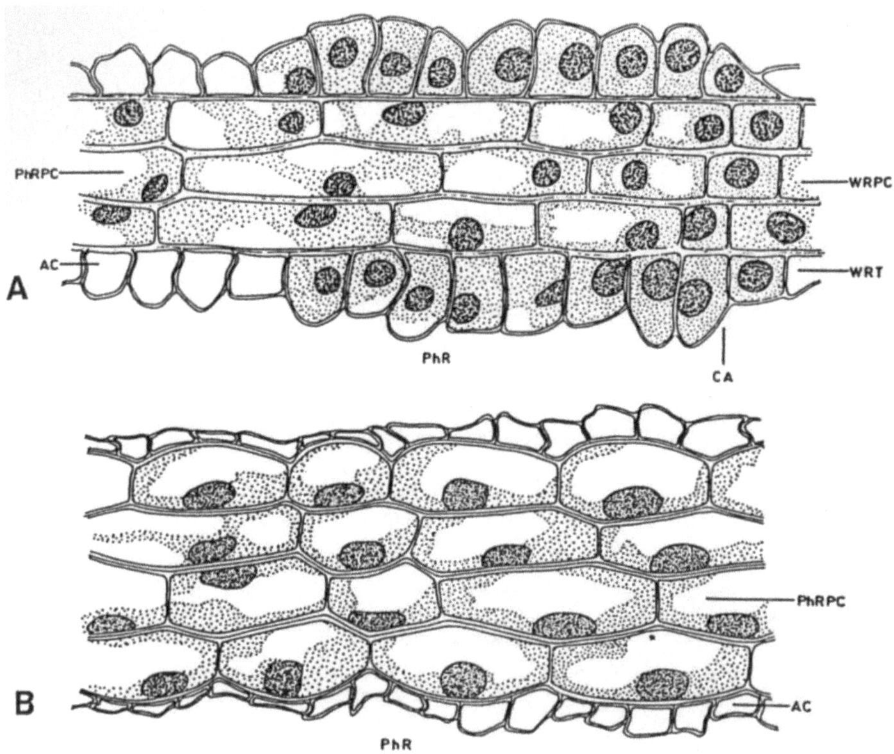

Fig. 8.54A,B. Phloem rays of *Pseudotsuga menziesii* in radial section. **A** Near the cambial zone showing functional albuminous cells (*AC*) with large nuclei and abundant protoplasm. **B** Farther outward from the cambium showing nonfunctional, collapsed albuminous cells that have lost their nuclei and protoplasm. *CA* Cambium; *PhR* phloem ray; *PhRPC* phloem-ray parenchyma cell; *WRT* wood ray tracheid; *WRPC* wood ray parenchyma cell. (den Outer 1967)

the phloem side. The resulting ray cells are at first taller than the procumbent cells of the ray and their shapes conform closely to those of the fusiform initials from which they are derived. Because of their positions on the margins of rays, they are often called sheath cells (Metcalfe and Chalk 1983) (Sect. 8.3.9). Erect cells, radial plates, and vertical albuminous cells have all been described as being present in the phloem of *Pseudotsuga menzesii* by Grillos and Smith (1959).

A radial plate is formed when the elongated erect cells on the lower margin of one phloem ray merge with those on the upper margin of another ray (Chrysler 1913). According to observations of Barghoorn (1940a), radial plates in conifers originated primarily from lateral divisions and secondarily from the ends or tips of fusiform initials. The radial plate initials generally divided for a short period after

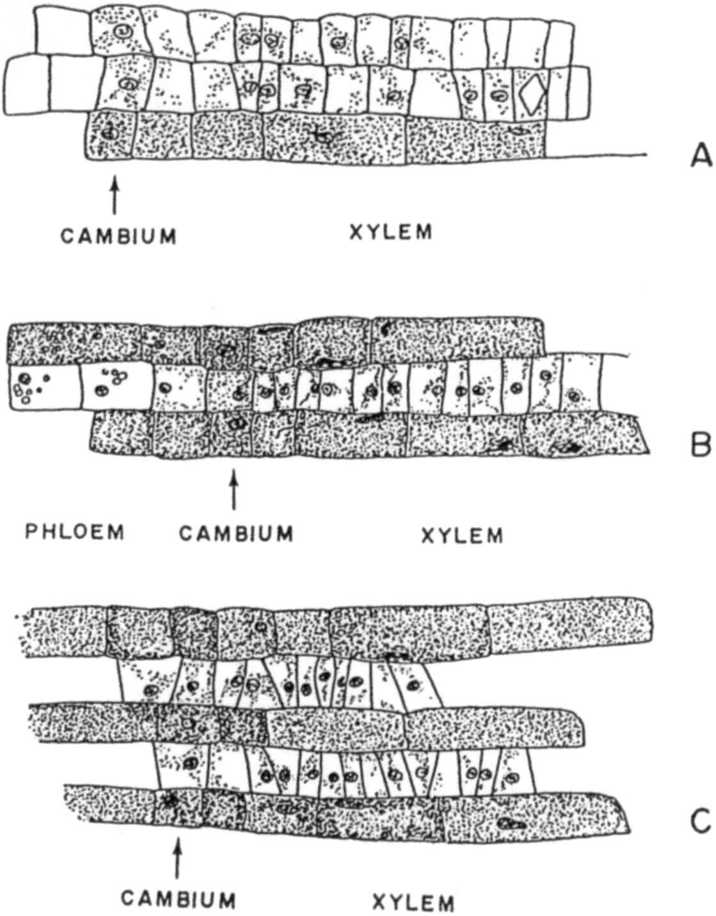

Fig. 8.55A-C. Radial sections of tile cells in rays of *Guazuma tomentosa* at three stages of development. **A** Procumbent (*shaded*) and tile cells originate from similar-shaped ray cell initials. **B** Daughter tile cells redivide, whereas daughter procumbent cells elongate. **C** Eight to ten mature tile cells occupy the space of one procumbent cell. (Chattaway 1933b)

formation but meristematic activity soon subsided. However, the radial plate initials that bordered on the marginal ray initials persisted and produced erect cells on the phloem side of the ray and either marginal tracheids or parenchyma cells on the xylem side. Whether a ray tracheid or marginal ray parenchyma cell was formed depended entirely on secondary differentiation processes because either type could arise from the same cambial initial. The erect cell initials that produced these border cells on the xylem side divided less frequently than those of the normal ray cell initials with the result that their progeny were scattered along the margins of the ray (Fig. 8.53).

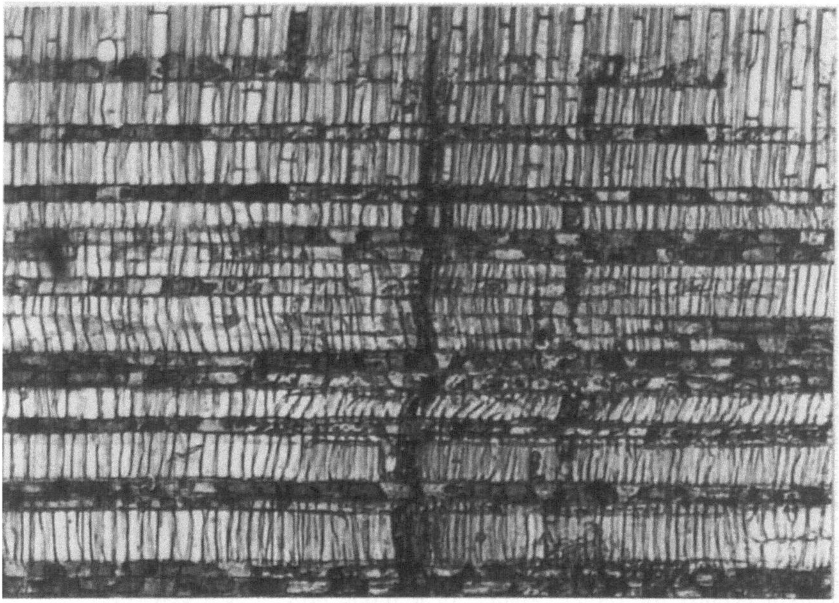

Fig. 8.56. Tile cells in *Pterospermum heterophyllum* shown in tangential section. 100x. (Manchester and Miller 1978)

Srivastava (1963a) disagreed with the idea of radial plate initials as described by Barghoorn (1940a). He was of the opinion that these initials were nothing more than declining fusiform initials. Segments of the declining initials that occurred close to the margins of existing rays generally persisted in the cambium and became attached to the margins as new initials (Fig. 8.10). In contrast, those segments that occurred near the middle positions of existing rays generally failed to become established. In general agreement with Srivastava, den Outer (1967) observed that most phloem rays in gymnosperms originated either from declining radial plates of initials one cell in width or from gradually declining fusiform initials. He found that declining initials contributed elements mostly on the phloem side. Consequently, declining radial plates on the xylem side were shorter in height than those on the phloem side.

According to den Outer (1967), a typical, newly formed phloem ray in *Pseudotsuga menziesii* one cell high consisted of a radial row of albuminous cells. After division, the two-celled ray still consisted exclusively of albuminous cells. Only when the ray became three or more cells high did a parenchymatous cell occur in the center of the ray. Thereafter, albuminous cells were confined to the ray borders (Fig. 8.54A). These albuminous cells were rich in protoplasm and contained large nuclei, but farther from the cambium the cells lost their protoplasm and collapsed (Fig. 8.54B). In places where albuminous cells occurred on the phloem side of a ray, narrow, mostly dead tracheids occurred on the xylem side (Fig. 5.11), and where parenchymatous cells occurred on the phloem side of a ray, parenchyma cells occurred on the xylem side.

As the erect ray cell initials lose their viability, or degenerate, they often give rise to the so-called "ghost" tracheids described by Thompson (1912b), Chrysler (1915), and Bannan (1936a) (Fig. 8.29). The ghost cells which occurred intermittently as erect cells on the margins of rays differed from those that occurred at the beginnings of new secondary rays (Bannan 1934a, 1951a, 1953) (Sect. 8.3.7). Gregory and Romberger (1975) interpreted these cells as "members of intermittent files" rather than degenerate ray cell initials. In *Abies balsamea*, these isolated cells invariably occurred on the xylem ray margins at the boundary between latewood and earlywood.

A special type of erect cell, called a "tile cell," is found in some species of the Bombacaceae, Sterculiaceae, and Tiliaceae (Chattaway 1933a,b). Unlike the marginal rows of erect cells that occur as described above, radial rows of tile cells are interspersed among the rows of procumbent cells. In one species, *Guazuma tomentosa*, both the tile and procumbent cells were found to originate from similar-shaped cambial initials by periclinal divisions (Fig. 8.55A). Subsequently, however, the daughter procumbent cells elongated radially, whereas the daughter tile cells kept pace by repeated divisions (Fig. 8.55B). At maturity, eight to ten tile cells might occupy the radial expanse of a single procumbent cell (Fig. 8.55C). The formation of tile cells obviously depends on the capacity of the daughter cells to subdivide and that of the procumbent daughter cells to elongate radially.

Chattaway suggested that this differential capacity might reside in the cambial ray cell initials, the contents of which appeared transparent in the initials of tile cells and extremely dark in those of procumbent cells. That is, biochemical constituents in the ray cell initials in some way determined the ultimate shapes of the daughter cells. However, in some species the differences in cell constituents presumably originated in the daughter cell and not in the initial cell (Chattaway 1951b). A confirmation and a critique of Chattaway's work were later issued by Milanez (1938) and Webber (1934), respectively. The comments of the latter author related primarily to the cell contents and they had little or no bearing on the cambial origin of these cells. Well-developed tile cells have been observed in other species of the Malvaceae (Manchester and Miller 1978) (Fig. 8.56). Extremely large tile cells have been reported in a fossil specimen of *Chattawaya*, believed to be an extinct relative of *Pterospermum* (Manchester 1980).

8.6 Effect of Eccentricity

Stems and branches are seldom perfectly cylindrical, and a wide variety of external conditions can contribute to eccentric diameter growth. As a general rule, eccentric growth is accompanied by the formation of reaction wood, tension wood in

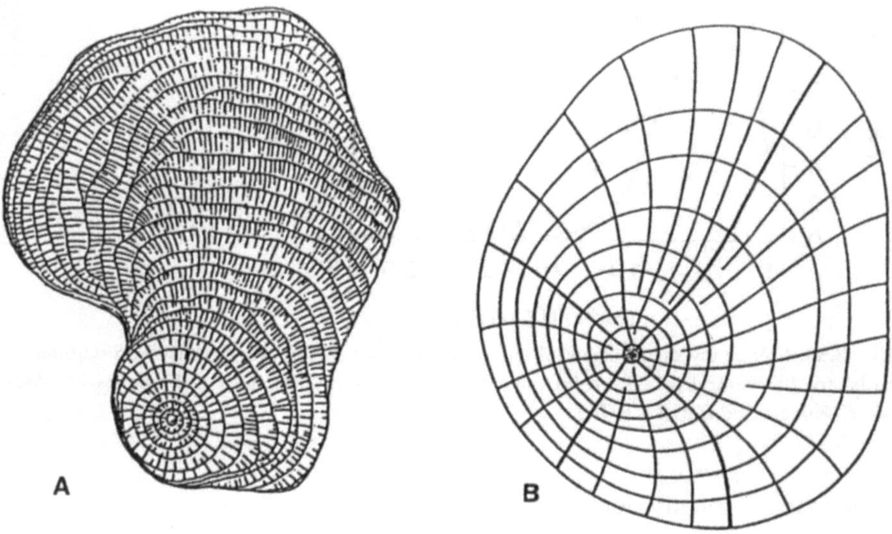

Fig. 8.57A,B. Eccentric root of *Cissampelos pareira* (**A**) and stem of *Tilia heterophylla* (**B**) in transverse sections. **A** Note extreme eccentricity of growth rings and the nonparallelism of rays in succesive rings. (Wigand 1856). **B** Note how the rays tend to intersect growth rings at right angles irrespective of the degree of eccentricity. (Sachs 1879)

dicotyledons and compression wood in conifers. Eccentric growth also results in changes in the patterns of ray development.

Some of the earliest observations of these patterns were made by Wigand (1856) on roots of *Cissampelos pareira,* a woody shrub from Mexico. In the eccentric growth rings of this species, the rays were always oriented perpendicular to the growth ring. As the direction of the rings progressively varied, the rays did likewise (Fig. 8.57A). Consequently, the rays lacked the "usual radiating structure" and the rays of one growth ring did not necessarily coincide with those of another. Similar radiating ray patterns were observed in the eccentric roots of *Ononis spinosa.*

In later years, a rather extensive literature developed regarding the trajectories produced by rays on transverse sections of tree stems. Sachs (1879), for example,

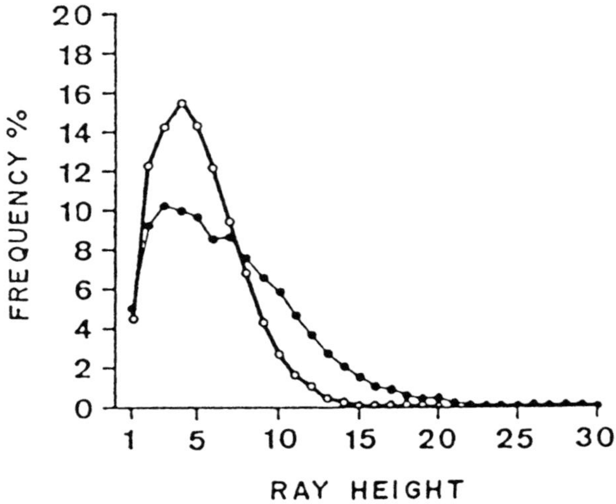

Fig. 8.58. Frequency distribution (%) of rays according to ray height (based on number of cells) for the ventral (*open circles*) and dorsal (*closed circles*) sides of branches in *Abies alba*. (Kucera and Bariska 1972)

in his study of cell division planes, observed that rays tended to intersect the cambium at right angles, or orthogonal trajectories. However, this perfect relationship was disturbed in very eccentric growth rings. Sachs therefore established a rule stating that the intersection of a growth ring was by concave rays on the short radius and by convex rays on the long radius (Fig. 8.57B). The cause of this growth shift was presumed to be related to a tendency of the cambium to expand on the side where bark pressure was least. Further examination of microsections showed that deflection of the rays responded passively to reorientation of xylem elements, as demonstrated by later workers (Chap. 6.6.2.1).

Schwendener (1880) disagreed with Sachs' idea that the orientation of cambial cell walls during anticlinal divisions was always right-angled and thus determined the direction of the rays. Rather, he believed that cell division was a phenomenon in itself and in no way dependent on prior causes. His interpretation, which was entirely theoretical and based on mechanistic principles, attempted to prove that bark tension was greatest on the side of maximum growth. Consequently, the rays were "pulled" rather than being literally "pushed" apart as Sachs assumed.

Detlefsen (1882) agreed in principle with Sachs (1879), as did Kny (1882) and Hoffmann (1885). La Rivière (1921), on the other hand, leaned toward the view of Schwendener (1880) by claiming that lateral tension, but not bark tension, might account for the curvature of rays in *Vitis lanceolaria*. In this species, an expanding fascicular sector often arched over an adjacent, less-developed sector. La Rivière assumed that ray cells on the convex part of the fascicular sector were subjected to tension which caused the rays to elongate. Krabbe (1882), a student of Schwendener, demonstrated that neither bark tension nor bark pressure were in any way responsible for the deviation of rays in eccentric stems. He concluded that ray deviation must therefore be due to some other as yet unknown causal mechanism. Küster (1899) concurred with Krabbe in his study of root grafting.

Kny (1902) attempted to simulate the effect of bark pressure by installing pinch clamps on the stems of young *Salix purpurea* and *Aesculus hippocastanum* stems. Both species produce uniseriate rays almost exclusively and Kny's objective was to induce the formation of multiseriate rays. However, despite the extreme stem eccentricities and irregularities in ray structure, most rays remained uniseriate. According to Kny, the few biseriate rays produced in *A. hippocastanum* arose in the newly formed xylem and not in the cambium.

Other workers were concerned with the distributional patterns of rays in eccentric stems. For example, Essner (1882) examined the possibility of using ray structures as diagnostic criteria for the identification of coniferous woods. As part of this investigation, he looked at ray distribution in eccentric stems of several conifers. However, due to the paucity of his data, he could detect no differences in either the numbers or sizes of rays on opposite stem sides.

Table 8.9. Numbers and sizes of rays in branches at different heights in *Sequoia sempervirens* stems (After Jaccard 1915b)

Height in (m)	Branch diameter (mm)	Number of rays[a]		Length of rays[b] stem	
		Upper side	Lower side	Upper side	Lower side
3.6	52	59	66	10	11
4.5	73	67	64	11	10
4.9	42	69	88	8	8
5.6	38	76	81	11	7
7.4	48	60	63	10	9
9.0	38	44	47	16	12
10.0	50	67	85	10	11
11.0	52	61	58	11	10
12.0	36	81	82	11	9
13.0	21	81	84	8	8
15.0	22	79	89	16	11
16.0	17	70	72	13	13

[a] Average number of rays in microscopic field 0.75 mm^2.
[b] Length of largest ray in mm x 80.

In a more thorough study, Jaccard (1915b) found that rays on the ventral sides of branches of *Sequoia sempervirens* were slightly taller than those on the dorsal sides (ratio 1.05:1.00), and these differences were related to position of the branch in the crown (Table 8.9). This observation was confirmed by Bannan (1937) in *Tsuga canadensis* and several other conifers, by Kucera and Bariska (1972) in *Abies alba*, and Chung and Lee (1989) in *Pinus densiflora*. The ratio of ventral to dorsal ray heights (1.35:1.00), was referred to by Kucera and Bariska (1972) as the "quotient of dorsiventrality of ray distribution." The frequency distributions of ray heights on the ventral sides of branches showed only one narrow peak skewed toward the lower ray heights (Fig. 8.58). On the other hand, frequency distributions of ray heights on the dorsal sides were flatter and showed several small peaks. These results were confirmed, in general, by Kucera and Necesany (1970) in *Abies alba*. Although dorsal branch sides contained a higher number of wood rays and ray cells per unit area than the ventral sides, the average ray heights were smaller. The authors suggested that higher concentrations of hormones, specifically indole-acetic acid, suppressed formation of xylem rays on the ventral sides. On the contrary, Timell (1972) suggested that the rapid growth characteristics of compression wood were associated not only with larger but more numerous rays.

As shown in Timell's (1986) comprehensive survey, eccentric growth can result from a wide variety of external conditions. One of the more interesting is the growth proliferation and induction of compression wood by certain insect infestations such as the balsam wooly aphid, *Adelges piceae*. The anatomical changes induced by aphid infestations were examined by Doerksen and Mitchell (1965) in three true firs, *Abies grandis*, *A. amabilis*, and *A. lasiocarpa*. These workers observed that the number of rays encountered on a traverse along the cambium-xylem boundary was almost double that of normal wood (Table 8.10). In contrast to normal wood where only uniseriate rays were observed, the infested wood contained some biseriate and occasional triseriate rays.

Table 8.10. Relative number of rays in wood from noninfested trees and trees infested with *Adelges piceae*[a] (Doerksen and Mitchell 1965)

Host species	Noninfested trees	Infested trees: Before infestation	Infested trees: During infestation	Ratio of abnormal to normal
Abies amabilis	8.3	8.8	16.5	1.99
Abies grandis	10.3	9.4	19.4	1.88
Abies lasiocarpa	10.1	10.6	19.6	1.94

[a] Numbers based on counts of rays in transverse section along 1.55 mm of cambium-xylem boundary.

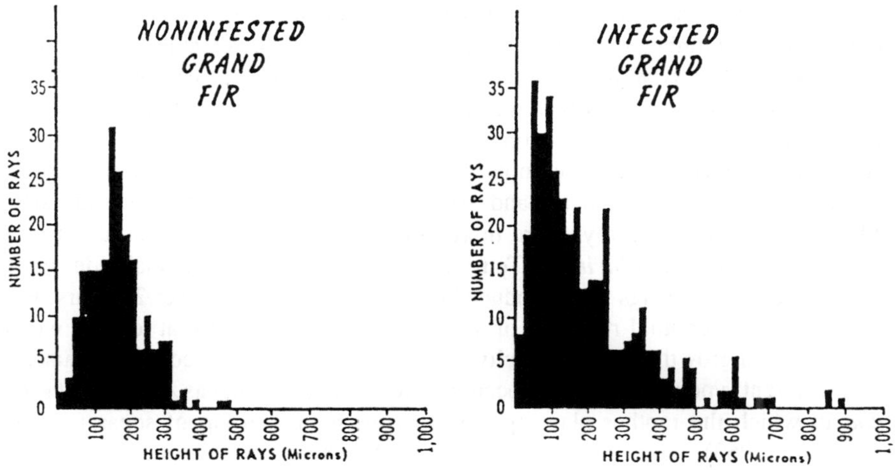

Fig. 8.59. Distribution of rays according to height in the wood of noninfested (*left*) and aphid-infested (*right*) *Abies grandis* trees. (Mitchell 1967)

A more detailed study by Mitchell (1967) on the same species revealed that the rays in infested wood were not only more numerous but they also differed in size and structure when compared to those in normal wood. Rays in the infested wood possessed fewer cells but because these cells were larger than normal, the rays tended to be higher. For example, in *A. grandis* the aphid infestation appeared to stimulate the production of many tall rays (Fig. 8.59). The combined effect of all these factors contributed to ray volumes ranging from 150 to 180% greater in the infested trees. Similar results were obtained by Smith (1967) in a companion study of *A. grandis*.

On the basis of available information, Mitchell (1967) concluded that the stimulus produced by the aphids must in some way interact directly with cambial division because only through division of the cambial initials could the number of rays and the number of cells within rays be increased. Research by others, such as Balch et al. (1964) suggested that the stimulus secreted by the aphids was hormonal in nature. Ray heights in the infested, compression wood region of *Abies* spp. (Fig. 8.59) corresponded more closely to the dorsal than to the ventral side of *A. alba* (Fig. 8.58). However, whereas Balch et al. (1964), Mitchell (1967), and others suggested a stimulating effect of growth hormones on ray cell formation and structure on the dorsal side, Kucera and Necesany (1970) suggested a suppressing effect on the ventral side. For a complete review of the effects of *Adelges* spp. on the growth and wood formation in *Abies*, the interested reader should consult Timell (1986).

Studies on the distribution of rays within eccentric stems of dicotyledonous trees have led to inconsistent conclusions. Linnemann (1953) could find no difference in the number of rays between the upper and lower sides of eccentric branches in *Fagus sylvatica*. Neither could Ollinmaa (1956) in eccentric stems of *Betula pubescens* and *B. verrucosa*, although rays on the upper, tension wood sides tended to be narrower. In contrast, Kucera and Necesany (1970) found that although rays were fewer in number on the dorsal sides of *Populus monilifera*, both the number of cells per ray and average ray heights were greater. Similarly, White (1962) found that rays were more numerous on the lower, ventral sides of branches in *Sassafras officinale*. Tension wood formed on the dorsal side while much wider growth rings were produced on the ventral side for over 20 years. Because of the differential response of species, White concluded that tension wood formation and growth ring eccentricity were not attributable to a common cause.

A different type of eccentricity occurs in desert shrubs, such as *Artemisia herba alba* in which the mother plant appears to consist of a group of separate units (Ginzburg 1963). This condition has been shown to result from exaggerated growth of aggregate rays that undergo suberization and eventual splitting of the axis. The eccentric stems of *Ambrosia dumosa*, consisting of flutes and sinuses, have been found to result from splits that appear near the base of two diverging branches (Jones and Lord 1982). These splits frequently appeared to be initiated within a ray.

Many tropical species, particularly lianas, have extremely eccentric stems in which vascular bundles are separated by wide rays. Pfeiffer (1926) exhaustively reviewed early work on this subject and pointed out that eccentricity in these species is often due to deep furrowing of the stem periphery. He distinguished among what he called furrow, strip, and projection cambia in the formation of these structures and he described their relation to the separating rays.

The eccentric growth of *Bauhinia surinamensis*, a woody vine, was analyzed in detail by Basson and Bierhorst (1967). They found that periclinal cambial divisions produced localized, excessive amounts of xylem that contributed to the formation of wings. Areas of excessive anticlinal cambial divisions were localized in a zone on either side of, but close to, the edge of a wing. The latter divisions tended to remove portions of the cambium from the periclinally active zone which resulted in a pronounced curvature to the rays. Additional information of these so-called cambial variants can be found in Carlquist (1988a).

Recently, Dobbins (1990) examined the causal mechanism contributing to formation of these segmented cambia. He hypothesized that "... the ray is acting as a physiological barrier allowing the growth factors regulating cambial activity to be very different on one side of the multiseriate ray than on the other side, allowing the induction to be broken and the cambium cylinder fragmented." It is apparent that the ray cell initials interact with the fusiform cambial initials both physically and physiologically in the formation of eccentric stems and branches.

8.7 Miscellaneous Features

A peculiar type of ray development was reported by Chattaway (1948) in *Banksia*. When the bark was peeled from fresh material, the rays projected outward from the xylem surface like small spikes. The rays of these species possess vascular strands with fibers and pitted vessels. Chattaway suggested that the ray projections were caused by radial expansion of the vascular tissue that extended outward into the undifferentiated cells of the cambial zone, thus disturbing cell orientation.

Similar protuberances occur in clones of *Populus euramericana*, but the causal mechanism is quite different (Luxová and Lux 1981). In this species, root primordia are initiated as a small group of cells associated with a biseriate ray in the secondary phloem of a stem internode that has just stopped elongating. Development of the root primordium in the phloem stimulates growth of the xylem immediately beneath the primordium site (Luxová 1985). Disturbances to the cambium caused by the rays and associated xylem elements lead to formation of the protuberances (Fig. 8.60A). The structures continue to enlarge in subsequent years and they can be easily seen and felt on debarked stems (Fig. 8.60B).

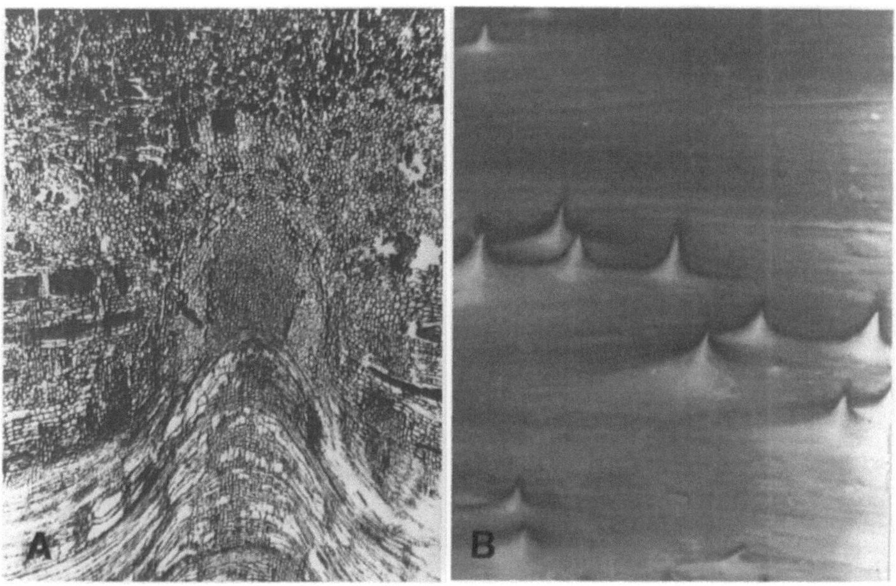

Fig. 8.60A,B. Root primordia in 4-year-old stems of *Populus euramericana*. **A** Transverse section showing an older root primordium associated with a secondary phloem ray. 20x. **B** Xylem protuberances under root primordia after decortication of an older stem. (Luxová and Lux 1981)

Van Vliet (1976) reported the presence of radial vessels in *Combretum* and *Quisqualis*. Rays of these genera are exclusively uni- or biseriate, consisting of homogeneous procumbent cells except for some erect cells in juvenile wood near the pith. The radial vessels formed relatively long radial tubes and their walls contained perforations and vestured pits characteristic of true vessel members. Van Vliet did not doubt that these radial vessel members arose from ray cell initials. Radial vessels were also found associated with multiseriate rays in the Annonaceae by Botosso and Gomes (1982). They were generally found to link two radially displaced axial vessel segments by means of simple perforations in the tangential or radial walls of terminal ray cells. Based on the characteristics of the radial vessels, the authors also concluded that they originated from cambial ray cell initials.

An abnormal type of ray development in *Fagus sylvatica* was noted by Kucera and Bonsen (1991). When viewed on transverse sections, the affected rays, which extended across several growth rings, appeared dark brown in color. When viewed on tangential sections, the rays were found to contain oval-shaped, dark flecks whose long axes paralleled those of the fibers. These flecks contributed to wound-like cell tissues and distorted ray structures that were obviously caused by a disruption of cambial activity and ray cell metabolism. The authors attributed this abnormality to bacteria that had invaded the cambium.

Chapter 9

Intercellular Spaces

A wide variety of intercellular spaces can be found in plants. Many of which occur in the xylem (IAWA 1964). The following simplified classification was suggested by Stern (1954).

A. Nonsecretory (interstitial) intercellular spaces
B. Secretory intercellular spaces
  I. Intercellular cavities
    a. Gum cavity
    b. Resin cavity
  II. Intercellular canals
    a. Gum canal (duct)
    b. Resin canal (duct)

A similar scheme was followed by Kucera (1985a,b).
Formation of these intercellular spaces can in many cases be related to events occurring within the cambial zone, and they are therefore relevant to the theme of this treatise.

9.1 Nonsecretory

A network of nonsecretory intercellular spaces, or interstitial spaces (IAWA 1964), can be found in both conifers and dicotyledons. In many woody plants a radial system, associated with ray parencyhma, is interconnected and in communication with a longitudinal system (McGinnes and Phelps 1972). The two systems presumably communicate through slit-like blind pits, i.e., pits on the ray parenchyma side which lack matching pits in the intercellular space (Panshin and de Zeeuw 1980).

A rather thorough investigation of the intercellular space system was conducted by Russow (1882b, 1883b) in both coniferous and dicotyledonous species. When the rays were uniseriate, Russow found that these air-filled, intercellular spaces occurred between the ray and nonray cells, and when the rays were multiseriate, intercellular spaces also occurred within the rays. The xylem and phloem systems were not separate but were in continuity across the cambium, and on occasion they could be followed outward to lenticels in the bark.

Kny (1890) extended this work to several dicotyledons and found that the intercellular spaces in rays were always associated with the "Merenchym" (procumbent) cells and never with the "Palissaden" (upright) cells. In radial sections of fresh wood, the interconnected spaces appeared as black lines between tiers of procumbent cells. He later (Kny 1896) attributed these schizogenously formed intercellular spaces in *Liriodendron* and *Tilia* to stretching of the radial walls of the ray cells induced by tangential expansion of the phloem, as had Koeppen (1889) a few years earlier in *Carya, Juglans,* and *Tilia.* Nicoloff (1911) also found that the intercellular spaces occurred most often between procumbent ray cells in the many species of dicotyledons he examined, although in some species they were present to a lesser degree between the tiers of procumbent and erect cells. Hartig (1893) mentioned the occurrence of intercellular spaces in rays of *Quercus,* but he did not elaborate on their formation. Kalberlah (1898) could find no evidence of intercellular spaces between ray cells in the liana *Tetrastigma scariosum,* although he did observe them in *Vitis.*

As noted by Russow (1883b), intercellular spaces might also be present in the uniseriate rays of conifers. Nyrén and Back (1959) found that they comprised a small proportion of the ray volume in *Pinus sylvestris,* and on tangential sections of *Picea abies* these same authors (Nyrén and Back 1960a) observed them as minute intercellular channels between some of the parenchymatous rows. Close examinations by Nyrén and Back (1960a) and others (Laming 1974, Mio and Matsumoto 1979a, 1981) proved that blind pits, lacking both perforations and plasmodesmata, were present in the walls of ray parenchyma cells bordering the intercellular channels.

Some authors had suggested that intercellular spaces might result from collapse of the thin-walled ray parenchyma cells as a result of water loss. To test for this possibility, Bamber (1973) forced India ink through sapwood of several *Pinus* species treated to produce collapse. The ink penetrated deeply in the wood by following the rays and fusiform resin canals, thus providing indirect evidence of collapse of the interstitial spaces.

Like the intercellular spaces in the rays of conifers, those in dicotyledons are associated with ray parenchyma cells, the walls of which contain blind pits (Back 1969, Kucera 1985b). The most detailed and comprehensive study of these structures was conducted by Kucera (1985b) in *Fagus sylvatica, Quercus petraea,* and *Q. robur.* Formation of the intercellular space system began with division of ray cell initials and was followed by subsequent growth of their derivatives in the xylem. Close examination showed that within the broad rays there existed regions with and without intercellular spaces. Although formation and distribution of these regions differed in *Fagus* and *Quercus,* they nonetheless were comparable in

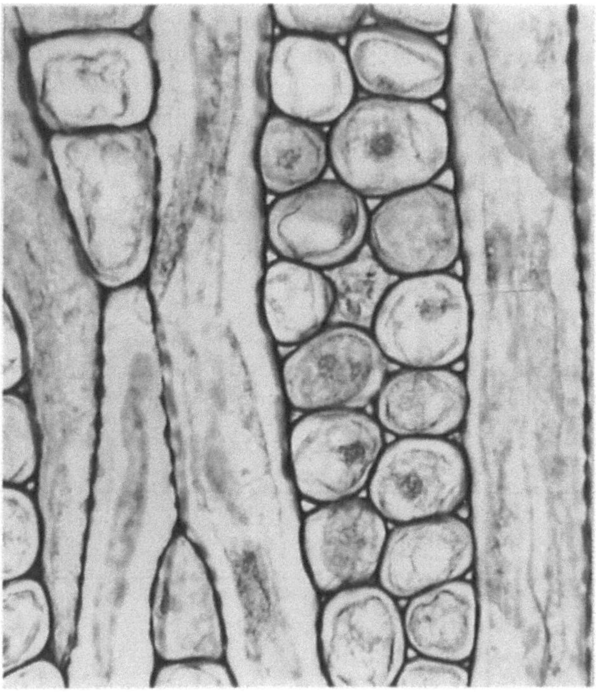

Fig. 9.1. Tangential section of *Nyssa sylvatica* showing well-defined intercellular spaces among cambial ray cell initials. (Hook and Brown 1972)

one important way. That is, the intercellular-free regions within the xylem part of a ray were developmentally correlated with sclerified regions in the phloem part. This correlation suggested that the sclerified region served as a barrier to gas exchange in the phloem. During subsequent development, phloem expansion created a cavity bordered by the sclerified cells. These events suggested, in turn, that although the initiation of new intercellular spaces was directly dependent on cambial activity, it was also secondarily dependent on differentiation processes in the phloem, as perhaps visualized earlier by Kny (1896). Formation of these intercellular spaces appeared to be entirely schizogenous, a process in which the walls of differentiating parenchymatous ray cells simply separate, although some might conceivably have formed by rhexigenesis, a process in which intercellular spaces form by mechanical rupture.

The intercellular network has long been presumed to be a system for the exchange of gases among living cells (Sifton 1945, 1957). An effective gas exchange system must maintain continuity between xylem and phloem. Consequently, there has been a long-standing debate as to whether or not the intercellular spaces continue radially through the cambium as suggested by Russow (1882b). Srivastava and O'Brien (1966a) could find no evidence of "intercellular air spaces in normal cambium" of *Pinus strobus*; presumably this statement included the rays. However, others, such as Mio and Matsumoto (1979a) and Kucera (1985b), claimed to have observed continuity across the cambium. In a more detailed study, den Outer (1967) described how intercellular spaces in the wood rays of many gymnosperms traversed radially through the ray cell initials and continued in the phloem rays. Esau (1938) also observed that "... prominent intercellular spaces pass in radial direction through the rays" of *Nicotiana tabacum*. Although she did not specifically mention intercellular spaces in the cambium, the above statement referred to her Fig. 12, which showed the phloem rays, and presumably the intercellular spaces, passing through the cambium.

Table 9.1. Average size of intercellular spaces among cambial ray cell initials and in ray cells of the differentiating and conducting phloem and maturing xylem of several bottomland species (Hook and Brown 1972)

Species	Ray cell initials µm	Phloem µm	Xylem µm
Fraxinus pennsylvanica	1-3	2-4	2-4
Liquidambar styraciflua	Absent or <1	2-6	2-4
Liriodendron tulipifera	Absent	2-4[a]	1-3
Nyssa aquatica	3-5	3-6	3-4
Platanus occidentalis	<1	2-5	2-5
Populus deltoides	<1	2-4	1-3

[a] Intercellular spaces as large as 6-10 µm are common in the rays of the conducting and nonconducting phloem in this species.

The most convincing evidence was obtained by Hook and Brown (1972), who not only presented photomicrographs of intercellular spaces in the cambium (Fig. 9.1), but also demonstrated that air could be physically pulled across the cambium of several dicotyledonous species. Intercellular spaces were large and prominent in species occupying hydrophytic sites, but they were small and sometimes absent in species growing on more mesic sites (Table 9.1).

Some workers have demonstrated that a remarkable degree of continuity exists between the radial and axial intercellular space systems (Bolton et al. 1975, Mio and Matsumoto 1979b, Kucera 1985b). A typical example is shown in Fig. 9.2. As noted earlier, it has been generally accepted that these spaces serve as a gas exchange system. Preusser et al. (1961) and Fengel (1966) questioned whether this dual system did indeed function in gas exchange. They suggested instead that

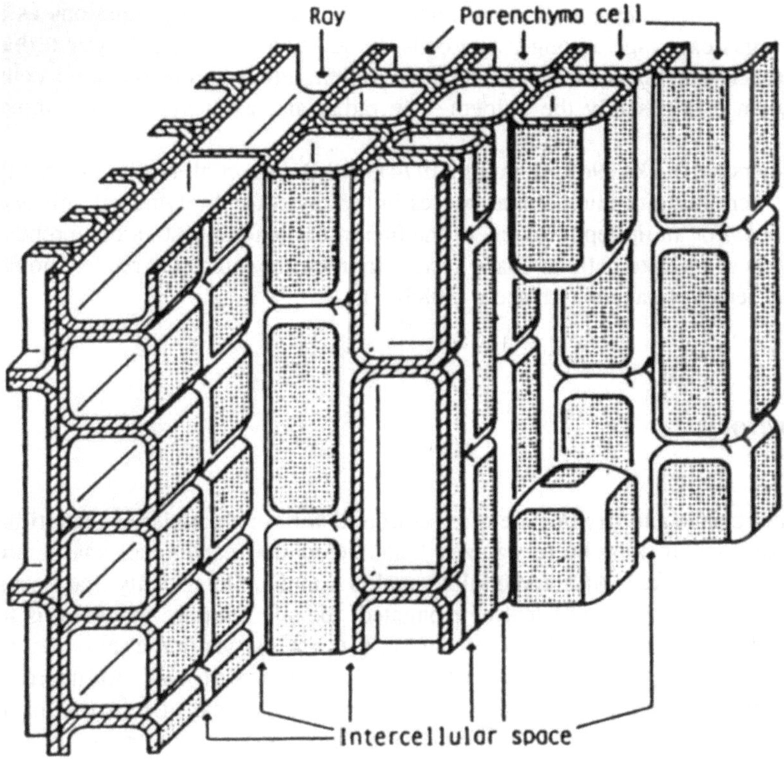

Fig. 9.2. Schematic three-dimensional drawing of the intercellular space system in a dicotyledon. (Mio and Matsumoto 1981)

it might serve as a storage reservoir because the channels sometimes appeared blocked and often filled with phenolic substances.

Mio and Matsumoto (1979a) listed 109 species in which such spaces were observed. In many of these species an interconnected network of axial and radial intercellular spaces was demonstrated, strongly suggesting a functional role for gas exchange. The system, although extensive, does not comprise a large volume and it seldom exceeds 0.5% of the tangential xylem surface (Back 1969, Mio and Matsumoto 1979a, Kucera 1985b).

Based on the foregoing review, one might conclude that intercellular spaces are of widespread occurrence in most woody species. These spaces are especially prominent in parenchymatous cells of the rays, where they communicate via blind pits with adjacent cells in both the xylem and phloem as well as with other parenchymatous cells within large multicellular rays. The intercellular spaces apparently arise in the cambium and they then expand during cellular differentiation. Thus, continuity of the intercellular space system is maintained across the cambium.

Evidence to date suggests that the intercellular space network functions as a system for gas exchange among living cells. In addition to the radial system that unites parenchymatous cells of rays, an axial system unites parenchymatous cells of the xylem and possibly the phloem. The radial and axial systems are interconnected.

The papers by Back (1969), Laming (1974), and Bolton et al. (1975) cite many references pertaining to intercellular spaces in the rays of both gymnosperms and dicotyledons. For an in-depth review of the literature, in addition to a comprehensive analysis of the intercellular space system in trees, the interested reader should consult the series of papers by Kucera (1985a,b).

9.2 Secretory

Intercellular canals and cavities are often confused with intercellular or interstitial spaces. However, it must be recognized that intercellular canals and spaces are fundamentally different. An intercellular canal is a multicellular entity that forms an internal lumen, either spherical or elongated, the sole purpose of which is to store what the surrounding cells secrete into it. It is not a space in the sense that in the living plant it is probably always filled with a secretory product. An intercellular or interstitial space, on the other hand, is an area external to cells that have no special relationship other than being living cells with a requirement for gas exchange (N. R. Lersten 1992, personal communication).

Intercellular cavities are structurally similar to intercellular canals but shorter in length (Webber 1938, IAWA 1964) (Sect. 9.2.2.6). Both intercellular canals and cavities might be formed either normally or pathologically and by either schizogenous, lysigenous, or schizolysigenous processes. Schizogenous refers to intercellular canals created by the physical separation of cells, lysigenous to the dissolution of cells, and schizolysigenous to a combination of the two processes. Elongated secretory intercellular spaces are referred to as canals by some authors and ducts by others. In the present context, the terms are considered synonymous and will be used interchangeably.

Intercellular canals occur either axially (vertical) or radially (horizontal). Although the resin canals of conifers often exhibit both orientations, it is rare to find both types occurring concurrently in dicotyledonous woods (Record 1918b, 1921, 1925). Radial rays of conifers that contain resin canals are commonly called fusiform rays, although any ray that appears spindle-shaped in tangential section might be so called (IAWA 1964).

Secretory canals are tubular structures that store gum and/or resin in the intercellular cavity. They are entirely different from laticifers in which latex is stored intracellularly within individual cells or articulated cell series (Metcalfe and Chalk 1983, Carlquist 1988a). Nonetheless, laticifers might also arise as a direct result of cambial activity. For example, in *Hevea brasilensis* the most important rubber yielding species, laticifers are located in the inner bark and they are derived from the cambium (Fahn 1979). Also, in another member of the Euphorbiaceae, *Croton conduplicatus*, laticifers originate in the vascular cambium (Rudall 1989). These laticifers then branch and intrude between cambial cells to penetrate the secondary xylem.

Lists of dicotyledonous families and genera in which various types of intercellular canals are present have been compiled by Record (1918b, 1921, 1925), Metcalfe and Chalk (1950), and Carlquist (1988a). According to Carlquist (1988a), the canals probably contain secondary plant products that function as defense mechanisms against herbivorous insects and vertebrates. Additional information on these products can be found in Smith and Montgomery (1959), Ehlers and Sandermann (1961), Dell and McComb (1978), and Hillis (1987).

Secretory canals can contain either resins or gums. The distinction between the two exudates is based on chemical composition. They are often distinguished by their relative solubilities in various solvents (Howes 1949, Smith and Montgomery 1959, Dell and McComb 1978, Hillis 1987). Most gums are found in dicotyledonous species, whereas resins are found in both gymnosperms and dicotyledons. When the distinction is unclear, some authors simply consider the exudate as a gum/resin. Resin is a generic term that came into use before rigorous chemical tests were available (Dell and McComb 1978). Though not an all-inclusive rule, the resins of conifers are commonly regarded as oleoresins and those of dicotyledons as natural resins (Hillis 1987).

9.2.1 Dicotyledons

Resins and gums are produced not only by a wide spectrum of plant species but by many different organs and tissues within these species (Howes 1949, Dell and McComb 1978, Fahn 1979). Because of this diversity, only a few representative families in which the cambium has been shown to be more or less directly involved in formation of secretory canals will be discussed.

Unlike the gymnosperms (Sect. 9.2.2), gum/resin canal formation in the dicotyledons is extremely variable. Depending on the species, canals can arise either in primary or secondary tissues and they can be confined to either the xylem or the phloem. Moreover, a radial system can be present or absent, and the canals can be of normal origin, traumatic origin, or both. However, in all cases, the canals in the secondary body arise in either xylem or phloem mother cells, but rarely, if ever, directly from fusiform cambial initials.

Anacardiaceae: Secretory canals in the Anacardiaceae occur both in the primary body and in the secondary phloem. Those occurring in the xylem are commonly radial canals (Metcalfe and Chalk 1950). Many investigators have studied the resin canals in this family and all modes of formation - lysigenous, schizogenous,

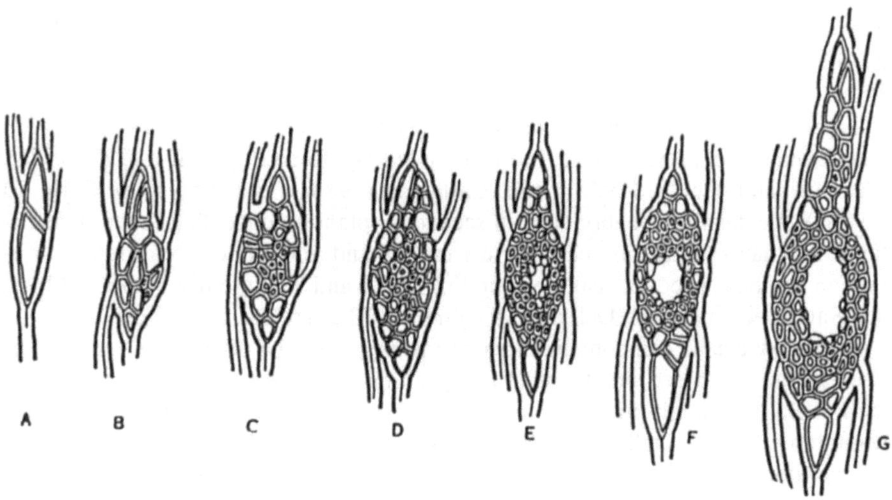

Fig. 9.3A-G. Development of a radial secretory canal extending through approximately 2.5 cm of xylem in *Euroschinus falcatus*. 175X (Chattaway 1951a)

and schizolysigenous - have been reported (Fahn 1979, Jöel and Fahn 1980, Bhat and Mohan Ram 1992).

Secretory canals in the secondary phloem of *Rhus glabra* are arranged in tangential bands oriented parallel to the longitudinal axis of the stem. Fahn and Evert (1974) observed that differentiation of a canal began at the phloem edge of the cambial zone when the middle lamellae separating a group of cells began to swell and to exhibit many electron-dense globular structures. Cavity formation proceeded rapidly by schizogeny and completely open canals were observed very close to the cambium. The canal lumen was lined with tangentially flattened epithelial cells and these in turn were surrounded by layers of sheath cells. As stem size increased by continued cambial activity, the older canals enlarged by repeated divisions and stretching of the epithelial cells. Cell stretching often led to anastomosing of adjacent canals in the tangential plane.

In contrast to *R. glabra*, the primary secretory canals of *Mangifera indica* were found to arise by lysigeny (Jöel and Fahn 1980) and those in *Semecarpus anacardium* by schizolysigeny (Bhat and Mohan Ram 1992). In the latter species, the gum-resin ducts originated in the phloic procambium of the young stem by formation of an intercellular space among a group of densely stained cells with conspicuous nuclei and nucleoli. Subsequent widening of the duct cavity occurred in several stages as follows: (1) longitudinal divisions of the epithelial cells; (2) partial or complete separation of the epithelial cells along their radial walls; (3) subsequent tangential flattening of the epithelial cells; and (4) lysis of some of the epithelial cells.

Development of the horizontal secretory canals in several tropical species of this family (*Euroschinus falcatus*, *Microstemon velutina*, *Pistacia lentiscus*, and *Pleiogynium cerasiferum*) was followed by Chattaway (1951a). *E. falcatus* was studied most thoroughly and was considered typical of the group. The canals originated in rays of the phloem but only in association with a vertical canal. That is, a horizontal canal could only arise in a phloem ray when the ray made contact with a vertical canal in the same way that horizontal canals arise in the xylem of conifers (Sect. 9.2.2.3). Once initiated, the horizontal canal was perpetuated in both the xylem and phloem by periclinal divisions of ray cell initials. In her investigation, Chattaway followed canal development through the xylem because canals were better preserved than in the crushed and distorted tissues of the old phloem.

In primary xylem near the pith of *E. falcatus*, Chattaway found that all rays were at first uniseriate and very tall; see Chapter 8.2.1 concerning primary rays. These tall primary rays quickly subdivided to form uniseriate secondary rays about 30 cells in height. At about 3 mm from the pith biseriate rays began to form accompanied by the inception of resin canals in some rays. Canals formed when a group of small cells in the ray center began separating schizogenously to form a cavity (Fig. 9.3), followed by the formation of a sheath of thin-walled, secretory cells that lined the cavity. In all species examined, the canals arose through changes in the ray cell initials that resulted in permanent alteration of cell types within the ray (Chattaway 1951b).

Dipterocarpaceae: Resin canals are commonly found in the Dipterocarpaceae, a family of large resinous trees widely dispersed throughout tropical Asia. Several commercially important damar resins are obtained from members of this family, the most widely recognized being "damar penak" from *Balanocarpus heimii* (Howes 1949). In the process described by Howes, trees were tapped by making incisions at intervals up the stem. The resin flowing from these wounds gradually hardened and it was collected about 3 months later when the stem was again wounded to induce further flow. The best resin flow occurred from the upper stem. Because some trees were 12 m in circumference and over 1000 years old, various types of ladders were constructed by the tappers.

A damar resin is also collected from *Shorea javanica*. In this species, Messer (1990) described how traumatic resin ducts were first induced by drilling a series of small holes in the stem. These holes presumably rendered the tree susceptible to fungal attack and resin production was induced as a defensive mechanism. Larger holes were then opened for commercial resin production. Shiokura (1989) also induced traumatic resin canals in several *Shorea* species by wounding the stems.

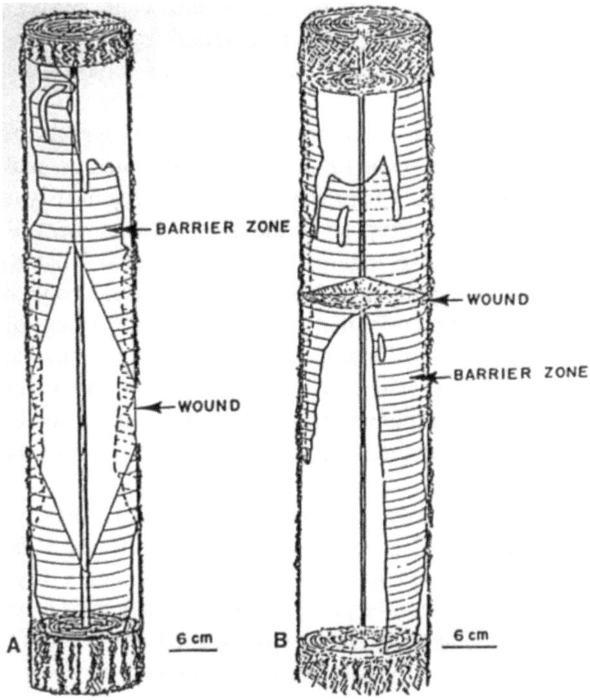

Fig. 9.4A,B. Schematic diagrams showing surface of cambium of *Liquidambar styraciflua* stems that reacted to wounding by formation of gum canals and associated barrier zone tissues. **A** Superficial wound. **B** Deep wound. (Moore 1978)

Tangential bands of traumatic resin canals formed both above and below wound sites created by driving nails into the trees.

The investigations of Deutschmann (1954) on herbarium and seedling material from 76 species representing 14 genera in the family showed that the resin canals arose first in the primary wood and not in the pith as previous workers believed. These primary canals were situated on the xylem side of the boundary between primary xylem and pith. Examinations of *Dipterocarpus subalpinus* and *Shorea robusta* clearly showed that the primary resin canals developed from a procambial ring in the former and in association with the first xylem elements of the procambial ring in the latter. A similar origin was found in all genera investigated with the exception of *Dryobalanops*, in which they "probably" arose in the procambium, and *Monotes*, in which primary canals were absent. In addition to the circle of primary canals situated on the periphery of the pith, canals were also associated with primary vascular bundles in the cortex. In many species, secondary canals also occurred in the wood. These canals, surrounded by parenchymatous cells, were arranged in a circle around the stem.

Hamamelidaceae: Gum ducts in *Liquidambar styraciflua* are of traumatic origin. Hepting and Blaisdell (1936) described how a protective zone consisting of a gum zone proliferating from the ray cells formed following a fire. This gum zone, usually associated with a barrier zone (Chap. 10.2), is formed by cells of the cambium adjacent to a wound (Moore 1978). The first response to injury that Moore observed in this species was the production of traumatic gum canals which often extended a meter or more above and below the wound site (Fig. 9.4). Gum was produced by epithelial cells lining the duct cavity. The gum, known as storax, has been shown to inhibit the growth of decay fungi (Shortle and Cowling 1978, Shortle 1979).

Observations by Gerry (1921) on *L. styraciflua* indicated that wounding induced formation of tangential rows of gum canals that filled the available spaces between rays. The rays, as in many other wounded tissues, increased in width but otherwise maintained continuity through the wound tissue. In some specimens, more than one tangential row of ducts developed. Although ducts formed in abundance above the wound, those induced lateral to it were smaller and less numerous.

The storax gum produced by *L. styraciflua* was at one time widely used in pharmaceuticals and specialty chemicals. When trees were tapped during the active growing season by horizontal cuts that injured the cambium, traumatic ducts were induced to form (Gerry 1921). The exudate was absorbed by cloth strips wrapped over the wound site.

Leguminosae: Of the gum-producing species in this large family, *Acacia senegal* is perhaps best known. This species, growing mainly in the Sudan region of North Africa, is the major source of gum arabic. Gum is collected by removing vertical strips of bark, one to two inches wide, from the tree during the dry season after the leaves have died (Howes 1949, Smith and Montgomery 1959). In the rainy season, when trees are in full leaf, tapping does not produce gum. Gum forms along the length of the wound and continues to form for an entire season.

Despite the commercial importance of this gum, little was known concerning its source within the tree for many years. Most prevailing ideas were extrapolations based on gummosis in fruit trees. Joseleau and Ullmann (1990) attempted to determine the site of synthesis chemically by analyzing tissues from the inner bark, cambium, and xylem of non-producing and producing branches that had been previously tapped for gum. Because they could find no evidence of either the existence or formation of gum ducts in the tissues examined, they suggested that gum was synthesized in "gummiferous cambial tissues" and transported to adjacent tissues. Gum production was presumed to be in competition with tree growth, because gum in reasonable amounts could not be obtained until the tree attained a certain stage of maturity.

Myrtaceae; *Eucalyptus* gum is commonly referred to as "kino" and the pockets in which it occurs are called "veins" (Hillis 1987). The veins form tangential layers of anastomosing passages embedded in a sheet of traumatic parenchyma (Fig. 9.5A). The passages contain kino, a polyphenol, which flows from puncture wounds in the veins but solidifies on exposure to the atmosphere (Skene 1965). Kino veins form in response to cambial injury, with the type and extent of response varying with species. The veins might arise either in the xylem or the phloem, depending on the species, although the developmental processes are identical in both tissues (Tippett 1986).

A critical examination of kino vein formation in *Eucalyptus obliqua* in response to wounding was conducted by Skene (1965). About 5 days after wounding of the cambium, the frequency of transverse and pseudotransverse divisions of the fusiform initials increased dramatically. Derivatives of these divisions developed as bands of traumatic parenchyma cells and polyphenols began to accumulate within them. The cells at this stage retained their radial alignment. Kino veins originated when certain parenchymatous cells destined to be the central cells of the future vein collapsed to form a lumen. Concomitantly, a peripheral cambium was initiated surrounding the lumen. The peripheral cambium continued to divide actively, producing callus-like layers of tissue by both anticlinal and periclinal divisions (Fig. 9.5B). Although the lumen formed initially by collapse of the central cells, further development was lysigenous. The secretory mechanism involved destruction of the secretory cells and continued secretory activity required constant cell replacement by the peripheral meristem.

In some *Eucalyptus* species, cambial activity associated with strands of traumatic parenchyma results in deformation of the radial and tangential cell alignment (Hillis 1987). Further activity of the cambium, together with breakdown of cells in the strand center, creates cavities. This process results in interconnected cavities which are either bridged by rows of parenchyma as in a vein or absent as in a pocket. Veins are 2-5 mm thick radially and of varying widths tangentially. They are vertically oriented, but tangential anastomoses form networks that often completely encircle the tree (Fig. 9.5C). In *E. maculata*, these bands of traumatic parenchyma and kino veins forming after cambial injury have been associated with ring shake (Wilkes 1986), a defect in which growth rings separate tangentially and longitudinally at the ring boundaries.

Rutaceae: In *Citrus* plants, the gum secretion resulting from injury is known as gummosis (Fahn 1979). It has been suggested that the gum serves to block the progress of invading pathogens, possibly by creation of a barrier zone.

Gedalovich and Fahn (1985a) examined gum duct formation in *Citrus aurantium* and *C. volkameriana* caused by the parasitic pathogen *Phytophthora citrophthora*. The fungus killed the cambium and phloem in the vicinity of the wound but not the xylem mother cells. First evidence of duct formation was swelling and then dissolution of the middle lamella of certain xylem mother cells. The narrow intercellular space created by this process occurred between two, occasionally several, cell layers (Fig. 9.6A-B). During the next phase of development, epithelial cells rich in cytoplasm and organelles were evident surrounding the duct

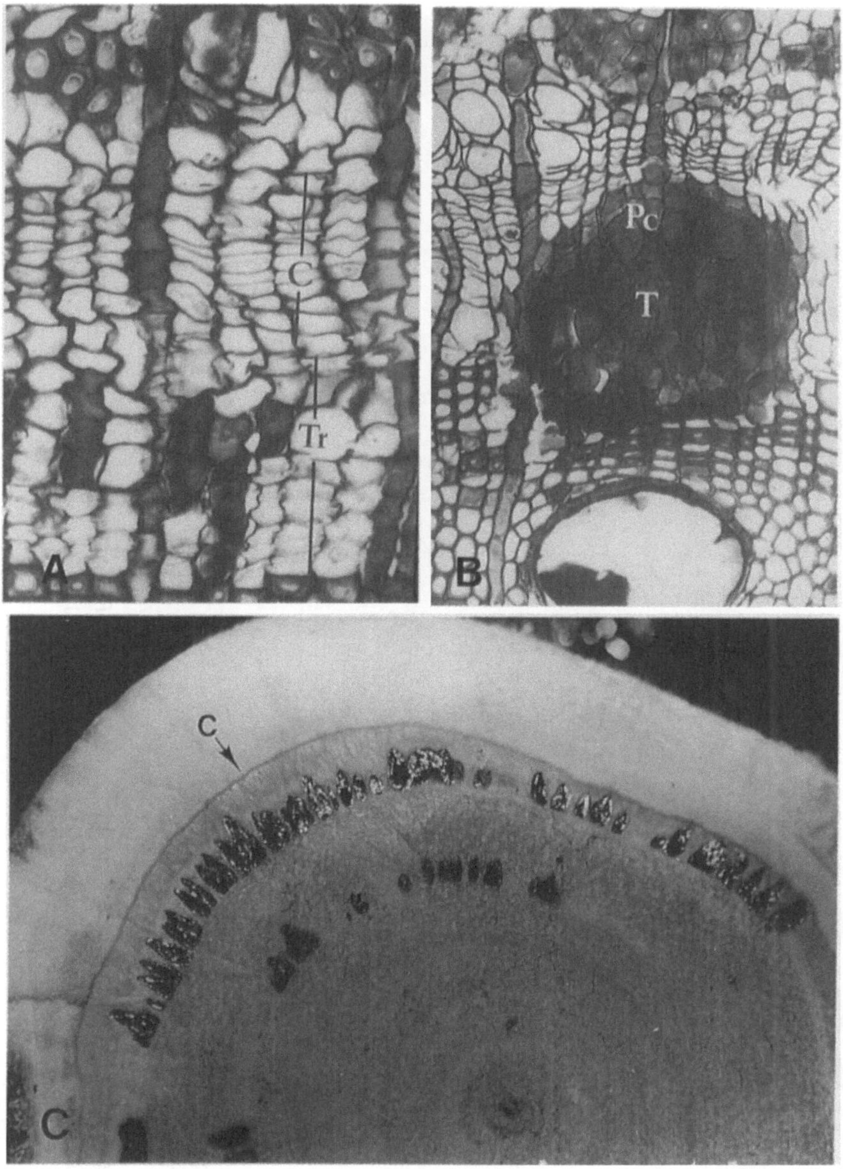

Fig. 9.5A-C. Development of kino veins in *Eucalyptus obliqua* (**A,B**) and *E. `camaldulensis'* (**C**) shown in transverse sections. **A** Section through cambial region (*C*) and developing network of traumatic parenchyma (*Tr*) 20 days after wounding. 241x. **B** Section through a kino vein at an early stage of development, 35 days after wounding. Note the peripheral cambium (*PC*), the small vein lumen, and torn cells (*T*). 181x. (Skene 1965). **C** Section of stem showing tangential bands of kino veins included in the xylem of a stem 10 cm in diameter. *C* cambium. (Tippett 1986)

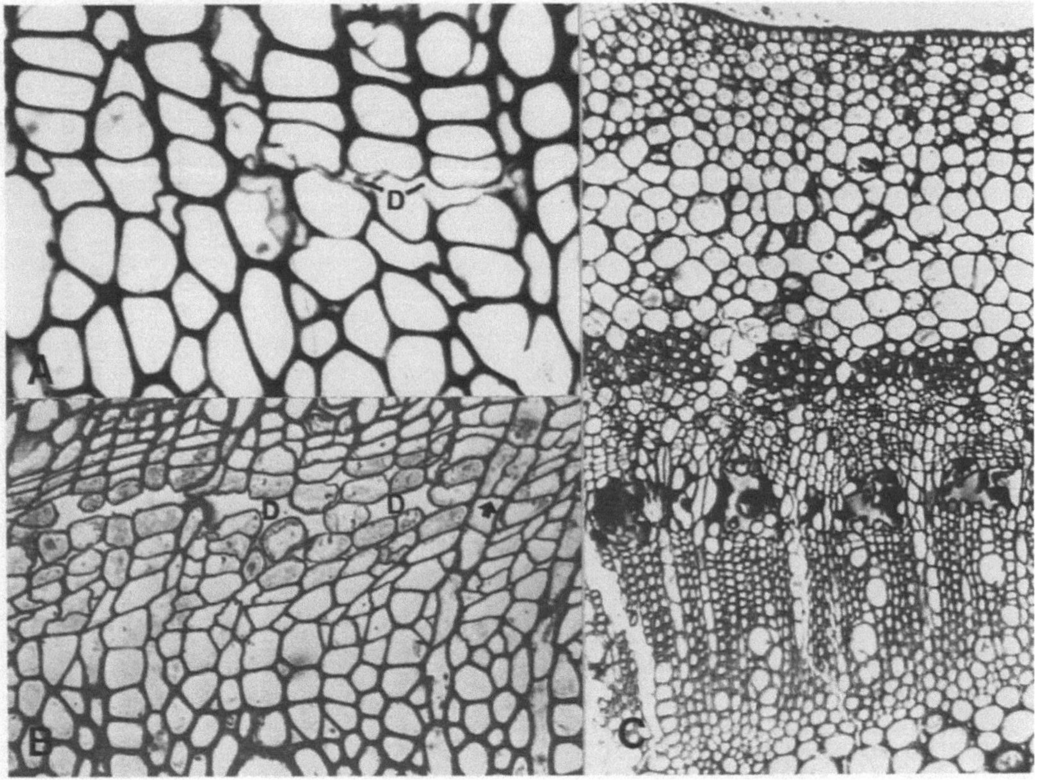

Fig. 9.6A-C. Formation of gum ducts in *Citrus* sp. invaded by the parasitic pathogen *Phytopthora citrophthora*. **A,B** Transverse sections of the cambium showing two developmental stages of a duct lumen (*D*). In **B** a ray cell initial has divided anticlinally (arrow). **C** Transverse section of stem with mature gum ducts. **A** 924x; **B** 547x; **C** 131x. (Gedalovich and Fahn 1985a)

cavity. Despite the presence of a few dead cells in the duct cavity resulting from the fungal infection, the cavity was considered to have developed schizogenously. Further enlargement of the duct cavity was due to expansion of epithelial cells. The short lengths of these cells, as well as their shapes, suggested that they were derived from fusiform initials. Mature ducts were filled with polysaccharides which exuded into the surrounding tissues, suggesting epithelial cell breakdown after the secretory phase was completed (Fig. 9.6C).

Applications of ethrel (2-chloroethane phosphonic acid) to *Citrus* stems induced rapid formation of gum ducts that mimicked the effect of the fungal pathogen *Phytophthora citrophthora* (Gedalovich and Fahn 1985b). Results of other studies had indicated that gum duct formation was caused by the release of ethylene from the infected tissues. The cambium of inoculated trees responded within 2 to 3 days by switching from the production of normal xylem elements to the production of gum secreting cells. The authors suggested that ethylene might act by inducing the synthesis of hydrolytic enzymes in the xylem mother cells. These enzymes, in turn, would dissolve the middle lamellae between the cells, thus leading to formation of the duct lumen.

9.2.1.1 Traumatic Canals

Most dicotyledonous species that form normal gum/resin canals also produce accessory traumatic canals when wounded. Other species that do not normally produce gum canals do so when wounded. For example, canals in the Hamamelidaceae, Myrtaceae, and Rutaceae, described in the foregoing sections, are all of traumatic origin. To cite another example, *Azadirachta indica* only forms traumatic resin ducts following injury to the cambium (Nair et al. 1983). In the specimen illustrated by the authors, three radial files of cambial derivatives were involved in canal formation in the xylem. Widening of the duct lumen was lysigenous, with epithelial cells first becoming secretory and then entering a phase of senescence and disintegration. Thus, the duct was filled with cell wall and cytoplasmic debris mixed with secreted gum.

Traumatic canal formation, particularly in commercially important gum/resin yielding species, is often enhanced by the application of growth hormones or other chemicals (Hillis 1987). In one study, the hormone ethylene (ethrel) was applied to wounds of *Ailanthus excelsa* (Babu et al. 1987). Ethrel induced the cambium to produce several layers of traumatic xylem parenchyma consisting of groups of cells that divided anticlinally and transversely to form an incipient gum cavity. The cavity, which resembled a canal, was initiated when a centrally situated cell underwent lysis, and the surrounding cells reoriented to form an epithelium. Traumatic parenchyma adjacent to the epithelial cells then became meristematic and divided. As new epithelial cells were added to the periphery, the inner cells underwent lysis, the cavity enlarged and eventually anastomosed with tangentially adjacent cavities. Interestingly, the multiseriate rays remained unaltered and continued radially through the anastomosing network of cavities. Meanwhile, the epithelial cells secreted gum into the cavities. Traumatic gum cavities developed in a similar manner in *Bombax ceiba* (Babu and Shah 1987), *Sterculia urens*, and several other species (Menon and Babu 1989, Babu and Menon 1990). The gum collected from *S. urens* is known in commerce as karaya gum (Smith and Montgomery 1959, Hillis 1987).

9.2.2 Gymnosperms

In the living conifers, normal resin canals are found in the secondary xylem of only a few members of the Pinaceae, more specifically in *Larix*, *Picea*, *Pinus*, and *Pseudotsuga*. Canal diameters are largest in the genus *Pinus* (60-300 µm), and smaller in *Larix* (40-80 µm), *Picea* (40-70 µm), and *Pseudotsuga* (40-45 µm) (Greguss 1955). Both axial and radial resin canals are found in these genera, and both types occur in traumatic as well as normal canal systems. As a general rule, epithelial cells in *Pinus* remain thin-walled in the sapwood, whereas those of the other three genera become thick-walled as they mature (Mayr 1893, Thomson and Sifton 1925, Bannan 1936b). The epithelial cells of *Pinus* can remain viable for many years. Peters (1970), for example, observed mitosis in a epithelial cell of a horizontal resin canal 11 growth rings inward from the cambium in *Pinus elliottii* var. *elliottii*.

Species represented by *Abies*, *Cedrus*, *Pseudolarix*, and *Tsuga* do not normally possess resin canals. However, traumatic axial canals have been reported in *Abies* (Dippel 1863, Bailey 1925a,b, Monange 1961), *Cedrus* (Fahn et al. 1979), *Pseudolarix* (Bannan 1936b), and *Tsuga* (Penhallow 1907, Bannan 1933a, 1934b, Monange 1961, Leney and Moore 1977), as well as in *Keteleeria* (Bannan 1936b) and *Sequoia* (Penhallow 1907, Hemenway 1926, Bailey and Faull 1934). Radial resin canals rarely occur in these genera.

9.2.2.1 Primary Resin Canals

The primary resin canal system in conifers is not a product of the cambium and connections between the primary and the secondary canal systems rarely, if ever, exist in above-ground parts (Mayr 1893, Thomson and Sifton 1925, Hanes 1927, Pauly 1963). Nonetheless, a brief summary of the primary canal system helps to place the secondary system in perspective.

In some conifers, resin canals are present in the hypocotyl of the embryo within the seed (Pauly 1963, Werker and Fahn 1969, Charon et al. 1990). A special type of resin-containing cell, called a secretory cell, also develops in embryos and seedlings of conifers (Schopf 1943, Allen 1947, Spurr 1949, 1950, Werker 1970). When these extremely long secretory cells are oriented longitudinally, they resemble laticifers (Werker 1970).

Primary resin canals are of common occurrence in many conifers, even in some species that do not normally produce secondary resin canals (Hanes 1927). Over the years, a number of workers have examined various aspects of these primary canal systems. In two of the earliest papers, both Frank (1867) and Müller (1867)

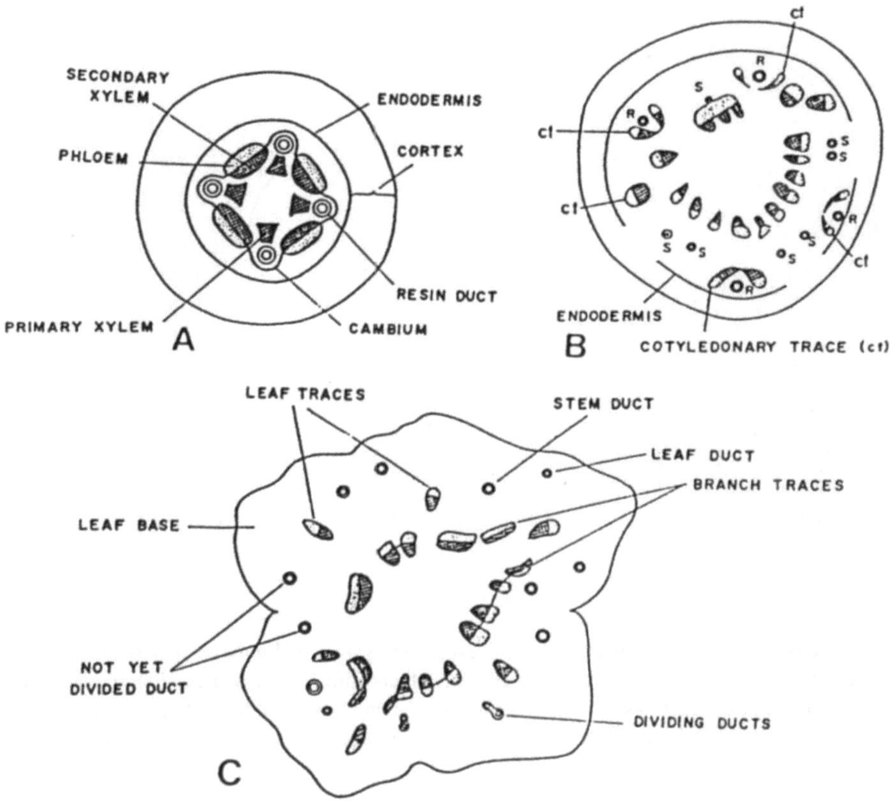

Fig. 9.7A-C. Schematic drawings of transverse sections of *Pinus halepensis* seedlings. **A** Tetrarch root showing four primary resin ducts. **B** Upper part of hypocotyl showing resin ducts of both the root-hypocotyl (*R*) and the stem (*S*). **C** Stem section above **B** showing departure of leaf traces and associated stem ducts, one upward in the stem and the other outward in the leaf. Resin ducts indicated by *circles*; xylem is *cross-hatched*; phloem is *stippled*. (Werker and Fahn 1969)

described how the primary canals arose from "cambial mother cells" that first divided to produce a rosette of four daughter cells and then separated to form the duct cavity. More recently, in a thoroughly detailed as well as beautifully illustrated paper, Suzuki (1979) described the development and distribution of primary

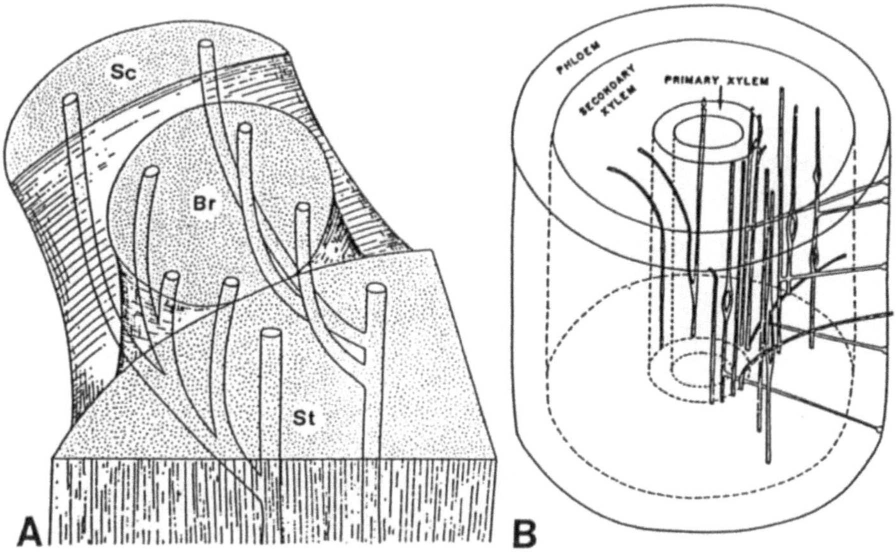

Fig. 9.8A,B. Schematic reconstructions of resin duct system in *Pinus halepensis* seedlings. **A** Part of stem (*St*) showing branching of resin ducts into a brachyblast (*Br*) and a scale (*Sc*). **B** Part of resin duct system of a 1-year-old branch showing ducts (*stippled*) extending from the inner circle of secondary xylem into the lateral branches. (Werker and Fahn 1969)

resin canals in the shoot and cortex of species representing all genera of living conifers.

Detailed investigations of the primary resin canal system have been conducted by Pauly (1963) and Werker and Fahn (1969) in *Pinus pinaster* and *P. halepensis*, respectively. In both species, evidence of canal initial cells could be recognized in embryos still within the seed. These initial cells first appeared near the root apex in the region of procambial differentiation, with the number of canals corresponding with the number of xylem poles in the root (Fig. 9.7A). The canals developed acropetally toward both the shoot and root apices adjacent to the protoxylem strands. Some canals ended at the upper end of the hypocotyl, whereas others entered the base of the cotyledons before terminating (Fig. 9.7B). Formation of a canal began when certain cells divided to produce a rosette of four cells, and the duct cavity was initiated when two cells in the middle of the rosette began to separate. Further divisions of the rosette and surrounding cells, in combination with expansion of these cells, enlarged the duct cavity. Cavity formation was entirely schizogenous.

A second system of new resin canals was initiated in the region of the cotyledonary nodes. These canals had no connections with those of the root-hypocotyl system. They developed in close association with procambial traces serving juvenile leaves of the epicotyl and they closely conformed to phyllotaxy. Each leaf trace was bounded by two so-called stem canals or ducts at the level where the trace departed the vascular cylinder (Fig. 9.7C). Each of these canals divided with one branch passing into the leaf and the other continuing upward in the stem. Thus, each juvenile leaf was served by two longitudinal resin canals situated along its abaxial side. The stem canals continued their acropetal development in close association with the procambial system. Farther down in the primary stem, Werker and Fahn could not determine whether the procambial leaf trace and resin duct systems arose from common initial cells. However, it was clear that common factors controlled both patterns of development and both were correlated with phyllotaxy.

The foregoing pattern, in which both the primary resin canal and primary vascular systems developed acropetally in close association, prevailed in the primary body of the shoot of both species investigated from seedling to mature tree. This meant that the primary resin canal system of the shoot was continuous throughout the plant except for the needles of the brachyblasts and possibly the radicle-hypocotyl system. The brachyblasts are actually short shoots, and the two canals that diverge from the stem canal system serve the brachyblast and its scale (Fig. 9.8A). As in other conifers (Mayr 1884, 1893, Hanes 1927, Charon and Vindt-Balguerie 1986), the needles of *P. halepensis* and *P. pinaster* are provided with an independent resin duct system. Both the primary stem and the needle canal systems are entirely independent of the secondary resin canal system that forms later in the xylem. Thus, in these two species of *Pinus*, and perhaps other *Pinus* species as well, four independent resin canal systems could be identified, namely: (1) the primary radicle-hypocotyl system; (2) the primary cortical-brachyblast system; (3) the needle system; and (4) the secondary xylem system. Earlier descriptions of the primary resin canal system by Mayr (1884) and Hanes (1927) were in general agreement with those of Pauly (1963) and Werker and Fahn (1969). However, in the species studied by the former authors, the cortical canals arose independently in each new shoot. That is, canals formed in one year's shoot were not continuous with those of either the previous or succeeding year's shoots.

Development of the primary resin canal system in *Pinus pinaster* was described in a slightly different manner by Charon et al. (1990). Again, incipient resin canals could be detected in embryos still within the seed. At germination, canals were situated at the periphery of the central cylinder immediately outside each of the xylem poles of the radicle. However, a second center of canal initiation was observed in the hypocotyl. The radicle-hypocotyl canal system then formed by a

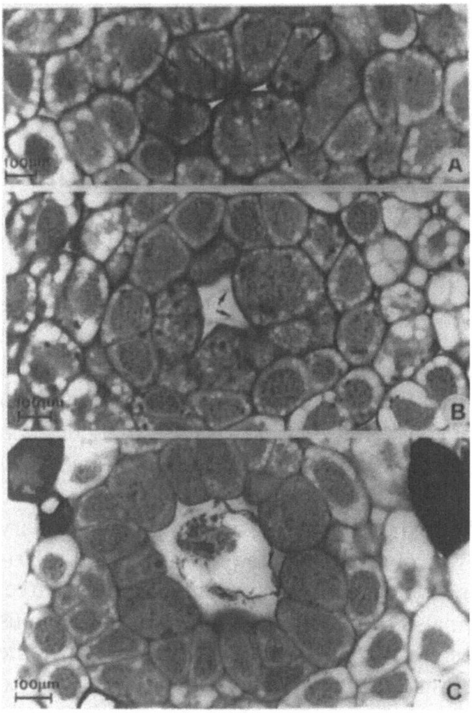

Fig. 9.9A-C. Transverse sections of *Pinus pinaster* showing development of primary resin canals. **A** Cell arrangement at early stage of canal formation. **B** Torn Cell fragments (*arrows*) in center of canal lumen. Two rows of epithelial cells encircle the canal. **C** Lysigenous cell residue in expanded canal lumen. A single row of epithelial cells now encircles the canal lumen. (Charon et al. 1986)

merger of these two parts. The acropetal extension of this canal system did not enter the cotyledons but terminated just below the level of their insertion.

Development of the epicotyl bud gave rise to the juvenile leaves. Resin canals did not begin to differentiate, however, until these leaves were about 160 μm long. At this time, two canals were initiated at the base of each juvenile leaf and development extended basipetally in the hypocotyl and acropetally in the epicotyl. As in *P. halepensis*, the canals that developed basipetally from the epicotyl did not merge with the radicle-hypocotyl canal system. Thus, the primary resin canals of *P. pinaster* also consisted of two autonomous systems.

The primary resin canal system in elongating shoots of *P. pinaster* was also studied by Charon et al. (1986). Their interpretation of canal development was somewhat different than that of either Pauly (1963) or Werker and Fahn (1969). Whereas the latter authors viewed canal development as being acropetal and continuous in the elongating shoots, Charon et al. (1986) considered it to be discontinuous. That is, the two canals serving a brachyblast were initiated independently

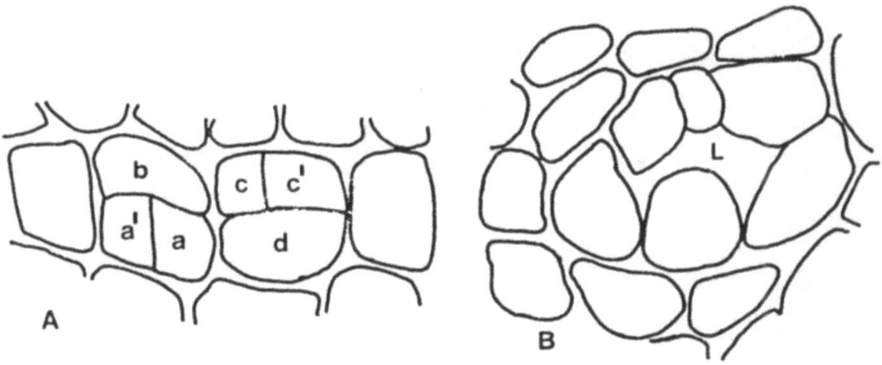

Fig. 9.10A,B. Schematic drawings of secondary resin canal formation in *Pinus sylvestris*. **A** The first anticlinal divisions occurred in two adjacent cambial cells (*a,c*). **B** The canal lumen (*L*) developed by radial growth and tangential expansion of the resulting epithelial cells. 300x. (After Sanio 1873)

at its base in close association with the procambium, and they then differentiated basipetally to eventually merge with the acropetally developing stem canals. Thus, the primary resin canal system differentiated bidirectionally just as the primary xylem elements did. Moreover, the primary canal system conformed precisely to phyllotaxis, a characteristic that appears to be common to all conifers (Suzuki 1979). In contrast to most previous workers, Charon et al. (1986) also presented evidence suggesting that all primary resin canals in *P. pinaster* were formed by schizolysigenous processes, i.e., in part by separation and in part by enzymic digestion of the cells (Fig. 9.9).

9.2.2.2 Secondary Resin Canals - Axial

Axial resin canals are found in first-year growth rings but their occurrence is not always consistent (Bannan 1936b). As observed by Thomson and Sifton (1925), in some trees of *Picea* the first six growth rings were entirely devoid of axial canals, whereas in other trees these rings were densely populated with canals. Radial resin canals were frequently initiated in the first-year ring, and these canals always arose in association with a secondary, but never a primary, axial canal. First-year growth rings of *Larix decidua* contained many axial resin canals but radial canals were absent (Hug 1979). Radial canals were initiated in second and later years in this species, again always in association with axial canals.

The earliest stages of axial resin canal formation were examined by Werker and Fahn (1969) in *Pinus halepensis*. In first-year growth rings, canal formation began shortly after the first periclinal divisions of the newly formed cambium. The canal initials, usually four in number, began to enlarge while the tracheid initials were still tangentially flattened. As these canal cells continued to enlarge, an intercellular space gradually formed. Later, the number of canal cells increased to six or

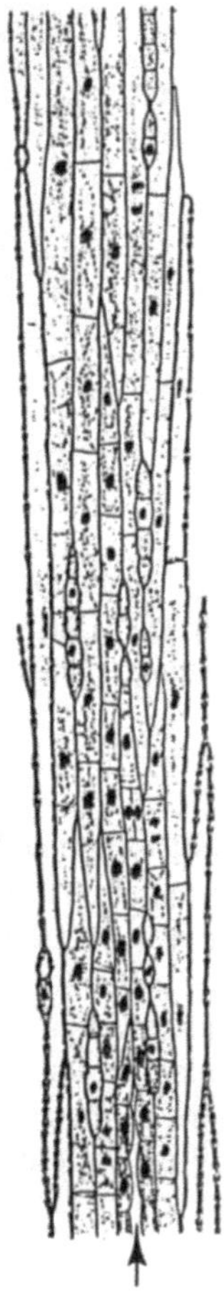

Fig. 9.11. Resin canal formation in *Pinus strobus*. Tangential section cut on bias. The upper segment, very close to the cambial zone, shows early stages in the partitioning of fusiform elements. The lower segment, deeper in the xylem, shows later stages of partitioning and first appearance of the canal (*arrow*). (Bannan 1936b)

more and the intercellular space expanded to contact all of them. By dividing tangentially, the axial canals increased in number, with some canals continuing upward in the stem and others diverging to serve lateral branches. The inner circle of canals was supplemented by other axial canals scattered throughout the first-year growth ring (Fig. 9.8B), a distributional pattern unique to the first ring. In subsequent growth rings no connections were observed between axial canals situated at different levels in the stem. Also, rather than being scattered throughout the growth ring, the axial canals tended to concentrate in the last-formed earlywood and first-formed latewood.

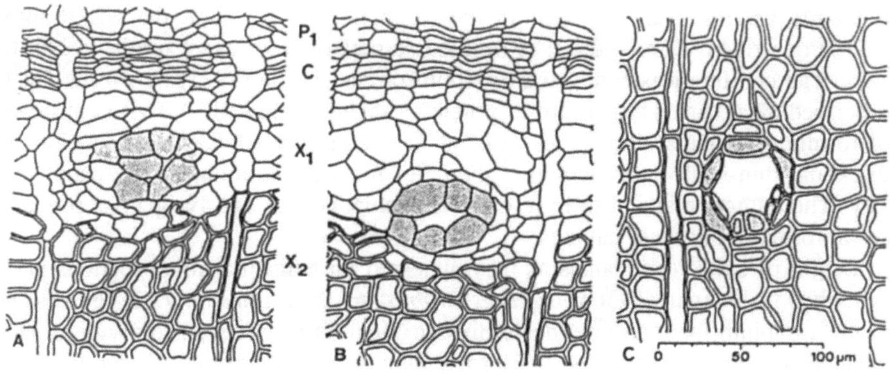

Fig. 9.12A-C. Three stages in the differentiation of a axial resin canal in transverse sections of *Larix decidua*. Epithelial cells are shaded. Double-lined epithelial cells in **C** are thick-walled and lignified. P_1 Differentiating phloem; **C** cambium; X_1 and X_2, differentiating and differentiated xylem, respectively. (Hug 1979)

Axial Canal Origin: Several views as to the origin of secondary resin canals prevailed among early workers. According to Sanio (1873), Karsten and Wigand had asssumed that resin canals arose by liquifaction and transformation of cell membranes. Frank (1867) and Müller (1867), on the contrary, showed that they arose by the separation of cells while Dipple (1863) invoked both phenomena. Frank (1867) claimed that the axial resin canals in *Pinus sylvestris* arose from a single radial series of cambial cells, one of which expanded and then divided; Mayr (1884) later confirmed this type of origin in *Larix* and *Picea*. Further divisions gave rise to a rosette of four daughter cells, which then separated to form the canal. Müller (1867) described the process in *Picea abies* in much the same way, with the exception that two radial series of cambial cells, rather than one, provided the mother cells. After separating to produce the duct cavity, the four daughter cells forming the rosette divided in various orientations as the canal enlarged. Sanio (1873) demonstrated that both Frank and Müller were correct in describing the process of canal formation, although Frank erred in assuming that a single cambial cell was involved.

In his own investigations of *Pinus sylvestris*, Sanio (1873) showed that the first longitudinal divisions occurred in two adjacent cambial cells (Fig. 9.10A). These divisions might occur in various configurations. By cambial cell, Sanio presumably meant immediate derivatives, not the initial cell. Transverse divisions immediately followed these first longitudinal divisions, producing parenchymatous cells that appeared as elongated rectangles in radial section. Formation of the canal cavity was by cell separation, and it was believed to result from increased growth and tangential expansion of the wood ring (Fig. 9.10B). Sanio did not investigate further development of the resin canal system.

Numerous later investigators verified that resin canals, both of normal and traumatic origins, were initiated in the cambial zone in much the way Sanio (1873) described. One of the first detailed studies of resin canal origin in North American conifers was conducted by Thomson and Sifton (1925) on *Picea*. They described axial canals arising from "potential tracheary elements." That is, instead of forming tracheids, derivatives of cambial initials formed "septate tracheids or thick and thin-walled parenchyma," the latter forming the lining of the canal cavity. The authors claimed to have observed schizogenous and lysigenous processes in formation of these cavities.

Origin of axial resin canals as they appeared on tangential microsections was described by Bannan (1936b) in *Pinus strobus*. As shown in Fig. 9.11, cells close to the cambium that had previously divided longitudinally began to divide transversely. Segmentation of these fusiform elements proceeded rapidly providing the epithelial cells that lined the duct cavity as well as the surrounding parenchymatous layers. The duct cavity formed schizogenously. According to this view, an axial canal developed both acropetally and basipetally from its point of origin.

Hug (1979) also described how differentiation of axial canals began with transverse divisions of strand parenchyma cells which were, in turn, derived from cambial initials in *Larix decidua* (Fig 9.12). The epithelial cells continued dividing post-cambially, both longitudinally and transversely, to form the lining of the canal. The canal opening was formed by schizogenous separation of the epithelial

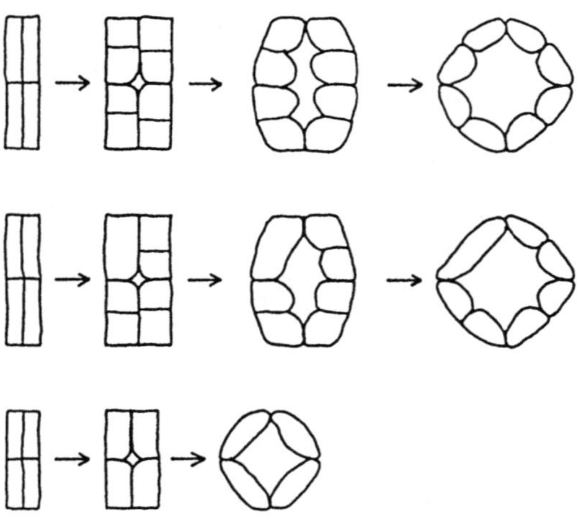

Fig. 9.13. Schematic, transverse drawings of three ways in which "epithelial initials" might divide to produce a resin canal in *Larix leptolepis*. (Sato and Ishida 1983b)

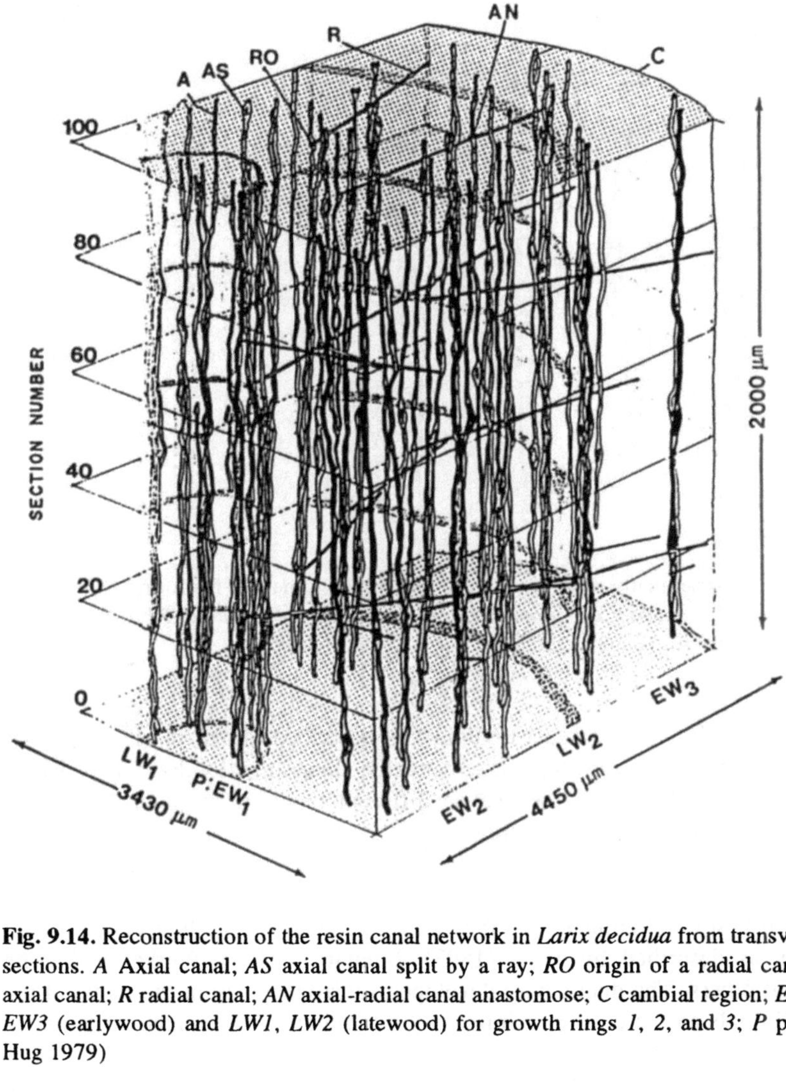

Fig. 9.14. Reconstruction of the resin canal network in *Larix decidua* from transverse serial sections. *A* Axial canal; *AS* axial canal split by a ray; *RO* origin of a radial canal from a axial canal; *R* radial canal; *AN* axial-radial canal anastomose; *C* cambial region; *EW1*, *EW2*, *EW3* (earlywood) and *LW1*, *LW2* (latewood) for growth rings *1*, *2*, and *3*; *P* pith. (After Hug 1979)

cells. Thus, the initiation of resin canals from cambial derivatives in the secondary body is essentially the same as their initiation from procambial derivatives in the primary body.

In addition to the epithelial cells that line the duct cavity, an unusual tracheid-type element has recently been reported in branches of *Pinus aristata* by Süss and Schirarend (1990). These tracheid-like elements, with ring-shaped and spiral thickenings, were interspersed among the parenchymatous epithelial cells. Like the epithelial cells, the tracheid-like elements arose very close to the cambial initials, presumably from xylem mother cells.

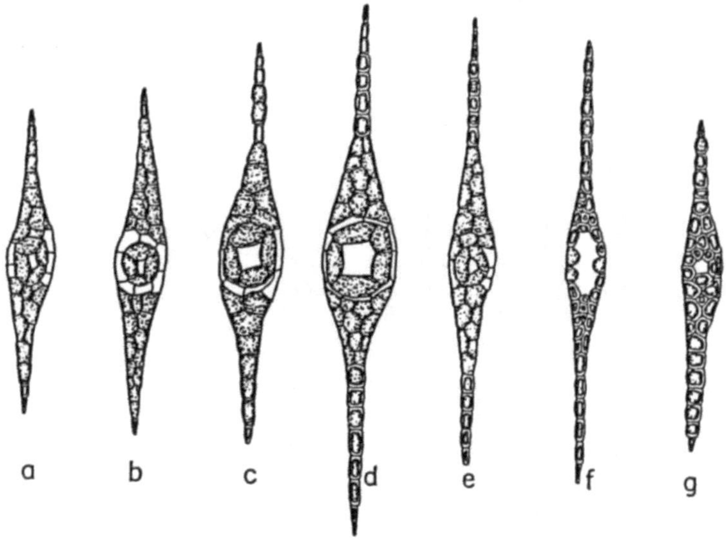

Fig. 9.15a-g. Diagrammatic representation of radial resin canals in fusiform rays of: **a** *Pinus resinosa*; **b** *P. contorta*; **c** *P. ponderosa*; **d** *P. strobus*; **e** *P aristata*; **f** *Picea glauca*; **g** *Pseudotsuga menziesii*. (Bannan 1965b)

Axial resin canals are initiated essentially as described above in all coniferous species. A typical process was schematically illustrated by Sato and Ishida (1983b). These authors described seven different types of origin in *Larix leptolepis* that differed only in the number and arrangement of the precursor cells. The simplest type, shown in Fig. 9.13, involved four so-called "epithelial initials" in two radial files. The epithelial initials were derived from cambial derivatives that divided first longitudinally and then transversely to form a four-celled group, or rosette, as described by Mayr (1893), Bannan (1936b), Werker and Fahn (1969), and others. The duct cavity developed schizogenously in the midst of the epithelial initials and in close proximity to the cambium. Subsequent variations in canal size and configuration depended on the number and orientation of divisions in the epithelial initials and surrounding parenchyma cells as shown in Fig. 9.13. Because differentiation took place either in immediate derivatives of cambial initials or in xylem mother cells, it was considered to be post-cambial. A recently formed axial canal initiated in xylem mother cells in close proximity to the cambium in *Picea abies* is shown in Fig. 9.21C.

Axial resin canals normally do not occur in the secondary phloem (Mayr 1884, Thomson and Sifton 1925, Chattaway 1951a, Alfieri and Evert 1968, Werker and Fahn 1969). Discussions concerning the ultrastructure of resin canal cells can be found in Fahn and Benayoun (1976), Fahn (1979), and Cheniclet (1987).

Axial Canal Lengths: As a rule, axial resin canals are relatively short (von Tubeuf 1918, Bannan 1936b). For example, in one seedling of *Larix decidua* Hug (1979) found that 80% of the axial canals were less than 0.5 cm long and in another seedling 65% were less than 1.0 cm long. In the upper internode of a mature tree, 35% of the canals were less than 0.5 cm long, whereas in the xylem of the mature trunk only a few were less than 10 cm long. Similar positional differences were observed by Mayr (1893). He found that canals in the upper and lower stems of *Picea* averaged 40 cm and 70 cm, respectively, and in *Abies* 15 cm and 30 cm, respectively.

The shortest canals occurred in the vicinity of branches. Canals in branches of various ages from *Pinus halepensis* trees averaged 4 to 10 cm in length (Werker and Fahn 1969). Axial canal lengths also increased with age of the tree. In *Pinus contorta*, Reid and Watson (1966) calculated an average canal length of 5.2 cm in a 30-year-old compared to 18.1 cm in a 75-year-old tree. However, traumatic resin canals induced in this species by the bark beetle *Dendroctonus ponderosae* often extended 90 to 120 cm above and below the site of injury (Reid et al. 1967).

Axial resin canal lengths were measured in several *Pinus* species by LaPasha and Wheeler (1990). Fresh stem bolts were peeled, the surfaces scraped, and canal lengths determined directly from the stem surfaces. For *P. taeda* the average canal length for a large number of samples was 10.5 cm with a range of approximately 5 to 20 cm. Mean values for 10- and 20-year-old *P. elliottii* trees were 20.7 and 49.8 cm, respectively. Mean values for single stems of *P. contorta*, *P. monticola*, and *P. ponderosa* were 7.5, 18.5, and 5.9 cm, respectively.

Although axial canals often end blindly, they do not necessarily end abruptly. The ends of axial canals in *Larix decidua* were characterized by the small diameters and triangular forms of their epithelial cells (Bosshard 1976b).

Axial Canal Meanders: Axial canals, although relatively short, do not always develop in a strictly vertical direction. Rather, they often migrate or meander both tangentially and radially, but not necessarily simultaneously. As noted earlier, Werker and Fahn (1969) described how axial canals dichotomized to give rise to new axial canals in first-year growth rings of *Pinus halepensis*. Because of their relatively short lengths and scattered distributions, tangential mergers of axial canals in older growth rings, while not frequent, do exist. The most commonly observed phenomenon was the bifurcation of an axial canal and the almost immediate reunification of its parts (Hug 1979) (Fig. 9.14). Some of these temporary separations occurred without apparent cause, but in many others the axial canal was pierced by a ray.

An axial canal commonly contacts one or more uniseriate and/or fusiform rays in its vertical traverse. This contact can occur in several different ways (Thomson and Sifton 1925, Turner et al. 1979, Sato and Ishida 1982, LaPasha and Wheeler 1990). Hug (1979), for example, found that 55% of the axial canals ended in a ray, 11% in another axial canal, 9% in both (presumably as in Fig. 9.20), and 25% in parenchymatous tissues. Some reports suggest that axial canals actually bend to

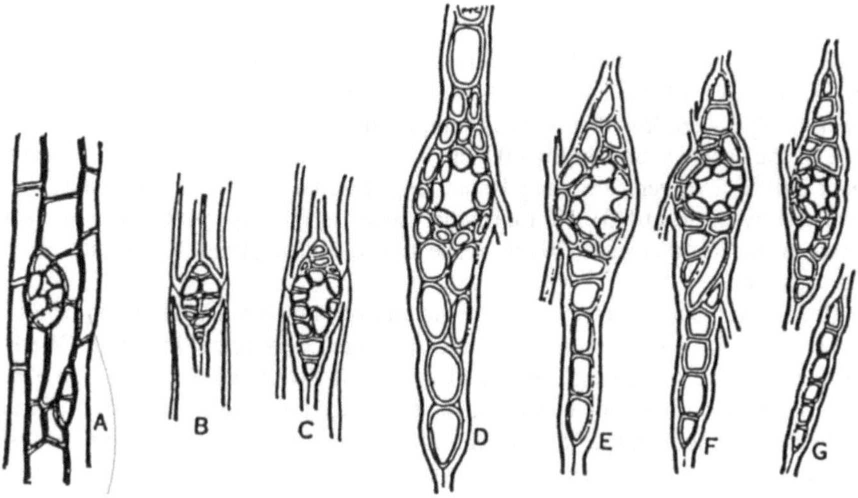

Fig. 9.16A-G. Tangential sections illustrating origin of a radial resin canal in *Pinus radiata*. See text for details. (Chattaway 1951a)

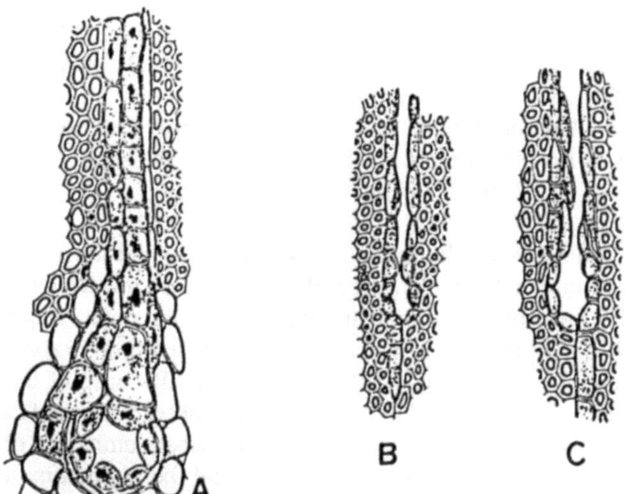

Fig. 9.17A-C. Transverse sections showing origin of radial resin canals in association with axial canals. **A** *Pinus radiata*; **B** *Picea smithiana*; **C** *Pseudotsuga menziesii*. (Chattaway 1951a)

intercept a ray. More likely, the axial canal simply intercepted the ray in its vertical meander. Werker and Fahn (1969) suggested that in some rare instances the radial canal might have been present before the axial canal formed.

Tangential meanders suggest that different radial files of xylem mother cells are involved in the formation of an axial resin canal. Evidence indicates that axial canals develop bidirectionally from a point of origin (Mayr 1893, Bannan 1936b, Reid and Watson 1966, Hug 1979), which means that a tangential meander would require the step-wise stimulation of contiguous cambial zone cells.

Radial meanders also present some interesting growth relationships involving the cambium. Mayr (1893), for example, observed in *Pinus sylvestris* that the central part of a canal was embedded more deeply in the xylem than its ends. This unusual orientation suggested that the canal developed bidirectionally, with the extremities forming last. The ends terminated blindly in a group of parenchymatous cells and they never crossed a growth ring boundary. Such a developmental pattern meant that cambial zone cells at different levels in the stem and at different times were activated to divide and differentiate as canal precursor cells.

Reid and Watson (1966) observed similar canal formation in *Pinus contorta* as well as the reverse condition. That is, the midsection of the canal was situated at the periphery of the growth ring and both extremities were buried 30 to 40 cells deeper in the xylem. Such an orientation suggests that the ends of the canal formed well in advance of the midsection. It is difficult to visualize a system whereby these undirectionally developing ends might join, considering the normal processes of resin canal formation and xylem cell differentiation.

A somewhat different pattern was described by Hug (1979) in *Larix decidua*. Unlike Mayr (1893), Hug found that canals situated at the growth ring boundary were not confined to a single growth ring. That is, canals that were initiated in one growth ring often completed their development in the next growth ring. This meant that the midsection of the axial canal was a product of cambial zone cells in the latewood of one ring, whereas the extremities were a product of the earlywood of the succeeding ring. Bannan (1936b) reported similar cross-overs of growth ring boundaries for traumatic resin canals in *Pinus strobus*.

Axial Canal Distribution: Axial resin canals tend to be more numerous per unit transverse area in young growth rings near the pith as opposed to old ones, especially in *Pinus* (Mayr 1893, Reid and Watson 1966). Yet as Hug (1979) found in *Larix decidua*, the wood volume occupied by axial canals was relatively small, ranging from only 0.18 to 0.36% in the young growth rings. Although most investigators have reported the presence of axial canals in first-year growth rings, their distribution in rings near the pith can be sporadic as noted by Thomson and Sifton (1925).

When present, axial canals are more or less uniformly dispersed in growth rings nearest the pith (Pauly 1963, Werker and Fahn 1969, Hug 1979). In growth rings formed later, the canals tend to concentrate either in the region of the earlywood-latewood transition (Chalk 1927, Rees 1929, Pauly 1963, Werker and Fahn 1969,

Fahn 1979), or in the latewood (Anderson 1896, Hobert 1913, von Tubeuf 1918, Reid and Watson 1966, Stephan 1967, Alfieri and Evert 1968, Zamski 1972, Sato and Ishida 1983a). Some workers could discern no consistent pattern (Jaccard 1939, Hug 1979). As a general statement for all conifers, Greguss (1955) concluded that axial canals were usually concentrated in the latewood, but in some situations they were dispersed more uniformly within the growth ring. The latter condition would be more likely to occur in wide growth rings of fast-growing trees. A review of the pertinent literature leaves one with the impression that axial canals are more likely to be confined to the latewood in *Pinus* than in *Picea* and *Larix*. In all conifers, this tendency increases with increasing growth ring age from the pith, but it can be modified toward greater dispersal with increasing vigor of the tree.

The resin canal system of roots, which has been less thoroughly studied, is in general the same as that in the above-ground stem. The secondary ducts of the stem and the root comprise a continuous, interconnected system (Werker and Fahn 1969).

Variation with Growth Rate: Attempts have been made to correlate the formation of axial resin canals with rate of growth, or ring width. Mayr (1893) outlined the many sources of variation and the difficulties involved in such a determination. However, over the years, investigators conducting quantitative studies arrived at either positive (Hobert 1913, von Tubeuf 1918, Bannan 1933a, 1936b, Wagg 1948, Mergen and Echols 1955, Fahn and Zamski 1970, Zamski 1972), negative (Thomson and Sifton 1925, Reid and Watson 1966), or inconclusive (Jaccard 1939) results.

Wagg (1948) claimed that the number of axial resin canals in *Pseudotsuga menziesii* formed a straight-line relationship with growth increment. According to Stephan (1967), the number of axial resin canals increased with increasing ring width in *Pinus sylvestris*, but the density of the canals per cm^2 of xylem surface was greater in trees with narrow growth rings. Although Reid and Watson (1966) could detect no relationship between the number of axial canals and ring width, they did observe that the number of canals appeared to vary in alternate seasons. This biannual pattern suggested to them that canals formed in one year might exert some influence on those formed the next season (Chap. 6.5.6).

Growth rings containing compression wood as a rule are wider than normal growth rings. After summarizing the available literature, Timell (1986) concluded that resin canals were of less frequent occurrence in compression wood than in normal wood, a view recently confirmed by Chung and Lee (1989) in *Pinus densiflora*. From the evidence available, it appears that a correlation exists between the number of axial resin canals and ring width in normal wood. It is not known whether a similar correlation exists in compression wood.

9.2.2.3 Secondary Resin Canals - Radial

Radial resin canals in conifers always occur in the midst of an elongated ray; hence the name fusiform ray. As pointed out long ago by Hartig (1848), these are

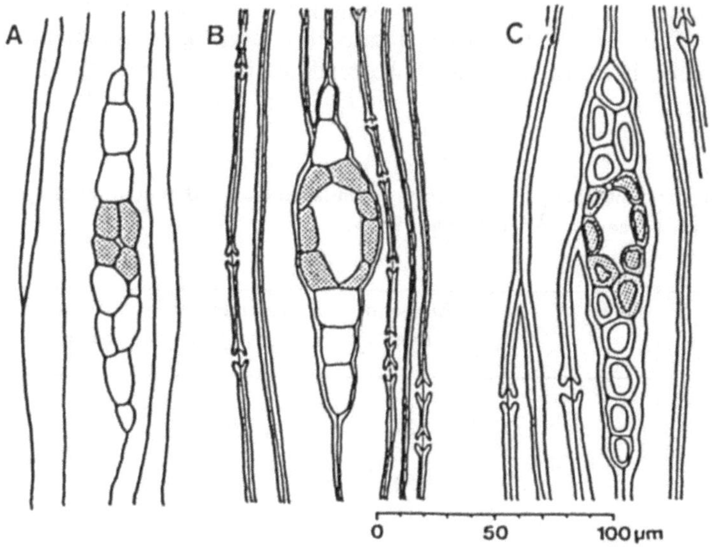

Fig. 9.18A-C. Differentiation of radial resin canals during three stages of development in *Larix decidua*. Epithelial cells *shaded*. Double-lined cells in **C** are thick-walled and lignified. (Hug 1979)

the only broad rays that normally occur in most conifers. Fusiform rays are present to varying extents in species of *Larix*, *Picea*, *Pinus*, and *Pseudotsuga* (Fig. 9.15), and absent in the normal wood of other conifers. Radial canals never arise independently, but always in association with an axial canal (Mayr 1893, Thomson and Sifton 1925, Bannan 1936b, Pauly 1963, Werker and Fahn 1969, Fahn et al. 1979, Sato and Ishida 1982, 1983b). Savidge and Farrar (1984) observed that traumatic radial canals did not always originate in association with axial canals in *Pinus contorta*. These are the only authors that I am aware of who have made such a claim, although Werker and Fahn (1969) suggested such a possibility for *P. halepensis*. The unusual response obtained by Savidge and Farrar might be related to the diagonal bridge they employed to induce traumatic resin canal formation.

Radial Canal Origin: Chattaway (1951a) investigated the earliest stages of fusiform ray initiation in *Pinus radiata*. By examining tangential sections from the pith outwards in young stems, she was able to demonstrate that the "primary medullary rays" developed from "loosely organized masses of parenchyma." These masses gradually subdivided into smaller, more compact masses that soon became recognizable as the small uniseriate rays characteristic of secondary wood (Fig. 9.16A). The radial resin canals originated near, but not in, the pith. They were not present in the parenchymatous tissue mass of primary medullary rays, but began to appear very soon after the secondary rays became recognizable (Fig. 9.16B,C).

During these later stages of development, the loosely arranged ray became more compact (Fig. 9.16D,E), and in the particular case illustrated, the ray subdivided to form a new uniseriate ray from the lower end of the fusiform ray (Fig. 9.16F,G).

Table 9.2. Frequency and total height of fusiform rays in the outer wood of coniferous species in trees 25-50 cm in diameter (After Bannan 1965b)

Species	Number of trees	Number of rays (cm^2)	Total ray height (μm)
Larix laricina	17	66	470
Picea engelmannii	26	58	385
Picea glauca	54	65	415
Picea mariana	18	62	365
Pinus aristata	10	54	475
Pinus banksiana	25	54	320
Pinus contorta	46	42	335
Pinus edulis	10	150	450
Pinus flexilis	11	33	520
Pinus lambertiana	16	53	600
Pinus ponderosa	39	41	420
Pinus resinosa	51	48	295
Pinus strobus	46	35	560
Pseudotsuga menziesii	35	75	320

According to Chattaway (1951a), new fusiform rays invariably arose in connection with an axial resin canal. That is, a fusiform ray formed when the initial cells of a uniseriate ray became contiguous with the initial cells that formed a axial resin canal (Fig. 9.17A). Similar fusiform ray origins were observed in *Picea smithiana* and *Pseudotsuga menziesii* (Fig. 9.17B,C). Werker and Fahn (1969) included in their paper a photomicrograph showing a very early stage of the axial-radial canal complex in *Pinus halepensis*. In this species, contact formed between the lumina of the axial and radial canals when the epithelial cells on one side of the radial canal separated and merged with those cells bordering a similar opening in the axial canal.

Like Chattaway (1951a), Hug (1979) found that all radial resin canals in *Larix decidua* were initiated in conjunction with axial canals. Radial canals arose by periclinal divisions of ray cell initials, derivatives of which formed a small group of epithelial cells in a biseriate ray that might vary from 5 to 11 cells in height (Fig. 9.18). These cells, called "epithelial cell initials," continued dividing post-cambi-

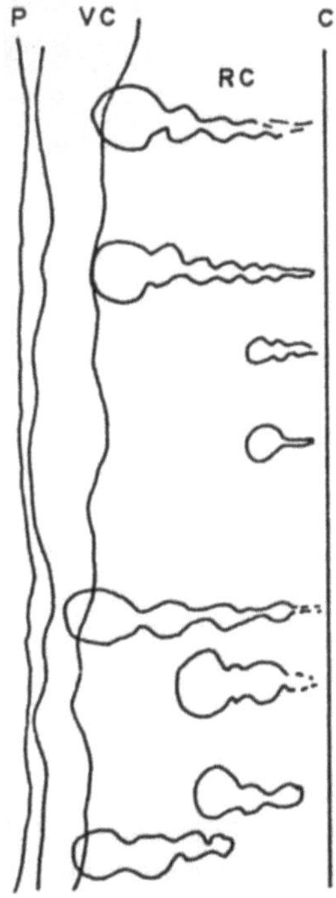

Fig. 9.19. Diagrams of radial resin canals (*RC*) in the phloem of *Picea glauca* showing development of bulb-like cavities. *C* Cambium; *VC* vertical resin canal of primary cortex; *P* periderm. (After Thomson and Sifton 1925)

ally to increase the canal surface area. The canal itself formed schizogenously. In the mature canal the epithelial cells were thick-walled and lignified, but, like those lining the axial canal, they retained living nuclei.

No radial canals were found in the first-year growth rings of *L. decidua*, but many were initiated in the second-year ring. Hug (1979) speculated that ray heights in the first-year ring might have been below the five-cell minimum for canal formation. However, radial canals formed rapidly in the second-year ring. In the innermost growth rings they were often ten times as numerous as the axial canals, suggesting that each axial canal was associated with several to many radial canals at different levels in its vertical traverse.

Bannan (1936b) had advanced the idea that every resin canal was initiated as a reaction to a wound stimulus. Hug (1979) suggested that such a stimulus might be present in *Larix decidua* because the arrangement of axial and radial canal systems resembled in many respects the wound callus tissue described by Kucera (1972) in *Taxus baccata*. Furthermore, Hug suggested that the stimulus for canal formation might reside in derivatives of the fusiform cambial initials because

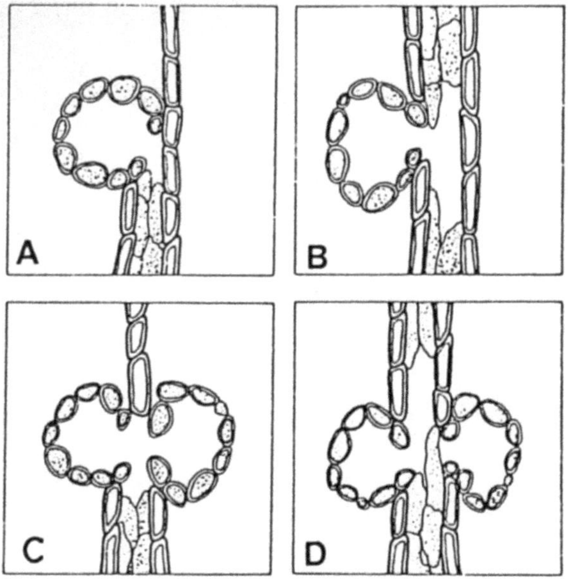

Fig. 9.20A-D. Four types of resin canal anastomoses in *Larix decidua, Picea abies,* and *Pinus sylvestris* as viewed on transverse sections. Cambium toward the bottom. Uniseriate (**A,C**) and multiseriate (**B,D**) rays either intersect (**A,B**) or bisect (**C,D**) an axial resin canal. In **A** and **C** the uniseriate ray is transformed to a multiseriate fusiform ray at the anastomosis, and in **B** and **D** the multiseriate fusiform ray is continuous through the anastomosis. (Bosshard and Hug 1980)

radial canals always arose in association with axial canals. As to the source of the stimulus, he suggested that lateral tension exerted on the cambium might mimic a wound stimulus. That is, new cells continue to differentiate as epithelial cells as long as the associated cambial cells divide periclinally.

Bosshard and Bangerter (1984) found that, unlike the axial canals that arise from xylem mother cells, radial resin canals arise directly from ray cell initials and continue to extend with radial growth. A similar observation had been made earlier by Werker and Fahn (1969). These authors advanced the hypothesis that the stimulus for radial canal formation first affects the ray cell initials. That is, the stimulus which affects the ray cell initials is conducted horizontally inward by the ray. It then spreads vertically for a certain distance in the fusiform xylem mother cells causing them to differentiate to duct cells.

Radial Canal Lengths: Once initiated, radial canals usually continue indefinitely from one growth ring to another through the xylem. Unlike axial canals, they seldom meander from their straight-line traverse. During passage through the xylem, a radial canal might come in contact with several to many axial canals (Werker and Fahn 1969, Hug 1979).

Radial Canal Distribution: The number of fusiform rays that occur in the xylem of different conifers is quite variable. As a general average for all conifers, Brown and Panshin (1940) estimated the ratio of fusiform to uniseriate rays to be 1:20. Kärkkäinen (1973) derived a ratio of 1:40 for his counts in *Pinus sylvestris*, and Jaccard (1939) about 1:35 in *Larix*. Nyrén and Back (1959) found that fusiform rays comprised only 0.3% of the wood volume and less than 10% of the ray volume in *Pinus sylvestris*. In *Larix decidua*, Hug (1979) estimated that fusiform rays comprised about 0.9% of the ray volume in young growth rings or about 0.53 rays/mm^2 of tangential surface. Bannan (1965b) compiled data on both the frequency of occurrence and total heights of fusiform rays in some North American conifers (Table 9.2). In this publication, he also presented data on the variation in fusiform ray heights with rate of growth and growing region.

Radial resin canals are usually more numerous than axial canals in normal wood (Hug 1979). Jaccard (1939) claimed that the radial canals outnumbered axial canals 20:1 in the innermost growth rings and 3:1 in the older rings of *Larix*. Wagg (1948) reported that radial canals were many times more numerous in *Pseudotsuga menziesii*, but no data were provided in the cited abstract. In *Pinus koraiensis*, radial resin canals were more numerous in compression wood than in opposite wood (Lee and Eom 1988).

The number of radial resin canals per unit of tangential surface appears to decrease with growth ring age (Mayr 1893). According to Mergen and Echols (1955), the number of radial canals in growth ring five was twice that in growth ring 25 in *Pinus elliottii*. Similarly, in *Pinus taeda*, De Angelis et al. (1986) reported that radial canals occupied approximately 0.16% of the tangential wood surface of inner growth rings near the pith, but this value decreased logarithmically with age to 0.08% in the older growth rings of 27-to 77-year-old trees. Takizawa et al. (1980), on the contrary, observed an increasing trend in the number of radial resin canals from pith to bark. The number of epithelial cells lining the duct cavity, which affects the diameter of the radial canal, was shown by von Diannelidis (1953) to increase with growth ring age from the pith in *Larix* and *Picea*.

As in the axial canals, reports on the relation between number of radial resin canals and growth ring width are variable and inconclusive. Some workers have found a positive relationship (Mergen and Echols 1955, De Angelis et al. 1986) and others a negative one (Hobert 1913, Jaccard 1939).

Radial Canals in Phloem: Although the radial canals are normally closed at the cambium, they again appear open in the phloem. In contrast to the xylem part of a canal, the phloem part in *Larix decidua* was lined with parenchyma cells that arose by post-cambial divisions of phloem ray-cell derivatives (Hug 1979). The

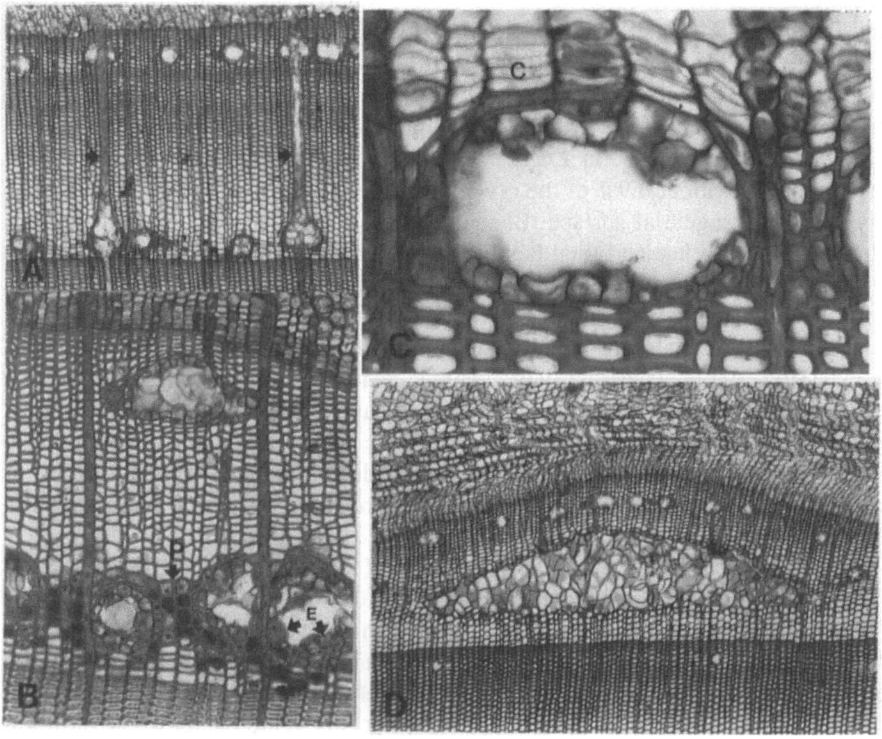

Fig. 9.21A-D. Traumatic resin canals in *Picea abies*. **A** Tangential series of resin canals in earlywood and latewood of a single growth ring. Two radial ducts (*arrows*) intersect axial ducts. 23x. **B** A similar series showing duct structure. *P* Strand parenchyma; *E* epithelial cells. 58x. **C** A recently formed duct in close proximity to the cambium (*C*). 232x. **D** A traumatic resin pocket. (Lo 1987)

phloem canals also formed schizogenously, but they were short and terminated at varying distances within the phloem tissues.

Thomson and Sifton (1925) described how the radial canals in *Picea* produced a series of spherical, bulb-like chambers (Fig. 9.19). Each chamber increased in size outward from the cambium by cell divisions that not only increased the size of the cavity but also the number of cell layers that enclosed it. The oldest chambers, situated farthest outward in the cortex, were often enclosed by six or more layers of thin-walled, compact cells. The authors demonstrated how transverse sections through these bulb-like cavities had been misinterpreted by earlier workers as axial, cortical canals. Similar structures were observed by Hanes (1927), Pauly (1963), and Werker and Fahn (1969) in the phloem rays of *Pinus australis*, *P. pinaster*, and *P. halepensis*, respectively.

Xylem-Phloem Continuity: The condition of the radial resin canals in passing through the cambial region was difficult to determine by early workers because of the tender cells in this region. Thus, the question was often raised whether the ducts were open or closed. Numerous observations have demonstrated that they are in fact closed (Mayr 1884, Thomson and Sifton 1925, Werker and Fahn 1969, Hug 1979). According to Thomson and Sifton (1925), the cambial part of a fusiform ray appeared as an ordinary biseriate ray. However, according to Wodzicki and Brown (1973), in those genera that normally produce thick-walled epithelial cells in the xylem - *Larix, Picea, Pseudotsuga* - the resin canals were open across the cambial zone "permitting free movement of resin between the xylem and phloem portion of existing canals." In contrast, the resin canals of *Pinus* with thin-walled epithelial cells were always closed in the cambial initial region. In an unconfirmed observation not mentioned by others, Thomson and Sifton (1925) claimed that the radial canals in roots of *Picea*, as opposed to those in stems, were open across the cambium.

9.2.2.4 Axial-Radial Canal Anastomoses

Over the years, the question has often been raised as to whether or not the radial and vertical resin canal systems actually formed an anastomosing system, i.e., a true inosculation of the duct cavities. Some workers maintained that although the two canals might merge, the duct cavities between them remained closed. For example, Thomson and Sifton (1925) expressed doubt as to the existence of true anastomoses of duct cavities. They always observed a separating layer(s) of thin-walled epithelial cells. More recently, Turner et al. (1979) could find no evidence of direct communication between axial and radial ducts in *Picea abies*. They claimed that no one had ever convincingly demonstrated actual anastomoses.

On the contrary, it has been generally accepted that open communication exists between the two systems (Greguss 1955). While some workers have simply described these anastomoses, others have provided either camera lucida drawings and/or photomicrographs (Chattaway 1951a, Pauly 1963, Werker and Fahn 1969, Hug 1979, Bosshard and Hug 1980, Sato and Ishida 1982, LaPasha and Wheeler 1990) (Fig. 9.17). Bosshard and Hug (1980) not only confirmed the presence of open anastomoses, but described four different types shown in Fig. 9.20. These anastomoses resulted in a fully open network of resin canals in the secondary xylem. The network was described as two-dimensional because the axial canals intersected radial canals but not other axial canals. Axial canals rarely anastomose among themselves, with the exception of traumatic canals situated in closely contiguous tangential bands (Bosshard and Bangerter 1984).

Open anastomoses between axial and radial resin canals are extremely difficult to detect, and those who denied their existence might simply have overlooked them in their preparations. According to Werker and Fahn (1969), a radial canal coming in contact with an axial canal might be subject to one of three possibilities. First, the radial canal might pass between the axial canal cells without either forming an open union or undergoing a noticeable change in its own cells. Secondly, the radial canal might split the axial canal, again without forming an open union. However, because of cellular changes, in this case the two canals would be separated by only one to several layers of thin-walled epithelial cells. Finally, contact between the two canals would result in a complete anastomosis. Similar radial-axial canal contacts were described and illustrated by Sato and Ishida (1982).

9.2.2.5 Traumatic Resin Canals

Traumatic, or pathological resin canals, are produced in response to cambial injury (Fig. 9.21). Hanes (1927) was unable to induce traumatic resin canals in conifer seedlings and concluded that they did not occur in primary resin canal systems. However, secondary traumatic canals can be induced in many genera that do not normally produce resin canals as well as in those that do. Research has shown that they are initiated in much the same way as normal canals.

Because of the wide variation in response to wounding among genera, Penhallow (1907) differentiated between resin canals and cysts. The former term was applied to the scattered tube-like structures and the latter to the more sac-like structures that occurred in compact masses. The two types intergraded not only when different species were compared, but also in response to different types of wounds within a species. It was for these reasons that Bannan (1936b) adopted the term resin duct for all traumatically formed structures in conifers. Greguss (1955) differentiated between resin ducts and resin cysts in that the latter did not "conduct resin," only stored it. Years earlier, Dippel (1863) had made reference to "resin reservoirs" in *Abies alba* as opposed to resin canals in *Larix*, *Picea*, and *Pinus*. Resin cysts are more commonly found in species that do not normally produce resin canals.

Initiation and development of traumatic resin canals comprised a part of the investigation of *Picea* by Thomson and Sifton (1925). These workers found that pressure wounds to stems of seedling plants, either by squeezing or pinching with spring-clip clothes pegs, induced tangential bands of axial canals. These canals were produced when true canals branched and fused vertically, forming an irregular series of resin cysts. If the cambium was not severely injured, three or four tangential bands of axial canals might be produced. Interestingly, uniseriate rays penetrated these bands without change. That is, despite the presence of bands of traumatic resin canals, the rays continued their radial traverse relatively undisturbed.

Thomson and Sifton (1925) also compared artificially induced and naturally occurring resin canals in seedlings and saplings. In almost every case they were

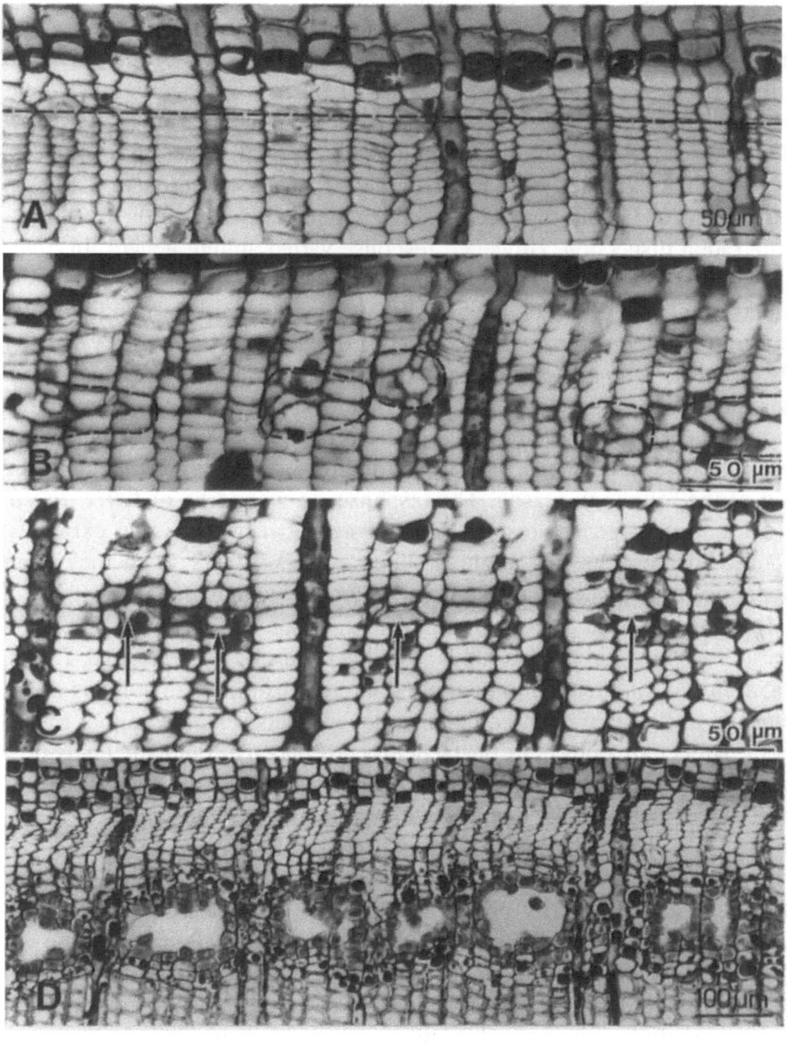

Fig. 9.22A-D. Transverse sections of *Tsuga sieboldii* showing formation of traumatic axial resin canals in days after wounding by inserting metal needles in the cambium. **A** Cambial zone before wounding. **B** Six days. Small intercellular spaces have formed among the swollen cells within the zone of xylem mother cells (*circled*). **C** Ten days. Intercellular spaces have enlarged (*arrows*) and epithelial cell division is underway. **D** Thirty-six days. Tangential band of mature, traumatic resin canals has formed. (Kuroda and Shimaji 1983)

able to relate the presence of resin canals to some type of prior injury. Many normal appearing canals were found to be the persistent ends of a traumatic series.

Based on these findings, the authors concluded that "... all resin canals in the secondary tissues of *Picea* are caused by injury to, or irritation of, the cambium."

Bannan (1933a,b, 1934b) extended the work of Thomson and Sifton (1925) to other conifers. He wounded stems and branches of *Larix larcina, Tsuga canadensis*, and several other conifers, either by small needle punctures or larger chisel wounds. Open wounds, in which the cambium was exposed, generally resulted in a series of ducts that extended both laterally and vertically from the wound area. Typically, this type of wound produced a single tangential series of ducts with relatively little radial dispersion. However, response varied among genera. In *Abies* and *Tsuga* (Bannan 1933a, 1934b), genera that did not normally produce resin canals, the ducts tended to be aggregated in cyst-like masses in the wound vicinity. The ducts were relatively short, and they rarely extended very far either vertically or laterally from the wound. In contrast, similar wounds in *Pinus* produced true resin canals, never cyst-like ducts. The canals were widely scattered, they extended great distances from the wound area both vertically and laterally, and those formed late in the season of one year might complete their development in the growth ring of the following season. The responses to wounding of *Larix, Picea, Pseudotsuga*, and other genera resided somewhere between these two extremes. Severe wounds in these species might produce cyst-like ducts when they were tangentially concentrated but canal-like farther from the wound area.

Cambial wounds induced by pressure, pinching, or some other means elicited a different response because such wounds remained closed. In *Abies* and *Tsuga*, ducts induced in this way tended to be more scattered than in open wounds and it was not unusual to find two or three radial series of ducts separated either by several rows of parenchymatous cells or tracheids. In *Pinus* the traumatic ducts produced in response to pressure were rarely tangentially aligned even at the center of injury. They were instead widely dispersed throughout the growth ring.

Bannan (1933a,b, 1934b) found that wounds caused by natural agents induced similar responses, depending on whether the wound was open or closed as well as other factors such as the severity of the wound, tree vigor, position on the tree, time of year, and age of the cambium at the time of wounding. Because of the variable responses to wounding, traumatic resin ducts might occur quite some distance from the wound area and appear isolated even in genera such as *Abies* and *Tsuga*. Such a distribution made it difficult to relate the presence of isolated or scattered resin ducts in older, internal growth rings to some past injury. Nonetheless, Bannan cautioned that the mere presence of isolated, unaccounted-for resin ducts did not warrant the conclusion that such ducts were of normal occurrence. Like Thomson and Sifton (1925), he found that nearly all vertical canals in seedlings and branch wood could be related to prior wounds. Many canals in older trees could be related to wounds, but for others it was difficult to establish such a connection because of the great lengths of the canals and their wide dispersal (Bannan 1933b). Similar precautions in establishing causality were cited by Nylinder (1951).

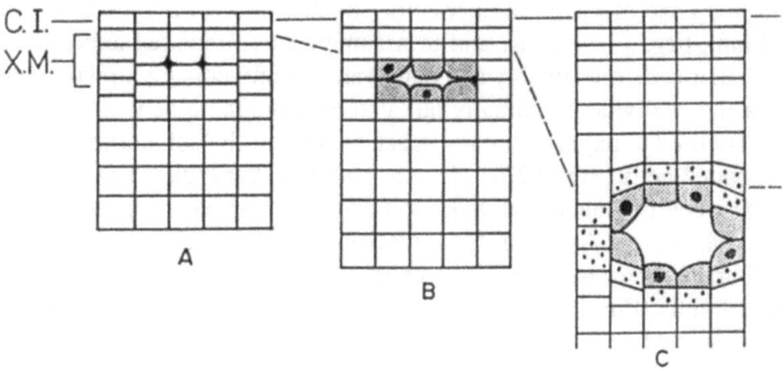

Fig. 9.23A-C. Interpretive diagram of Fig. 9.22 showing the process of traumatic resin canal formation in relation to the assumed site of the cambial initials at 3 (**A**), 10 (**B**), and 30 (**C**) days after wounding. *CI* Cambial initials (*solid lines*); *XM* xylem mother cells (*dashed lines*). (Kuroda 1986)

The distinction between normal and traumatic resin canals was also a matter of concern to Fahn and Zamski (1970). They, too, found it difficult to distinguish between the two types of canals in samples taken from wounded stems. Their experimental results showed that the same factors affected both normal canals and those formed after injury and, for this reason, they suggested that the term traumatic should be avoided. Although it is true that traumatic resin canals might not be correctly identified retrospectively, they could be identified when associated with recent injuries, for example insect attacks (Reid et al. 1967, Cerezke 1972). In a contrary report, Grosser (1986) observed extensive tangential bands of traumatic resin canals in very narrow growth rings of *Cedrus atlantica*. The resin canals occurred in successive bands with no apparent source of injury, suggesting some form of cambial stimulus. As a further anomaly, an above-average frequency of trabeculae originated within or near the traumatic parenchyma in which the resin canals were embedded.

Bosshard and co-workers (Bosshard 1976a, Bosshard and Bangerter 1984) related formation of traumatic resin canals to the concept of "functional tropism," i.e., the adaptation of structure to functional necessity. Traumatic resin canals conformed to this concept because in some cases the epithelial cells underwent hypertrophy, whereas in other cases the walls became lignified. These events were correlated with the completion of wound occlusion and the consequent reduction in metabolic rate to normal levels.

Development of traumatic resin canals was examined in *Tsuga sieboldii* by Kuroda and Shimaji (1983, 1984) and Kuroda (1986). Wounds were induced by inserting metal pins 250 μm in diameter into the stems and immediately withdrawing them (Chap. 10.7). The wound response was then studied by microscopic examination of tissue samples harvested at periodic intervals beginning with the unwounded cambium (Fig. 9.22A). Three days after wounding a group of four to six cells within the zone of xylem mother cells, situated three to five cells inward from the cambium, began to enlarge and to assume a more rounded appearance. These xylem mother cells presumably responded almost immediately to wounding not only by ceasing to divide periclinally but also by initiating a new direction of differentiation. By the 6th day, small gaps began to form at the cell corners within the group of rounded cells (Fig. 9.22B). By the 10th day, these gaps had enlarged and extended between the tangential walls to form the incipient resin canal. The cells around the gap were filled with dense cytoplasm and contained conspicuous nuclei. These characteristics were associated with septation of the cells destined to become epithelial cells (Fig. 9.22C). By the 20th day, the gaps had enlarged further and the thin-walled cells lining the canal contained resin-like droplets. The wound response was not confined to formation of a single traumatic resin canal, but to an irregular line of canals extending laterally about 0.5 cm on either side of the wound. By the 36th day, the tangential band of resin canals was mature (Fig. 9.22D). The process is depicted schematically in Fig. 9.23.

Some conifers, such as *Tsuga heterophylla* (Leney and Moore 1977), produce traumatic, axial resin canals but not radial ones. Other conifers that produce neither normal nor traumatic resin canals in the xylem might produce traumatic axial resin canals in the functional phloem (Lo 1985). For example, Yamanaka (1984) was able to induce traumatic resin canals in the phloem of certain species of *Chamaecyparis*, *Cryptomeria*, *Cunninghamia*, *Juniperus*, and *Thujopsis*. These canals arose schizogenously from phloem parenchyma cells in close proximity to the cambium, but not from the phloem mother cells (Yamanaka 1989).

As noted, traumatic resin canals can be induced by a large number of both physical and environmental factors that cause either disturbance or injury to the cambium. Three examples have been selected to illustrate the variable and often diverse occurrence of traumatic resin canals, namely, false rings, frost rings, and the effect of tapping for commercial gum production.

In False Rings: Disordered and widely dispersed tangential bands of traumatic resin canals are often found in association with false growth rings (Bailey 1925a). In *Picea*, Thomson and Sifton (1925) frequently observed the concurrent formation of false growth rings and tangential bands of axial resin canals emanating from wound sites. The wound response was usually localized in a tangential arc of tissue and, in the case of slight wounds, thick-walled resinous tracheids were produced instead of resin canals. Close examination showed that false rings were reliable indicators of concealed wounds. The authors concluded that the false rings and traumatic resin canals were correlated phenomena, both due to a common disturbance to the cambium. A similar relationship was reported by Bannan (1933b, 1936b) for *Larix*, *Picea*, and *Pinus*. In addition to the concealed wounds described by Thomson and Sifton (1925), false rings and bands of traumatic resin canals at times could be directly related to an obvious injury such as regrowth following destruction of a terminal shoot or bud. Cown (1973) suggested that the

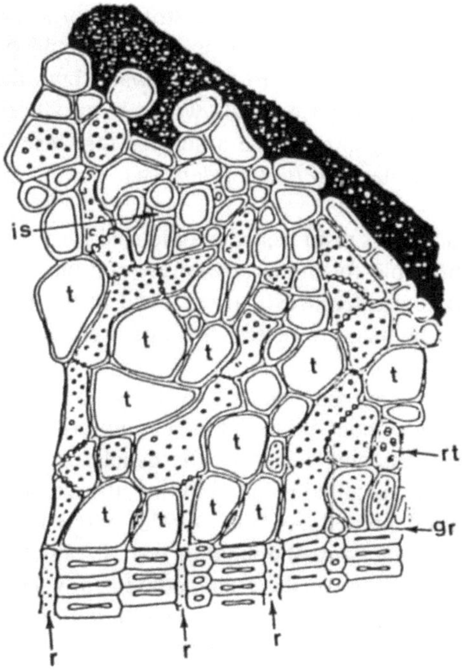

Fig. 9.24. Wound tissue formed in conjunction with a resin pocket in *Larix decidua* in transverse section. *gr* Growth ring boundary; is intercellular space; *m* ray; *rt* ray tracheid; *t* tracheid. (Frey-Wyssling 1938)

false rings containing traumatic resin canals in *Pinus radiata* were zones of weakness possibly associated with a period of water stress to the cambium.

In Frost Rings: A second factor that might cause wide dispersal of traumatic resin canals is frost injury to the cambium (Anderson 1896, Bannan 1933b). Although Mayr (1893) and Hartig (1895a) claimed that traumatic resin canals were often associated with frost rings, Rhoads (1923) disagreed, and suggested they were due to mechanical injury. Depending on the circumstances and extent of frost injury, the traumatic response might be either localized or extensive, i.e., it might extend around the entire circumference (Bailey 1925a, Bannan 1936b). In such cases the frost ring is often associated with false ring formation (Bailey 1925a). On the contrary, if the frost injury is slight or occurs during the dormant period, parenchymatous wound tissue entirely devoid of resin ducts might be produced (Bannan 1933b). Wound callus in frost rings of *Pinus rigida* and *P. virginiana* consisted of proliferation of uniseriate ray tissues and cell collapse (Hemenway 1926).

Monange (1961) examined several exotic species of *Abies* and *Tsuga* damaged by frost in the Jouéou Arboretum in France. Three types of reaction were observed. If frost injury was severe, cambial activity might be arrested for an entire growing season. If frost injury was moderate, resulting in the destruction of a few cambial initials, abnormal tissues and series of traumatic resin canals would be produced in the first-formed part of the growth ring; if frost injury was slight, the cambium might function normally but traumatic resin canals would be produced later in the season. Fahn and Zamski (1970) also observed a delayed response in resin canal formation in artifically wounded stems of *Pinus halepensis*.

Resin Extraction: The commercial extraction of resin has been practiced for many years and its importance has been responsible for much of the research on resin canal formation in conifers. Many pertinent references can be found in the previously cited papers and no attempt will be made to review the vast literature on the subject.

Resin extraction, or tapping, usually involves removal of a narrow strip of bark and/or wood extending in a shallow diagonal about 1/4 to 1/3 the circumference of the stem. This so-called "streaking" or "chipping" exposes the canals, causing them to exude resin. In one type of extraction, called wood-chipping, the streak is cut to a depth of about 0.6 to 1.2 cm into the wood. As Gerry (1923) pointed out, once resin tapping by this method is well underway, the yield of resin is obtained almost exclusively from the axial traumatic resin canals. This method is now considered obsolete, at least in the United States, and it has given way to bark-chipping. Bark-chipping exposes the wood surface and taps the radial resin canals. However, because the resin crystalizes rapidly, sulfuric acid must be sprayed on the freshly cut surface. The acid destroys the epithelial cells and prolongs resin flow for up to 2 weeks.

In both types of chipping the process is repeated at regular intervals during the growing season, with each new streak exposing fresh xylem upward in the stem. Like any other wound, resin tapping induces the formation of traumatic resin canals. However, unlike most wounds, tapping is repetitive and over time it induces a phenomenal increase in the number of axial canals that extend a considerable distance upward in the stem. Although bark chipping exposes the radial canal system, the resin is derived primarily from the axial traumatic system because of the radial/axial anastomoses.

Various chemicals have been applied to the exposed wound surface in addition to sulfuric acid in attempts to stimulate the flow of resin (Mergen and Echols 1955, Fahn and Zamski 1970). More recently, chemicals such as Paraquat have been applied to conifer stems in attempts to induce traumatic resin duct formation (Kiatgrajai et al. 1976, Stone 1977). Such chemical stimulation of *Pinus elliottii* var. *elliottii* results in extensive resin-soaked wood known as light-wood and the accumulated resin is then extracted during processing for pulpwood (Roberts et al. 1973).

Attempts have also been made to induce resin flow by mimicking the effects of bark-beetle infestations. Cheniclet (1987) investigated these effects in *Pinus pinaster* by inoculating bore-holes with sporulating mycelium from cultures of

Verticicladiella sp., fungi isolated from the bark-beetle *Tomicus pinipeda*. Although the resin canals were at first activated, they soon underwent lysis and no longer exuded resin. However, like Paraquat applications, terpene accumulation in the xylem was extensive and continued for a long time.

9.2.2.6 Resin Cysts and Pockets

As defined by Penhallow (1907), resin cysts are closed, sac-like structures that often occur in compact masses. Such cyst-like masses are typical of the traumatic resin canals that are found in the immediate vicinity of wounds. They are especially prominent in those species that do not normally produce resin canals (Bannan 1933a, 1934b, 1936b, Werker and Fahn 1969, Fahn et al. 1979, Sato and Ishida 1982). However, when defined simply on the basis of clustering, cyst-like masses have also been observed in *Abies* (Bailey 1925), *Larix* (Bannan 1933b, Sato and Ishida 1982), *Picea* (Thomson and Sifton 1925), *Sequoia* (Penhallow 1907), and occasionally in *Pseudotsuga* (Bannan 1936b). In *Pinus*, traumatic ducts are more scattered and they never occur as cysts (Bannan 1936b).

Greguss (1955) considered resin cysts to be part of a canal system that stored resin but "did not conduct it." That is, resin cysts often lacked epithelial cells. Cyst-like structures fitting this description sometimes occur in the radial system. For example, Hanes (1927) found that the canals in *Cedrus* formed without prior cell division and suggested that the duct opening was caused by pressure of exuding resin. The traumatic radial canals in *Cedrus* are also unique in that they are not always associated with the axial system (Record 1934).

Resin cysts differ from radial streaks which are overly large radial resin canals (Gerry 1942) and resinous tracheids which are often associated with rays (Record 1918a). The latter have been described as normal tracheids into which resinous material has diffused from the adjoining ray (Bailey 1925a). An example of overly enlarged radial canals was described by Prince (1934) in *Picea engelmannii*. In contrast to the normal canals which averaged 50 μm in diameter, some traumatic radial canals attained diameters of 800 μm. These abnormal canals, which were presumed to be of fungal origin, originated in the cambium and they were accompanied by frequent trabeculae in adjacent tracheids.

Resin pockets or pitch pockets are tangential openings in the cambial region into which resin from normal and/or traumatic canals has exuded (Fig. 9.21D). The size of the opening can vary considerably and in some cases the wound area is recognizable as a minute resin pocket and in others simply as a pitch ring. Resin pockets are not traversed by the parenchymatous bridges that separate tangential arrays of closely spaced resin ducts (Fig. 9.21B).

Resin pockets have often been associated with wind-induced stress (Frey-Wyssling 1938), but they might be caused by other natural agents (von Tubeuf 1918, Jaccard 1939, Cown 1973). Large resin pockets formed in *Pseudotsuga menziesii* as a consequence of repeated attacks by *Dendroctonus* bark beetles (Belluschi et al. 1965). According to Frey-Wyssling (1938), resin pockets in *Larix decidua* originated as a small cambial rupture between the xylem and phloem. The gap filled not only with exuded resin but also with callus tissue that proliferated into what the author described as a completely "verwildertes," or wild, tissue consisting of hypertrophied ray cells and transformed tracheids (Fig. 9.24). Interestingly, similarly induced wounds in *Juniperus virginiana*, a species that does not produce resin canals, were completely devoid of callus tissue.

Three distinct types of resin pockets were recognized in *Pinus radiata* in New Zealand by Somerville (1980). Type 1 was the typical oval, lens-shaped defect described earlier by Clifton (1969), Cown (1973), and Harris and Barnett (1975). Initial rupture, occurring in the cambial zone, created a tangential discontinuity that expanded under pressure of exuded resin from the radial canals. The resin pocket later became filled with callus nodules that developed from proliferating ray and resin canal parenchyma. Nodules consisted of several layers of phellem and phelloderm cells surrounding a tracheid-containing woody tissue. Origin of a pocket was usually associated with presumed planes of weakness, either along the earlywood/latewood boundary or in bands of closely spaced vertical resin canals. The cambium apparently resumed normal development rather quickly, as the subsequently formed growth ring was seldom affected.

Type 2 resin pockets also originated, as in Type 1, along a plane of weakness, and the tangential separation in the cambial zone became filled with resin under pressure. However, in this case, the developing pocket ruptured through the cambium causing resin to exude. The original pocket then flattened to a thin line of callus, and dry resin covered the wound surface. If the adjacent cambium died, some bark tissues were often occluded over by the healthy surrounding cambium. Wound healing proceeded as in other wounded tissues (Chap. 10.3.2.1). The resin pocket consisted of both a tangential and an expanded radial component that was filled with resin, callus, and included bark.

Type 3 resin pockets were narrow, longitudinally oriented breaks in the continuity of xylem that became filled with dry resin, parenchymatous tissues, and sometimes cork cells. The defect apparently arose as a lesion in the cambial zone in the vicinity of closely spaced axial resin canals. Like the Type 2 defect, the cambium ruptured and resin exuded from the wound. However, a Type 3 pocket differed primarily in shape and size of the wound occlusion, which was longitudinally oriented and more confined tangentially.

Chapter 10

Cambial Wounding

10.1 Introduction

Records of wound closure and wound treatment on trees date to ancient times (Bartlett 1935). Wound closure in trees not only intrigued early botanists, but it was also one of the methods they used to examine the origin of cambium. Lengthy quotes from some of these papers were included in the reviews of Trécul (1852, 1853b). For example, Duhamel in 1758 concluded that the wood of girdled stems was capable of producing new bark and this bark in turn produced new layers of wood. Meyen in 1839 wounded stems of *Corylus, Salix, Syringa,* and *Viburnum* by removing strips of bark and covering the wounds with mastic. He noted "gelatinous-formed drops" exuding from the medullary rays on the wood surface. This liquid initiated "corticoïde" tissues, or false bark, but no new wood was formed. Dalbert in 1830 also removed patches of bark on stems of *Fraxinus* and covered them with glass plates sealed over the wound. He, too, observed drops of a gelatinous substance covering the bare wood surface. These drops multiplied and eventually gave rise to new bark similar to that described by Duhamel. Dutrochet, another contemporary, was not sure how the new bark formed after wounding. He suggested that it might arise from a metamorphosis of the "médulle centrale," or pith. Although pith tissue does not extend to the secondary wood surface, he might have been referring to medullary ray tissue. Thus, the general view at this time was that a gelatinous sap exuded from the rays. According to Meyen, the sap was "a mucilage without organization," although it presumably contained "the principle of ulterior organization" as a tissue (Trécul 1853b). This mucilagenous sap was, of course, an early stage of callus formation.

Trécul (1853b), after reviewing the literature, debated whether the new tissue arose from the interior or from the outer surface of the exposed wood. As a result of his many experiments on various dicotyledonous trees with both open and covered wounds, he concluded that the tissue did not arise from unorganized, mucilagenous sap. Rather, young fibrous cells, cells of medullary rays, and sometimes very young vessels contributed to production of cellular tissues which then gave rise to new bark and wood. In some cases only the woody fibers gave rise to new elements and in others only the rays, as Brongniart (1852) had also demonstrated. Trécul obviously examined late stages of wound closure in which vascular elements were differentiating within the callus mass.

In a response to Trécul's observations, Hartig (1853) stated that he had never observed new fiber tissue forming directly on an exposed wound surface. Hartig (1845) had taken the precaution of enclosing girdled stems of *Quercus pedunculata* with cylindrical, glass lamp chimneys and sealing the edges with grafting wax. Under these air-tight, humid conditions, ray cells contributed to new growth on the exposed wood surface, first forming a new layer of green bark to the outside and then new wood to the inside. Hartig performed this test about 30 times on both young and old stems and on different species, always with the same results. He concluded that it was proof of the "spontaneous regeneration" of rind and wood on the exposed wood surface without the participation of cambium. According to Hartig, the wound surface was first covered by cell tissue derived from the rays and this tissue served as mother cells for both wood and rind. New tissues did not form on uncovered, exposed surfaces because of rapid drying of the delicate rays.

In the ensuing years, investigators were particularly ingenious and creative in the kinds of wounds they inflicted on trees in attempts to mimic natural causes. Many of these wounds involved injury to the cambium in some way or another. Observed responses of the cambium to these wounds were neither consistent nor predictable, although at some point in the process of wound closure typical cambial division events usually prevailed. For this reason, an attempt has been made in the present review to recognize the most commonly encountered types of wounds on woody plants.

Within the category of wounding, one might include girdling and grafting, both of which unnaturally disrupt the normal patterns of growth and cambial activity. Also included are wounds of biological origin, and growth disturbances due to environmental factors such as frost, drought, and lightning. Although not usually classified as wounds in the narrow sense, these growth disturbances often induce cambial responses similar to mechanical wounding. Other wound responses, such as overgrown stumps and decapitation, are discussed later in this chapter. The extensive literature on pathogenic injury to the cambium has not been included. Most pathogens are secondary invaders of established wounds, although some, such as *Phytophtora cinnamomi* (Robin et al. 1992), can colonize the cortical tissues and locally kill the cambium. Nonetheless, it must be recognized that the invasion of pathogens not only prolongs wound closure but also induces responses in the host's defense mechanism, such as accentuated barrier zone formation.

A voluminous literature exists on various types of wounding and the resultant effects on activity and reorientation of the cambium. Because it is impossible to even begin a thorough coverage, a few references have been selected for discussion in each category to illustrate the type of wound and involvement of the cambium in the processes of wound closure.

The response to wounding is not confined to the immediate vicinity of the wound. It can extend quite some distance both radially and longitudinally from the wound site. Attempts to understand the process have contributed to a fertile field of research from the standpoint of determining both the extent of the response (Wodzicki and Zajaczkowski 1970, Shigo and Dudzik 1985) and the nature of the traumatic signals (Bloch 1952, Sachs 1975, 1984, Warren Wilson 1978, Davies and Schuster 1981, Walker-Simmons et al. 1984, Bostock and Stermer 1989). An indication of the distances to which the signal might be propagated in some species can be found in the discussion of traumatic resin canals (Chap. 9.2.2.5). Because of the traumatic response, a distinction must be made between closure of the wound itself and recovery of the traumatized tissues surrounding the wound. These two regions will be referred to as the exposed surface and the barrier zone. The latter term, introduced in the pathological literature (Tippett and Shigo 1981), is now being used to describe analogous responses to wounding from many causes. For an excellent pictorial as well as descriptive analysis of tree wounds and types of wound closure from a wide variety of external causes, *A New Tree Biology* by Shigo (1986) is highly recommended.

10.2 Barrier Zone

Barrier zone is a term applied to the traumatized tissue adjacent to a wound in which the cambium is induced to respond but is not destroyed (Fig. 9.4). As defined by Shigo (1984b): "A barrier zone is a protective tissue consisting of unique cells and formed by the living cambium of a tree in response to mechanical wounding or infection, or both."

The perceived function of the barrier zone is to seal off internal tissues from the invasion of pathogens. Thus, in its broadest sense, the barrier zone includes not only the formation of regenerated vascular tissues but also the deposition of biochemical constituents in a process known as compartmentalization (Shortle 1979, Shigo 1984b). As generally interpreted in the forest pathology literature, a barrier zone refers to the defensive tissue derived from anomalous sapwood formed after wounding. A protective zone (Sect. 10.3.1), in contrast, refers to the defensive tissue derived from normal sapwood formed before wounding (Shortle 1992, personal communication). The term compartmentalization has been rather loosely applied, at times to include just the barrier zone and at other times to include other wound-altered tissues along with the barrier zone. Barrier zone development might be viewed either on the micro scale, stressing cellular events, or the macro-

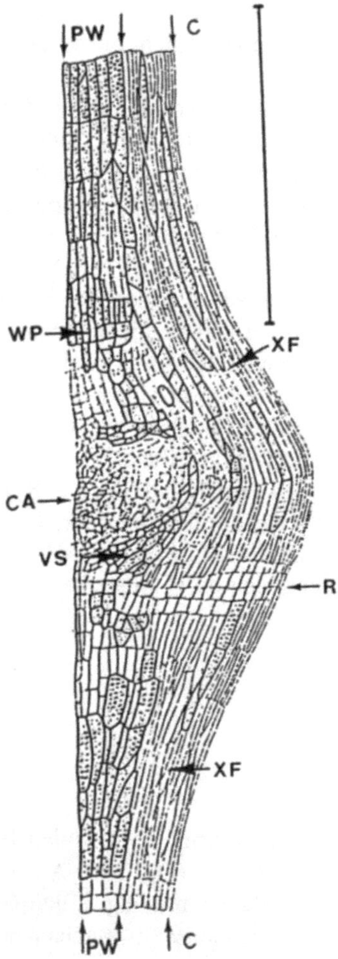

Fig. 10.1. Schematic representation of short-celled wound wood in radial view in *Viburnum opulus*. Wound callus (*CA*) is immediately surrounded by short vessel strands (*VS*), wood parenchyma cells (*WP*), and xylem cells comprising primary wound wood (*PW*). Xylem fibers (*XF*) in the short-cell zone are considerably shorter than those of normal wood represented by the length of the vertical line. **C** Inner border of cambial zone; *R* ray. (After de Vries 1876)

scale, stressing tissue changes extending over many years. The former approach will be emphasized here because of the role of the cambium in these cellular events.

Although it is only recently that the barrier zone as a defensive tissue has been described in detail and its significance appreciated, its presence and formation have long been known. Interestingly, one of the most comprehensive early studies of this phenomenon, made by de Vries (1876), has been only briefly mentioned in the barrier zone literature.

In reviewing the early literature, de Vries (1876) described how previous workers had interpreted wound callus as a homogeneous tissue which formed first a cork layer to the outside and later a meristematic layer of isodiametric cells to the inside. The meristematic layer then gave rise to "callus wood" and "callus bast." However, up to that time, a comprehensive examination of wound wood had never been made. De Vries applied numerous types of wounds to over 50 woody species, although most of his tests were conducted on *Caragana arborescens* and *Salix purpurea*. He noted that the effect of wounding was greatest in the immediate vicinity of the wound and decreased with distance from the wound. The abnormal wood produced by the cambium in the wound vicinity, which he called "wound wood", was believed to be of primary concern. The tissue that developed on the wound surface, which he simply called "callus," was considered of lesser importance.

De Vries recognized "primary wound wood" and "secondary wound wood" in addition to callus (Fig. 10.1). Wound wood represented "all wood that arises in the vicinity of a wound and deviates from normal wood of a species." Primary wound wood was the innermost layer that arose first from the cambium after wounding. Here, transverse divisions were numerous resulting in a "short-cell zone" of poorly differentiated cells. The short-cell zone graded gradually and imperceptibly into the "long-cell zone" of secondary wound wood where cell differentiation began to approach normality. Thus, primary wound wood graded into secondary wound wood not only radially in the immediate vicinity of the wound, but also centrifugally with increasing distance from the wound center. For example, above a ring-girdle in *Caragana arborescens*, cambial cells in the short-cell zone divided transversely an average of three times whereas those in the long-cell zone divided only once. Callus, which developed on the exposed wound surface, differed from primary wound wood. The wound wood that eventually differentiated within the callus was considered secondary wound wood.

Some interesting observations relating to wounding were also made by Bertrand (1884) in his treatise on the *Law of Free Surfaces*. One type of free surface was created by wounding, in which case the inactivated wound surface was surrounded by active cambial tissues. Rather than callus formation, Bertrand spoke of "cambiform" tissue, which he defined as a generative tissue that closed tangentially but did not produce secondary xylem or phloem. When the cambiform tissue invaded the free surface of a wound, a cork layer was produced between the cambiform tissue and the free surface which was now externally exposed. Fundamental and vascular tissues then differentiated internal to the cork layer. Bertrand claimed that all cambial zones dependent on a free surface would develop in the same way. That is, phloem always developed nearest the free surface, then cambium, and xylem to the inside.

Wounds of all types have been found to form a barrier zone (Bostock and Stermer 1989). The extent of barrier zone formation varies with the type and de-

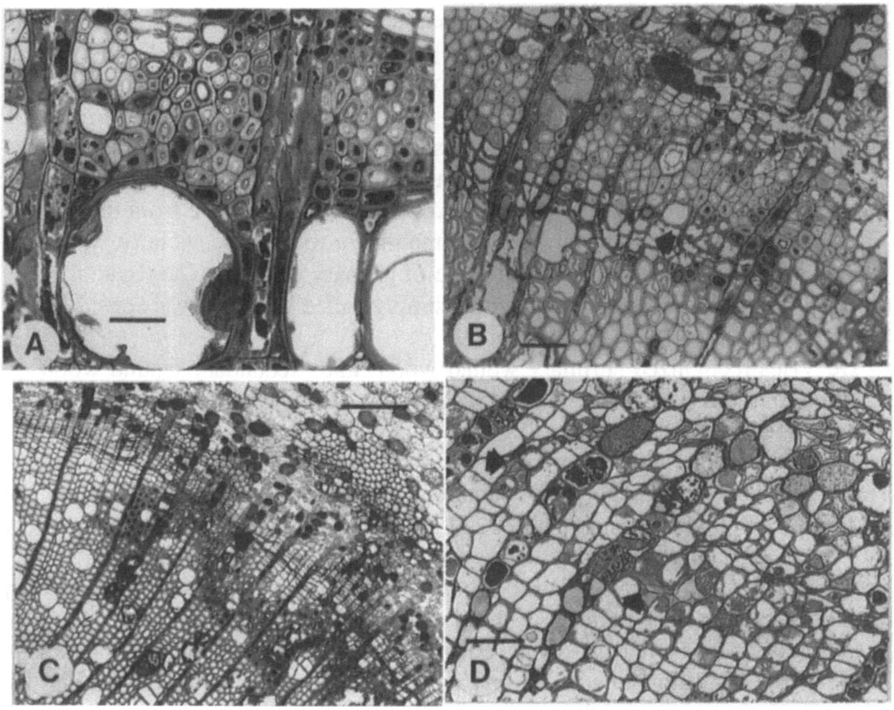

Fig. 10.2A-D. A, B Barrier zone formation in *Ulmus americana*, **C** *Populus balsamifera*, and **D** *Prunus pensylvanica*. **A** Cells located in the cambial zone at top arch outward in the vicinity of rays. Fibers form a barrier zone over the partially occluded vessels. *Bar* 25 µm. **B** Barrier zone cells (*arrow*) consist mainly of empty parenchymatous cells and fibers with thin, distorted walls. *Bar* 40 µm. **C** Recently differentiated barrier zone cells (*arrow*) contrast with adjacent cambial cells. Barrier zone formation appears to be asynchronous in adjacent sectors of the cambial zone. *Bar* 100 µm. **D** Early stages of two gum canals (*arrows*) in the cambial zone. *Bar* 25 µm. (Rioux and Ouellette 1991)

gree of wounding in addition to many other contributing factors. As noted later, secondary causes, such as the invasion of pathogens, can often accentuate the size and extent of the barrier zone. This effect was demonstrated by Rioux and Ouellette (1991) in branches of several tree species inoculated with the vascular pathogen *Ophiostoma ulmi*. At the onset of barrier zone formation in *Ulmus americana*, masses of cambial cells and adjacent barrier zone cells extended outward,

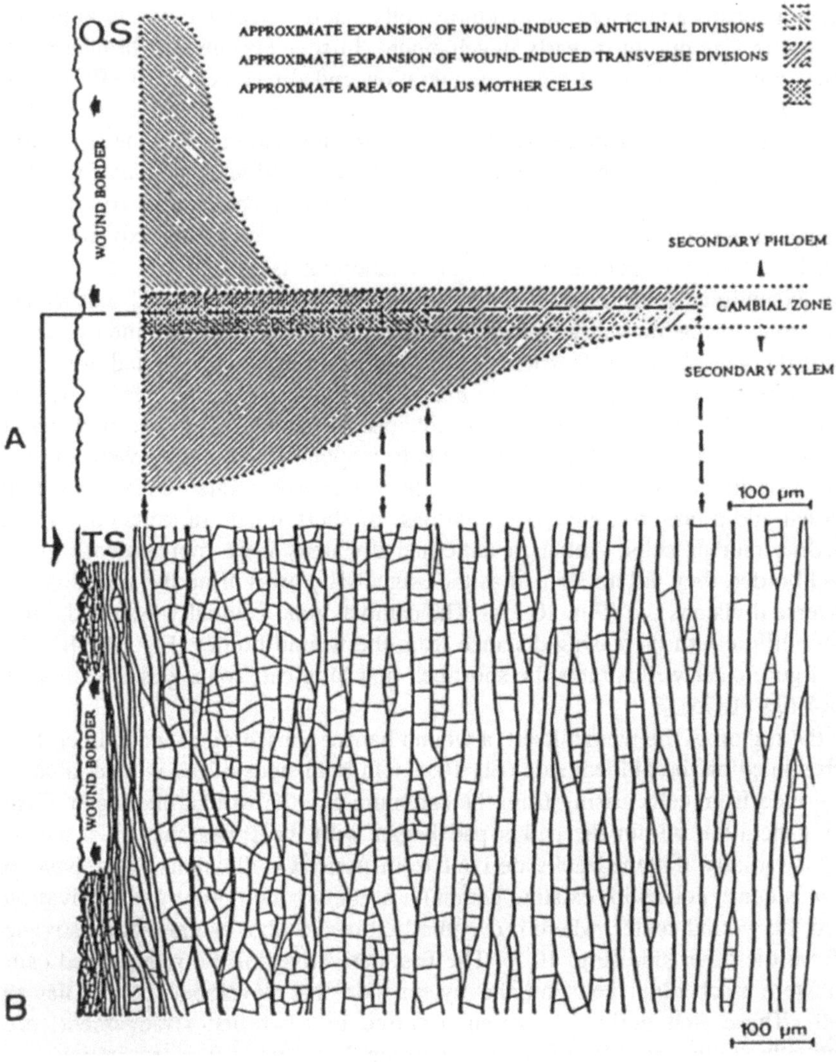

Fig. 10.3 A,B. First wound-induced histological changes in the cambial wound border 7 days after wounding a 4-year-old *Larix decidua* stem. **A** Idealized drawing of the mode and dimensions of wound-induced anticlinal and transverse divisions in the cambial zone and recent derivatives. **B** Simplified drawing of cambial zone tissue altered by anticlinal and transverse divisions. The frequency of wound-induced divisions decreases with increasing distance from the wound border, i.e., to the *right*. Drawn from SEM sections; *QS* Transverse; *TS* tangential. 115x. (Bangerter 1984)

forming a convex mass of tissue that contrasted with the contiguous normal xylem elements and cambial tissues (Fig. 10.2A). Barrier zone tissue consisted mainly of parenchyma cells and fibers, although ray cells that traversed the barrier zone also increased in size during its early development. In severely infected branches, the cambial zone cells showed signs of alteration and disrupted organization (Fig. 10.2B). *Populus balsamifera* responded in a similar manner; the arrow in Fig. 10.2C indicates the outward extension of the cambium and barrier zone at an early stage of development in this species. *Prunus pensylvanica* responded by formation of gum canals in the wound area (Fig. 10.2D). The gum canals arose in the cambial zone in the manner described in Chapter 9.2.1.6 and they later expanded lysigenously to form a tangential band of gum-containing tissue.

An exceptionally clear and detailed account of both deep wound closure and barrier zone formation was provided by Bangerter (1984). He examined injury to cells of the wound border (barrier zone) as well as those of the wound area itself in stems of *Larix decidua* and *Picea abies*. Following wounding, water loss and shrinkage at the wound border caused collapse of cambial zone and undifferentiated vascular cells. The first border cells to respond to wounding were ray cell initials that divided anticlinally and expanded transversely. Intense divisions, both anticlinal and transverse, gave rise to a mass of short, quadratic cells comprising the callus mother cells. Although anticlinal divisions were more intense at the wound border, they did not extend as far tangentially away from the wound as the transverse divisions did (Fig. 10.3A). The cambial zone gradually assumed a normal condition with increasing distance from the wound border (Fig. 10.3B). The latter figure clearly illustrates the short-cell and long-cell zones described earlier by de Vries (1876).

In the region of the wound itself, a wound barrier consisting of wound periderm differentiated on the phloem side (Fig.10.4). On both sides of the wound, a callus pad formed from cells arising from the cambium and the wound phellogen. Cambial regeneration within the callus pad began with the appearance of a narrow band of radially flattened cells derived from periclinal divisions. This band of cells was contiguous with existing cambium at the wound border and it advanced toward the wound center, where it eventually joined with a cambial band advancing from the other side (Fig. 10.5). The first derivatives of the regenerated cambium were tracheoidal cells on the xylem side that developed by bipolar tip growth. These first derivatives often occurred in a whirled arrangement, presumably due to the rapid intrusive growth of the first-formed fusiform initials.

Although anatomical descriptions of barrier zone formation by other investigators have been less detailed than that of Bangerter (1984), their observations are similar in many respects. The process has been described to varying degrees in *Acer rubrum* (Mulhern et al. 1979, Bauch et al. 1980, Smith and Shortle 1990), *Acer saccharum* (Sharon 1973), *Betula papyrifera* (Bauch et al. 1980), *B. pendula* (Schmitt and Liese 1990), *Juglans nigra* (Armstrong et al. 1981), *Liquidambar styraciflua* (Moore 1978, Shortle and Cowling 1978), *Liriodendron tulipifera* (Shortle and Cowling 1978, Lowerts et al. 1986), *Populus deltoides* (Shigo et al. 1977), *Prunus persica* (Wisniewski et al. 1985), *Ulmus americana* (Rioux and Ouellette 1989), and many other species. The potential for compartmentalization

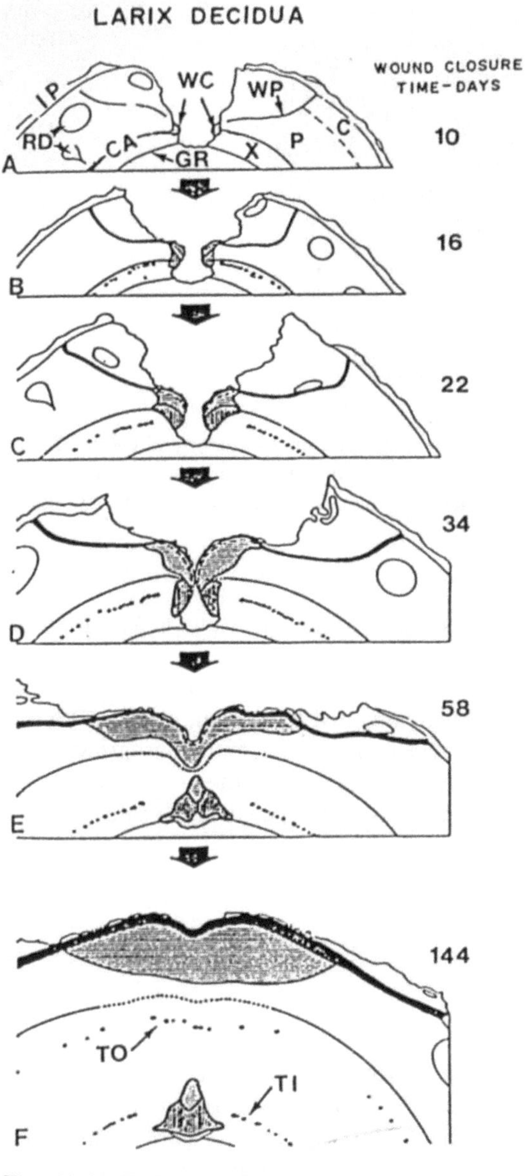

Fig. 10.4A-F. Histology of the wound occlusion process in a 4-year-old *Larix decidua* stem. *Vertical* and *horizontal hatching* represent callus originating from cambium and wound phellogen, respectively; *bold lines* wound periderm; *dashed lines* callus-producing wound phellogen; *dotted lines* regenerating cambium; *C* cortex; *CA* cambium; *GR* growth ring boundary; *IP* initial periderm; *P* secondary phloem; *RD* resin ducts; *TI* and *TO* inner and outer traumatic, axial resin ducts, respectively; *WC* wound callus; *WP* wound periderm, *X* secondary xylem. Transverse sections. 68x. (Bangerter 1984)

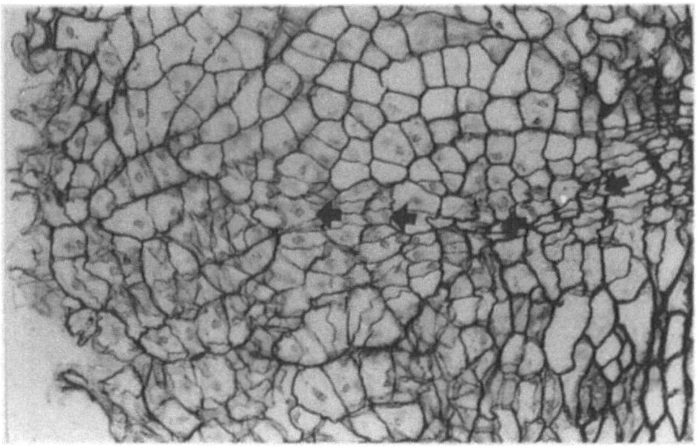

Fig. 10.5. *Larix decidua*. First stage of cambial regeneration within wound callus 22 days after wounding. Regeneration began in pre-existing cambium (*right*) and advanced in direction of the *arrows* by successive periclinal divisions. Transverse section 90x. (Bangerter 1984)

of the discolored and decayed wood associated with stem wounds appears to be genetically controlled (Santamour 1984).

A healed-over barrier zone embedded in the xylem of *Acer saccharum* is shown in Fig. 10.6, a photomicrographic montage prepared by Dudzik (1988). The barrier zone appears as an arc of tissue resembling a discontinuous growth ring (Chap. 11.8). It consists of typical wound tissues that proliferated in the immediate vicinity of the wound and their gradual disappearance to either side.

10.3 Mechanical Wounds

10.3.1 Shallow Wounds

Shallow wounds can be defined as those in which cortical tissues are removed in such a way that the cambium is either directly injured or destroyed but the underlying xylem remains essentially untouched. They differ from deep wounds that penetrate varying distances in the xylem. Shallow wounds also differ in their rates of closure, depending on whether they are left exposed to the atmosphere or sealed in some way to prevent desiccation and exposure to air.

10.3.1.1 Exposed Surfaces

Callus formation following wounding was for many years believed to be derived directly from the cambium. Although a few workers had recognized the role of ray

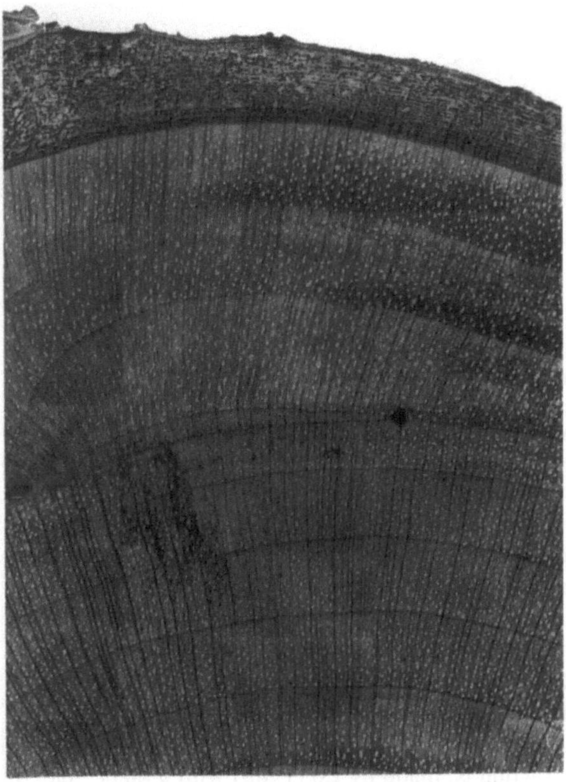

Fig. 10.6. Healed-over barrier zone in *Acer saccharum (arrow)*. (Dudzik 1988)

parenchyma, the true significance of these cells was not fully appreciated. Sharples and Gunnery (1933) were perhaps first to critically examine the anatomical processes involved in callus formation and wound closure. They made rectangular wounds in stems of *Hibiscus rosa-sinensis* by carefully removing bark patches and exposing the underlying xylem surface. Callus formed by division and proliferation of xylem ray cells. As the callus pad covered the wound area, phellogen developed from the wound edges and advanced inward, completely restoring the phellogen as a protective layer. Cambial activity followed and, like the phellogen, new cambium began where the severed ends of older cambium impinged on the callus as shown in Fig. 10.5. As described by the authors, the new cambium "...like a slowly closing diaphragm sweeps gradually inwards until the opposing edges meet and the cambial cylinder is thus restored."

The two main points described by Sharples and Gunnery, that the callus arose from xylem ray cells and that the new cambium differentiated centripetally inward in continuity with pre-existing cambium, have been confirmed by many other

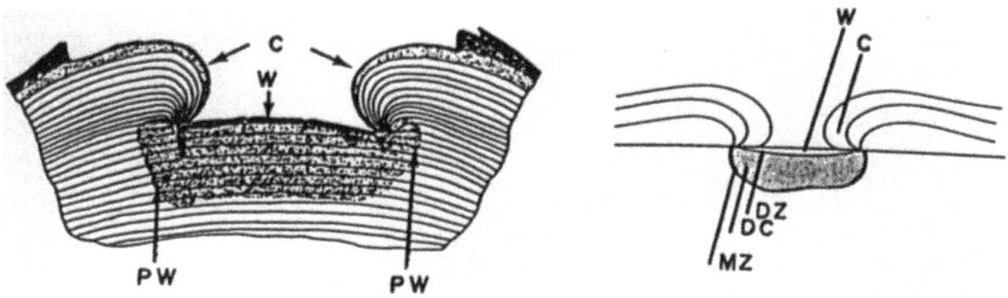

Fig. 10.7A,B. Schematic representation of protection wood in transverse view. **A** *Betula* showing the shallow wound surface (*W*) with overgrown callus (*C*) and dark protection wood (*PW*) extending inward across several annual rings. **B** Interpretation of protection wood in **A** consisting of a desiccation zone (*DZ*), a discoloration zone (*DC*), and a surrounding marginal zone (*MZ*). (Grosser et al. 1991)

workers. The former conclusion was tested by Barker (1954) on *Tilia americana*. When bark segments were removed, as in bark peeling, callus arose from all recent cambial derivatives but primarily from ray cells. However, when the bark segments removed about 1 mm of xylem, the ray cells did not contribute to callus formation except for a few that were recent cambial derivatives. In this case, rather than callus, small nodules analogous to pith flecks arose by proliferation of mature ray cells.

Vertically oriented V-shaped wounds, made by Lo (1988) with a razor blade in stems of *Chamaecyparis formosensis*, also closed by callus arising from the invasion of ray cell parenchyma. Again, Lo found that wound cambium within the callus was produced by extension from existing cambium on both sides of the wound. New cambial initials were connected by radial walls, and cambial divisions advanced cell by cell toward the wound center.

The process of wound closure described by Sharples and Gunnery (1933) is typical of that for many types of wounds. The distinctive characteristic is the continuity of the new cambial layer as it advances toward the wound center under the protective layer of phellogen. However, because the authors carefully removed bark patches from actively growing stems, the uninjured xylem permitted the ray cells to proliferate. In most exposed wounds, rapid drying of the xylem surface inhibits callus formation. Rates of wound closure can differ greatly in exposed wounds, varying with tree species, type and size of wound, and growth rate of the tree (Neely 1970, 1988).

The exposed surface of a shallow wound also responds by the formation of "protection wood," a concept introduced by Frank in 1884 (Grosser et al. 1991). Frank (1884) observed colored tissues in this region of gum deposition beneath the wound in ***Prunus avium*** and similar depositions beneath wounds in ***Gleditsia triacanthos, Juglans nigra, Quercus pedunculata,*** and ***Pyrus malus***. The gum masses filled the vessels in a manner analogous to tyloses. Frank also noted that although gum formation and tyloses arose by different processes, they induced the same physiological response, i.e., the exclusion of air and the prevention of desiccation of the underlying tissues.

Grosser et al. (1991) reexamined and reinterpreted Frank's concept of protection wood in seven dicotyledonous species (Fig. 10.7A). They recognized three zones in protection wood, namely: (1) a desiccation zone; (2) a discoloration zone; and (3) a surrounding marginal zone (Fig. 10.7B). Thus, in contrast to the barrier zone, which applies to the wound response in tissues adjacent to the wound, the concept of protection wood applies to the wound response of woody tissues previously formed beneath the wound surface.

Years earlier, Hepting (1935) had described a similar response to fire scarring which he referred to as a "protective zone." He noted a "marked difference between the sapwood extant at the time of scarring and the sapwood laid down later as to their susceptibility to decay." In *Liquidambar styraciflua*, the protective zone usually comprised several of the outer layers of sapwood cells heavily infiltrated with a brown substance that appeared to be gum. The infiltrated gum, presumably produced by the ray and xylem parenchyma cells, caused the affected sapwood to resemble heartwood (Hepting and Blaisdell 1936). A photograph through a partially closed fire scar, reproduced by Hepting (1935), appears strikingly similar to Fig. 10.7A. Protective zones were artificially induced by superficial wounding stems of *L. styraciflua* and *Liriodendron tulipifera* by Shortle and Cowling (1978). However, in almost all cases the protective zone was ineffective in preventing decay due to physical disruption either by boring insects or cracks that developed during drying.

10.3.1.2 Enclosed Surfaces

Wound closure proceeds somewhat differently when the wound surface is covered to reduce desiccation and exclude air. In covered wounds not only can the origin of callus differ, but the formation of a cambial layer can either be greatly delayed or arise independent of the pre-existing cambium.

The work of Noel (1968) on *Trema orientalis* growing in Rhodesia provides a good example of wound closure under a protective cover. Bark patches 5 cm^2 were removed, the cambial surface was scraped, and the wound covered with polyethylene film and aluminum foil. All immature xylem cells, with the exception of vessels, contributed to formation of callus by dividing both periclinally and anticlinally. The protruding callus clumps gradually thickened and spread laterally to cover the wound surface. Later, a phellogen began to differentiate in the outer region of the callus and a cambium differentiated internal to it. Unlike exposed wounds, the new cambium did not originate in continuity with pre-existing cambium. Rather, it was initiated discontinuously in isolated patches. Because these islands of cambia lacked functional vascular connections, they died and the wound eventually closed by overgrowth from the wound edges as in exposed wounds. Similar wounds, in which the wound surface was not scraped, responded dif-

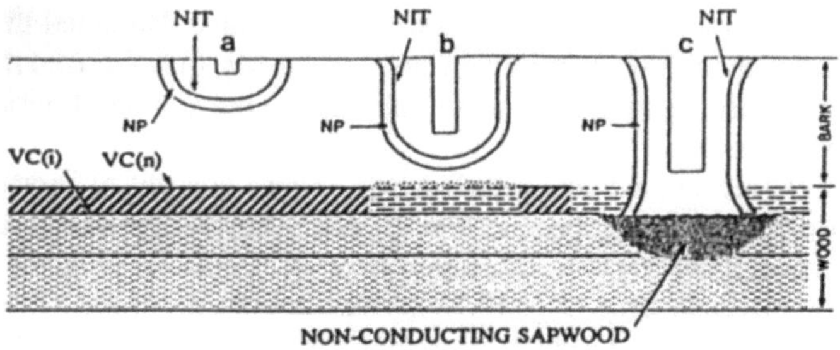

NP = NECROPHYLACTIC PERIDERM

NIT = NON-SUBERIZED IMPERVIOUS TISSUE

VC(n) = POSITION OF VASCULAR CAMBIUM AT TIME OF PHELLOGEN RESTORATION

VC(i) = POSITION OF VASCULAR CAMBIUM AT TIME OF INJURY

= CONDUCTING SAPWOOD FORMED AFTER INJURY

= CONDUCTING SAPWOOD EXTANT AT TIME OF INJURY

= TRANSFORMED CAMBIAL-PHLOIC ZONE

= ZONE OF NEWLY RESTORED VC

= ZONE OF NON-CONDUCTING SAPWOOD

Fig. 10.8. *Pinus contorta.* Diagrammatic view of the nonspecific nature of a tree's defense from wounding during: **a** Penetration of bark surface; **b** penetration to vascular cambium; **c** penetration of sapwood. See text for details. (Mullick 1977)

ferently. The cambium increased in thickness by periclinal divisions and it soon merged with the normal vascular tissues both above and below the debarked square.

The failure of cambia to regenerate in the expected manner in covered wounds led Warren Wilson and Warren Wilson (1961, 1984) and Warren Wilson (1978) to formulate a "gradient induction hypothesis." They proposed that wound closure was facilitated by the concentration gradient of an unknown "diffusable morphogen," the concentration of which was in some way maintained from its source to the wound. Covered wounds might be considered atypical, as they are unlikely to occur in nature.

One type of shallow wound, closely analogous to a covered wound, was described by Mullick (1977). In this case, the bark was removed from *Pinus contorta* seedlings down to and including some living tissues but leaving intact the cambial zone and its derivatives. In addition to phellogen restoration, the vascular cambium was affected by this procedure. The cambium surrounding the injury continued producing xylem and phloem. However, the cambium immediately beneath the point of injury stopped producing xylem, and both the cambium and the preexisting phloem essentially ceased enlargement. After about 10 days, when a zone of nonsuberized impervious tissue (NIT) had formed, the cambium resumed activity in cells external to the previous cambium as shown in Fig. 10.8. At this time an impervious necrophylactic periderm (NP) had formed internal to the NIT. The three levels of wounding shown in Fig. 10.8 illustrate the three major points of weakness for potential pathogens: namely, phellogen, vascular cambium, and apoplastic tissue of the functional xylem.

10.3.2 Deep Wounds

Deep wounds, which Mäule (1896) referred to as "Kerbwund," lack the viable xylem surface from which a callus pad might develop. Consequently, wound healing involves both filling of the cavity and regeneration of the cambial surface. Trécul (1852) examined healing of bore-holes made with an augur in stems of *Nyssa angulisans*. Cells of the "couche régénératrice" and other living cells bordering the wound were the first to elongate and divide two or more times. These new cells enlarged and divided gradually filling the cavity with "utricles." Eventually, fibro-vascular tissues were produced by those cells demonstrating "a special memory" for the transformation to cellular tissue.

Years earlier, Schultz (1831) had also pointed out how deep wounds were healed by cells derived from the cambium, which eventually produced new wood and bark. He remarked that this would be impossible if the cambium consisted of a sap (Chap. 3.2). He also mentioned that wherever one observed an overgrowth after wounding it was always associated with swelling of the underlying cambium. The vessels and fibers in these gnarled tissues were frequently deflected from their normal, vertical orientation in various directions. Other workers also described how tissues on the borders of deep wounds were deflected so that they appeared to bend around the wound (de Vries 1876, Mäule 1896).

The foregoing workers described, according to knowledge at the time, how deep wounds were at first filled with parenchymatous callus tissues. Later, the surface of the cavity was invaded by callus from the wound margins. As this callus pad enveloped the wound cavity, phellogen and cambium extended circumferentially inward in continuity with pre-existing tissues. Thus, other than filling the xylem cavity and proceeding more slowly, closure of deep wounds is simply a modification of that found in surface wounds.

The wound plug that eventually covered bore-holes made in early spring in stems of *Juglans nigra* was derived from the vascular cambium adjacent to the wound according to Armstrong et al. (1981). Parenchymatous-like callus arose in the disrupted cambial zone from cambial initials, immature cambial derivatives, and phloem parenchyma cells. Numerous anticlinal, longitudinal divisions of these cells increased the number of derivative files that expanded laterally as an overgrowth into the wound opening (Fig. 10.9). Wound closure was considered anatomically complete when continuity of the vascular cambium was reestablished following lateral contact of the wound-plug tissue. In this species, callus did not form on the internal surfaces of the wound and the wound cavity remained more or less open.

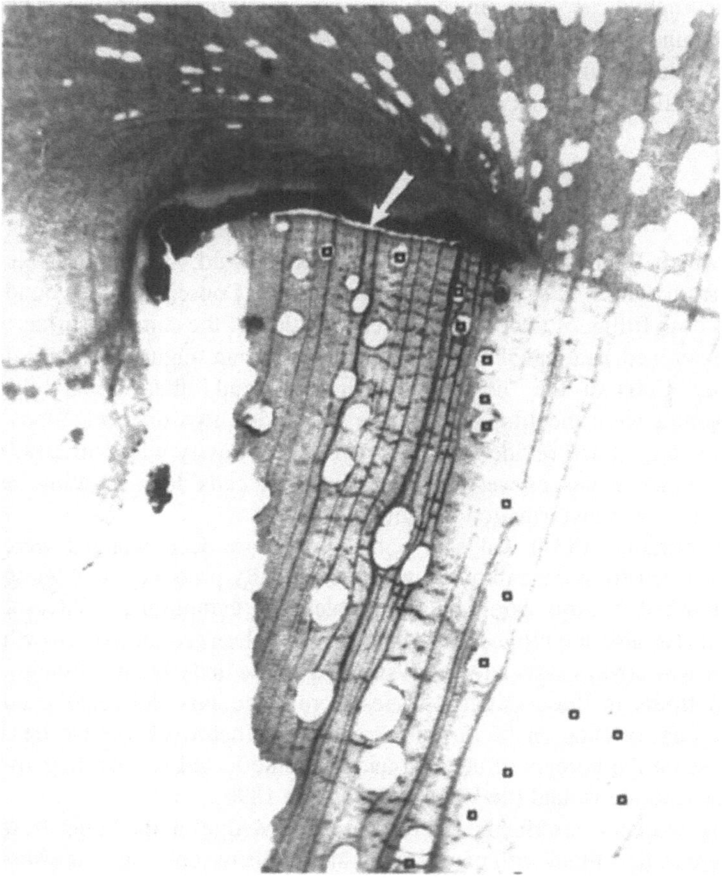

Fig. 10.9. Transverse section through a 15 mm wide, partially closed stem wound made in early spring in *Juglans nigra*. Both the periderm and cambium of the wound plug are in continuity with pre-existing tissues. The *arrow* indicates area where cambium was torn at the time of wounding. The lateral compartmental wall is delimited by ray and axial parenchyma with dark-colored contents and vessels containing tyloses (*squares*). (Armstrong et al. 1981)

Similar results were obtained in an earlier study conducted on the same species by Smith (1980). However, in this case, the boreholes were completely filled with a callus pad consisting of parenchymatous cells interspersed with elongated tracheary-like cells. The abnormal tissue following fall wounding was more intensive and extensive than that following spring wounding. For example, in wounds made in November, dieback of the vascular cambium extended about 9 mm both above and below the circular boreholes, whereas in wounds made in March, dieback averaged only about 4.5 mm in both directions. According to data obtained by Martin and Sydnor (1987), both tree vigor and radial growth were poorly correlated with rates of wound closure of boreholes made in stems of 12 dicotyledonous species. The greatest source of variability was among species.

Most studies of deep wound closure in woody plants with a well-developed cambium have been conducted more or less retrospectively when the process was either well underway or completed. Few studies have critically examined the process from the time of wound inception. The reasons are obvious, primarily because of the difficulties involved in examining the cambium and its disruption in stems of mature trees. Kuroda and Shimaji (1984, 1985) and Kuroda (1986) attempted to overcome these difficulties by creating deep wounds that disrupted relatively small regions of the cambium. Wounds were made in various patterns and at different times by inserting metal needles 250 to 400 μm in diameter into tree stems and then extracting them immediately (Sect. 10.7). Tissue blocks were removed from the wound sites at periodic intervals and prepared for microscopic examination.

Response varied according to the state of development of the various tissues at the time of wounding in 12- to 15-year-old *Pinus taeda* (Kuroda and Shimaji 1984, Kuroda 1986). In the mature xylem and the zone of wall thickening, ray parenchyma cells were highly activated and they proliferated randomly to eventually fill the wound gap (Fig. 10.44A,B). Ray parenchyma also proliferated in the wound cavity in the zone of cell enlargement (Fig. 10.44C,D). In neither zone did ray tracheids participate to any appreciable extent. Occasional fusiform cells immediately adjacent to the wound gap divided and on rare occasions one became septate. Because the wound was small, only about eight radial rows of cambial cells were crushed or pushed aside. This region closed over as new cambial initials gradually developed laterally in continuity with existing cambial initials on both sides of the wound in much the same manner as interfascicular cambium develops from fascicular cambium.

Reports of tests with 400 μm needles on stems of *Cedrus deodara, Cryptomeria japonica,* and *Pinus taeda* by Yoshimura et al. (1981a,b) yielded similar results. Radial rows of proliferating tracheids arose from the cambium by anticlinal divisions on both sides of the wound. Immediately beneath these cells, abnormally large and disfigured tracheids arose from cells in the early stages of differentiation. The wound cavities were filled by proliferating ray parenchyma cells, some of which retained the potential for dividing after having differentiated. As in many conifers, tangential bands of traumatic resin canals were common wound responses (Fig. 9.22D).

The response to similar wounds in *Populus euramericana* differed in several significant ways (Kuroda and Shimaji 1985, Kuroda 1986). In addition to reactivation of ray parenchyma cells, conspicuous abnormalities occurred in the fusiform cells. Vessel formation was suppressed in cells recently derived from the cambial initials. However, in the xylem mother cell zone, fusiform derivatives frequently divided transversely and they showed various transitional patterns ranging from vessel element-like through tracheid-like, and axial parenchyma cell-like, to fiber-like. In the cambial region, wound closure proceeded as it did in *P. taeda*. These observations suggested to the authors that the direction of cytodifferentiation was determined either in the cambial initials or in the neighboring xylem mother cells, and that it might be controlled by wound-stimulated substances.

10.3.2.1 Radial Cracks and Fissures

Filling of the wound cavity and closure of the wound gap do not always proceed in the manner just described. In addition to differences in species response, closure also varies with depth and severity of the wound. Woody lianas, for example, are often subjected to twisting and other natural forces that cause longitudinal splits and fissures in their stems. An attempt was made by Fisher and Ewers (1989) to duplicate these wounds and to induce stem splitting by twisting liana stems around pipes of various sizes. In *Jasminum fluminens*, a species with typical vascular cambium and normally oriented xylem and phloem, large cracks or fissures often extended inward to the pith. Neither the pith nor the xylem rays responded to these injuries. Callus was produced by tissues at the ruptured edges of the cambial zone that grew into these xylem cracks. New cambia, developing inward along both margins of the crack, eventually merged at some point deep within the crack and in this way a continuous cambium was reestablished.

Not all woody lianas responded in this manner. In some, the wound healed by overgrowths across the gap. In others, with deep fissures, the inward developing cambia failed to merge and the isolated parts became independently stemlike. Species with internal phloem and anomalous cambia often responded in unexpected and unique ways, depending on the extent of injury and the structures of their cambia.

Closure of radial splits in the sapwood of *Anthocephalus cadamba* was studied by Bamber and Lanyon (1966). Cause of the split was not known, but it was presumed to be mechanical in nature. The fissure was filled by a mass of parenchymatous callus cells, some with thin primary walls and others with thick, lignified walls. All evidence suggested that the callus originated from transformed ray cells in the sapwood. Interestingly, the sapwood ray cells had lignified, thickened walls but still living cytoplasm at the time of their transformations.

Radial splits or fissures are common in woody stems and they have been attributed to many causes. Occasionally, these fissures are confined to the xylem without rupturing the cambium. Filling of the cavity, however, proceeds in much the same way. A particularly interesting type of radial fissure in *Pinus radiata* was attributed to adverse growth conditions during latewood formation that caused partial collapse of the thin-walled earlywood cells (Amos 1954). The fissure was filled by proliferation of the still-living, thin-walled ray cells which were presumably stimulated to divide in the humid environment of the newly created cavity.

In a review of radial cracks in trees, Kubler (1983) pointed out that the wounded tree does not as a general rule really heal. It only covers the wound with wood and bark callus tissues that grow over the wound from the edges. Therefore, it is the bark that finally closes the wound when the callus of opposing flanks touch each other (Fig. 10.10). In this case, neither the opposing bark surfaces nor the two wood surfaces grow together. The cambial layers of the opposing flanks are forced together later when the growth of the callus wood crowds out the callus bark. It is at this time that cambial continuity is reestablished and normal wood formation resumed. Whereas Kubler attributed most of the radial cracks in his paper to frost injury, Dietrichson et al. (1985) found that similar appearing radial cracks in *Picea abies* were due to drought (Sect. 10.6.2). In addition to bark inclusions, Göppert (1846) and Seidel (1879) mentioned how stones, nails, and other foreign objects often became enclosed within deep wounds. References to many oddities of this kind can be found in the literature (Studhalter 1955).

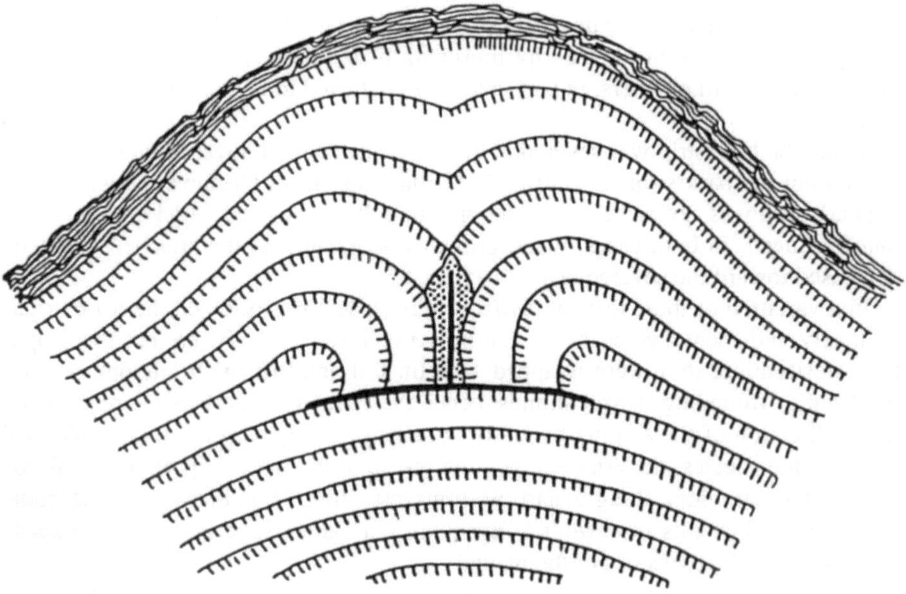

Fig. 10.10. Partial transverse section of stem with a closed wound, showing the resulting ring shake and wound crack. (Kubler 1983)

10.3.3 Bark Strips

One of the earliest anatomical descriptions of a bark-strip trial was that of Hartig (1845). He cut a bark strip so that it separated from the wood but remained attached at the upper end. Separation occurred through the wide-lumened wood cells and both the wood and bast mother cells remained on the bast side; Hartig referred to cambial initials as mother cells and he believed that wood and bast each possessed their own mother cells (Chap. 4.1.1). The new wood ring did not form on the exposed wood surface but on the inside of the bark strip, where a complete new stem developed with an included pith and a peripheral layer of bark. A few years later, Trécul (1852) obtained somewhat similar results with bark strips attached at the upper end in *Paulownia imperialis* stems.

De Vries' (1876) classical paper on wound wood contained a wide variety of different experiments. Among these were some unique trials in which two parallel, longitudinal incisions 20 to 40 cm long were made in the bark of *Magnolia cordata*. The bark strip was then gently loosened from the wood and either the two ends left intact or one end severed. When either both ends or the upper end were left intact, a thin callus layer was produced by the cambium and phloem that remained on the inner side of the bark strip. The callus soon gave way to a relatively thick layer of tissue that approached normal wood in structure except that vessels were more numerous and fibers were significantly wider in radial diameter. De Vries attributed the deviation from normal wood structure to the fact that bark pressure had been reduced by loosening of the bark strip. Reduction in the number and length of fibers, in particular, was believed to be the result of reduced bark pressure.

When the bark strip was severed at the lower end, a thick layer of wood formed throughout the bark strip, as Hartig (1845) had described. On the contrary, when the bark strip was severed at the upper end, both the callus and wood growth were much reduced. In both cases, wood growth was greatest at the attached end and decreased toward the severed end.

In a variation of these tests on *Salix purpurea*, de Vries oriented the bark strips obliquely and then severed either the upper or lower end. To protect the strips from desiccation, they were wrapped with tin foil and then with oilcloth. A few weeks after wounding, vessel strands, hence the cambium, had become reoriented in the direction of the sloping bark strip attached at its upper end. In another test on the same species, de Vries scraped the inside of the bark strip with a knife to destroy the cambium along a narrow, transverse line. Wound wood arose both above and below this new wound. Eventually, a callus ridge formed and vessel strands united the two sides of the wound.

Similar, but far more sophisticated, tests were conducted many years later by Brown and Sax (1962); these authors apparently overlooked the earlier work by Hartig (1845) and de Vries (1876). Longitudinal bark strips of *Pinus strobus* and *Populus trichocarpa* were separated from the bole wood during early spring and maintained in a humid environment while attached at the upper end to the parent tree. Bark separation occurred in the xylem mother cell zone, leaving the cambium intact along the inner surface of the bark strip. As in other exposed wounds, callus on the bark surface was derived almost entirely from proliferation of ray cells. As the callus pad increased in thickness and mutual cell pressures increased, radial rows of periclinally dividing cells began to form. First a phellogen and shortly thereafter a cambium developed within the callus pad. Both meristems differentiated tangentially from the margins of the pre-existing meristems in the bark strip. The bark strip now possessed a complete circle of cambium consisting of its original cambium and the "new" cambium extending through the callus pad. Cambial closure isolated a callus-derived "pith" region in the center. Subsequent cambial activity produced a typical stemlike structure in the bark strip (Fig. 10.11A). The new cambium developed in continuity with pre-existing cambium just as it did in other types of exposed wounds.

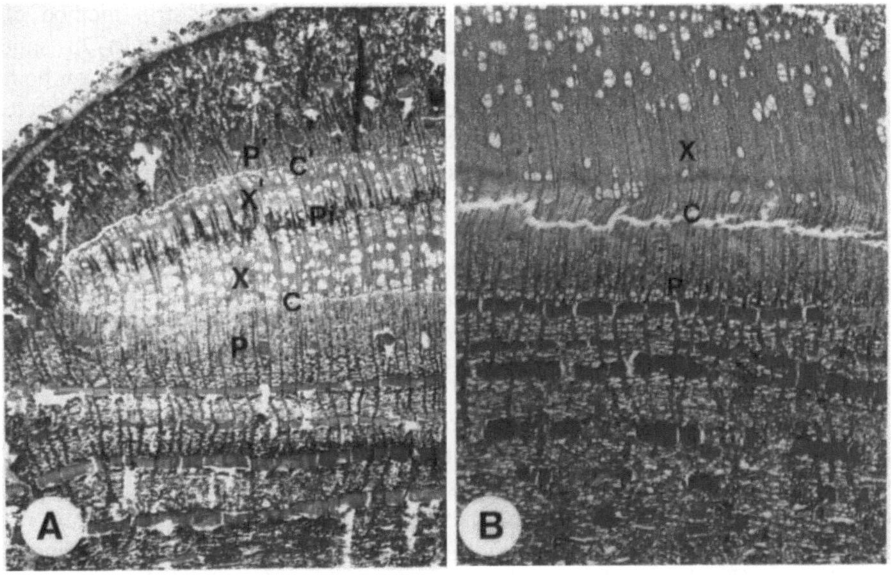

Fig. 10.11A,B. Transverse sections of *Populus trichocarpa* bark strips 60 days after treatment. **A** A complete cambium has formed surrounding the "pith" (*Pi*) area. 10x. **B** As in **A** except that the bark strip was enclosed in polyethylene film and replaced on the stem under applied pressure. A normally oriented cambium has formed. *C* and *C1* Original and later formed cambia, respectively; *X* and *X1*, *P* and *P1* xylem and phloem, respectively, produced by the original and newly differentiated cambia. 13x. (Brown and Sax 1962)

In another test, the bark strip was enclosed with polyethylene film to prevent cellular contact, and carefully replaced on the stem. The stem was next wrapped with rubber grafting bands to provide both physical confinement and applied pressure on the bark strip. Under these conditions, proliferating ray cells again filled the entire inner wound surface. However, under the relatively uniform mutual pressure of the new environment, parenchyma cells began to differentiate as xylem elements with highly lignified and pitted secondary walls. Somewhat later, the cambium began to function normally, producing new xylem and phloem derivatives that matured into typically elongated, functional elements (Fig. 10.11B).

Cambial explants of *Populus deltoides* grown in vitro responded similarly to applied pressure (Brown 1964), as did those of *Cryptomeria japonica* (Makino et al. 1983). That is, cambial explants to which external pressure was applied continued producing apparently normal xylem derivatives for a short period of time.

10.3.4 Split Stems

Split stems might be considered a modification of the bark-strip method of wounding. In an attempt to duplicate the effect of gall insects, Kny (1877b) split young stems longitudinally with a scalpel. Cell division was very active on both exposed surfaces, and after a time new cork and vascular cambia were produced. The vascular cambia produced normal conducting tissues, xylem to the inside and phloem to the outside, in each of the split stem parts. Results were essentially identical in all woody plants examined, such as *Acer, Catalpa, Lonicera, Sambucus, Salix,* and *Syringa.*

This experimental approach was repeated by Snow (1942) in hypocotyls of *Helianthus annuus* and epicotyls of *Phaseolus multiflorus.* After 3 to 4 weeks, new cambia formed in each half-stem by extension from the cut edges. When the split halves were pressed together, the cambial edges joined and reunited the old cambial ring.

A somewhat similar regeneration process occurred in rootstock cuttings of *Malus* that were longitudinally split at the basal ends (MacKenzie et al. 1986). Cambial callus expanded in two opposing directions, gradually occupying the deep gap between the two separated halves of the woody cylinder. Cambium regenerated within the callus in continuity with the original cambium and extended into the gap. However, rather than joining with cambium from the other stem half, each cambium remained independent. In contrast, the cambial callus produced from the wound edges of shallower incisions pushed outward from the wound gap instead of inward. In this case, the regenerating cambia extending across the callus bridge joined to reunite the cambial cylinder.

The split-stem experiments of Kny (1877b) and Snow (1942) were repeated by Warren Wilson and Grange (1984) in an attempt to quantify the effects. Internodes of young *Capsicum annuum, Datura stramonium,* and *Lycopersicon esculentum* were split and held apart by small pieces of either nylon or glass tubing. Three phases of wound regeneration were recognized. The initial lag phase lasted about 2 days, during which time neither cell division nor enlargement were detected but mitotic figures were observed within about 300 μm of the cut surface. Throughout the second division phase, from about days 2 to 10, both cell division and enlargement produced a callus mass with new cell walls parallel to the surface. During the latter part of the division phase, mitotic figures within the callus became concentrated in the regions of the future cork and vascular cambia and ceased in the regions of the future cortex and xylem (Fig. 10.12). At this time, the position of the future cambium became defined. During the third differentiation phase, cell divison was restricted to the cambial zone and derivatives differentiated into cork, phloem, and xylem according to position. The vascular cambium differentiated in continuity with pre-existing cambium from the wound edges and from common parent cells in the callus interior. Thus, each half-stem became independently stemlike.

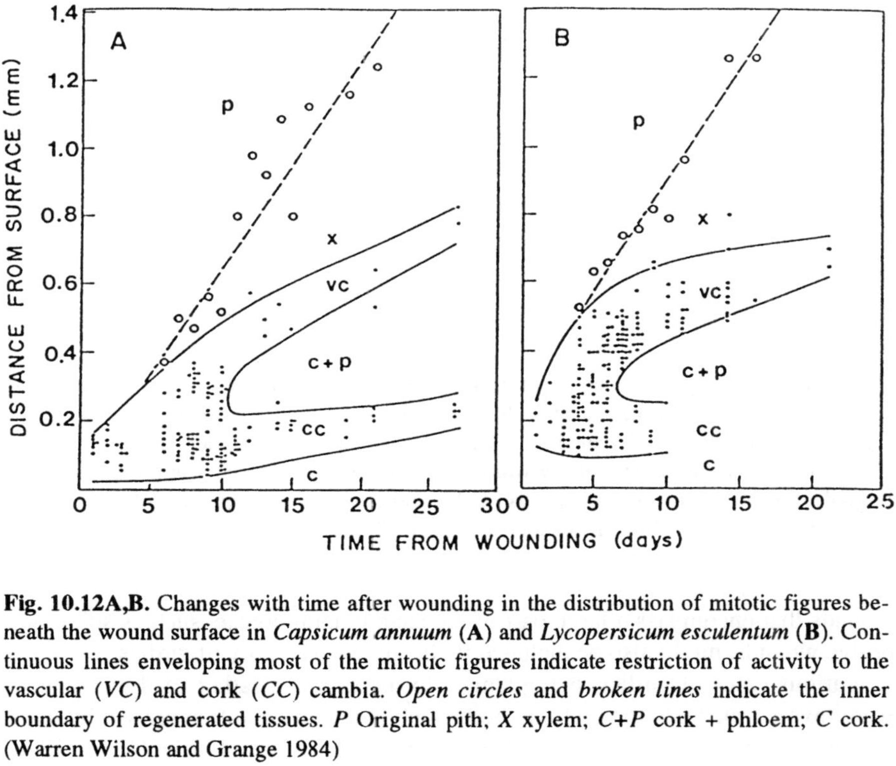

Fig. 10.12A,B. Changes with time after wounding in the distribution of mitotic figures beneath the wound surface in *Capsicum annuum* (**A**) and *Lycopersicum esculentum* (**B**). Continuous lines enveloping most of the mitotic figures indicate restriction of activity to the vascular (*VC*) and cork (*CC*) cambia. *Open circles* and *broken lines* indicate the inner boundary of regenerated tissues. *P* Original pith; *X* xylem; *C+P* cork + phloem; *C* cork. (Warren Wilson and Grange 1984)

10.3.5 Girdling

One of the most commonly practiced mechanical injuries is stem girdling, whereby translocation pathways are interrupted by transverse wounds perpendicular to the tree's axis. Stem girdles have been used in a wide variety of ways to study a myriad of translocation and growth processes as well as more practical objectives such as the induction of flowering, rooting of cuttings, and killing of unwanted stems (Noel 1970). As defined by Noel (1970), "... a girdled tree is one in which the phloem is completely severed either by a narrow incision or by the removal from the trunk of a more or less wide cylinder of bark, with or without damage to the underlying tissues." Although not true girdles, other types of transverse

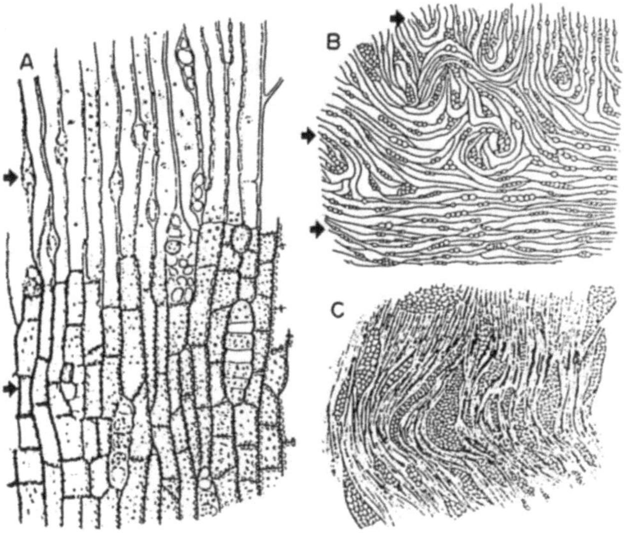

Fig. 10.13A-C. Tangential sections of *Abies cephalonica* (**A,B**) and *Rosa centifolia* (**C**). **A** Above a girdle showing the transition from wound parenchyma (*lower arrow*) to tracheids (*upper arrow*). 126x. **B** Tracheid orientation along the wound border of a spiral girdle (*lower arrow*), in the middle of the whirled zone (*middle arrow*), and in the transition to normal orientation (*upper arrow*). 70x. **C** Wound tissue of a completely closed, narrow girdle. Wound to *right*. Note dissected rays. 21x. (Mäule 1896)

wounds that encompass a large part of the stem circumference, such as spiral and bridge wounds, might also be included in the general category of girdles.

Various types of girdles, being simple to perform and rather easily evaluated, were favored methods for studying both cambial activity and tree growth by early botanists. Schultz (1831), Trécul (1852, 1853b, 1855), T. Hartig (1845, 1858, 1859b, 1862, 1863a,b), R. Hartig (1889, 1890), Schumann (1873), and many others used stem girdles to great advantage in their research. Girdling disrupts the downward translocation stream and the usual response is an increase in tissue growth. This increased growth, in the form of a swollen cicatrice at the wound margin, is invariably much larger above the girdle than below. The development and structure of these tissues are of interest from the anatomical point of view. Unfortunately, most investigators simply described the wound tissue in broad terms rather than its development. De Vries (1876), for example, interpreted the wound response in the same general terms of primary and secondary wound wood that he used to describe other types of wounds. However, Brongniart (1852) wrote of "excrescences" from the medullary rays that formed a kind of "couche génératrice" over the debarked region of the girdle.

One of the better descriptions of girdling was by Mäule (1896), who studied under Vöchting. He examined the wound response to ring girdling in 28 coniferous and dicotlyedonous species. Above the girdle, he observed a wedge-shaped mass of wound callus. This callus consisted of isodiametric, irregularly arranged wound parenchyma elements that differed from normal parenchyma cells; in wide girdles, wound callus below the girdle was negligible. In *Abies cephalonica*, wound callus arose not only from cambial cells but also from the youngest phloem and thin-walled, unlignified xylem elements. With increasing distance from the wound border, wound parenchyma cells gradually became intermixed with other xylem elements (Fig. 10.13A). The nature of this transition region varied considerably among the species examined. A new cambium developed within the callus by extension from the uninjured cambium distant from the wound site. This cambium functioned normally and, radially from within outward, Mäule could recognize old wood, wound parenchyma, unlignified callus tissue, new wound wood, cambium, and rind or cortex. Although the newly formed cambium functioned normally in the sense that it produced new cell files, the xylem elements were by no means normal. They were often distorted and convoluted in various configurations (Fig. 10.13B). As a student of Vöchting, Mäule considered this structure to be an expression of cell polarity. That is, the wound elements in "striving" to attain their normal lengths did so in the direction of least resistance.

In the case of fairly narrow girdles, the callus overgrowths from the upper and lower borders sometimes merged. Each callus ridge was covered by a cork layer, and upon meeting, these cork layers became flattened by mutual pressure. According to Mäule's observations, walls of the intervening cork cells became thinner and they were somehow "resorbed." The cambium soon bridged the parenchymatous gap in the wound callus and continuity was reestablished (Fig. 10.13C). During the bridging process, the large multiseriate rays of some species were dissected. In more recent work, Lev-Yadun and Aloni (1991b) also demonstrated how multiseriate rays were dissected in the regions of stem girdles.

Among the many papers reviewed by Noel (1970), relatively few of the authors actually studied the wound response in girdled stems. Descriptions of those that did differed little from the report by Mäule (1896). One of the major differences expressed by authors relates to the tissues that give rise to callus. In addition to the rays, callus might arise from the cambium, xylem and phloem parenchyma, or any combination of these tissues. Apparently, variations exist among species and possibly to the time and degree of wounding in the source of callus tissues.

Although the cambium can seldom regenerate on an efficiently girdled surface (Noel 1970), the exposed surface can produce callus and even completely regenerate in some species and under certain conditions. As an example, a variety of locations and types of living tissues were involved in wound closure of the debarked regions in lianas studied by Dobbins and Fisher (1986). In addition to margins of the girdles, which proliferated in the form of callus in all species, the exposed surface of the xylem proliferated in a few species. These cells most often were derived primarily from living xylem ray cells. In species with supernumerary cambia, protected segments of the internal cambium proliferated profusely, thereby contributing to wound closure. In still other species, callus on the exposed surface arose from either internal xylem parenchyma or protected regions of internal and included phloem.

An exceptional example of regeneration after girdling was described by Zhengli and Keming (1988) in *Eucommia ulmoides*. Stems of this species are rountinely girdled for 1 to 2 m lengths and the bark removed for medicinal purposes. After girdling, ray cells of the immature xylem swell, proliferate, and form callus. A cork layer forms to the outside and a vascular cambium differentiates deep within the callus. Cambial development is completed within 3 to 4 months after wounding. Thus, in this species, the debarked region can regenerate providing that treatment is applied when many immature xylem elements are present and the exposed xylem is covered with plastic film to prevent desiccation (Cui et al. 1989).

Based on the foregoing results, Cui et al. (1989) questioned whether the tree could regenerate its stem if the xylem was removed in the form of an inverse girdle. They removed the central xylem cylinder and pith from 1- and 2-year-old stems of *Broussonetia papyrifera* through a longitudinal slit in the bark and immediately wrapped the stem with plastic film. Callus formed from the residual cambium and immature xylem and phloem cells. Meristematic regions differentiated within the callus and began to produce new xylem and phloem. These regions, which first appeared as patches or islands, eventually joined laterally to form a continuous cambial cylinder.

In another type of girdling, employed by both de Vries (1876) and Mäule (1896), a narrow incision was made circumferentially around the stem without removal of any bark. This "einfacher Ringelschnitt," or simple ring incision, usually closed by formation of a callus pad through which the two severed cambial parts soon reunited. In some species, such as *Abies cephalonica* described by Mäule, the response was somewhat more complicated.

This process, also called scoring by horticulturists, was later studied by Soe (1959) on *Acer saccharinum* and *Populus deltoides* stems. In both species, callus formation was derived mainly from living cells of the vascular rays. Activity of the new cork, which always preceded the vascular cambium, began as soon as the callus pad was well developed. Closure of these narrow wounds was rapid and new cambial formation occurred on the 6th and 9th days after wounding in *P. deltoides* and *A. saccharinum*, respectively. Cambium was initiated where the callus tissue met the normal, undisturbed cambial zone. Cambial formation was independent of both the rate and amount of callus production. Xylem elements differentiated before phloem elements from the newly formed cambium, but they were not abundant and all were abnormal and distorted. Normal xylem and phloem elements began to appear about 20 days after wounding.

Girdling by ringing with an encircling cord or wire, a ligation, has been frequently employed in place of scoring. As described by Grill and Gigler (1985) on *Thuja occidentalis*, the stem above the ligature thickened primarily by an increase in cell number, not cell size. Tracheids were shorter, often with transverse septae, and many were S-formed and otherwise distorted. In tight girdles, cambial union usually occurred in the callus pad that grew over the ligature.

Eom and Lee (1989) described in detail the anatomical changes induced in a stem of *Ailanthus altissima* girdled with a 19-mm wide plastic band for a full year. They referred to the swollen callus growth that formed above the band as "tumor wood." Although structurally similar to tumorous wood described by others (Sect. 10.5.1) as well as that produced by girdling, the wound tissue was artificially induced and it should not be equated with tumors produced in nature.

A report by Wilson (1968) contained several items of interest. Girdles 2 cm wide were made on *Pinus strobus* stems and the wounds were then filled with petroleum jelly to prevent desiccation. Essentially all cambial division and radial expansion stopped below the girdle, although wall thickening continued slowly in xylem elements that were still differentiating. Above the girdle, in the region where cell lengths and cell diameters decreased markedly, the local effect extended upward from 2 to 10 cm. A long-distance effect in the form of increased cell division extended upward for at least another 310 cm (Table 10.1). This report by Wilson is one of the few that has attempted to measure the long-distance effect of wounding.

Table 10.1. Effect of low girdling on cell production above girdles in *Pinus strobus* trees[a]. (Wilson 1968)

Position	1966 cell number/1965 cell number (%)			
	Control		Low girdle	
	2	7	4	8
310 cm above girdle point	79	125	105	138
210 cm " " "	91	73	107	123
110 cm " " "	98	59	108	103
10 cm " " "	68	51	175	129

[a] Average 1965 cell numbers at all heights for control trees 2 and 7 were 44 and 32, respectively, and for girdled trees 4 and 8 they were 17 and 47, respectively.

Cambial activity does not necessarily stop below the girdle in all species or under all growth conditions. Brown (1937), for example, observed feeble cambial activity at the lower edge of the girdle in *Populus balsamifera*, and others have reported similar responses. Such responses might have resulted from root grafts with neighboring living trees as described by Bormann (1966) in *Pinus strobus*. In these root-grafted trees, radial growth beneath the girdle was extensive over a se-

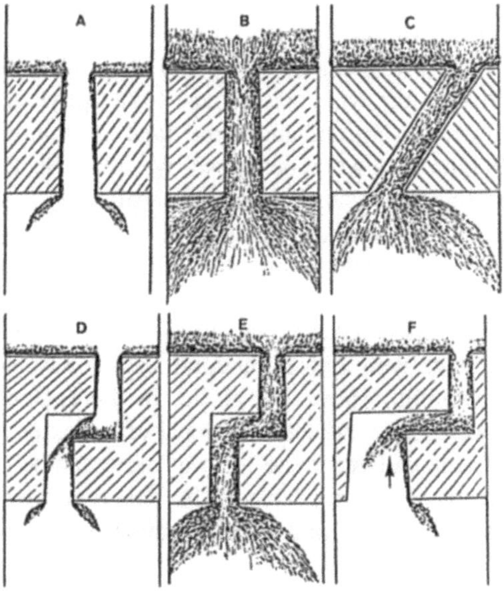

Fig. 10.14A-F. Phloem bridges on *Populus balsamifera. Dashed lines* indicate stem tissue, and *shaded areas* the extent and direction of renewed xylem activity after removal of the bark. *Vertical dashed lines* below the *girdle* in **B** and the *arrow* in **F** indicate areas where cambial activity was either weak or failed to occur, respectively. See text for details. (Brown 1937)

veral-year period. The wood produced immediately below the girdle exhibited grain reorientation just like that in wood above the girdle (Fig. 10.13B,C).

10.3.5.1 Phloem Bridges

Descriptions of phloem or bark bridges of many kinds can be found in the literature and they appear to be limited in structure only by the objectives and imaginations of the investigators. Phloem bridges, within which the cambium remains undisturbed, differ from, and should not be confused with, the tissue bridges that develop after severing vascular bundles in herbaceous stems (Simon 1902, Jost 1942, Bloch 1952, Sachs 1968, 1969, 1975, Benayoun et al. 1975, Jacobs 1979). Research in the latter area has expanded rapidly in recent years, and the findings have much relevance to cambial wounding in woody stems (Hush et al. 1990).

A variety of phloem bridges were installed during the winter on 1- to 4-year-old stems of *Populus balsamifera* and *P. tremuloides* by Brown (1936, 1937). In one type, a longitudinal bridge was left connecting the upper and lower parts of an otherwise complete girdle (Fig. 10.14A). At the resumption of spring growth, callus developed above the girdle but none below. Reactivation of the cambium in spring normally progresses downward in the stem. Brown judged its progress by the extent and vigor of xylem development, mainly by the vessel patterns that became visible on the xylem surface when he periodically debarked phloem bridges for examination. Interestingly, the initial burst of cambial activity did not continue vertically downward but flared obliquely outward below the girdle (Fig. 10.14A). Within the phloem bridge, cambial activity was at first confined to the lateral edges, but with time it spread not only throughout the entire bridge but also more extensively below the girdle (Fig. 10.14B). Feeble callus developed below the lower edge of the girdle but neither fibers nor vessels differentiated within it. A similar pattern of cambial activity and xylem development was observed in oblique phloem bridges (Fig. 10.14C).

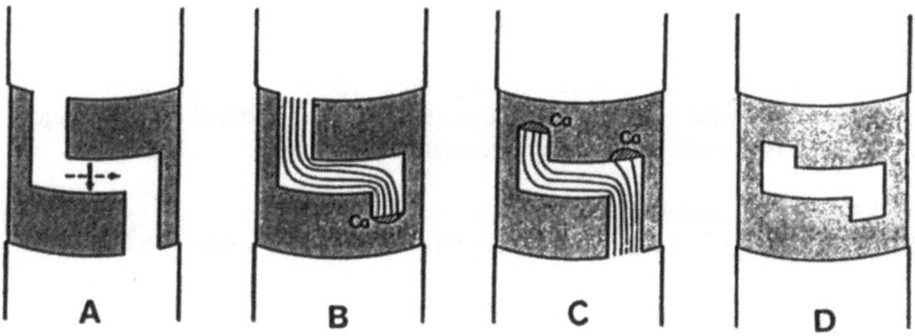

Fig. 10.15A-D. Diagrams showing the reorientation of xylem in phloem bridges on *Robinia pseudoacacia* stems. *Shaded areas* indicate regions from which the bark and cambium were removed, *thin lines* the general course of vessel regeneration, and *Ca* the regions of callus growth. See text for details. (Kirschner et al. 1971)

In a variant of the longitudinal phloem bridge, a short horizontal diversion was made between the two vertical ends producing a so-called zig-zag bridge (Fig. 10.14D). Brown (1937) noted that the first indications of cambial activity in these bridges occurred simultaneously above the girdle and on the horizontal bar of the bridge. Cambial activity within the vertical parts of the bridge and below the bridge were similar to those in the longitudinal bridge. At a later stage of development a marked "piling up" of the new xylem occurred not only above the girdle but also at the lower margin of the horizontal bridge (Fig. 10.14E). In narrow bridges, as in Fig. 10.14E, new xylem development spread obliquely to both sides below the girdle. However, if the bridge was very wide, as in Fig. 10.14F, the initial basipetal progression of cambial activity separated in two distinct paths. One adhered closely to the inner margin of the bark strip and the other diverged outward, leaving an area in which there was no cambial activity. Similar results were obtained by Collins (1930) in *Castanea*, Tupper-Carey (1930) in *Acer pseudoplatanus* and *Laburnum vulgare*, and by many other workers.

Kirschner et al. (1971) applied modified zig-zag phloem bridges to stems of *Robinia pseudoacacia*. A typical zig-zag bridge (Fig. 10.15A) resulted in xylem reorientation similar to that described by Brown (1937). One of the first effects of girdling was the almost immediate differentiation of vessel elements from fusiform initials and reorientation of these vessel strands. When the bridge was isolated from the roots but remained in contact with the shoot (Fig. 10.15B), cambial activity was much reduced and did not proceed very far. The new vessels connected the shoot vasculature with the callus tissue that developed on the lower side of the bridge. When the bridge was isolated from the shoot but remained in contact with the roots (Fig. 10.15C), a few vessels regenerated upward in the bridge. These new vessels connected tissues leading to the root with callus that developed on the upper side of the bridge. In bridges that were isolated from both the shoot and root (Fig. 10.15D), cambial activity ceased almost entirely. However, small patches of immature xylem cells differentiated as disorganized, circular vessels (Chap. 6.6.2.5). The authors concluded that vascular reorientation was due to a hormonal stimulus arising in the growing shoot. The effect of this inductive stimulus in the reorientation of xylem elements was not immediate, but was subject to oscillations in its direction before eventually stabilizing.

10.3.5.2 Spiral Girdles

Spiral or helical girdles are of two main types. Either a bark strip is left spiraling downward around the stem through a girdled region, or a bark strip is removed in a downward spiral around the stem. Response to the first type, applied by Brown (1937) to stems of *Populus balsamifera*, was similar to that of the oblique phloem bridge (Fig. 10.14C). New xylem differentiation was at first confined to the lower side of the bark strip and gradually spread to the upper side. New vessels were oriented in the direction of the spiral.

The second type of spiral girdle is most common, and the general growth response has been known for many years (Büsgen and Münch 1929). Hartig (1858), for example, described how growth essentially stopped on the lower side of the spiral girdle and increased in a swollen pad on the upper side. In the course of the year, all new vascular elements became reoriented in the direction of the spiral. Hartig obviously looked at late stages of development. De Vries (1876), and especially Mäule (1896), described in greater detail how new vascular elements developing in the callus gradually changed their orientations.

The minute anatomy was examined by MacDaniels and Curtis (1930) in spirally girdled stems of *Malus*. First accommodation to the changed conditions for conduction was not the orientation of newly formed cells, but the position of pore openings between vessel elements. Instead of the pore openings being end to end in a vertical series of vessel segments, they were oriented from side to side in a staggered series of elements. These changes occurred very rapidly in new cambial derivatives within the callus pad. Reorientation of the vessel elements occurred more gradually. In time they became reoriented 45° in the direction of the spiral and their pore openings were again end to end. Vessel reorientation was, of course, a reflection of cambial reorientation and all cambial derivatives were similarly reoriented 45° to the stem axis.

Wound response in the phloem bridges applied to stems of *Robinia pseudoacacia* by Kirschner et al. (1971) was almost identical to that of the spiral girdles described above. One of the first responses was the division of fusiform initials into short cells, some of which assumed the characteristics of callus. Other fusiform cells did not divide but underwent considerable enlargement and differentiated as vessel elements (Fig. 10.16A). These vessel elements, with perforations in their radial walls, formed networks in the direction of the girdle. Short cells associated with the vessels grew intrusively in the general direction of the girdle. In later-formed wood, derivatives of the short cells differentiated as fibers. The direction of elongation of these fibers was at first inconsistent and many fibers crossed one another (Fig. 10.16B). Eventually, however, both fibers and vessel elements became completely reoriented.

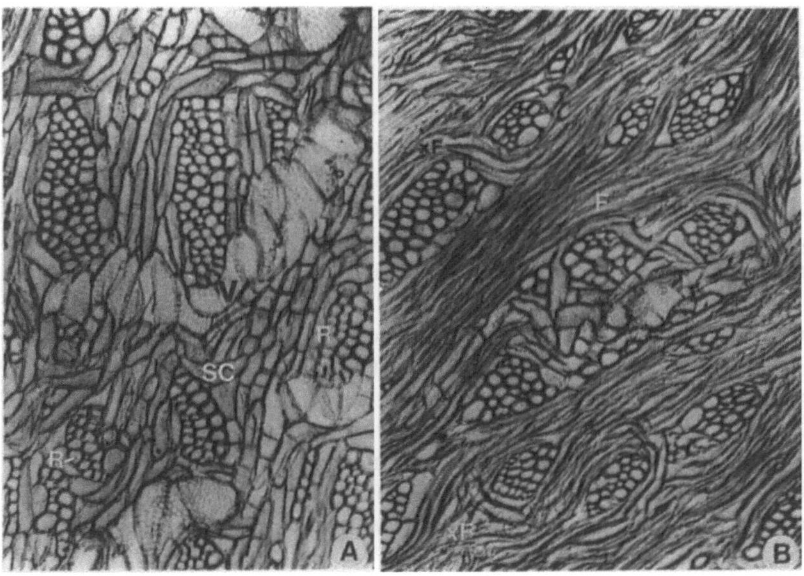

Fig. 10.16A,B. Tangential sections of *Robinia pseudoacacia* showing late stages of xylem reorientation in phloem bridges. **A** Vessel elements (*V*) have formed a network in the direction of the phloem bridge. Short cells (*SC*) have grown intrusively, but their direction of growth is not consistent. Rays (*R*) are heterogeneous and they have irregular shapes. **B** At a later stage, short cells have elongated and differentiated as fibers (*F*). Again, the direction of intrusive growth is not always consistent, and fibers often cross one another in the vicinity of rays (*xF*). 123x. (Kirschner et al. 1971)

Although conifers do not produce vessels, the general response to spiral girdles is similar. The change in xylem cell reorientation in *Pinus radiata* was followed by Harris (1969, 1973). Again, cambial initials in the region above the girdle divided frequently both periclinally and anticlinally, the latter producing many short cells. The xylem derivatives elongated by intrusive growth preferentially in the direction of the girdle. Rays responded more slowly with the result that many tracheids tended to bend around the more vertically oriented rays. Harris attempted to show that the reorientation of xylem tissues after spiral girdling was analogous to that experienced by xylem tissues during spiral grain formation (Chap. 6.6.2.3).

Preferential orientation of the dividing wall during anticlinal division of fusiform initials is apparently a decisive factor in reestablishing conduction channels in conifers. Savidge and Farrar (1984) examined these events in 45° oblique bridges in *Pinus contorta*. Although not true spiral girdles, the wound responses were closely related to those described by Harris (1973).

Savidge & Farrar (1984) analyzed 1800 μm of wound-response tissue in 20 μm thick, serial, tangential sections to determine the rate and degree of cell reorientation. Section 0 μm represented xylem that differentiated immediately before construction of the diagonal bridge. At Section 500 μm, anticlinal divisions had occurred in both fusiform and ray cells, the latter indicating the formative stages of a radial resin canal. At Section 1000 μm, obvious changes were evident in the radial files of tracheids, particularly a substantial decline in their lengths. The post-girdling increase in the frequency of anticlinal divisions did not elicit an immediate change in tracheid orientation. First evidence of a change in orientation occurred about Section 1600 μm, when the tips of tracheids began shifting in the direction of the diagonal bridge. Most tracheids at this time were about one-half the length of those in Section 0 μm. Additional anticlinal divisions occurring in a preferred direction, together with either cell shortening or the loss of some radial files, contributed to the further reorientation of tracheids in subsequent sections. However, Savidge and Farrar (1984) found no evidence that intrusive growth of the tracheid tips might have contributed to their inclination as suggested by Harris (1973). Rather, they suggested that cell shortening might have been an essential mechanism for transposing tracheid tips from one side of a ray to the other. As in *P. radiata* and other species, the rays of *P. contorta* resisted the change in grain inclination.

A different conclusion was arrived at by Zagórska-Marek and Little (1986), who made spiral girdles 5 mm wide extending downward at a 60° angle for 1.5 turns on stems of *Abies balsamea*. At the end of the growing season, the radial width of new xylem increased from the top to bottom ends of the girdle. Tracheid realignment was not complete at season's end, averaging only about 23° in the direction of the spiral girdle. Girdling induced the same wound responses found by others in conifers, such as a decrease in fusiform initial length, increased frequency of anticlinal divisions, reduction in ray heights, conversion of many rays from uniseriate to biseriate, and formation of traumatic resin canals. Rapid fusiform initial reorientation was facilitated in two ways. First, one anticlinal division often was associated with a second one in a tangentially adjacent initial, thereby allowing oriented intrusive growth to begin promptly (Fig. 10.17A). Second, in-

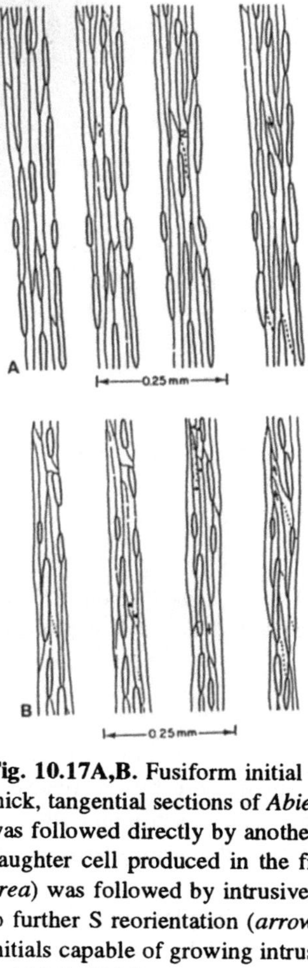

Fig. 10.17A,B. Fusiform initial reorientation in an S-oriented bridge as reflected in 30 μm thick, tangential sections of *Abies balsamea* xylem. **A** An anticlinal division (*broken line 1*) was followed directly by another (2), enabling immediate S-oriented intrusive growth by a daughter cell produced in the first division (*arrow*). **B** Elimination of an initial (*stippled area*) was followed by intrusive growth in the end of the initial positioned most favorably to further S reorientation (*arrows*). *Broken lines* Anticlinal divisions; *dark circles* ends of initials capable of growing intrusively into spaces created by elimination. (Zagórska-Marek and Little 1986)

trusive growth of one initial into the space created by the elimination of another initial typically occurred in the end of the initial that was best located to promote realignment (Fig. 10.17B). Thus, Zagórska-Marek and Little agreed with Harris (1973), but not with Savidge and Farrar (1984), with regard to the role of intrusive growth in inducing tracheid reorientation.

10.3.6 Grafting

Another commonly practiced mechanical injury dating back to ancient times is grafting (Roberts 1949, Dormling 1963). Grafts are made in many ways. Conse-

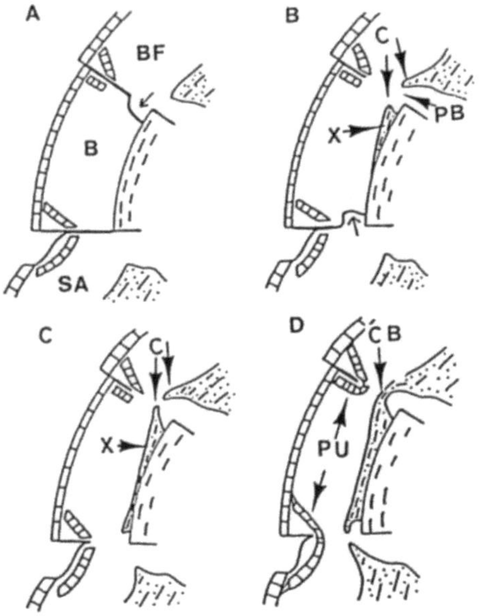

Fig. 10.18A-D. Closure of side-split grafts in *Populus*. *SA* Stock axis; *B* scion; *BF* bark flap; *C* cambium; *PB* parenchyma bridge of advancing cambium; *X* new xylem; *CB* cambial bridge; *PU* periderm union. (Braun 1958)

quently, the wound response will depend on the type of graft, although in every case the first response is callus formation. Early botanists were by no means in agreement as to the source of the callus, whether from stock or scion and whether from cambium or rays. The general belief, however, was that callus arose from cambium of the stock plant.

In cleft-grafts of *Hibiscus rosa-sinensis*, made by Sharples and Gunnery (1933), the bulk of the callus was derived from phloem ray cells of the stock. Derivatives of the phloem rays proliferated to fill the interstices between stock and scion with parenchymatous callus, and there was no evidence that the cambium participated during these early stages. When the callus cushion was fully formed, both phellogen and vascular cambium advanced peripherally through the callus. All evidence at this time indicated that new cambial elements developed in complete continuity with pre-existing cambium.

Several types of graft union on 1- to 4-year-old *Populus* stems were investigated by Braun (1958, 1959, 1960). Because of the vigor of these young stems, many tissues contributed to initial callus formation. In side-slit grafts (Braun 1958), in which the scion was inserted between the xylem and a bark flap on the stock, the first cell divisions were confined to the stock. Callus formation was derived from the stock cambium, undifferentiated xylem cells, and the rays. Parenchymatous cells of the primary and secondary cortex became active somewhat later. This callus cushion, called intermediary tissue, developed more rapidly on the wood side than on the bark flap side, forcing a tight union between stock and scion (Fig. 10.18A). It was at this time that tissues of the scion became activated. Tissues produced by cambia of both the scion and bark flap of the stock first formed a parenchymatous tissue bridge (Fig. 10.18B) and, quickly thereafter, a cambial bridge and cambial union (Fig. 10.18C,D). The gap between the scion and underlying wood of the stock was bridged in the same way.

Braun (1958) invoked the concept of "homogenetic induction" to account for advance of the cambium (Fig. 10.19). According to this concept, a stimulus originating in pre-existing cambium induces its progressive advance by dedifferentiation of cells across the parenchymatous tissue bridge. Thus, the cambium advanced in the graft union in much the same way that others had observed it advancing during wound closure (Fig. 10.5). Although Warren Wilson and Warren Wilson (1981) described the progressive advance of the cambium in graft unions in much the same way as Braun (1958), they referred to the inductive process as "gradient induction."

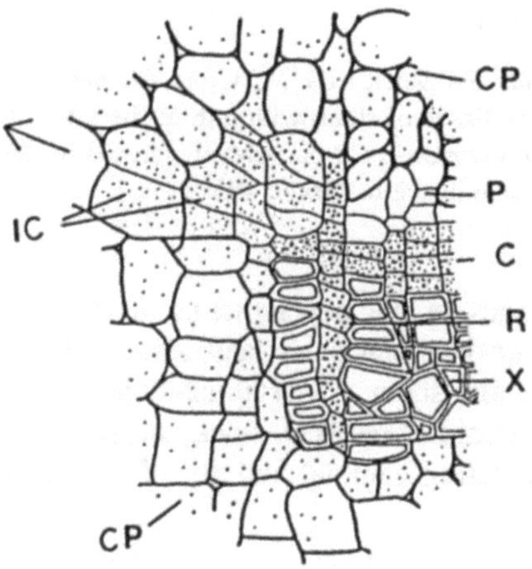

Fig. 10.19. Homogenetic induction of cambial development progressing from existing cambium (*C*) in direction of the *arrow* by dedifferentiation of intermediary cortical parenchyma (*IC*). *CP* Cortical parenchyma; *P* phloem parenchyma; *R* ray; *X* xylem fibers. (Braun 1958)

In veneer side grafts described by Braun (1959), intermediary tissue was derived entirely from the cambium, undifferentiated xylem cells, and inner phloem of the stock. Like the side-slit grafts, build-up of this callus cushion forced a tight union between stock and scion. Shortly thereafter tissues of the scion, which had remained passive until this time, now became activated. Cambial bridges developed through the intermediary tissue uniting stock and scion (Fig. 10.20).

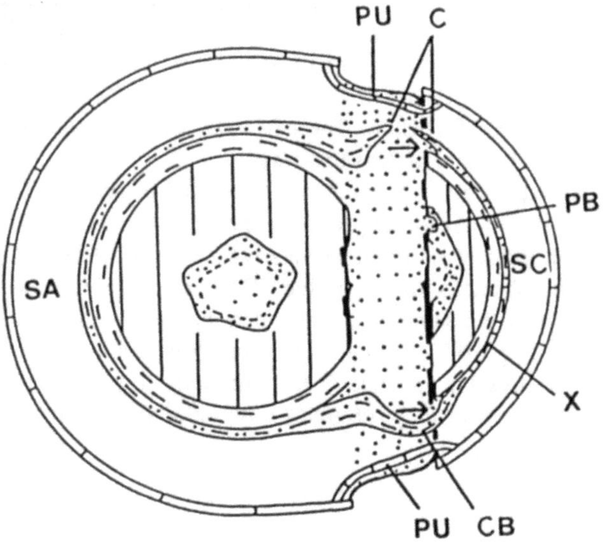

Fig. 10.20. Closure of side-veneer graft in *Populus*. *SA* Stock axis; *SC* scion; *PU* periderm union; *PB* parenchyma bridge uniting intermediary tissue of stock with pith of the scion; *C* cambium of stock developing in direction of scion (*arrow*); *CB* completed cambial bridge between stock and scion; *X* renewed xylem formation in scion. (Braun 1959)

Braun (1959) recognized three main stages in the development of all graft unions: (1) establishment of primary contact between stock and scion; (2) increasing build-up of pressure between stock and scion; (3) establishment of normal tissue contact. In addition to good initial contact between stock and scion, he stressed the importance of pressure in stimulating cambial activity of the scion and promoting a successful union.

An extremely detailed anatomical investigation of several types of graft union in *Pinus sylvestris* and *Picea abies* was conducted by Dormling (1963). She found that cell divisions initiating callus were most vigorous in regions best suited for storage and conduction, i.e., phloem ray tissues (Table 10.2). Although both stock and scion might contribute to the callus, contributions from the stock usually were most vigorous. In some veneer grafts of *P. sylvestris*, but not *P. abies*, wood and pith parenchyma cells contributed to some extent to callus formation. Cambial fusiform initials and mother cells were conspicuously inactive during the initial stages of callus formation. These cells would have to undergo repeated transverse divisions before they could contribute to callus, and in well matched grafts, the scion and stock were united before such divisions were completed.

According to Dormling (1963), the first cells to unite in the cambial regions were always short and isodiametric, and they consequently differed from normal fusiform cambial cells. Cambial cells arose by dedifferentiation of the parenchy-

Table 10.2. Intensity of cell division in different tissues of grafts of *Pinus sylvestris* and *Picea abies* (Dormling 1963)

Tissue	*Pinus*	*Picea*
Periderm	-	-
Cortex, ground tissue	+++	+++[a]
Cortex, resin ducts	++++	++++[a]
Phloem, rays	++++	++++[a]
Phloem, vertical parenchyma	++	++[a]
Phloem, completely differentiated sieve cells	-	-
Cambial region, rays	+++	+++
Cambial region, other undifferentiated or incompletely differentiated cells	++	++
Xylem, rays	+	-(+)
Xylem, resin ducts	+	-
Xylem, completely differentiated tracheids	-	-
Pith, leaf and branch gaps	++	-

a Tissues influenced by leaf and branch traces show greater activity than others.
++++ Very high intensity of cell division.
+++ High intensity of cell division.
++ Intensity of cell division varies; in pith, leaf, and branch gaps of *Pinus* divisions occur at an early stage, otherwise not until neighboring tissues have started to divide.
\+ Divisions in variable amounts can occur, but often fail to appear.
(+) When late grafted, ray cells in the newly formed xylem may divide.
\- No divisions.

matous callus cells, progressing upward from stock to scion and downward from scion to stock. The new cambium arose by a process of induction, always in continuity with pre-existing cambium in both stock and scion. Xylem differentiation invariably preceded that of phloem in the graft unions.

Mergen (1954) also found that phloem rays of the stock plant, and other parenchymatous tissues to a lesser extent, were the primary contributors to callus formation in side-slit grafts of *Pinus elliottii*. He, too, considered the contribution of cambial cells to the initial stages of graft union, but not the later stages, to be overemphasized by previous investigators. Origin and development of callus and differentiation of cambium proceeded in much the same way in cleft grafts of *Pseudotsuga menziesii* (Copes 1969).

Contrary to the foregoing investigators, Weatherhead and Barnett (1986) and Barnett et al. (1990) reported that recent cambial derivatives as well as phloem parenchyma contributed to callus in graft unions of *Picea sitchensis*. At the edges of graft interfaces in this species, the cambia of scion and stock continued to proliferate until they joined.

10.3.6.1 Graft Inversions

Cell polarity plays a prominent role in all types of wound closure. It accounts for the distorted elements that often form above transverse wounds, as well as the reorientation of elements in spiral girdles. Mäule (1896) included detailed descriptions of polarity phenomena in his study of wounding.

Sax and Dickson (1956) examined polarity by grafting a ring of bark from a *Malus* tree with red wood and red bark onto a *Malus* tree with white wood and green bark. At season's end, new wood that formed beneath the ring of red bark was red, but the wood formed at the vertical seam was white. The red wood beneath the graft was derived from the cambium adhering to the bark patch, whereas the white wood at the seam was derived from regeneration of the underlying xylem of the host tree. When the bark ring was grafted in an inverted position, the result was the same except that very little wood formed under the grafted red bark because of reversed polarity of the phloem.

In a further test of this phenomenon, one-inch squares of bark were removed from *Malus* trees with white wood and replaced with similar bark squares from trees with red wood. A smaller square of bark was removed from the center of the red square leaving the white wood beneath exposed. The grafts were covered with polyethylene film to prevent desiccation. The grafts of red wood and bark were oriented in the normal position and also at 90° and 180° orientations from normal on the white wood. Although xylem growth beneath the inverted and laterally oriented grafts was much reduced, the regenerated xylem elements were oriented in a normal vertical position at season's end. In all grafts, the new wood beneath the graft patch was derived from cambium of the scion.

Additional studies using these "patch grafts" were conducted by Thair and Steeves (1976) on *Cornus stolonifera, Sorbus aucuparia, Thuja occidentalis, Malus baccata* (with white wood), and two varieties of *Malus sylvestris*, one with pink and one with red wood. All tests were made in spring and early summer when the bark separated easily and the cambium remained on the isolated bark patches. Most tests were performed with *Malus* spp.

Bark patches oriented normally recovered rapidly and they appeared to be completely functional when examined 62 days after wounding. Early development of callus at the interface between stock and scion of bark patches oriented 90° from normal was similar to that of the normally oriented grafts. In both cases, callus arose mainly by proliferation of ray cells from both stock and scion. However, the cambium of the bark patch remained in its original morphological position at right angles to the stock axis throughout the first growing season. This position was verified by the fact that all new xylem beneath the graft retained the color of the scion. Consequently, it was derived from cambium of the scion; but the new vessels were considerably shortened and their perforation plates were radially rather than terminally oriented. In the second and third seasons, vessel element lengths gradually increased and perforation plates resumed their normal, terminal positions despite the fact that neither the scion cambium nor its derivatives deviated from the original 90° orientation. Conduction between stock and scion was presumably facilitated by the reorientation of vascular tissues at the lateral edges of the bark patches.

Thair and Steeves (1976) found that inverted bark patches oriented 180° to the tree's axis were slower to heal and secondary growth was drastically reduced. The few vessel elements that were produced had perforation plates located predominantly on the radial walls. On the basis of the results and observations obtained from these tests, the authors concluded that the cambium possesses a high degree of autonomy in its organization. Moreover, it probably possesses an inherent polarity as suggested by Vöchting (1906).

10.3.6.2 Natural Root Grafts

Root grafts are natural grafts resulting from the pressure of intertwined roots. Although intraspecific root grafts are commonly encountered in forest stands, interspecific root grafts are infrequent. Detopped trees (Hartig 1890) and so-called "living stumps" (Graham and Bormann 1966, Liphschitz et al. 1987) (Chap. 6.5.4.5) have survived for years because of root grafts with neighboring living trees.

Bertog (1895) encountered "missing" growth rings at the mid-stem level in his investigation of suppressed *Picea* trees. The rings were distinct and complete in the upper stem and they reappeared at the stem base. Bertog suspected that enhanced growth at the stem base was due to root grafts with neighboring trees. A comprehensive review of the prevalence of natural root grafting has been prepared by Graham and Bormann (1966). Despite extensive research on the subject, relatively few studies have dealt with the anatomy of the graft union.

One of the earliest anatomical investigations of root grafting was by Göppert (1846). He attributed Reum in 1835 with being the first to relate living stumps to root grafting. Göppert considered root grafting to be the rule in *Abies*, the exception in *Picea*, and never to occur in *Pinus*. Bark grafts, but never cambium or wood grafts, occurred in *Pinus*. He also claimed to have observed a wood graft between *Abies* and *Picea* roots. Root grafts were presumed to result from mutual pressure of the intertwining roots. The union was at first mechanical, but later the inner bark tissue was believed to be "resorbed" and carried away in the sap stream. Göppert cited many reports by his contemporaries who had observed similar instances of root grafting and living stumps.

Early work on root grafting was reviewed by Franke (1883). In his own research he was unable to find suitable root grafts in the forest, so he used the aerial roots of *Tecoma radicans, Hedera helix,* and *Hoya carnosa*. On the basis of his observations, he agreed with Göppert and others regarding the three main requirements for successful root grafting: (1) mutual root pressure; (2) nature of the plants; only individuals of the same species were capable of root grafting; (3) cha-

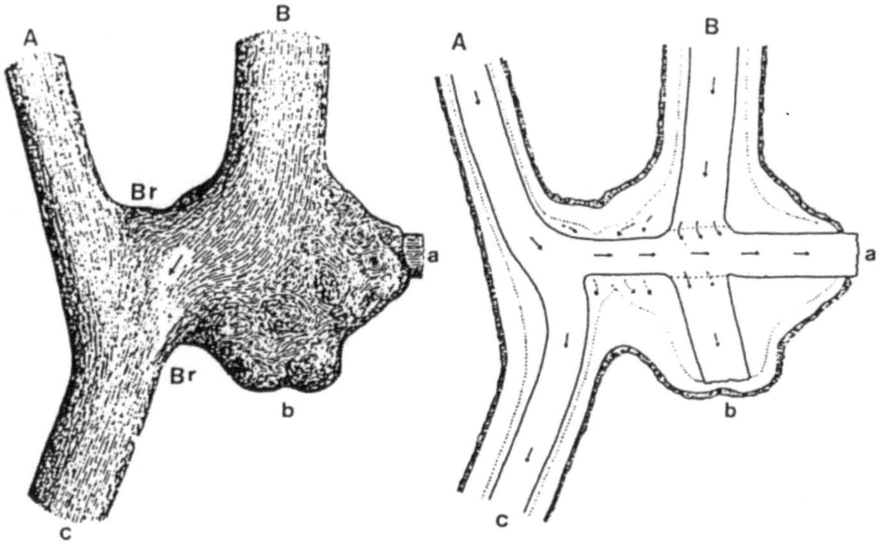

Fig. 10.21. Root graft in *Fagus sylvatica*. *Left* Debarked root showing grain patterns. *Arrow* through tissue bridge (*Br*) points in direction of root apex. *Right* Schematic drawing of internal root body. *Arrows* indicate the direction of growth of roots *A* and *B*. (Neeff 1922)

racteristics of the tissues; only young tissues capable of dividing could form grafts (cambium, soft bast, and rind parenchyma).

Franke was able to observe root grafting in young roots of *Fagus sylvatica*. At the contact area where mutual pressures increased, the bark "verottet," or disintegrated, and formed brown, isolated patches between the two roots. Initial union occurred throughout the rays, which by active divisions broadened fan-like in all directions. The rays formed a meristematic tissue that enclosed and displaced the dead bark elements. These dead elements were presumably resorbed by the rays of one root which eventually united with rays of the other root. Meanwhile, the cambia of the two roots united to form a common cambial cylinder.

A mature root graft in *Fagus sylvatica* is shown in Fig. 10.21. Root grafts, like many exposed wounds, display regions of distorted and convoluted vascular elements in the region of the graft union. Neeff (1922) attributed these distortions to a disruption of polarity. That is, the normal polarity along the root was disrupted by the bridge of tissue resulting from the root graft.

Another study of root grafting in aerial roots of *Ficus globosa* was made by Rao (1966), although no mention was made of the earlier work of Göppert (1846) and Franke (1883). As described by Rao, first contact between two aerial roots was by the fusion of epidermal hairs which established cortical connections. As secondary growth continued, the two roots became firmly adpressed, the cortical cells became narrower, and they gradually lost their identities. Further pressure

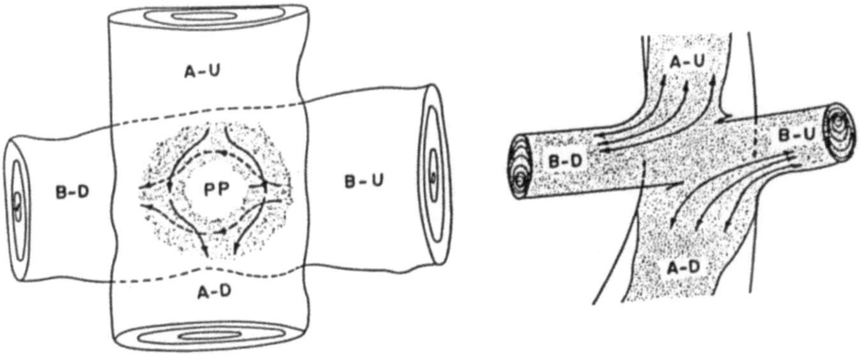

Fig. 10.22. Root grafting in *Pinus strobus*. *Left* Two nongrafted roots (*A*,*B*) in the early stage of graft union. The letters *U* and *D* represent root parts that will be upgraft and downgraft, respectively. *PP* represents the pressure point and the *stippled area* the doughnut-shaped tissue area. *Right* Diagram of an excised, functional root graft. *Arrows* indicate the grain orientation through the graft region. (Bormann 1966)

was exerted on the cortical tissues by the xylem rays that enlarged and produced abundant parenchyma in both roots. Ultimately, the cortex became completely indistinguishable in the contact region and the enlarged ray tissues merged to establish a continuous parenchymatous zone between the two steles. Meristematic cell layers formed by further differentiation of some of these parenchymatous cells. A new cambium was then established by dedifferentiation of the parenchymatous cells as pre-existing cambia from the two roots advanced across the fused zone. Cambial fusion completed formation of a cambial ring around the two roots. In this union, no callus was produced. Rather, new cambial cells arose by dedifferentiation of parenchymatous tissues derived from xylem ray cells.

The most intensive study of root grafting was made by Bormann (1966) in *Pinus strobus*. A typical graft formed at a "pressure point" where two roots overlapped (Fig. 10.22). When cambial continuity was established, grain reorientation began in the doughnut-shaped ring surrounding the pressure point. Development of new vascular tissue did not proceed equally in all areas of the ring. Strongest grain development occurred between A-U and B-D, and B-U and A-D, where U was the "upgraft" part of the root nearest the bole and D the "downgraft" part nearest the root tip. Vascular connections between A-U and B-U, and A-D and B-D not only were very weak but contained highly convoluted grain patterns. Analysis of grain alignment led Bormann to conclude that orientation of cambial deri-

Fig. 10.23. Natural stem graft between two *Fagus sylvatica* stems. (From Neeff 1922, who reproduced it from a 1885 paper by Ziegler)

vatives was determined by substances moving upward in the xylem rather than downward in the phloem.

10.3.6.3 Natural Stem Grafts

Natural stem grafts are less common than root grafts. When they do occur, stem grafts most often unite stems of the same species. Seidel (1879) described how a 60- to 80-year-old *Pinus sylvestris* stem had become united with a *Fagus sylvatica* stem of similar age to form a single, rounded stem. He did not examine these stems to determine whether vascular union had occurred. However, based on results of other investigators, we might assume that the stems were simply intertwined rather than vascularly intergrown. As a more plausible example, Seidel described two *F. sylvatica* stems, 10.0 and 12.5 cm in diameter, that had become united at the stem base to form a single 20 cm stem extending upward 10 to 12 m.

Like root grafts, stem grafts occur by the mutual pressure of interlocking stems or branches (Fig. 10.23). A particularly clear description of the process was provided by Küster (1899) who examined *Fagus sylvatica, Ficus stipularis, Hedera helix, Platanus* sp., and *Quercus* sp. The first effect of pressure was a flattening of the stem and deviation of the rays. The former was brought about by a decrease in cambial activity at the point of contact and an increase on the flanks where a bulge of tissue was produced, particularly in *F. stipularis*.

In thin-barked species, such as *F. stipularis* and *H. helix*, Küster described how the outer bark between the two stems often became abraded, exposing the tender cortical cells beneath. Walls of the phloem ray cells thickened, lignified, and became pitted. Cell division in deeper layers resulted in a tissue consisting of thin-walled, unlignified parenchyma interspersed with thick-walled, lignified cells that united cortical tissues of the two stems. Union of the underlying tissues was more complicated, and here the cambium played the principal role. Complications were due to the included bark and cortical tissues that were invariably present. With increased mutual pressure, a new cambial zone formed between the two stems, but regions of included bark remained.

In older stems of all species, included bark at the point of contact restricted growth of the underlying tissues. In no case was Küster able to confirm the resorption of bark by ray tissues as described by Franke (1883) in root grafts. The bark inclusions, in effect, forced the cambium to develop around them and in so doing the cambium produced abnormal tissues. For example, in *Fagus sylvatica*, a homogeneous mass of thick-walled, lignified and pitted parenchymatous tissue was produced. Therefore, before the cambium could begin producing normal tissues, it must first be transformed to a parenchymatous meristem. Küster believed that this transformation was the result of increased pressure because true parenchyma with cells arranged in radial series formed in regions where pressure was presumed to be greatest; but pressure was not the cause, it simply provided the stimulus.

In *Fagus* it was difficult for Küster to establish the transition boundary between two stems because in some areas prosenchymatous and in closely adjacent areas parenchymatous cells were being produced. The transition to normal wood was very diffuse. In contrast, the transition to normal wood formation was rapid and very abrupt in *Platanus*. The newly formed libriform fibers and vessels of *Fagus, Quercus*, and *Platanus* stem grafts were often inclined 90° from normal and distorted in C- and S-forms with all gradations in between. These abnormal cells reflected the condition of cells in the cambium, which presumably was responding to increased physical pressure of the stem union.

10.3.7 Pruning Wounds

Green pruning of forest trees has been widely practiced over the years in attempts to improve both stem form and wood quality. Past experience suggests that

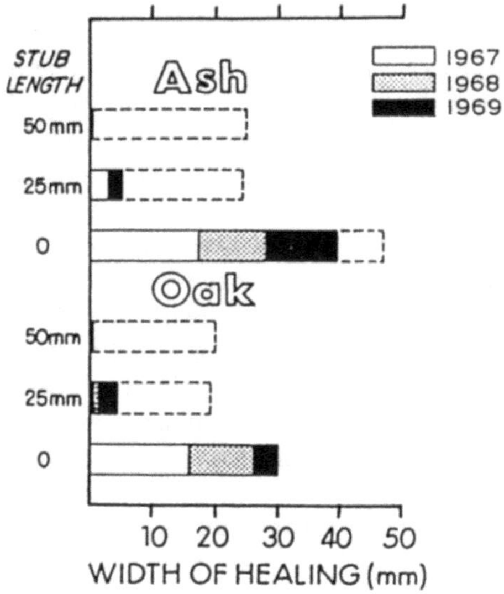

Fig. 10.24. Healing of pruned branches of *Fraxinus americana* and *Quercus palustris* with residual branch stubs of 0, 25, and 50 mm. Each value is the average of 18 branches, 16 to 25 mm in diameter, pruned during May 1967. *Dotted lines* indicate the width of wood still exposed in September 1969. (Neely 1970)

pruning wounds made flush with the stem, or with the branch collar, close more rapidly than those leaving a protruding branch stub (Skilling 1958, Neely 1970, Shigo 1986) (Fig. 10.24). In fact, callus rarely or ever forms on the exposed stub (Mayer-Wegelin 1952). Failure of the exposed surface to form callus has been attributed to maturation of the ray and xylem parenchyma cells of the branch wood. These mature cells do not proliferate even when kept moist (Brown 1971).

Wound closure, or healing, has been described as a two-stage process in those wounds created by pruning flush with the branch collar (Romell 1940). In the first stage, stem wood grows around the residual branch stub until the stub has "healed in," i.e., radial growth has advanced to the cut surface. In the second stage, "healing over proper," new wood begins to cover the surface end of the stub. It has often been claimed that the closure rate of pruning wounds is related to tree vigor. In testing this assumption, Neely (1970) found that wound closure was not necessarily related to tree vigor but to radial growth at the wound site.

Some workers have suggested that wound closure could be stimulated by cutting through the branch collar (Curtis 1936, Roth and Hepting 1943, Herring et al. 1958). Neely (1970) provided data indicating that cuts through the branch collar closed more rapidly and consistently even though the wound areas were considerably enlarged. Mayer-Wegelin (1952) suggested that this procedure exposed the longitudinally oriented fibers (cambial cells) and these fibers would then divide periclinally. Shigo (1984a), however, vociferously deplored this practice, claiming that it not only delayed wound closure but created entrance channels for microorganisms. He recommended pruning flush with the branch collar, but not through it. In earlier work, he (Shigo et al. 1979) found that pruning cuts through the branch collar of *Juglans nigra* often resulted in ring shakes, which in turn created planes of weakness for radial separation. When a radial separation extended to the cambium, a seam called a "frost crack" formed.

According to Mayer-Wegelin (1952), wound callus is produced by the stem cambium, not by the branch cambium. In pruning wounds on *Carya* stems examined by Brown (1971), callus formed at the cambial zone around the wound periphery and it grew centripetally until the branch scar was covered. Callus growth was not uniform, however. Callus on the lateral flanks grew more rapidly than that at the top and bottom resulting in a spindle-shaped scar. Such a wound-closure pattern has been commonly observed following green pruning (Romell 1940, Mayer-Wegelin 1952, Neely 1970). Roth (1948), for example, observed that pruning wounds in *Quercus* stems remained at approximately their original vertical lengths for a variable number of years after they had completely closed by lateral callus growth.

Tissue organization above pruned branches resembles that above exposed wounds. For example, Kienitz (1878) examined wound closure following live branch pruning in both conifers and dicotyledons. In the region immediately above the wound, the elements were twisted and distorted resembling whirled grain (Chap. 6.6.2.4). Kienitz attributed this pattern to the fact that the elements were normally oriented parallel to the stem axis, but pruning caused first disorientation and then reorientation of the cambium around the wound.

Closure of the so-called branch hole following pruning was considered from a mechanistic standpoint by Mattheck et al. (1990). They postulated that "notch stresses" at the branch hole might be responsible for the stimulation and regulation of the wound-closure process. Highest stresses were found in regions of branch holes, i.e., pruning wounds without protruding branch stubs. In those with protruding branch stubs, stresses increased with increasing amounts of rot in the branch stub. Stresses were also high at the lateral edges of branch holes where callus formation was rapid. To further test their hypothesis, the authors cut circular windows in the stem to simulate a branch hole. The observed callus formation compared favorably with the notch-stress distributions calculated numerically.

A unique type of branch-wound closure was observed in *Adansonia digitata* by Fisher (1981). When large branches were transversely cut, the entire surface of secondary xylem and pith responded by producing wound tissue. Pith, axial parenchyma, and xylem ray parenchyma cells that had differentiated 16 to 21 years earlier were still capable of dividing and contributing to formation of callus-like tissue that sealed the exposed, transverse surface. In large, cut branches, a cambium still had not differentiated within this parenchymatous tissue after 15 years. The protective, bark-like layer was maintained by continued cell divisions below the wounded surface and within the callus-like tissue.

10.3.8 Decapitation

Removal of a growing bud or shoot by decapitation is usually followed by a subordinate axillary bud growing out to assume the role of a new terminal. Provision of new vascular pathways to this new terminal shoot often requires significant changes in cellular organization. Jost (cited by Neeff 1914), for example, had observed in *Aesculus* that wood fibers were oriented parallel to the stem axis in intact shoots but inclined in the direction of a new side shoot after decapitation. Neeff (1914), one of Jost's students, re-examined this effect in nine woody species in which side shoots were allowed to grow for varying periods of time after decapitation of the terminal. In all tested plants, the cambia became reoriented in the direction of the side shoots, with the time of change occurring fastest in the most vigorous side shoots and in those nearest the site of decapitation.

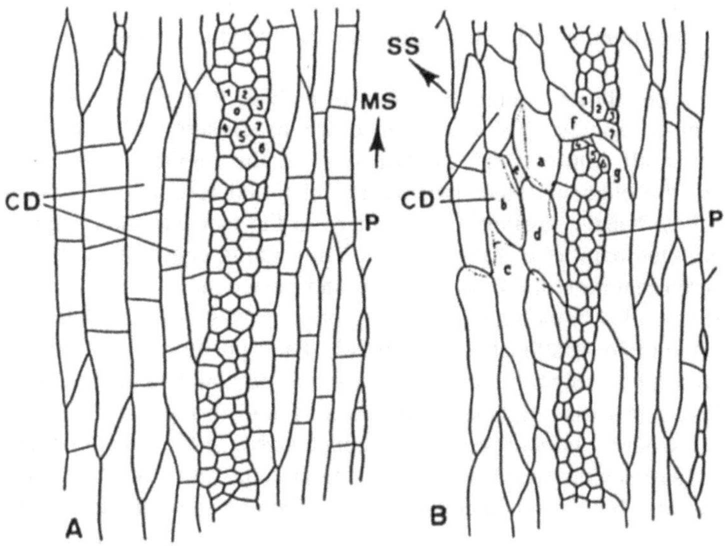

Fig. 10.25A,B. Cambial reorientation in *Tilia americana* shoots after decapitation. **A** Cambial cells have divided transversely to produce short daughter cells (*CD*) oriented in the direction of the main shoot (*MS*). **B** Cambial daughter cells have become more rounded and inclined in the direction of the new side shoot (*SS*). Cells *f* and *g* penetrated the pith during their reorientation. (Neeff 1914)

The first stage in the reorientation process was an intense increase in the number of transverse divisions at which time cambial cells divided repeatedly to produce short cells (Fig. 10.25A). During the next stage, the cambial cells became more rounded and inclined toward the side shoot. Occasionally, the elongating tips of reorienting cells penetrated between ray cells breaking them into smaller units and also through the pith (Fig. 10.25B). Crowding of cells during the reorientation process also resulted in the interpenetration of elongating tips between the walls of neighboring cells (Fig. 6.14). In subsequent stages, the cells were completely reoriented in the direction of the side shoot, and the previous transverse walls became strongly oblique. The same reorientation process and splitting of rays occurred in *Tilia tomentosa* roots that had been decapitated (Fig. 10.26).

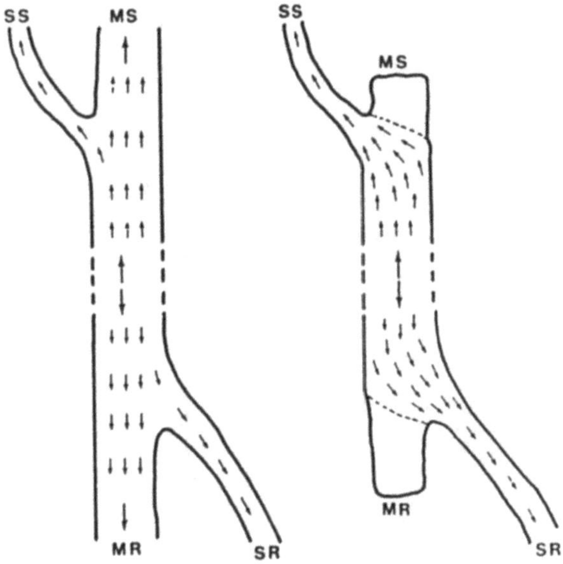

Fig. 10.26. Schematic representation of fiber orientation in *Tilia tomentosa* before (*left*) and after (*right*) shoot and root decapitation. *MS* Main stem; *MR* main root; *SS* side shoot; *SR* side root. (After Neeff 1922)

10.3.8.1 Cladoptosis

Cladoptosis is the natural shedding of branches. It involves the separation of branches from the parent plant through well-defined cleavage zones that are produced by tissue weakening and periderm formation. In trees with pendulant or weeping branches, such as *Picea abies*, growth rings are often missing in the branch extremities and xylem cells that are formed are structurally atypical as described by Rubner (1910b). In other species, such as *Salix babylonica* and especially *Sophora japonica* var. *pendula*, certain of the pendant branch parts are lost through "natural decapitation," or cladoptosis (Neeff 1914). Neeff examined cell reorientation following growth of new side shoots in the latter species. The process was much the same as that following artificial decapitation in *Aesculus*. Cambial cells first divided anticlinally and then elongated in the direction of the side shoot, splitting the rays in the process. The first derivatives of the inclined cambial cells were predominantly short, parenchymatous cells. With increasing inclination, the paren-

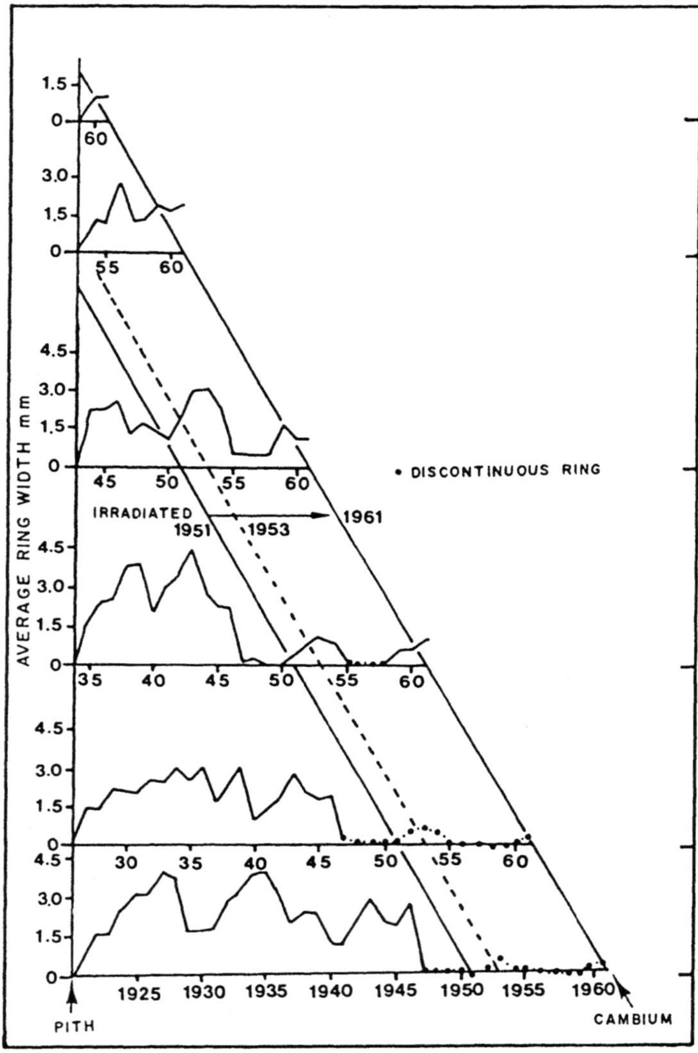

Fig. 10.27. Long-term effect of ionizing radiation (average 3 R/day) on diameter growth at different heights in a *Pinus rigida* tree. Growth reduction increased progressively downward resulting in discontinuous rings at the stem base. (After Sparrow et al. 1965)

chyma cells became longer as did the vessel elements included among them. Xylem fibers, however, were distorted, often S-shaped, with gabled and forked tips. These fibers were only 0.2 to 0.6 mm long, as opposed to a normal length of 0.8 to 1.0 mm. Neeff observed the same processes in sympodial formation of the branch systems of *Populus, Sambucus,* and *Ulmus,* genera that also exhibited cladoptosis.

Jacquiot (1977) attempted to demonstrate in *Castanea vesca* that tissue organization depended on vegetative organs in the crown. That is, when the vegetative organs were removed by decapitation, the cambial-produced tissues of the axis became completely disorganized. His experiments demonstrated that normal organization could be completely restored by grafting buds to the decapitated stems.

Wound closure following leaf and twig abscission and cladoptosis have been thoroughly reviewed by Addicott (1981) and Millington and Chaney (1973), respectively.

10.4 Ionizing Radiation

The effects of ionizing radiation by gamma rays on forest trees are, in general, unfavorable (Lynn 1967). Sensitivity to irradiation varies not only with dosage received but also with species, season of year, and a host of growth conditions. Terminal shoots and buds are more radiation-sensitive than the cambium (Brandenburg et al. 1962), possibly because of protective effects of the bark. For example, dosages required to inhibit terminal and radial growth of *Pinus taeda* were 1000 and 3000 rads[1], respectively (Pedigo 1963).

Long-term exposure of *Pinus rigida* to low dosages of about 3 roentgen/day showed the effects of cumulative exposure on radial growth (Woodwell and Miller 1963, Sparrow et al. 1965). Radial growth was reduced at all stem levels in irradiated trees, but to a greater degree at the stem base (Fig. 10.27). This effect was accentuated during unfavorable growing seasons when growth rings were often completely absent at the stem base although present, but reduced in width, at higher levels (Chap. 11.8). The authors questioned whether the cambium at the stem base could remain inactive for a period of years. They speculated that cambial initials might continue dividing on the phloem side and then resume divisions on the xylem side when growth conditions improved.

The results of a number of studies suggest that the effects of nonlethal radiation on the cambium might be indirect with the primary site of injury residing in the vegetative meristems (Woodwell and Miller 1963). Some support for this idea can be found in the data of Cordero and Gunckel (1982a,b) and Cordero (1982) in their work on *Lupinus albus*. One of the main responses to gamma irradiation was the suppression of primary vascularization in shoot apices. In stem tissues, the area of xylem decreased while the area of phloem increased with increasing dosage rates. Although the total number of cells within the secondary phloem increased by 50% in plants irradiated at 25 rads/h, the proportions of each cell type

---

[1] A rad is a unit of absorbed dose of ionizing radiation. One rad = 100 ergs/gram. A roentgen is a unit of exposure. One R = 2.58×10^{-4} coulomb of ions/kilogram of air under standard conditions (Vozzo 1978).

were not appreciably changed. The increase in phloem cell number and area lend a degree of credence to the suggestion of Sparrow et al. (1965) that cambial division and phloem differentiation continue in the absence of xylem cell production.

Additional data on these effects were recently provided by Chandorkar and Dengler (1987) on *Pinus sylvestris* seedlings. Continuous low-level exposure to gamma irradiation reduced height growth about 21%. Concomitantly, however, it stimulated mitotic activity of the cambium and increased the number of xylem and phloem cells over that of control plants. As suggested by the authors, stimulation of cambial activity was probably caused by an increased supply of nutrients not used in stem elongation together with an alteration in the normal supply of growth regulating substances.

Cambial activity of woody herbaceous plants is often affected in more striking ways by ionizing radiation. Seeds of *Linum usitatissimum* were exposed to acute dosages of gamma rays ranging from 25 000 to 100 000 rads in a study by Ghouse et al. (1981). Seedlings developing from these seeds had shorter fibers and vessel members with thicker and more heavily lignified walls and the xylem rays were taller and broader than in control plants.

The effects of ionizing radiation on the cambium are reflected in structure of xylem derivatives in addition to reductions in radial increment. An 8-year exposure of *Pinus radiata* to high, but nonlethal, levels of radiation resulted in a 17.9% reduction in tracheid length, 10.0% reduction in lumen diameter, and 17.0% reduction in wall thickness (Hamilton 1963). Similar responses were observed in *P. echinata* exposed to a sudden high-level burst of radiation over a 1-month period. In this case, cambial activity and xylem structure returned to normal at the termination of treatment. These results were confirmed in a later study of *P. echinata* (Clark and Hamilton 1968).

Morphological changes in xylem derivatives were relatively slight in *Liquidambar styraciflua* and *Quercus alba* exposed to 14 900 and 35 000 rads, respectively, over a 2-year period (Hamilton and Chesser 1969). Both fiber length and wall thickness were reduced to some extent, but normal tissues were produced at the cessation of treatment. *Acer rubrum* trees exposed to cumulative dosages of 13 000 rads for one growing season showed no changes in radial increment (Romberger et al. 1980). However, intrusive growth of xylem elements was severalfold greater than the pretreatment level resulting in a temporary suppression of the local domain pattern (Chap. 6.6.2).

10.5 Biological Wounds

10.5.1 Tumorous Growths

Tumorous growths of unknown cause often arise on stems of forest trees. Von Tubeuf (1930a,b) described a tumor-like hypertrophy that occurred on stems and branches of *Pinus sylvestris* from certain forest regions in Germany. He could find no visible evidence of infection, and suggested that the growth resulted from mutations (Chap. 8.4.1.5). More detailed investigations of this abnormality on both *Pinus* and *Picea* were conducted by Sprengel (1936). In trees of both genera he was able to trace the origin of the tumor to either a single cell or a small group of cells. The origin was usually in the first growth ring but occasionally in an older ring. Once initiated, the diseased tissue spread not only vertically but also laterally by anticlinal cambial divisions. However, laterally the wedge of diseased tissue remained sharply demarcated from normal tissues.

Sprengel's most detailed studies were conducted on *Quercus*. Tumorous growth always began in one to several cells that appeared as a microscopic dark spot in the first growth ring next to the pith. These spots were of very regular occurrence and they showed no relation to either leaves or axillary buds. As growth proceeded, the diseased cells divided, producing a radial wedge of tissue confined between two broad rays. Diseased tissue was characterized by rapid growth both tangentially and radially and by the absence of broad rays. To all appearances, the cambium remained normal and the derivatives diseased. Like *Picea* and *Pinus*, the tumorous tissue in *Quercus* remained rigidly confined with a sharply demarcated boundary separating diseased and normal cells on both lateral margins. However, unlike the tumors on coniferous stems which were relatively long-lived, those on *Quercus* stems rarely lived for more than 5 years. In its final stages, the tumor produced a vortex of distorted fibers at its terminus. Adjacent tissues produced a callus which often failed to close over the dead tumor for many years. Eventually, however, the callus did grow over the vertically oriented notch of dead tissue and included bark and united the cambial cylinder.

Sprengel was never able to find any microscopic evidence of a stimulator in the diseased tissues nor was he able to isolate any biological agent in cultures. He agreed with von Tubeuf (1930a,b) that the tumorous cells probably arose by mutation. Consequently, the abnormality was not transmitted to other cells and the tumorous tissue remained laterally confined. He tentatively compared these tumors to the analogous origin of faciations that de Vries had described arising from primary meristems.

A unique type of tumorous growth occurring on *Picea glauca* was examined by White and Millington (1954a,b). These tumors, presumably caused by an unknown biological agent, were confined to stems of trees growing on exposed rocky shores of New England islands off the Atlantic coast of North America (Fig. 10.28A). They appeared to be similar to those growing on wet, exposed sites described by Sprengel (1936) in Germany. The tumors originated from either a single cell or a small group of cells in the first-year growth ring (Fig. 10.28B). For example, on a transverse section a tumor one-cell wide the first year was 7 and 20 cells wide the third and fourth years, respectively. Despite the tangential expansion of the tumorous region, the tumor boundaries remained intact and nontumorous cells never invaded the affected area. Tumors grew in all three directions producing large, bulbous growths on the stem surface. Tangential expansion was by anticlinal divisions and cell enlargement. Radial expansion for the first few years was entirely by cell enlargement, but in later years periclinal divisions predominated.

Fig. 10.28. A *Picea glauca* tree from Mt. Desert Island, Maine, heavily infected with tumorous growths. **B** Transverse section of a 13-year-old *P. glauca* stem showing tumorous growth beginning in the first growth ring. Although the tumor expanded tangentially in subsequent growth rings, it remained confined within a wedge of tissue (*arrows*). 14x. (White and Millington 1954b)

In older tumors, periclinal divisions in the tumorous region exceeded those in the nontumorous region by a 10:1 ratio. Thus, the most striking difference between normal and tumorous wood was the wider growth rings in the latter, with some rings consisting of 800 or more cells in a radial file. This massive increase in radial growth was brought about by an acceleration of periclinal cell divisions and not by prolongation of growth in the fall as suspected in earlier work (White et al. 1967). Longitudinal expansion of the tumors was due to intrusive growth of fusiform initials accompanied by occasional anticlinal divisions. On tangential sections tumorous wood often exhibited complex erratic whorls of tracheids. The authors speculated that these whorls might be caused by occasional stimulation of anticlinal divisions together with concomitant suppression of divisions in the nor-

Fig. 10.29. A cambial excrescence on *Castanea dentata* caused by chestnut blight (*Endothia parasitica*) (*Cryphonectria parasitica*). After the fungus killed a patch of bark, some still-living portions of the cambium "exploded" forming a tumor-like growth of wood and bark. (Shigo 1983)

mal wood without the former invading the latter. Tumors from the same region were later re-examined by Tsoumis (1965), who reached similar conclusions.

Tumors almost identical in development and structure have been found on *Picea sitchensis* trees growing in coastal forests of the American Pacific Northwest (Torok and White 1960, Rickey et al. 1974), and also on *Pinus contorta* trees growing in the Rocky Mountains (Peterson 1961). Anatomical examination of the latter species showed that it shared most of the characteristics described by White and Millington (1954a,b) on *Picea glauca*. Tumors extended from the pith to the cambium in a steadily enlarging conical zone sharply separated from normal tissues. However, contrary to the later finding of White et al. (1967), Peterson noted that late-summer cambial activity continued in tumorous tissues beyond that for healthy tissues. He suggested that the tumor stimulus might have prolonged the season of growth rather than accelerating the rate of cell division.

Cultures of tumorous tissues conducted by Torok and White (1960) showed that the tumorous cells were extremely unstable. Mitoses appeared more numerous than normal, and a high percentage of the mitoses were aneuploid. Causally, these tumors might arise epigenetically in the absence of a pathogen and, once established, be perpetuated via habituation processes analogous to those described by Gaspar et al. (1991).

In their investigations of tumorous growths, all of the foregoing investigators observed distorted vascular elements similar to those described by Hejnowicz (1971) and Wloch (1976) in tumors on *Picea abies* stems (Chap. 6.6.2).

A tumor-like outgrowth, called briarwood, occurs between the root and stem of *Erica arborea* (Tsoumis et al. 1988). As in the tumorous growths on *Picea* stems, the disturbance, of unknown origin, first appears in 1-year-old plants and continues to develop for 40 to 50 years. The xylem consists of short, distorted fibers that often display circular patterns. Structures resembling tumors also have been observed in old *Ginkgo biloba* trees growing in humid environments in China (Li and Lin 1991). These structures emerge as tumor-like protuberances and develop as slender, hanging branches more than one meter long. If these "stalactite-like" branches succeed in reaching the ground, they often develop adventitious roots. These branches resemble tumorous growth at emergence and, like other tree tumors, the wood contains zones of transversely oriented, whirled tracheids and short, irregularly arranged rays. The causal mechanism is unknown.

Other structures that resemble tumors are actually cambial excrescences. Shigo (1983) described and illustrated such an occurrence in *Castanea dentata* infected with chestnut blight (*Endothia parasitica*) (*Cryphonectria parasitica*). Small sectors of the cambium resisted the fungal invasion by the formation of barrier zones. These viable cambial sectors often produced tumor-like outgrowths consisting essentially of fibrous tissue richly permeated with rays (Fig. 10.29).

10.5.2 Insects

Many insects such as scales, aphids, bark beetles, and cambial miners either directly or indirectly injure the cambium. In addition, many insects serve as vectors for diseases that invade the tree and cause cambial damage.

10.5.2.1 Scales

One scale insect, *Matsucossus josephi*, is a parenchymatous feeder that infects *Pinus halepensis*. Liphschitz and Mendel (1987) described how prolonged feeding by this insect caused necrotic spots and pronounced narrowing of the annual ring at the attack site. The xylem tissue was disorganized, and abnormal lignified parenchyma cells were produced instead of tracheids. The abnormal xylem extended for some distance both above and below the injured spot, and this tissue was accompanied by traumatic resin canals. Tangential microsections of these regions revealed abnormal tracheids with twisted walls, reflecting similar disorganization in the cambial layer.

10.5.2.2 Aphids

Most aphids attack the leaves of trees where their feeding induces gall formation (Philipson 1964). Some aphids, however, feed on buds and small branches in which case the gall induces changes in the underlying tissues. One aphid, *Chermes abietis*, was found to feed exclusively on cortical parenchyma of young *Picea abies* buds (Rohfritsch 1977). Feeding stimulated differentiation of a highly specialized nutritive tissue that strongly resembled the phloem sieve cells. Another gall-forming aphid, *Slavum wertheimae*, induced differentiation changes in the xylem below the gall in *Pistacia atlantica* (Aloni et al. 1989). The affected xylem

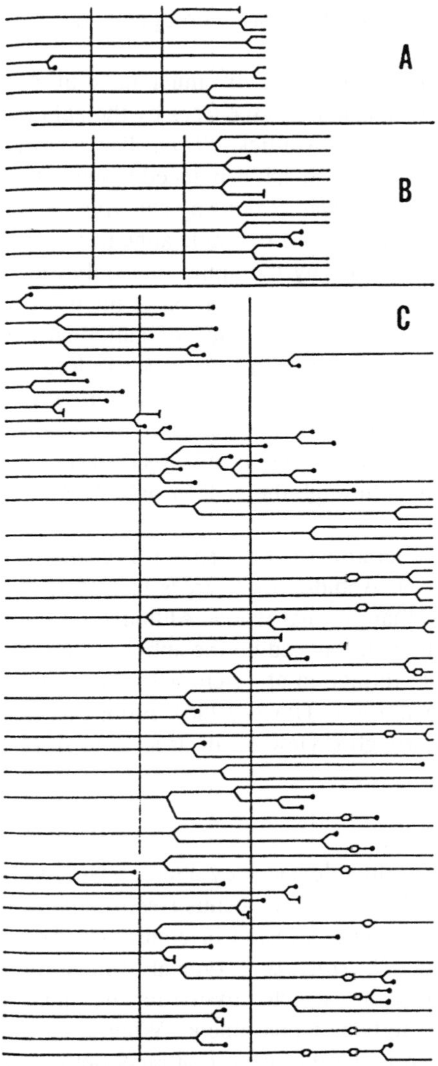

Fig. 10.30. Diagrammatic representation of the anticlinal divisions in fusiform initials of *Abies grandis* during one season's growth before infestation by *Adelges piceae* (**A**), the first year of appreciable infestation (**B**), and after infestation was well established (**C**). A *dot* indicates the initial declined and disappeared at that point, and a *vertical bar* indicates the initial shortened and withdrew from the study area. Anticlinal divisions in xylem mother cells occurred where a line is branched and rejoined. Diagrams were based on numbered, serial sections, but the *vertical lines* divide the annual increments into thirds by micrometers. (Smith 1967)

was characterized by wide vessels that continued to form throughout the entire season even in the latewood.

The balsam wooly aphid, *Adelges piceae*, feeds on the inner bark of *Abies* twigs and branches. During feeding, saliva injected into the trees by the aphid stimulates abnormal growth of the xylem tissues (Mullick 1977). The abnormal xylem tissues reflect changes in the cambium by the aphid-induced wound stimulus. Smith (1967) investigated these changes in *Abies grandis* by studying the tiers of tracheids and rays in serial, tangential microsections of single annual rings both before and after infestation. In branches attacked by the aphid, not only did the frequency of anticlinal divisions increase but a greater proportion of the divisions were transverse rather than pseudotransverse. The frequency of anticlinal divisions both before and after aphid infestation are shown in Fig. 10.30. Before infestation, only 13.3% of the initials divided anticlinally during the year's growth, but after infestation 58% divided anticlinally at least once and some two to four times. Daughter cells of divisions in the infested branches were also subjected to a high rate of decline. Both daughter cells survived through the year in only 32% of the divisions.

The aphid also caused an abnormal proliferation of ray tissue in *A. amabilis*, *A. grandis*, and *A. lasiocarpa* (Mitchell 1967). Aphid-infested trees had 35 to 73% more rays per mm^2 of tangential surface than normal wood. Although aphid infestation is a form of wounding, the stimulus introduced by the aphid resulted in a marked increase in both rate of growth and formation of xylem tissues resembling compression wood (Doerksen and Mitchell 1965, Timell 1986). From this latter viewpoint, Timell (1986) conducted a thorough review of all aspects of *Adelges* infestations.

10.5.2.3 Bark Beetles

Bark beetles (Scolytidae) are distributed worldwide and they attack a wide range of woody plants (Heliövaara et al. 1991). Members of the genus *Dendroctonus* are especially injurious and are responsible for serious losses in coniferous forests. The life history of the southern pine beetle, *D. frontalis*, is typical of most members of the genus (Payne, no date). Host trees are selected by the female, possibly on the basis of olfactory cues such as the terpene alpha-pinene. When a suitable host tree is found, the female emits a pheromone that in turn attracts both males and other females. The female initiates boring and creates a nuptial chamber in the inner bark where mating takes place upon arrival of the male. The female then mines a gallery in the cambial region that extends diagonally across the tree's axis. Eggs are laid in niches cut in the walls of the gallery as depicted by Wilson (1977) (Fig. 10.31). Upon hatching, the first instar larvae feed on the cambium, usually by making lateral passages perpendicular to the adult gallery. As the larvae molt to successive stages their galleries enlarge. Fully mature larvae emerge from the tree and form pupal cells in the outer bark. Although individual bark beetle attacks are relatively insignificant, mass attacks by many insects can severely weaken a tree and very often result in its death.

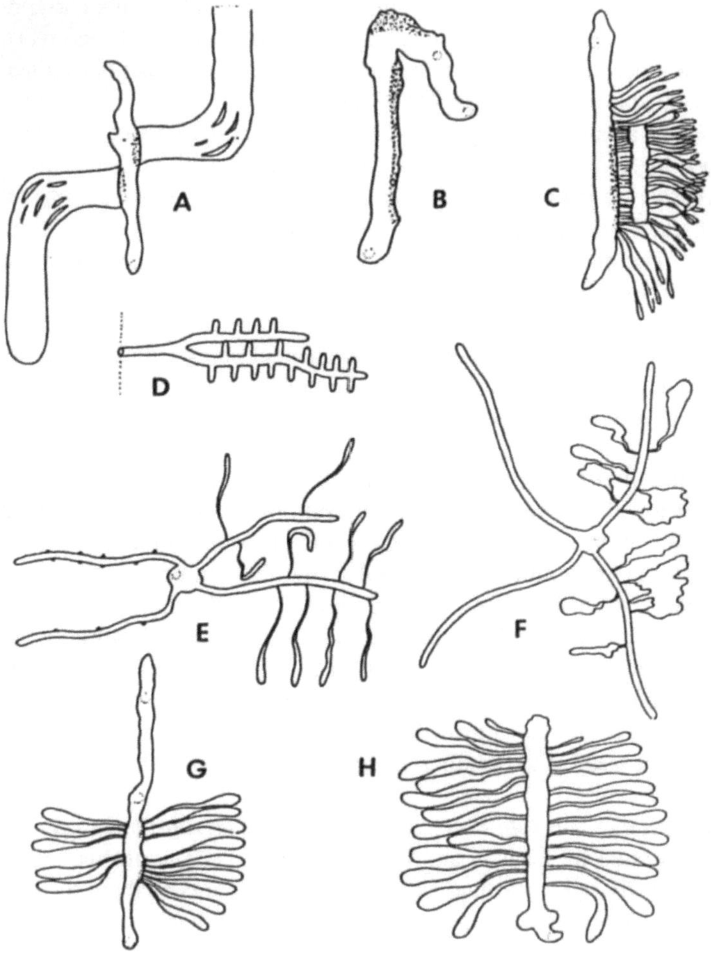

Fig. 10.31A-H. Typical bark and ambrosia beetle galleries in the cambial region of conifer stems. Larvae, either individually or in congress, mine tunnels away from the main gallery excavated by the adult beetles. Pupation occurs in cells at the ends of the tunnels or mines. Galleries are not drawn to scale. **A** *Dendroctonus murranae,* **B** *D. valens,* **C** *D. rufipennis,* **D** *Trypodendron sp.,* **E** *Pityokteines sparsus,* **F** *Ips pini,* **G** *D. simplex,* **H** *Phloeosinus sp.* (Wilson 1977)

10.5.2.4 Cambial Miners

Cambial miners are the larvae of certain Lepidoptera insects that create mines or feeding channels in the cambium of the host tree. These mines, usually oriented longitudinally in the stem axis, appear as small, brown, half-moon shaped patches on transverse sections in the xylem of many woods, especially in the Aceraceae, Betulaceae, Rosaceae, and Salicaceae (Record 1911). On tangential and radial surfaces they appear as brown streaks. The mines are commonly known as pith flecks, but they are also referred to as medullary spots, parenchyma flecks, worm tracks, and several other terms. For a review of early research on cambial miners the reader should consult Grossenbacher (1910) and Record (1911).

The life history of a cambial miner that attacked *Prunus avium* and *P. domestica* was described by Grossenbacher (1915a). The adult, a small black fly, provisionally called *Agromyza pruni (Phytobia pruni)* (Kulman 1964), laid its eggs in lenticles of the bark. Upon hatching, the larvae invaded the cambial zone where they began to feed. The mines, or feeding channels, appeared random with no preferred orientation. They were of uniform depth radially but of variable width tangentially. The tangential diameter of a feeding channel averaged about one to four times the diameter of the cambial miner, with the channel being widest where the larvae paused to feed from side to side. The portion of the cambium most desired by the larvae for food was observed to be a sheath of cells only a few cells in thickness.

Closure of a mine often began shortly after the larva had proceeded onward. A new cambium developed over the channel by division of the first to third phloem cells external to the mine. Because the new cambium often formed over the wound before the true cambium closed laterally, the "substitute cambium" appeared to arch over the channel. Cell proliferations from phloem rays that had been disrupted by the larvae filled the channel with enlarged cells having the general appearance of pith, creating the so-called pith flecks. However, during the processes of occlusion and wound closure, the channels decreased in size, possibly because of compression due to growth of neighboring cells. In *Prunus serotina*, the pith flecks caused by the cambial miner were often associated with gum spots (Kulman 1964).

Another cambial miner, *Phytobia setosa*, that attacks *Acer rubrum* and *A. saccharum*, was investigated by Hanson and Benjamin (1967). Eggs of this species were laid by the adult fly in or near lenticles on 1-year-old twigs at about the time of spring bud break. Newly hatched larvae bored through the bark and began mining toward the buds, then reversed course and proceeded about 2 m down the main stem during the first instar. After molting, the second instar larvae mined downward approximately another 4 m. The third instar larvae mined to the stem base and then continued mining upward and downward in a zig-zag fashion. With each instar the mines became larger as the larvae increased in size. To pupate, the larvae mined downward in a root and bored out through a slit cut in the bark. They emerged in late July when cambial activity declined and the bark began to tighten. Wounds made by these larvae appeared as pith flecks in the xylem.

A prominent feature of the wood of *Dactylocladus stenostachys* examined by Chalk and Murthy (1963) was a characteristic pattern of small holes or flecks on tangential surfaces. Cambium at the site of the flecks had been killed, and the callus appeared to consist of phloem strands that grew inward to fill the voids. Although the cause was unknown, the authors suggested that it was probably due to insects. Buczacki (1973) suggested that "xylem flecks" in *Larix decidua* and *L.*

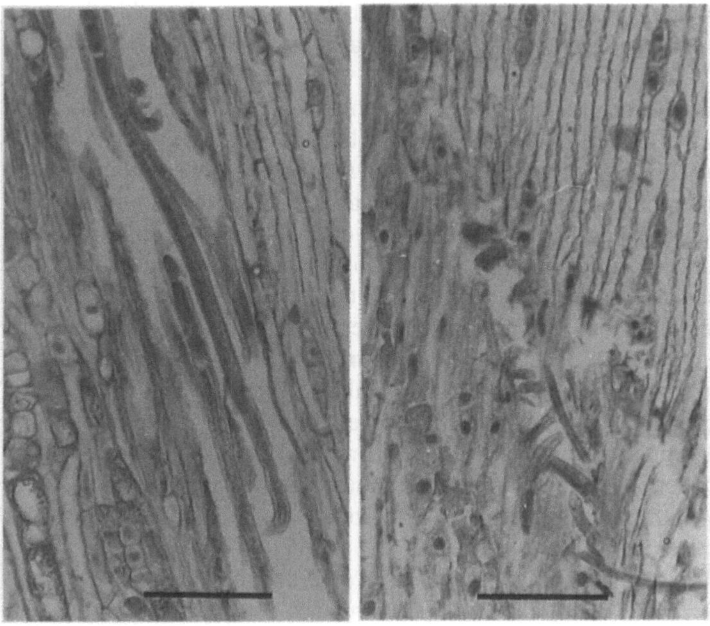

Fig. 10.32. Pinewood nematode *(Bursaphelenchus xylophilus)* feeding in cambial regions of *Pinus resinosa (left)* and *P. strobus (right)*. *Bars* 100 µm. (Myers 1986)

leptolepis might be due to environmental stress, although he, too, did not rule out possible association with Adelges infestations. Injuries recognized as pith flecks are not necessarily confined to insect attacks. Similar appearing traumatic tissues have been attributed to frost and drought (Carlquist 1988a).

Although pith flecks do not significantly affect the strength of wood products, they do detract from the aesthetic appeal of certain finished products such as veneers and paneling. On the other hand, the wavy grained wood in *Grewia tilifolia*, presumably induced by a species of Lyctidae, is often highly prized for its ornamental properties (Bhat and Bhat 1984).

10.5.3 Nematodes

Nematodes are primarily root and lower stem feeders that attack many tree species. Injury caused by the pinewood nematode, *Bursaphelenchus xylophilus*, in seedlings of ten coniferous species was studied by Myers (1986). The nematode migrated from the cortex either through the rays or directly across the sieve cells to the cambium (Fig.10.32). While feeding, it moved both vertically and circumferentially, destroying the fusiform and ray cell initials and their immediate derivatives. Destruction of the cambium ultimately resulted in seedling death, primarily by disruption of ray transport and associated processes such as blockage of tracheids by gas embolism, oleoresin exudates, and metabolites from dying ray tissues.

More intensive research on the pinewood nematode, *B. xylophilus*, has been conducted by Kuroda et al. (1988) and Kuroda (1989, 1991). The nematode was found to be widely distributed throughout *Pinus thunbergii* stems within 1 week after infection; one report suggested that the nematodes might migrate from 20 to 50 cm/day. Among the first obvious anatomical changes was the appearance of white streaks in the xylem indicating cavitation of tracheids. Tracheid cavitation, in turn, caused blockage of xylem sap ascent and this effect was noticeable within 2 weeks after infection. About 4 weeks after infection, the cavitated areas had enlarged drastically and finally reached the cambium. At this time, water content of the xylem was only about 30 to 40% that of healthy trees, and necrosis of cambium, phloem, and parenchyma cells had become conspicuous in the trunk. Further observations supported the hypothesis that tracheid cavitation resulted from the excessive production of volatile terpenes released from the infected cells. As cavitation spread throughout the tree, the symptoms of pine wilt intensified and heavily infested trees died of water shortage.

Another member of the same genus, *B. lignicolus*, invades both seedling and mature stems of *Pinus densiflora* and *P. thunbergii* in Japan (Mamiya and Kiyohara 1972). It is believed to be one of the few nematodes that actually reside in the wood. These nematodes move through both the axial and radial resin canals while feeding on the epithelial and canal-associated parenchyma cells.

An earlier report of a wood-inhabitating nematode, *Aphelenchoides xylophilus*, was made by Steiner and Buhrer (1934) in *Pinus palustris* and *P. echinata*. The authors suggested that the nematodes used the bark-beetles, *Dendroctonus frontalis*, as carriers to invade the wood. This nematode did not feed directly on woody tissues but on secondary invaders of the wood, blue-stain fungi.

10.5.4 Birds

Birds inflict many types of wounds on trees that either directly or indirectly affect the cambium. Cavity-nesting birds, as a rule, are not the primary cause of injury to living trees (Pearce 1947). When new cavities are excavated in living trees by woodpeckers, the selected site most often is either an old wound or a decayed branch stub (Scott et al. 1977). Such trees also are highly susceptible to heart rot, which also favors cavity-nest building.

One of the most commonly encountered direct effects on the cambium is caused by sapsuckers, *Sphyrapicus* spp. (Ziller and Stirling 1961, Oliver 1968, 1970, Erdman and Oberg 1974). The birds feed on sap, phloem, and cambium of a large number of tree species. Cambium and phloem tissues sometimes comprise 12% or more of the diet during summer, whereas sap and cambium are often the main source of food in winter when insects are scarce.

An analysis of the seasonal food preferences of *Sphyrapicus varius varius*, the yellow-bellied sapsucker, compiled by Tate (1973), is shown in Fig. 10.33. It is obvious that tree bast is a preferred food throughout the year, but especially so during early spring and summer. A study by Beal (1911, cited by Tate) showed that 49% of the stomach contents in April consisted of tree bast. Because the data of Fig. 10.33 were based on stomach analyses, bast included all tissues of the soft inner bark and cambium that were consumed by the birds along with the sap.

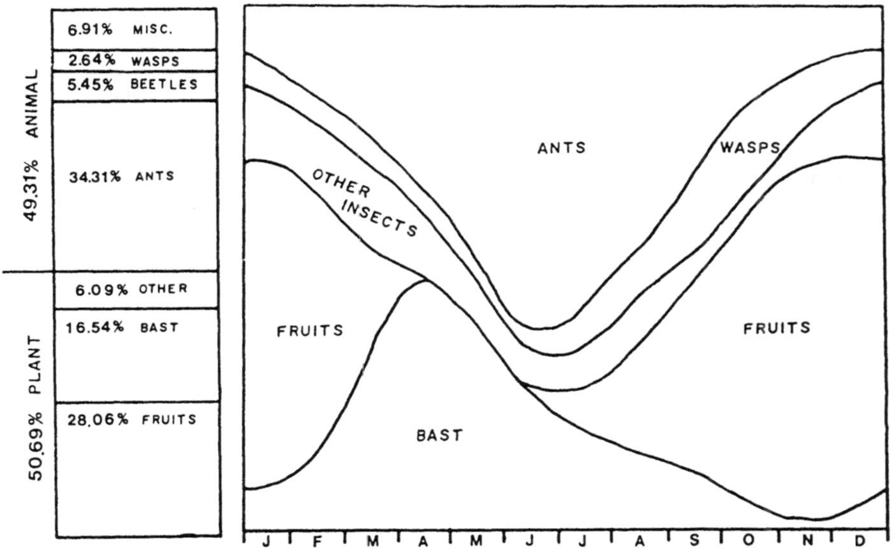

Fig. 10.33. Stomach contents of 313 sapsuckers (*Sphyrapicus varius varius* and *S. v. nuchalis*). Tree bast, the preferred food in early spring and summer, included soft bast and cambium consumed along with the sap. (After Tate 1973)

Bird feeding sites or sap wells are selected by test borings, small circular punctures through the bark. If sap flow, bark thickness, and other conditions appear favorable, the bird will continue feeding. Initial feeding holes are spaced about one half-inch apart in horizontal bands and sometimes in vertical rows. Although their feeding preferences vary to some degree with season, the birds tend to concentrate on a single species. Certain trees within range of the nest, called sap orchards, are attacked repeatedly not only during the season but also in subsequent years. Later feedings often remove the bark between the original holes, or possibly the callus tissue formed after wounding.

Tate (1973) categorized the types of food tap holes according to season of the year and other characteristics (Table 10.3). Sap bands and columns are drilled mainly in spring and summer when sap flows readily from the wound. Bast bands and columns, in contrast, are drilled mainly in fall and winter when the feeding sites are tapped for solid tissues such as cambium and inner bark. Spiral bands are produced when the birds drill holes in pencil-thin branches to obtain the early spring sap. As noted in Table 10.3, the type of tap hole varied not only with species preference of the birds but also with season of the year in northern Michigan.

Table 10.3 Types of food tap holes drilled by the yellow-bellied sapsucker and their characteristics (Tate 1973)

	Primary bands			Columns	
Type	Sap bands	Bast bands	Spiral bands	Sap	Bast
Season					
Summer	Rare	Rare	None	Abundant	As drilled
Fall	Common	Rare-common	None	Rare	Rare
Winter	Rare-common	Rare-common	None	Rare-common	Common
Spring	Common	Common	Common	As available	Abundant
Preferred plants	Pines Hemlocks Maples Many others	Maples Beeches Elms Few others	Aspens Others?	Birches Aspens Beeches Many others: nonsummer	
Location on plant	Most on bole Angiosperms: major limbs	Bole	Uppermost branches	Bole Angiosperms: limbs	
Group Configuration	Line	Line	Spiral	Band below columns	
Shape of hole	Round	Ragged	Rectangular	Round	
Cross-section	"V"-shaped	Inverted "V"	Squared	Inverted "V"	

Light to moderate feeding by sapsuckers results in cambial wounds that usually close, although not without contributing to stem defects. Upper stem parts and sometimes entire trees are killed when the horizontal feeding bands are sufficiently concentrated to effectively girdle the stem. Concentrated small injuries also weaken the fibers along newly formed growth rings resulting in ring shakes (Shigo 1963). Secondary injuries are often caused by other birds and squirrels feeding on the sapsucker holes (Kilman 1958). The feeding sites also provide ac-

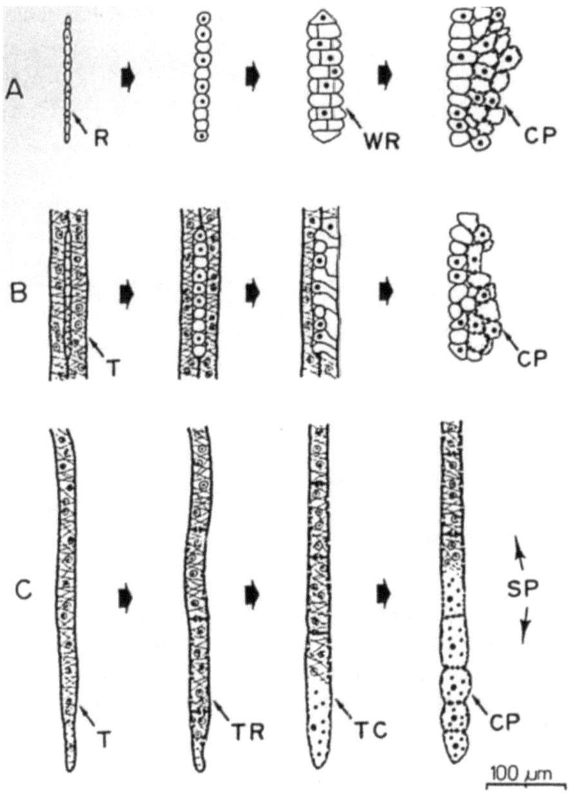

Fig. 10.34A-C. Stages in wound closure of *Taxus baccata* trees injured by woodpeckers *(Dendrocopus major)*. **A** Uniseriate rays (*R*) broadened to form wound rays (*WR*). **B** Segmentation of fusiform initials and differentiation to callus parenchyma (*CP*). **C** Segmentation of fusiform initials and differentiation to either strand parenchyma (*SP*) or tracheids (*T*). *TC* Transition cells; *TR*, tracheoidal cells. (Kucera 1972)

cess channels for pathogens. Ohman and Kessler (1964), for example, described how the sooty-black bark of *Acer saccharum* stems was caused by fungi feeding on the exuded sap from sapsucker holes. In this species, feeding occurred on the lower stem. Feeding was moderate during any one season, but the same trees were repeatedly attacked for many years leaving buried pockets of wound tissue embedded in the xylem.

Wounds in *Taxus baccata* caused by woodpeckers (*Dendrocopus major*) were investigated by Kucera (1972). The wounds consisted of a tangential series of holes each of which extended inward through the cambium. Wound closure by xylem-derived callus formation involved three processes. In the first, the uniseriate rays broadened to form wound rays (Fig. 10.34A). As in most surface wounds, the rays were found to be the most important contributors to callus for-

Fig. 10.35A-C. Feeding injury to *Pinus contorta* trees. **A** Snowshoe hare (*Lepus americanus*). Note traces of inner bark still attached to the sapwood. **B** Red squirrel (*Tamiasciurus hudsonicus*). *Arrow* indicates accumulation of bark strips on the ground. **C** Porcupine (*Erethizon dorsatum*). Note the prominent incisor marks on the exposed sapwood. (Sullivan and Sullivan 1982)

mation. Reaction to cambial injury not only began with increased activity of the ray cell initials but it also continued longest in the ray tissues. In the second process, the fusiform initials adjacent to the rays and nearest the callus mass were obliterated by the hypertorphy of the neighboring ray cells (Fig. 10.34B). In the third process, fusiform initials segmented to form either strand parenchyma or short tracheids (Fig. 10.34C). A typical callus therefore consisted of parenchymatous callus followed by small bands of transition cells, then a zone of tracheoidal cells, and finally a broad zone of short tracheids. Wound wood terminated when tracheids reattained their normal lengths.

10.5.5 Mammals

Mammals damage the cambium of trees in a variety of ways (Pearce 1947, Hawthorne 1980). In addition to the clipping of seedlings and young shoots, many rodents such as deer mice (*Peromyscus*), woodchucks (*Marmota*), and wood rats (*Neotoma*) will girdle or partially girdle small trees as a winter food source. Voles (*Microtus*) often cause extensive damage to forests by girdling seedlings and mature trees during the fall and winter months (O'Brien 1983). Even such an unlikely candidate as the pocket gopher (*Geomys*) has been known to partially debark trees to a height of 3.5 m by tunneling to the trees under deep snow cover (Case 1983, Hawthorne 1983). Other rodents, such as beavers (*Castor canadensis*) not only fell trees up to 25 cm in diameter for food and dam building, but also girdle or partially girdle larger trees. In some regions, beavers show a preference for stems of *Pinus* spp. and *Liquidambar styraciflua*, apparently to consume the resin and storax, respectively, that exude from the girdled stems (Miller 1983).

Porcupines (*Erethizon*) are recognized widely for the extensive damage they inflict in forest stands (Timell 1986). Damage is caused by basal gnawing, branch clipping, and girdling of the upper stems where they often feed for days in a single tree (Fig. 10.35C). Secondary effects of feeding result from exposing the xylem to attacks by pathogens, insects, and birds (Schemnitz 1983). In plantations of *Pinus resinosa* and *P. sylvestris*, porcupines fed on all sizes of trees but they showed a preference for trees at least 7.5 cm in diameter and 4 m high (Rudolf 1949). Although porcupines feed on many tree species, they do show preferences as indicated in Table 10.4. In compiling these data, Brander and Stearns (1963) examined tree tops in a logging operation for past and current porcupine damage. Fifty-eight percent of the trees had been gnawed on at some time during the past 20 years. Fifty percent of those gnawed showed a high degree of activity, defined as girdling or stripping of large patches of bark. Porcupines also show a preference for mistletoe-infected wood (Schemnitz 1983). Apparently, the nutrient-rich tissues, like those of *Cronartium* infections (Hirt 1939), are particularly appealing to rodents.

Table 10.4. Porcupine (*Erethizon*) activity by tree species at Golden Lake, Michigan (Brander and Stearns 1963)

Species	Number of tops in sample	Percent gnawed in each species	Species preference[a]
Acer rubrum	2	50.0	0.86
Acer saccharum	170	51.8	0.89
Betula alleghaniensis	55	50.9	0.87
Ostrya virginia	2	0.0	0.0
Picea glauca	1	0.0	0.0
Prunus serotina	1	0.0	0.0
Tilia americana	8	75.0	1.29
Tsuga canadensis	32	90.6	1.55
Ulmus americana	17	94.1	1.61

[a] Expressed as a ratio representing the percentage of trees gnawed within a species divided by the percentage of all trees gnawed.

Like porcupines, squirrels (*Sciurus, Tamiasciurus*) usually, but not always, confine their feeding primarily to upper stem parts. However, rather than gnawing, they strip the bark from the stem by biting and pulling and then eating the soft, inner phloem and cambial tissues (Fritz and Averill 1924, Sullivan and Sullivan 1982, Kenward and Parish 1986, Sullivan and Vyse 1987) (Fig. 10.35B). At the other extreme, rabbit damage is confined to lower stem parts. However, whereas squirrels feed on tree tissues mainly in times of food stress, rabbits and hares are voracious feeders at most all times of the year. In addition to clipping seedlings and shoots, they, too, will strip bark from small trees (Sullivan and Sullivan 1982, Hawthorne 1983). The eastern cottontail (*Sylvilagus floridanus*) is especially inclined to girdle small, thin-barked trees when winter snow is deep (Craven 1983). Like the cottontail, the snowshoe hare (*Lepus americana*) feeds on a wide variety of woody plants but shows a stronger preference for coniferous species (Pearce 1947) (Fig. 10.35A).

Damage by large mammals, such as deer (*Odocoileus*), elk (*Cervus*), and moose (*Alces*) is primarily confined to shoot damage by browsing (Heikkilä and Mikkonen 1992). However, when populations are high, deer often strip trees and consume the soft inner bark (Michael 1987). Cambial damage to sapling trees also results from buck deer rubbing velvet from their antlers, a process known as scoring (Pearce 1947). As noted by Hawthorne (1983), any antlered animal might scar tree trunks during the rutting season. It appears that European deer are far more inclined than American deer to peel and eat the bark of trees (Timell 1986).

A somewhat similar type of cambial injury is caused by black bears (*Ursus americanus*). Bears not only claw off bark to leave a territorial marker, but also feed on the inner bark (Boddicker 1983). In addition to claws, teeth are used to pull ragged strips of bark from pole-sized trees. After the bark is pulled away, the bears scrape off the residual cambial and outer xylem layers with their incisor teeth leaving vertical tooth marks on the exposed wood surface.

10.5.6 Parasitic Plants

Many species of parasitic plants are found on forest trees throughout the world. The biology of these plants has been exhaustively reviewed by Ozenda (1965) and in a richly illustrated text by Kuijt (1969). Most plant parasites invade aboveground stem tissues but a surprisingly large number invade the roots. Mann and Musselman (1979), for example, described 25 species in five families of root parasites that damaged forest trees in the southern United States. The haustoria of all species, irrespective of the site of attack, invariably either invade or penetrate the cambia of the host species.

10.5.6.1 *Exocarpus bidwillii*

Exocarpus bidwillii is a small, procumbent woody plant of the Santalaceae family that parasitizes the roots of a number of tree species in New Zealand (Fineran

1962, 1963a,b,c, 1965a,b). *E. bidwillii* is considered a semiparasite as the above-ground vegetative organs are nourished by a well-developed root system in addition to the parasitic haustoria. Haustoria develop as lateral appendages from a parent root, and their contact with a host root is more or less by chance. Once contact is established, the haustorium produces a pressure zone followed by digestive action of the apical sucker cells. The suckers penetrate the cortex and phloem until their tips enter the cambial zone (Fig. 10.36A). As the host root grows radially, the sucker tissues become more deeply embedded in the host xylem. Tissues of the sucker and host root are clearly separated and lateral expansion of the sucker is accommodated to that of the host's cambium. Thus, the widest part of a sucker is in the cambial zone, where the sucker not only enlarges by cambial activity but also sends out new branches that extend along the zone of the host's cambium (Fig. 10.36B). The parasite derives nutrients via vessels that make direct contact with those of the host.

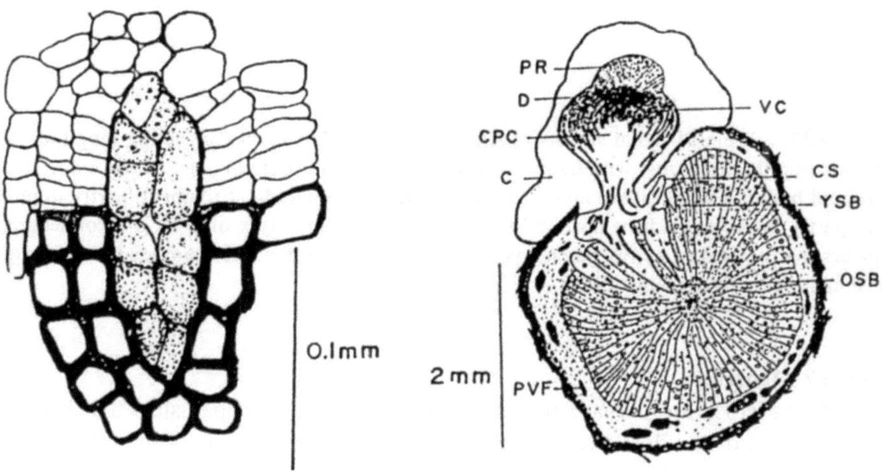

Fig. 10.36. *Exocarpus bidwillii* in transverse section. *Left* A young sucker branch near its distal end showing recent radial growth in the phloic region. The host cambium appears relatively undisturbed at this time. ***Right*** A mature haustorium with a deeply embedded sucker branch and a body which is woody in structure. Vascular tissues appear discontinuous due to the plane of sectioning. Tissues of the sucker and the host root remain clearly separated. *C* Cortex of haustorium body; *CPC* central parenchymatous core of body; *CS* cortical strand; *D* duramen; *OSB* an old, deeply embedded sucker branch; *PR* parent root of haustorium; *PVF* perivascular fibers of host root; *VC* vascular core region; *YSB* young branch of sucker. (Fineran 1963c)

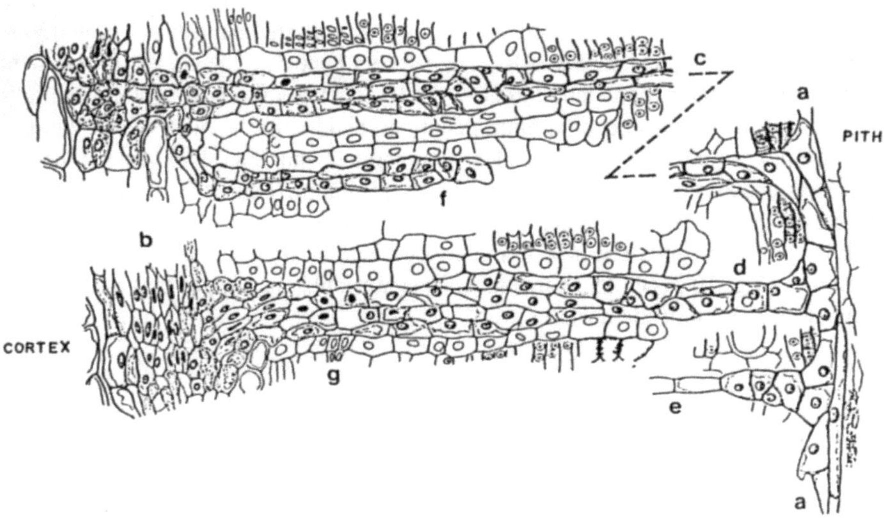

Fig. 10.37. System of radial sinkers in *Arceuthobium americanum* connecting longitudinal strands in pith of the host with those in the cortex. Area illustrated is just below a dormant terminal bud of a witches' broom on *Pinus contorta*. *a* Longitudinal filament, partly in pith and partly in protoxylem of host; *b* longitudinal strands in cortex of host; *c-e* connecting sinkers, partly flanked by ray parenchyma of host; *f* young sinker with single apical cells; *g* host cambium. 175x. (Kuijt 1969)

10.5.6.2 Mistletoes

Dwarf mistletoes are perhaps the most widely known plant parasites that invade the cambium of the host tree. The part that develops within the host is called the endophytic system. External symptoms are initial hypertrophy of the host branch due to an abnormal increase in phloem and xylem tissues followed in some species by the development of witches' brooms (Kuijt 1955, 1969). Witches' brooms are typically masses of twiggy growth that vary greatly in size and weight. Brooms are initiated when the parasite invades the procambium of a dormant bud. As branch diameter increases, parasitic tissues also divide, producing radial sinkers that extend from pith to cortex of the host plant (Fig. 10.37). When infection occurs on the trunk of the tree, hypertrophies usually develop without brooming. These hypertrophies can be extensive, due to stimulation of the host cambium for some distance from the primary infection site. Bole infections affect wood structure in the regions of hypertrophy. In conifers, the tracheids are short and distorted and the proportions of ray tissues are greatly increased (Hawksworth 1979). Some

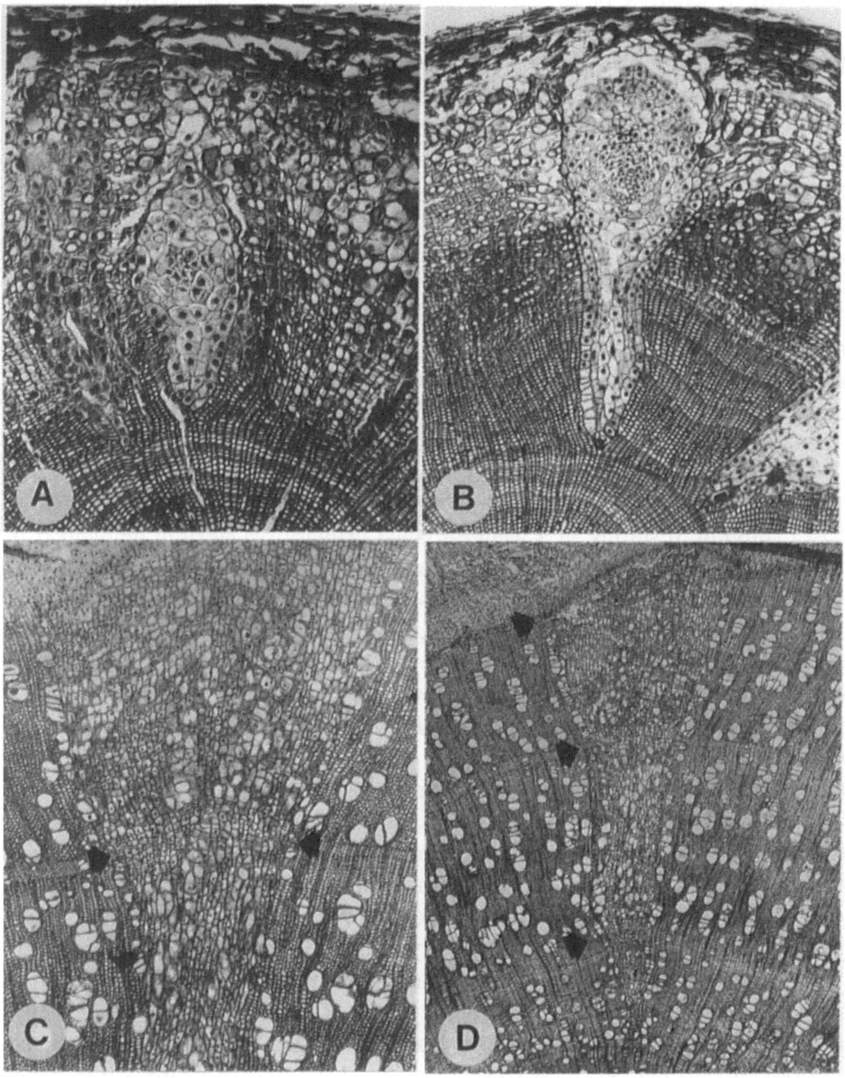

Fig. 10.38A-D. Transverse sections illustrating various stages in sinker development of *Phoradendron juniperinum* on branches of *Juniperus occidentalis* (**A,B**) and *P. flavescens* on branches of *Juglans hindsii* (**C,D**). **A** Young longitudinal strand initiating a sinker. 32x. **B** An older strand with mature tracheary elements and a sinker deeply embedded in the host xylem. 25x. (Calvin et al. 1991). **C** The sinker is widest at the cambium as a result of radial anticlinal divisions that occurred at the margins of the sinker. Note the correlation of the sinker and host cambia and also the pronounced differences in cambial cell structure (*arrows*). 29x. **D** Tangential width of the sinker increased markedly during its 4 years of growth. Note how the annual ring boundaries of the sinker correspond with those of the host (*arrows*). 18x. (Calvin 1967)

trees are killed, whereas others are weakened and become susceptible to secondary attacks by insects and fungi.

Mistletoes infect both coniferous and dicotyledonous trees. In North America alone, Hawksworth (1979) listed six species of *Phoradendron* that invaded dicotyledonous trees, and four species of *Phoradendron* and 16 species of *Arceuthobium* that invaded coniferous trees. The vast majority of the parasitic plants are dicotyledons. Only one parasitic gymnosperm, *Parasitaxus ustus*, a member of the Podocarpaceae, has been discovered (Wilkinson and Musselman 1979). Interestingly, its primary host, *Falcatifolium taxodes*, is also a member of the Podocarpaceae.

Phoradendron: Development of the mistletoes, *Phoradendron flavescens* on *Juglans hindsii* and *P. juniperinum* on *Juniperus occidentalis*, were investigated by Calvin (1967) and Calvin et al. (1991), respectively. A germinating seed formed an attachment disc on the host branch, and from the lower side of the disc a primary haustorium grew through host tissue until it reached the vascular cambium. Meristematic activity of the parasite tissues became correlated with that of the host cambium, resulting in a wedge-shaped primary haustorium embedded in the host tissue. Cortical haustoria, growing more or less parallel to the long axis of the branch, gave rise to new protuberances that also established contact with the host cambium in the form of sinkers (Fig. 10.38A,B). As the cortical haustoria continued to grow along the branch, new sinkers became embedded in the cambium of the host. The new sinkers superficially resembled rays, although they were much larger than rays of the host. In *P. juniperinum* infecting *Juniperus occidentalis*, the sinkers formed in close proximity to leaf traces of the host and sometimes completely surrounded the leaf trace vasculature.

Correlation of the host cambium, as noted above, referred to the fact that once a sinker made contact with the host cambium, a common cambial zone developed between parasite and host (Fig. 10.38C). Cambial activity of the sinker duplicated that of the host, producing xylem tissues and forming annual ring boundaries (Fig. 10.38D). Most sinkers enlarged tangentially with age. Tangential expansion was particularly prominent at the beginning of a new growth ring where the anticlinal divisions appeared to be restricted to border cells of the sinker, suggesting a correlation with cambial reactivation in the host. The cortical haustorium associated with an old sinker enlarged by activity of cambial tissue that encircled the haustorium but maintained continuity with cambium of both the sinker and host. In *J. hindsii*, vessels of the parasite and host were connected by common perforation plates and the tracheary elements by pits.

Phoradendron was once thought to be a "water parasite," acquiring little or no food from its host (Hull and Leonard 1964). However, more recent investigations, based on carbon isotope ratios, showed that these xylem-tapping mistletoes extract substantial quantities of organic carbon and nitrogen dissolved in the xylem sap of the tree (Marshall and Ehleringer 1990).

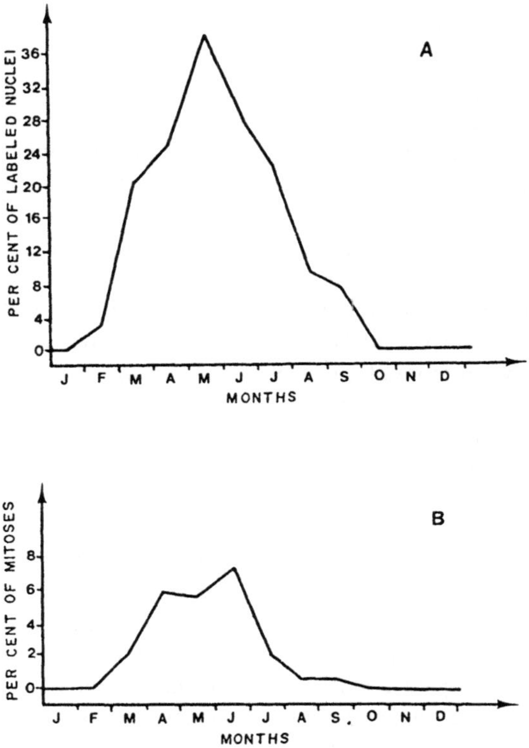

Fig. 10.39A,B. Variations of the DNA (**A**) and mitotic (**B**) indices in the intercalary meristem zone of sinkers of *Viscum album* growing on *Pyrus malus* during a 12-month period. (After Sallé 1979)

Arceuthobium: The genus *Arceuthobium* is an important parasite in the coniferous forests of the United States (Hawksworth and Scharpf 1984). The many species within this genus differ both in their morphology and in their host's response (Alosi and Calvin 1984). Anatomy and development of the sinkers of *Arceuthobium* in seven species of conifers was examined by Srivastava and Esau (1961a,b). General morphology was similar to that of *Phoradendron* described above. The sinkers merged more or less completely with rays of the host wood, with some remaining uniseriate and others multiseriate. The multiseriate rays of the host apparently broadened with age because they were observed to be wider near the cambium than deeper in the wood. Sinkers were composed either of parenchyma cells only or of parenchyma cells and tracheary elements, including vessels. Direct contact between cells of the sinkers and those of the host tracheids occurred in both the axial and radial systems. Although the sinkers were usually associated with the host rays, the inner parenchymatous ends were often wedged in radially among axial tracheids of the host xylem. As in *Phoradendron*, growth appeared to be coordinated with that of the host cambium, so that over time the sinkers became more deeply embedded in the host tissues. At the level of the cambium, the sinkers expanded tangentially by anticlinal divisions that occurred concomitant with those of the host cambium. Despite the coordinated growth of parasite and host, xylem elements produced by the sinkers were often stretched and ruptured during the spring renewal of cambial activity. Continuity of the parasite xylem was only temporarily disrupted, however, as new tracheary elements differentiated and replaced the old ones.

The most pronounced changes in the host tissue occurred in the rays, which not only assumed a hypertrophied appearance but fused to form large composite structures (Fig. 8.16). Ray union involved the intrusive growth of ray cells as well as the displacement and loss of fusiform initials. Some compound rays were formed when a sinker fused with two or more of the host's rays. *Arceuthobium* apparently acquires food from its host via the rays (Alosi and Calvin 1984). Although it has no phloem, it is nonetheless a highly specialized complete parasite withdrawing significant amounts of carbohydrate from its host (Cohen 1963).

Viscum: A few additional features of mistletoe morphology were revealed in a study of *Viscum album* infecting *Pyrus malus* (Sallé 1979). Like other species of mistletoe, when a primary haustorium reached the host cambium its further growth was taken over by a secondary meristem, or intercalary meristem, located in line with the cambial zone of the host. Also, as in other mistletoes, cortical strands gave rise to sinkers which also made contact with the cambial zone of the host by the formation of intercalary meristematic zones. Seasonal activity of the intercalary meristems of the sinkers was correlated with, and functioned in complete harmony with, that of the host cambium as determined by both DNA synthesis and mitotic activity (Fig. 10.39). Densitometric studies of DNA contents of the intercalary meristem cells showed that they were blocked in the G_1 phase of the cell cycle during winter rest. In a similar way, cambial cells have been shown to be arrested in the G_1 phase during dormancy (Chap. 5.3.1.4). At the resumption of growth, new vessel members were initiated simultaneously by intercalary meristems of haustoria and the cambium of the host. Although many direct contacts between vessel members of the host and parasite were observed, this species also posessed true phloem tissue with enucleated sieve elements and densely cytoplasmic nucleated companion cells.

10.6 Environmental Factors

The effects of environmental factors on the cambium are not strictly speaking wounds. Nonetheless, disturbances to the cambium often resemble wounds and, when they are severe, the recovery processes are similar.

10.6.1 Frost

The effect of frost or freezing temperatures on the cambia of trees has been of interest to foresters and horticulturists for many years. Some early experiments were conducted by Caspary (1857), who attempted to relate tangential frost splits to radial temperature differentials in the wood. He concluded that these splits were not due to freezing of the sap stream but to the wood itself that contracted more circumferentially than radially.

According to Ratzeburg (1869, 1871), the relation between frost damage and false ring formation was still unclear as late as 1870. He examined *Abies alba* and *Fagus sylvatica* stems subjected to frost damage in 1866 and described in great detail how he verified the presence of a frost ring. The two most prominent characteristics were disorganized cell series and sometimes unrecognizable rays. Ratzeburg arrived at two conclusions or laws. First, ring doubling, or false rings, never resulted from "Johannistrieben," i.e., from environmentally produced second flushes of shoot growth. Secondly, ring doubling only resulted from "Ersatztrieben," or substitute shoots, that grew out immediately after suppression or death of the May shoot as a result of frost injury or rapid defoliation by insects. Years later, Mayr (1893) also recognized zones of abnormal tissue in conifers and associated them with frost damage.

Hartig (1895a) examined frost rings formed by a late June frost in 1894 that severely injured the new growth of several young coniferous species. In older internodes below the injured shoots, a frost ring was created by damage not only to cambial cells but also to the differentiating xylem cells. The first response to a moderate frost was the expansion and division of ray cells (Fig. 10.40A). New cells, originating on the phloem side of the cambial zone, produced a distorted tissue that was parenchymatous in character, followed by a short-celled tracheid tissue, and finally normal tracheids. Thus, the new cambium appeared to arise on the phloem side. The rays, although distorted, continued through the wound tissue without change in number. The response to a light frost in *Picea abies* was similar (Fig. 10.40B).

The cambium is the tissue most readily damaged by frost during periods of active growth. A study of frost damage to fruit trees by Pisek and Eggarter (1959) showed that the cambium was most resistant to frost during winter rest and it maintained this resistance until its cells began dividing during spring bud break. In some frost-damaged buds the cambium remained viable even though most other tissues had been killed. The living cambium was then capable of regenerating a new bud, although the process often required 2 years. A severe frost sometimes damaged the cambium to such an extent that the residual cambium appeared as isolated "cell nests" comprising only one-quarter of the original cambial circumference. Such cambial strips often regenerated, provided that contact was maintained with cambium of the main axis.

Horticultural varieties of peach (*Prunus persica*) growing in the southern United States are frequently injured by late winter and early spring frosts. Jensen

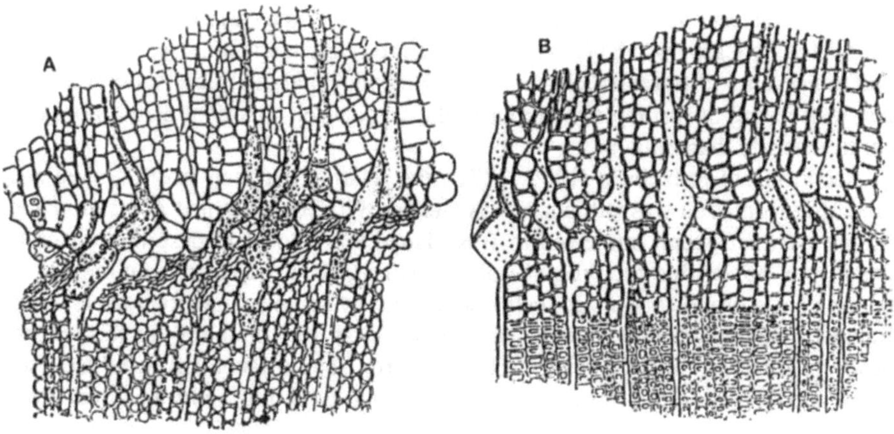

Fig. 10.40A,B. Responses registered in the xylem to a moderate frost in *Pinus sylvestris* (**A**) and to a light frost in *Picea abies* (**B**). Note the distorted rays which in this case were first to respond to the effects of frost. (Hartig 1895a)

et al. (1970) inserted thermistor probes in the trunk cambia to monitor temperature changes. They found that damage usually occurred when high cambial temperatures during the day were followed by a rapid fall to freezing temperatures at night. The high day temperatures during early spring were due to solar radiation impinging on the bark of the trunk which, in turn, caused a premature loss of hardiness after the winter rest. Under these conditions, cambial injury was often concentrated on the trunk and branch-crotch areas of the trees.

In contrast to the foregoing, frost damage to many indigenous trees in the mild climate of Japan occurred at the ground line (Sakai 1968). Bark temperatures varied considerably at different heights above the snow surface. Sakai found that the basal part of the stem near the snow surface was subjected to an especially sharp temperature drop due to radiation on clear nights. The cambium and adjacent cortical cells were most frost sensitive in *Castanea crenata*, whereas the cambium remained undamaged but the cortical cells were killed in *Abies sachalinensis*. Grossenbacher (1915b) described similar basal frost damage on fruit trees growing in New York.

The cambium and newly formed xylem cells of *Castanea vesca* were found by Holdheide (1940) to be injured most severely by late spring frosts although phloem tissues were also injured to some extent. Late spring frosts are also common in alpine forests as described by Lenz (1967). Again, damage was most severe when the cambial initials and mother cells were dividing. Early fall frost can

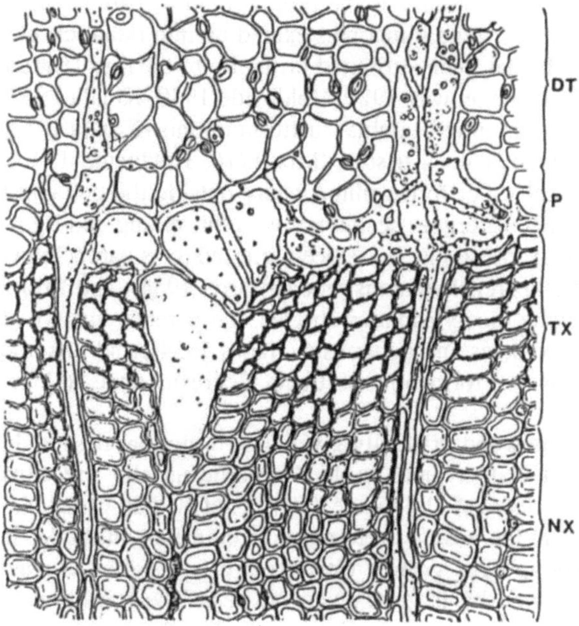

Fig. 10.41. Transverse section of the 1939-40 growth ring boundary showing frost injury to the cambial region of *Larix gmelini var. japonica*. The disturbed tissue arose primarily from the rays and parenchymatous cells. *DT* Distorted tracheids; *P* pathological parenchymatous tissues; *TX* thin-walled tracheids; *NX* normal xylem. 360x. (Kujala 1952)

be equally damaging to still-growing shoots and cambial tissues. Grossenbacher (1915b), for example, mentioned instances where the callus associated with pruning wounds was delayed in entering dormancy and was injured by late fall frosts. Day and Peace (1934) described how a fall frost might result in a frost ring extending from the latewood of one year into the earlywood of the next year.

Neger (1916) related his observations of an early frost that occurred in 1912 during a cold, wet fall. The following spring, the cambia in injured *Picea* stems produced bands of distorted parenchyma cells that abutted directly on the previous season's latewood. Some workers had mistakenly associated this abnormal xylem tissue with pathogens that secondarily invaded the injured stems. According to Day (1950), frost had often been overlooked as the causal mechanism and it was not uncommon to attribute the cambial injury to some other factor. General reviews of the factors related to frost resistance and frost hardiness have been prepared by Parker (1963) and Skre (1988) and a review of the phenomena associated with frost cracks can be found in Kubler (1983).

Damage to the cambium and associated tissues results in a zone of compressed or collapsed cells bordering the xylem. The xylem cells might also show signs of collapse (Fig. 10.40A) if they were only partially differentiated at the time of the frost. First response to freezing temperatures is the proliferation of parenchymatous tissues along the entire cambial zone sometimes extending into radial checks (Fig. 10.41). Origin of these cells has been attributed primarily to the rays (Hartig 1895a, Mix 1916, Bailey 1925a, Hemenway 1926, Holdheide 1940, Kujala 1952, Lenz 1967), but also to isolated, uninjured cambial fusiform initials (Kujala 1952, Lenz 1967), and to other parenchymatous tissues (Mix 1916, Holdheide 1940). The parenchymatous band of cells that forms immediately outside the collapsed zone at the resumption of growth undergoes secondary wall thickening and lignification. Because of the nature of this tissue, it has been referred to as parenchyma wood (Mix 1916, Kujala 1952), parenchymatous wood (Harris 1934), and filling tissue (Holdheide 1940). In some dicotyledonous species, the parenchyma wood has been associated with cell exudates and gummosis (Harris 1934), and in some conifers it is accompanied by the formation of traumatic resin ducts (Chap. 9.2.2.5).

The disorganized parenchyma wood gradually gives way to normal tissues. In conifers, Rhodes (1923) found that atypical tracheidal tissue usually preceded normal tracheids, and in dicotyledons, Holdheide (1940) noted that vessel development was at first weak and the vessels were small in diameter. In the latter case, rather than a normal ring-porous growth ring, the frost ring of *Castanea vesca* consisted of small- followed by large-diameter vessels. The transition to normal wood might also be very irregular, depending on the localized severity of injury to the cambial tissue (Harris 1934).

As pointed out by Hartig (1895a), the rays usually continue through a frost ring without change in their numbers. That is, although ray cells might proliferate and the rays might become distorted, ray numbers neither increase nor decrease. This is a common characteristic in frost-injured wood (Rhodes 1923, Day and Peace 1934, Harris 1934, Holdheide 1940, Kujala 1952), and it is obvious in the many photomicrographs that have appeared in the literature. Templeton (1923), however, found that the rays in frost-injured tissues of *Acer pseudoplatanus* were discontinuous and they did not complete their traverse through the abnormal wood. In more recent work, Cinotti (1991) found that ray size and number were among the factors contributing to radial frost cracks in *Quercus* stems. Apparently the rays served as radial planes of weakness.

One of the most detailed anatomical investigations of frost damage, conducted by Glerum and Farrar (1966), involved the artificial induction of frost rings in seedlings of six coniferous species at different times during the year. Frost rings were divided into two parts; an inner part composed of cells killed by frost, and an outer part formed by cells that developed abnormally after the frost. Freezing caused contraction and deformation of cambial zone cells. Differentiating tracheids appeared to be most susceptible to freezing temperatures, although xylem mother cells and cambial initials might also be killed. The frost ring was formed first by expansion and deformation of immature tracheids followed by callus-like parenchyma cells. The latter arose primarily from xylem rays, but also from uninjured fusiform initials, when present. The authors were unable to determine the exact means by which the cambium became re-established. Because moderate frosts seldom affect the phloem, the phloem mother cells were considered to be the logical source of these new cells.

Frost rings are formed when the cambium resumes production of normal xylem and phloem. The degree of frost injury varies among species, the severity of the frost and its duration, and the time of occurrence. As Bailey (1925a) found, frost rings usually occurred only in the youngest wood unprotected by thick layers of cortex and bark, However, in small trees the frost ring sometimes extended to the stem base (Rhodes 1923). Although frost rings sometimes encircle the entire stem, at other times they are confined to part of a growth ring as "moon rings" (Rhodes 1923, Paclt 1989). In alpine forests, Lenz (1967) reported that frost rings were often present as arcs either on the stem side exposed to the sun or on the compression-wood side where the cambium was most active. Frost rings are not necessarily confined to northern latitudes. Stone (1940) described typical frost rings in seedlings of *Pinus palustris* injured by a mid-winter frost in 1932 in southern Mississippi.

In addition to formation of a frost ring resulting from cambial injury, severe frosts might also induce formation of a false growth ring (Ratzeburg 1869, 1871). As observed by Glock (1951), such rings occurred as extra rings following the frost ring in the radial increment. Although Glock did not ascribe a causal factor, it might be assumed that the frost killed the terminal bud or shoot, thus inducing a second flush of growth. The presence of frost rings has been a valuable aid in dating growth rings in dendrochronological studies (Studhalter 1955, Glock et al. 1963, Fritts 1976).

10.6.2 Drought

Drought rings resemble frost rings in some respects, although cambial injury is usually less serious and recovery is more rapid. The only evidence of cambial disturbance in 2-year-old *Picea glauca* and *Pinus resinosa* seedlings exposed to artificial drought by Glerum (1970) was the presence of distorted tangential walls in undifferentiated xylem cells in fragile tissues near the stem tips. Exposure of *Pinus rigida* seedlings to artificial drought by Barnett (1976) induced collapse of the thin-walled, unlignified tracheids and formation of a ring of crushed cells.

Some examples of more severe drought damage in *Picea abies* were presented by Dietrichson et al. (1985). Radial stem cracks, sometimes several meters in length and extending inward to the pith, were attributed by these authors to drought stress. In one case, an internal drought crack that formed at the end of the growing season produced a very narrow false ring and a tangential band of traumatic resin canals. The radial crack closed the following season, although production of traumatic tissues continued.

Yamanaka (1985) described and illustrated an abnormal ring of tissue in swollen stemwood of *Chamaecyparis obtusa*. Cause of the disturbance was unknown, but evidence suggested that moisture stress induced collapse of the tracheids during secondary wall thickening. Minute radial fissures were filled with callus derived from the proliferation of ray cells. Again, the growth ring of the following year was normal, with the exception of the radial files of cells associated with the fissures.

Unlike the false rings resulting from frost, those resulting from drought are rarely preceded by either distorted or collapsed cells. As shown in Fig. 10.42, most false rings are produced by an extreme narrowing of the xylem elements. This response to drought is common in conifers (Foote 1954, Larson 1963a, Shepherd 1964, Glerum 1970). At the onset of drought, cambial cell division first subsides, followed by declines in cell expansion and maturation. At the expiration of the drought, cambial activity resumes and the only evidence of drought is the false growth ring.

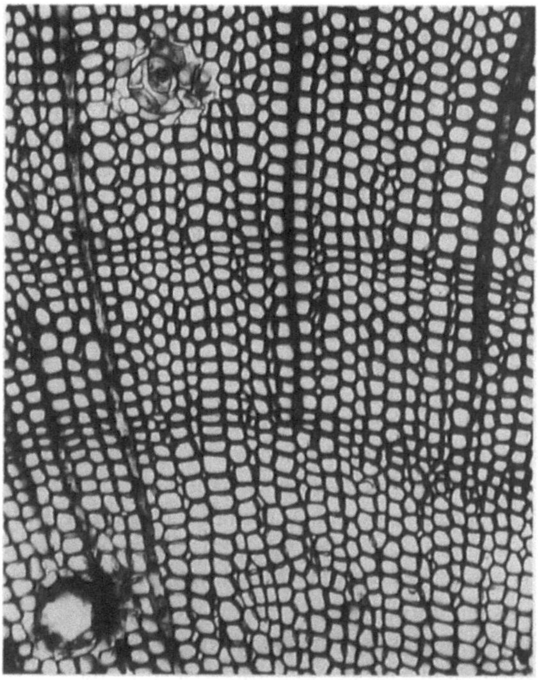

Fig. 10.42. A double false ring in a *Pinus resinosa* seedling produced by two artificially imposed drought periods during the growing season. 60x. (Larson 1963a)

Seedlings of a dicotyledon, *Liriodendron tulipifera*, exposed to a simulated drought responded quite differently (Doley 1970). During the drought period, the radial files of xylem elements became more regular and the cells assumed rectangular outlines. The proportion of xylem occupied by fibers decreased by almost 50% while the proportion occupied by vertical parenchyma increased from very sparse to about 25%. Although the number of vessels did not change, vessel element diameters were drastically reduced. The ray cells were shorter and broader in diameter, and rays became wider by an almost doubling of the number of cell files within a ray. All these structural changes quickly returned to normal at termination of the drought period. As in conifers, the only outwardly visible evidence of the drought was the appearance of a false ring on transverse stem sections.

10.6.3 Flooding

Many species are adapted to prolonged water-immersion (Worbes 1985), but many others can tolerate only temporary submergence of their roots. Although physiological and growth responses of woody plants to flooding of soil have been studied in considerable detail, relatively few studies have been concerned with the effects of water immersion on cambial activity and the anatomy of wood and bark tissues.

The cambial response to prolonged flooding varies widely according to species, duration of water immersion, and other factors. In one example, Smith and McGinnes (1984) examined *Quercus* trees, the lower stems of which had been submerged in water during the dormant season. Rather than normal earlywood vessels, a darkened callus layer formed above and below the mean dormant-season water line. Later, a new cambium developed from the innermost radial layers of the undifferentiated callus mass. This new cambium rapidly merged with existing active cambium, thus isolating the xylem core.

An example of a differential response to flooding was provided by Angeles (1992) who examined adult trees as well as seedlings from the mangrove vegetation of Veracruz, Mexico. In *Rhizophora mangle*, development of secondary xylem and phloem was similar in both the flooded and nonflooded stem parts. However, in *Pachira aquatica*, the vascular cambium was more active in the flooded stem parts. The cambial zone was wider, the phloem produced thicker gelatinous fibers, and the vessels were radially more elongated than in the non-flooded stem parts. These differences in cambial activity and development were attributed to the action of ethylene, which has been found to accumulate in flooded stem parts.

Tree seedlings are often subjected to temporary flooding at different times of the year. As a test of flood-tolerance, Yamamoto et al. (1987) grew seedlings of *Pinus halepensis* for 43 days in flooded soil and then examined them for anatomical changes associated with cambial activity. Flooded seedlings contained proportionally more parenchymatous tissues in the form of abundant xylem rays, enlarged ray cells, numerous epithelial cells associated with an increased number of resin ducts, and more phloem parenchyma cells than did stems of unflooded seedlings. The abnormal xylem of flooded seedlings had several, but not all, the cha-

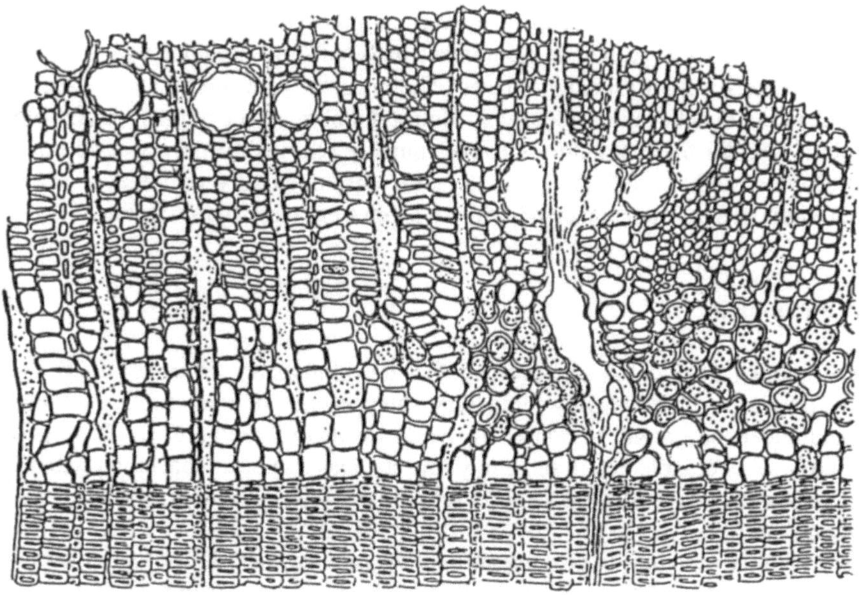

Fig. 10.43. Response to moderate lightning injury in a *Picea abies* stem shown in transverse section. In this region of the stem the bark remained firm but gaps in the xylem were filled by proliferating ray and parenchymatous cells. 100x. (Hartig 1897b)

racteristics of well-developed compression wood. Similar, but somewhat less striking, changes were also induced in flooded seedlings of *Thuja orientalis* (Yamamoto and Kozlowski 1987a), *Cryptomeria japonica* (Yamamoto and Kozlowski 1987b), and *Taxodium distichum* (Yamamoto 1992).

10.6.4 Lightning

Lightning damage is common in forest stands, particularly in trees growing in exposed locations or on ridge tops. Typical lightning damage is a jagged wound conforming to the path of the electrical charge downward on the stem. However, not all trees are equally susceptible to lightning injury. Büsgen and Münch (1929), for example, cited the work of Stahl, who in 1912 demonstrated how susceptibility varied with tree form, bark thickness, and other factors that influenced conduction of the electrical charge.

An extensive study of injuries to trees caused by lightning strikes was made by Hartig (1897b). Wound closure progressed as in other types of mechanical wounds, and the rate of recovery depended on wound severity. Traumatic resin canals, extending tangentially to varying distances on both sides of the wound, invariably formed in *Picea abies*. These canals did not form immediately following injury but some time later, occasionally in the next season's growth ring. In some cases, the bark was detached from the wood adjacent to the primary wound site. In other cases, where the bark appeared to be firm, parenchymatous callus-like tissue filled gaps that formed between wood and bark. A normal type of xylem gradually formed during subsequent cambial activity (Fig. 10.43). In one interesting example, a tree growing on an exposed alpine site had been struck repeatedly by lightning producing a so-called "Blitzfichte", or lightning spruce.

Occasionally, an area of die-off spreads outward from the central point of lightning impact creating a "lightning gap" (Kucera et al. 1985). Von Tubeuf (1905, 1906) investigated a lightning gap in a *Picea* stand a year after the lightning strike and described one tree that had survived. Other than a dead top, it showed no external evidence of injury. At a 4-m level below the stem tip, the phloem had been killed and had become encrusted with resin. The injured xylem was distorted and compressed. The renewed xylem consisted of large, disorganized parenchymatous cells and traumatic resin canals followed by normal xylem. In this case, the new xylem resembled frost damage. Peace (1940) and Kucera et al. (1985) described similar lightning gaps in stands of *Pseudotsuga menziesii* and *Picea abies*, respectively. The latter authors reviewed the various causes of this phenomenon suggested by previous investigators. They concluded that the lightning gap resulted from insufficient water supply due to lightning damage to the roots and/or crowns of the affected trees. Their conclusions were in general agreement with those of Bosshard and Meier (1969) who performed simulated lightning strikes on *Picea* trees. Many years earlier, von Tubeuf and Zehnder (1903) had attempted to experimentally relate shoot death of *Picea* seedlings to atmospheric electrical discharges.

10.7 Cambial Marking

Investigators of cambial activity often find it necessary to precisely mark the cambium so that subsequent events can be studied. Various methods of cross-dating have been used advantageously in dendrochronological studies (Studhalter 1955, Fritts 1976) and in studying natural stand disturbances (Telewski and Lynch 1991), but they are seldom of value in studying specific events in the cambium.

Many techniques for marking the cambium have been tried with varying degrees of success. Some techniques simply induce changes in the structure of the growth ring by mimicking natural causes. For example, Glock and Studhalter (1941) artificially induced frost injury on branches of *Cupressus arizonica*. Not only did the injury exactly duplicate that produced by natural frost, but the procedure made it possible to control the time, position, and intensity of the injury. Artificial drought might be used in a similar way as the cambial response is rapid. Kennedy and Farrar (1965) suggested tilting seedlings of *Larix laricina* to induce

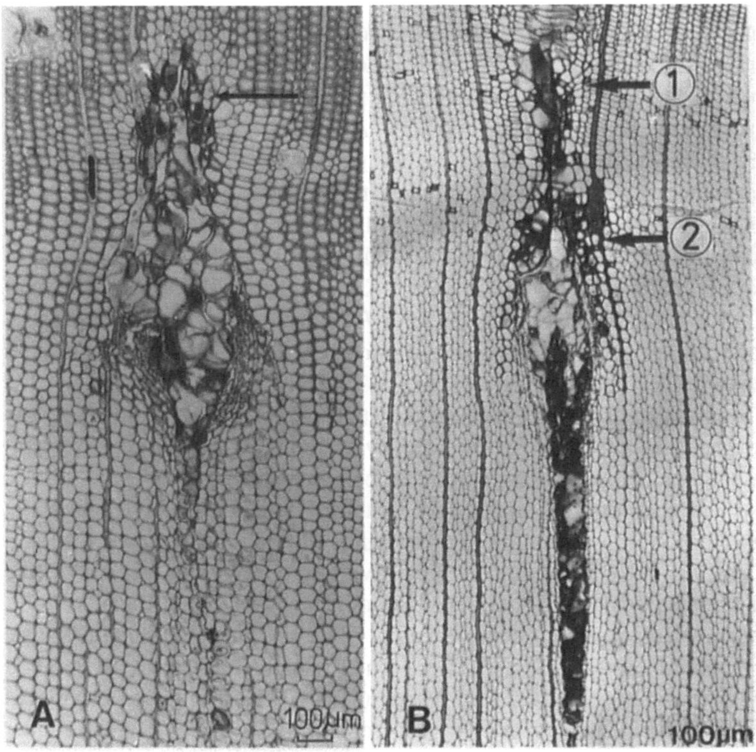

Fig. 10.44A,B. Wound response induced by inserting metal needles in the cambium. **A** Transverse section of *Pinus taeda* wounded in July. *Arrow* indicates the "assumed" site of the cambial initials at the time of wounding. **B** Transverse section of *Cryptomeria japonica*. *Arrows* at *1* and *2* indicate the deduced site of the cambial initials and the initiation of secondary wall thickening, respectively, at the time of wounding. (Kuroda 1986)

bands of compression wood. Although each of these techniques might have limited application in studies of seedling material or branches, they are not readily adaptable to large trees.

A noninvasive technique employed by Waisel and Fahn (1965) involved labeling cells of several woody species with ^{14}C. Radioactive $^{14}CO_2$ was fed photosynthetically to foliage of potted plants and the xylem examined a week later. Tests with *Eucalyptus camaldulensis* and *Zygophyllum dumosum* showed a zone of newly formed, unlabeled xylem elements between the cambial zone and the ^{14}C-labeled zone. However, the interface between labeled and unlabeled elements was not sharp.

Mechanical injuries of various types have received the greatest attention and have been tested most thoroughly. In attempts to determine the rates of cambial activity in tropical species of Africa with indistinct growth rings, Mariaux (1967,

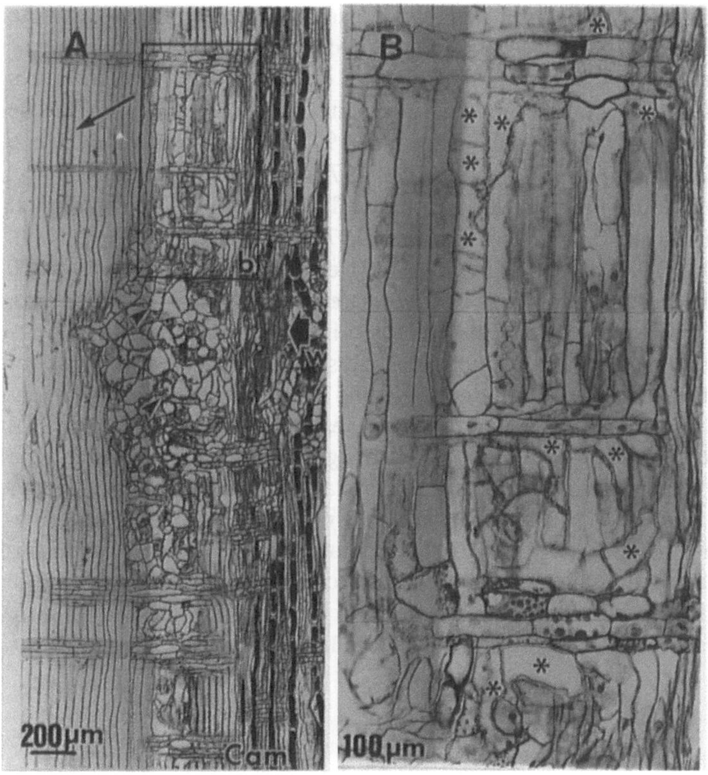

Fig. 10.45A,B. Radial sections of a *P. taeda* stem 26 days after wounding by inserting metal needles in the cambium. **A** Ray cell derivatives, some of which contain droplets (*arrowhead*), completely fill the wound gap. Xylem tracheids (*arrow*) are dividing by transverse walls. *W* Wound site. **B** Rectangular area (*b*) in **A**. Disfigured ray tracheids and derivatives (*) after wall thickening. (Kuroda 1986)

1969) wounded stems by removing bark patches at various times of the year. These small wounds provided fairly reliable markers for evaluating seasonal growth ring development. The technique has been used by others, such as Détienne (1989) and Vetter and Botosso (1989), in their investigations of tropical species.

10.7.1 Pinning Method

A far less destructive method of cambial marking was introduced by Wolter (1968). A small needle was inserted through the bark and into the cambial zone. The needle was either left in place for future identification or removed and the point of insertion marked. At a later date, wood samples were removed from the tree and sectioned for analysis. Needles 300 μm in diameter injured an average of 10-12 cells tangentially in the cambial zone. Wolter estimated that the method was capable of identifying the cells responding to wounding within two cells. Needle wounds in *Pinus radiata* (Jenkins and Shepherd 1972), *Picea sitchensis* (Denne 1977), and *Pseudotsuga menziesii* (Emmingham 1977) responded similarly.

This so-called pinning method was adapted and thoroughly tested by several research groups in Japan (Kawana et al. 1973, 1974, Yoshimura et al. 1981a,b, Kuroda and Shimaji 1984, 1985, Kuroda 1986). Needles with 250, 400, and 700 μm diameters were inserted into stems of *Cedrus deodara, Cryptomeria japonica,* and *Pinus taeda*. The wound responses, described in detail in Section 10.3.2, varied considerably among species, wound shape and size, season of the year, rate of growth, and needle size. By examining the wound response in periodically analyzed samples, it was found possible to accurately estimate the site of the cambial initials at the time of pinning within a few cells (Fig. 10.44A,B). This point could be recognized by the proliferation of tracheid rows and the wall residue of the destroyed cambial zone cells (Fig. 10.45A,B). The outer border of the gap (Fig. 10.46), which Kawana et al. (1973, 1974) had assumed to be the site of the cambial initials, was actually the site of initiation of secondary wall thickening at the time of pinning. More recent research by Nobuchi et al. (1989) on *Cryptomeria japonica* further clarified the probable site of cambial initials at the time of pin wounding.

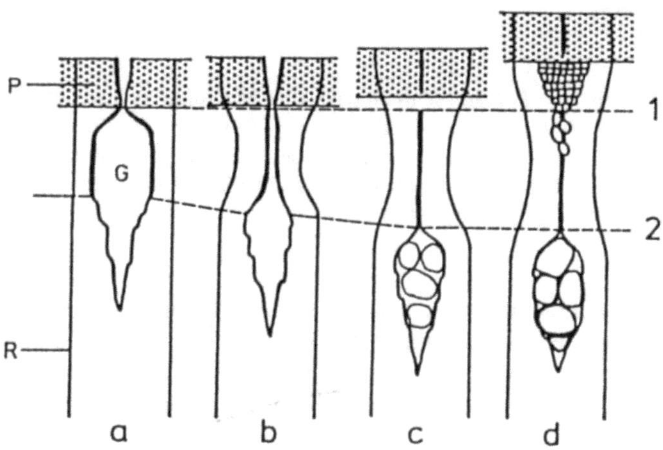

Fig. 10.46a-d. Diagrammatic representation of the process of abnormal tissue formation in *Pinus taeda* at (**a**) immediately, (**b**) several days, (**c**) about 10 days, and (**d**) about a month after pinning. *1* and *2*, as in Fig. 10.44B, indicate the deduced sites of the cambial initials and the initiation of secondary-wall thickening, respectively, at the time of pinning. *P* Phloem; *R* ray parenchyma; *G* cavity formed by the pin. (Kuroda 1986)

A variant of the pin method was used on the thick-barked Dipterocarpaceae trees in Indonesia, where the pin method was impractical (Shiokura 1989). Nails 1 mm in diameter were driven into the trees at periodic intervals to create a series of barrier zones similar to the wounds induced by Mariaux (1967, 1969). In Shorea spp. the cambium formed tangential bands of gum canals in addition to the barrier zones.

10.7.2 Electrical Stimulation

A more recent innovation is electrical stimulation to induce localized wounding in cambial zone cells. In the studies of Imagawa and Ishida (1981b, 1982, 1983a,b), electrical stimulation was applied by direct-current, dry batteries at various voltages and times. Electrodes were inserted in stems spaced vertically 5 cm apart. The response of *Abies sachalinensis* varied with voltage and duration of the applied current. Stimulation at 1.5 Volts for 10 min resulted in crushing of undifferentiated tracheids. Stimulation at 12 Volts for 1 h caused formation of tangential rows of traumatic resin canals, some extending more than one-half of the stem circumference. Results of these preliminary trials with *A. sachalinensis* were not satisfactory for cambial marking. Further trials were made by Imagawa (1985) on the same species at different voltage levels in an attempt to evaluate the wound response in a manner similar to that of Yoshimura et al. (1981b). While more promising than the earlier trials, it was still not possible to accurately identify former cambial cells in the wound area.

Additional tests with *Larix leptolepis* (Imagawa and Ishida 1983b) at 220 Volts through electrodes spaced vertically 5 cm apart also resulted in a zone of crushed cells. Microscopic examination showed that these crushed cells were differentiating tracheids and the cambial cells appeared to be unaffected by the electrical stimulation. Interestingly, tracheids that were in the process of forming the S2 wall layer were crushed to a greater degree than those that had just completed formation of the S1 layer. Intracellular layers were fully lignified in the former cells, whereas lignification had just begun in the cell corners of the latter. The authors speculated that the electric current flowed through the cytoplasmic-rich cambial and undifferentiated xylem cells. However, the differentiating xylem cells with S2 walls were in a suitable state for the generation of Joule's heat. Therefore, they were killed and collapsed. The method was judged inadequate for cambial marking because the cambial cells were unaffected.

Electrical stimulation of two dicotyledons, *Alnus japonica* and *Cinnamomum camphora*, induced similar, but somewhat more complex, responses (Imagawa and Ishida 1983a). The crushed cells of *A. japonica* appeared to be both differentiating fibers and vessels, but in *C. camphora* only fibers were crushed. The crushed fibers had S_1 wall layers whereas the vessels had relatively thick walls but had not yet initiated S_2 layers. Rays in the region of crushed cells increased in both height and width by new cell divisions, and abundant deposition of some material had occurred in the ray parenchyma cells. Unlike the conifers, the crushed cells were not killed by the voltages applied. However, as in the conifers, the cambium appeared to be unaffected by electrical stimulation.

10.8 Cambial Electrical Resistance

Cambial marking is just one of the many ways in which electric currents have been used to either examine or influence cambial activity. The resistance of cambial cells to pulsed electric currents has received considerable research attention (Fensom 1963, Glerum and Krenciglowa 1970). The technique has been proposed as a means of evaluating tree vigor, rate of wound closure, and certain pathological conditions. Tests by Shortle et al. (1977) on *Acer rubrum* and several *Populus* hybrids showed that resistance to a pulsed electric current was inversely proportional to rate of growth, and in *Populus* inversely proportional to wound closure. The lower electrical resistance in cambia of vigorous trees was believed due to a greater concentration of mobile cations, especially potassium. Similar relationships were obtained on *Abies balsamea* trees growing on 26 sites in New England (Blanchard et al. 1983) (Fig. 10.47). Further tests on *A. balsamea* demonstrated that cambial electrical resistance was inversely related to the number of cells in the active cambial zone (Smith et al. 1984).

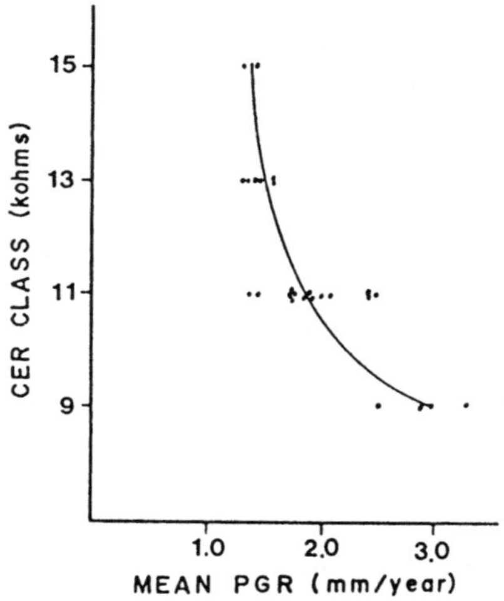

Fig. 10.47. Relationship between cambial electrical resistance class (*CER*) and mean periodic growth rate (*PGR*) in *Abies balsamea* trees from 26 different sites. (Blanchard et al. 1983)

Tests by many workers have shown that cambial electrical resistance can be used to evaluate changes in tree vigor brought about either by silvicultural practices or environmental conditions (Wargo and Skutt 1975, Smith et al. 1976, Blanchard et al. 1983, Gagnon et al. 1988). The technique has also proven useful in evaluating injuries and defects in trees, both mechanically and pathologically induced (Shortle et al. 1977, Imagawa and Ishida 1982, Shortle and Ostrofsky 1983). Tests of tree vigor based on cambial electrical resistance are widespread, having been conducted on *Picea abies* (Garrec et al. 1989, Lindberg and Johansson 1989, Hüttl et al. 1990), *Pinus densiflora* (Ko and Morimoto 1985), *Prunus persica* (Wisniewski et al. 1985) and several species indigenous to South Africa (van Daalen 1988). Wilhelmi (1969) examined the effect of electrical potential differences on diameter growth of *Pinus peuce*, and Zhuravleva (1972) measured electrical resistance to determine the status of cambial activity. In the latter study, the transition to cambial dormancy was accompanied by an increase in electrical resistance due to a decrease in cell hydration and electrolyte content.

A slightly different approach, stem electrical capacitance, was used by Lekas et al. (1990) to examine tree vigor in *Abies balsamea, Picea glauca, P.mariana,* and *P. rubens*. Changes in cambial zone growth rates could be largely accounted for by the seasonal patterns of capacitance. The first production of new xylem cells in spring was associated with an increase in electrical capacitance and the highest values coincided with maximum cambial cell production in June. Thus, the seasonal pattern of electrical capacitance paralleled the seasonal cycle of cambial activity.

10.9 Reconciliation

In most woody species, wound closure conforms to a more or less predictable pattern irrespective of the type and severity of the wound. The first response to wounding is invariably formation of a mass of undifferentiated callus cells that tend to shield the exposed meristematic tissues. In shallow wounds, protected from desiccation, callus might arise directly on the wound surface from living ray and xylem parenchyma. However, in most wounds callus formation proceeds peripherally inward from the wound margins. This marginal callus might be derived from any living tissue bordering the wound, but usually from the rays, vascular parenchyma, and undifferentiated xylem and phloem derivatives. Seldom does the cambium contribute significantly to the initial formation of callus. Callus development is most vigorous on the upper margins of transverse wounds, considerably less so on lateral margins, and weak to nonexistent on the lower margins. Irrespective of its derivative cells, callus tissue is invariably parenchymatous in nature (Lo 1985).

The rates of callus development and wound closure are influenced by many factors, with tree vigor or rate of growth, season of wounding, and intensity of pathogenic invasion being of primary importance. The time required for wound closure will then depend on the depth and severity of the wound. As the callus pad or cushion progresses peripherally inward to occlude the wound, a phellogen develops in the outer layers of callus and a cambium in deeper layers to the inside. Phellogen formation always precedes that of cambium. Both the phellogen and the cambium that develop in the callus pad do so in complete continuity with pre-existing phellogen and cambium in uninjured border tissues. Moreover, both phellogen and cambium are preceded in the callus pad by what might be called the "premeristematic" organization of certain callus cells. That is, although wound callus is often classified as a mass of undifferentiated cells, these cells eventually differentiate (Gaspar et al. 1991). Thus, phellogen and cambium that cover the wound site result from the dedifferentiation of parenchymatous callus tissues (Kuroda 1986). Incipient wound closure occurs when the advancing callus encounters either the opposite wound border or advancing callus from that border. Complete wound closure occurs when the phellogen and cambium have regained circumferential continuity and normal secondary tissue production is underway.

Rarely does a new cambium form independently within the callus mass and advance toward the wound borders. More commonly, the cambium advances peripherally inward in the callus mass "like a slowly closing diaphragm" in the analogy of Sharples and Gunnery (1933). Some workers (Kuroda 1986) have suggested that wound cambium advances in much the same way that the interfascicular cambium develops at the onset of secondary differentiation. This comparison is logical because numerous investigators have demonstrated that the periclinal divisions contributing to the interfascicular cambium originate within the fascicular bundles. These divisions then proceed from the edges of the fascicular bundles uninterruptedly across the interfascicular regions (Esau 1938, 1965, 1977, Engard 1944, Gunckel and Wetmore 1946, Dittmer and Spensley 1947, Eames and MacDaniels 1947, Bünning 1956, Resch 1959, Thompson and Heimsch 1964, Lang 1965, Neubauer 1969, Steeves and Sussex 1972, Larson 1975, Schnettker 1977, Yoshizawa et al. 1986, Soh 1992, and many others).

Neither the wound response nor callus formation is confined to the immediate wound border. Because cells at the immediate wound border are usually killed, callus tissue proliferates both centripetally to occlude the wound surface and centrifugally beneath the injured, but still intact, necrotic phloem and bark. Farther removed from the wound border, the callus gradually gives way to uninjured, normal tissues. However, the release of biochemical products such as resins, gums, phenols, and other constituents often contributes to formation of a protective barrier zone that can extend well beyond that of callus development. Thus, the barrier zone, while protecting the tree from microorganisms, might also reduce the rate of wound closure.

Chapter 11

Cambial Zone Characteristics

11.1 Cambial Zonation

In Chapter 4.4, several different schemes were presented for distinguishing cambial initials from their derivatives (Figs. 4.15, 4.16). According to these schemes, the cambial initials and their immediate derivatives, the xylem and phloem mother cells, were collectively referred to as the cambial zone. However, as pointed out in previous chapters, it is seldom possible to precisely identify cambial initials in histological preparations. Consequently, characteristics of the initials must, by necessity, be determined retrospectively by examining cambial derivatives in various states of differentiation.

The numbers of cells in these states of differentiation also vary seasonally, which adds to the difficulty in identifying cells that comprise the cambial zone. One attempt to describe and standardize the seasonal zonation patterns was made by Whitmore and Zahner (1966) as shown in Fig. 11.1. In this scheme the cambial zone, as defined above, is not clearly demarcated. It might include a few cells in the cytoplasm zone in addition to the primary wall zone. Whitmore and Zahner's interpretation of latewood cells was based on Mork's[1] (1928) definition of latewood in *Picea abies*. Thus, all tracheids in which the common cell wall between two cell cavities multiplied by two was greater than the width of the lumen were considered latewood. The zonation model shown in Fig. 11.1 has been widely adopted by many subsequent investigators in their work on cambial activity in conifers. When these criteria were applied to 25-year-old *Pinus resinosa* trees, the authors arrived at the growth model shown in Fig. 11.2. The seasonal course of cambial zone development in *Larix leptolepis* in Japan, based on slightly different criteria, is shown in Fig. 11.3 from Imagawa and Ishida (1970).

The present chapter consists of a brief overview of certain cambial zone characteristics that both complement and supplement those presented in previous chapters. In many of the papers reviewed, the cambial zone has been ill-defined, at

[1] For a discussion of Mork's (1928) paper, see Denne (1988).

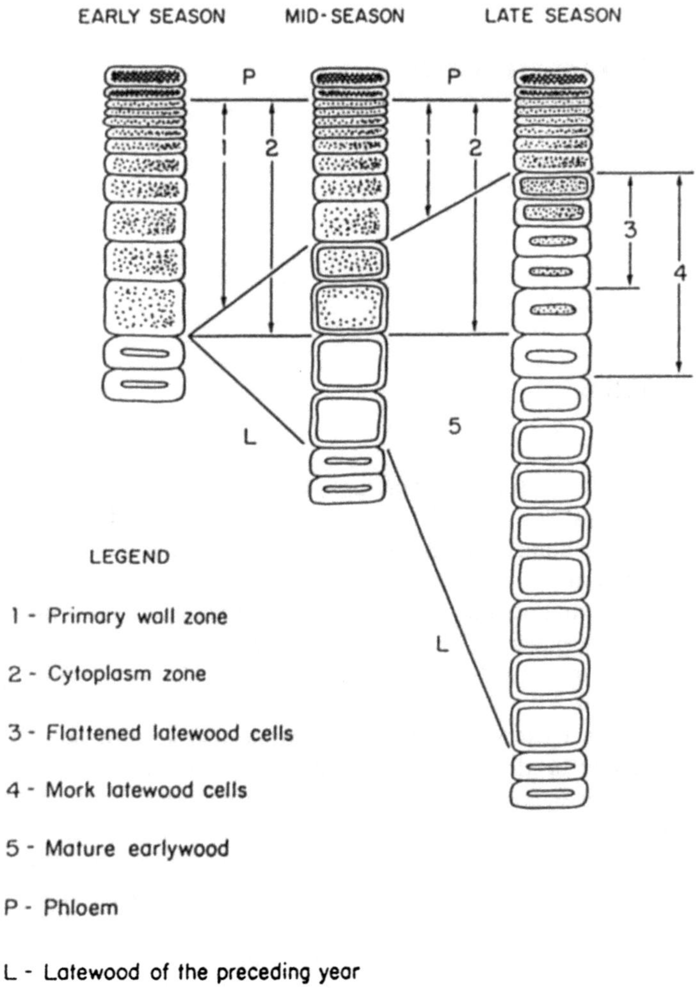

Fig. 11.1. Proposed nomenclature for the cambial tissue zones in conifers. Data were derived from transverse microsections collected at three times during the growing season from 25-year-old *Pinus resinosa* trees. (Whitmore and Zahner 1966)

times conforming more closely to the definitions of Figs. 4.15 and 4.16 and at other times to that of Fig. 11.1. Moreover, like so many other topics relating to the cambium, these, too, have a long history and an enormous literature so that it is impractical to survey it comprehensively.

11.2 Radial Cell Wall Thickness

Tangential wall thickness was one of the criteria used by Sanio (1873) and others to aid in the identification of cambial initial cells (Chap. 4.2.1). Although radial wall thickness has not been found suitable for this purpose, variation in this dimension among cells of the cambial zone has intrigued botanists over the years. For example, Unger (1847) noted that the radial walls of cambial zone cells tended to be thicker than the tangential walls. This relationship comprised an integral part of the concept of Sanio's four (Sanio 1873, von Höhnel 1884b, Mischke 1890, Schoute 1902, Gustin and De Sloover 1955, Newman 1956, Mahmood 1968, 1971, Murmanis 1970,1971, Timell 1973, 1979, 1980). Radial walls tended to be

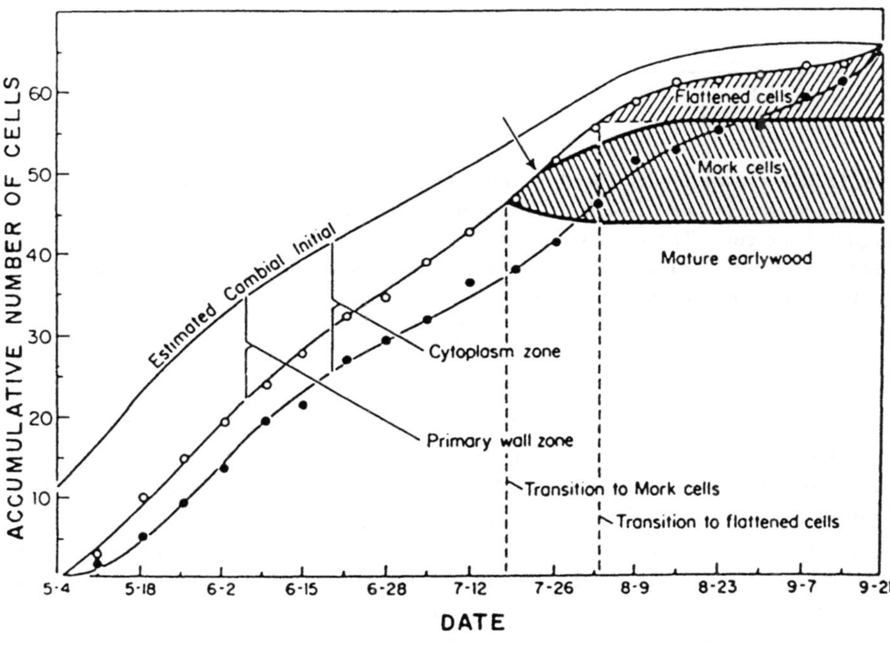

Fig. 11.2. A growth model, based on criteria in Fig. 11.1, applied to 25-year-old *Pinus resinosa* trees. Points plotted along the primary wall zone and cytoplasmic zone boundaries represent the actual mean number of cells in these zones, determined weekly. They were located with reference to the upper curve representing the estimated positions of cambial initials. (Whitmore and Zahner 1966)

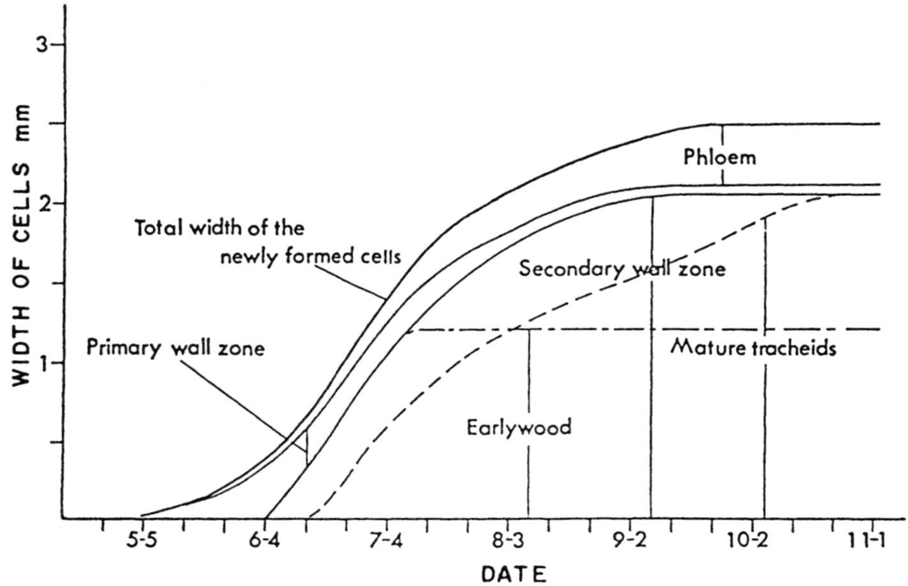

Fig. 11.3. A growth model, based on cell counts made at 10-day intervals throughout the growing season in *Larix leptolepis*. The primary wall zone consists of cambium and enlarging tissue cells. The secondary wall zone consists of secondary-wall thickening cells and mature tracheids. The total number of newly formed cells is the sum of the cells in the primary wall zone, secondary wall zone, and phloem in a given radial file of the current growth ring. (Imagawa and Ishida 1970)

thicker than tangential walls despite the occasional extra-thick tangential wall associated with a group of four (Chap. 4.5.1). Even workers such as Raatz (1892), who did not fully accept the concept, nonetheless observed the relationship. Rothert (1913) accepted the fact that radial walls of cambial zone cells were thicker than the tangential walls, but he added that wall thickness also decreased outward to both sides from the middle of the cambial zone. Rather than relating these changes to cell division, he suggested that the thick walls contained some kind of reserve food necessary for further cell expansion.

The aforementioned relationship has been observed by many other workers without reference to Sanio's four in both conifers (Russow 1883a, Dodd 1948) and dicotyledons (De Bary 1877, Krabbe 1884, Nordhausen 1898, Kuhla 1900, Cockerham 1930, Priestley et al. 1933, Kubella 1985, Ajmal and Iqbal 1987a). In *Populus canescens*, the radial walls of cambial zone cells were two to three times as thick as the tangential walls (Roland 1978).

A peculiar aspect of resting cambial zone cells is the fact that the differences between radial and tangential walls often become accentuated at this time of year.

The effect appears to be more or less universal in both conifers (Russow 1881, Grillos and Smith 1959, Bannan 1962b, Alfieri and Evert 1968, Murmanis 1971, Meier 1973, Timell 1973, Riding and Little 1984) (Figs. 11.4, 11.13) and dicotyledons (Krüger 1892, Haberlandt 1914, Priestley et al. 1933, Wight 1933, Elliott 1935, Evert 1960, Catesson 1961, Robards and Kidwai 1969, Paliwal and Prasad 1970, Paliwal et al. 1975, 1976, Farooqui and Robards 1979, Imagawa

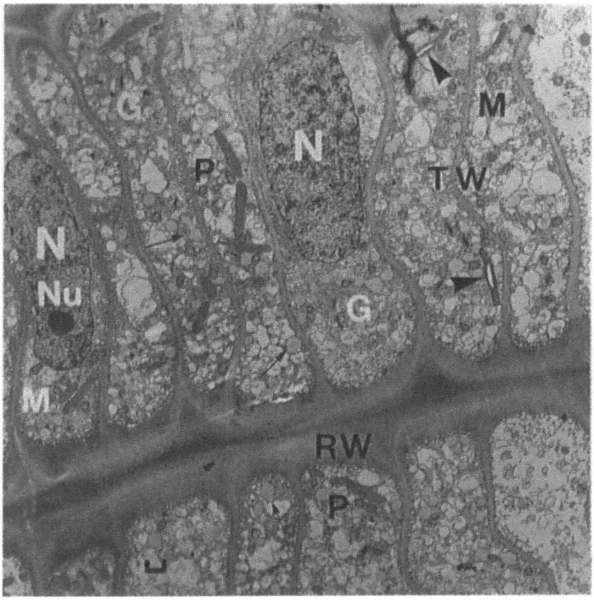

Fig. 11.4. Transverse section of the dormant cambial zone in *Picea abies* illustrating the thick radial walls (*RW*) and the uniformly thin tangential walls (*TW*). *G* Golgi bodies; *M* mitochondria; *N* nuclei; *Nu* nucleolus; *P* plastids, those identified by *darts* contain starch; *arrow* points to lipid droplets. (Timell 1973)

1981, Rao and Dave 1981, 1983c, 1985b, Dave and Rao 1982a, Sennerby-Forsee et al. 1984, Iqbal and Ghouse 1985b, 1987, Rao 1985, Kurczynska 1986, Rao et al. 1989) (Figs. 11.5, 11.6).

In *Aesculus hippocastanum*, maximum wall thickness of fusiform initials increased from 0.15 to 0.5 μm for tangential walls and from 0.7 to 1.7 μm for radial walls as the plants progressed from active growth to dormancy (Rao and Catesson 1987). A similar increase in radial wall thickness during dormancy induction was observed by Riding and Little (1984) in *Abies balsamea* but tangential wall thickness remained unaltered.

Wall thickness was also measured in cambial cells of *Salix fragilis* by Robards and Kidwai (1969). During active growth, cambial cell walls were relatively thin, averaging 0.06 to 0.30 μm for tangential and 0.30 to 1.0 μm for radial walls. In contrast, the walls of resting cambial zone cells were comparatively thick, averaging 0.17 to 0.70 μm for tangential and 2.0 to 3.0 μm for radial walls. Because of these thick radial walls, the first periclinal divisions at the beginning of the growing season should produce an odd, unequally thickened cell. The presence of such a cell was used by some early workers to estimate the beginning of cambial activity (Chap. 4.5.1).

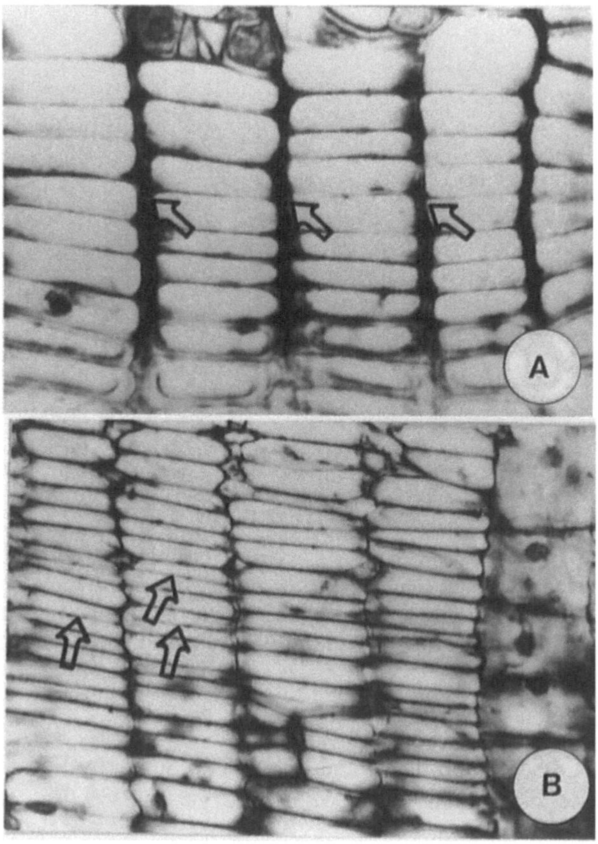

Fig. 11.5A,B. Transverse sections of the dormant (**A**) and active (**B**) cambial zones in *Tectona grandis*. Note how the thick radial walls in **A** (arrows) have become thinner during active growth in **B**. Arrows in **B** indicate newly formed thin tangential walls. **A** 443x; **B** 338x. (Rao and Dave 1981)

According to Catesson's (1981) observations on *Acer pseudoplatanus*, cambial wall thickening coincided with the cessation of mitoses in the fall. It was brought about by persistent activity of the Golgi vesicles that fused with the plasmalemma, thereby increasing the surface of the latter even though cell volume remained constant. Later work, also on *A. pseudoplatanus* (Catesson 1987), indicated that polysaccharides involved in wall synthesis were transported to the Golgi vesicles by the intervention of dictyosomes.

As to causality, Catesson (1990) suggested that radial wall thickness in cambial zone cells entering dormancy might be related to an increase in apoplastic translocation through the cambial zone when storage materials accumulate in xylem and phloem parenchyma cells. In this paper, Catesson described in detail the

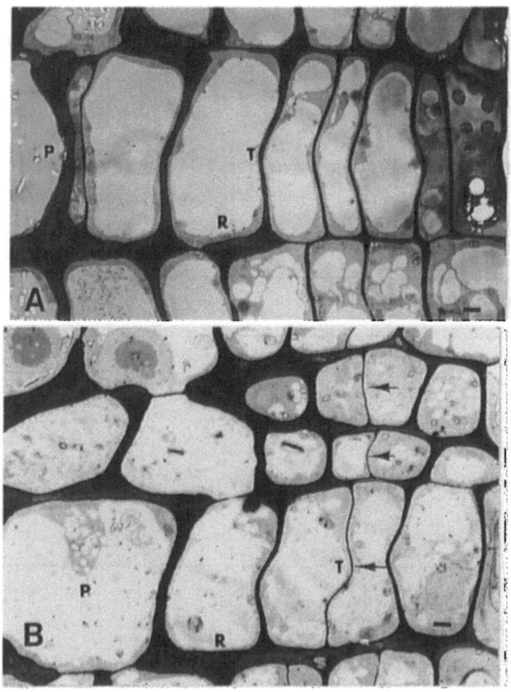

Fig. 11.6A,B. Transverse sections of dormant (**A**) and active (**B**) cambia in *Fraxinus excelsior*. **A** Note the thick radial walls (*R*) compared to the relatively thin tangential walls (*T*). **B** Arrows indicate the first periclinal divisions in the cambial zone. *P* Phloem; *X* xylem. *Bars* 1 µm. (Funada and Catesson 1991)

differences in structure and composition between radial and tangential walls of both resting and actively dividing cambial zone cells. In keeping with this idea, Rao (1985) had suggested that a partial hydrolysis of the thick radial walls during spring growth activation might contribute to the radial growth of cambial initial cells and their derivatives. As noted, Rothert (1913) made a similar suggestion years earlier.

The possibility of some form of cell wall lysis was investigated by Funada and Catesson (1991) during the resumption of cambial activity in *Fraxinus excelsior.* They postulated that the decrease in radial wall thickness probably resulted from localized changes in wall composition and structure facilitating radial expansion under the force of turgor pressure (Fig. 11.6). Cytochemical tests revealed partial autolysis in the radial walls, especially in the junction zones with tangential walls where the cellulosic skeleton was poorly developed. Autolysis of wall material began simultaneously at different levels along a radial wall. It was therefore incomplete, leaving a loose polysaccharide network. This pattern of autolysis was closely associated with a localized decrease in the level of cell wall-bound calcium ions (Fig. 11.7). The authors suggested that wall lysis might be due, in part, to breaking down of the acidic pectin network, thus increasing wall plasticity and allowing radial expansion of the cambial cells.

Table 11.1. Winter condition of the cambium in a *Pinus strobus* tree 17 m in height (After Brown 1915)

Position	Cambium thickness (μm)	Cell rows in cambium
2-year twig	5-9	2-4
6-year twig	5-9	2-4
3.7 m from apex	9-31	6-7
7.6 m from apex	32-40	7-9
12.6 m from apex	29-35	7-10
15.8 m from apex	29-35	7-9

11.3 Number of Cambial Zone Cells

11.3.1 Dormant Cambium[2]

Because of the difficulties in detecting initial cells in active cambia, many investigators have relied on observations of dormant cambia in which the cambial zone is greatly reduced. The initial cell often has been assumed to be an exceptionally

[2] The term "dormant cambium" will be used in a broad sense to include all states of dormancy in which cambial cells that are predisposed to either enlarge or divide fail to do so (Romberger 1963).

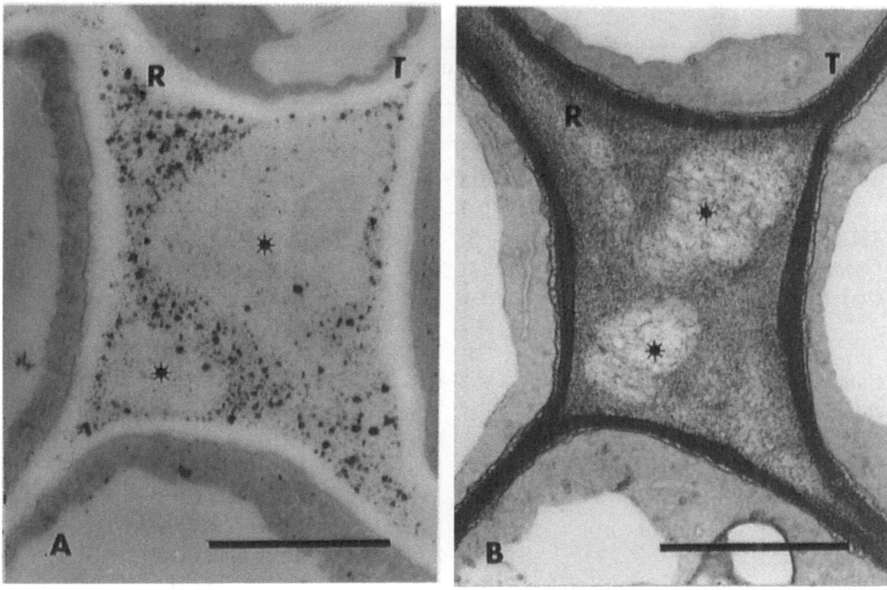

Fig. 11.7A,B. Transverse sections of recently activated cambia in *Fraxinus excelsior*. Localization of calcium in junction between radial (*R*) and tangential (*T*) walls was by the antimonate precipitation technique. **A** Precipitate is less dense in two areas (*asterisks*) indicating reduced calcium deposits. **B** Staining to visualize polysaccharides intensified the two regions (*asterisks*) suggesting lysis of the acidic pectin wall network. *Bars* 1 µm. (Funada and Catesson 1991)

narrow cell situated in the midst of slightly larger ones. In more recent work to be described later it appears that the initial cell is most often located on the xylem side of the resting cambium.

Dormant cambial cells are usually narrow in radial diameter, although this dimension can be quite variable even within a single sample. Bannan (1955) provided values of 5-6 µm for resting cambial cells in *Thuja occidentalis*. In three *Pinus* species examined by Alfieri and Evert (1968), radial diameters averaged 4-6 µm with some cells ranging up to 8-10 µm. The latter cells were found most often on the phloem and xylem sides of the resting cambium. Dormant and active cambia can be compared in two dicotyledons in Fig. 11.5 (*Tectona grandis*) and Fig. 11.6 (*Fraxinus excelsior*), and in Figure 11.13 in a conifer (*Larix leptolepis*).

Table 11.2. Number of cells in resting cambia of gymnosperms

Species	Number of cells	Reference
Abies alba	1-3	Fink (1986)
Abies balsamea	7-8	Kutscha et al. (1975)
" "	2-7	Riding and Little (1984)
Abies firma	5	Tsuda (1975a)
Chamaecyparis obtusa	4	Tsuda (1975a)
Cryptomeria japonica	3-6	Itoh et al. (1968)
" "	4-5	Itoh (1971)
" "	4	Tsuda (1975a)
Cunninghamia lanceolata	2-3	Lo (1985)
Juniperus californica	1	Alfieri and Kemp (1983)
Juniperus communis	1	De Bary (1877)
Larix decidua	4-7	Meier (1973)
Larix laricina	5-6	Knudson (1913)
Picea abies	5-7	Meier (1973)
" "	6-8	Timell (1973)
" "	5-6[a]	Höster (1974)
" "	4-8[a]	Timell (1980)
" "	7-9	Raatz (1892)
Picea glauca	2-8	Gregory (1971)
Picea rubens	1-2	Wilcox et al. (1956)
Picea rubra	2-8	Rees (1929)
Pinus banksiana	4-11	Alfieri and Evert (1968)
Pinus densiflora	4-6	Tsuda (1975a)
Pinus elliottii	6-10	Nix and Villiers (1985)
Pinus patula	4-8	Mahmood (1971)
Pinus pseudostrobus	6-10	Mahmood (1971)
Pinus radiata[b]	12	Barnett (1973)
Pinus resinosa	4-11	Alfieri and Evert (1968)
Pinus strobus	2-10	Brown (1915)
" "	8	Fraser (1952)
" "	5-6	Srivastava and O'Brien (1966)a
" "	8	Wilson (1966)
" "	4-11	Alfieri and Evert (1968)
" "	6-10	Murmanis (1971)
" "	6-10	Panshin and de Zeeuw (1980)
Pinus sylvestris	4	Unger (1847)
" "	3-4	Mischke (1890)
" "	7-9	Raatz (1892)
" "	4-6	Wight (1933)
" "	5-9	Meier (1973)
Pinus taeda	6-10	Nix and Villiers (1985)

Pinus thunbergii	6-8	Itoh (1971)
Pseudotsuga menzesii	4-6	Grillos and Smith (1959)
Thuja occidentalis	1-4	Bannan (1955, 1964b)

[a] Compression wood side
[b] This species is seldom completely dormant in the New Zealand climate.

Schacht (1856) was unable to determine how many cells overwintered in the cambia of species he examined. Wieler (1886a) simply stated that the cambium was "very broad" in *Cytisus*, Nordhausen (1898) that it was "few-layered" in *Quercus*, and Artschwager (1945), and Artschwager (1950) that it was composed of "several rows of cells" in *Parthenium* and *Carya*, respectively. Hartig (1891) made the general statement that the dormant cambium consisted of one to three cells. In an even more restrictive sense, Priestley et al. (1933), without verification, claimed that "... the dormant cambial layer on the surface of the old wood is usually about one cell deep."

As expected, many workers reported that width of the dormant cambium varied not only with growth conditions but also among trees of a stand and with the sampling position within individual trees. One example of the latter source of variation in *Pinus strobus* by Brown (1915) is presented in Table 11.1. A similar trend was found by Riding and Little (1986) in *Abies balsamea*, with the dormant cambium averaging 5.2 cells near the top and 7.6 cells at the base of a 19-year-old tree.

Reports on widths of dormant cambia in terms of cell numbers vary widely among species as indicated in Tables 11.2 and 11.3. Investigators not only have interpreted the cambial zone differently, but conditions under which the sample trees were growing also varied greatly. The values appearing in Tables 11.2 and 11.3 must therefore be accepted as best estimates.

11.3.2 Active Cambium

It has been estimated that the cambium of a large tree might possess 500 million dividing cells (Clowes and Juniper 1968). Such a large number of cells is difficult both to visualize and to verify. A more easily comprehended value is the number of cells comprising radial cell files in the active cambial zone. The difficulty here, however, lies in determining the cambial zone boundaries, and criteria for doing so are often quite arbitrary. A host of factors can influence the number of cells in the cambial zone in different species. Foremost among these factors is rate of growth, which is largely controlled by environmental conditions. In general, the number of cells in the active cambial zone parallels the pattern of radial growth. Thus, cell number usually increases from a low during dormancy, to a more or less constant level during the main growth period, and then decreases again with the approach of dormancy (Bannan 1955, Wilson 1964, 1966). For example, the

Table 11.3. Number of cells in resting cambia of dicotyledons

Species	Number of cells	Reference
Acacia caven	2	Aljaro et al. (1972)
Acer negundo	3-6	Tucker and Evert (1969)
Acer pseudoplatanus	2-3	Catesson (1962a), (1964)
Acer saccharinum	3-5	Hanson and Brenke (1926)
Acer saccharum	6	Evert et al. (1972)
Aesculus hippocastanum	3-4	Nordhausen (1898)
" "	2-4	Rao (1985)
" "	4-6	Barnett (1992)
Albizzia lebbeck	3-5	Venugopal and Krishnamurthy (1989)
Arbutus unedo	1-2	Arianoutsou-Faraggitaki et al. (1984)
Betula alleghaniensis	2-4	Wilcox et al. (1956)
Callophyllum inophyllum	2-4	Venugopal and Krishnamurthy (1989)
Celastrus scandens	5-8	Davis and Evert (1970)
Dalbergia sissoo	2-4	Paliwal and Prasad (1970)
Diospyros lotus	3-5	Archer and Cameron (1939)
Fagus sylvatica	2-3	Kidwai and Roberts (1969)
" "	2-3	Farooqui and Roberts (1979)
Ficus rumphii	6-7	Sagheer and Iqbal (1987)
" "	6-7	Ajmal and Iqbal (1987b)
Fraxinus americana	3-4	Srivastava (1966)
" "	3-4	Tepper and Hollis (1967)
Fraxinus campestris	3-6	Hanson and Brenke (1926)
Fraxinus excelsior	4-5	Kurczynska (1986)
Gmelina arborea	4-9	Dave and Rao (1982a)
Holoptelia integrifolia	3-6	Rao and Dave (1983a)
" "	3-8	Rao and Dave (1985b)
Liriodendron tulipifera	6	Panshin and de Zeeuw (1980)
Malus spp.	4-6	Swarbrick (1927)
Mangifera indica	4-9	Dave and Rao (1982b)
" "	3-5	Venugopal and Krishnamurthy (1989)
Morinda tinctoria	2-4	Venugopal and Krishnamurthy (1989)
Parthenium argentatum	3-5	Joseph et al. (1989)
Parthenocissus inserta	4-9	Davis and Evert (1970)
Polyalthia longifolia	4-7	Paliwal et al. (1975)
Populus euramericana	4-6	Stahel (1968)
Populus tremuloides	5-8	Raphael (1954)
" "	5-8	Davis and Evert (1968)
Prosopis flexuosa	3-4	Villalba (1985)
Prosopis juliflora	3	Meyer et al. (1971)
Prosopis spicigera	6-7	Iqbal and Ghouse (1985b)
Proustia cunefolia	2-4	Aljaro et al. (1972)

Psidium guajava	1-2	Chou and Chiang (1973)
Pyrus communis	5-10	Evert (1960)
Pyrus malus	5-10	Evert (1960, 1963a)
Quercus robur	4-5	Kubella (1985)
Quercus rubra	3-4	Murmanis (1977)
" "	4-6	Zasada and Zahner (1969)
Robinia pseudoacacia	4-9	Wareing and Roberts (1956)
" "	4-9	Derr and Evert (1967)
Salix dasyclados	2-3	Sennerby-Forsse and von Fricks (1987)
Salix fragilis	2-4	Robards and Kidwai (1969)
Salix viminalis	3-4	Sennerby-Forsse (1986)
Salvadora persica	2	Paliwal et al. (1976)
Sambucus nigra	6	Nordhausen (1898)
Streblus asper	5-6	Ajmal and Iqbal (1987a)
Tectona grandis	5-10	Rao and Dave (1981)
" "	3-4	Venugopal and Krishnamurthy (1989)
Terminalia crenulata	3-4	Venugopal and Krishnamurthy (1989)
Tilia americana	2-3	Mia (1970)
Vitis riparia	1-2	Davis and Evert (1970)

number of cells in the cambial zone of *Abies balsamea* in New Brunswick, Canada, varied from 5-9 in April, to 3-5 in July, and to 4-5 in September (Mellerowicz et al. 1992a).

Different approaches have been used to estimate the mean cell number in active cambia. Imagawa and Ishida (1970, 1972), for example, analyzed the seasonal trend in cambial cell number in *Larix leptolepis* stems in Japan. They defined cambial cells as those with a R/T ratio of less than 0.3, where R and T were the radial and tangential cell diameters, respectively. By applying this ratio, they demonstrated how width of the cambial zone varied seasonally (Fig. 11.8). A similar pattern, based on different criteria, was derived for *Pinus radiata* by Jenkins (1975) (Fig. 11.9). In this case, however, note that the winter minimum occurred during June to August in Australia. Another similar pattern for *Populus euramericana* cv. '*robusta*', grown in the northern hemisphere, is shown in Fig. 11.10. In some species, such as *Cryptomeria japonica* shown in Fig. 11.20, the active cambium is interrupted by a mid-season growth pause.

A computer simulation of cambial activity for *Pinus resinosa*, developed by Wilson and Howard (1968), showed a relatively stable cambial zone width over a 140-day time span (Fig. 11.11). The cambial zone included the initial cell and all xylem and phloem mother cells still capable of dividing. The model illustrates how the zones of enlargement and wall thickening disappear when no new cells enter their respective phases. When read vertically, the graph shows the number of cells in each zone of differentiation for each day of the season. For instance, at day 40 two cells were in the zone of wall thickening, six cells in the zone of enlargement, and seven cells in the cambial zone. When read horizontally, the graph shows how long each cell remained in the phase of enlargement and cell wall

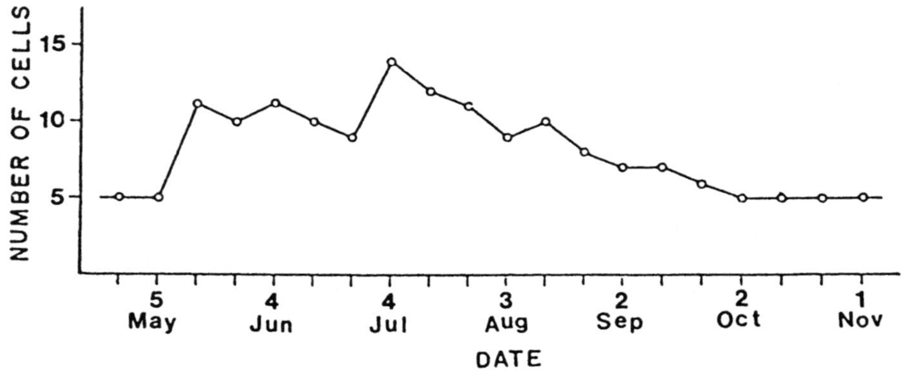

Fig. 11.8. Seasonal variation in the number of cells defined as cambial cells by means of the R/T ratio in *Larix leptolepis*, where R and T were the radial and tangential cell diameters, respectively. See text for details. (Imagawa and Ishida 1970)

thickening. For instance, cell number 50 spent 2 days in the phase of enlargement and 6 days in the phase of wall thickening. The authors noted that the number of cells in the cambial zone of adjacent radial files did not vary greatly over a period of days. In fact, as Wilson (1968) found, cell number in the cambial zone hardly changed even above a girdle where the rate of cell production doubled or tripled.

Reports of the number of cells in the active cambial zone are exceedingly variable and must be accepted provisionally. As noted, the extent of the cambial zone is difficult to determine, and authors are not consistent in how they define it. When defined as the number of cells capable of dividing periclinally, then the main determinants are the number of xylem and phloem mother cells. In some species, the width of the active zone does not differ greatly from that of the dormant cambial zone (Srivastava 1966, Murmanis 1977). In other species, it might be quite broad. A few representative values for the active cambial zone are given in Table 11.4.

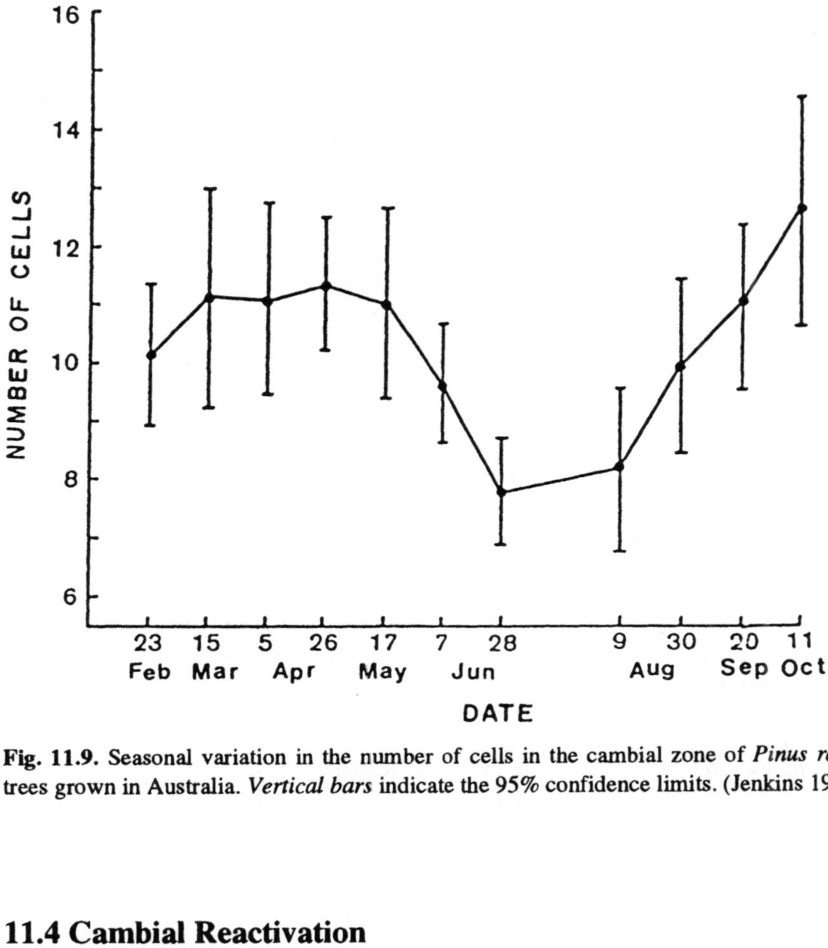

Fig. 11.9. Seasonal variation in the number of cells in the cambial zone of *Pinus radiata* trees grown in Australia. *Vertical bars* indicate the 95% confidence limits. (Jenkins 1975)

11.4 Cambial Reactivation

11.4.1 Bark Slippage and Cambial Swelling

Reactivation of the cambium in the spring results in a condition known as cell or bark slippage. It can be experienced by the ease with which bark can be peeled from a stem or branch, and it has been developed as a common procedure for evaluating the status of cambial activity (Cooper et al. 1955, Husain and Cooper 1958, Madden 1977). This condition has been known, although not fully understood, since ancient times (Grew 1682, Studhalter 1955). For example, Trécul (1853a), in attempting to describe bark slippage, stated that it occurred before multiplication of the "utricles" had begun, i.e., while the previous season's wood and bark were still in "two parts." Hartig (1837b) believed the cambium to be a fluid that not only filled the intercellular spaces between wood and bark but also the wood cells themselves. Therefore, if one stripped the bark from a tree in

Table 11.4. Number of cells in the active cambial zone

Species	Number of cells	Reference
Gymnosperms		
Abies alba	5-10	Fink (1986)
Abies balsamea	15-16	Kutscha et al. (1975)
" "	12-14	Timell (1979)
Cryptomeria japonica	6-11	Itoh et al. (1968)
Juniperus californica	8	Alfieri and Kemp (1983)
Larix leptolepis	10	Imagawa and Ishida (1970)
Picea glauca	12-16	Gregory (1971)
" "	14-20	Yumoto et al. (1982)
Pinus strobus	15-20	Murmanis (1971)
Thuja occidentalis	10-15	Bannan (1962b)
Dicotyledons		
Dalbergia sissoo	8-10	Venugopal and Krishnamurthy (1989)
Ficus rumphii	7-13	Sagheer and Iqbal (1987)
Fraxinus americana	4-6	Srivastava (1966)
Fraxinus angustifolia	3-4	Matovic (1980)
Holoptelia integrifolia	6-14	Rao and Dave (1985b)
Parthenium argentatum	7-9	Joseph et al. (1989)
Populus tremuloides	5-15	Raphael (1954)
Prosopis flexuosa	9-16	Villalba (1985)
Quercus rubra	3-4	Murmanis (1977)
Streblus asper	7-13	Ajmal and Iqbal (1987a)
Tectona grandis	15-24	Rao and Dave (1981)
Tilia americana	8-10	Mia (1970)
Tilia cordata	8-12	Wloch and Zagórska-Marek (1982)

early spring, the tender cells would rupture and release their cambial sap stream. When he examined the separated regions, he found that membranes of the new wood layer occurred on the inner rind surface as well as on the outer wood surface.

Some workers found that bark slippage preceded actual cell division in the cambium. Thus, von Mohl (1862) described how the bark could be easily peeled from twigs of *Fraxinus excelsior* and how the underlying bare wood felt slimy even though microscopic examination showed no new cell growth. Elliott (1935) reported similar conditions in *Acer pseudoplatanus*, and Wight (1933) established criteria for judging the time of cambial reactivation. This phase of cambial reactivation is often associated with the vernal ascent of water from the roots and the phenomenon of "bleeding sap" from fresh wounds in the xylem (Hartig 1862,

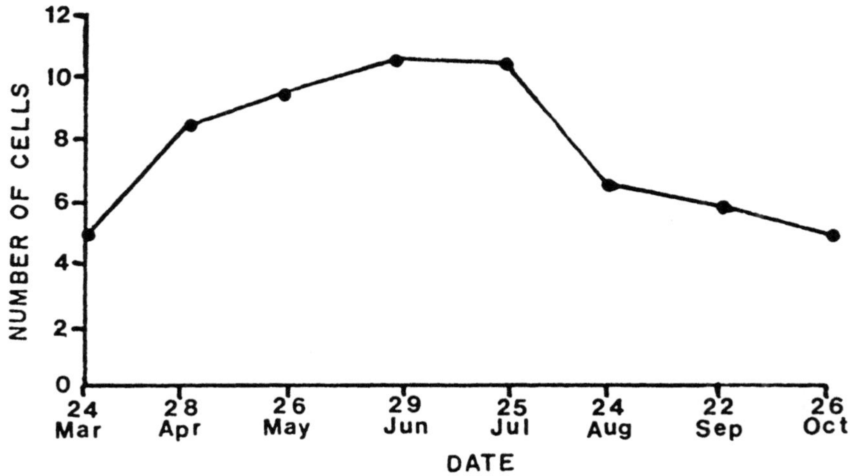

Fig. 11.10. Seasonal variation in the number of cells in the cambial zone of a 77-year-old *Populus euramericana* cv. robusta tree grown in Europe. (After Stahel 1968)

Burström and Krogh 1947). It is brought about by the continuous increase in the movement of water and spring sap, the latter forming first in the cambial zone of the roots and lower trunk and then extending upward in the stem and branches (Essiamah and Eschrich 1986). The latter authors suggested that the sap was derived from storage products in the sapwood, and that it might be transported in cells of the cambial zone as they became progressively reactivated.

The most commonly reported pattern of bark slippage places separation in the newly formed xylem. However, whether ease of separation precedes or immediately follows active cambial cell division reflects not only when during the spring season the observations are made but also the species examined. Most early workers were not overly concerned with accuracy and they simply assumed that separation occurred in the cambium. Tupper-Carey (1930), for example, stated that bark strips of *Acer pseudoplatanus* "split naturally at the cambium" when collected in the spring, and MacDaniels and Curtis (1930) claimed that removal of bark strips "extended only inward to the cambium." This view was sharply criticized by Bailey (1943), who claimed it to be without foundation. He found that

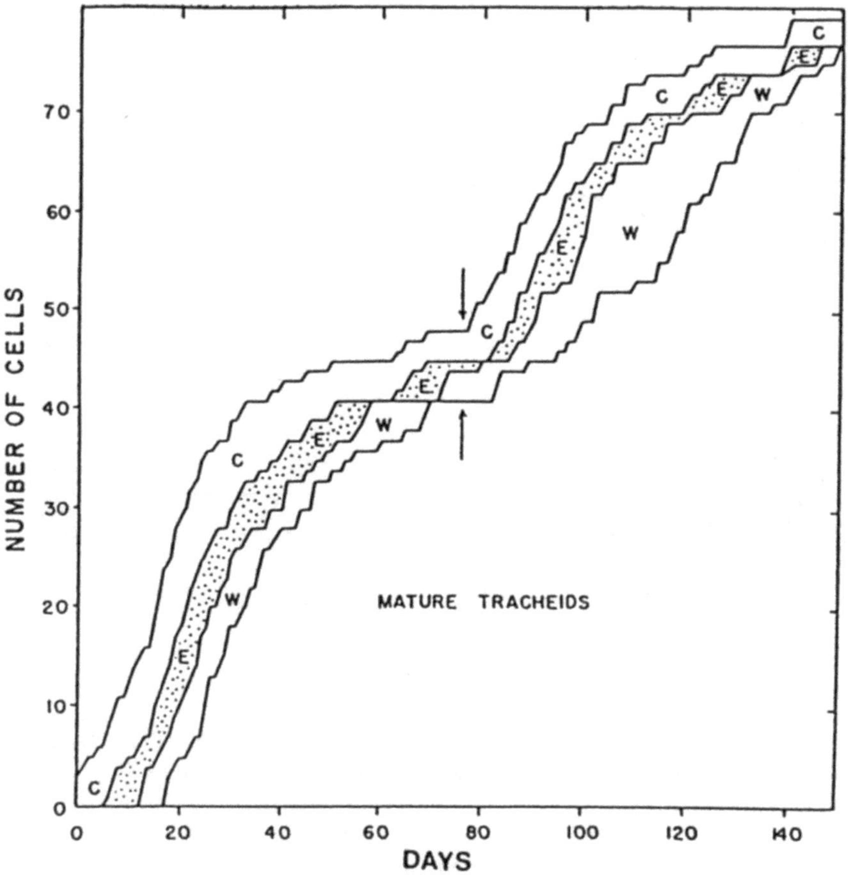

Fig. 11.11. Simulated development of a *Pinus resinosa* annual growth ring. *C* Cambial zone; *E* cell enlarging zone; *W* cell wall thickening zone. *Arrows* indicate the end of a simulated mid-season drought. (Wilson and Howard 1968)

the zone of excessive weakness was well within the maturing xylem. A report by Russow (1883a) placed it in still older cells in certain dicotyledonous trees. He observed that the cambial layer, together with a thin layer of wood cells, adhered to the separated bark strip.

In contrast to the foregoing, Noel (1968) observed that bark separation occurred within the cambial zone in *Julbernardia globiflora* and *Trema orientalis* growing in Rhodesia. Microscopic examination showed that the cambium and newly formed phloem cells adhered to the debarked surfaces and not to the de-

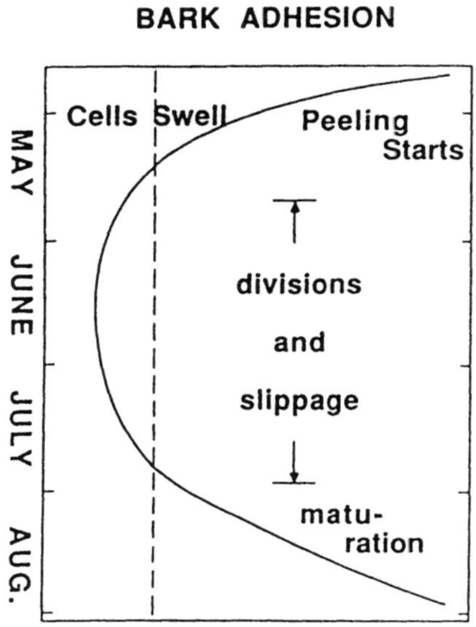

Fig. 11.12. Change in bark adhesion on *Acer saccharum* stems during the growing season in the Adirondack Mountains of New York. *Base line* Adhesion of loose bark; *dashed line* adhesion up to which stems could be debarked with hand tools. (Kubler 1990)

tached bark. However, the zone of young wood cells in which separation occurs has been interpreted by most workers as occurring in either xylem mother cells or immature tracheids (Artschwager 1950, Schneider 1952, Bannan 1955, Evert 1960, 1963b, Thair and Steeves 1976). The ease with which these tender-walled cells within the cambial zone separated comprised the basis for the "bark-strip" method of analyzing cambial activity developed by Priestley et al. (1933).

Bark slippage has practical significance, particularly in the ease of bark peeling for pulpwood, poles, and other commercial products (Huber 1948, Raphael 1954, Wilcox et al. 1956, Kubler 1990) (Fig. 11.12).

Bark slippage has been found to either precede or to be accompanied by the well-known phenomenon of cambial swelling in association with thinning of the radial walls (Swarbrick 1927, Cockerham 1930, Priestley 1930c, Priestley et al. 1933, Esau 1948, Artschwager 1950, Ladefoged 1952, Bannan 1955, 1962b, Wilcox et al. 1956, Wareing 1958, Catesson 1964, Derr and Evert 1967, Philipson et al. 1971, Essiamah and Eschrich 1985, Iqbal and Ghouse 1985a, Siddiqui 1991). Like so many other terms, the term swelling, with reference to the cambium, lacks specificity. For example, Evert (1960) found that although swelling occurred in the cambial zone of *Pyrus communis*, it was confined to the differentiating sieve elements and did not occur in the cambial initials. Consequently, he and his co-workers (Evert 1963b, Derr and Evert 1967, Davis and Evert 1968, Tucker and Evert 1969) questioned the earlier more generalized reports of cambial swelling, implying active swelling of the cambial initial.

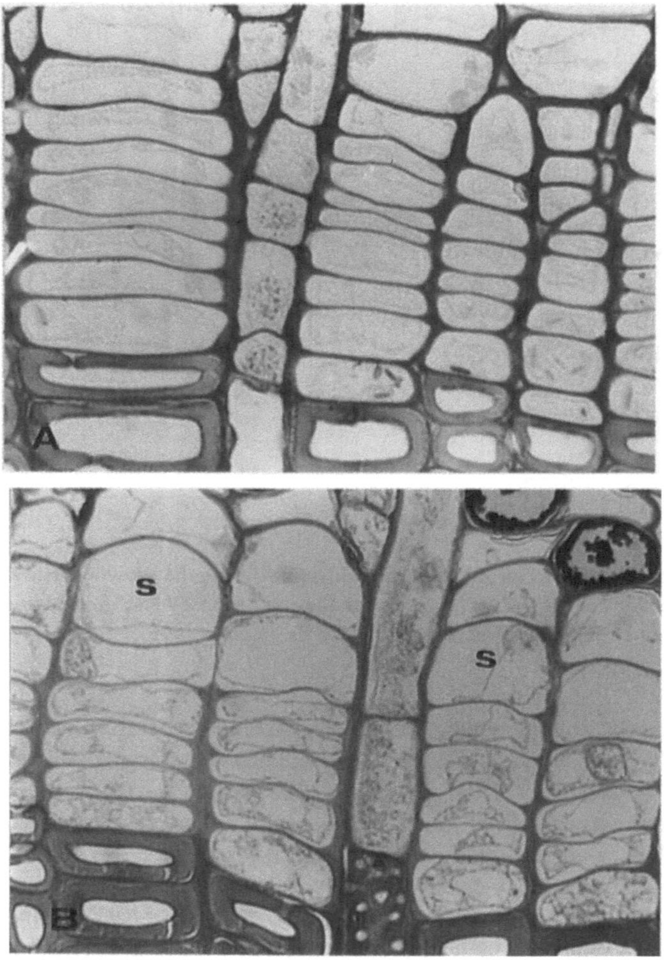

Fig. 11.13A,B. Cambia of *Larix leptolepis* in transverse view. **A** Dormant cambium. Mature tracheids abut directly on the tangentially flattened, dormant cambial cells. Note the relatively thick radial walls. **B** Transitional stage from dormant to active. Cambial zone cells have enlarged radially, but cell division has not begun. The overwintering sieve cells (S) are much enlarged. (Imagawa and Ishida 1981a)

Cambial cell swelling has been demonstrated in several other species. Particularly clear photomicrographs illustrating this phenomenon in the cambial zone of *Larix leptolepis* were prepared by Imagawa and Ishida (1981a). Dormant cambial zone cells, abutting directly on the mature tracheids in the middle of Fig. 11.13A, appear thick-walled and flattened. Cells adjacent to the mature tracheids have radial diameters slightly larger than the others. These cells appear to have lost the capacity for cell division and will probably differentiate immediately in the spring. Just before the initiation of cell division, shown in Fig. 11.13B, the cambial zone cells appear thinner-walled and radially enlarged. As Evert and coworkers observed, the sieve cells are much enlarged. The general phenomenon of cambial swelling has been attributed to hydration of the cambial zone cells, thinning of their walls, and separation of the primary pit fields (Chap. 6.7).

11.4.2 Xylem vs. Phloem Reactivation

Another lack of specificity closely related to the foregoing concerns the nature of the reactivation phenomenon. Some workers regarded cambial reactivation to mean cambial cell division, whereas others interpreted it to include the maturation of overwintering, immature cambial zone cells. For many years, workers simply assumed that xylem cell division preceded that of phloem because they either confined their examinations to the xylem or found the phloem too difficult to evaluate. In a typical case, Priestley et al. (1933) concluded that cambial cell divison on the xylem side preceded that on the phloem side in both conifers and dicotyledons, but their examination of phloem was entirely cursory. Several workers claimed that the first-formed earlywood tracheids in conifers were derived from xylem mother cells that were reactivated in the spring (Mischke 1890, Bannan 1955, Tepper and Hollis 1967, Höster 1974). Bannan (1955), for example, claimed that late-season anticlinal divisions would provide the initial cells for rapid tracheid production during the following spring surge of growth (Chap. 6.5.3).

Doley and Leyton (1968) and Zasada and Zahner (1969) suggested that the first vessel elements in *Fraxinus excelsior* and *Quercus rubra*, respectively, might have overwintered in an undifferentiated state, i.e., with no visible evidence of prior differentiation. The report by Matovic (1980) on *Fraxinus angustifolia* tended to confirm this observation.

The possibility that vessels might overwinter in an undifferentiated state was examined in five ring-porous species by Imagawa and Ishida (1972). During the resumption of cambial activity in all five species, the frequency of periclinal divisions was lower near the previous year's latewood than in the middle of the cambial zone. Moreover, a considerable number of cells in the cambial zone did not divide and these cells had thicker tangential walls than the newly divided cells. The authors arrived at this conclusion by calculating frequency ratios of cell division for each cell position radially outward from the previous season's xylem. Frequency (F) was calculated by the formula F(%) = 100N/MF, where N = total number of cell divisions in each cell position and MF = total number of radial files measured on the transverse sections. Data for four species are shown in Fig. 11.14. On the basis of these observations, the authors concluded that many cells near the previous year's latewood were not newly formed but had overwintered as

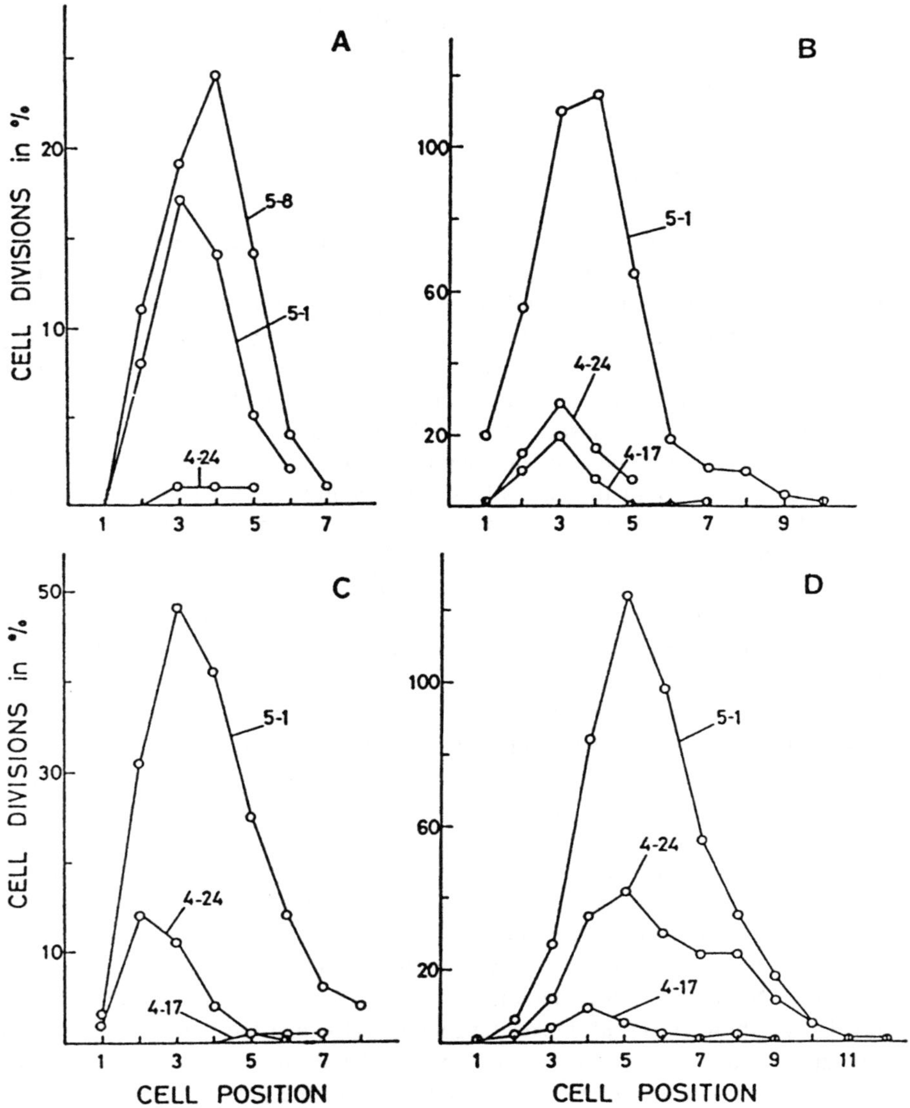

Fig. 11.14A-D. Frequency (%) of the number of cell divisions in each cell position radially removed from the annual ring boundary at different collection dates. **A** *Fraxinus mandshurica* var. *japonica*. **B** *Quercus mongolica* var. *grosseserrata*. **C** *Robinia pseudoacacia*. **D** *Ulmus davidiana* var. *japonica*. (Imagawa and Ishida 1972)

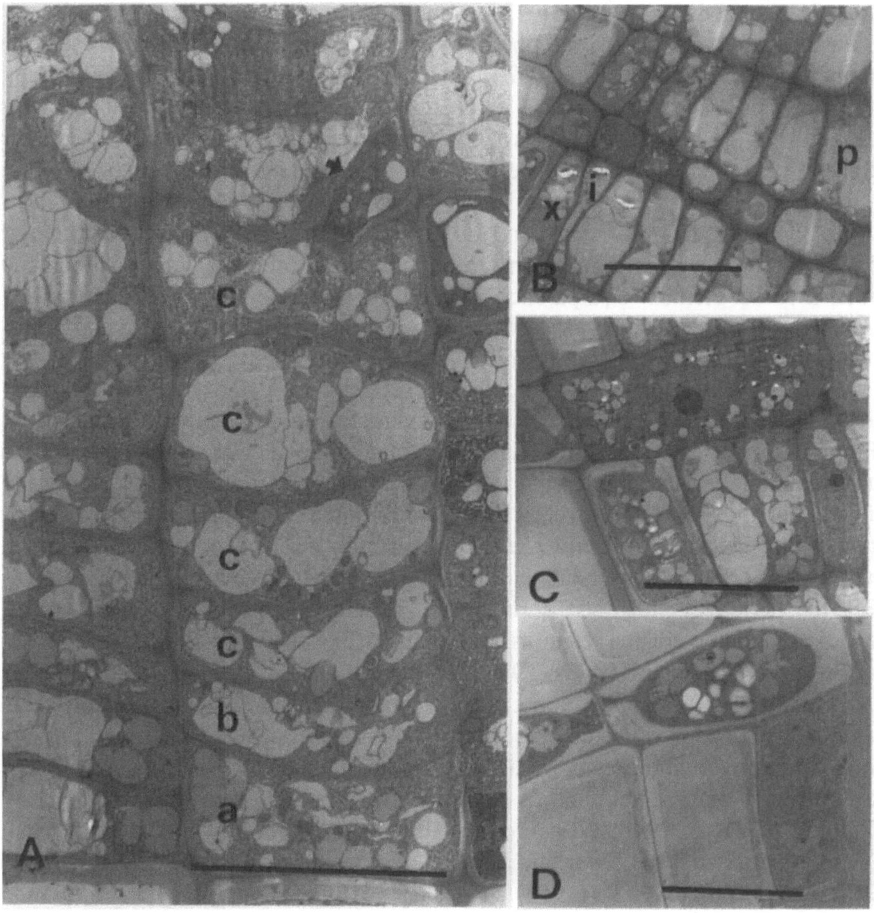

Fig. 11.15A-D. Transverse sections of cambial zone cells in *Aesculus hippocastanum*. **A** Dormant cambium showing the boundary-layer cell (*a*), the cell adjacent to the boundary-layer cell deduced to be the initial cell (*b*), and presumptive phloem precursor cells (*c*). Division of a sieve element giving rise to a companion cell is indicated by the arrow. **B** Radially enlarging cells on the phloem side (*p*) in early spring. Boundary-layer cells (*x*) and cells adjacent (*i*) have not changed in size. **C-D** Dormant cambial zone ray cells with incomplete (**C**) and complete (**D**) secondary walls protruding into the fusiform initial zone. **A-B** Bars 10 μm; **C-D** Bars 1 μm. (Barnett 1992)

xylem derivatives. This conclusion was especially applicable to the first earlywood vessels, which often differentiated in close proximity to the previous season's latewood.

Further confirmation for the possible overwintering of prospective vessel elements was provided by Barnett (1992) in *Aesculus hippocastanum*. By examining both dormant and newly activated cambia, he determined that the boundary-layer cell abutting the previous season's xylem was destined to become either a vessel element or a parenchymatous cell without undergoing further division (Fig. 11.15A). He suggested that the boundary-layer cell was the last-formed daughter cell on the xylem side of the fusiform initial. No ultrastructural features were found to indicate that such cells were preprogrammed to become either a vessel element or a parenchyma cell. However, Barnett reasoned that because cells on the phloem side of the fusiform initial appeared to be preprogrammed, similar events were likely to occur on the xylem side.

Most illustrations and photomicrographs of dormant cambia clearly show undifferentiated cambial zone cells abutting directly on fully mature xylem cells (Figs.11.6, 11.13, 11.15A). Although these cambial zone cells might very well be undifferentiated xylem mother cells, it is unlikely that they would be in a partially differentiated state except perhaps in very mild climates or under unusual circumstances, as, for example, *Pinus radiata* in Australia where Skene (1969) observed partially differentiated tracheids in the cambial zone at all times of the year including the period of presumed dormancy. It should be noted that a partially differentiated cell is not the same as a committed, but undifferentiated, cell as described by Imagawa and Ishida (1972) and Barnett (1992).

An entirely different situation exists on the phloem side of resting dicotyledon cambia, where it is quite common to find immature sieve elements or phloem parenchyma overwintering in partially differentiated states. These cells are often first to renew enlargement and to complete their maturation the following spring (Elliott 1935, Esau 1948, Wilcox et al. 1956, Evert 1960, 1963b, Catesson 1964, Cheadle and Esau 1964, Derr and Evert 1967, Davis and Evert 1968, Tucker and Evert 1969). Wilcox et al. (1956) suggested that precocious phloem reactivation might account for the beginning of early bark loosening prior to cambial activity in some species.

As noted previously, Barnett (1992) deduced that cells in the dormant cambium on the phloem side of the fusiform initial in *Aesculus hippocastanum* appeared to be preprogrammed and committed to differentiate as phloem derivatives (Fig. 11.15A). During reactivation, cambial zone cells farther from the initial cell began radial enlargement well in advance of either the initial cell or the boundary-layer cell (Fig. 11.15B). The first periclinal divisions also occurred in cambial zone cells nearest the previous year's phloem followed by those situated inward toward the initial cell. The cell adjacent to the boundary-layer cell was last to divide, and by the time it did so all cambial zone cells external to it were committed to phloem formation. Barnett was unable to determine whether development of these phloem cells was arrested at the end of the previous growing season, or whether their development proceeded slowly throughout the winter season.

Evidence for developmental arrest was found in ray cells of the cambial zone. Whereas some fully enlarged ray cells with unthickened walls protruded between mature fusiform xylem elements (Fig. 11.15C), others with completed secondary walls protruded between fusiform cells of the cambial zone (Fig. 11.15D). Barnett reasoned that once secondary wall formation began in a ray cell, the process went on to completion even though cell enlargement and other developmental processes had stopped. This evidence from ray cells, together with the fact that tangential wall thickness increased across the dormant cambial zone from the boundary-layer cell toward the phloem, suggested a similar process of arrest for the phloem precursor cells.

Similar reports of either uncommitted or partially differentiated cells on the phloem side of resting cambia have been made for a number of conifers (Brown 1915, Abbe and Crafts 1939, Bannan 1955, Srivastava and O'Brien 1966a, Wilson 1966). On the contrary, Alfieri and Evert (1968) could find no evidence for partially differentiated sieve elements in dormant cambia of the three species (*Pinus banksiana, P. resinosa, P. strobus*) they examined. These authors suggested that although mature sieve cells in conifers might resume functioning the following spring, it was unlikely that such elements would overwinter in a partially differentiated state. However, more recent work by Kutscha et al. (1975) and Riding and Little (1984) has indicated that immature sieve cells might possibly overwinter in *Abies balsamea.*

Further confirmation for overwintering sieve cells in a conifer can be found in the work of Imagawa (1981) on *Larix leptolepis*. Phloem elements situated immediately outside the dormant, flattened cambial zone cells in Fig. 11.13A were produced the previous season. Just before the initiation of cell division in spring, shown in Fig. 11.13B, the radial diameters of several sieve cells increased dramatically. Imagawa suggested that these sieve cells were produced near the end of the previous growth period, they overwintered in an immature state, and began their radial enlargement prior to the initiation of cell division in early spring. More thorough discussions of these various aspects of phloem reactivation can be found in Evert (1960, 1963a,b) and particularly in Esau (1969).

The completion of either xylem or phloem cell maturation in the spring is not necessarily the same as cambial reactivation. As pointed out by Esau (1969), cell maturation and cell division might or might not coincide. For example, Rees (1929) reported that in *Picea rubra*, differentiation of overwintering phloem parenchyma coincided with the inception of cambial cell division on the xylem side. Esau (1948) found in *Vitis vinifera* that although immature phloem cells were the first to resume development, cambial cell divisions began more or less simultaneously on the xylem and phloem sides. In *Abies balsamea*, Kutscha et al. (1975) and Riding and Little (1984) observed that even though overwintering phloem cells enlarged and matured the following spring, the first cambial divisions occurred on the xylem side.

Table 11.5. Reactivation of cambial cell division after dormancy

Species	Reference
Xylem reactivated first	
Abies alba	Ermich (1963)
Abies balsamea	Kutscha et al. (1975)
" "	Riding and Little (1984)
Acer pseudoplatanus	Elliott (1935)
Betula alleghaniensis	Wilcox et al. (1956)
Chamaecyparis pisifera	Tsuda (1975a)
Citrus sinensis	Schneider (1952)
Cytisus laburnum	Wieler (1886a)
Delonix regia	Ghouse and Hashmi (1982)
Fagus sylvatica	Ermich (1963)
Fraxinus americana	Lodewick (1925)
" "	Tepper and Hollis (1967)
Malus spp.	Swarbrick (1927), Knudson (1913)
Mimusops elengi	Ghouse and Hashmi (1983)
Parthenium argentatum	Artschwager (1945)
Picea excelsa	Raatz 1892, Ermich (1962)
Picea rubens	Wilcox et al. (1956)
Picea rubra	Rees (1929)
Pinus halepensis	Lipschitz et al. (1984)
Pinus rigida	Brown (1912)
Pinus sylvestris	Mischke (1890)
Pseudotsuga menzesii	Grillos and Smith (1959)
Quercus robur	Kubella (1985)
Quercus rubra	Zasada and Zahner (1969)
Sambucus nigra	Nordhausen (1898)
Strebulus asper	Ajmal and Iqbal (1987a)
Thuja occidentalis	Bannan (1955)
Vitis vinifera	De Bary (1877)
Phloem reactivated first	
Acer negundo	Tucker and Evert (1969)
Acer pseudoplatanus	Cockerham (1930), Catesson (1962a, 1964)
" "	Lawton (1976)
Aesculus hippocastanum	Rao (1985)
" "	Barnett (1992)
Betula verrucosa	Boutin (1985)
Cryptomeria japonica	Itoh et al. (1968)
Ficus rumphii	Ajmal and Iqbal (1987b)
" "	Sagheer and Iqbal (1987)
Holoptelia integrifolia	Rao and Dave (1985b)

Juglans regia	Schaad and Wilson (1970)
Juniperus californica	Alfieri and Kemp (1983)
Larix decidua	Ladefoged (1952), Meier (1973)
Larix laricina	Knudson (1913)
Malus sylvestris	Algan and Toker (1981)
Parthenocissus inserta	Davis and Evert (1970)
Picea abies	Meier (1973), Höster (1974)
Pinus banksiana	Alfieri and Evert (1968)
Pinus resinosa	"
Pinus sylvestris	Wight (1933), Meier (1973)
Pinus strobus	Abbe and Crafts (1939)
" "	Brown (1915), Wilson (1966)
" "	Murmanis and Sachs (1969), Murmanis (1971)
Polyalthia longifolia	Ghouse and Hashmi (1978)
Populus euramericana	Stahel (1968)
Populus serotina	Gill (1933)
Populus tremuloides	Evert and Kozlowski (1967)
" "	Davis and Evert (1968)
Prosopis spicigera	Iqbal and Ghouse (1985b)
Pyrus communis	Evert (1960)
Pyrus malus	Evert (1963b)
Quercus rubra	Murmanis (1977)
Salix caprea	Gill (1933)
Salix dasyclados	Sennerby-Forsse and von Fircks (1987)
Salix fragilis	Lawton (1976)
Salix viminalis	Sennerby-Forsse (1986)
Vitis vinifera	Esau (1948)
Xylem and phloem reactivated simultaneously	
Carya illinoensis	Artschwager (1950)
Celastrus scandens	Davis and Evert (1970)
Fraxinus angustifolia	Matovic (1980)
Picea glauca	Gregory (1971)
Tectona grandis	Rao and Dave (1981)
Tilia americana	Evert (1962)
Vitis labrusca	Knudson (1916)
" *riparia*	Davis and Evert (1970)
Many spp.	Lodewick (1928)

Despite the occasional report of xylem reactivation preceding that of phloem, an increasing number of studies are showing just the opposite in both conifers and dicotyledons (Table 11.5). These studies raise the question as to whether this latter pattern might be more prevalent in temperate zone species than previously thought. However, the data in Table 11.5 should be accepted with reservations. Although reactivation usually referred to periclinal divisions, in some papers it was difficult to determine whether it referred to cell enlargement or cell division.

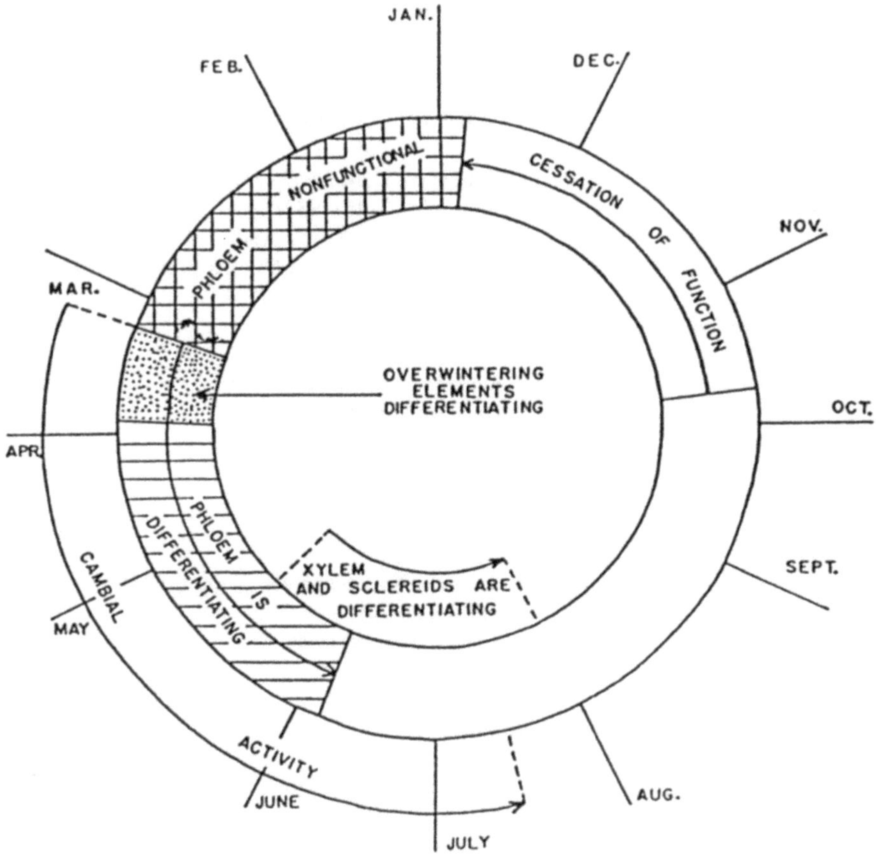

Fig. 11.16. Diagram illustrating the seasonal changes that occur during one year of secondary growth in a branch of *Pyrus communis*. (Evert 1960)

The diagram shown in Fig. 11.16 illustrates the relation between cambial activity and xylem and phloem differentiation in *Pyrus communis*.

In an early report by Huber and Rouschal (1938), the authors stated that the first sieve elements in several dicotyledonous trees differentiated shortly before leaf opening. In a more recent study in *Pinus banksiana, P. resinosa,* and *P. strobus*, Alfieri and Evert (1968) found that cambial reactivation began when one or two layers of cells on the phloem side began to expand and differentiate directly into sieve cells without dividing. At about the same time, some cambial cells began to expand radially, followed a few days later by the appearance of mitoses and

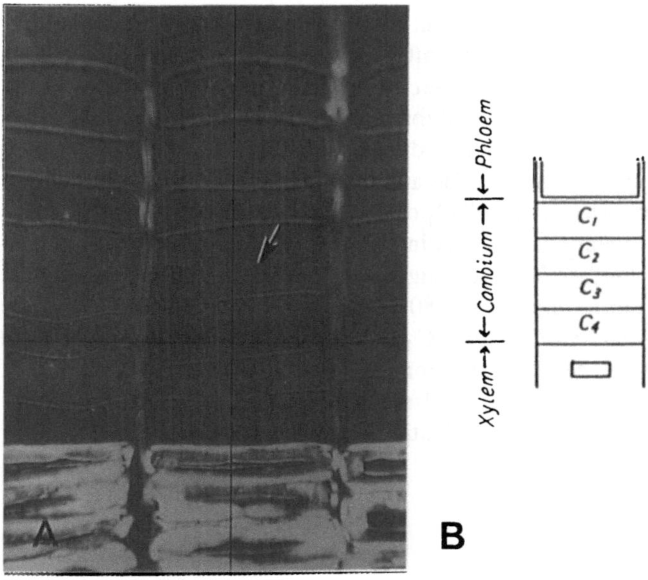

Fig. 11.17A,B. First cell divisions in the cambial zone of *Cryptomeria japonica*. **A** First periclinal division occurred in the cambial initial. *Arrow* indicates new cell wall. 384x. **B** Model of four cambial zone cells as in **A**. C_1 was judged to be the initial cell. 306x. (Itoh et al. 1968)

phragmoplasts in cambial zone cells. The first fusiform cambial cells to divide periclinally were on the phloem side. New phloem differentiation preceded that of xylem by approximately 6 weeks, although cell division on the xylem side occurred much earlier.

Recent work by Barnett (1992) has shown that an almost identical pattern of development occurs in *Aesculus hippocastanum*. Activity in the form of vesicle production of dictyosomes and endoplasmic reticulum, and coated vesicle formation at the plasmalemma was observed in cambial samples collected in mid-February. The first cell divisions, involved exclusively in phloem production, began between March 9 and 16, 5 weeks before xylem development began. The fusiform initial was identified as the cell adjacent to the xylem-boundary-layer of parenchyma cells (Fig. 11.15A). It was the last cell in the cambial zone to divide, and it did so 2 weeks after the first divisions in the phloem precursor cells.

A somewhat similar reactivation pattern in *Cryptomeria japonica* was not only followed by Itoh et al. (1968), but it was also clearly illustrated by a series of photomicrographs. In a typical tree, the cambial zone consisted of four cells (Fig. 11.17A). The first periclinal division occurred in cell C_1 in about 80% of the trees examined. Cell C_1 lay adjacent to the overwintering phloem (Fig. 11.17B). As cell division proceeded, derivatives could be distinguished from each other by the birefringence of their parent walls when viewed with polarization optics (Fig. 4.20A). Cell C_4 generally matured without dividing, occasionally after one division producing two derivatives, or rarely after two divisions producing three derivatives (Fig. 4.20B). Cell C_3 usually differentiated after only one division, sometimes two, rarely three. Cell C_2 produced three to seven cells after two to six divisions, most commonly six cells after five divisions, and cell C_1 produced four to ten cells after three to nine divisions. The authors rationalized that the initial cell in the dormant cambium was either cell C_1 or C_2 (Fig. 11.17B). However, because they could detect no phloem mother cells in dormant cambia, cell C_1 was judged to be the initial cell. Further analyses suggested that of the approximately 60 xylem cells produced during the season, 80% were derived from cell C_1 and the remaining 20% from cells C_2, C_3, and C_4. All of the new phloem cells were derivatives of the mother cell derived from the first division(s) of cell C_1. A somewhat similar scheme of cambial cell reactivation was presented by Höster (1974) for *Picea abies*. A cambial reactivation pattern is shown in Fig. 11.18 for *Pinus strobus* by Wilson (1966).

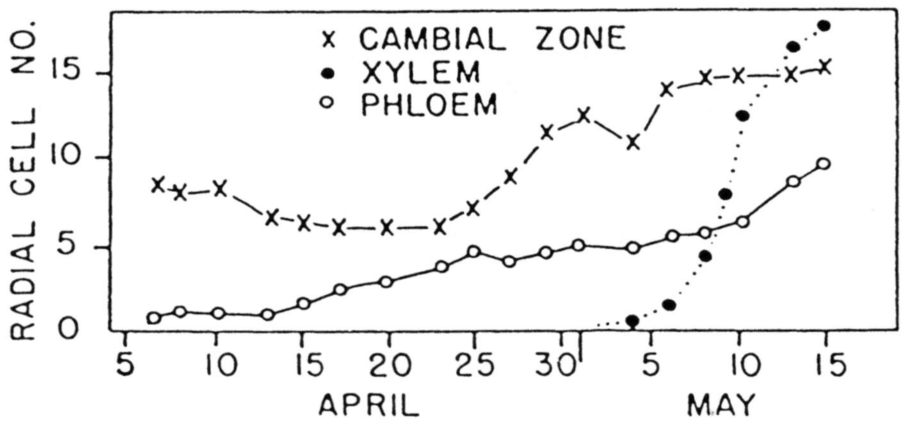

Fig. 11.18. Changes in radial cell numbers associated with the initiation of cambial activity in *Pinus strobus*. Each *point* is the average count for nine radial files. (Wilson 1966)

Cambial reactivation in tropical species often exhibits patterns quite different than those found in temperate zone trees. In *Polyalthia longifolia*, for example, over-wintering phloem mother cells differentiated as a narrow strip of "precursor phloem"[3] (Ghouse and Hashmi 1978, 1979). As spring temperatures rose, more phloem differentiated from derivatives of cambial cell division. With a further rise in air temperature and commencement of the monsoon rains, phloem production stopped and xylem production began. In autumn, when air temperatures began to decline, xylem production stopped and phloem production resumed. Thus, the cambium alternated from phloem to xylem and back to phloem production each season. A somewhat similar pattern was found in *Delonix regia* (Ghouse and Hashmi 1982) and *Mimusops elengi* (Ghouse and Hashmi 1983). However, in these species, differentiation of "precursor phloem" was not followed by development of new phloem from the cambium and the first cambial derivatives were on the xylem side. Rao and Dave (1981) were unable to observe the second phloem phase in *Tectona grandis* in India.

In several tropical species growing in Nigeria, Amobi (1973) observed that the resumption of cambial activity was positively correlated with bud break and the unfolding of new leaves which usually occurred before the end of the dry season. Typically, the dormant cambial zone consisted of one row each of xylem and phloem mother cells in addition to several cambial cells. Upon the resumption of cambial activity, these mother cells expanded without further divisions. Frequently, the xylem mother cells abutted on incompletely differentiated xylem cells, suggesting an abrupt cessation of cambial activity the previous season.

Acacia caven, an evergreen species from Chile, possesses what appears to be a dormant cambium. Yet, lying next to the two layers of flattened cambial cells, Aljaro et al. (1972) found turgid, immature xylem cells that occurred in patches and suggested a state of incomplete dormancy.

With regard to the initiation of cambial cell division, fusiform initials usually divide before ray cell initials (Rao 1988a).

11.5 Alternation of Cambial Divisions

Sanio (1873) assumed that divisions of the cambial initial cell regularly alternated between xylem and phloem. He referred to this process as the "principal law of cell formation by the cambium" and stated it as follows: "Of the two daughter cells that arise by tangential division of a cambial mother cell, the upper (outer) either remains as the cambial mother cell while the lower (inner) divides once tangentially and the twin pair passes over to wood, or of the two daughter cells that arise by division of the cambial mother cell the lower (inner) remains as the cambial mother cell while the upper (outer) divides once and the twin pair passes over to bast. In both cases they alternate with one another, resulting in a twin pair

---

[3] According to Esau (1969), the term "precursory phloem" should be restricted to the seedling stage, where it has a specific meaning.

to the outside for the bast and a twin pair to the inside for the wood." One daughter cell always remained as a cambial mother cell. However, each daughter cell divided to form a twin pair, the cells of which often divided a second time to form the so-called group of four (Chap. 4.5.1).

Sanio's assumption of a regularly alternating pattern of cambial mother cell division was soon proven false. Although the initial cell did alternate in producing cells toward the xylem and phloem, it did not do so with regularity. As pointed out by Newman (1956) many years later, "... the destination of the products of the cambial initial's daughter cell destined to tissue formation is not xylem and phloem alternately for successive 'initial' divisions, but is xylem and phloem alternately for successive *sequences* of 'initial' divisions."

Once the cambium is reactivated, cell divisions usually predominate on the xylem side, although precocious phloem production can be appreciable in some species (Catesson 1962a). For example, in *Juniperus californica* about one-half the phloem increment is produced before xylem differentiation has begun (Alfieri and Kemp 1983). Space does not permit a discussion of the numerous and variable patterns and the relative rates of cell production in the xylem and phloem. Several typical cell production curves are shown in Figs. 11.19, 11.20, and 11.21. As indicated in these figures, xylem production far exceeds that of phloem and in some species it appears to be proportional (Fig. 11.22).

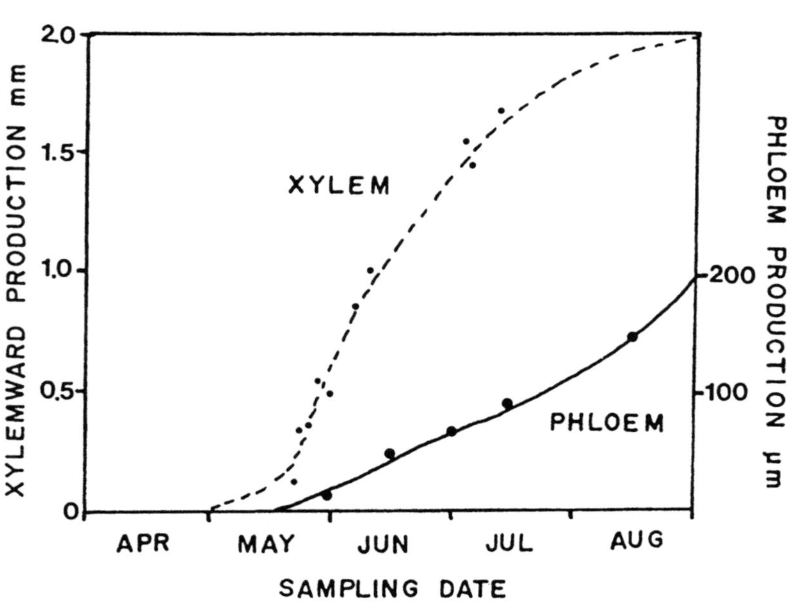

Fig. 11.19. Seasonal production of xylem and phloem cells in a hypothetical *Thuja occidentalis* tree having an annual xylem increment of 2 mm. (After Bannan 1955)

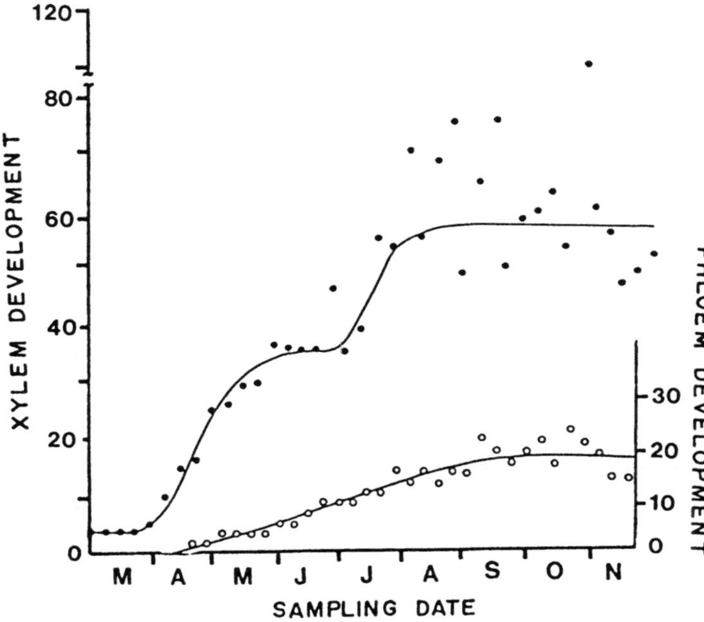

Fig. 11.20. Seasonal production of xylem and phloem based on number of cells in *Cryptomeria japonica.* Although the first periclinal division occurred on the phloem side, the number of xylem divisions was overwhelming large. The temporary midseason growth pause in the xylem did not occur in the phloem. (After Itoh et al. 1968)

A few representative ratios of seasonal xylem to phloem cell production are presented in Table 11.6. It must be recognized, however, that a host of circumstances might influence these ratios. For example, in *Abies concolor*, the ratio might vary from 14:1 to 1:1 in fast-growing versus slow-growing trees, respectively (Wilson 1964). Ratios of about 1:1 were also found in slow-growing trees of *Pseudotsuga menziesii* and *Thuja occidentalis* by Grillos and Smith (1959) and Bannan (1955), respectively. These low ratios are presumably brought about because phloem production does not increase appreciably with increasing cambial activity; that is, within certain limits, it remains about the same irrespective of the rate of growth. In *Thuja occidentalis*, a quadrupling in width of xylem increment from 0.5 to 2.0 mm was accompanied by only a doubling in expansion of the phloem (Bannan 1962b). A similar relationship prevailed in compression wood of *Abies balsamea*, where xylem production exceeded that of phloem by a ratio of 21:1 (Kutscha et al. 1975). In *Eucalyptus camaldulensis*, application of the growth hormone gibberellic acid increased the xylem/phloem ratio from a normal 4:1 to about 10:1 (Waisel et al. 1966). The treated wood resembled reaction wood.

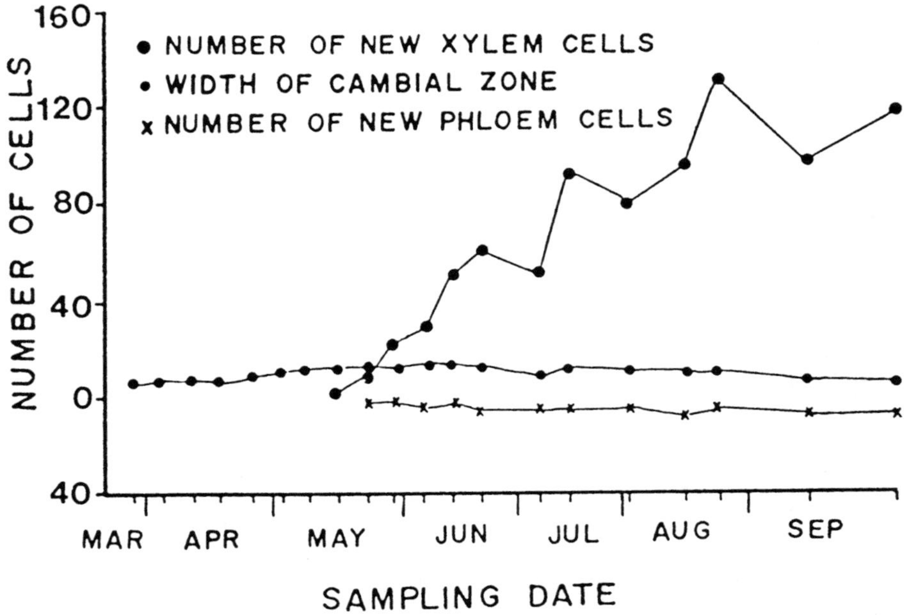

Fig. 11.21. Seasonal production of xylem and phloem cells and width of the cambial zone in *Abies balsamea*. (After Kutscha et al. 1975)

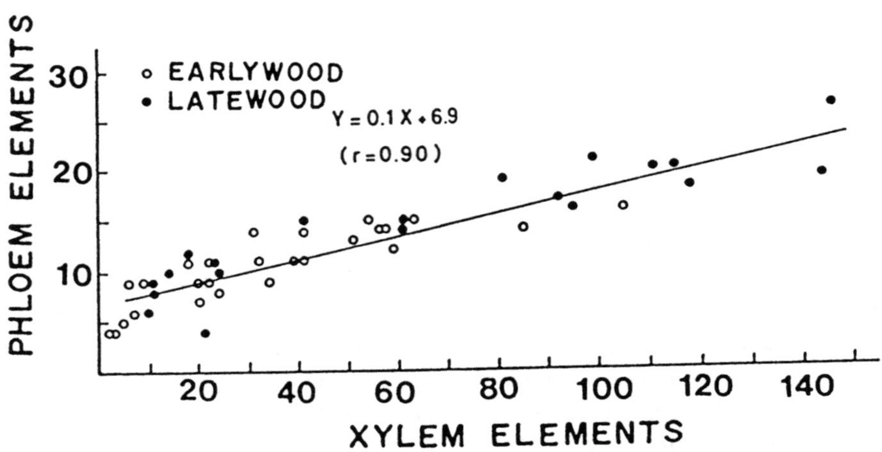

Fig. 11.22. Relation between the number of phloem and xylem elements produced throughout three growth periods in *Larix leptolepis*. (Imagawa 1981)

Table 11.6. Ratio of xylem to phloem cell production in a few representative species

Species	Xylem:phloem ratio	Reference
Abies balsamea	14:1	Kutscha et al. (1975)
Abies concolor	14:1	Wilson (1963)
Eucalyptus camaldulensis	4:1	Waisel et al. (1966)
Juniperus californica	1:1	Alfieri and Kemp (1983)
Larix leptolepis	10:1	Imagawa (1981)
Picea abies	10:1	Raatz (1892)
Picea glauca	7:1	Gregory (1971)
Pinus halepensis	6:1	Lipschitz et al. (1984)
Pinus sylvestris	10:1	Raatz (1892)
Populus euramericana	10:1	Stahel (1968)
Pseudotsuga menzesii	10:1	Grillos and Smith (1959)
Quercus robur	7:1	Kubella (1985)
Salix nigra	3:1	Nordhausen (1898)
Thuja occidentalis	15:1	Bannan (1955)

Calculations of the relative proportions of xylem and phloem in a number of species by Huber (1939) showed that phloem comprised about 10-20% of the annual cell production in dicotyledons and 20-30% in conifers.

Some species produce a second burst of phloem production late in the season as reported in *Acer pseudoplatanus* (Cockerham 1930, Elliott 1935, Catesson 1962a). Such a condition was reported by Lawton (1972) in several tropical species, *Albizia adianthifolia*, *Ricinodendron heudelotii*, and *Tectona grandis*, growing in Nigeria. The largest amount of phloem production coincided with the rainy season (Lawton and Lawton 1971). Rao and Dave (1981), however, could find no evidence of the second period in *T. grandis* growing in India.

11.6 Mid-Season Growth Pause

Studies of the seasonal progression of cambial activity by early workers very often revealed a mid-season growth pause in temperate-zone species (Schröder 1879, Mischke 1890, Reuss 1893), which Christison (1893) called the "law of retardation." This growth pause was believed due to some unknown endogenous factor that influenced the growth rhythm of the tree. For example, Korstian (1921) concluded that intermittent rest periods of short duration were essential for

"...proper health and optimum efficiency of vital activities of the tree." Wieler (1898) re-examined Mischke's (1890) results and conducted extensive investigations of his own on these "growth periodicities." Although he observed occasional mid-season diminutions of growth, they were of short duration and variable in occurrence in both conifers and dicotyledons. Wieler concluded that these growth pauses were not endogenous but were due to some unfavorable growth condition.

Other workers suggested that the mid-season growth pause might signify the transition from earlywood to latewood (Lodewick 1925, 1928, Chalk 1927, Kienholz 1934, Schober 1951, Wodzicki 1962, Wodzicki and Peda 1963). In *Citrus sinensis*, mid-season growth stoppages coincided with formation of bands of metatracheal xylem parenchyma and phloem fibers, suggesting to Schneider (1952) the possibility of some endogenous factor. Most workers, however, related the growth pause, when present, to external growth conditions (Friedrich 1897, Jost 1892), the most common being a mid-season drought of varying duration (Friesner and Walden 1946, Daubenmier 1955, Itoh et al. 1968) (Figs. 11.20, 11.23). A search of many papers on tree growth by this reviewer has provided no evidence for a mid-season growth pause due to endogenous factors. Almost all such growth pauses can be accounted for by external growth conditions which result in growth ring aberrations such as false rings, frost rings, discontinuous rings, and the like. False rings are commonly associated with the periodic growth flushes that occur seasonally in many north-temperate zone trees (Grossenbacher 1915b, Studhalter 1955), but these growth flushes are seldom preceded by significant reductions in cambial cell division.

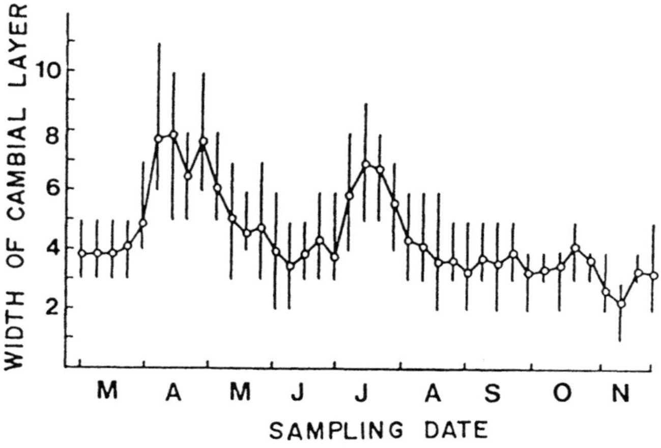

Fig. 11.23. Variation in width of the cambial zone based on number of cells in *Cryptomeria japonica*. The midseason growth pause in late May was attributed to lack of rainfall. *Vertical lines* indicate the minimum and maximum values and the *open circles* the mean values for 40 observations at each sampling date (Itoh et al. 1968)

Mid-season growth pauses are common in tropical species exhibiting long periods of cambial activity (Fahn 1982, Rao and Dave 1985b). For example, the cambium of *Holoptelea integrifolia* exhibited three growth flushes during the season with short periods of inactivity in between (Fig. 11.24). Reactivation of phloem preceded that of xylem in the first two flushes, but reactivation of the two tissues occurred simultaneously in the third flush. Cessation of xylem and phloem occurred simultaneously in the first two flushes but xylem ceased first in the third flush.

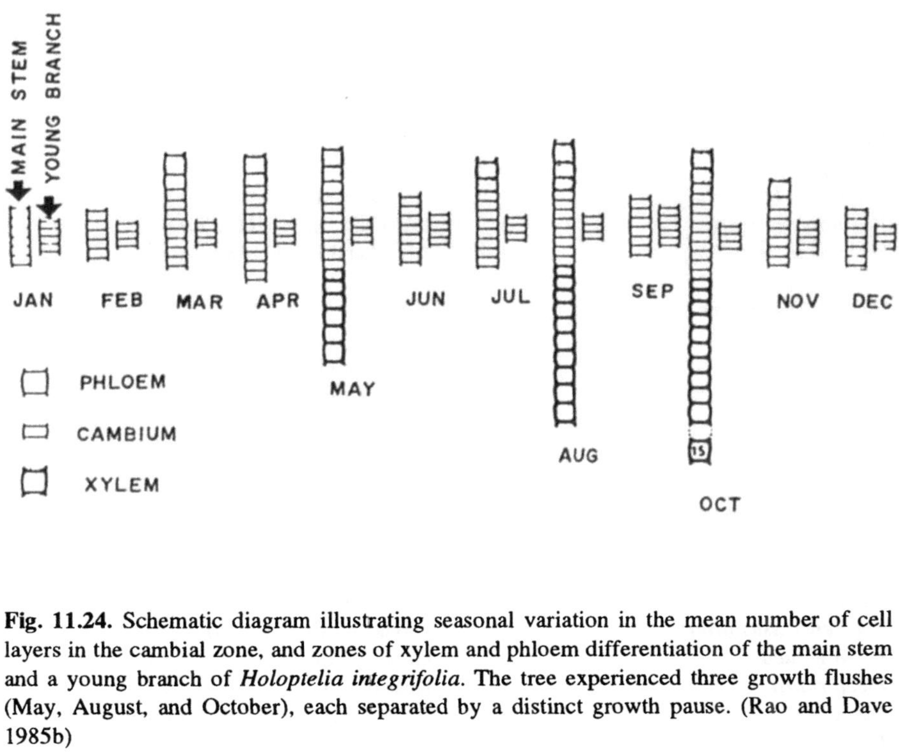

Fig. 11.24. Schematic diagram illustrating seasonal variation in the mean number of cell layers in the cambial zone, and zones of xylem and phloem differentiation of the main stem and a young branch of *Holoptelia integrifolia*. The tree experienced three growth flushes (May, August, and October), each separated by a distinct growth pause. (Rao and Dave 1985b)

11.7 Cambial Growth Cessation

Like cambial reactivation, the cessation of cambial activity lends itself to several different interpretations, depending on whether the investigator considered cessation to refer to cell division or to cell differentiation. Cessation of cambial cell

Table 11.7. Cessation of cambial activity in the autumn

Species	Reference
Xylem ceases first	
Acer negundo	Brown (1915)
Acer pseudoplatanus	Cockerham (1930), Elliott (1935)
" "	Catesson (1962a)
Aesculus hippocastanum	Rao (1985)
Betula alleghaniensis	Wilcox et al. (1956
Delonix regia	Ghouse and Hashmi (1982
Fraxinus americana	Lodewick (1925)
Mimusops elengi	Ghouse and Hashmi (1983
Parthenium argentatum	Artschwager (1945)
Picea rubens	Wilcox et al. (1956)
Pinus strobus	Brown (1915)
Polyalthia longifolia	Ghouse and Hashmi (1978)
Prosopis spicigera	Iqbal and Ghouse (1985b)
Streblus asper	Ajmal and Iqbal (1987a)
Phloem ceases first	
Abies balsamea	Kutscha et al. (1975)
Ficus rumphii	Ajmal and Iqbal (1987b)
Quercus robur	Kubella (1985)
Pyrus communis	Evert (1960)
Salix fragilis	Lawton (1976)
Xylem and phloem cease simultaneously	
Celastrus scandens	Davis and Evert (1970)
Juniperus californica	Alfieri and Kemp (1983)
Malus sp.	Knudson (1913)
Parthenocissus inserta	Davis and Evert (1970)
Picea glauca	Gregory (1971)
Picea rubra	Rees (1929)
Pinus spp.	Alfieri and Evert (1968), Lodewick (1928)
Populus tremuloides	Davis and Evert (1970)
Pyrus malus	Evert (1963b)
Robinia pseudoacacia	Derr and Evert (1967)
Salix viminalis	Sennerby-Forsee (1986)
Tectona grandis	Rao and Dave (1981)
Vitis riparia	Davis and Evert (1970)

division is extremely difficult to determine unless one collects samples frequently at season's end. Consequently, the evaluations in Table 11.7, which presumably refer to cessation of cambial division, must be accepted as best estimates by most of the cited authors. Moreover, they are generalizations in the sense that the termination of cambial activity varies widely due to prevailing growth conditions. It is perhaps for these reasons that many workers cautiously concluded that both xylem and phloem ceased activity simultaneously.

Several workers were unable to arrive at definite conclusions. Bannan (1955), for example, found no pattern in the cessation of cambial activity in *Thuja occidentalis* because "varied circumstances obtain in neighboring sectors." That is, at season's end some radial files were dividing on the phloem side and some on the xylem side while still others had apparently stopped dividing. In a more obvious situation, Barnett (1972) could detect no truly dormant period for phloem of *Pinus radiata* in New Zealand where winters are mild. Although mature latewood tracheids abutted directly on undifferentiated cambial cells, phloem cells were still differentiating. Antevs (1917) stated, as a general conclusion, that phloem cell production continued later in the fall than that of xylem.

Cambial cell derivatives continue to differentiate long after the final cell divisions. On the phloem side, this fact has been well documented by the many reports of late phloem maturation (Esau 1969) as well as the precocious reactivation of undifferentiated phloem elements in the spring. Although it is uncommon, as noted previously, for xylem derivatives to overwinter in a partially differentiated state, differentiation does continue into late fall and winter. Von Mohl (1862), for example, noted that the cambial zone appeared extremely irregular in the fall. He found it impossible to arrive at a predictable pattern for cambial growth cessation because xylem cells continued their maturation into the winter months.

Another report is that of Sigmond (1936), who examined the seasonal progression of cambial activity in *Picea abies* growing at different altitudinal sites. With regard to the cessation of cambial activity, which he interpreted to include cell differentiation, he concluded that it continued longer than previously thought. His observations showed that although cambial cell division usually ceased in late August or September, tracheid cell wall development continued until October and in some cases was resumed the following spring. Therefore, according to his report, xylem differentiation continued much later in the fall than that of phloem.

11.8 Discontinuous and Missing Growth Rings

Missing rings are those in which the annual increment is entirely absent around the stem periphery at some level within the stem or branch. In a purely technical sense, the term missing ring is a misnomer, for it results from failure of a ring to form rather than to the loss of a ring. Discontinuous rings are those in which the annual increment is absent on only part of the stem periphery at some level within the stem or branch.

Missing and discontinuous rings are not the same as reduced growth rings. Wiesner (1872), for instance, reported growth rings consisting of only three to five cells in 230- to 240-year-old *Larix russica* driftwood collected from the North Polar Sea. The narrow three-celled rings often contained a single latewood tracheid. Outer growth rings in old trees of *Sequoia sempervirens* also were found by Bailey and Faull (1934) to be impoverished of latewood. The earlywood to latewood ratio was often 20:1 and some old growth rings attained the minimum width of two tracheids, one earlywood and one latewood. In both of these cases, the growth rings were simply reduced in width but apparently present around the entire stem periphery.

11.8.1 Discontinuous Growth Rings

Reports of discontinuous growth rings appeared in the early literature. Hartig (1862) claimed to have been first to observe a partial growth layer in trees as a result of his disbudding and defoliation studies. This claim is of doubtful validity. Studhalter et al. (1963), in reviewing the early literature on cross-dating of tree stems, cited several earlier reports of partial growth rings. The text by Treviranus published in 1811, for example, included a figure showing two growth rings running together. One confirmed report is that of Wigand (1856), who illustrated discontinuous growth rings in a number of eccentric stems and roots (Fig. 8.57A).

Grossenbacher (1915b) and Nägeli (1935) mentioned three principal types of ring failure in trees: (1) earlywood fades out, leaving a broad band of fused latewood zones; (2) latewood fades out, leaving a broad band of fused earlywood zones; and (3) both latewood and earlywood fade out, leading to fusion of nonconsecutive growth rings. Nägeli (1935) never observed the third type in his investigations.

11.8.1.1 Fused Latewood Zones

Sanio (1863b) described many anomalous features of growth rings in the different species he examined. In a dwarf, 14-year-old *Fraxinus excelsior* tree growing in a swamp, he observed that vessels appeared to be missing in some growth rings. Sanio suggested that earlywood had failed to form, allowing the latewood of consecutive growth rings to merge. A similar merger of latewood bands was described by Ursprung (1904) in *Tectona grandis*. On one part of the stem two distinct rings were present, whereas 90° removed from this point they merged to form a single, dark band.

One of the most comprehensive studies of discontinuous growth rings, mostly of the fused latewood type, was conducted by Rubner (1910a). He found that *Carpinus* spp. provided exceptionally good study material because of their pronounced tendencies for asymmetric growth. Undulations in circumferential growth were associated with the presence of "Scheinstrahlen," or false rays. In places where small rays aggregated to form these false rays, wood growth was

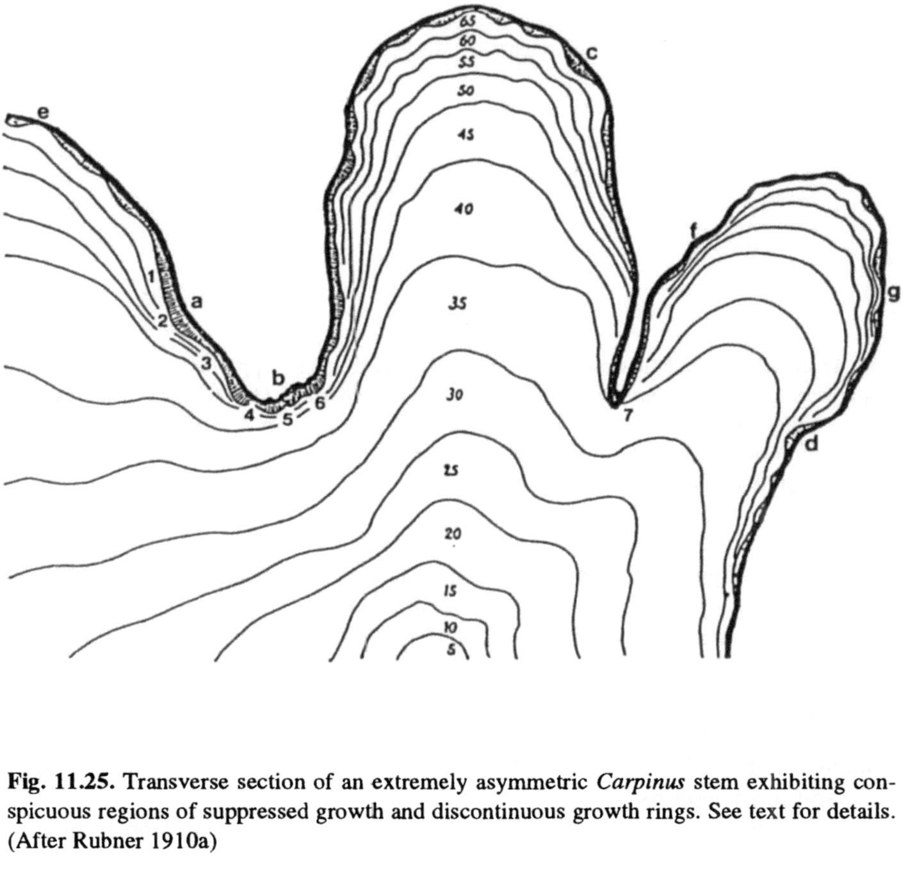

Fig. 11.25. Transverse section of an extremely asymmetric *Carpinus* stem exhibiting conspicuous regions of suppressed growth and discontinuous growth rings. See text for details. (After Rubner 1910a)

suppressed and phloem growth enhanced (Chap. 8.4.2.2). Although formation of false rays began in first-year wood, macroscopically visible undulations in stem growth usually did not appear until about the 10th to 12th year. However, once established, the growth differential often resulted in extremely asymmetric stems (Fig. 11.25). This precocious manifestation of false ray formation also was observed by Bailey (1911) who referred to them as dips in the cambium (Chap. 8.4.2.2).

Rubner (1910a) examined tissue samples microscopically at positions numbered 1 through 7 in Fig. 11.25 representing varying degrees of growth suppression. As an example, in the extremely recessed region at position number 7, approximately 30 growth rings have apparently terminated. Termination of these growth rings was brought about by the merger of several latewood bands with no

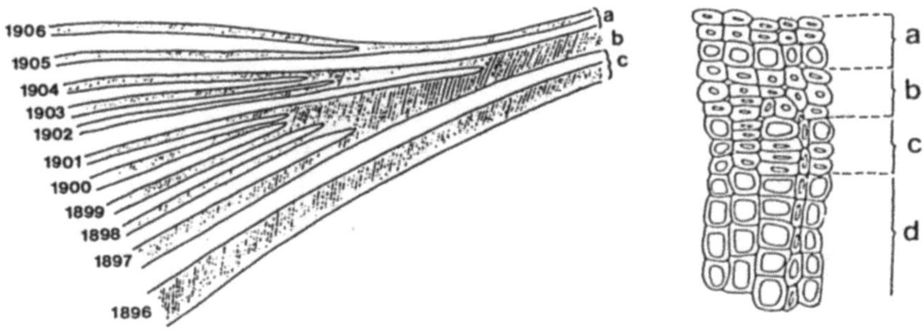

Fig. 11.26A,B. Schematic drawings of growth rings from the upper side of a 37-year-old suppressed branch of ***Juniperus nana***. **A** Eight growth rings (1897-1904) have fused to form a single latewood band (*b*). **B** Enlargement of region *a-c* in **A**. The latewood band (*b*) consisted of only three thick-walled cells. The 1896 growth ring is shown in (*c*) and part of a normal growth ring in (*d*). (Rubner 1910a)

intervening earlywood. Rubner commented on the fact that he had never observed a growth ring consisting of earlywood without accompanying latewood, i.e., earlywood always faded out first. These composite latewood bands then merged to form a single, broad latewood band in the recess at position 7. When viewed microscopically, this latewood band consisted of only eight cells, although it represented approximately the last 30 growth rings. According to Rubner, the cambium appeared to be dead because the cells were thick-walled and filled with reddish-brown deposits.

In regions where xylem growth was suppressed to varying degrees but the cambium remained active, phloem growth was promoted (Fig. 11.25, Positions a,b,c,d). Within these broad phloem bands, the numerous false rays serving them were dispersed among rings of stone cells. On the contrary, on the minor ridges where xylem growth was more strongly promoted (Positions e,f,g), phloem development was often suppressed. In some places the phloem apparently remained quiescent for many years. This tendency for xylem to be suppressed in regions where phloem is promoted and vice versa is apparently characteristic of stems with extremely undulating surfaces. Rubner cited similar cases in *Populus*, *Alnus*, *Betula*, *Fagus*, and *Aesculus*, species with pronounced epitropic branching, where deep depressions occurred immediately beneath branch insertions.

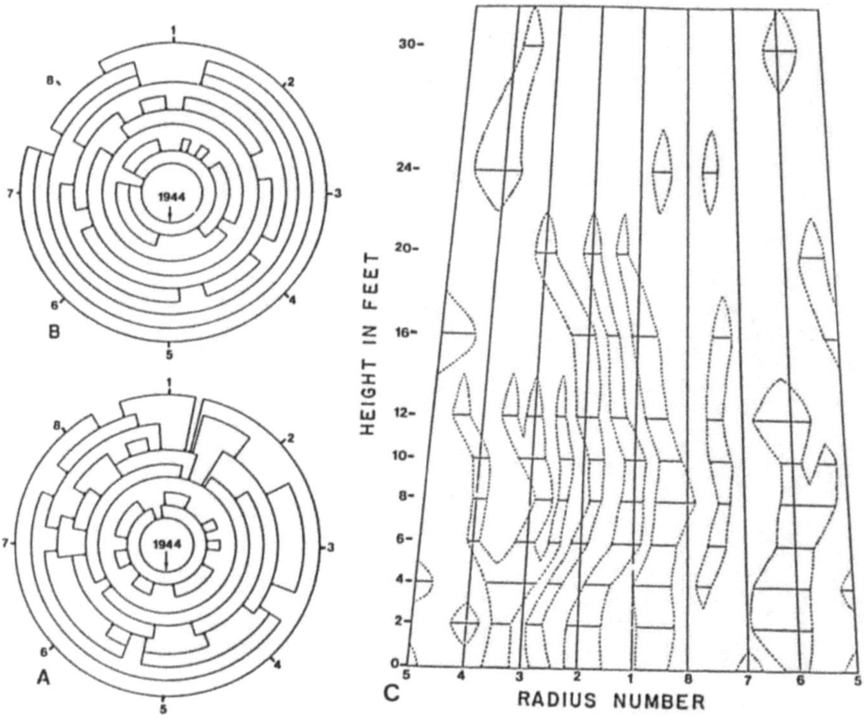

Fig. 11.27A-C. Annual ring diagrams for the years 1944-1953 inclusive, showing the positions of ring discontinuities on transverse sections of the 0-foot (**A**) and the 16-foot (**B**) levels in a suppressed *Pinus elliottii* tree. The *inward-pointing arrows* indicate that the ring from which the arrow originated merged with the ring to which the arrow points. **C** Vertical distribution of discontinuities for the 1946 growth ring shown in **A** and **B**. The *solid lines* indicate the actual positions on each transverse section, shown on the ordinate, where the 1946 earlywood failed to form resulting in fusions of the 1945-1946 latewood bands. The *dashed lines* were arbitrarily drawn to illustrate an apparent vertical pattern. All rings were complete above the 30-foot level. (Larson 1956)

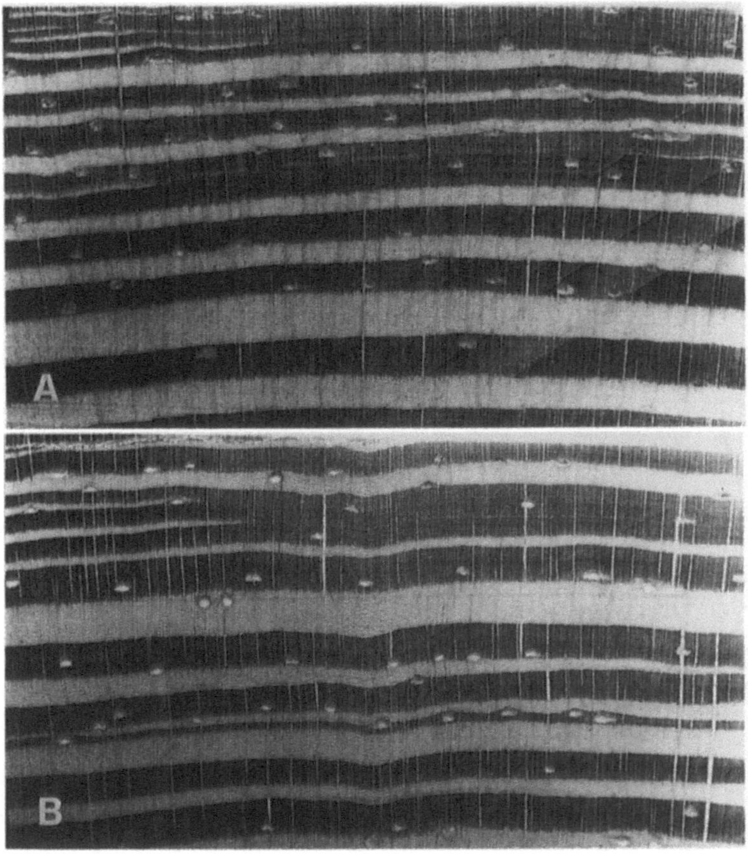

Fig. 11.28A, B. Discontinuous growth rings in *Pinus elliottii.* The 1953 latewood is visible at the top of each photomicrograph. **A** Triple fusion of the 1949-1951 latewood bands. Note the abrupt fading out of the intervening earlywood zones. **B** Five distinct growth rings, 1949-1953, have fused into a single inseparable latewood band. The four uppermost latewood bands on the *extreme right* represent the total radial growth for 10 years, 1944-1953 inclusive. 6x. (Larson 1956)

Whereas xylem development was greatly suppressed and rings were often missing in the depressions, it was strongly promoted in the "Stutzleisten," or supporting ridges on both sides.

In another example, Rubner found that the 16 last-formed growth rings were essentially absent from the upper side of a 37-year-old suppressed branch of *Juniperus nana.* As shown in Fig. 11.26A, eight growth rings (1897-1904) fused to form a single latewood band (b) consisting of only three cells (Fig. 11.26B). In this case, the cambium apparently remained alive and capable of further division. Rubner (1910b) also found missing rings to be the rule in pendant branches of *Pi-*

cea abies. In one 16-year-old branch only 12 rings were present at its point of attachment with the parent stem.

Many examples of fused latewood bands can be found in the dendrochronology literature. Glock et al. (1960), for example, described thin latewood bands they referred to as "stringers" in *Cupressus arizonica*. Some growth rings were reduced to a single earlywood and a single latewood cell on parts of their circumference. Occasionally, the earlywood cell faded out, in which case the latewood cell lay immediately on the outer margin of the previous season's growth ring. The authors described stringers of many types and configurations and with varying numbers of cells.

Fused latewood bands are commonly encountered in both conifers and dicotyledons. Stem analyses usually reveal that all growth rings are present and distinct in the upper stem and the prevalence of discontinuities increases toward the stem base. Detailed stem analyses, conducted by Larson (1956) on several suppressed *Pinus elliottii* trees, illustrate this vertical pattern. In transverse sections of one stem, shown in Fig. 11.27A,B, growth ring discontinuities were all of the fused latewood type, and they occurred in a wide variety of configurations (Fig. 11.28A,B). In one case, five distinct growth rings, 1949-1953, fused into a single, broad latewood band in a circumferential passage of just a few centimeters (Fig. 11.28A). All the fused growth zones separated into distinct, complete annual rings at progressively higher levels in the stem. The vertical distribution of these discontinuities for a single growth ring (1946) is shown in Fig. 11.27C. The seven discontinuities present at the four-foot level conform to those appearing in the 1946 growth ring of Fig. 11.27A.

11.8.1.2 Fused Earlywood Zones

Fused earlywood bands are far less common than fused latewood. Schacht (1856) claimed to have observed fused earlywood in suppressed *Pinus sylvestris* trees growing on wet, moor soils. However, rather than earlywood fusion, these broad earlywood bands were more likely due to the presence of thin-walled latewood cells that resembled earlywood. Such was the case with the fused earlywood in *Pinus palustris* seedlings reported by Pessin (1933). As long as these seedlings remained in the "grass" stage, all tracheids resembled earlywood and they failed to produce distinguishable growth rings. Failure to form typical latewood-like cells also accounts for much of the difficulty in distinguishing growth rings in many tropical species (Christison 1890, Coster 1927, 1928).

11.8.1.3 Fused Growth Rings

A fused growth ring, in which both earlywood and latewood fail to form on a part of the stem or branch circumference, is probably the most frequently encountered type of discontinuous ring. These growth-ring fusions often appear in the form of lenses (Fig. 11.29). Dendrochronologists usually examine entire stem sections in their cross-dating, a procedure that permits ready detection of both the lenses and the ring fusions. Consequently, numerous reports of these ring structures occur in

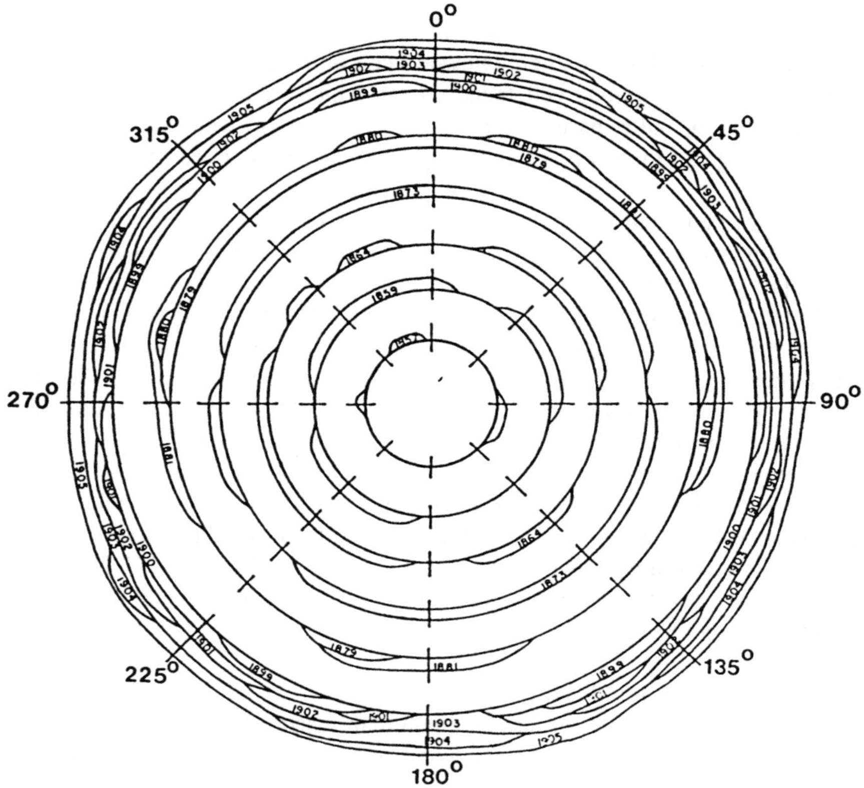

Fig. 11.29. Partial growth layers, or lenses, in a *Pinus ponderosa* tree. The diagram is a composite of transverse sections from several levels in the lower stem. (Glock et al. 1963)

the dendrochronology literature (Antevs 1917, Glock 1941, 1955, Glock et al. 1960, 1963, Studhalter et al. 1963, Fritts 1976). However, as noted later, many other investigators have misinterpreted these fused rings as missing rings because of incomplete stem analyses.

Fused growth rings are not confined to temperate zone trees. Such ring structures have been encountered with varying degrees of frequency in many tropical tree species, particularly very slow-growing specimens like those reported from the Ivory Coast and Cameroons regions of Africa (Détienne 1975a, b, 1976, Détienne and Mariaux 1976, 1977).

Tropical gymnosperms often display aberrant growth ring patterns as described in *Agathis robusta* and *A. vitiensis* by Ash (1983, 1985). Zones of cambial activity apparently migrate around the trunks, producing a distinctive wave pattern in some trees. The cambium might be virtually inactive for several years on some parts of the trunk while active on other parts. The zones of inactive cambium produce no growth and there is a gradual transition to adjacent regions where growth is active. The zones of inactive cambium shift periodically so that the trunk maintains a nearly cylindrical form. However, discontinuous growth rings occur where regions of active cambium gradually merge with inactive regions.

11.8.2 Missing Rings

Reference has been made in many publications to missing growth rings, usually with insufficient verification. Actual missing rings are difficult to verify, and the only certain methods are comprehensive stem analyses, precise cross-dating, and comparison with trees of known age. Although some reports of missing growth rings are perhaps valid, most others are more likely to represent undetected discontinuous rings of one type or another. Whether or not a growth ring is missing then becomes a matter of terminology, with some authors claiming that the ring was indeed missing at the point of sampling. In cases of extreme growth suppression where large parts of the cambial circumference have died, growth rings are obviously missing. Kelly et al. (1992), for example, described *Thuja occidentalis* trees over 1000 years old with only 2 to 4% of the potential cambial circumference still viable.

Examples of suspected growth ring discontinuities, pronounced as missing rings, abound in the literature. For instance, Flury (1907) examined a suppressed *Picea* tree in which eight rings were assumed to be completely missing at the stem base. The missing rings could not be accounted for at the stem base but all were present in the upper stem. Complete rings appeared successively at the 0.13, 0.25, 0.39, 2.08, 7.27, 8.59, 9.07, and 9.49 m levels in the stem. Similar stem analyses in another *Picea* stem revealed 15 rings missing in the outer sheath of a slightly eccentric stem (Flury 1926). In this case, despite the author's careful stem analyses, he admitted that the missing rings were probably present on the other stem side. Lakari (1915) also suggested that the missing rings he encountered might be present as incomplete rings despite his detailed microscopic analyses. Cursory examination of a 40-year-old drought-stricken *Abies alba* stem by Nägeli (1935) indicated that one ring was missing. Microscopic analysis, however, revealed a narrow ring consisting of only three to four latewood cells, the inner two of which were thin-walled. Farther down the stem it was impossible to distinguish the growth ring, although Nägeli could not conclude that the ring was missing. Many previous claims of missing growth rings might have been invalidated by similar detailed stem analyses.

Like Flury's (1907) report of eight missing rings at the base of a suppressed *Picea* stem, many reports of missing growth rings can be neither rejected outright nor validated. However, most such observations of multiple missing rings have been made on either extremely suppressed trees or those suffering retardation due to defoliation or serious injury. Like dying branches, in which growth rings disappear first at the branch base (Andrews and Gill 1939, Romell 1940, Labyak and Schumacher 1954, Reukema 1959), many cases of missing rings at the stem base undoubtedly are a prelude to the tree's impending death.

11.8.3 Causality

The proximate cause of discontinuous and missing growth rings is some form of severe growth retardation or suppression. Suppressed and stunted growth due to stand conditions is the most frequently reported factor (Hartig 1869, 1871, Flury 1907, 1926, Turberville and Hough 1939, Harris 1952, Larson 1956, Bormann 1965, Löhr 1969, Nogler 1981), although these conditions are often confounded with factors associated with site (Schacht 1856, Lakari 1915) or geographic location (Wiesner 1872, Douglass 1946, Fritts et al. 1965). Growth might also be retarded simply as a consequence of old age (Wiesner 1872, Bailey and Faull 1934, Connor and Lanner 1990) and in a great many cases by eccentric stem growth (Fritz and Averill 1924, Flury 1926, Gassner and Christiansen-Weniger 1942).

Many natural environmental factors that result in defoliation, root and shoot death, and general growth suppression can also induce discontinuous and missing rings. Contributing factors of this type are drought (Nägeli 1935, Glock et al. 1963), frost (Glock and Studhalter 1941, Glock et al. 1960, Paclt 1989), fire (Craighead 1927, Wade and Johansen 1986, Johansen and Wade 1987), insect defoliation (Evenden 1940, O'Neil 1963), and dieback (Marchand 1984).

Human intervention in the tree's growth processes can induce similar responses as, for example, artificial defoliation (Hartig 1892, Harper 1913, Coster 1927, Craighead 1940), heavy green pruning (Hartig 1869, 1871, 1872, Nördlinger 1871), stem girdling (Hartig 1889, 1890, Coster 1927), ionizing radiation (Woodwell and Miller 1963, Sparrow et al. 1965) and, of more immediate concern at the present time, atmospheric pollution (Hartig 1897a, Vins 1961, de Kort 1986, Torelli et al. 1986, Sutherland and Martin 1990).

Discontinuous and missing growth rings present a dilemma to researchers. Fused and missing growth rings imply that the cambium has remained quiescent for an extended period of time. Fabricius (1905), for example, concluded that the cambium could remain in a latent state for many years in *Picea abies* roots. In one eccentric root, he found that the broad side, measuring 12 cm, consisted of 55 growth rings whereas the narrow side, measuring 3.5 cm, consisted of only 37 rings. In the last 20 and 15 rings, respectively, the rays were starch filled and the entire cambium appeared healthy when the tree was felled on February 19, 1904. In the eccentric root of another tree, 70 growth rings were present on one side and only 20 on the other. Rays in the last 20 and 10 growth rings, respectively, were starch filled when the trees were felled on October 28, 1903. Again the cambial cells appeared viable.

Antevs (1917) questioned whether a tree could survive an entire season with no diameter growth. Glock (1955), after a lengthy discussion of the subject, also questioned previous reports of extended cambial inactivity. Rarely have previous investigators examined the status of the cambium associated with presumed missing growth rings, although several reported that the cambium "appeared healthy." Aloni (1987b) has presented evidence that xylem does not differentiate in the absence of phloem, although phloem often develops in the absence of xylem. That some trees do survive for a number of years without producing new xylem cells appears incontrovertible. However, whether this inactivity also applies to the phloem remains unanswered. Without such information, it is impossible to evaluate either the duration or the status of cambial dormancy associated with discontinuous and missing growth rings.

11.9 Complete Maturation of the Cambium

During the normal course of development, cambial initial cells periodically contribute new derivatives either to the xylem or the phloem. As noted in the previous section, this process can be interrupted under extreme circumstances in which case growth rings might be either incomplete or possibly absent. In such cases, the cambium is presumed to be temporarily quiescent and it might resume activity when and if more favorable growth conditions permit.

In some dicotyledons the cambium actually matures to the point where all cells of the cambial zone differentiate as vascular elements (Artschwager 1918). This condition is found more often in stems subtending flowering or fruiting organs. For example, Boodle (1906) described how cambium in the upper stem of *Helianthus tuberosus* differentiated in late fall after maturation of the flower head. Lignification of phloem fusiform elements proceeded centripetally toward the xylem, while that of phloem rays began adjacent to the xylem and proceeded centrifugally. Lignification generally began adjacent to previously lignified elements. Consequently, groups of sclerified cells were connected to the xylem by bridges of lignified tissue and some areas of phloem were bypassed leaving small islands of unaltered phloem. Alexandrov (1929) and Alexandrov and Alexandrova (1929) described a similar condition in *Helianthus*. In some stems, phloem elements with sieve plates lay immediately adjacent to vessels at the completion of cambial maturation.

Cessation of cambial activity in flowering stems is apparently quite common. In a wide variety of species examined by Wilton and Roberts (1936), the status of the cambium varied with the degree of flowering. The cambia of plants with most

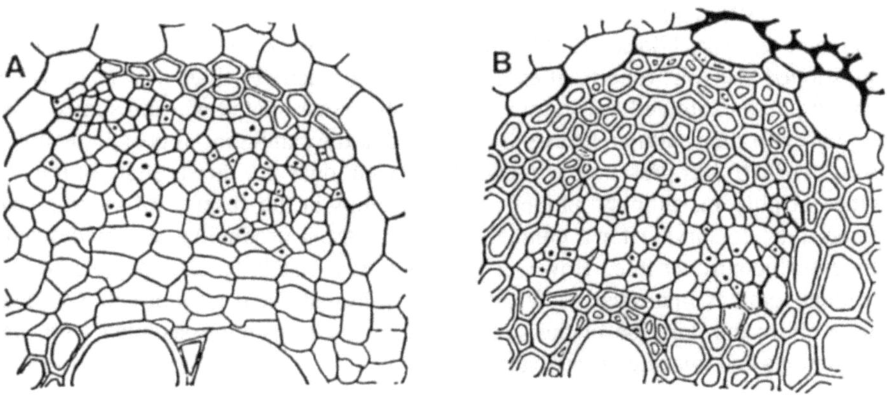

Fig. 11.30A,B. Camera lucida drawings in transverse view of vascular bundles of *Chrysanthemum morifolium*. **A** Nonflowering plant in which the cambium has remained active. **B** Flowering plant with no apparent meristematic cells in the cambial region. All phloem cells, except for a few sieve elements and companion cells, have become thick-walled. The last-formed secondary xylem consists entirely of small, thick-walled elements. (Wilton and Roberts 1936)

of their primordia in a reproductive state showed either a disorganized structure or complete absence of meristematic cells, whereas the cambia of plants with a few vegetative apices remained active (Fig. 11.30). Complete maturation of the cambium usually occurred in species that died at the completion of a single flowering cycle.

Cessation of cambial activity was found to accompany the progressive development of the flowering process. For example, in *Salvia scandens*, a reduction in cambial activity was noted after exposure to only five short-day photoperiods (Struckmeyer 1941). Additional short-day exposures deepened the flowering induction process and further reduced cambial activity. Plants that had received 39 short days showed no active cambia. On the contrary, if the inductive process was interrupted by conditions unfavorable to flowering but before flower primordia formed, then a renewal of vegetative growth would be accompanied by a renewal of cambial activity (Struckmeyer 1941, Roberts and Struckmeyer 1948). As a rule, the last xylem elements produced by the cambium in a flowering stem were thick-walled and narrow in diameter. On the phloem side, parenchyma cells were produced instead of sieve elements and companion cells (Struckmeyer and Roberts

1939). Other studies (Wilton 1938) showed that the cessation of cambial activity progressed basipetally from the region of the inflorescence toward the plant base.

Complete maturation of the cambium was described in quite different circumstances by Evert et al. (1972) in *Acer saccharum*. Circular patches of bark were isolated on stems by cutting two concentric circles 10.0 and 11.25 cm in diameter, one within the other. In the isolated regions, the fusiform initials subdivided into strands of cells which ultimately lapsed into maturity, leaving only strands of parenchymatous elements between the xylem and phloem. Subdivision of the fusiform initials is a typical wound response, but in this case further cambial cell division was inhibited by the isolation procedure.

References

Abbe LB, Crafts AS 1939 Phloem of white pine and other coniferous species. Bot Gaz 100: 695-722

Abromeit J 1884 Ueber die Anatomie des Eichenholzes. Jahrb Wiss Bot 15: 209-281

Addicott FT 1981 Abscission. University of California Press, Berkeley, 369 pp

Ajmal S, Iqbal M 1987a Annual rhythm of cambial activity in Streblus asper. IAWA Bull 8: 275-283

Ajmal S, Iqbal M 1987b Seasonal rhythms of structure and behaviour of vascular cambium in Ficus rumphii. Ann Bot 60: 649-656

Ajmal S, Khan R, Iqbal M 1985 Dimensional relationship between fusiform initials and their derivative xylem elements in Eucalyptus. J Tree Sci 4: 1-6

Ajmal S, Khan MIH, Khan R, Mahmooduzzafar, Iqbal M 1986a Structure of vascular cambium in certain exotic eucalypts. J Tree Sci 5: 80-87

Ajmal S, Khan R, Iqbal M 1986b Cambial structure of Holoptelea integrifolia Planch. in relation to age. Flora 178: 197-202

Aldaba VC 1927 The structure and development of the cell wall in plants I. Bast fibers of Boehmeria and Linum. Am J Bot 14: 16-24

Alexandrov WG 1929 Beiträge zur Kenntnis des Gefässbündels der dikotylen Krautpflanze. Ber Deutsch Bot Ges 47: 451-460

Alexandrov WG, Alexandrova OG 1929 Über die Struktur verschiedener Abschnitte ein und desselben Bündels und den Bau von Bündeln verschiedener Internodien des Sonnenblumenstengels. Planta 8: 465-486

Alfieri FJ, Evert RF 1968 Seasonal development of the secondary phloem in Pinus. Am J Bot 55: 518-528

Alfieri FJ, Kemp RI 1983 The seasonal cycle of phloem development in Juniperus californica. Am J Bot 70: 891-896

Algan G, Toker C 1981 The relationship between the cambial reactivation and bud swelling in two agricultural forms of Malus silvestris. Commun Fac Sci Univ Ankara Ser C2 Bot 24: 1-13

Aljaro ME, Avila G, Hoffmann A, Kummerow J 1972 The annual rhythm of cambial activity in two woody species of the Chilean "Matorral". Am J Bot 59: 879-885

Allen GS 1947 Embryogeny and the development of the apical meristems of Pseudotsuga. II. Late embryogeny. Am J Bot 34: 73-80

Aloni R 1987a Differentiation of vascular tissues. Ann Rev Plant Physiol 38: 179-204

Aloni R 1987b The induction of vascular tissues by auxin. In: Davis PJ (ed) Plant hormones and their role in plant growth and development. Martinus Nijhoff, Dordrecht, 363-374

Aloni R 1991 Wood formation in deciduous hardwood trees. In: Raghavendra AS (ed) Physiology of trees. John Wiley, New York, 175-197

Aloni R, Wolf A 1984 Suppressed buds embedded in the bark across the bole and the occurrence of their circular vessels in Ficus religiosa. Am J Bot 71: 1060-1066

Aloni R, Katz DA, Wool D 1989 Effect of the gall-forming aphid Slavum wertheimae on the differentiation of xylem in branches of Pistacia atlantica. Ann Bot 63: 373-375

Alosi MC, Calvin CL 1984 The morphology of the endophytic system of Arceuthobium spp. In: Biology of the dwarf mistletoes. U S For Serv Gen Tech Rep RM-111, 40-52

Amer M 1977 Interactions contrôlant la formation et le fonctionnement du cambium caulinaire chez Gleditsia triacanthos L. Thèse Univ d'Aix-Marseille III France, 147 pp

Amobi CC 1973 Periodicity of wood formation in some trees of lowland rainforest in Nigeria. Ann Bot 37: 211-218

Amos GL 1954 Radial fissures in the earlywood of conifers. Aust J Bot 2: 22-34

Amos GL, Bisset IJW, Dadswell HE 1950 Wood structure in relation to growth in Eucalyptus gigantea Hook. F. Aust J Sci Res Ser B 3:393-413

Anand SK, Sajwan VS, Paliwal GS 1978 Size correlations among the cambium and its derivatives in Dalbergia sissoo. J Indian Bot Soc 57: 16-24

Anand SK, Singh DP, Paliwal SP, Paliwal GS 1981 Size correlations among cambial initials and their derivatives in Jacaranda minosifolia D. Don and Putranjiva roxburghii Wall. Indian J For 4: 179-185

Anderson AP 1896 Über abnorme Bildung von Harzbehälter und andere zugleich auftretende anatomische Veränderungen im Holz erkrankter Coniferen. Forst-Naturwiss Z 5: 439-454, 461-482

Andrews SR, Gill LS 1939 Determining the time branches on living trees have been dead. J For 37: 930-935

Angeles G 1992 Cambial activity and development of xylem and phloem in flooded trees. IAWA Bull 13: 251-252

Antevs E 1917 Die Jahresringe der Holzgewächse und die Bedeutung derselben als klimatischer Indikator. Prog Rei Bot 5: 285-386

Anucin NP 1960 Opredelenie tekuscego prirosta nasazdenij po bokovoj poverhnosti derev'ev. (Determining the current increment of stands by the lateral {cambial} surface area of the trees). Voprosy Lesovedeni Lesovodstva, Moscow, pp 346-361 (Cited from For Abst 22: 2258, 1961)

Arber A 1946 Gothe's botany. Chron Bot 10: 63-124

Archer CJ, Cameron SH 1939 Cambial activity in the Hachiya persimmon. Proc Am Soc Hortic Sci 37: 127-129

Arianoutsou-Faraggitiki M, Psaras G, Christodoulakis N 1984 The annual rhythm of cambial activity in two woody species of the Greek "Maquis". Flora 175: 221-229

Armstrong JE 1982 Polyploidy and wood anatomy of mature white ash, Fraxinus americana. Wood Fiber 14: 331-339

Armstrong JE, Shigo AL, Funk DT, McGinnes EA, Smith DE 1981 A macroscopic and microscopic study of compartmentalization and wound closure after mechanical wounding of black walnut trees. Wood Fiber 13: 275-291

Artschwager E 1918 Anatomy of the potato plant, with special reference to the ontogeny of the vascular system. J Agr Res 14: 221-252

Artschwager E 1945 Growth studies on guayule (Parthenium argentatum). U S Dep Agr Tech Bull 885, 19 pp

Artschwager E 1950 The time factor in the differentiation of secondary xylem and phloem in pecan. Am J Bot 37: 15-24

Arzee T 1953 Morphology and ontogeny of foliar sclereids in Olea europaea II. Ontogeny. Am J Bot 40: 745-752

Ash J 1983 Growth rings in Agathis robusta and Araucaria cunninghamii from tropical Australia. Aust J Bot 31: 269-275

Ash J 1985 Growth rings and longevity of Agathis vitiensis (Seemann) Benth. & Hook. f. ex Drake in Fiji. Aust J Bot 33: 81-88

Aymard M 1968 Le xylème secondaire chez Lysimachia punctata L. Bull Soc Bot Fr 115: 187-196

Baas P 1986 Ecological patterns in xylem anatomy. In: Givnish TJ (ed) On the economy of plant form and function. Cambridge University Press, 327-352

Baas P, Chenglee L, Xinying Z, Keming C, Yuefen D 1984 Some effects of dwarf growth on wood structure. IAWA Bull 5: 45-63

Baas P, Werker E, Fahn A 1983 Some ecological trends in vessel characters. IAWA Bull 4: 141-159

Baas P, Schmid R, van Heuven BJ 1986 Wood anatomy of Pinus longaeva (Bristlecone pine) and the sustained length-on-age increase of its tracheids. IAWA Bull 7: 221-228

Babu AM, Menon ARS 1990 Distribution of gum and gum-resin ducts in plant body: certain familiar features and their significance. Flora 184: 257-261

Babu AM, Shah JJ 1987 Unusual tissue complexes formed in association with traumatic gum cavities in the stem of Bombax ceiba L. Ann Bot 59: 293-299

Babu AM, Nair GM, Shah JJ 1987 Traumatic gum-resin cavities in the stem of Ailanthus excelsa Roxb. IAWA Bull 8: 167-174

Back EL 1969 Intercellular spaces along the ray parenchyma - the gas canal system of living wood? Wood Sci 2: 31-34

Bailey IW 1910a Notes on the wood structure of the Betulaceae and Fagaceae. For Q 8: 178-185

Bailey IW 1910b Reversionary characters of traumatic oak woods. Bot Gaz 50: 374-380

Bailey IW 1911 The relation of the leaf-trace to the formation of compound rays in the lower dicotyledons. Ann Bot 25: 225-241

Bailey IW 1912 The evolutionary history of the foliar ray in the wood of the dicotyledons: and its phylogenetic significance. Ann Bot 26: 647-661

Bailey IW 1919a Structure, development, and distribution of so-called Rims or Bars of Sanio. Bot Gaz 67: 449-468

Bailey IW 1919b Phenomena of cell division in the cambium of arborescent gymnosperms and their cytological significance. Proc Nat Acad Sci USA 5: 283-285

Bailey IW 1920a The cambium and its derivative tissues. II. Size variations of cambial initials in gymnosperms and angiosperms. Am J Bot 7: 355-367

Bailey IW 1920b The cambium and its derivative tissues. III. A reconnaissance of cytological phenomena in the cambium. Am J Bot 7: 417-434

Bailey IW 1920c The formation of the cell plate in the cambium of the higher plants. Proc Nat Acad Sci USA 6: 197-200

Bailey IW 1920d The significance of the cambium in the study of certain physiological problems. J Gen Physiol 2: 519-534

Bailey IW 1923 The cambium and its derivative tissues. IV. The increase in girth of the cambium. Am J Bot 10: 499-509

Bailey IW 1925a The "Spruce Budworm" biocoenose. I. Frost rings as indicators of the chronology of specific biological events. Bot Gaz 80: 93-101

Bailey IW 1925b The "Spruce Budworm" biocoenose. II. Structural abnormalities in Abies balsamea. Bot Gaz 80: 300-310

Bailey IW 1930 The cambium and its derivative tissues. V. A reconnaissance of the vacuome of living cells. Z Zellforsch Mikros Anat 10: 651-682

Bailey IW 1943 Some misleading terminologies in the literature of "plant tissue culture". Science 98: 539

Bailey IW 1944a The development of vessels in angiosperms and its significance in morphological research. Am J Bot 31: 421-428

Bailey IW 1944b The comparative morphology of the Winteraceae. III. Wood. J Arnold Arbor 25: 97-103

Bailey IW 1952 Biological processes in the formation of wood. Science 115: 255-259

Bailey IW 1953 Evolution of the tracheary tissue of land plants. Am J Bot 40: 4-8

Bailey IW 1957 Additional notes on the vesselless dicotyledon, Amborella trichopoda Baill. J Arnold Arbor 38: 374-380

Bailey IW, Faull AF 1934 The cambium and its derivative tissues. IX. Structural variability in the redwood, Sequoia sempervirens, and its significance in the identification of fossil woods. J Arnold Arbor 15: 233-254

Bailey IW, Shepard HB 1915 Sanio's laws for the variation in size of coniferous tracheids. Bot Gaz 60: 66-71

Bailey IW, Srivastava LM 1962 Comparative anatomy of the leaf-bearing Cactaceae. IV. The fusiform initials of the cambium and the form and structure of their derivatives. J Arnold Arbor 43: 187-202

Bailey IW, Tupper WW 1918 Size variation in tracheary cells: I. A comparison between the secondary xylems of vascular cryptogams, gymnosperms and angiosperms. Proc Am Acad Arts Sci 54: 149-204

Bailey IW, Zirkle C 1931 The cambium and its derivative tissues. VI. The effects of hydrogen ion concentration in vital staining. J Gen Physiol 14: 363-383

Baker JR 1948-55 The cell theory: a restatement, history and critique. Q J Micros Sci 89: 103-125; 90: 87-108; 93: 157-190; 94: 407-440; 96: 449-481

Bakhuizen R, Goosen-de Roo L, van Spronsen PC, Libbenga KR 1985 Polarly organized membrane maturation of Golgi vesicles "en route" to the plane of division in phragmoplasts of periclinally dividing fusiform cambial cells of Fraxinus excelsior L. Acta Bot Neerl 34: 134

Balch RE, Clark J, Bonga JM 1964 Hormonal action in production of tumours and compression wood by an aphid. Nature 202(4903): 721-722

Bamber RK 1973 The formation and permeability of interstitial spaces in the sapwood of some Pinus species. J Inst Wood Sci 6: 36-38

Bamber RK, Lanyon JW 1966 Callus-like growth in radial splits in the wood of Anthocephalus cadamba (Roxb.) Miq. J Inst Wood Sci 16: 43-48

Bangerter UM 1984 Der Verschlussmechanismus von Längswunden am Stamm von Larix decidua Mill. und Picea abies (L.) Karst.). Vierteljahrsch Naturforsch Ges Zürich 129: 339-398

Bannan MW 1933a Vigour of growth and traumatic resin cyst development in the hemlock Tsuga canadensis (L.) Carr. Trans R Soc Can 27 (Sect V): 197-202

Bannan MW 1933b Factors influencing the distribution of vertical resin canals in the wood of the larch-Larix laricina (Du Roi) Koch. Trans R Soc Can 27 (Sect V): 203-217

Bannan MW 1934a Origin and cellular character of xylem rays in gymnosperms. Bot Gaz 96: 260-281

Bannan MW 1934b Seasonal wounding and resin-cyst production in the hemlock, Tsuga canadensis (L.) Carr. Ann Bot 48: 857-868
Bannan MW 1936a Comparison and the distribution of albuminous and tracheary ray cells in the gymnosperms. Am J Bot 23: 36-40
Bannan MW 1936b Vertical resin ducts in the secondary wood of the Abietineae. New Phytol 35: 11-46
Bannan MW 1937 Observations on the distribution of xylem-ray tissue in conifers. Ann Bot 1: 717-726
Bannan MW 1941 Vascular rays and adventitious root formation in Thuja occidentalis L. Am J Bot 28: 457-463
Bannan MW 1942 Wood structure of the native Ontario species of Juniperus. Am J Bot 29: 245-252
Bannan MW 1944 Wood structure of Libocedrus decurrens. Am J Bot 31: 346-351
Bannan MW 1950a Abnormal xylem rays in Chamaecyparis. Am J Bot 37: 232-237
Bannan MW 1950b The frequency of anticlinal divisions in fusiform cambial cells of Chamaecyparis. Am J Bot 37: 511-519
Bannan MW 1951a The reduction of fusiform cambial cells in Chamaecyparis and Thuja. Can J Bot 29: 57-67
Bannan MW 1951b The annual cycle of size changes in fusiform cambial cells of Chamaecyparis and Thuja. Can J Bot 29: 421-437
Bannan MW 1953 Further observations on the reduction of fusiform cambial cells in Thuja occidentalis L. Can J Bot 31: 63-74
Bannan MW 1954 Ring width, tracheid size, and ray volume in stem and wood of Thuja occidentalis. Can J Bot 32: 466-479
Bannan MW 1955 The vascular cambium and radial growth in Thuja occidentalis L. Can J Bot 33: 113-138
Bannan MW 1956 Some aspects of the elongation of fusiform cambial cells in Thuja occidentalis. Can J Bot 34: 175-196
Bannan MW 1957a Girth increase in white cedar stems of irregular form. Can J Bot 35: 425-434
Bannan MW 1957b The relative frequency of the different types of anticlinal divisions in conifer cambium. Can J Bot 35: 875-884
Bannan MW 1957c The structure and growth of the cambium. Tappi 40: 220-225
Bannan MW 1960a Cambial behavior with reference to cell length and ring width in Thuja occidentalis L. Can J Bot 38: 177-183
Bannan MW 1960b Ontogenetic trends in conifer cambium with respect to frequency of anticlinal division and cell length. Can J Bot 38: 795-802
Bannan MW 1961 Anticlinal division and cell length in the cambium of conifers. IAWA Bull 1961/1: 2-6
Bannan MW 1962a Cambial behavior with reference to cell length and ring width in Pinus strobus L. Can J Bot 40: 1057-1062
Bannan MW 1962b The vascular cambium and tree-ring development. In: Kozlowski TT (ed) Tree Growth. Ronald Press, New York, 3-21
Bannan MW 1963a Cambial behavior with reference to cell length and ring width in Picea. Can J Bot 41: 811-822
Bannan MW 1963b Tracheid size and rate of anticlinal division in the cambium of Cupressus. Can J Bot 41: 1187-1197

Bannan MW 1964a Tracheid size and anticlinal divisions in the cambium of Pseudotsuga. Can J Bot 42: 603-631

Bannan MW 1964b Tracheid size and anticlinal divisions in the cambium of lodgepole pine. Can J Bot 42: 1105-1118

Bannan MW 1965a The rate of elongation of fusiform initials in the cambium of Pinaceae. Can J Bot 43: 429-435

Bannan MW 1965b Ray contacts and rate of anticlinal division in fusiform cambial cells of some Pinaceae. Can J Bot 43: 487-507

Bannan MW 1965c The length, tangential diameter, and length/width ratio of conifer tracheids. Can J Bot 43: 967-984

Bannan MW 1966a Cell length and rate of anticlinal division in the cambium of the Sequoias. Can J Bot 44: 209-218

Bannan MW 1966b Spiral grain and anticlinal divisions in the cambium of conifers. Can J Bot 44: 1515-1538

Bannan MW 1967a Anticlinal divisions and cell length in conifer cambium. For Prod J 17: 63-69

Bannan MW 1967b Sequential changes in rate of anticlinal division, cambial cell length, and ring width in the growth of coniferous trees. Can J Bot 45: 1359-1369

Bannan MW 1968a Polarity in the survival and elongation of fusiform initials in conifer cambium. Can J Bot 46: 1005-1008

Bannan MW 1968b Anticlinal divisions and the organization of conifer cambium. Bot Gaz 129: 107-113

Bannan MW 1968c The problem of sampling in studies of tracheid length in conifers. For Sci 14: 140-147

Bannan MW 1970 A survey of cell length and frequency of multiplicative divisions in the cambium of conifers. Can J Bot 48: 1585-1589

Bannan MW, Bayly IL 1956 Cell size and survival in conifer cambium. Can J Bot 34: 769-776

Bannan MW, Bindra M 1970a The influence of wind on ring width and cell length in conifer stems. Can J Bot 48: 255-259

Bannan MW, Bindra M 1970b Variations in cell length and frequency of anticlinal division in the vascular cambium throughout a white spruce tree. Can J Bot 48: 1363-1371

Bannan MW, Whalley BE 1950 The elongation of fusiform cambial cells in Chamaecyparis. Can J Res 28C: 341-355

Barghoorn ES Jr 1940a Origin and development of the uniseriate ray in the Coniferae. Bull Torrey Bot Club 67: 303-328

Barghoorn ES Jr 1940b The ontogenetic development and phylogenetic specialization of rays in the xylem of dicotyledons. I. The primitive ray structure. Am J Bot 27: 918-928

Barghoorn ES Jr 1941a II. Modification of the multiseriate and uniseriate rays. Am J Bot 28: 273-282

Barghoorn ES Jr 1941b III. The elimination of rays. Bull Torrey Bot Club 68: 317-325

Barghoorn ES Jr 1964 Evolution of cambium in geologic time. In: Zimmermann MH (ed) The formation of wood in forest trees. Academic Press, New York, 3-17

Barker WG 1954 A contribution to the concept of wound repair in woody stems. Can J Bot 32: 486-490

Barlow PW 1978 Endopolyploidy: Towards an understanding of its biological significance. Acta Theor 27: 1-18

Barlow PW 1991 From cell wall networks to algorithms. The simulation and cytology of cell division patterns in plants. Protoplasma 162: 69-85

Barnett JR 1972 Winter activity in the cambium of Pinus radiata. N Z J For Sci 1: 208-222

Barnett JR 1973 Seasonal variation in the ultrastructure of the cambium in New Zealand grown Pinus radiata D. Don. Ann Bot 37: 1005-1111

Barnett JR 1975 Seasonal variation of organelle numbers in sections of fusiform cambium cells of Pinus radiata D. Don. N Z J Bot 13: 325-332

Barnett JR 1976 Rings of collapsed cells in Pinus radiata stemwood from lysimeter-grown trees subjected to drought. N Z J For Sci 6: 461-465

Barnett JR 1979 Current research into tracheary element formation. Curr Adv Plant Sci 11: 33.1-33.13

Barnett JR 1981 Secondary xylem cell development. In: Barnett JR (ed) Xylem cell development. Castle House, Turnbridge Wells, 47-95

Barnett JR 1987a The development of fibre-tracheid pit membranes in Pyrus communis L. IAWA Bull 8: 134-142

Barnett JR 1987b Changes in the distribution of plasmodesmata in developing fibre-tracheid pit membranes of Sorbus aucuparia L. Ann Bot 59: 269-279

Barnett JR 1991 Microtubules in interphase nuclei of Aesculus hippocastanum L. Ann Bot 68: 159-165

Barnett JR 1992 Reactivation of the cambium in Aesculus hippocastanum L.: A transmission electron microscope study. Ann Bot 70: 169-177

Barnett JR, Harris JM 1975 Early stages of bordered pit formation in radiata pine. Wood Sci Tech 9: 233-241

Barnett JR, Weatherhead I, Miller H 1990 Anatomical studies of the developing graft union in Picea sitchensis (Bong) Carr. IAWA Bull 11: 116-117

Bartlett FA 1935 Historical background of cavity work. Proc Natl Shade Tree Conf 11: 39-51

Bartwal BS, Siddiqui F, Iqbal M 1983 Cambium structure in some Indian fruit trees. Kalikasan 12: 61-69

Baskin TI, Cande WZ 1990 The structure and function of the mitotic spindle in flowering plants. Ann Rev Plant Physiol Mol Biol 41: 277-315

Basson PW, Bierhorst DW 1967 An analysis of differential lateral growth in the stem of Bauhinia surinamensis. Bull Torrey Bot Club 94: 404-411

Bauch J, Shigo AL, Starck M 1980 Wound effects in the xylem of Acer and Betula species. Holzforschung 34: 153-160

Beebe DU, Turgeon R 1991 Current perspectives on plasmodesmata: Structure and function. Physiol Plant 83: 194-199

Beijer JJ 1927 Die Vermehrung der Radialen Reihen im Cambium. Rec Trav Bot Neerl 24: 631-786

Belluschi PG, Johnson NE, Heikkenen HJ 1965 Douglas-fir defects caused by the Douglas-fir beetle. J For 63: 252-256

Benayoun J, Aloni R, Sachs T 1975 Regeneration around wounds and the control of vascular differentiation. Ann Bot 39: 447-454

Bergman SI 1949 Lengths of hardwood fibers and vessel segments. A statistical analysis of forty-nine hardwoods indigenous to the United States. Tappi 32: 494-498

Berlyn GP 1979 Physiological control of differentiation of xylem elements. Wood Fiber 11: 109-126

Berlyn GP 1982 Morphogenetic factors in wood formation and differentiation. In: Bass P (ed) New perspectives in wood anatomy. Nijhoff/Junk, The Hague, 123-150
Bertog H 1895 Untersuchungen über den Wuchs und das Holz der Weisstanne und Fichte. Forst-Naturwiss Z 4: 97-112, 177-216
Bertrand CE 1884 Loi des surfaces libres. C R Acad Sci 98: 48-51
Bethel JS 1941 The effect of position within the bole upon fiber length of loblolly pine (Pinus taeda L.). J For 39: 30-33
Bhat KM 1983 A note on aggregate rays in Betula species. IAWA Bull 4: 183-185
Bhat KV, Bhat KM 1983 Anatomical changes associated with interlocked grain in Anacardium occidentale L. IAWA Bull 4: 179-182
Bhat KV, Bhat KM 1984 Wavy grain in Grewia tilifolia Vahl. IAWA Bull 5: 249-252
Bhat JR, Mohan Ram HY 1992 Development and ultrastructure of primary secretory ducts in the stem of Semecarpus anacardium (Anacardiaceae). IAWA Bull 13: 173-185
Bisset IJW, Dadswell HE 1949 The variation of fibre length within one tree of Eucalyptus regnans, F.v.M. Aust For 13: 86-96
Bisset IJW, Dadswell HE 1950 The variation in cell length within one growth ring of certain angiosperms and gymnosperms. Aust For 14: 17-29
Bisset IJW, Dadswell HE, Wardrop AB 1951 Factors influencing tracheid length in conifer stems. Aust For 15: 17-30
Blanchard RO, Shortle WC, Davis W 1983 Mechanism relating cambial electrical resistance to periodic growth rate of balsam fir. Can J For Res 13: 472-480
Bloch R 1952 Wound healing in higher plants. II. Bot Rev 18: 655-679
Boddicker ML 1983 Black bears. In: Timm RM (ed) Prevention and control of wildlife damage, Great Plains Agr Council, University of Nebraska, Lincoln, C5-C23
Böhlmann D 1970a Anatomisch-histologische Untersuchungen im Bereich der Astabzweigung bei Nadel- und Laubbäumen. I. Die Verhältnisse im Abzweigungsbereich der Langtriebe von Nadelbäumen. Allg Forst-Jagdztg 141: 134-140
Böhlmann D 1970b II. Die Verhältnisse im Abzweigungsbereich der Kurztriebe von Larix und Pinus. Allg Forst-Jagdztg 141: 189-193
Böhlmann D 1970c III. Die Abzweigungsverhältnisse bei Quercus robur L. und Populus Sektion Aigeiros. Allg Forst-Jagdztg 141: 224-230
Böhlmann D 1970d IV. Die Abzweigungsverhältnisse bei Juglans, Fraxinus, Betula und Fagus und ihre Zuordnung zu Abzweigungstypen. Allg Forst-Jagdztg 141: 245-250
Bolton AJ, Jardine P, Jones GL 1975 Interstitial spaces. A review and observations on some Araucariaceae. IAWA Bull 1975/1: 3-12
Boodle LA 1906 Lignification of phloem in Helianthus. Ann Bot 20: 319-321
Bormann FH 1965 Changes in the growth pattern of white pine trees undergoing suppression. Ecology 46: 269-277
Bormann FH 1966 The structure, function, and ecological significance of root grafts in Pinus strobus L. Ecol Monogr 36: 1-26
Bosshard HH 1951 Variabilität der Elemente des Eschenholzes in Funktion von der Kambiumtätigkeit. Schweiz Z Forstwes 102: 648-665
Bosshard HH 1952 Elektronenmikroskopische Untersuchungen im Holz von Fraxinus excelsior L. Ber Schweiz Bot Ges 62: 482-508
Bosshard HH 1965 Aspects of the aging process in cambium and xylem. Holzforschung 19: 65-69
Bosshard HH 1976a Funktioneller Tropismus. Holztechnologie 17: 107-112

Bosshard HH 1976b Differenzierungsmechanismus von Wundkambium in Koniferen. Schweiz Z Forstwes Beih 57: 54-65

Bosshard HH, Bangerter UM 1984 Functional tropism in traumatic excretory tissue of Larix decidua Mill. and Picea abies (L.) Karst. In: Sudo S (ed) Proc Pac Region Wood Anat Conf, Tsukuba, 46-48

Bosshard HH, Hug UE 1980 The anastomoses of the resin canal system in Picea abies (L.) Karst., Larix decidua Mill. and Pinus sylvestris L. Holz Roh-Werkst 38: 325-328

Bosshard HH, Meier B 1969 Über den Einfluss von Blitzeinwirkungen auf Fichten (Picea abies L.). Schweiz Z Forstwes 120: 476-485

Bosshard HH, Stahel J 1969 Modifikationen in der sekundären Rinde von Populus robusta. Holzforsch Holzverwert 21: 109-113

Bostock RM, Stermer BA 1989 Perspectives on wound healing in resistance to pathogens. Ann Rev Phytopathol 27: 343-371

Botosso PC, Gomes AV 1982 Radial vessels and series of perforated ray cells in Annonaceae. IAWA Bull 3: 39-44

Boutin B 1985 Étude de la réactivation cambiale chez un arbre ayant un bois à zones poreuses (Castanea sativa) et deux autres au bois à pores diffus (Betula verrucosa, Acer campestre). Can J Bot 63: 1335-1343

Bower FO 1930 Size and form in plants with special reference to the primary conducting tracts. MacMillan, London, 232 pp

Brandenburg MK, Mills HL, Rickard WH, Shields LM 1962 Effects of acute gamma radiation on growth and morphology in Pinus monophylla Torr. and Frem. (Pinyon pine). Radiat Bot 2: 251-263

Brander R, Stearns FW 1963 Porcupine winter feeding activity in merchantable stands of northern hardwood-hemlock. U S For Serv Res Note LS-16, 2 pp

Branski S 1970 (Relations between anticlinal divisions, intrusive growth and loss of fusiform initials in the cambium of Pinus silvestris.) Acta Soc Bot Pol 39: 593-615

Braun A 1854 Über den schiefen Verlauf der Holzfaser und die dadurch bedingte Drehung der Bäume. Monatsber Berl Akad 1854: 432-484

Braun HJ 1955 Beiträge zur Entwicklungsgeschichte der Markstrahlen. Bot Stud Jena 4: 74-131

Braun HJ 1958 Die normalen Verwachsungsvorgänge nach Pfropfung von Laubbäumen I. Das Verfahren des seitlichen Einspitzens. Z Bot 46: 309-338

Braun HJ 1959 II. Die Verfahren des seitlichen Anplattens und der Kopulation. Z Bot 47: 145-166

Braun HJ 1960 III. Das seitliche T-Schnitt-Verfahren, die Geissfuss- und Keilpfropfung, sowie das Ablaktieren. Z Bot 48: 58-65

Braun HJ 1963 Die Organisation des Stammes von Bäumen und Sträuchern. Wissenschaftliche Verlagsgesellschaft MBH, Stuttgart, 162 pp

Braun HJ 1970 Funktionelle Histologie der sekundären Sprossachse I. Des Holzes. Handb Pflanzenanat 9(1): 1-190

Braun HJ, den Outer RW 1964 Die unterschiedlichen Beziehungen der Holzstrahlen zum Hydrosystem als wesentliches Differenzierungsprinzip I. Die einschichtigen Strahlen und die durch Verschmelzung aus ihnen entstehenden mehrschichtigen Strahlen. Z Bot 52: 539-571

Brister GH 1973 A biometrical study of the sequential development of secondary xylem in Douglas fir. Diss Abst Intl B 33(12): 5596

Brongniart A 1852 Note sur la formation de nouvelles couches ligneuses dans les tiges des arbres dicotylédonés. Ann Sci Nat Bot Ser 3, 18: 56-64

Brown AB 1936 Cambial activity in poplar with particular reference to polarity phenomena. Can J Res 14C: 74-83

Brown AB 1937 Activity of the vascular cambium in relation to wounding in the balsam poplar, Populus balsamifera L. Can J Res 15C: 7-31

Brown CL 1964 The influence of external pressure on the differentiation of cells and tissues cultured in vitro. In: Zimmermann MH (ed) The formation of wood in forest trees. Academic Press, New York, 389-404

Brown CL 1970 Physiology of wood formation in conifers. Wood Sci 3: 8-22

Brown CL 1971 Secondary growth. In: Zimmermann MH, Brown CL (eds) Trees structure and function. Springer Berlin, Heidelberg, New York, 67-123

Brown CL, Sax K 1962 The influence of pressure on the differentiation of secondary tissues. Am J Bot 49: 683-691

Brown HP 1912 Growth studies in forest trees 1. Pinus rigida Mill. Bot Gaz 54: 386-403

Brown HP 1915 2. Pinus strobus L. Bot Gaz 59: 197-241

Brown HP, Panshin AJ 1940 Commercial timbers of the United States. Their structure, identification, properties and uses. 1st edn McGraw-Hill, New York, 554 pp

Buczacki ST 1973 Xylem flecks: a note on some unexplained abnormalities in the wood of European larch and other trees. New Phytol 72: 825-831

Bünning E 1952 Morphogenesis in plants. Surv Biol Prog 2: 105-140

Bünning E 1956 General processes of differentiation. In: Milthorpe FL (ed) The growth of leaves. Butterworth, London, 18-30

Bünning E 1965 Die Entstehung von Mustern in der Entwicklung von Pflanzen. Handb Pflanzenphysiol 15: 383-408

Burggraaf PD 1974 Cell divisions in the cambial zone of Fraxinus excelsior L. and a model for cambial activity. Acta Bot Neerl 23: 355-356

Burström H, Krogh A 1947 Bleeding and bud development in Carpinus. Svensk Bot Tidskr 41: 17-44

Büsgen M, Münch E 1929 The structure and life of forest trees. 3rd edn. Thomson T (transl) John Wiley, New York, 436 pp

Butterfield BG 1972 Developmental changes in the vascular cambium of Aeschynomene hispida Willd. N Z J Bot 10: 373-386

Butterfield BG 1973 Variation in the size of fusiform cambial initials and vessel members in Hoheria angustifolia Raoul. N Z J Bot 11: 391-410

Butterfield BG 1975 Terminology used for describing the cambium. IAWA Bull 1975/1: 13-14

Butterfield BG 1976 The ontogeny of the vascular cambium in Hoheria angustifolia Raoul. New Phytol 77: 409-420

Butterfield BG, Meylan BA 1972 Trabeculae in a hardwood. IAWA Bull 1972/1: 3-9

Buvat R 1956a Variations saisonnières du chondriome dans le cambium de Robinia pseudoacacia. CR Acad Sci 243: 1908-1911

Buvat R 1956b Variations saisonnières du chrondiome dans les cellules parenchymateuses du phloéme de Robinia pseudoacacia. CR Acad Sci 243: 2127-2130

Calvin CL 1967 Anatomy of the endophytic system of the mistletoe, Phoradendron flavescens. Bot Gaz 128: 117-137

Calvin CL, Wilson CA, Varughese G 1991 Growth of longitudinal strands of Phoradendron juniperinum (Viscaceae) in shoots of Juniperus occidentalis. Ann Bot 67: 153-161

Carlquist S 1962 A theory of paedomorphosis in dicotyledonous woods. Phytomorphology 12: 30-45

Carlquist S 1970 Wood anatomy of insular species of Plantago and the problem of raylessness. Bull Torrey Bot Club 97: 353-361

Carlquist S 1975 Ecological strategies of xylem evolution. University of California Press, Berkeley, 259 pp

Carlquist S 1978 Wood anatomy of Bruniaceae: correlations with ecology, phylogeny, and organography. Aliso 9: 323-364

Carlquist S 1981 Types of cambial activity and wood anatomy of Stylidium (Stylidiaceae). Am J Bot 68: 778-785

Carlquist S 1985a Wood and stem anatomy of Misodendraceae: Systematic and ecological conclusions. Brittonia 37: 58-75

Carlquist S 1985b Wood anatomy of Coriariaceae: Phylogenetic and ecological implications. Syst Bot 10: 174-183

Carlquist S 1988a Comparative wood anatomy. Systematic, ecological, and evolutionary aspects of dicotyledon wood. Springer, Berlin, Heidelberg, New York, 460 pp

Carlquist S 1988b Near-vesselessness in Ephedra and its significance. Am J Bot 75: 598-601

Carlquist S 1989 Wood and bark anatomy of Empetraceae: comments on paedomorphosis in woods of certain small shrubs. Aliso 12: 497-515

Carlquist S, Hoekman DA 1985 Wood anatomy of Staphyleaceae: Ecology, statistical correlations, and systematics. Flora 177: 195-216

Carmi A, Sachs T, Fahn A 1972 The relation of ray spacing to cambial growth. New Phytol 71: 349-353

Carr DJ 1976 Plasmodesmata in growth and development. In: Gunning BES, Robards AW (eds) Intercellular communications in plants: studies in plasmodesmata. Springer Berlin, Heidelberg, New York, 243-289

Case RM 1983 Pocket gophers. In: Timm RM (ed) Prevention and control of wildlife damage, Great Plains Agr Council, University of Nebraska, Lincoln B13-B26

Caspary R 1857 Neue Untersuchungen über Frostspalten. Bot Ztg 15: 329-335, 345-350, 361-371

Caspary R 1862 Untersuchungen über die Gefässbündel der Pflanzen. Monatsber Akad Wiss Berl 1862: 448-483

Catesson AM 1961 Variations saisonnières du chondriome dans le cambium d`Acer pseudoplatanus (Acéracées). C R Acad Sci 252: 2588-2590

Catesson AM 1962a Nodalités de l'activité proliferatrice du cambium d'Acer pseudoplatanus au cours du cycle annuel. C R Acad Sci 255: 3462-3464

Catesson AM 1962b Modifications saisonnières des vacuoles et variations de la pression osmotique dans le cambium d'Acer pseudoplatanus. C R Acad Sci 254: 3887-3889

Catesson AM 1964 Origine, fonctionnement et variations cytologiques saisonnières du cambium de l'Acer pseudoplatanus L. (Acéracées). Ann Sci Nat Bot Ser 12, 5: 229-456

Catesson AM 1966 Présence de phytoferritine dans le cambium et les tissues conducteurs de la tige de sycomore, Acer pseudoplatanus. C R Acad Sci 262: 1070-1073

Catesson AM 1974 Cambial cells. In: Robards AW (ed) Dynamic aspects of plant ultrastructure. McGraw Hill, New York, 358-390

Catesson AM 1980 The vascular cambium. In: Little CA (ed) Control of shoot growth in trees. Proc IUFRO Workshop July 20-24, 1980 Fredericton NB, 12-40

Catesson AM 1981 Le cycle saisonnièr des cellules cambiales chez quelques feuillus. Bull Soc Bot Fr Act Bot 128: 43-51

Catesson AM 1984 La dynamique cambiale. Ann Sci Nat Bot Ser 13, 6: 23-43

Catesson Am 1987 Characteristics of radial cell walls in the cambial zone. A means to locate the so-called initials? IAWA Bull 8: 309

Catesson AM 1990 Cambial cytology and biochemistry. In: Iqbal M (ed) The vascular cambium. Research Studies Press, Taunton, England, 63-112

Catesson AM 1992 Ultrastructural and biochemical changes in the cambial zone in relationship with cell differentiation and seasonal activity. IAWA Bull 13: 253

Catesson AM, Roland JC 1981 Sequential changes associated with cell wall formation. IAWA Bull 2: 151-162

Cerezke HF 1972 Effects of weevil feeding on resin duct density and radial increment in lodgepole pine. Can J For Res 2: 11-15

Chalk L 1927 The growth of the wood of ash (Fraxinus excelsior L. and F. oxycarpa Willd.) and Douglas fir (Pseudotsuga Douglasii Carr.). Q J For 21: 102-122

Chalk L 1930 Tracheid length, with special reference to Sitka spruce. Forestry 4: 7-14

Chalk L 1955 Ray volumes in hardwoods. Trop Woods 101: 1-10

Chalk L 1970 Short fibres with clearly defined intrusive growth, with special reference to Fraxinus. In: Robson NKB, Cutler DF, Gregory M (eds) New research in plant anatomy. Academic Press, New York, 163-168

Chalk L, Chattaway MM 1934 Measuring the length of vessel members. Trop Woods 40: 19-26

Chalk L, Chattaway MM 1935 Factors affecting dimensional variations of vessel members. Trop Woods 41: 17-37

Chalk L, Murthy LSV 1963 Radial strands of phloem in the wood of Jongkong, Dactylocladus stenostachys Oliv. in Sarawak. Commonw For Rev 42: 290-292

Chalk L, Ortiz CM 1961 Variation in tracheid length within the ring of Pinus radiata D. Don. Forestry 34: 119-124

Chalk L, Marstrand EB, Walsh JP de C 1955 Fibre length in storeyed hardwoods. Acta Bot Neerl 4: 339-347

Chandorkar, KR, Dengler NG 1987 Effect of low level continuous gamma irradiation on vascular cambium activity in Scotch pine (Pinus sylvestris L.). Environ Exp Bot 27: 165-175

Charon J, Launay J, Vindt-Balquerie E 1986 Ontogenèse des canaux sécréteurs d'origine primaire dans le bourgeon de Pin maritime. Can J Bot 64: 2955-2964

Charon J, Vindt-Balquerie E, Launay J 1990 Ontogenèse des canaux sécréteurs d'origine primaire dans la jeune plante de Pinus pinaster. Can J Bot 68: 2119-2126

Chattaway MM 1933a Ray development in the Sterculiaceae. Forestry 7: 93-108

Chattaway MM 1933b Tile-cells in the rays of the Malvales. New Phytol 32: 261-273

Chattaway MM 1936 Relation between fibre and cambial initial length in dicotyledonous woods. Trop Woods 46: 16-20

Chattaway MM 1948 Note on the vascular tissue in the rays of Banksia. J Council Sci Ind Res Aust 21: 275-278

Chattaway MM 1951a The development of horizontal canals in rays. Aust J Sci Res 4B: 1-11

Chattaway MM 1951b Morphological and functional variations in the rays of pored timbers. Aust J Sci Res 4B: 12-27

Cheadle VI, Esau K 1964 Secondary phloem of Liriodendron tulipifera. Univ Calif Publ Bot 36: 143-252

Cheniclet C 1987 Effects of wounding and fungus inoculation on terpene producing systems of maritime pine. J Exp Bot 38: 1557-1572

Chou T, Chiang SHT 1973 Seasonal changes of cambial activity in the young branch of Psidium guajava Linn. Taiwania 18: 35-41

Chowdhury KA 1936 Terminal and initial parenchyma cells in the wood of Terminalia tomentosa W. & A. New Phytol 35: 351-358

Chowdhury KA 1953 The role of initial parenchyma in the transformation of the structure diffuse-porous to ring-porous in the secondary xylem of the genus Gmelina Linn. Proc Nat Inst Sci India 19: 361-369

Chowdhury KA, Purkayastha SK, Ghosh SS, Sond JS 1956 Septate fibres in angiosperms and their length. Nature 177: 1243-1244

Christison D 1890 On the difficulty of ascertaining the age of certain species of trees in Uruguay, from the number of rings. Trans Proc Bot Soc Edinburgh 18: 447-455

Christison D 1893 The weekly rate of girth-increase in certain trees, and its relation to the growth of leaves and twigs. Trans Proc Bot Soc Edinburgh 19: 101-120

Chrysler MA 1913 The origin of the erect cells in the phloem of the Abietineae. Bot Gaz 56: 36-50

Chrysler MA 1915 The medullary rays of Cedrus. Bot Gaz 59: 387-396

Chung YJ, Lee Pw 1989 (Anatomical studies on the features of rays in compression wood of Korean red pine (Pinus densiflora S. et Z.). J Korean For Soc 78: 119-131

Cichan MA 1985a Vascular cambium and wood development in Carboniferous plants. I. Lepidodendrales. Am J Bot 72: 1163-1176

Cichan MA 1985b II. Sphenophyllum plurifoliatum Williamson and Scott (Sphenophyllales. Bot Gaz 146: 395-403

Cichan MA 1986a III. Arthropitys (Equisetales; Calamitaceae). Can J Bot 64: 688-695

Cichan MA 1986b IV. Seed plants. Bot Gaz 147: 227-235

Cichan MA, Taylor TN 1990 Evolution of cambium in geologic time - a reappraisal. In: Iqbal M (ed) The vascular cambium. Research Studies Press, Taunton, England, 213-228

Cinotti B 1991 Recherche de propriétés intrinsèques du bois pouvant expliquer la sensibilité à la gélivure de Quercus petraea (Liebl.) et Q. robur (L.). Ann Sci For 48: 453-468

Clark A III, Hamilton JR 1968 Effects of ionizing irradiation on the xylem derivatives of Pinus echinata Mill. W Va Univ Agr Exp Stn Bull 567T, 16 pp

Clifton NC 1969 Resin pockets in Canterbury radiata pine. N Z J For 14: 38-49

Clowes FAL, Juniper BE 1968 Plant cells. Blackwell, Oxford, 546 pp

Cockerham G 1930 Some observations on cambial activity and seasonal starch content in sycamore (Acer pseudoplatanus). Proc Leeds Philos Lit Soc Sect 2: 64-80

Cohen LI 1963 Studies on the ontogeny of the dwarf mistletoes, Arceuthobium. I. Embryogeny and histogenesis. Am J Bot 50: 400-407

Cohn F 1857 Bericht über die Verhandlungen der botanischen Sektion im Jahre 1857. Schlesische Ges Vaterlandische Cultur Breslau Bot Sect 1857: 43-44

Collins JF 1930 On changing the direction of sap conducting tissues. Rhodora 32: 145-146

Connor KF, Lanner RM 1990 Effects of tree age on secondary xylem and phloem anatomy in stems of Great Basin bristlecone pine (Pinus longaeva). Am J Bot 77: 1070-1077

Cooper WC, Tayloe S, Maxwell N 1955 Preliminary studies on cold hardiness in Citrus as related to cambial activity and bud growth. Proc Rio Grande Valley Hortic Soc 9: 1-14

Copes D 1969 Graft union formation in Douglas-fir. Am J Bot 56: 285-289
Cordero RE 1982 The effects of acute and chronic gamma irradiation on Lupinus albus L. - III. Chronic effects. Environ Exp Bot 22: 359-372
Cordero RE, Gunckel JE 1982a I. Effects of acute irradiation on the vegetative shoot apex and general morphology. Environ Exp Bot 22: 105-126
Cordero RE, Gunckel JE 1982b II. Effects of acute irradiation on floral development. Environ Exp Bot 22: 127-137
Core HA, Côté WA, Day AC 1976 Wood structure and identification. Syracuse University Press, 168 pp
Coster C 1927 Zur Anatomie und Physiologie der Zuwachszonen- und Jahresringbildung in den Tropen. Pt I. Ann Jard Bot Buitenzorg 37: 49-160
Coster C 1928 Pt II. Ann Jard Bot Buitenzorg 38: 1-114
Cown DJ 1973 Resin pockets: their occurrence and formation in New Zealand forests. N Z J For 18: 233-251
Cown DJ 1975 Variation in tracheid dimensions in the stem of a 26-year-old Radiata pine tree. Appita 28: 237-245
Craighead FC 1927 Abnormalities in annual rings resulting from fires. J For 25: 840-842
Craighead FC 1940 Some effects of artificial defoliation on pine and larch. J For 38: 885-888
Craven SR 1983 Cottontail rabbits. In: Timm RM (ed) Prevention and control of wildlife damage, Great Plains Agr Council, University of Nebraska Lincoln D69-D74
Creber GT, Chaloner WG 1984 Climatic indications from growth rings in fossil woods. In: Brenchley P (ed) Fossils and climate. John Wiley, London, 49-74
Cronshaw J 1965 The organization of cytoplasmic components during the phase of cell wall thickening in differentiating cambial derivatives of Acer rubrum. Can J Bot 43: 1401-1407
Cui K, Lu P, Liu Q, Li Z 1989 Regeneration of vascular tissues in Broussonetia papyrifera stems after removal of the xylem. IAWA Bull 10: 193-199
Cumbie BG 1960 Anatomical studies in the Leguminosae. Trop Woods 113: 1-47
Cumbie BG 1963 The vascular cambium and xylem development in Hibiscus lasiocarpus. Am J Bot 50: 944-951
Cumbie BG 1967a Development and structure of the xylem in Canavalia (Leguminosae). Bull Torrey Bot Club 94: 162-175
Cumbie BG 1967b Developmental changes in the vascular cambium of Leitneria Floridana. Am J Bot 54: 414-424
Cumbie BG 1969a Developmental changes in the xylem and vascular cambium of Apocynum sibiricum. Bull Torrey Bot Club 96: 629-640
Cumbie BG 1969b Developmental changes in the vascular cambium of Polygonum Lapathifolium. Am J Bot 56: 139-146
Cumbie BG 1983 Developmental changes in the wood of Bocconia vulcania Donn. Smith. IAWA Bull 4: 131-140
Cumbie BG 1984 Origin and development of the vascular cambium in Aeschynomene virginica. Bull Torrey Bot Club 111: 42-50
Cumbie BG, Mertz D 1962 Xylem anatomy of Sophora (Leguminosae) in relation to habit. Am J Bot 49: 33-40
Currey DR 1965 An ancient bristlecone pine stand in eastern Nevada. Ecology 46: 564-566
Curtis JD 1936 A method of pruning dead branches. For Chron 12: 291-299
Cusset G 1982 The conceptual basis of plant morphology. Acta BioTheor 31A: 8-86

Czaninski Y 1964 Variations saisonnières du chondriome dans les cellules du parenchyme ligneux vertical du Robinia pseudo-acacia. C R Acad Sci 258: 679-682

Czaninski Y 1968 Étude du parenchyme ligneux du Robinier (parenchyme à réserves et cellules associées aux vaisseaux) au cours du cycle annuel. J Microsc 7: 145-164

Czaninski Y 1970 Étude cytologique de la différenciation du bois de Robinier II. Différenciation des cellules du parenchyme (cellules à réserves et cellules associées aux vaisseaux). J Microsc 9: 389-406

Czaninski Y 1972 Mise en évidence de cellules associées aux vaisseaux dans le xyléme du sycomore. J Microsc 13: 137-140

d'Arbaumont M 1881 La tige des Ampélidées. Ann Sci Nat Bot Ser 6, 11: 186-255

Daubenmire R 1955 Xylem layers of trees as related to weather and altitude in the northern Rocky Mountains. Ecology 36: 456-463

Dave YS, Rao KS 1981 Seasonal nuclear behavior in fusiform cambial initials of Tectona grandis L.f. Flora 171: 299-305

Dave YS, Rao KS 1982a Seasonal activity of the vascular cambium in Gmelina arborea Roxb. IAWA Bull 3: 59-65

Dave YS, Rao KS 1982b Cambial activity in Mangifera indica L. Acta Bot Sci Hung 28: 73-79

Dave YS, Rao KS 1982c Plastid ultrastructure in the cambium of teak (Tectona grandis L.f.). Ann Bot 49: 425-427

Davies E, Schuster A 1981 Intercellular communication in plants: evidence for a rapidly generated, bidirectionally transmitted wound signal. Proc Nat Acad Sci USA 78: 2422-2426

Davis JD, Evert RF 1968 Seasonal development of the secondary phloem in Populus tremuloides. Bot Gaz 129: 1-8

Davis JD and Evert RF 1970 Seasonal cycle of phloem development in woody vines. Bot Gaz 131: 128-138

Day WR 1950 Cambial injuries in a pruned stand of Norway spruce. For Commonw For Record 4 HMSO London, 11 pp

Day WR, Peace TR 1934 The experimental production and the diagnosis of frost injury on forest trees. Oxford For Mem 16, 60 pp

De Angelis JD, Nebeker TE, Hodges JD 1986 Influence of tree age and growth rate on the radial resin duct system in loblolly pine (Pinus taeda). Can J Bot 64: 1046-1049

De Bary A 1877 Vergleichende Anatomie der Vegetationsorgane der Phanerogamen und Farne. Wilhelm Engelmann, Leipzig, 663 pp

De Bary A 1884 Comparative anatomy of the vegetative organs of the phanerogams and ferns. Bower FO, Scott DH (transl), Clarendon Press Oxford, 659 pp

de Kort I 1986 Wood structure and growth ring width of vital and non-vital Douglas fir (Pseudotsuga menziesii) from a single stand in the Netherlands. IAWA Bull 7: 309-318

de Mirbel CFB 1816 Note sur le cambium et le liber. Soc Philom Bull Sci Paris 1816: 107-108

de Mirbel CFB 1827 Mémoire sur l'origine, le développement et l'organisation du liber et du bois. Mem Mus Hist Nat Paris 8: 139-164

de Mirbel CFB 1837 Observations sur le cambium et sur quelques modes de formations utriculaires ou vasculaires dans les végétaux. C R Acad Sci 5: 295-301

de Mirbel CFB 1839 Nouvelles notes sur le cambium, extraites d'un travail sur la racine du dattier. C R Acad Sci 8: 645-658. Same title, 1839 Ann Sci Nat Bot Ser 2, 11: 321-336

de Mirbel CFB, Payen A 1843 Sur la composition du cambium et sur le rôle qu'il joue dans l'organogénie végétale. C R Acad Sci 16: 98-101
Dell B, McComb AJ 1978 Plant resins - their formation, secretion and possible functions. Adv Bot Res 6: 277-316
Denne MP 1977 Some effects of wounding on tracheid differentiation in Picea sitchensis. IAWA Bull 1977/3: 49-50
Denne MP 1988 Definition of latewood according to Mork (1928). IAWA Bull 10: 59-62
Denne MP, Dodd RS 1981 The environmental control of xylem differentiation. In: Barnett JR (ed) Xylem cell development Castle House Turnbridge Wells, 236-255
den Outer RW 1967 Histological investigations of the secondary phloem of gymnosperms. Meded Landbou Wageningen 67(7): 1-119
den Outer RW 1986 Storied structure of the secondary phloem. IAWA Bull 7: 47-51
Derr WF, Evert RF 1967 The cambium and seasonal development of the phloem in Robinia pseudoacacia. Am J Bot 54: 147-153
Deshpande BP, Vishwakarma AK 1992 Calcium oxlate crystals in the fusiform cells of the cambium of Gmelina arborea. IAWA Bull 13: 297-300
De Smit WJ 1922 Studies of the distribution and volume of the wood rays in slippery elm (Ulmus fulva Michx.). J For 20: 352-362
Détienne P 1975a Nature et périodicité des cernes dans le bois de Doussié (Afzelia sp.). Rapp Rech Cent Tech For Trop Nogent-sur-Marne, 17 pp
Détienne P 1975b Nature et périodicité des cernes dans le bois de bété (Mansonia altissima A. Chev.). Rapp Rech Cent Tech For Trop Nogent-sur-Marne, 19 pp
Détienne P 1976 Nature et périodicité des cernes dans les bois d'iroko (Chlorophora excelsa Benth. & Hook. f.). Rapp Rech Cent Tech For Trop Nogent-sur-Marne, 20 pp
Détienne P 1989 Appearance and periodicity of growth rings in some tropical woods. IAWA Bull 10: 123-132
Détienne P, Mariaux A 1976 Nature et périodicité des cernes dans le bois de Samba. Bois For Trop 169: 29-35
Détienne P, Mariaux A 1977 Nature et périodicité des cernes dans les bois rouges de Méliacées Africaines. Bois For Trop 175: 52-61
Detlefsen E 1878 Ueber Dickenwachsthum cylindrischer Organe. Arbeit Bot Inst Würzburg 2: 18-45
Detlefsen E 1882 Versuch einer mechanischen Erklärung des excentrischen Dickenwachsthums verholtzter Aschen und Wurzeln. Arbeit Bot Inst Würzburg 2: 670-688
de Torok D, White PR 1960 Cytological instability in tumors on Picea glauca. Science 131: 730-732
Deutschmann F 1954 Anatomische Untersuchungen über die Harzgänge einiger Dipterocarpaceae. Ber Deut Bot Ges 67: 381-384
de Vries H 1876 Über Wundholz. Flora 59: 2-8, 17-25, 38-45, 81-88, 97-108, 113-121, 129-139
de Vries H 1885 Ueber die Bedeutung der Circulation und der Rotation des Protoplasma für den Stofftransport in der Pflanze. Bot Ztg 43: 1-6
Dietrichson J, Rognerud PA, Haveraaen O, Skrøppa T 1985 Stem cracks in Norway spruce (Picea abies (L.) Karst.). Medd Norske Skogforsøksves 38(2): 1-32
Dinwoodie JM 1961 Tracheid and fibre length in timber: a review of literature. Forestry 34: 125-144
Dinwoodie JM 1963 Variation in tracheid length in Picea sitchensis Carr. Dept Sci Indust Res, For Prod Res Spec Rep 16 London, 55 pp

Dippel L 1863 Zur Histologie der Coniferen. II. Die Harzbehälter der Weisstanne und die Entstehung des Harzes in denselben. Bot Ztg 21: 253-259
Dittmer HJ, Spensley RD 1947 The developmental anatomy of Descurainia pinnata ochroleuca (Woot). Detling. Univ New Mexico Publ Biol 3: 1-47
Dobbins DR 1990 The role of rays in anomalous cambial activity. IAWA Bull 11: 122-123
Dobbins DR, Fisher JB 1986 Wound responses in girdled stems of lianas. Bot Gaz 147: 278-289
Dodd JD 1948 On the shapes of cells in the cambial zone of Pinus silvestris L. Am J Bot 35: 666-682
Dodds JH 1981 The role of the cell cycle and cell division in xylem differentiation. In: Barnett JR (ed) Xylem cell differentiation. Castle House Turnbridge Wells, 153-167
Doerksen AH, Mitchell RG 1965 Effects of the balsam wooly aphid upon wood anatomy of some western true firs. For Sci 11: 181-188
Doley D 1970 Effects of simulated drought on shoot development in Liriodendron seedlings. New Phytol 69: 655-673
Doley D, Leyton L 1968 Effects of growth regulating substances and water potential on the development of secondary xylem in Fraxinus. New Phytol 67: 579-594
Dormer KJ 1980 Fundamental tissue geometry for biologists. Cambridge University Press, 149 pp
Dormling I 1963 Anatomical and histological examinations of the union of scion and stock in grafts of Scots pine (Pinus silvestris L.) and Norway spruce (Picea abies (L) Karst.). Studia For Suec 13: 1-136
Dostál R 1967 On integration in plants. Kiely JM (transl) Harvard University Press, 218 pp
Douglass AE 1946 Precision of ring dating in tree-ring chronologies. Univ Ariz Bull 17: 1-21
Dudzik KR 1988 Macro-microscopic anatomy: obtaining a composite view of barrier zone formation in Acer saccharum. IAWA Bull 9: 183-186
Eames AJ 1910 On the origin of the broad ray in Quercus. Bot Gaz 49: 161-167
Eames AJ 1911 On the origin of the herbaceous type in the angiosperms. Ann Bot 25: 215-224
Eames AJ, MacDaniels LH 1925 An introduction to plant anatomy. 1st edn. McGraw-Hill, New York, 364 pp
Eames AJ, MacDaniels LH 1947 An introduction to plant anatomy. 2nd edn. McGraw-Hill, New York, 427 pp
Edelstein L 1982 A molecular switching mechanism in differentiation. Induction of polarity in cambial cells. Differentiation 23: 1-9
Ehlers RC, Sandermann W 1961 Chemische Studien an Tropenhölzern - sechste Mitteilung: Untersushungen über harzhaltige Tropenhölzer. Holz Roh Werkst 19: 187-195
Eichhorn E 1895 Untersuchungen über das Holz der Roteiche. Forst-Naturwiss Z 4: 233-264, 281-296
Elliot E 1951 Formation of new cell walls in cell division. Nature 168: 1089
Elliott JH 1935 Seasonal changes in the development of the phloem of the sycamore, Acer pseudo platanus L. Proc Leeds Phil Lit Soc, Sci Sect 3: 55-67
Emmingham WH 1977 Comparison of selected Douglas-fir seed sources for cambial and leader growth patterns in four western environments. Can J For Res 7: 154-164
Engard CJ 1944 Organogenesis in Rubus. Univ Hawaii Res Publ 21: 1-234

Eom YG, Lee PW 1989 Anatomical studies on tumorous tissue formed in a stem of Ailanthus altissima Swingle by artificial banding and its subsequent removing treatment: characters of individual elements. J Korean For Soc 78: 287-301

Erdmann GC, Oberg RR 1974 Sapsucker feeding damages crown-released yellow birch trees. J For 72: 760-763

Ermich K 1963 The inception and the end of the annual tree ring formation in Fagus silvatica L., Abies alba Mill. and Picea excelsa Lk. in the Tatra mountains. Ekol Pol 11: 311-336

Errera L 1888 Ueber Zellformen und Seifenblasen. Bot Cbl 34: 395-398

Esau K 1938 Ontogeny and structure of the phloem of tobacco. Hilgardia 11: 343-424

Esau K 1942 Vascular differentiation in the vegetative shoot of Linum. I. The procambium. Am J Bot 29: 738-747

Esau K 1943 II. The first phloem and xylem. Am J Bot 30: 248-255

Esau K 1948 Phloem structure in the grapevine, and its seasonal changes. Hilgardia 18: 217-296

Esau K 1954 Primary vascular differentiation in plants. Biol Rev 29: 46-86

Esau K 1965 Plant anatomy. John Wiley, New York, 767 pp

Esau K 1969 The phloem. Encl Plant Anat Vol 5,Pt 2, Gebrüder Borntraeger, Berlin, 505 pp

Esau K 1977 Anatomy of seed plants. 2nd edn John Wiley, New York, 550 pp

Esau K, Cheadle VI 1955 Significance of cell divisions in differentiating secondary phloem. Acta Bot Neerl 4: 348-357

Esau K, Cheadle VI 1969 Secondary growth in Bougainvillea. Ann Bot 33: 807-819

Esau K, Gill RH 1965 Observations on cytokinesis. Planta 67: 168-181

Esau K, Gill RH 1991 Distribution of vacuoles and some other organelles in dividing cells. Bot Gaz 152: 397-407

Essiamah S, Eschrich W 1985 Changes in starch content in the storage tissues of deciduous trees during winter and spring. IAWA Bull 6: 97-106

Essiamah S, Eschrich W 1986 Water uptake in deciduous trees during winter and the role of conducting tissues in spring reactivation. IAWA Bull 7: 31-38

Essner B 1882 Ueber den diagnostischen Werth der Anzahl und Höhe der Markstrahlen bei den Coniferen. Abh Nat Ges Halle 16: 1-33

Evenden JC 1940 Effects of defoliation by the pine butterfly upon ponderosa pine. J For 38: 949-955

Evert RF 1960 Phloem structure in Pyrus communis L. and its seasonal changes. Univ Calif Publ Bot 32: 127-194

Evert RF 1961 Some aspects of cambial development in Pyrus communis. Am J Bot 48: 479-488

Evert RF 1962 Some aspects of phloem development in Tilia americana. Am J Bot 49: 659

Evert RF 1963a Ontogeny and structure of the secondary phloem in Pyrus malus. Am J Bot 50: 8-37

Evert RF 1963b The cambium and seasonal development of the phloem in Pyrus malus. Am J Bot 50: 149-159

Evert RF, Deshpande BP 1970 An ultrastructural study of cell division in the cambium. Am J Bot 57: 942-961

Evert RF, Kozlowski TT 1967 Effect of isolation of bark on cambial activity and development of xylem and phloem in trembling aspen. Am J Bot 54: 1045-1055

Evert RF, Kozlowski TT, Davis JD 1972 Influence of phloem blockage on cambial growth of sugar maple. Am J Bot 59: 632-641

Fabricius L 1905 Untersuchungen über den Stärke- und Fettgehalt der Fichte auf der oberbayerischen Hochebene. Naturwiss Z Forst-Landwiss 3: 137-176

Fahn A 1974 Plant anatomy. 2nd edn Pergamon, New York, 611 pp

Fahn A 1979 Secretory tissues in plants. Academic Press, New York, 302 pp

Fahn A 1982 Plant anatomy. 3rd edn Pergamon, Oxford, 544 pp

Fahn A, Arnon N 1963 The living wood fibers of Tamarix aphylla and the changes occurring in them in transition from sapwood to heartwood. New Phytol 62: 99-104

Fahn A, Benayoun J 1976 Ultrastructure of resin ducts in Pinus halepensis development. Possible sites of resin synthesis, and mode of its elimination from the protoplasts. Ann Bot 40: 857-863

Fahn A, Ben-Sasson R, Sachs T 1972 The relation between the procambium and the cambium. In: Ghouse AKM (ed) Research trends in plant anatomy. Tata McGraw Hill, New Delhi, 161-170

Fahn A, Evert RF 1974 Ultrastructure of the secretory ducts of Rhus glabra L. Am J Bot 61: 1-14

Fahn A, Leshem B 1963 Wood fibres with living protoplasts. New Phytol 62: 91-98

Fahn A, Werker E 1990 Seasonal cambial activity. In: Iqbal M (ed) The vascular cambium. Research Studies Press, Taunton, England, 139-157

Fahn A, Zamski E 1970 The influence of pressure, wind, wounding and growth substances on the rate of resin duct formation in Pinus halepensis wood. Isr J Bot 19: 429-446

Fahn A, Werker E, Ben-Tzur P 1979 Seasonal effects of wounding and growth substances on development of traumatic resin ducts in Cedrus libani. New Phytol 82: 537-544

Falkenberg P 1876 Ueber das sekundäre Dickenwachsthum der Mesembryanthemum. Bot Ztg 34: 317-319, 324-326

Farooqui P, Robards AW 1979 Seasonal changes in the ultrastructure of cambium of Fagus sylvatica L. Proc Indian Acad Sci 88B, Pt 2 No 6: 463-472

Fengel D 1966 Elektronenmikroskopische Beiträge zum Feinbau des Buchenholzes (Fagus sylvatica L.)-zweite Mitteilung: weitere Beobachtungen an Markstrahlenzellen der Buche. Holz Roh- Werkst 24: 177-185

Fensom DS 1963 The bioelectric potentials of plants and their functional significance V. Some daily and seasonal changes in the electrical potential and resistance of living trees. Can J Bot 41: 831-851

Ferreirinha MP 1965 Appraisal of the variation in the micrographic fibre characteristics within and between trees of tropical hardwood species. Garcia Orta 13: 383-390

Feucht W, Schmid PPS, Christ E 1986 Distribution of flavanols in meristematic and mature tissues of Prunus avium shoots. J Plant Physiol 125: 1-8

Fineran BA 1962 Studies on the root parasitism of Exocarpus bidwillii Hook. F. I. Ecology and root structures of the parasite. Phytomorphology 12: 339-355

Fineran BA 1963a II. External morphology, distribution and arrangement of haustoria. Phytomorphology 13: 30-41

Fineran BA 1963b III. Primary structure of the haustoria. Phytomorphology 13: 42-54

Fineran BA 1963c IV. Structure of the mature haustorium. Phytomorphology 13: 249-267

Fineran BA 1965a V. Early development of the haustorium. Phytomorphology 15: 10-25

Fineran BA 1965b VI. Haustorial attachment to non-living objects and the phenomenon of self-parasitism. Phytomorphology 15: 387-399

Fink S 1982 Adventitious root primordia - the cause of abnormally broad xylem rays in hard- and softwoods. IAWA Bull 3: 31-38
Fink S 1986 Microscopical investigations of wood formation and function in diseased trees. IAWA Bull 7: 351-355
Fischer A 1891 Beiträge zur Physiologie der Holzgewächse. Jahrb Wiss Bot 22: 73-160
Fischer H 1885 Ein Beitrag zur vergleichenden Anatomie des Markstrahlgewebes und der jährlichen Zuwachszonen im Holzkörper von Stamm, Wurzel und Aesten bei Pinus Abies L. Flora 68: 263-278, 279-294, 302-309, 313-324
Fisher JB 1981 Wound healing by exposed secondary xylem in Adansonia (Bombaceaceae). IAWA Bull 2: 193-199
Fisher JB, Ewers FW 1989 Wound healing in stems of lianas after twisting and girdling injuries. Bot Gaz 150: 251-265
Flury P 1907 Ertragstafeln für die Fichte und Buche der Schweiz. Mitt Schweiz Centralanst Forst Versuchswes 9: 1-290
Flury P 1926 Über gewisse Störungen in der Jahrringbildung. Schweiz Z Forstwes 77: 141-146
Foote TK 1954 False rings in larch, Larix decidua Mill. J Oxford Univ For Soc Ser 4, 2: 5-8
Ford ED 1981 Can we model xylem production by conifers? Studia For Suec 160: 19-29
Ford ED, Robards AW, Piney MD 1978 Influence of environmental factors on cell production and differentiation in the early wood of Picea sitchensis. Ann Bot 42: 683-692
Forsaith CC 1920 Anatomical reduction in some alpine plants. Ecology 1: 124-135
Fosket DE, Morejohn LC 1992 Structural and functional organization of tubulin. Ann Rev Plant Physiol Plant Mol Biol 43: 201-240
Foster AS 1947 Structure and ontogeny of the terminal sclereids in the leaf of Mouriria huberi Cogn. Am J Bot 34: 501-514
Frank AB 1864 Ein Beitrag zur Kenntnis der Gefässbündel. Bot Ztg 22: 149-154, 157-162, 165-172, 177-182, 185-188, 377-384, 385-395, 397-403, 405-414
Frank AB 1867 Ueber die Entstehung der Intercellularräume der Pflanzen. Bot Ztg 25: 182-184
Frank B 1884 Ueber die Gummibildung im Holze und deren physiologische Bedeutung. Ber Deutsch Bot Ges 2: 321-332
Franke M 1883 Beiträge zur Kenntniss der Wurzelverwachsungen. Beitr Biol Pflanz 3: 307-334
Fraser DA 1952 Initiation of cambial activity in some forest trees in Ontario. Ecology 33: 259-273
Frey-Wyssling A 1938 Über die Entstehung von Harztaschen. Holz Roh- Werkst 1: 329-332
Frey-Wyssling A 1950 Physiology of cell wall growth. Ann Rev Plant Physiol 1: 169-182
Frey-Wyssling A 1959 Die pflanzliche Zellwand. Springer Berlin, Heidelberg, New York, 367 pp
Frey-Wyssling A, Lopez-Saez JF, Mühlethaler K 1965 Formation and development of the cell plate. IAWA Bull 1965/1: 4-12
Friedrich J 1897 Ueber den Einfluss der Witterung auf den Baumzuwachs. Cbl Ges Forstwes 23: 471-495
Friesner RC, Walden G 1946 A five-year dendrometer record in two trees of Pinus strobus. Butler Univ Bot Stud 8: 1-23
Fritts HC 1976 Tree rings and climate. Academic Press, New York, 567 pp

Fritts HC, Smith DG, Cardis JW, Budelsky CA 1965 Tree-ring characteristics along a vegetation gradient in northern Arizona. Ecology 46: 393-401

Fritz E, Averill JL 1924 Discontinuous growth rings in California redwood. J For 22: 31-38

Fujita M, Tohyama M, Harada H 1984 Three-dimensional analysis of the cambium and the derivatives in Robinia pseudoacacia L. In: Sudo S (ed) Proc Pac Reg Wood Anat Conf Oct 1-7, 1984 Ibaraki Japan, 91-93

Fujiwara S, Nakayama Y 1978 (Distribution of rays and total number of ray cells in stem of Sugi (Cryptomeria japonica D. Don). J Jpn Wood Res Soc 24: 841-845

Fukazawa K, Ohtani J 1984 Indented rings in Sitka spruce. In: Sudo S (ed) Proc Pac Reg Wood Anat Conf Oct 1-7, 1984 Ibaraki Japan, 28-30

Fukuda T, Mori I, Terashima N, Kanada T 1969 (Studies on the enzyme in cambial zone. II. Seasonal changes in the activities of phenylalanine ammonia-lyase and peroxidase, rate of growth and rate of lignification.) J Jpn Wood Res Soc 15: 182-187

Funada R, Catesson AM 1991 Partial cell wall lysis and the resumption of meristematic activity in Fraxinus excelsior cambium. IAWA Bull 12: 439-444

Furqan M, Ahmad Z 1981 Ratio of ray and fusiform initials in the tree axis of Sterculia urens Roxb. at various height levels. Indian J For 4: 293-295

Furuya M 1984 Cell division patterns in multicellular plants. Ann Rev Plant Physiol 35: 349-373

Gagnon RR, Chabot M, Pineau M 1988 Effets de l'élagage sur la résistance électrique cambiale de la tige de jeunes Sapins baumiers. Can J For Res 18: 1655-1658

Gahan PB 1989 How stable are cambial initials? Bot J Linn Soc 100: 319-321

Gao Q, Pitt RE 1991 Mechanics of parenchyma tissue based on cell orientation and microstructure. Trans ASAE 34: 232-238

Gardiner W 1907 On the mode of formation of the initial cell-wall, the genesis and neogenesis of the connecting threads, and the method of connection of living tissue cells. Proc Cambridge Phil Soc 14: 209-210

Garrec JP, Heymes C, Rose C, deLegge R, Toulemonde JC 1989 Etude de la résistance électrique du cambium dans les éspicéas (Picea abies L.) dépérissants. Ann Sci For 46: 373-385

Gaspar T, Hagége D, Kevers C, Penel C, Crèvecoeur M, Engelmann I, Greppin H, Foidart JM 1991 When plant teratomas turn into cancers in the absence of pathogens. Physiol Plant 83: 696-701

Gassner G, Christiansen-Weniger F 1942 Dendroklimatologische Untersuchungen über die Jahresringentwicklung der Kiefern in Anatolien. Nova Acta Leopold 12: 1-137

Gaudichaud C 1852 Remarques générales sur un rapport relatif au mémoire de M. Trécul. Ann Sci Nat Bot Ser 3, 18: 24-55

Gedalovich E, Fahn A 1985a The development and ultrastructure of gum ducts in Citrus plants formed as a result of brown-rot gummosis. Protoplasma 127: 73-81

Gedalovich E, Fahn A 1985b Ethylene and gum duct formation in Citrus. Ann Bot 56: 571-577

Geiger F 1915 Anatomische Untersuchungen über die Jahresringbildung von Tectona grandis. Jahrb Wiss Bot 55: 521-607

Gerry E 1910 The distribution of the "Bars of Sanio" in the Coniferales. Ann Bot 24: 119-123

Gerry E 1915 Fiber measurement studies; length variations: where they occur and their relation to the strength and uses of wood. Science 41: 179

Gerry E 1921 American storax production: results of different methods of tapping red gum trees. J For 19: 15-24

Gerry E 1923 Recent observations on the effects of turpentining on the structure of second-growth slash and longleaf pines. J For 21: 236-241

Gerry E 1942 Radial streak (red) and giant resin ducts in spruce. U S For Prod Lab Rpt 1391, 2 pp

Ghouse AKM, Hashmi S 1977a Ray and fusiform initials in the vascular cambium of some Indian tropical trees. Geobios 4: 149-150

Ghouse AKM, Hashmi S 1977b Comparative study of cambial structure of five Indian tropical trees. Phytomorphology 27: 33-39

Ghouse AKM, Hashmi S 1978 Seasonal cycle of vascular differentiation in Polyalthia longifolia (Annonaceae). Beitr Biol Pflanz 54: 375-380

Ghouse AKM, Hashmi S 1979 Cambium periodicity in Polyalthia longifolia. Phytomorphology 29: 64-67

Ghouse AKM, Hashmi S 1980 Changes in the vascular cambium of Polyalthia longifolia Benth. et Hook. (Annonaceae) in relation to the girth of the tree. Flora 170: 135-143

Ghouse AKM, Hashmi S 1981 Developmental changes in the vascular cambium of Delonix regia Rafin. Proc Indian Acad Sci (Plant Sci) 90: 143-151

Ghouse AKM, Hashmi S 1982 Impact of extension growth and flowering on the cambial activity of Delonix regia Rafin. Proc Indian Acad Sci (Plant Sci) 91: 201-209

Ghouse AKM, Hashmi S 1983 Periodicity of cambium and the formation of xylem and phloem in Mimusops elengi L., an evergreen member of tropical India. Flora 173: 479-488

Ghouse AKM, Iqbal M 1975 A comparative study on the cambial structure of some arid zone species of Acacia and Prosopis. Bot Not 128: 327-331

Ghouse AKM, Iqbal M 1977a Variation trends in the cambial structure of Prosopis spicigera L. in relation to the girth of the tree axis. Bull Torrey Bot Club 104: 197-201

Ghouse AKM, Iqbal M 1977b Trends of size variation in phloem fibres and sieve-tube cells within the bark of some arid-zone trees. Flora 166: 517-521

Ghouse AKM, Iqbal M 1979 Intrusive growth in the secondary phloem of Acacia and Prosopis. New Bot 6: 91-96

Ghouse AKM, Khan MIH 1977 Seasonal variation in the nuclear number of fusiform cambial initials in Psidium guajava L. Caryologia 30: 441-444

Ghouse AKM, Sabir D 1974 Intrusive growth in the phloem fibers of Erythrina indica and Pongamia glabra. Isr J Bot 23: 223-225

Ghouse AKM, Yunus M 1972 An example of the stratified cambium among the indigenous tropical trees. Curr Sci 41: 569-570

Ghouse AKM, Yunus M 1973 Some aspects of cambial development in the shoots of Dalbergia sisoo Roxb. Flora 162: 549-558

Ghouse AKM, Yunus M 1974a The ratio of ray and fusiform initials in some woody species of the Ranalian complex. Bull Torrey Bot Club 101: 363-366

Ghouse AKM, Yunus M 1974b Cambial structure in Dalbergia. Phytomorphology 24: 152-158

Ghouse AKM, Yunus M 1974c A new record in the occurrence of stratified cambium in the family Mimosaceae. Geobios 1: 138

Ghouse AKM, Yunus M 1975 Intrusive growth in the phloem of Dalbergia. Bull Torrey Bot Club 102: 14-17

Ghouse AKM, Yunus M 1976 Ratio of ray and fusiform initials in the vascular cambium of certain leguminous trees. Flora 165: 23-28

Ghouse AKM, Yunus M, Iqbal M 1975a Ratio of ray and fusiform initials in the vascular cambium of some arid-zone plants. Curr Sci 44: 361-362

Ghouse AKM, Yunus M, Iqbal M 1975b Ray and fusiform initials in Bauhinia species. Curr Sci 44: 442-443

Ghouse AKM, Yunus M, Iqbal M 1976 A comparative study on the cambial structure of some Bauhinia species. Bot Jahrb Syst 95: 411-417

Ghouse AKM, Khan MIH, Khan S, Khan AH 1979 Occurrence of polynucleate condition in the fusiform cambial initials of some Verbenaceae. Chrom Inf Serv 26: 16-17

Ghouse AKM, Khan MIH, Khan S, Khan AH 1980 Comparative study on the structure of vascular cambium in some Verbenaceae. Phytomorphology 30: 32-40

Ghouse AKM, Abidi SH, Khan PR, Reshma P 1981 The cambial activity of Linum usitatissimum var. T397 as influenced by gamma rays. In: Verma SC (ed) Contemporary trends in plant science. Kalyani, New Delhi, 263-266

Gibson AC 1978 Rayless secondary xylem of Halophytum. Bull Torrey Bot Club 105: 39-44

Giesenhagen K 1909 Die Richtung der Teilungswand im Pflanzenzellen. Flora 99: 355-369

Gill N 1933 The relation of flowering and cambial activity. Observations on vascular differentiation and dryweight changes in the catkins of some early flowering catkin-bearing dicotyledons. New Phytol 32: 1-12

Giltay E 1882 Sur le collencyhme. Arch Neerl 17: 432-459

Ginzburg C 1963 Some anatomic features of splitting of desert shrubs. Phytomorphology 13: 92-97

Giraud B 1980 Correlations between wood anatomical characters in Entandrophragma utile (Meliaceae). IAWA Bull 1: 73-75

Glerum C 1970 Drought ring formation in conifers. For Sci 16: 246-248

Glerum C, Farrar JL 1966 Frost ring formation in the stems of some coniferous species. Can J Bot 44: 879-886

Glerum C, Krenciglowa EM 1970 The dependence of electrical impedance of woody stems on various frequencies and tissues. Can J Bot 48: 2187-2192

Glock WS 1941 Growth rings and climate. Bot Rev 7: 649-713

Glock WS 1951 Cambial frost injuries and multiple growth layers at Lubbock, Texas. Ecology 32: 28-36

Glock WS 1955 Tree growth II. Growth rings and climate. Bot Rev 21: 73-188

Glock WS, Studhalter RA 1941 Frost injury artificially induced in Arizona cypress. Am J Bot 28(suppl): 3s

Glock WS, Studhalter RA, Agerter SR 1960 Classification and multiplicity of growth layers in the branches of trees at the extreme lower forest border. Smithsonian Misc Coll Publ 4421 Vol 140 (1), 292 pp

Glock WS, Germann PJ, Agerter SR 1963 Uniformity among growth layers in three ponderosa pines. Smithsonian Misc Coll Publ 4508, Vol 145 (4), 374 pp

Goffinet MC, Larson PR 1981 Structural changes in Populus deltoides terminal buds and in the vascular transition zone of the stems during dormancy induction. Am J Bot 118-129

Goosen-de Roo L, Creyghton-Schouten EAM 1981 Plasmodesmata in cell walls in the cambial zone of Fraxinus excelsior L. Proc 2nd Cell Wall Mtg, Göttingen, 46 pp

Goosen-de Roo L, van Spronsen PC 1978 Electron microscopy of the active cambial zone of Fraxinus excelsior L. IAWA Bull 4: 59-64

Goosen-de Roo L, Venverloo CJ, Burggraaf PD 1980 Cell division in highly vacuolated plant cells. In: Brederoo P, de Priester W (eds) Electron microscopy, 7th Eur Cong Elec Micro Found Leiden Vol 2: 232-233

Goosen-de Roo L, Burggraaf PD, Libbenga KR 1983 Microfilament bundles associated with tubular endoplasmic reticulum in fusiform cells in the active cambial zone of Fraxinus excelsior L. Protoplasma 116: 204-208

Goosen-de Roo L, Bakhuizen LR, van Spronsen PC, Libbenga KR 1984 The presence of extended phragmosomes containing cytoskeletal elements in fusiform cambial cells of Fraxinus excelsior L. Protoplasma 122: 145-152

Göppert HR 1842 Recherches sur la structure anatomique des tiges des Casuarina. Ann Sci Nat Bot Ser 2, 18: 5-12

Göppert HR 1846 Ueber die Ueberwallung von Tannenstöcke. Bot Ztg 4: 505-514

Govindarajalu E, Swamy BGL 1955 Size variations of septate wood fibres in Pithecolobium dulce Benth. Nature 176: 315

Graham BF, Bormann FH 1966 Natural root grafts. Bot Rev 32: 255-292

Gray RL 1973 Multiseriate rays in redwood Sequoia sempervirens (D.Don.) Endl. IAWA Bull 1973/1: 7-8

Green JR 1909 A history of botany 1860-1900, Being a continuation of Sachs' `History of botany, 1530-1860'. Reprint of 1909 edn, Russell & Russell, New York 1967, 543 pp

Greenwood JS, Demmers C, Wetzel S 1990 Seasonally dependent formation of protein-storage vacuoles in the inner bark tissues of Salix microstachya. Can J Bot 68: 1747-1755

Gregory RA 1971 Cambial activity in Alaskan white spruce. Am J Bot 58: 160-171

Gregory RA 1977 Cambial activity and ray cell abundance in Acer saccharum. Can J Bot 55: 2559-2564

Gregory RA 1978 Living elements of the conducting xylem of sugar maple (Acer saccharum Marsh.). IAWA Bull 1978/4: 65-69

Gregory RA 1981 A rapid method of estimating the relative amount of vascular ray tissue. IAWA Bull 2: 77-78

Gregory RA, Romberger JA 1975 Cambial activity and height of uniseriate vascular rays in conifers. Bot Gaz 136: 246-253

Gregory RA, Wilson BF 1968 A comparison of cambial activity of white spruce in Alaska and New England. Can J Bot 46: 733-734

Greguss P 1955 Identification of living gymnosperms on the basis of xylotomy. Jóesik L (transl) Akad Budapest, 263 pp

Grew N 1682 The anatomy of plants. Rawlings, London, 1965 Johnson Reprint, New York, 304 pp

Grill D, Gigler A 1985 Holzanatomische Untersuchungen an einer geringelten Thuja occidentalis L. Phyton 25: 293-296

Grillos SJ, Smith FH 1959 The secondary phloem of Douglas-fir. For Sci 5: 377-388

Groom P, Rushton W 1913 The structure of the wood of East Indian species of Pinus. J Linn Soc Bot Lond 41: 457-490

Grossenbacher JG 1910 Medullary spots: a contribution to the life history of some cambium miners. N Y Agr Exp Sta Tech Bull 15: 49-65

Grossenbacher JG 1914 Gliding growth and the bars of Sanio. Am J Bot 1: 522-530

Grossenbacher JG 1915a Medullary spots and their causes. Bull Torrey Bot Club 42: 227-239

Grossenbacher JG 1915b The periodicity and distribution of radial growth in trees and their relation to the development of "annual rings". Trans Wisc Acad Sci Arts Lett 18: 1-77

Grosser D 1986 On the occurrence of trabeculae with special consideration of diseased trees. IAWA Bull 7: 319-341

Grosser D, Schulz H, Utschig H 1985 Mögliche anatomische Veränderungen in erkrankten Nadelbäumen. Holz Roh- Werkst 43: 315-323

Grosser D, Lesnino G, Schulz H 1991 Histologische Untersuchungen über das Schutzholz einheimischer Laubbäume. Holz Roh- Werkst 49: 65-73

Gunckel JE, Wetmore RH 1946 Studies of development in long shoots and short shoots of Ginkgo biloba L. I. The origin and pattern of development of the cortex, pith and procambium. Am J Bot 33: 285-295

Gunning BES, Wick SM 1985 Preprophase bands, phragmoplasts, and spatial control of cytokinesis. J Cell Sci 2(suppl): 157-179

Gustin R, De Sloover J 1955 Recherches sur l'histogénèse des tissus conducteurs- I. Problèms posés et donnés acquises. La Cellule 57: 97-128

Haberlandt G 1886 Ueber das Markstrahlmeristem von Cytisus laburnum. Ber Deutsch Bot Ges 4: 144-150

Haberlandt G 1914 Physiological plant anatomy. 4th edn Drummond M (transl) Macmillan, London, 777 pp

Hale JD 1923 The bars or rims of Sanio. Bot Gaz 76: 241-256

Hamilton JR 1963 Characteristics of tracheids produced in a gamma and gamma-neutron environment. For Prod J 13: 62-67

Hamilton JR, Chesser AH 1969 Effects of acute ionizing radiation on the xylem fibers of Quercus alba and Liquidambar styraciflua. Rad Bot 9: 331-339

Hanes CS 1927 Resin canals in seedling conifers. J Linn Soc Bot Lond 47: 613-636

Hanson HC, Brenke B 1926 Seasonal development of growth layers in Fraxinus campestris and Acer saccharinum. Bot Gaz 82: 286-304

Hanson JB, Benjamin DM 1967 Biology of Phytobia setosa, a cambium miner of sugar maple. J Econ Entomol 60: 1351-1355

Harper AG 1913 Defoliation: its effects upon the growth and structure of the wood of Larix. Ann Bot 27: 621-642

Harris HA 1934 Frost ring formation in some winter-injured deciduous trees and shrubs. Am J Bot 21: 485-498

Harris JM 1952 Discontinuous growth layers in Pinus radiata. N Z For Prod Res Notes 1(4): 1-10

Harris JM 1969 On the causes of spiral grain in corewood of Radiata pine. N Z J Bot 7: 189-213

Harris JM 1973 Spiral grain and xylem polarity in radiata pine: microscopy of cambial reorientation. N Z J For Sci 3: 363-378

Harris JM 1981 Spiral grain development. In: Barnett JR (ed) Xylem cell development. Castle House, Turnbridge Wells, 256-274

Harris JM 1989 Spiral grain and wave phenomena in wood formation. Springer Berlin, Heidelberg, New York, 214 pp

Harris JM, Barnett JR 1975 Differentiated callus nodules in resin pockets of Pinus ponderosa Laws. N Z J For Sci 5: 226-229

Hartig R 1869 Das Aussetzen der Jahresringe bei unterdrückten Stämmen. Z Forst- Jagdwes 1: 471-476

Hartig R 1871 Ueber das Dickenwachsthum der Waldbäume. Z Forst- Jagdwes 3: 66-104

Hartig R 1872 Einfluss verscheiden starker Ausästung und Entnadelung auf den Zuwachs der Weymouthskiefer und gemeinen Kiefer. Z Forst- Jagdwes 4: 240-254
Hartig R 1889 Ein Ringlungsversuch. Allg Forst- Jadgztg 65: 365-373, 401-410
Hartig R 1890 Ueber die Folgen der Baumringelung. Bot Cbl 41: 251-253, 283-285
Hartig R 1891 Lehrbuch der Anatomie und Physiologie der Pflanzen; unter besonderer Berücksichtigung der Forstgewächse. Julius Springer, Berlin, 308 pp
Hartig R 1892 Ueber den Entwicklungsgang der Fichte im geschlossenen Bestande nach Höhe, Form und Inhalt. Forst-Naturwiss Z 1: 169-185
Hartig R 1893 Ueber neuere Untersuchungen zur Physiologie der Eiche. Bot Cbl 56: 357-358
Hartig R 1894 Untersuchungen über die Entstehung und die Eigenschaften des Eichenholzes. Forst-Naturwiss Z 3: 1-13, 49-68, 172-191, 193-203
Hartig R 1895a Doppelringe als Folge von Spätfrost. Forst-Naturwiss Z 4: 1-8
Hartig R 1895b Untersuchungen des Baues und der technischen Eigenschaften des Eichenholzes. Forst-Naturwiss Z 4: 49-82
Hartig R 1895c Ueber den Drehwuchs der Kiefer. Forst-Naturwiss Z 4: 313-326
Hartig R 1897a Ueber den Einfluss des Hütten- und Steinkohlenrauches auf den Zuwachs der Nadelholzbäume. Forst-Naturwiss Z 6: 49-60
Hartig R 1897b Untersuchungen über Blitzschläge in Waldbäumen. Forst-Naturwiss Z 6: 97-120
Hartig R 1901 Ueber die Ursachen des Wimmerwuchses. Cbl Ges Forstwes 27: 145-150
Hartig T 1837a Vergleichende Untersuchungen über die Organisation des Stammes der einheimischen Waldbäume. Jahrb Fortschr Forstwiss Forst Naturk 1: 125-168
Hartig T 1837b Ueber die Vegetationsperioden der Waldbäume und die sie begleitenden Erscheinungen. Jahrb Fortschr Forstwiss Forst Naturk 1: 601-637
Hartig T 1845 Ueber künstliche Erzeugung neuer Holz- und Rindenschichten auf entrindeten Holzflächen durch Fensterung. Allg Forst-Jagdztg 11: 165-169
Hartig T 1848 Beiträge zur Geschichte der Pflanzen und zur Kenntniss der norddeutschen Braunkohlen-Flora. Bot Ztg 6: 122-128, 137-141, 166-172, 185-190
Hartig T 1853 Ueber die Entwickelung des Jahrringes der Holzpflanzen. Bot Ztg 11: 553-560, 569-579
Hartig T 1854a Ueber Bildung und Entwickelung der sogenannten Knospenwurzeln. Bot Ztg 12: 1-7, 27-34
Hartig T 1854b Ueber die Querscheidewände zwischen den einzelnen Gliedern der Siebröhren in Cucurbita pepo. Bot Ztg 12: 51-54
Hartig T 1854c Ueber das Verhalten des Zellkerns bei der Zellentheilung. Bot Ztg 12: 893-902
Hartig T 1855a Ueber die Entstehung der Markstrahlen. Bot Ztg 13: 217-221
Hartig T 1855b Ueber die Wirkung verdünnter Schwefelsäure auf die Ablagerungsschichten der Zellwand in deren jugendlichstem Zustande. Bot Ztg 13: 222-223
Hartig T 1856 Beiträge zur physiologischen Forstbotanik. Über den Vegetations-Zyklus der Holz-pflanzen und die Bedeutung der Reserve-Stoffe für denselben. Allg Forst-Jagdztg 32: 361-372
Hartig T 1858 Ueber die Bewegung des Saftes in den Holzpflanzen. Bot Ztg 16: 329-335, 337-342
Hartig T 1859a Beiträge zur vergleichenden Anatomie der Holzpflanzen. Bot Ztg 17: 93-101, 105-112

Hartig T 1859b Beiträge zur physiologischen Forstbotanik. Ueber den Lauf der Wander-Säfte in den Holzpflanzen. Allg Forst-Jagdztg 35: 129-138

Hartig T 1862 Ueber die Bewegung des Saftes in den Holzpflanzen. Bot Ztg 20: 73-76, 81-87, 89-94, 97-100, 105-109

Hartig T 1863a Folgen der Ringelung an einer Linde. Bot Ztg 21: 286

Hartig T 1863b Folgen der Ringelung an Nadelholz-Aesten. Bot Ztg 21: 286-287

Hartig T 1863c Ueber die Thätigkeit des Siebfasergewebes bei Rückleitung der Bildungssäfte. Bot Ztg 21: 287-288

Hartig T 1863d Ueber den Zucker und über einem dem Salicin ähnlichen Körper aus den Cambialsäften der Nadelhölzer. Bot Ztg 21: 413-414

Hartig T 1878 Anatomie und Physiologie der Holzpflanzen. Julius Springer Berlin, 404 pp

Hartley WR 1960 Nutrients and tracheid length in seedlings of Pinus radiata D. Don. Emp For Rev 39: 474-482

Hashmi S, Ghouse AKM 1978 On the nuclear number of the fusiform initials of Polyalthia longifolia Benth. and Hook. J Indian Bot Soc 57: 24

Hawksworth FG 1979 Mistletoes and their role in North American forestry. In: Musselman LD, Worsham AD, Eplee RE (eds) Proc 2nd Intl Symp on Parasitic Weeds. N C State Univ Raleigh, 13-23

Hawksworth FG, Scharpf RF (eds) 1984 Biology of dwarf mistletoes: Proc Symp U S For Serv Genl Tech Rpt RM-111 Fort Collins CO, 131 pp

Hawthorne DW 1980 Wildlife damage and control techniques. In: Schemnitz SD (ed) Wildlife management techniques manual. 4th edn, 411-439

Hawthorne DW 1983 Identifying wildlife damage. In: Timm RM (ed) Prevention and control of wildlife damage. Great Plains Agr Council, University of Nebraska, Lincoln pp A1-A18

Haygreen JG, Bowyer JL 1982 Forest products and wood science. Iowa State University Press, Ames, 495 pp

Heikkilä R, Mikkonen T 1992 Effects of density of young Scots pine (Pinus sylvestris) stand on moose (Alces alces) browsing. Acta For Fenn 231: 1-14

Hejnowicz A, Hejnowicz Z 1958 Variations of length of vessel members and fibres in the trunk of Populus tremula L. Acta Soc Bot Pol 27: 131-159

Hejnowicz A, Hejnowicz Z 1959 Variations of length of vessel members and fibres in the trunk of Robinia pseudoacacia. Acta Soc Bot Pol 28: 453-460

Hejnowicz Z 1961 Anticlinal divisions, intrusive growth, and loss of fusiform initials in nonstoried cambium. Acta Soc Bot Pol 30: 729-748

Hejnowicz Z 1963a Intrusive growth, transverse and pseudotransverse divisions in fusiform initials of wounded cambium of Larix europaea. Acta Soc Bot Pol 32: 493-503

Hejnowicz Z 1963b Significance of intrusive growth in establishment of cambial union after transverse incision in Larix. Acta Soc Bot Pol 32: 625-630

Hejnowicz Z 1964 Orientation of the partition in pseudotransverse division in cambium of some conifers. Can J Bot 42: 1685-1691

Hejnowicz Z 1967 Interrelationship between mean length, rate of intrusive elongation, frequency of anticlinal divisions and survival of fusiform initials in cambium. Acta Soc Bot Pol 36: 367-378

Hejnowicz Z 1968 The structural mechanism involved in the changes of grain in timber. Acta Soc Bot Pol 37: 347-365

Hejnowicz Z 1971 Upward movement of the domain pattern in the cambium producing wavy grain in Picea excelsa. Acta Soc Bot Pol 40: 499-512

Hejnowicz Z 1973 Morphogenetic waves in cambia of trees. Plant Sci Lett 1: 359-369
Hejnowicz Z 1974a Pulsations of domain length as support for the hypothesis of morphogenetic waves in the cambium. Acta Soc Bot Pol 43: 261-271
Hejnowicz Z 1974b How genetic information can be translated into the spatial pattern of cellular differentiation. 19th Hortic Cong Warsaw Sept 11-18 1974, 37-48
Hejnowicz Z 1975 A model for morphogenetic map and clock. J Theor Biol 54: 345-362
Hejnowicz Z 1980 Tensional stress in the cambium and its developmental significance. Am J Bot 67: 1-5
Hejnowicz Z 1990 Phenomena of orientation in the cambium. In: Iqbal M (ed) The vascular cambium. Research Studies Press, Taunton, England, 127-137
Hejnowicz Z, Branski S 1966 Quantitative analysis of cambium growth in Thuja. Acta Soc Bot Pol 35: 395-400
Hejnowicz Z, Krawczyszyn J 1969 Oriented morphogenetic phenomena in cambium of broadleaved trees. Acta Soc Bot Pol 38: 547-560
Hejnowicz Z, Kurczynska EU 1987 Occurrence of circular vessels above axillary buds in stems of woody plants. Acta Soc Bot Pol 56: 415-419
Hejnowicz Z, Romberger JA 1973 Migrating cambial domains and the origin of wavy grain in xylem of broadleaved trees. Am J Bot 60: 209-222
Hejnowicz Z, Romberger JA 1979 The common bases of wood grain figures is the systematically changing orientation of cambial fusiform cells. Wood Sci Tech 13: 89-96
Hejnowicz Z, Zagórska-Marek B 1974 Mechanism of changes in grain inclination in wood produced by storeyed cambium. Acta Soc Bot Pol 43: 381-398
Heliövaara K, Väisänen R, Immonen A 1991 Quantitative biogeography of the bark beetles (Coleoptera, Scolytidae) in northern Europe. Acta For Fenn 219: 1-35
Hemenway AF 1926 Late frost injury to some trees in central Kentucky. Am J Bot 13: 364-366
Hepting GH 1935 Decay following fire in young Mississippi delta hardwoods. USDA Tech Bull 494, 32 pp
Hepting GH, Blaisdell DJ 1936 A protective zone in red gum fire scars. Phytopathology 26: 62-67
Herbst A 1894 Beiträge zur Kenntniss der Markstrahlen dicotyler Kräute und Stauden. Bot Cbl 57: 257-266
Herring HG, Rudolph VJ, Lemmien WA 1958 Wound area as influenced by closeness of pruning. J For 56: 219-220
Hess RW 1950 Classification of wood parenchyma in dicotyledons. Trop Woods 96: 1-20
Hill AW 1901 The histology of the sieve-tubes of Pinus. Ann Bot 15: 575-611
Hill AW 1908 The histology of the sieve-tubes in Angiosperms. Ann Bot 22: 245-290
Hillis WE 1955 Formation of leucoanthocyanins in Eucalypt tissues. Nature 175: 597-598
Hillis WE 1987 Heartwood and tree exudates. Springer, Berlin, Heidelberg, New York, 268 pp
Hillis WE, Carle A 1958 The polyphenols and shikimic acid of eucalypt cambium and wood. Holzforschung 12: 136-141
Hirakawa Y, Ishida S, Ohtani J 1979 A SEM observation of organelles in the cambial and living xylem cells in Todomatsu (Abies sachaliensis). Res Bull Hokkaido Univ Exp For 36: 459-468
Hirt RR 1939 Canker development by Cronartium ribicola on young Pinus strobus. Phytopathology 29: 1067-1076
Hobert VC 1913 Number and distribution of resin passages in slash pine. For Worker 8: 8-9

Hoffmann R 1885 Untersuchungen über die Wirkung mechanischer Kräfte auf die Theilung, Anordnung und Ausbildung der Zellen beim Aufbau des Stammes der Laub- und Nadelhölzer. Inaug Diss Berlin Rev: Bot Cbl 25: 359-360 1886

Hohl HR 1960 Über die submikroskopische Struktur normaler und hyperplastischer Gewebe von Datura stramonium L. I. Normalgewebe. Ber Schweiz Bot Ges 70: 395-439

Holden R 1912 Reduction and reversion in the North American Salicales. Ann Bot 26: 165-173

Holdheide W 1940 Über Kambialfrostschäden an Edelkastanien und Eschen. Thar Forst Jahrb 91: 582-590

Holdheide W 1955 Über das abnorme Dickenwachstum der Hainbuche (Carpinus betulus L.) und die Rolle der falschen Markstrahlen. Bot Stud Jena 4: 132-164

Hook DD, Brown CL 1972 Permeability of the cambium to air in trees adapted to wet habitats. Bot Gaz 133: 304-310

Höster HR 1974 On the nature of the first-formed tracheids in compression wood. IAWA Bull 1974/1: 3-9

Howard RA, Wilson BF 1972 A stochastic model for cambial activity. Bot Gaz 133: 410-414

Howes FN 1949 Vegetable gums and resins. Chronica Botanica, Waltham MA, 188 pp

Huber B 1939 Das Siebröhrensystem unserer Bäume und seine jahreszeitlichen Veränderungen. Jahrb Wiss Bot 88: 176-242

Huber B 1948 Physiologie der Rindenschälung bei Fichte und Eiche. Forstwiss Cbl 67: 129-164

Huber B 1949 Zur Frage der anatomischen Unterscheidbarkeit des Holzes von Pinus sylvestris L. und Pinus nigricans Host. mit Betrachtungen über heterogene Markstrahlen. Forstwiss Cbl 68: 456-468

Huber B, Prütz G 1938 Über den Anteil von Fasern, Gefässen und Parenchym am Aufbau verscheidener Hölzer. Holz- Roh Werkst 1: 377-381

Huber B, Rouschal E 1938 Anatomische und zellphysiologische Beobachtungen am Siebröhrensystem der Bäume. Ber Deutsch Bot Ges 56: 380-391

Huber B, Schmidt E 1938 Botanik und Koloniale Holzforschung. Kolonial Forst Mitt 1: 163-169

Hueber FM, Nambudiri EMF, Tidwell WD, Wheeler EF 1991 An Eocene fossil tree with cambial variant wood structure. Rev Paleobot Palynology 68: 257-267

Hug UE 1979 Das Harzkanalsystem im juvenilen Stammholz von Larix decidua Mill. Z Schweiz Forst Beih 61: 1-127

Hugentobler UH 1965 Zur Cytologie der Kernholzbildung. Vierteljahrsch Naturforsch Ges Zürich 110: 321-342

Hull RJ, Leonard OA 1964 Physiological aspects of parasitism in mistletoes (Arceuthobium and Phoradendron) I. The carbohydrate nutrition of mistletoe. Plant Physiol 39: 996-1007

Husain A, Cooper WC 1958 Cambial activity in grapefruit trees as influenced by winter irrigation. Am J Bot 45: 517-519

Hush JM, Hawes CR, Overall RL 1990 Interphase microtubule re-orientation predicts a new cell polarity in wounded pea roots. J Cell Sci 96: 47-61

Hüttl RF, Orlovius K, von Schenck B 1990 Messung elektrischer Widerstände im Kambialbereich gedüngter und ungedüngter Fichten. Forstwiss Cbl 109: 190-199

IAWA 1964 Multilingual glossary of terms used in wood anatomy. Buchdruckerei Konkordia, Winterthur, 186 pp

Imagawa H 1981 Study on the seasonal development of the secondary phloem in Larix leptolepis. Res Bull Coll Exp For Hokkaido Univ 38: 31-44

Imagawa H 1985a (Study on the ray formation in Japanese larch). Res Bull Coll Exp For Hokkaido Univ 42: 133-147

Imagawa H 1985b (Study on the seasonal sequence of the radial growth in forest trees). Res Bull Coll Exp For Hokkaido Univ 42: 149-178

Imagawa H 1985c Marking method by electrical stimulation for studying xylem formation IV Occurring area of crushed tracheids. Res Bull Coll Exp For Hokkaido Univ 42: 585-594

Imagawa H, Ishida S 1970 (Study on the wood formation in trees Report 1. Seasonal development of the xylem ring of Japanese larch stem, Larix leptolepis Gordon). Res Bull Coll Exp For Hokkaido Univ 27: 373-394

Imagawa H, Ishida S 1972 (Study on the wood formation in trees Report III. Occurrence of the overwintering cells in cambial zone in several ring-porous trees). Res Bull Coll Exp For Hokkaido Univ 29: 207-221

Imagawa H, Ishida S 1981a An observation of the cambial cells in Larix leptolepis by semi-ultrathin sections. Res Bull Coll Exp For Hokkaido Univ 38: 45-54

Imagawa H, Ishida S 1981b New marking method by electrical stimulation for studying xylem formation. Res Bull Coll Exp For Hokkaido Univ 38: 241-248

Imagawa H, Ishida S 1982 Preliminary experiment of relationship between electrical resistance and state of cells in Pinus luchuensis stems. Res Bull Coll Exp For Hokkaido Univ 39: 127-136

Imagawa H, Ishida S 1983a New marking method by electrical stimulation for studying xylem formation II Application to broad-leaved trees. Res Bull Coll Exp For Hokkaido Univ 40: 387-396

Imagawa H, Ishida S 1983b New marking method by electrical stimulation for studying xylem formation III Identification of the influenced tracheids. Res Bull Coll Exp For Hokkaido Univ 40: 397-408

Imamura Y 1978 (Abnormal ray tissues in excrescence featured wood of Sugi (Cryptomeria japonica D. Don). J Jap Wood Res Soc 24: 71-74

Iqbal M, Ghouse AKM 1977 Ontogenetic size variation of sieve-tube elements in Prosopis spicigera L. Bull Soc Bot Fr 124: 445-450

Iqbal M, Ghouse AKM 1985a Cell events of radial growth with special reference to cambium of tropical trees. In: Malik CP (ed) Widening horizons of plant sciences. Cosmo, New Delhi, 218-252

Iqbal M, Ghouse AKM 1985b Impact of climatic variation on the structure and activity of vascular cambium in Prosopis spicigera. Flora 177: 147-156

Iqbal M, Ghouse AKM 1987 Anatomy of the vascular cambium of Acacia nilotica (L.) Del. var. telia Troup (Mimosaceae) in relation to age and season. Bot J Linn Soc Lond 94: 385-397

Iqbal M, Ghouse AKM 1990 Cambial concept and organisation. In: Iqbal M (ed) The vascular cambium. Research Studies Press, Taunton, England, 1-36

Isebrands JG, Larson PR 1973 Some observations on the cambial zone in cottonwood. IAWA Bull 1973/3: 3-11

Itoh T 1971 On the ultrastructure of dormant and active cambium of conifers. Wood Res Kyoto 51: 33-45

Itoh T, Hayashi S, Kishima T 1968 Cambial activity and radial growth in Sugi trees (Japanese Cryptomeria). Wood Res Kyoto 45: 23-35

Jaccard P 1910a Etude anatomique de bois comprimés. II. Bois comprimés naturellement. 1° Epicéa plissé (Wellenholz). Mitt Schweiz Centralanst Forst Versuchswes 10: 71-77
Jaccard P 1910b Etude anatomique de bois comprimés. 2° Bois des aisseles (Anstansatz) comprimé par la croissance des rameaux. Mitt Schweiz Centralanst Forst Versuchswes 10: 78-84
Jaccard P 1915a Neue Untersuchungen über die Ursachen des Dickenwachstums der Bäume. Naturwiss Z Forst-Landwirtsch 13: 321-360
Jaccard P 1915b Über die Verteilung der Markstrahlen bei den Coniferen. Ber Deutsch Bot Ges 33: 492-498
Jaccard P 1939 Harzgänge und mechanische Eigenschaften des Holzes. Schweiz Z Forstwes 90: 122-126
Jackson BD 1928 A glossary of botanic terms with their derivation and accent. 4th edn Gerald Duckworth, London, 481 pp
Jacobs WP 1979 Plant hormones and plant development. Cambridge University Press, 339 pp
Jacquiot C 1966 Plant tissues and excised organs cultures and their significance in forest research. J Inst Wood Sci 16: 22-34
Jacquiot C 1977 Fonctionnement du cambium et corrélations. Soc Bot Fr Coll Mult Veget 124: 129-133
Jacquiot C 1978a Nouvelles recherches sur l'action de bourgeons greffés sur du tissu cambial cultivé in vitro. 103e Cong Nat Soc Savantes Nancy I:87-91
Jacquiot C 1978b Étude cytologique d'une souche ancienne de tissu cambial de Fagus silvatica L. cultivé in vitro. 103e Cong Nat Soc Savantes Nancy I: 93-96
Jacquiot C 1984 Sur la physiologie du cambium des arbres forestiers. Rev For Fr 36: 113-121
Jamal A, Ghouse AKM 1979 Proportion of ray and fusiform initials in the cambium of Cassia. Geophytology 9: 160-161
Jamal A, Ghouse AKM 1981 Interspecific variation in the cambium of Cassia. New Bot 8: 1-4
Jane FW 1934 Terminal and initial parenchyma in wood. Nature 133: 534
Janse JM 1887 Die Mitwirkung der Markstrahlen bei der Wasserbewegung im Holze. Jahrb Wiss Bot 18: 1-69
Janse JM 1914 Les sections annulaires de l'écorce et le suc descendant. Ann Jard Bot Buitenzorg 28: 1-89
Janse JM 1921 La polarité des cellules cambiennes. Ann Jard Bot Buitenzorg 31: 167-180
Janssonius HH 1931 Die Verteilung des stockwerkartigen Aufbaues im Holz der Dikotylen. Rec Trav Bot Neerl 28: 97-106
Jeffrey EC 1917 The anatomy of woody plants. University of Chicago Press, 478 pp
Jeffrey EC, Torrey RE 1921a Transitional herbaceous dicotyledons. Ann Bot 35: 227-249
Jeffrey EC, Torrey RE 1921b Physiological and morphological correlations in herbaceous angiosperms. Bot Gaz 71: 1-31
Jenkins PA 1975 Seasonal trends in translocation of ^{14}C photosynthates and their association with wood formation in Radiata pine seedlings. N Z J For Sci 5: 62-73
Jenkins PA, Shepherd KR 1972 Influence of temperature on cambial activity and cell diameter in Pinus radiata D. Don. J Inst Wood Sci 16: 36-39
Jensen RE, Savage EE, Hayden RA 1970 The effect of certain environmental factors on cambium temperatures of peach trees. J Am Soc Hortic Sci 95: 286-292

Jöel DM, Fahn A 1980 Ultrastructure of the resin ducts of Mangifera indica L. (Anacardiaceae). I. Differentiation and senescence of the shoot ducts. Ann Bot 46: 225-233

Johansen RW, Wade DD 1987 Effects of crown scorch on survival and diameter growth of slash pine. S J Appl For 11: 180-184

Jones BE 1963 Cell adjustments accompanying the development of spiral grain in a specimen of Pseudotsuga taxifolia Brit. Commonw For Rev 42: 151-158

Jones CS, Lord EM 1982 The development of split axes in Ambrosia dumosa (Gray) Payne (Asteraceae). Bot Gaz 143: 446-453

Jones MGK 1976 The origin and development of plasmodesmata. In: Gunning BES, Robards AW (eds) Intercellular communications in plants: studies in plasmodesmata. Springer, Berlin, Heidelberg, New York, 81-105

Joseleau JP, Ullmann G 1990 Biochemical evidence for the site of formation of gum arabic in Acacia senegal. Phytochemistry 29: 3401-3405

Joseph JP, Shah JJ, Inamdar JA 1989 Cambial activity and distribution of rubber in guayule (Parthenium argentatum). Ann Bot 63: 339-347

Jost L 1892 Beobachtungen über den zeitlichen Verlauf des secundären Dickenwachsthums der Bäume. Ber Deutsch Bot Ges 10: 587-605

Jost L 1901 Ueber einige Eigenthümlichkeiten das Cambiums der Baüme. Bot Ztg 59: 1-24

Jost L 1942 Über Gefässbrücken. Z Bot 38: 161-215

Kaeiser M 1964 Vascular cambial initials in eastern cottonwood in relation to mature wood cells derived from them. Trans Ill State Acad Sci 57: 182-184

Kalberlah A 1898 Der Bau von Tetrastigma scariosum Pl. Ein Beitrag zur Kenntniss der Lianenstructur. Z Naturwiss 71: 161-218

Kang KD, Soh WY 1988 Origin of the vascular cambium in developing hypocotyl of Glycine max seedling. Korean J Bot 31: 289-298

Kärkkäinen M 1973 Mäntyrunkojen ydinsäteiden määrä ja koko. (Amount and size of rays in pine stems). Silva Fenn 7: 69-95

Kaussmann B 1963 Pflanzenanatomie. Gustav Fischer, Jena, 624 pp

Kawana A, Doi M, Motoyama Y 1973 (The seasonal pattern of xylem formation denoted by pin marking method. I. The stem growth in fertilized Sugi (Cryptomeria japonica) seedlings). J Jap For Soc 55: 202-206

Kawana A, Doi M, Motoyama Y 1974 (II. The morphological study of denoted date and healed wounds after pin markings on Sugi (Cryptomeria japonica) trunks). J Jap For Soc 56: 16-19

Keith CT 1971 Observations on the anatomy and fine structure of the trabeculae of Sanio. IAWA Bull 1971/3: 3-11

Kelly PE, Cook ER, Larson DW 1992 Constrained growth, cambial mortality, and dendrochronology of ancient Thuja occidentalis on cliffs of the Niagara escarpment: an eastern version of bristlecone pine? Int J Plant Sci 153: 117-127

Kennedy RW, Farrar JL 1965 Tracheid development in tilted seedlings. In: Côté WA (ed) Cellular ultrastructure of woody plants. Syracuse University Press, 419-453

Kenward RE, Parish T 1986 Bark-stripping by grey squirrels (Sciurus carolinensis). J Zool Lond Ser A 210(Pt3): 473-481

Kerr T, Bailey IW 1934 The cambium and its derivative tissues. No. X. Structure, optical properties and chemical composition of the so-called middle lamella. J Arnold Arbor 15: 327-349

Khan KK, Ahmad Z, Iqbal M 1981 Trends of ontogenetic size variation of cambial initials and their derivatives in the stem of Bauhinia parviflora Vahl. Bull Soc Bot Fr Lett Bot 128: 165-175

Khan MIH, Khan AH, Siddiqui TQ 1979a Ratio of ray and fusiform initials in some Eucalyptus species. Geobios 6: 280-281

Khan MIH, Khan S, Khan AH, Iqbal S 1979b A comparative study on the variation in cell size of wood fibres and their mother initials in some species of Eucalyptus. Aust For Res 9: 143-145

Khan MIH, Siddiqui TO, Khan AH 1983 Ontogenetic changes in the cambial structure of Citrus sinensis L. Flora 173: 151-158

Khan S, Khan MIH 1983 The extent of apical growth of the bast fibres over their mother initials in some trees of Myrtaceae. Bull Soc Bot Fr Lett Bot 130: 103-107

Kiatgrajai P, Conner AH, Rowe JW, Peters W, Roberts DR 1976 Attempt to induce light wood in balsam fir and tamarack by treating with Paraquat. Wood Sci 9: 31-36

Kidwai P, Robards AW 1969 On the ultrastructure of resting cambium of Fagus sylvatica L. Planta 89: 361-368

Kienholz R 1934 Leader, needle, cambial and root growth of certain conifers and their interrelations. Bot Gaz 96: 73-92

Kienitz M 1878 Ueber die Aufastung der Waldbäume. Forst Jagd 10(suppl): 58-82

Kienitz-Gerloff F 1891 Die Protoplasmaverbindungen zwischen benachbarten Gewebselementen in der Pflanze. Bot Ztg 49: 1-10, 17-26, 33-46, 49-60, 64-74

Kilman L 1958 Red squirrels feeding at sapsucker holes. J Mammal 39: 596-597

Kimmerer TW, Stringer MA 1988 Alcoholdehydrogenase and ethanol in the stems of trees. Evidence for anaerobic metabolism in the vascular cambium. Plant Physiol 87: 693-697

Kirschner H, Sachs T, Fahn A 1971 Secondary xylem reorientation as a special case of vascular tissue differentiation. Isr J Bot 20: 184-198

Kisser JG 1967 History of wood anatomy. Wood Sci Tech 1: 161-164

Klauditz W 1942 Über den Markstrahlanteil einiger Laubhölzer unter Berücksichtigung papiertechnischer Fragen. Holz Roh- Werkst 5: 145-148

Kleinmann A 1923 Ueber Kern- und Zellteilungen im Cambium. Bot Arch 4: 113-147

Klinken J 1914 Ueber das gleitende Wachstum der Initialen im Kambium der Koniferen und den Markstrahlenverlauf in ihrer secundären Rinde. Bibl Bot 19 (84): 1-37

Knudson L 1913 Observations on the inception, season, and duration of cambium development in the American larch [Larix laricinia (Du Roi) Koch.]. Bull Torrey Bot Club 40: 271-293

Knudson L 1916 Cambial activity in certain horticultural plants. Bull Torrey Bot Club 43: 533-537

Kny L 1877a Über das Dickenwachsthum des Holzkörpers an beblätterten Sprossen und Wurzeln und seine Abängigkeit von äusseren Einflüssen, insbesondere von Schwerkraft und Druck. Sitzungsber Ges Naturforsch Freunde Berlin 17: 23-50

Kny L 1877b Über künstliche Verdoppelung des Leitbündel Kreises im Stamme der Dicotyledonen. Sitzungsber Ges Naturforsch Freunde Berlin 17: 189-192

Kny L 1882 Ueber das Dickenwachsthum des Holzkoerpers in seiner Abhaengigkeit von aeusseren Einfluessen. Parey, Berlin, 136 pp

Kny L 1886 Ein Beitrag zur Entwickelungsgeschichte der "Tracheïden". Ber Deutsch Bot Ges 4: 267-276

Kny L 1890 Ein Beitrag zur Kenntniss der Markstrahlen dicotyler Holzgewächse. Ber Deutsch Bot Ges 8: 176-188

Kny L 1896 Ueber den Einfluss von Zug und Druck auf die Richtung der Scheidewände in sich theilenden Pflanzenzellen. Ber Deutsch Bot Ges 14: 378-391

Kny L 1902 Ueber den Einfluss von Zug und Druck auf die Richtung der Scheidewände in sich theilenden Pflanzenzellen. Jahrb Wiss Bot 37: 55-98

Ko JH, Morimoto K 1985 Loss of tree vigor and role of boring insects in red pine stands heavily infested by the pine needle gall midge in Korea. Esakia 23: 151-158 (Cited from For Abstr 47: 3038, 1986)

Koeppen M 1889 Ueber das Verhalten der Rinde unserer Laubbäume während der Thätigkeit des Verdickungsringes. Nova Acta Ksl Leop-Carol Deutsch Akad Naturforsch 53:441-496

Kohl EJ 1933 An explanation of the cause of spiral grain in trees. Science 78: 58-59

Kollmann R, Glockmann C 1991 Studies on graft unions III. On the mechanism of secondary formation of plasmodesmata at the graft interface. Protoplasma 165: 71-85

Kollmann R, Schumacher W 1961 Über die Feinstruktur des Phloems von Metasequoia glyptostroboides und seine jahreszeitlichen Veranderüngen. I. Das Ruhenphloem. Planta 57: 583-607

Koltzenburg C 1969 Der Einfluss von Lichtgenuss, soziologischer Stellung und des Standortes auf Holzeigenschaften der Rotbuche (Fagus sylvatica L.). Holz Roh- Werkst 25: 465-473

Korovin VV, Zuikhina SP 1985 (Some regularities in the structure of abnormal wood of maple, birch and alder). Biol Nauki 8: 68-73

Korstian CF 1921 Diameter growth in box elder and blue spruce. Bot Gaz 71: 454-461

Krabbe G 1882 Über die Beziehungen der Rindenspannung zur Bildung der Jahrringe und zur Ablenkung der Markstrahlen. Sitzungsber Kgl Preuss Akad Wiss Berlin 60: 1093-1143

Krabbe G 1884 Über das Wachsthum des Verdickungsringes und der jungen Holzzellen in seiner Abhängigkeit von Druckwirkungen. Abh Kgl Preuss Akad Wiss Berlin 1: 1-83

Krabbe G 1886 Das gleitende Wachsthum bei der Gewebebildung der Gefässpflanzen. Gebrüder Borntraeger, Berlin, 100 pp

Krawczyszyn J 1971 Unidirectional splitting and uniting of rays in the cambium of Platanus accompanying the formation of interlocked grain in wood. Acta Soc Bot Polon 40: 57-79

Krawczyszyn J 1972 Movement of the cambial domain pattern and mechanism of formation of interlocked grain in Platanus. Acta Soc Bot Pol 41: 443-461

Krawczyszyn J 1973 Domain pattern in the cambium of young Platanus stems. Acta Soc Bot Pol 62: 637-648

Krawczyszyn J 1977a Cambial domain pattern in the root and root collar of Platanus. Acta Soc Bot Pol 46: 531-541

Krawczyszyn J 1977b The transition from nonstoried to storied cambium in Fraxinus excelsior. I. The occurrence of radial anticlinal divisions. Can J Bot 55: 3034-3041

Krawczyszyn J, Romberger JA 1979 Cyclical cell length changes in wood in relation to storied structure and interlocked grain. Can J Bot 57: 787-794

Krawczyszyn J, Romberger JA 1980 Interlocked grain, cambial domains, endogenous rhythms, and time relations, with emphasis on Nyssa sylvatica. Am J Bot 67: 228-236

Kribs DA 1928 Length of tracheids in jack pine in relation to their position in the vertical and horizontal axes of the tree. Univ Minn Agr Exp Tech Bull 54, 14 pp

Kribs DA 1935 Salient lines of structural specialization in the wood rays of dicotyledons. Bot Gaz 96: 547-557

Kribs DA 1937 Salient lines of structural specialization in the wood parenchyma of dicotyledons. Bull Torrey Bot Club 64: 177-187

Kriedemann PE, Buttrose MS 1971 Chlorophyll content and photosynthetic activity within woody shoots of Vitis vinifera (L.). Photosynthetica 5: 22-27

Krishnamurthy KV, Venugopal N 1984 Studies on the successive cambia of a few plants. Indian J Bot 7: 183-189

Krüger F 1892 Ueber die Wandverdickungen der Cambiumzellen. Bot Ztg 50: 633-640, 649-657, 665-670, 681-688, 702-708

Kubella J 1985 Die jahreszeitliche Entwicklung des Sekundären xylems und phloems im Stamm und Wurzel von Quercus robur L. Thesis Univ Hamburg (Mimeo), 71 pp

Kubler H 1983 Mechanism of frost crack formation in trees - a review and synthesis. For Sci 29: 559-568

Kubler H 1990 Natural loosening of the wood/bark bond: a review and synthesis. For Prod J 40: 25-31

Kubler H 1991 Function of spiral grain in trees. Trees Struct Funct 5: 125-135

Kucera L 1968 Anatomische Studie über die Entwicklung und Verteilung von Markstrahlen bei der Tanne (Abies alba Mill.) II. Verteilung von Markstrahlen im jungen Tannenstamm. Drev Vysk 4: 167-184

Kucera LJ 1972 Wundgewebe in der Eibe (Taxus baccata L.). Vierteljahrsch Naturforsch Ges Zürich 116: 445-470

Kucera LJ 1978 Vascular nodules in the pith of yew (Taxus baccata L.). IAWA Bull 1978/4: 81-85

Kucera LJ 1985a On the morphology of the intercellular spaces in the wood rays. Proc Symp For Prod Res Intl - achievments and the future. Pretoria S Africa Apr 22-26, 1985 Vol 1 Pap 16, 15 pp

Kucera LJ 1985b Zur Morphologie der Interzellularen in den Markstrahlen. Teil 1-4. Vierteljahrsch Naturforsch Ges Zürich 130: 35-74, 157-198, 251-310, 374-397

Kucera L, Bariska M 1972 Einfluss der Dorsiventralität des Astes auf die Markstrahlbildung bei der Tanne (Abies alba Mill.). Vierteljahrsch Naturforsch Ges Zürich 117: 305-313

Kucera L, Bonsen KJM 1991 Markstrahlenkrankheit der Buche (Fagus sylvatica): anatomische Beschreibung und mögliche Entstehungsursache. Eur J For Pathol 21: 172-178

Kucera LJ, Bosshard HH 1975 The presence of biseriate rays in fir (Abies alba Mill.). IAWA Bull 1975/4: 51-56

Kucera L, Kucera J 1967 Anatomische Studie über die Entwicklung und Verteilung der Markstrahlen bei der Tanne (Abies alba Mill.) I. Charakteristik der Anfangsentwicklung des Markstrahles. Drev Vysk 12: 179-189

Kucera L, Necesany V 1970 The effect of dorsiventrality on the amount of wood rays in the branch of fir (Abies alba Mill.) and poplar (Populus monilifera Henry). Part I. Some wood ray characteristics. Drev Vysk 15: 1-6

Kucera L, Bosshard HH, Katz E 1980 Über den Keilwuchs und den welligen Jahrringverlauf in Buche (Fagus sylvatica L.). Holz Roh- Werkst 38: 161-168

Kucera LJ, Eichenberger B, Stoll A 1985 Das Blitzloch - Ursache und Entstehungsweise. In: Kucera LJ (ed) Xylorama. Trends in wood research. Birkhäuser, Basel, 127-138

Kuhla F 1900 Die Plasmaverbindungen bei Viscum album. Bot Ztg 58: 29-58

Kuijt J 1955 Dwarf mistletoes. Bot Rev 21: 569-626

Kuijt J 1969 The biology of parasitic flowering plants. University of Calif Press, Berkeley, 246 pp

Kujala V 1952 Eräistä ruskettumisiimiv1öistä elävien puiden rungoissa. (Über einige Verbräunungserscheinungen in lebenden Baumstämmen in Süd- und Mittel-finnland). Commun Inst For Fenn 40(14): 1-24

Kulman HM 1964 Defects in black cherry caused by bark beetles and agromizid cambium miners. For Sci 10: 258-266

Kurczynska E 1986 Terminal vessel and early vessel arrangement in internodes of Fraxinus excelsior. Acta Soc Bot Pol 55: 3-10

Kurczynska EU, Hejnowicz Z 1991 Differentiation of circular vessels in isolated segments of Fraxinus excelsior. Physiol Plant 83: 275-280

Kuroda K 1986 Wound effects on cytodifferentiation in the secondary xylem of woody plants. Wood Res 72: 67-118

Kuroda K 1989 Terpenoids causing tracheid-cavitation in Pinus thunbergii infected by the pine wood nematode (Bursaphelenchus xylophilus). Ann Phytopathol Soc Jpn 55: 170-178

Kuroda K 1991 Mechanism of cavitation development in the pine wilt disease. Eur J For Pathol 21: 82-89

Kuroda K, Shimaji K 1983 Traumatic resin canal formation as a marker of xylem growth. For Sci 29: 653-659

Kuroda K, Shimaji K 1984 Wound effects on xylem cell differentiation in a conifer. IAWA Bull 5: 295-305

Kuroda K, Shimaji K 1985 Wound effects on cytodifferentiation in hardwood xylem. IAWA Bull 6: 107-118

Kuroda K, Yamada T, Mineo K, Tamura H 1988 Effects of cavitation on the development of pine wilt disease caused by Bursaphelenchus xylophilus. Ann Phytopathol Soc Jpn 54: 606-615

Küster E 1899 Ueber Stammverwachsungen. Jahrb Wiss Bot 33: 487-512

Kutscha NP, Hyland F, Schwarzmann JM 1975 Certain seasonal changes in balsam fir cambium and its derivatives. Wood Sci Tech 9: 175-188

Labyak LF, Schumacher FX 1954 The contribution of its branches to the main-stem growth of loblolly pine. J For 52: 333-337

Lachaud S 1989 Participation of auxin and abscisic acid in regulation of seasonal variations in cambial activity and xylogenesis. Tree Struct Funct 3: 125-137

Ladefoged K 1952 The periodicity of wood formation. Dansk Biol Skr 7(3):1-98

Ladell JL 1963 Needle density, pith size and tracheid length in pine. Commonw For Inst Univ Oxford Inst Pap 36, 76 pp

Lakari OJ 1915 Studien über die Samenjahre und Altersklassenverhältnisse der Kiefernwälder auf den nordfinnischen Heideboden. Acta Forest Fenn 5: 1-211

Laming PB 1974 On intercellular spaces in the xylem ray parenchyma of Picea abies. Acta Bot Neerl 23: 217-233

Lang A 1965 Progressiveness and contagiousness in plant differentiation and development. Handb Pflanzenphysiol 15(1): 409-423

Lanner RM 1961 Living stumps in the Sierra Nevada. Ecology 42: 170-173

LaPasha CA, Wheeler EA 1990 Resin canals in Pinus taeda. Longitudinal canal lengths and interconnections between longitudinal and radial canals. IAWA Bull 11: 227-238

La Rivière HCC 1916 Sur l'anatomie et l'épaississement des tiges du Gnetum moluccense Karst. Ann Jard Bot Buitenzorg 30: 23-58

La Rivière HCC 1921 L'épaississement des tiges du Vitis lanceolaria Wall. Ann Jard Bot Buitenzorg 31: 141-166

Larson PR 1956 Discontinuous growth rings in suppressed slash pine. Trop Woods 104: 80-99

Larson PR 1963a The indirect effect of drought on tracheid diameter in red pine. For Sci 9: 52-62

Larson PR 1963b Microscopic wood characteristics and their variations with tree growth. Proc IUFRO Sect 41, Madison WI Sept 11-13,1963, 24 pp

Larson PR 1975 Development and organization of the primary vascular system in Populus deltoides according to phyllotaxy. Am J Bot 62: 1084-1099

Larson PR 1976a Procambium vs. cambium and protoxylem vs. metaxylem in Populus deltoides seedlings. Am J Bot 63: 1332-1348

Larson PR 1976b Development and organization of the secondary vessel system in Populus grandidentata. Am J Bot 63: 369-381

Larson PR 1979 Establishment of the vascular system in seedlings of Populus deltoides Bartr. Am J Bot 66: 452-462

Larson PR 1980a Interrelations between phyllotaxis, leaf development and the primary-secondary vascular transition in Populus deltoides. Ann Bot 46: 757-769

Larson PR 1980b Control of vascularization by developing leaves. In: Little CHA (ed) Control of shoot growth in trees. Proc IUFRO Workshop July 20-24, 1980 Fredericton NB, 157-172

Larson PR 1982 The concept of cambium. In: Baas (ed) New perspectives in wood anatomy. Nijhoff/Junk, The Hague, 85-92

Larson PR 1984 The role of subsidiary bundles in stem and leaf development of the Dicotyledoneae. In: White RA, Dickison WC (eds) Contemporary problems in plant anatomy. Academic Press, New York, 109-143

Larson PR, Fisher DG 1983 Xylary union between elongating lateral branches and the main stem in Populus deltoides. Can J Bot 61: 1040-1051

Larson PR, Isebrands JG 1974 Anatomy of the primary-secondary transition zone in stems of Populus deltoides. Wood Sci Tech 8: 11-26

Larson PR, Richards JH 1981 Lateral branch vascularization: its circularity and its relation to anisophylly. Can J Bot 59: 2577-2591

Lawton JR 1972 Seasonal variations in the secondary phloem of some forest trees from Nigeria II Structure of the phloem. New Phytol 71: 335-348

Lawton JR 1976 Seasonal variation in the secondary phloem from the main trunk of willow and sycamore trees. New Phytol 77: 761-777

Lawton JR, Lawton JRS 1971 Seasonal variations in the secondary phloem of some forest trees from Nigeria. New Phytol 70: 187-196

le compte de Tristan 1840 Études phytologiques, Premier Mémoire. De la nature des tissus végétaux. Ann Sci Nat Bot Ser 2, 14: 16-47

Lee HN, Smith EM 1916 Douglas fir fiber, with special reference to length. For Q 14: 671-695

Lee PW, Eom YG 1988 Anatomical comparison between compression wood and opposite wood in a branch of Korean pine (Pinus koraiensis). IAWA Bull 9: 275-284

Lekas TM, MacDougall RG, MacLean DA, Thompson RG 1990 Seasonal trends and effects of temperature and rainfall on stem electrical capitance of spruce and fir trees. Can J For Res 20: 970-977

Leney L, Moore LD 1977 Traumatic resin canals in western hemlock, Tsuga heterophylla (Raf.) Sarg. IAWA Bull 1977/2: 23-24

Lenz O 1967 Action de la neige et du gel sur les arbres de montagne, en particulier sur leur forme et l'anatomie de la tige. Schweiz Centralanst Forst Versuchswes 43: 293-316

Lescaffette J 1951 Une propriétéi des arbres et des peuplements-la surface génératrice. Bull Soc For Fr-Comte 25: 712-725 (Cited from For Abstr 13: 530, 1951)

Levin ED, Rubchevskaya LP 1985 Composition of triglyderides in the cambial zone of Siberian larch. Khim Drev 1: 104-109 (Cited from For Abstr 46: 5765, 1985)

Levin ED, Cherepanova VE, Zimovtseva IA, Sedlova TO 1988 Prostaglandins: identification and content in higher plants. Phytochemistry 27: 3241-3243

Lev-Yadun S, Aloni R 1990 Vascular differentiation in branch junctions of trees: circular patterns and functional significance. Trees Struct Funct 4: 49-54

Lev-Yadun S, Aloni R 1991a Polycentric vascular rays in Saueda monoica and the control of ray initiation and spacing. Trees Struct Funct 5: 22-29

Lev-Yadun S, Aloni R 1991b Natural and experimentally induced dispersion of aggregate rays in shoots of Quercus ithaburensis Decne. and Q. calliprinos Webb. Ann Bot 68: 85-91

Lev-Yadun S, Aloni R 1991c An experimental method of inducing 'Hazel' wood in Pinus halepensis (Pinaceae). IAWA Bull 12: 445-451

Lev-Yadun S, Aloni R 1992a Experimental induction of dilatation meristems in Melia azedarach L. Ann Bot 70: 379-386

Lev-Yadun S, Aloni R 1992b The role of wounding and partial girdling in differentiation of vascular rays. Int J Plant Sci 153: 348-357

Lewis FT 1935 The shape of the tracheid in the pine. Am J Bot 22: 741-762

Li Z, Lin J 1991 Wood anatomy of the stalactite-like branches of Ginkgo. IAWA Bull 12: 251-255

Liang SC 1948 Variation in tracheid length from the pith outwards in the wood of the genus Larix, with a note on variation in other anatomical features. Forestry 22: 222-237

Liese W, Ammer U 1958 Untersuchungen über die Länge der Holzfaser bei der Pappel. Holzforschung 11: 169-174

Lindberg M, Johansson M 1989 The use of electrical resistance of cambium and phloem as a measure of tree vigor. Scand J For Res 4: 175-185

Linnemann G 1953 Untersuchungen über den Markstrahlanteil am Holz der Buche. Ber Deutsch Bot Ges 66: 37-63

Liphschitz N, Mendel Z 1987 Histological studies of Pinus halepensis stem xylem affected by Matsucoccus josephi (Homoptera: Margarodidae). IAWA Bull 8: 369-378

Lipschitz N, Lev-Yadun S, Rosen E, Waisel Y 1984 The annual rhythm of activity of the lateral meristems (cambial and phellogen) in Pinus halepensis Mill. and Pinus pinea L. IAWA Bull 5: 263-274

Liphschitz N, Bonneh O, Mendel Z 1987 Living stumps - circumstantial evidence for root grafting in Pinus halepensis and P. brutia plantations in Israel. Isr J Bot 36: 41-43

Little JR, Jones CE 1980 A dictionary of botany. Van Nostrand Reinhold, New York, 400 pp

Livingston LG 1935 The nature and distribution of plasmodesmata in the tobacco plant. Am J Bot 22: 75-87

Livingston LG 1964 The nature of plasmodesmata in normal (living) plant tissue. Am J Bot 51: 950-957

Lloyd CW 1987 The plant cytoskeleton: the impact of fluorescence microscopy. Ann Rev Plant Physiol 38: 119-139

Lo HC 1985 (Anatomical studies on the trunk of Luan-ta fir and peacock pine with special reference to the wound healing). Mem Coll Agr Natl Taiwan Univ 25(2): 121-123

Lo HC 1987 (Traumatic resin ducts in European spruces [Picea abies (L.) Karst.]). Mem Coll Agr Natl Taiwan Univ 27(2): 9-28

Lo HC 1988 Development of new vascular cambium from trunk callus in Chamaecyparis formosensis Mats. Mem Coll Agr Natl Taiwan Univ 28(1): 54-67

Lo HC 1989 (Dynamics of cambium and pattern of wood grain in camphor tree (Cinnamomum camphora (L.) Sieb.). Mem Coll Agr Natl Taiwan Univ 29(1): 120-140

Lodewick JE 1925 Growth studies in forest trees. III. Experiments with the dendrograph on Fraxinus americana. Bot Gaz 89: 311-323

Lodewick JE 1928 Seasonal activity of the cambium in some northeastern trees. NY State Coll For Syracuse Tech Bull 23, 87 pp

Löhr E 1969 Jahresringverlust bei Laubbäumen mit ringporigem Holz. Allg Forst-Jadgztg 140: 18-20

Lorch J 1967 The elusive cambium. Arch Int Hist Sci 80: 253-283

Lowerts G, Wheeler EA, Kellison RC 1986 Characteristics of wound-associated wood of yellow-poplar (Liriodendron tulipifera L.). Wood Fiber Sci 18: 537-552

Lundegårdh H 1922 Übersicht über die Geschichte der Pflanzenanatomie und der Zellenlehre. Handb Pflanzanat I Abt 1 Teil, 1:1-62

Lutz HJ 1943 Injuries to trees caused by Celastrus and Vitis. Bull Torrey Bot Club 70: 436-439

Luxová M 1985 Influence of the stem root primordia on the cambial activity in poplar stem. Acta Univ Agric Brno 33: 633-636

Luxová M, Lux A 1981 Latent root primordia in poplar stems. Biol Plant 23: 285-290

Lynn M 1967 Ionizing radiations in forests and forestry (Excluding the use of radio-active tracers). For Abstr 28: 1-18

MacDaniels LH 1923 The apple-tree crotch. Histological studies and practical considerations. Cornell Univ Agr Exp Stn Bull 419, 21 pp

MacDaniels LH, Curtis OF 1930 The effect of spiral ringing on solute translocation and the structure of the regenerated tissues of the apple. Cornell Univ Agr Exp Stn Mem 133, 31 pp

MacKenzie KAD, Howard BH, Harrison-Murray RS 1986 The anatomical relationship between cambial regeneration and root initiation in wounded winter cuttings of the apple rootstock M. 26. Ann Bot 58: 649-661

Madden JL 1977 Direct evaluation of cambial activity in trees. Aust For Res 7: 145-150

Mahmood A 1968 Cell grouping and primary wall generations in the cambial zone, xylem, and phloem in Pinus. Aust J Bot 16: 177-195

Mahmood A 1971 Numbers of initial-cell divisions as a measure of activity in the yearly cambial growth pattern in Pinus. Pakistan J For 21: 27-42

Mahmood A 1990 The parental cell walls. In: Iqbal M (ed) The vascular cambium. Research Studies Press, Taunton, England, 113-126

Mahmooduzzafar (-), Iqbal M 1986 Variation in size and amount of wood elements across and within the growth rings in Terminalia tomentosa. Flora 178: 191-196

Majumdar GP 1941 The sliding, gliding, symplastic or the intrusive growth of the cambium cells and their derivatives in higher vascular plants. J Indian Bot Soc 20: 161-171

Majumdar GP 1942 The organization of the shoot apex in Heracleum in the light of development. Ann Bot 6: 49-81

Makino R, Kuroda H, Shimaji K 1983 Callus formation, and effects of applied pressure to the cultured cambial explant of Sugi (Cryptomeria japonica D. Don.). Wood Res 69: 1-11

Makkonen O 1967 Ancient forestry. An historical study. I. Facts and information on trees. Acta For Fenn 82: 1-84

Mamiya Y, Kiyohara T 1972 Description of Bursaphelenchus lignicolus N. sp. (Nematoda: Aphelenchoididae) from pine wood and histopathology of nematode-infested trees. Nematologica 18: 120-124

Manchester SR 1980 Chattawaya (Sterculiaceae); a new genus of wood from the Eocene of Oregon and its implications for xylem evolution of the extant genus Pterospermum. Am J Bot 67: 59-67

Manchester SR, Miller RB 1978 Tile cells and their occurrence in Malvalean fossil woods. IAWA Bull 1978/2-3: 23-28

Mann WF Jr, Musselman LJ 1979 Native root parasites and forestry in southern United States: A review. In: Musselman LJ, Worsham AD, Eplee RE (eds) Proc 2nd Intl Symp Parasitic Weeds. N C State University, Raleigh, 24-33

Marchand PJ 1984 Dendrochronology of a fir wave. Can J For Res 14: 51-56

Margaris NS, Papadogianni P 1977 The ratio of ray and fusiform initials in some plants dominating Mediterranean formations in Greece. Flora 166: 219-222

Mariaux A 1967 Les cernes dans les bois tropicaux africains, nature et périodicité. Bois For Trop 113: 3-14; 114: 23-37

Mariaux A 1969 La périodicité des cernes dans le bois de Limba. Bois For Trop 128: 39-53

Marshall JD, Ehleringer JR 1990 Are xylem-tapping mistletoes partially heterotrophic? Oecologia 84: 244-248

Martin JM, Sydnor TD 1987 Differences in wound closure rates in twelve tree species. HortScience 22: 442-444

Matovic A 1980 Cambium activity and xylem differentiation in Fraxinus angustifolia Vahl. ssp. pannonica Soo et Simon. Drev Vysk 25: 25-46

Mattheck C, Huber-Betzer H, Keilen K 1990 Die Kerbspannungen am Astloch als Stimulanz der Wundheilung bei Bäumen. Allg Forst-Jagdztg 161: 47-53

Mäule C 1896 Der Faserverlauf im Wundholz. Eine anatomische Untersuchung. Bibl Bot 33: 1-32

Mayer-Wegelin H 1952 Das Aufästen der Waldbäume. M & H Schaper, Hannover, 92 pp

Mayr H 1884 Entstehung und Verteilung der Secretions-Organe der Fichte und Lärche. Bot Cbl 20: 23-25, 53-56, 86-91, 117-121, 148-153, 183-190, 213-219, 246-253, 278-283, 308-310

Mayr H 1893 Das Harz der Nadelhölzer. II. Verteilung des Harzes. Z Forst-Jagdwes 25: 389-417

McGinnes EA Jr, Phelps JE 1972 Intercellular spaces in eastern red cedar (Juniperus virginiana L.). Wood Sci 4: 225-229

Meeuse ADJ 1941 Plasmodesmata. Bot Rev 7: 249-262

Meeuse ADJ 1942 A study of intercellular relationships among vegetable cells with special reference to "sliding growth" and to cell shape. Rec Trav Bot Neerl 38: 18-140

Megraw RA 1985 Wood quality factors in loblolly pine. Tappi, Atlanta, 88 pp

Meier BA 1973 Über Kambiumtätigkeit und Jahrringentwicklung in Picea abies Karst., Larix decidua Mill. und Pinus silvestris L. an der oberen alpinen Baumgrenze. Vierteljahrschr Naturforsch Ges Zürich 118: 153-191

Mellerowicz EJ, Riding RT 1992 Does DNA endoreduplication occur during differentiation of secondary xylem and phloem in Abies balsamea? Int J Plant Sci 153: 26-30
Mellerowicz EJ, Riding RT, Little CHA 1989 Genomic variability in the vascular cambium of Abies balsamea. Can J Bot 67: 990-996
Mellerowicz EJ, Riding RT, Little CHA 1990 Nuclear size and shape changes in fusiform cambial cells of Abies balsamea during the annual cycle of activity and dormancy. Can J Bot 68: 1857-1863
Mellerowicz EJ, Coleman WK, Riding RT, Little CHA 1992a Periodicity of cambial activity in Abies balsamea. I. Effects of temperature and photoperiod on cambial dormancy and frost hardiness. Physiol Plant 85: 515-525
Mellerowicz EJ, Riding RT, Little CHA 1992b Periodicity of cambial activity in Abies balsamea. II. Effects of temperature and photoperiod on the size of the nuclear genome in fusiform cambial cells. Physiol Plant 85: 526-530
Mellerowicz EJ, Riding RT, Little CHA 1992c Cambial activity and dormancy in Abies balsamea is correlated with changes in nucleolar size and number. Am J Bot 79(Suppl): 41
Mendelsohn E 1963 Cell theory and the development of general physiology. Arch Int Hist Sci 16: 419-429
Menon ARS, Babu AM 1989 Structure and development of ethephon induced gum cavities in the stem of Sterculia urens Roxb. Phyton 29: 41-47
Mergen F 1954 Anatomical study of slash pine graft unions. Q J Fla Acad Sci 17: 237-245
Mergen F, Echols RM 1955 Number and size of radial resin ducts in slash pine. Science 121(3139): 306-307
Messer AC 1990 Traditional and chemical techniques for stimulation of Shorea javanica (Dipterocarpaceae) resin exudation in Sumatra. Econ Bot 44: 463-469
Metcalfe CR, Chalk L 1950 Anatomy of the dicotyledons. Leaves, stem, and wood in relation to taxonomy with notes on economic uses. Vol I & II 1st edn Clarendon, Oxford, 1500 pp
Metcalfe CR, Chalk L 1979 Anatomy of the dicotyledons. Systematic anatomy of the leaf and stem, with a brief history of the subject. Vol I 2nd edn Oxford University Press, 276 pp
Metcalfe CR, Chalk L 1983 Anatomy of the dicotyledons. Wood structure and conclusions of the general introduction. Vol II 2nd edn Oxford University Press, 297 pp
Meyer F 1961 Das Altern der Bäume. Mitt Deutsch Dendrol Ges 62: 59-65
Meyer KA 1949 Sprachliche und literarische Bemerkungen zum Problem <Drehwuchs>. Mitt Schweiz Centralanst Forst Versuchswes 26: 331-347
Meyer RE, Morton HL, Haas RH, Robison ED, Riley TE 1971 Morphology and anatomy of honey mesquite. USDA Tech Bull 1423, 186 pp
Meyer-Uhlenreid KH 1958 Holzanatomische Untersuchungen an der Pappel. Holzforschung 11: 150-157
Meylan BA, Butterfield BG 1973 A trabecula with a vestured pit. IAWA Bull 1973/3: 12-14
Mia AJ 1970 Fine structure of active, dormant, and aging cambial cells in Tilia americana. Wood Sci 3: 34-42
Michael ED 1987 Bark stripping by white-tailed deer in West Virginia. North J Appl For 4: 96-97
Middleton TM 1987 Aggregate rays in New Zealand Nothofagus blume (Fagacea) stem wood and their influence on vessel distribution. IAWA Bull 8: 53-57

Milanez FR 1938 Observacães sobre as celulas do lenho. (Observations on xylem cells). An Primeira Reuniaio Sul-Am Bot 3: 189-221
Miller JE 1983 Beavers. In: Timm RM (ed) Prevention and control of wildlife damage. Great Plains Agr Council, University of Nebraska, Lincoln, pp B1-B11
Millington WF, Chaney WR 1973 Shedding of shoots and branches. In: Kozlowski TT (ed) Shedding of plant parts. Academic Press, New York, 149-204
Mio S, Matsumoto T 1979a (Morpholgoical observations on intercellular spaces in wood ray). Bull Kyushu Univ For 51: 1-10
Mio S, Matsumoto T 1979b (Morphological observations on longitudinal intercellular spaces in normal softwoods). Bull Kyushu Univ For 51: 13-18
Mio S, Matsumoto T 1981 (Intercellular spaces in the axial parenchyma of hardwoods). Mokuzai Gakkaishi 27: 626-632
Mio S, Matsumoto T 1982 A note on parent cell walls in coniferous woods. IAWA Bull 3: 56-58
Mischke K 1890 Beobachtungen über das Dickenwachsthum der Coniferen. Bot Cbl 44: 39-43, 65-71, 97-102, 137-142, 169-175
Mitchell RG 1967 Abnormal ray tissue in three true firs infested by the balsam wooly aphid. For Sci 13: 327-332
Mix AJ 1916 The formation of parenchyma wood following winter injury to the cambium. Phytopathology 6: 279-283
Möbius M 1934 Über die Herkunft der Wörter Cambium und Protoplasma. Ber Deutsch Bot Ges 52: 154-161
Monange Y 1961 Le froid de l'hiver 1956 et l'anatomie du bois de quelques Gymnosperms. Bull Soc Hist Nat Toulouse 96: 175-180
Moorby J, Wareing PF 1963 Ageing in woody plants. Ann Bot 27: 291-308
Moore KE 1978 Barrier-zone formation in wounded stems of sweetgum. Can J For Res 8: 389-397
Mork E 1928 Die Qualität des Fichtenholzes unter besonderer Rücksichtnahme auf Schlief- und Papierholz. Papier-Fabr 26: 741-747
Morton AG 1981 History of botanical science. Academic Press, New York, 474 pp
Mühlethaler K 1950 Elektronenmikroskopische Untersuchungen über den Feinbau und das Wachstum der Zellmembranen in Mais- und Haferkoleoptilen. Ber Schweiz Bot Ges 60: 614-628
Mulhern J, Shortle WC, Shigo AL 1979 Barrier zones in red maple: an optical and scanning microscope examination. For Sci 25: 311-316
Müller C 1890 Ueber die Balken in den Holzelementen der Coniferen. Ber Deutsch Bot Ges 8: 17-46
Müller NJC 1867 Untersuchungen über die Vertheilung des Harze, ätherischen Oele, Gummi und Gummiharze, und die Stellung der secretionsbehälter im Pflanzenkörper. Jahrb Wiss Bot 5: 387-421
Müller NJC 1875-76 Über die Vertheilung der Molekularkräfte im Baume. 2. Der sogenannte absteigende Saftstrom. Bot Untersuch 1(IV): 85-300
Müller-Stoll WR 1965 Über intrazelluläre Stabbildungen (Trabeculae) im Holz als anatomische Eigenart bei Gehölzen exponierter Gerbirgslagen. Kulturpflanzen 13: 763-799
Müller-Stoll WR, Casperson G 1988 Elektronenmikroskopische Untersuchungen intrazellulärer Stäbe (Trabeculae) im Holz von Larix und Juniperus. Kulturpflanze 35: 441-452
Mullick DB 1977 The non-specific nature of defense in bark and wood during wounding, insect and pathogen attack. Rec Adv Phytochem 11: 395-441

Murmanis L 1970 Locating the initial in the vascular cambium of Pinus strobus L. by electron microscopy. Wood Sci Tech 4: 1-14
Murmanis L 1971 Structural changes in the vascular cambium of Pinus strobus L. during an annual cycle. Ann Bot 35: 133-141
Murmanis L 1977 Development of vascular cambium into secondary tissue of Quercus rubra L. Ann Bot 41: 617-620
Murmanis L, Sachs IB 1969 Seasonal development of secondary xylem in Pinus strobus L. Wood Sci Tech 3: 177-193
Myer JE 1922 Ray volumes of the commercial woods of the United States and their significance. J For 20: 337-351
Myers RF 1986 Cambium destruction in conifers caused by pinewood nematodes. J Nematol 18: 398-402
Nabuchi T, Takahara S, Harada H 1979 (Studies on the survival rate of ray parenchyma cells with ageing process in coniferous secondary xylem). Bull Kyoto Univ For 51: 239-246
Nägeli C 1858 Das Wachsthum der Stammes und der Wurzel bei den Gefässpflanzen und die Anordung der Gefässstränge in Stengel. Beitr Wiss Bot 1: 1-156
Nägeli C 1864 Dickenwachsthum des Stengels und Anordnung der Gefässstränge bei den Sapindaceen. Beitr Wiss Bot 4: 1-72
Nägeli W 1935 Aussetzende und auskeilende Jahrringe. Schweiz Z Forst 86: 209-215
Nair MNB, Shah JJ, Subramanyam SV 1983 Ultrastructure and histochemistry of traumatic gum ducts in the wood of Azadirachta indica A. Juss. IAWA Bull 4: 103-112
Nanko H, Saiki H, Harada H 1977 Development and structure of the phloem fibers in the secondary phloem of Populus euramericana. Jpn Pap Wood Res Soc 23: 267-272
Nathansohn A 1898 Beiträge zur Kenntniss des Wachsthums der trachealen Elemente. Jahrb Wiss Bot 32: 671-686
Neeff F 1914 Über Zellumlagerung. Ein Beitrag zur experimentellen Anatomie. Z Bot 6: 465-547
Neeff F 1920 Über die Umlagerung der Kambiumzellen beim Dickenwachstum der Dikotylen. Z Bot 12: 225-252
Neeff F 1922 Über polares Wachstum von Pflanzenzellen. Jahrb Wiss Bot 61: 205-283
Neely D 1970 Healing of wounds on trees. J Am Soc Hortic Sci 95: 536-540
Neely D 1988 Wound closure rates on trees. J Arbor 14: 250-254
Neger FW 1916 Über eine durch Frühfrost an Nectria cucurbitula Fr. und Dermatea encrita (Karst.) verursachte Gipfeldürre der Fichte. Nat Z Forst-Landw 14: 121-127
Neubauer HF 1969 Bemerkungen zum Stammbau einiger Pelargonien. Gartenbauwissenschaft 34: 365-374
Newman IV 1956 Pattern in meristems of vascular plants. 1. Cell partition in the living apices and in the cambial zone in relation to the concepts of initial cells and apical cells. Phytomorphology 6: 1-19
Newman IV 1961 II. A review of shoot apical meristems of gymnosperms, with comments on apical biology and taxonomy, and a statement of some fundamental concepts. Proc Linn Soc N S Wales 86: 9-59
Newman IV 1965 III. Pursuing the patterns in the apical meristem where no cell is a permanent cell. J Linn Soc 59: 185-214
Nicoloff T 1911 Contribution a l'histologie et a la physiologie des rayons médullaires chez les dicotylédones arborescentes. Rev Gen Bot 23: 369-403

Nix LE, Villiers K 1985 Tracheid differentiation in southern pines during the dormant season. Wood Fiber Sci 17: 397-403

Nobuchi T, Fujisawa T, Saiki H 1989 Marking of the differentiating zone by the pinning method and some characteristics of annual ring formation in sugi (Cryptomeria japonica D. Don). IAWA Bull 10: 341-342

Noel ARA 1968 Callus formation and differentiation at an exposed cambial surface. Ann Bot 32: 347-359

Noel ARA 1970 The girdled tree. Bot Rev 36: 162-195

Nogler P 1981 Auskeilende und fehlende Jahrringe in absterbenden Tannen (Abies alba Mill.). Allg Forstztg 28: 709-711

Nordhausen M 1898 Zur Kenntniss der Wachstumsvorgänge im Verdickungsringe der Dikotylen. Beitr Wiss Bot 2: 356-400

Nördlinger H 1871 Der Holzring als Grundlage des Baumkörpers. J G Gotta'schen, Stuttgart, 47 pp

Noskowiak AF 1978 Distribution of aggregate rays in red alder. Wood Fiber 10: 58-68

Nwankiti OC 1982 Phytoferritin in the cambium of Bougainvillea spectabilis Willd. (Nyctaginaceae). Z Pflanzenphysiol 105: 173-178

Nylinder P 1951 Om patologiska hartskanaler (On pathological resin canals). Medd Statens Skogsförsöksanst 40(7): 1-12

Nyrén V, Back E 1959 The ray volume in Pinus silvestris stemwood and its distribution between tracheidal and parenchymatous ray cells. Svensk Papperstidn 62: 681-686

Nyrén V, Back E 1960a Characteristics of parenchymatous cells and tracheidal ray cells in Picea abies Karst. Svensk Papperstidn 63: 501-509

Nyrén V, Back E 1960b The dimensions of tracheidal and parenchymatous ray cells in Picea abies (Karst.) pulpwood. Svensk Papperstidn 63: 619-624

O'Brien JM 1983 Voles. In: Timm RM (ed) Prevention and control of wildlife damage. Great Plains Agr Council, University of Nebraska, Lincoln, pp B147-B156

O'Brien TP 1974 Primary vascular tissues. In: Robards AW (ed) Dynamic aspects of plant ultrastructure. McGraw Hill, New York, 414-440

Ohman JH, Kessler KJ Jr 1964 Black bark as an indicator of bird peck defect in sugar maple. U S For Serv Res Pap LS-14, 8 pp

Ohtani J 1977 An observation of the trabeculae in some dicotyledonous woods using scanning electron microscopy. Res Bull Coll Exp For Hokkaido Univ 34: 69-78

Ohtani J 1985 SEM observations on trabeculae in Abies sachalinensis. IAWA Bull 6: 43-51

Ohtani J, Fukazawa K, Fukumorita T 1987 SEM observations on indented rings. IAWA Bull 8: 113-124

Olesen PO 1982 The effect of cyclophysis on tracheid width and basic density in Norway spruce. Arboretet Hørsholm, København For Tree Improv 15, 80 pp

Oliver WW 1968 Sapsucker damage to ponderosa pine. J For 66: 842-844

Oliver WW 1970 The feeding pattern of sapsuckers on ponderosa pine in northeastern California. Condor 72: 241

Ollinmaa PJ 1956 Koivun vetopuum anatomisesta rakenteesta ja ominaisuuksista. (On the anatomic structure and properties of the tension wood in birch). Acta For Fenn 64(3):1-263

O'Neil LC 1963 The suppression of growth rings in jack pine in relation to defoliation by the Swaine jack-pine sawfly. Can J Bot 41: 227-235

Ozenda P 1965 Recherches sur les phanérogames parasites. I. Revue des travaux récents. Phytomorphology 15: 311-338

Paclt J 1989 Mondringe: für eine einheitliche Erfassung ihrer Entstehungsjahre. Beitr Biol Pflanz 64: 17-22
Paliwal GS, Prasad NVSRK 1970 Seasonal activity of cambium in some tropical trees I. Dalbergia sissoo. Phytomorphology 20: 333-339
Paliwal GS, Prasad NVSRK 1971 Seasonal variations in the size of cambial initials in Dalbergia sissoo. Curr Sci 40: 114-115
Paliwal GS, Srivastava LM 1969 The cambium of Aleuosmia. Phytomorphology 19: 5-8
Paliwal GS, Prasad N, Sajwan VS, Aggarwal SK 1975 Seasonal activity of cambium in some tropical trees II Polyalthia longifolia. Phytomorphology 25: 478-484
Paliwal GS, Sajwan S, Aggarwal SK 1976 Seasonal activity of cambium in some tropical trees III Salvadora persica L. Acta Soc Bot Pol 45: 303-312
Paliwal SP, Singh DP, Badola HK 1984 Structure and ratio of fusiform and ray initials in the cambium of some Himalayan and Sub-Himalayan trees. In: Paliwal GS (ed) The vegetational wealth of the Himalayas. Puja Delhi, 301-310
Panshin AJ, de Zeeuw C 1980 Textbook of wood technology. 4th edn McGraw Hill, New York, 722 pp
Panshin AJ, de Zeeuw C, Brown HP 1964 Textbook of wood technology. Vol I. Structure, identification, uses, and properties of the commercial woods of the United States. McGraw Hill, New York, 643 pp
Parameswaran N 1964 Über die Länge der septierten Fasern bei Tectona grandis. Naturwissenschaften 51: 317-318
Parameswaran N, Liese W 1969 On the formation and fine structure of septate wood fibres of Ribes sanguineum. Wood Sci Tech 3: 372-386
Parameswaran N, Liese W 1974 Variation of cell length in bark and wood of tropical trees. Wood Sci Tech 8: 81-90
Parker J 1963 Cold resistance in woody plants. Bot Rev 29: 123-201
Parthasarathy MV, Pesacreta TC 1980 Microfilaments in plant vascular cells. Can J Bot 58: 807-815
Patel JD 1975 Occurrence of multinucleate fusiform initials in Solanum melongena L. Curr Sci 44: 516-517
Patel RN 1974 Wood anatomy of the dicotyledons indigenous to New Zealand. 4. Winteraceae. N Z J Bot 12: 19-32
Pauly G 1963 Les canaux sécréteurs du pin maritime. Univ Toulouse Lab For Trav 1(6):1-33
Payen A 1839 Mémoire sur la nutrition des plantes. C R Acad Sci 9: 509
Payne TL (No date) Life history and habits. In: Thatcher RC, Searcy JL, Coster JE, Hertel GD (eds) The southern pine beetle. U S For Serv Tech Bull 1631, 266 pp
Peace TR 1940 An interesting case of lightning damage to a group of trees. Q J For 34: 61-63
Pearce J 1947 Identifying injury by wildlife to trees and shrubs in northeastern forests. U S Fish Wildlife Serv Res Rep 13, 29 pp
Pedigo RA 1963 Effects of ionizing radiation on Pinus taeda L. In: Schultz V, Klement AW Jr (eds) Radioecology. Reinhold, New York, 295-299
Penhallow DP 1907 A manual of the North American gymnosperms. Ginn, Boston, 374 pp
Perrot E 1898 Anatomie comparée des gentianacées. Ann Sci Nat Bot Ser 8, 7: 105-292
Pessin LJ 1933 How old is a longleaf pine? U S For Serv For Work 9(1): 12-13
Peters WJ 1970 Mitosis in 11-year-old xylem of pine. Am Midl Nat 84: 255-256

Peterson RS 1961 Conifer tumors in the Central Rocky Mountains. Plant Dis Rep 45: 472-474

Petric B, Scukanec V 1973 Volume percentage of tissues in wood of conifers. IAWA Bull 1973/2: 3-7

Petric B, Scukanec V 1975 Ray tissue percentages in wood of Yugoslavian hardwoods. IAWA Bull 1975/3: 43-44

Pfeiffer H 1926 Das abnorme Dickenwachstum. Handb Pflanzenanat Bd IX Berlin, 272 pp

Philipson WR 1964 Morphogenesis as affected by associated organisms. Phytomorphology 14: 163-185

Philipson WR 1990 Anomalous cambia. In: Iqbal M (ed) The vascular cambium. Research Studies Press, Taunton, England, 210-212

Philipson WR, Butterfield BG 1967 A theory on the causes of size variation in wood elements. Phytomorphology 17: 155-159

Philipson WR, Ward JM 1965 The ontogeny of the vascular cambium in the stem of seed plants. Biol Rev 40: 534-579

Philipson WR, Ward JM, Butterfield BG 1971 The vascular cambium: its development and activity. Chapman & Hall, London, 182 pp

Phillips IDJ 1976 The cambium. In: Yeoman MM (ed) Cell division in higher plants. Academic Press, New York, 347-390

Phillips R 1980 Cytodifferentiation. In: Perspectives in plant cell and tissue culture. Int Rev Cytol 11A(suppl):55-70

Pisek A, Eggarter H 1959 Beobachtungen zur Überwindung von Frostschäden an Zweigen und Laubknospen von Apfel und Birne. Gartenbauwissenschaft 24: 446-456

Pomeroy MK, Siminovitch D 1971 Season cytological changes in secondary phloem parenchyma cells in Robinia pseudoacacia in relation to cold hardiness. Can J Bot 49: 787-795

Prat H 1948 Histo-physiological gradients and plant organogenesis. Bot Rev 14: 603-643

Preusser HJ, Dietrichs HH, Gottwald H 1961 Elektronenmikroskopische Untersuchungen an Ultradünnschnitten des Markstrahlparenchyms der Rotbuche-Fagus sylvatica L. Holzforschung 15: 65-75

Priestley JH 1930a Studies in the physiology of cambial activity I. Contrasted types of cambial activity. New Phytol 29: 56-73

Priestley JH 1930b II. The concept of sliding growth. New Phytol 29: 96-104

Priestley JH 1930c III. The seasonal activity of the cambium. New Phytol 29: 316-354

Priestley JH, Scott LI 1939 The formation of a new cell wall at cell division. Proc Leeds Philos Soc 3: 532-545

Priestley JH, Scott LI, Malins ME 1933 A new method of studying cambial activity. Proc Leeds Philos Lit Soc Sect 2 Pt 8: 365-374

Prince JB 1934 Pathological enlargement of resin canals in Picea. For Chron 10: 129-136

Pulawska Z 1972 General and peculiar features of vascular organization and development in shoots of Bougainvillaea glabra Choisy (Nyctaginaceae). Acta Soc Bot Pol 41: 39-70

Pulawska Z 1973 The parenchymo-vascular cambium and its derivative tissues in stems and roots of Bougainvillaea glabra Choisy (Nyctaginaceae). Acta Soc Bot Pol 42: 41-61

Pulawska Z 1982 Tissues development in stems of Aristolochia clemantitis L. in the point of view of multicellular complexes formation. Acta Soc Bot Pol 51: 107-125

Purkayastha SK 1958 Growth and development of septate and crystalliferous fibres in some Indian trees. Proc Natl Inst Sci India B Biol Sci: 24:239-244

Pyszynski W 1972 Downward movement of the domain pattern in Aesculus cambium producing wavy-grained xylem. Acta Soc Bot Pol 41: 511-517

Pyszynski W 1977a Complex wavy grain in the stem of Aesculus. Acta Soc Bot Pol 46: 231-249

Pyszynski W 1977b Mechanism of formation of spiral grain in Aesculus stems: dissymmetry of deformation of stems caused by cyclic torsion. Acta Soc Bot Pol 46: 501-522

Pyszynski W 1980 Pattern of ray arrangement on cross section of bark of Aesculus. Acta Soc Bot Pol 49: 415-422

Pyszynski W 1990 The relationship between the arrangement of rays in secondary phloem and grain inclination in the wood of Pinus silvestris L. and Picea abies Karst. Acta Soc Bot Pol 59: 9-24

Raatz W 1892 Die Stabbildungen im secundären Holzkörper der Bäume und die Initialentheorie. Jahrb Wiss Bot 23: 567-636

Radlkofer L 1858 Ueber das anomale Wachsthum des Stammes bei Menispermeen. Flora 41: 193-206. Also, same title and date: Sur l'accroissement anormal de la tige dans les ménispermées. Ann Sci Nat Bot Ser 4, 10: 164-180

Rao AN 1966 Developmental anatomy of natural root grafts in Ficus globosa. Aust J Bot 14: 269-276

Rao KS 1985 Seasonal ultrastructural changes in the cambium of Aesculus hippocastanum L. Ann Sci Nat Bot Ser 13, 7: 213-228

Rao KS 1988a Cambial activity and developmental changes in ray initials of some tropical trees. Flora 181: 425-434

Rao KS 1988b Fine structural details of tannin accumulations in non-dividing cambial cells. Ann Bot 62: 575-581

Rao KS, Catesson AM 1987 Changes in the membrane components of nondividing cambial cells. Can J Bot 65: 246-254

Rao KS, Dave YS 1981 Seasonal variations in the cambial anatomy of Tectona grandis (Verbenaceae). Nord J Bot 1: 535-542

Rao KS, Dave YS 1983a Ultrastructure of dormant cambium in Holoptelea integrifolia (Roxb.) Planch. Flora 174: 165-172

Rao KS, Dave YS 1983b Seasonal histochemical changes in the cambium of Tectona grandis Lif. and Gmelina arborea Roxb. Biol Plant 25: 241-245

Rao KS, Dave YS 1983c Ultrastructure of active and dormant cambial cells in teak (Tectona grandis L.f). New Phytol 93: 447-456

Rao KS, Dave YS 1984 Occurrence of crystals in vascular cambium. Protoplasma 119: 219-221

Rao KS, Dave YS 1985a Developmental changes in the fusiform cambial cells of some tropical trees. Flora 177: 187-194

Rao KS, Dave YS 1985b Seasonal variations in the vascular cambium of Holoptelea integrifolia (Ulmaceae). Beitr Biol Pflanz 59: 321-331

Rao KS, Menon ARS 1989 Histochemistry of cambium in Holoptelea integrifolia (Roxb.) Planch. and Mangifera indica L.: its seasonal changes. Pak J Bot 21: 81-87

Rao KS, Menon ARS, Dave Y 1989 Developmental and dimensional changes in the cambium fusiform initials of Holoptelea integrifolia (Roxb.) Planch. Biol Plant 31: 126-132

Raphael HJ 1954 Anatomical changes in trembling aspen (Populus tremuloides Michx.) induced by the application of chemical agents to the stem. Q Bull Mich Agr Exp Stn 37(2): 241-247

Ratzeburg T 1869 Neue Beobachtungen über Doppelringe. Krit Blätt Forst Jagd 51: 208-214
Ratzeburg T 1871 Ueber Verdoppelung des Jahrringes. Abh Bot Ver Prov Brandenburg 13: 71-76
Raven JA 1987 The role of vacuoles. New Phytol 106: 357-422
Record SJ 1911 Pith flecks or medullary spots in wood. For Q 9: 244-252
Record SJ 1918a Significance of resinous tracheids. Bot Gaz 66: 61-67
Record SJ 1918b Intercellular canals in dicotyledonous woods. J For 16: 429-441
Record SJ 1919 Storied or tier-like structure in certain dicotyledonous woods. Bull Torrey Bot Club 46: 253-273
Record SJ 1921 Further notes on intercellular canals in dicotyledonous woods. J For 19: 255-266
Record SJ 1925 Occurrence of intercellular canals in dicotyledonous woods. Trop Woods 4: 17-20
Record SJ 1934 Identification of the timbers of temperate North America. John Wiley, New York, 196 pp
Record SJ 1936 Classifications of various anatomical features of dicotyledonous woods. Trop Woods 47: 12-27
Record SJ 1943 Key to American woods X. Woods with storied structure. Trop Woods 76: 32-47
Rees LW 1929 Growth studies in forest trees. Picea rubra Link. J For 27: 384-403
Regnault M 1860 Recherches sur les affinités de structure des tiges des plantes du groupe des cyclospermées. Ann Sci Nat Bot Ser 4, 14: 73-166
Reid RW, Watson JA 1966 Sizes, distributions, and numbers of vertical resin ducts in lodgepole pine. Can J Bot 44: 519-525
Reid RW, Whitney HS, Watson JA 1967 Reactions of lodgepole pine to attack by Dendroctonus ponderosae Hopkins and blue stain fungi. Can J Bot 45: 1115-1126
Reinders-Gouwentak CA 1949 Homogenous and heterogeneous rays: their characteristics and a key for their identification. Meded LandHogesch Wageningen 49: 217-236
Rendle BJ, Clarke SH 1934a The problem of variation in the structure of wood. Trop Woods 38: 1-8
Rendle BJ, Clarke SH 1934b The diagnostic value of measurements in wood anatomy. Trop Woods 40: 27-37
Resch A 1959 Über Leptombündel und isolierte Siebröhren sowie deren Korrelationen zu den übrigen Leitungsbahnen in der Sprossachse. Planta 52: 467-489; 490-515
Reukema D 1959 Missing annual rings in branches of young-growth Douglas fir. Ecology 40: 480-482
Reuss H 1893 Beiträge zur Wachstumsthätigkeit des Baumes nach praktischen Beobachtungsdaten des laufenden Stärkenzuwachsganges an der Sommerlinde. Forst-Naturwiss Z 2: 145-158
Rhoads AS 1923 The formation and pathological anatomy of frost rings in conifers injured by late frosts. USDA Bull 1131, 15 pp
Richard A 1852 Rapport sur un mémoire de M. Trécul. Ann Sci Nat Bot Ser 3, 18: 14-24
Richards JH, Larson PR 1981 Morphology and development of Populus deltoides branches in different environments. Bot Gaz 142: 382-393
Rickey RG, Hamilton JK, Hergert HL 1974 Chemical and physical properties of tumor-affected Sitka spruce. Wood Fiber 6: 200-210

Riding RT, Greenwood MS, Mills EA 1989 Histochemistry of the vascular cambium in larch (Larix decidua). Am J Bot (suppl) 76(6): 53

Riding RT, Little CHA 1984 Anatomy and histochemistry of Abies balsamea cambial zone cells during the onset and breaking of dormancy. Can J Bot 62: 2570-2579

Riding RT, Little CHA 1986 Histochemistry of the dormant vascular cambium of Abies balsamea: Changes associated with tree age and crown position. Can J Bot 64: 2082-2087

Riedl H 1937 Bau und Leistungen des Wurzelholzes. Jahrb Wiss Bot 85: 1-75

Rioux D, Ouellette GB 1989 Light microscope observations of histological changes induced by Ophiostoma ulmi in various nonhost trees and shrubs. Can J Bot 67: 2335-2351

Rioux D, Ouellette GB 1991 Barrier zone formation in host and nonhost trees inoculated with Ophiostoma ulmi. I. Anatomy and histochemistry. Can J Bot 69: 2055-2073

Robards AW, Humpherson PG 1967 Phytoferritin in plastids of the cambial zone of willow. Planta 76: 169-178

Robards AW, Kidwai P 1969 A comparative study of the ultrastructure of resting and active cambium of Salix fragilis, L. Planta 84: 239-249

Robards AW, Lucas WJ 1990 Plasmodesmata. Ann Rev Plant Physiol Plant Mol Biol 41: 369-419

Roberts DR, Joye NM Jr, Proveaux AT, Peters WJ, Lawrence RV 1973 A new and more efficient method of naval stores production. Nav Stores Rev 83: 4-5

Roberts LW 1976 Cytodifferentiation in plants. Xylogenesis as a model system. Cambridge University Press, 160 pp

Roberts, LW, Gahan PB, Aloni R 1988 Vascular differentiation and plant growth regulators. Springer, Berlin, Heidelberg, New York, 154 pp

Roberts RH 1949 Theoretical aspects of graftage. Bot Rev 15: 423-463

Roberts RH, Struckmeyer BE 1948 Anatomical and histological changes in relation to vernalization and photoperiodism. In: Murneek AE, Whyte RO (eds) Vernalization and photoperiodism. Chronica Botanica, Waltham, 91-100

Robin C, Desprez-Loustau ML, Delatour C 1992 Spatial and temporal enlargement of trunk cankers of Phytophthora cinnamomi in red oak. Can J For Res 22: 362-366

Rodin RJ 1967 Ontogeny of foliage leaves of Gnetum. Phytomorphology 17: 118-128

Roelofsen PA 1959 The plant cell-wall. Enc Plant Anat 3(Pt 4): 1-335

Rohfritsch O 1977 Ultrastructure of the nutritive tissue of the Chermes abietis L. fundatrix on Picea excelsa L. Marcellia 40: 135-149

Roland JC 1978 Early differences between radial walls and tangential walls of actively growing cambial zone. IAWA Bull 1978/1: 7-10

Romberger JA 1963 Meristems, growth, and development in woody plants. USDA Agric Tech Bull 1293, 214 pp

Romberger JA, Hejnowicz Z, Wloch W 1980 Gamma irradiation suppresses intrusive growth in the Acer rubrum cambial region, but orientational domains are not destroyed. Bot Soc Am Misc Ser 158: 97

Romell LG 1940 Kvistningsstudier å tall och gran. (Studies on pruning in pine and spruce) Medd Statens Skogsförsöksanst 32: 143-194

Roth ER 1948 Healing and defects following oak pruning. J For 46: 500-504

Roth ER, Hepting GH 1943 Wounds and decay caused by removing large companion sprouts of oaks. J For 41: 190-195

Rothert W 1913 Das Gewebe der Pflanzen. Handwörterbuch Naturwiss 4: 1144-1284

Rothert W, Jost L 1934 Gewebe der Pflanzen. Handwörterbuch Naturwiss Aufl 2, 5: 1-139

Rubner K 1910a Das Hungern des Cambium und das Aussetzen der Jahrringe. Naturwiss Z Forst-Landwirtsch 8: 212-262

Rubner K 1910b Einiges über die Hängenzweige der Fichte. Mitt Bayer Bot Ges Erforsch Heim Flora 2: 307-308

Rudall P 1989 Laticifers in vascular cambium and wood of Croton spp. (Euphorbiaceae). IAWA Bull 10: 379-383

Rudolf P 1949 Porcupines' preferences in pine plantations. J For 47: 207-209

Russow E 1872 Vergleichende Untersuchungen betreffend die Histiologie (Histiographie und Histiogenie) der vegetativen und sporenbildenden Organe und die Entwickelung der Sporen der Leitbündel-Kryptogamen, mit Berücksichtigung der Histiologie der Phanerogamen, ausgehend von der Betrachtung der Marsiliaceen. Acad Imp Sci St. Pétersbourg Mem Ser 7, 19: 1-207

Russow E 1881 Über die Entwickelung des Hoftüpfels, der Membran der Holzzellen und des Jahresringes bei den Abietineen, in erster Linie von Pinus sylvestris L. Sitzungsber Naturforsch Ges Univ Dorpater 6: 109-158

Russow E 1882a Über den Bau und die Entwicklung der Siebröhren und Schloss daran Mittheilungen über Bau und Entwicklung der secundären Rinde der Dicotylen und Gymnospermen. Sitzungsber Naturforsch Ges Univ Dorpater 6: 257-327

Russow E 1882b Über den Inhalt der parenchymatischen Elemente der Rinde vor und während des Knospenaustriebes und Beginns der Cambiumthätigkeit im Stamm und Wurzel der einheimischen Lignosen. Sitzungsber Naturforsch Ges Univ Dorpater 6: 350-389

Russow E 1882c Gegenbemerkungen zu den Bemerkungen Dr. C. Sanio's zu meinem Aufsatz "Ueber die Entwicklung des Hoftüpfels, der Membran der Holzzellen und des Jahrringes bei den Abietineen, in erster Linie von Pinus silvestris". Bot Cbl 10: 62-70

Russow E 1883a Über die Perforation der Zellwand und den Zusammenhang der Protoplasmakörper benachbarter Zellen. Sitzungsber Naturforsch Ges Univ Dorpater 6: 562-582

Russow E 1883b Zur Kenntniss des Holzes, insonderheit des Coniferenholzes. Bot Cbl 13: 29-40; 60-68; 95-109; 134-144; 166-173

Sachs J 1878 Ueber die Anordnung der Zellen in jüngsten Pflanzentheilen. Arb Bot Inst Würzburg 2: 46-104

Sachs J 1879 Ueber Zellenanordnung und Wachsthum. Arb Bot Inst Würzburg 2: 185-208

Sachs J 1882 Text-book of botany, morphological and physiological. 1870 2nd edn Vines EH (transl) Clarendon, Oxford, 980 pp

Sachs J 1890 History of botany (1530-1860). Garnsey HEF (transl) Clarendon, Oxford, 568 pp

Sachs T 1968 On the determination of the pattern of vascular tissue in peas. Ann Bot 32: 781-790

Sachs T 1969 Polarity and the induction of organized vascular tissue. Ann Bot 33: 263-275

Sachs T 1975 The induction of transport channels by auxin. Planta 127: 201-206

Sachs T 1984 Axiality and polarity in vascular plants. In: Barlow PW, Carr DJ (eds) Positional controls in plant development. Cambridge University Press, 193-224

Sachs T 1986 Cellular patterns determined by polar transport. In: Bopp M (ed) Plant growth substances. Proc 12th Int Conf Pl Growth Subst, Heidelberg Aug 26-31, 1985 Springer, Berlin, Heidelberg, New York, 231-235

Sachs T, Cohen D 1982 Circular vessels and the control of vascular differentiation in plants. Differentiation 21: 22-26

Sagheer A, Iqbal M 1987 Seasonal rhythms of structure and behaviour of vascular cambium in Ficus rumphii. Ann Bot 60: 649-656
Sajwan VS, Paliwal GS 1976 On the occurrence of stratified cambium in Salvadora persica L. Curr Sci 45: 267-268
Sakai A 1968 Frost damage on basal stems in young trees. Cont Inst Low Temp Sci Ser B, 15: 1-14
Sallé GC 1979 The endophytic system of Viscum album: its anatomy, ultrastructure and relations with the host tissues. In: Musselman LJ, Worsham AD, Eplee RE (eds) Proc 2nd Int Symp Parasitic Weeds N C State Univ Raleigh, 115-127
Sanio C 1857a Über die in der Rinde dicotyler Holzgewächse vorkommenden krystallinischen Niederschläge und deren anatomische Verbreitung vor. Monatsber Akad Wiss Berl 1857: 252-272
Sanio C 1857b Untersuchungen über diejenigen Zellen des dicotylen Hölzkörpers, welche, ausser dem Markstrahlen, im Winter assimilierte Stoffe führen. Linnaea 29: 111-168
Sanio C 1860 Einige Bemerkungen über den Bau des Holzes. Bot Ztg 18: 193-198, 201-204, 209-217
Sanio C 1863a Vergleichende Untersuchungen über die Elementarorgane des Holzkörpers. Bot Ztg 21: 85-91; 93-98; 101-111; 113-118; 121-128
Sanio C 1863b Vergleichende Untersuchungen über die Zusammensetzung des Holzkörpers. Bot Ztg 21: 357-363; 364-375; 377-385; 389-399; 401-412
Sanio C 1872 Ueber die Grösse der Holzzellen bei der gemeinen Kiefer (Pinus silvestris). Jahrb Wiss Bot 8: 401-420
Sanio C 1873 Anatomie der gemeinen Kiefer (Pinus silvestris L.) II. Entwickelungsgeschichte der Holzzellen. Jahrb Wiss Bot 9: 50-128
Sanio C 1882 Bemerkungen zu dem Aufsatze "Ueber die Entwicklung des Hoftüpfels, der Membran der Holzzellen und des Jahrringes bei den Abietineen, in erster Linie von Pinus silvestris" von Prof. E. Russow. Bot Cbl 9: 316-320
Santamour FS Jr 1982 Cambial peroxidase isoenzymes in relation to systematics of Acer. Bull Torrey Bot Club 109: 152-161
Santamour FS Jr 1984 Early selection for wound compartmentalization potential in woody plants. J Environ Hort 2: 126-128
Sárkány A 1936 Über Entwicklung und Funktion des interfaszikularen Kambiums bei Ricinus communis. Jahrb Bot 82: 625-656
Sato K, Ishida S 1982 Resin canals in the wood of Larix leptolepis Gord. (II) Morphology of vertical resin canals. Res Bull Coll Exp For Hokkaido Univ 39: 297-316
Sato K, Ishida S 1983a (III) Morphology of vertical resin canals in compression wood. Res Bull Coll Exp For Hokkaido Univ 40: 455-462
Sato K, Ishida S 1983b V. Formation of vertical resin canals. Res Bull Coll Exp For Hokkaido Univ 40: 723-740
Sauter JJ 1980 The Strasburger cells - Equivalents of companion cells. Ber Deutsch Bot Ges 93: 29-42
Sauter JJ, Wellenkamp S 1988 Protein storage vacuoles in ray cells of willow wood (Salix caprea L.). IAWA Bull 9: 59-65
Savidge RA, Farrar JL 1984 Cellular adjustments in the vascular cambium leading to spiral grain formation in conifers. Can J Bot 62: 2872-2879
Sax K 1962 Aspects of aging in plants. Ann Rev Plant Physiol 13: 489-506
Sax K, Dickson AQ 1956 Phloem polarity in bark regeneration. J Arnold Arbor 37: 173-179

Schaad NW, Wilson EE 1970 Structure and seasonal development of secondary phloem of Juglans regia. Can J Bot 48: 1049-1053
Schacht H 1856 Lehrbuch der Anatomie und Physiologie der Gewächse. Theil 1, GWF Müller, Berlin, 446 pp
Schacht H 1860 Der Baum. Studien über Bau und Leben höheren Gewächse. 2nd edn GWF Müller, Berlin, 378 pp
Schemnitz SD 1983 Porcupines. In: Timm RM (ed) Prevention and control of wildlife damage, Great Plains Agr Council, University of Nebraska, Lincoln, pp B71-B84
Schenck H 1885 Ueber die Auskleidung der Intercellulargänge. Ber Deutsch Bot Ges 3: 217-225
Schmid R 1976 The elusive cambium - another terminological contribution. IAWA Bull 1976/4: 51-59
Schmidt E 1889 Ein Beitrag zur Kenntniss der secundären Markstrahlen. Ber Deutsch Bot Ges 7: 143-151
Schmitt U, Liese W 1990 Wound reaction of the parenchyma in Betula. IAWA Bull 11: 413-420
Schneider H 1952 The phloem of the sweet orange tree trunk and the seasonal production of xylem and phloem. Hilgardia 21: 331-366
Schneider H 1955 Ontogeny of lemon tree bark. Am J Bot 42: 893-905
Schnepf E 1986 Cellular polarity. Ann Rev Plant Physiol 37: 23-47
Schnettker M 1977 Zum Dickenwachstum bei Helianthus annuus L. (Compositae). Bot Jahrb Syst 98: 250-265
Schober R 1951 Zum jahreszeitlichen Ablauf des sekundären Dickenwachstums: Untersuchungen in Lehrforstamt Kattenbühl. Allg Forst-Jagdztg 122: 81-96
Schoch-Bodmer H 1945 Interpositionwachstum, symplastisches und gleitendes Wachstum. Ber Schweiz Bot Ges 55: 313-319
Schoch-Bodmer H, Huber P 1948 Fasergabelung durch Auftreten lokalisierter Wachstumsbezirke. Experientia 4: 146-147
Schoch-Bodmer H, Huber P 1949a Über Flächenwachstum, insbesondere über Fasergabelungen. Vierteljahrschr Naturforsch Ges Zürich 94: 188-193
Schoch-Bodmer H, Huber P 1949b Spitzenwachstum und Gabelbildung bei sekundären Fasern. Schweiz Z Forstwes 100: 551-567
Schoch-Bodmer H, Huber P 1951 Das Spitzenwachstum der Bastfasern bei Linum usitatissimum und Linum perenne. Ber Schweiz Bot Ges 61: 377-404
Schoch-Bodmer H, Huber P 1952 Local apical growth and forking in secondary fibres. Proc Leeds Philos Lit Soc Sci Sect 6 Pt 1: 25-32
Schopf JM 1943 The embryology of Larix. Ill Biol Monogr 19(4): 1-97
Schopfer M 1947 Les débuts de l'anatomie végétale moderne. L'historie du cambium. Archeion l: 270-279
Schoute JC 1902 Über Zellteilungsvorgänge im Cambium. Verhandl K Akad Wetensch Amsterdam Sec 2, 9: 1-60
Schröder R 1879 Zur Frage über die Verdickung der Bäume in verschiedenen Jahreszeiten. Mitt Land Forstw Acad Moskau 2: 1-4 (Cited from Just's Jahresber 7(1): 240-241, 1879)
Schüepp O 1926 Meristeme. Handb Pflanzanat Bd 4, Abt 1, T 2:Histologie. Gebrüder Borntraeger, Berlin, 114 pp
Schüepp O 1966 Meristeme; Wachstum und Formbildung in den Teilungsgeweben höherer Pflanzen. Birkhäuser, Basel, 253 pp
Schultz CH 1831 Ueber den Ursprung und die Natur des Cambiums. Flora 14: 752-767

Schultze-Dewitz G 1959 Einfluss der Kronenhöhe auf den Jahrringbau sowie auf die Faserlänge und den Markstrahlenanteil bei der Fichte verscheidener Altersklassen. Holzforsch Holzverwert 11: 67-74

Schultze-Dewitz G 1961 Notes concerning the influence of the position of a tree on its anatomical structure. Holzforschung 15: 89-91

Schultze-Dewitz G 1992 Frequency and distribution of wood rays in several conifers. IUFRO Div 5 Conf, Aug 23-28, 1992 Nancy, 9 pp mimeo

Schultze-Dewitz G, Götze H 1986 Abnorme Holzstrukturen. Teil 1: Haselwuchs und Wimmerwuchs. Drev Vysk 31: 1-10

Schulz H 1957 Der Anteil der einzelnen Zellarten an dem Holz der Rotbuche. Holz Roh-Werkst 15: 113-118

Schulz P 1883 Das Markstrahlengewebe und seine Beziehungen zu den leitenden Elementen des Holzes. Jahrb Kgl Bot Gart Bot Mus Berlin 2: 210-232

Schumann C 1873 Dickenwachsthum und Cambium. Historisch-experimentelle Untersuchungen. Inaug Diss Univ Breslau Görditz 40 pp (Cited from Just's Jahresber 1: 199, 1873)

Schwendener S 1880 Über die durch Wachsthum bedingte Verschiebung kleinster Theilchen in trajectorischen Curven. Akad Wiss Berlin Monatsber 1880: 408-433

Scott DH 1888 Das gleitende Wachsthum bei der Gewebebildung der Gefässpflanzen. (Review of Krabbe, G. 1886) Ann Bot 2: 127-136

Scott DH 1902 The old wood and the new. New Phytol 1: 25-30

Scott VE, Evans KE, Patton DR, Stone CP, Singer A 1977 Cavity-nesting birds of North American forests. USDA For Serv Agr Handb 511, 112 pp

Scukanec V, Petric B 1977 The relationship between wood ray shape and ray volume percentage in beech. IAWA Bull 1977/3: 57-59

Sebastine KM 1955 Studies on the variations in the structure and size of rays in the secondary wood. J Indian Bot Soc 34: 299-306

Seidel CF 1879 Ueber Verwachsungen von Stämmen und Zweigen von Holzgewächsen und ihren Einfluss auf das Dickenwachsthum der betreffenden Theile. Naturwiss Ges Isis Dresden Sitzber 1879: 161-168

Sennerby-Forsee L 1986 Seasonal variation in the ultrastructure of the cambium in young stems of willow (Salix viminalis) in relation to phenology. Physiol Plant 67: 529-537

Sennerby-Forsee L, von Fircks HA 1987 Ultrastructure of cells in the cambial region during winter hardening and spring dehardening in Salix dasyclados Wim. grown at two nutrient levels. Trees Struct Funct 1: 151-163

Sennerby-Forsee L, Berggren B, Brunkener L, Fjell I 1984 Growth behaviour and anatomy of meristems in Salix. Swed Agr Univ Dept Ecol Environ Res, Res Notes 15: 481-501

Seth MK, Bala M, Lal C 1988 Intra-increment radial variation in tracheid length in Cedrus deodara (Roxb.) Loud. Phytomorphology 38: 231-239

Seth MK, Veena, Sadhna 1989 Intra-increment tracheid length variation in branches and roots of blue pine. Indian For 115: 385-391

Sharma DD, Sharma HK, Paliwal GS 1979a Size correlations among cambial initials and their derivatives in Polyalthia longifolia Thw. Acta Soc Bot Pol 48: 93-98

Sharma DD, Sharma HK, Paliwal GS 1980 Annual rhythm of size variations in the cambial initials of Chorisia speciosa St. Hil. Sci Cult 46: 96-97

Sharma HK, Sharma DD, Paliwal GS 1979b Annual rhythm of size variations in cambial initials of Azadirachta indica A. Juss. Geobios 6: 127-129

Sharon EM 1973 Some histological features of Acer saccharum wood formed after wounding. Can J For Res 3: 83-89

Sharples A, Gunnery H 1933 Callus formation in Hibiscus Rosa-sinensis L. and Hevea brasiliensis Mull. Arg. Ann Bot 827-839

Shepard HB, Bailey IW 1914 Some observations on the variation in length of coniferous fibers. Proc Soc Am For 9: 522-527

Shepherd KR 1964 Some observations on the effect of drought on the growth of Pinus radiata D. Don. Aust For 28: 7-22

Shigo AL 1963 Ring shake associated with sapsucker injury. U S For Serv Res Pap NE-8, 10 pp

Shigo AL 1983 Tree defects: A photo guide. U S For Serv Gen Tech Rpt NE-82, 167 pp

Shigo AL 1984a Tree decay and pruning. Arbor J 8: 1-12

Shigo AL 1984b Compartmentalization: a conceptual framework for understanding how trees grow and defend themselves. Ann Rev Phytopathol 22: 189-214

Shigo AL 1985 How tree branches are attached to trunks. Can J Bot 63: 1391-1401

Shigo AL 1986 A new tree biology. Shigo & Trees Assoc, Durham, 618 pp

Shigo AL, Dudzik KR 1985 Response of uninjured cambium to xylem injury. Wood Sci Tech 19: 195-200

Shigo AL, Shortle WC, Garrett PW 1977 Genetic control suggested in compartmentalization of discolored wood associated with tree wounds. For Sci 23: 179-182

Shigo AL, McGinnes EA Jr, Funk DT, Rogers N 1979 Internal defects associated with pruned and nonpruned branch stubs in black walnut. U S For Serv Res Pap NE-440, 27 pp

Shimaji K 1950 (On the relation between fiber length of some dicotyledonous woods and the tension wrought on the cambium). J Jap For Soc 32: 371-372

Shimaji K, Nagatsuka Y 1971 Pursuit of the time sequence of annual ring formation in Japanese fir (Abies firma Sieb. et Zucc.). J Jap Wood Res Soc 17: 122-128

Shiokura T 1989 A method to measure radial increment in tropical trees. IAWA Bull 10: 147-154

Shortle WC 1979 Mechanisms of compartmentalization in living trees. Phytopathology 69: 1147-1151

Shortle WC, Cowling EB 1978 Development of discoloration, decay, and microorganisms following wounding of sweet-gum and yellow-poplar trees. Phytopathology 68: 609-616

Shortle WC, Ostrofsky A 1983 Decay susceptibility of wood in defoliated fir trees related to changing physical, chemical and biological properties. Eur J For Pathol 13: 1-11

Shortle WS, Shigo AL, Berry P, Abusamra J 1977 Electrical resistance in tree cambium zone: relationship to rates of growth and wound closure. For Sci 23: 326-329

Siddiqi TO 1991 Impact of seasonal variation on the structure and activity of vascular cambium in Ficus religiosa. IAWA Bull 12: 177-185

Sieber FW 1912 Über die physiologische Rolle von Kalk, Magnesia und Phosphorsäure im Kambium. Verh Phys Med Ges Würzburg 51: 215-270

Sieber M, Kucera LJ 1980 On the stem anatomy of Clematis vitalba L. IAWA Bull 1: 49-54

Sifton HB 1915 On the occurrence and significance of "Bars" or "Rims" of Sanio in the Cycads. Bot Gaz 60: 400-405

Sifton HB 1922 The bar of Sanio and primordial pits in the gymnosperms. Trans R Soc Can Ser 3, 16: 83-99

Sifton HB 1945 Air-space tissue in plants. Bot Rev 11: 108-143

Sifton HB 1957 Air-space tissue in plants. Bot Rev 23: 303-312
Sigmond J 1936 Kambiumtätigkeit u. Spätholzentwicklung bei der Fichte (Picea excelsa LK) and Standorten unterschiedlicher Höhenlage. Beih Bot Cbl 54-A: 531-568
Simon S 1902 Der Bau des Holzkörpers sommer- und wintergrüner Gewächse und seine biologische Bedeutung. Ber Deutsch Bot Ges 20: 229-249
Simson BW, Timell TE 1978 Polysaccharides in cambial tissues of Populus tremuloides and Tilia americana I. Isolation, fractionation, and chemical composition of the cambial tissues. Cellul Chem Tech 12: 39-50
Sinnott EW, Bailey IW 1914 Investigations on the phylogeny of the angiosperms. No. 4. The origin and dispersal of herbaceous angiosperms. Ann Bot 28: 547-600
Sinnott EW, Bailey IW 1922 The significance of the "Foliar Ray" in the evolution of herbaceous angiosperms. Ann Bot 36: 523-533
Sinnott EW, Bloch R 1939 Changes in intercellular relationships during the growth and differentiation of living plant tissues. Am J Bot 26: 625-634
Sinnott EW, Bloch R 1940 Cytoplasmic behavior during division of vacuolate plant cells. Proc Nat Acad Sci USA 26: 223-227
Sinnott EW, Bloch R 1941 Division in vacuolate plant cells. Am J Bot 28: 225-232
Sisson WE Jr 1968 Cambial divisions in Pseudotsuga menziesii. Am J Bot 55: 923-926
Skene DS 1965 The development of kino veins in Eucalyptus obliqua L'Hrit. Aust J Bot 13: 367-378
Skene DS 1969 The period of time taken by cambial derivatives to grow and differentiate into tracheids in Pinus radiata D. Don. Ann Bot 33: 253-262
Skene DS 1972 The kinetics of tracheid development in Tsuga canadensis Carr. and its relation to tree vigour. Ann Bot 36: 179-187
Skilling DD 1958 Wound healing and defects following northern hardwood pruning. J For 56: 19-22
Skre O 1988 Frost resistance in forest trees: a literature survey. Commun Norw For Res Inst 40.9: 1-35
Smith DE 1980 Abnormal wood formation following fall and spring injuries in black walnut. Wood Sci 243-251
Smith DE, McGinnes EA Jr 1984 Abnormal vascular cambial subdivision occurrence in Erythrobalanus oaks in response to water impoundment. Am J Bot 71(5) Pt2: 48
Smith DE, Shigo AL, Safford LO, Blanchard R 1976 Resistances to a pulsed electric current reveal differences between nonreleased, released, and released-fertilized paper birch trees. For Sci 22: 471-472
Smith F, Montgomery R 1959 The chemistry of plant gums and mucilages. Reinhold, New York, 627 pp
Smith FH 1967 Effects of balsam woolly aphid (Adelges piceae) infestation on cambial activity in Abies grandis. Am J Bot 54: 1215-1223
Smith KT, Shortle WC 1990 IAA oxidase, peroxidase, and barrier-zone formation in red maple. Eur J For Pathol 20: 241-246
Smith KT, Blanchard RO, Shortle WC 1984 Cambial electrical resistance related to the number of vascular cambial cells in balsam fir. Can J For Res 14: 950-952
Smoot EL 1984a Secondary phloem anatomy in Callistrophyton Boyssetii (Renault) Rothwell and histological changes in the outer phloem. Bot Gaz 145: 395-406
Smoot EL 1984b Phloem anatomy of the Carboniferous pteridosperm Medullosa and evolutionary trends in gymnosperm phloem. Bot Gaz 145: 550-564

Snow R 1942 On the causes of regeneration after longitudinal splits. New Phytol 41: 101-107

Soe K 1959 Anatomical studies of bark regeneration following scoring. J Arnold Arbor 40: 260-267

Soh WY 1990 Origin and development of cambial cells. In: Iqbal M (ed) The vascular cambium. Research Studies Press, Taunton, England, 37-62

Soh WY 1992 Origin and development of the interfascicular cambium from residual meristem in seed plants. Korean J Bot 35: 273-281

Somerville A 1980 Resin pockets and related defects of Pinus radiata grown in New Zealand. N Z J For Sci 10: 439-444

Sparrow AH, Schairer LA, Woodwell GM 1965 Tolerance of Pinus rigida trees to a ten-year exposure to chronic gamma irradiation from cobalt-60. Radiat Bot 5: 7-32

Sprengel F 1936 Ueber die Kropfkrankheit an Eiche, Kiefer und Fichte. Phytopathol Ztg 9: 583-635

Spurr AR 1949 Histogenesis and organization of the embryo in Pinus strobus L. Am J Bot 36: 629-641

Spurr AR 1950 Organization of the procambium and development of the secretory cells in the embryo of Pinus strobus L. Am J Bot 37: 183-197

Spurr SH, Hyvärinen MJ 1954 Wood fiber length as related to position in tree and growth. Bot Rev 20: 561-575

Srivastava LM 1963a Secondary phloem in the Pinaceae. University of Calif Publ Bot 36: 1-142

Srivastava LM 1963b Cambium and vascular derivatives of Ginkgo biloba. J Arnold Arbor 44: 165-192

Srivastava LM 1966 On the fine structure of the cambium of Fraxinus americana L. J Cell Biol 31: 79-93

Srivastava LM 1973 Cambial activity in trees. Arnoldia 33: 46-66

Srivastava LM, Esau K 1961a Relation of dwarf mistletoe (Arceuthobium) to the xylem tissue of conifers. I. Anatomy of parasite sinkers and their connection with host system. Am J Bot 48: 159-167

Srivastava LM, Esau K 1961b II. Effect of the parasite on the xylem anatomy of the host. Am J Bot 48: 209-215

Srivastava LM, O'Brien TP 1966a On the ultrastructure of cambium and its vascular derivatives I. Cambium of Pinus strobus L. Protoplasma 61: 257-276

Srivastava LM, O'Brien TP 1966b II. Secondary phloem of Pinus strobus L. Protoplasma 61: 277-293

Stahel J 1968 Quantitative und qualitative Alterungsphänomene in Pappeln (Populus x euramericana [Dode] Guinier cv. 'Robusta'). Holz Roh- Werkst 26: 418-427

Steeves TA, Sussex IM 1972 Patterns in plant development. Prentice-Hall, Englewood Cliffs, 302 pp

Steeves TA, Sussex IM 1989 Patterns of plant development. 2nd edn Cambridge University Press, 388 pp

Steiner G, Buhrer EM 1934 Aphelenchoides xylophilus N. sp., a nematode associated with blue-stain and other fungi in timber. J Agr Res 48: 949-951

Stephan G 1967 Untersuchungen über die Anzahl der Harzkanäle in Kiefern (Pinus silvestris). Arch Forstwes 16: 461-470

Sterling C 1946 Growth and vascular development in the shoot apex of Sequoia sempervirens (Lamb.) Endl. III. Cytological aspects of vascularization. Am J Bot 33: 35-45

Stern WL 1954 A suggested classification for intercellular spaces. Bull Torrey Bot Club 81: 234-235

Stern WL, Sweitzer EM, Phipps RE 1970 Comparative anatomy and systematics of woody Saxifragaceae. Ribes. In: Robson NKB, Cutter DF, Gregory M (eds) New research in plant anatomy. Academic Press, New York, 215-236

Stewart CM 1957 Status of cambial chemistry. Tappi 40: 244-256

Stewart CM 1966 The chemistry of secondary growth in trees. CSIRO Aust Div For Prod Tech Pap 43, 107 pp

Stieber J 1985 Wave nature and a theory of cambial activity. Can J Bot 63: 1942-1950

Stone EL Jr 1940 Frost rings in longleaf pine. Science 92: 478

Stone RN 1977 A critique of completed research on paraquat-induced lightwood. TAPPI For Biol Wood Chem Conf Pap Madison WI June 20-22, 1977, pp 57-62

Strain BR, Johnson PL 1963 Corticular photosynthesis and growth in Populus tremuloides. Ecology 44: 581-584

Strasburger E 1901 Ueber Plasmaverbindungen pflanzlicher Zellen. Jahrb Wiss Bot 36: 493-610

Strasburger E, Noll F, Schenck H, Karsten G 1908 Textbook of botany. Lang WH (transl) Macmillan, London, 746 pp

Struckmeyer BE 1941 Structure of stems in relation to differentiation and abortion of blossom buds. Bot Gaz 103: 182-191

Struckmeyer BE, Roberts RH 1939 Phloem development and flowering. Bot Gaz 100: 600-606

Studhalter RA 1955 Tree growth I. Some historical chapters. Bot Rev 21: 1-72

Studhalter RA, Glock WS, Agerter SR 1963 Tree growth. Some historical chapters in the study of diameter growth. Bot Rev 29: 245-365

Sugawa T, Fujii T 1992 Unusual multiseriate rays of Thujopsis dolabrata var. hondai (Cupressaceae). IAWA Bull 13: 248-249

Sullivan TP, Sullivan DS 1982 Barking damage by snowshoe hares and red squirrels in lodgepole pine stands in central British Columbia. Can J For Res 12: 443-448

Sullivan TP, Vyse A 1987 Impact of red squirrel feeding damage on spaced stands of lodgepole pine in the Cariboo Region of British Columbia. Can J For Res 17: 666-674

Sundvor T 1956 Noen undersıkelsea over når tykkelsestilveksten innstilles hos lauvtraerne om hysten på Vestlandet. (Some observations on the date of cessation of cambial activity in deciduous trees in Western Norway) Blyttia 14: 122-125

Süss H 1967 Über die Längenänderungen der Parenchymstränge, Holzfasern und Gefässglieder von Laubhölzern im Verlauf einer Zuwachsperiode. Holz Roh-Werkst 25: 369-377

Süss H 1989 Beitrage zur Holzanatomie der Pinaceae am Beispiel des Astholzes der Borstenkiefer, Pinus aristata Engelm. Feddes Repert 100: 617-624

Süss H, Müller-Stoll R 1970 Änderungen der Zellgrössen und des Anteils der Holzelemente in zerstreutporigen Hölzern innerhalb einer Zuwachsperiode. Holz Roh- Werkst 28: 309-317

Süss H, Müller-Stoll WR 1973 Zur Anatomie des Ast-, Stamm- und Wurzelholzes von Platanus x acerifolia (Ait.) Willd. Österr Bot Z 121: 227-249

Süss H, Müller-Stoll WR 1984 Längenänderungen einiger Holzelemente der Rotbuche (Fagus sylvatica L.) in Abhängigkeit von Stammhöhe und Himmelsrichtung. Holz Roh-Werkst 42: 409-414

Süss H, Schirarend C 1990 Ein neues tracheidales Element der Koniferen im Astholz von Pinus aristata Engelm. Feddes Repert 101: 11-12
Sutherland EK, Martin B 1990 Growth response of Pseudotsuga menziesii to air pollution from copper smelting. Can J For Res 20: 1020-1030
Suzuki M 1979 The cause of resin canals in the shoots of conifers. Parts I-III. Bot Mag Tokyo 92: 235-251, 253-274, 333-353
Swamy BGL, Krishnamurthy KV 1978 Certain conceptual aspects of meristems. III. A model. Phytomorphology 28: 1-7
Swamy BGL, Parameswaran N, Govindarajulu E 1960 Variation in vessel length within one growth ring of certain arborescent dicotyledons. Indian Bot Soc J 39: 163-170
Swarbrick T 1927 Studies in the physiology of fruit trees. I. The seasonal starch content and cambial activity in one- to five-year-old apple branches. J Pomol Hort Sci 6: 137-156
Swartz D 1971 Collegiate dictionary of botany. Ronald, New York, 520 pp
Takamatsu M 1928 On the arrangement of cambial cells in some woody plants. Tohoku Imp Univ Sci Rep 3: 615-624
Takizawa T, Takahashi M, Kawaguchi N 1980 A note on the distribution of radial resin canals in Larix leptolepis Gord. IAWA Bull 1: 111-112
Tate J Jr 1973 Methods and annual sequence of foraging by the sapsucker. Auk 90: 840-856
Taylor EL 1988 Secondary phloem anatomy in Cordaitean axes. Am J Bot 75: 1655-1666
Taylor FW 1977 Fiber length variation within growth rings of certain angiosperms. Wood Fiber 8: 116-119
Taylor FW, Moore JS 1981 A comparison of earlywood and latewood tracheid lengths of loblolly pine. Wood Fiber 13: 159-165
Telewski FW, Lynch AM 1991 Measuring growth and development of stems. In Lassoie JP, Hinckley TM (eds) Techniques and approaches in forest tree physiology. CRC, Boston, 503-555
Templeton J 1923 The effect of late frost on the wood of Acer pseudoplatanus. Linn R Bot Gard Edinburgh Notes 14: 9-12
Tepper HB, Hollis CA 1967 Mitotic reactivation of the terminal bud and cambium of white ash. Science 156: 1635-1636
Terrazas T 1991 Origin and activity of successive cambia in Cycas (Cycadales). Am J Bot 78: 1335-1344
Thair BW, Steeves TA 1976 Response of the vascular cambium to reorientation in patch grafts. Can J Bot 54: 361-373
Thimann KV, Kaufman D 1958 Cytoplasmic streaming in the cambium of white pine. In: Thimann KV (ed) The physiology of forest trees. Ronald, New York, 479-492
Thomas KI, Wall RE 1986 Stem pitting and annual ring discoloration in yellow birch. Can J Bot 64: 2165-2167
Thompson NP, Heimsch C 1964 Stem anatomy and aspects of development in tomato. Am J Bot 51: 7-19
Thompson WP 1910 The origin of ray tracheids in the Coniferae. Bot Gaz 50: 101-116
Thompson WP 1911 On the origin of the multiseriate ray of the dicotyledons. Ann Bot 25: 1005-1014
Thompson WP 1912a The anatomy and relationships of the Gnetales. I. The genus Ephedra. Ann Bot 26: 1077-1104
Thompson WP 1912b Ray tracheids in Abies. Bot Gaz 53: 331-339

Thomson RB, Sifton HB 1925 Resin canals in the Canadian spruce (Picea canadensis (Mill.) B.S.P.) - an anatomical study, especially in relation to traumatic effects and their bearing on phylogeny. Philol Trans R Soc Lond Ser B, 214: 63-111

Timell TE 1972 Beobachtungen an Holzstrahlen im Druckholz. Holz Roh- Werkst 30: 267-273

Timell TE 1973 Ultrastructure of the dormant and active cambial zones and the dormant phloem associated with the formation of normal and compression woods in Picea abies (L.) Karst. State Univ NY Coll For Tech Bull 96, 94 pp

Timell TE 1979 Formation of compression wood in Balsam fir (Abies balsamea) 1. Ultrastructure of the active cambial zone and its enlarging derivatives. Holzforschung 33: 137-143

Timell TE 1980 Organization and ultrastructure of the dormant cambial zone in compression wood of Picea abies. Wood Sci Tech 14: 161-179

Timell TE 1986 Compression wood in gymnosperms. Vol I-III Springer, Berlin, Heidelberg, New York, 2150 pp

Tippett JT 1986 Formation and fate of kino veins in Eucalyptus L'Hérit. IAWA Bull 7: 137-143

Tippett JT, Shigo AL 1981 Barrier zone formation: a mechanism of tree defense against vascular pathogens. IAWA Bull 2: 163-168

Torelli N, Cufar K, Robic D 1986 Some wood anatomical, physiological, and silvicultural aspects of silver fir dieback in Slovenia (NW Yugoslavia). IAWA Bull 7: 343-350

Trécul A 1852 Observations, relatives a l'accrôissement en diamètre des végétaux dicotyledonés ligneux. Ann Sci Nat Bot Ser 3, 17: 250-279

Trécul A 1853a Origine et développment des fibres ligneuses. Ann Sci Nat Bot Ser 3, 19: 63-74

Trécul A 1853b Accroissement des végétaux dicotylédonés ligneux, reproduction du bois et de l'ecorce par le bois décortiqué. Ann Sci Nat Bot Ser 3, 19: 157-192

Trécul A 1854 Mémoire sur les formations secondaires dans les cellules végétales, et sur les formations spirales, annulaires et réticulées, en particulier. Ann Sci Nat Bot Ser 4, 2: 273-356

Trécul A 1855 De l'influence des décortications annulaires sur la végétation des arbres dicotylédons. Ann Sci Nat Bot Ser 4, 3: 341-363

Trendelenburg R 1939 Das Holz als Rohstoff. JF Lehmanns, Berlin, 435 pp

Trendelenburg R, Mayer-Wegelin H 1955 Das Holz als Rohstoff. Carl Hanser, München, 541 pp

Tsoumis G 1965 Structural deformities in an epidemic tumor on white spruce, Picea glauca. Can J Bot 43: 176-181

Tsoumis G, Kezos N, Fanariotou (-), Voulgaridis E, Passialis C 1988 Characteristics of briarwood. Holzforschung 42: 71-77

Tsuda M 1972 Phytoferritin in plastids of the cambial zone and phloem of conifers. Mokuzai Gakkaishi 18: 367-371

Tsuda M 1975a The ultrastructure of the vascular cambium and its derivatives in coniferous species I. Cambial cells. Bull Tokyo Univ For 67: 158-226

Tsuda M 1975b II. Xylem cells and phloem parenchyma cells. Bull Tokyo Univ For 68: 25-89

Tucker MC, Evert RF 1969 Seasonal development of the secondary phloem in Acer negundo. Am J Bot 56: 275-284

Tupper-Carey RM 1930 Observations on the anatomical changes in tissue bridges across rings through the phloem of trees. Proc Leeds Philos Soc Sci Sect 2(2): 86-94

Turberville HW, Hough AF 1939 Errors in age counts of suppressed trees. J For 37: 417-418

Turner SH, Bush AP, Bolton AJ 1979 A note on the spatial relationship between rays and axial resin canals in Picea abies. IAWA Bull 1979/1: 19-21

Unger DF 1842 Mémoire sur l'origine des vaisseaux spiraux. Ann Sci Nat Bot Ser 2, 17: 227-243

Unger DF 1844 Ueber das Wachsthum der Internodien von anatomischer Seite beobachtet. Bot Ztg 2: 489-494, 506-511, 521-526

Unger DF 1847 Botanische Beobachtungen III. Die Intercellularsubstanz und ihr Verhältniss zur Zellmembran bei Pflanzen. Bot Ztg 5: 289-300

Ursprung A 1904 Zur Periodicität des Dickenwachsthums in den Tropen. Bot Ztg 62: 189-210

van Daalen JC 1988 Cambial electrical resistance as an indicator of tree growth in the Southern Cape indigenous forests. S Afr For J 146: 44-50

van der Graaff NA, Baas P 1974 Wood anatomical variation in relation to latitude and altitude. Blumea 22: 101-121

van Tieghem P 1871 Recherches sur la symmétrie de structure des plantes vasculaires. Ann Sci Nat Bot Ser 13, 5: 5-314

Vasiljevic S 1950 (O Sirini traka lignuma nekih vrsta roda Acer). The width of the xylem rays in some species of Acer. Glas Sum Fak Beograd 1: 116-151 (Cited from For Abst 12: 2991, 1951)

Velten W 1872 Ueber die Verbreitung der Protoplasmabewegungen im Pflanzenreiche. Bot Ztg 30: 645-653

Velten W 1875 Ueber die Entwickelung des Cambium und N. J. C. Müller's Ideen über diesen Gegenstand. Bot Ztg 33: 809-814, 825-829, 841-845

Venugopal N, Krishnamurthy KV 1988 Occurrence of multinucleate cambial initials in some tropical trees. Curr Sci 57: 1174-1175

Venugopal N, Krishnamurthy KV 1989 Organisation of vascular cambium during different seasons in some tropical timber trees. Nord J Bot 8: 631-638

Venverloo CJ, Libbenga KR 1987 Regulation of the plane of cell division in vacuolated cells. I. The function of nuclear positioning and phragmosome formation. J Plant Physiol 131: 267-284

Vetter RE, Botosso PC 1989 Remarks on age and growth rate determination of Amazonian trees. IAWA Bull 10: 133-145

Villalba R 1985 Xylem structure and cambial activity in Prosopis flexuosa DC. IAWA Bull 6: 119-130

Vincent L 1945 Remarque sur l'élaboration de chlorophylle par les tissus cambiaux d'arbres cultivés 'in vitro'. Bull Soc Bot Fr 92: 84-85

Vins B 1961 Störungen der Jahresringbildung durch Rauchschäden. Naturwissenschaften 48: 484-485

van Vliet GJCM 1976 Radial vessels in rays. IAWA Bull 1976/3: 35-37

Vöchting H 1873-74 Beiträge zur Morphologie und Anatomie der Rhipsalideen. Jahrb Wiss Bot 9: 327-484

Vöchting H 1875 Der Bau und die Entwicklung des Stammes der Melastomeen. Bot Abh 3: 1-92

Vöchting H 1906 Über Regeneration und Polarität bei höheren Pflanzen. Bot Ztg 64: 101-148

Vol EV, Levin ED 1984 Procedure for the isolation of phospholipids from the cambial zone of Larix sibirica and determination of their group content. Khim Drev No 1: 99-101 (Cited from For Abstr 45:4065, 1984)

von Diannelidis T 1953 Zur Frage der Unterscheidbarkeit des Holzes von Picea und Larix auf Grund der Markstrahl-Harzgänge. Forstwiss Cbl 72: 308-315

von Höhnel F 1884a Ueber den Einfluss des Rindendruckes auf die Beschaffenheit der Bastfasern der Dicotylen. Jahrb Wiss Bot 15: 311-326

von Höhnel F 1884b Ueber den etagenförmigen Aufbau einiger Holzkörper. Ber Deutsch Bot Ges 2: 2-5

von Höhnel F 1884c Über stockwerkartig aufgebaute Holzkörper. Ein Beitrag zur Holzanatomie. Sitzungsber Kaiser Acad Wiss Math-Natur Cl 89 (Pt 1): 30-47

von Mohl H 1844 Einige Bemerkungen über den Bau der vegetabilischen Zelle. Bot Ztg 2: 273-277, 289-294, 305-310, 321-326, 337-342

von Mohl H 1846 Ueber die Saftbewegung im Innern der Zellen. Bot Ztg 4: 73-78, 89-94

von Mohl H 1858 Ueber die Cambiumschicht des Stammes der Phanerogamen und ihr Verhältniss zum Dickenwachsthum desselben. Bot Ztg 16: 185-190, 193-198 (Also, Same title and date: Sur la couche de cambium de la tige des Phanérogames, et sur ses rapports avec l'accroisement en épaisseur de cette tige. Ann Sci Nat Bot Ser 4, 51-29; and, On the cambium-layer of the stem of the Phanerogamia, and on its relation to the increase of thickness. Henfrey A (transl) Ann Mag Nat Hist Ser 3, 2: 389-407

von Mohl H 1862 Einige anatomische und physiologische Bemerkungen über das Holz der Baumwurzeln. Bot Ztg 20: 225-230, 233-239, 269-278, 281-287, 289-295, 313-319, 321-327

von Tubeuf C 1905 Beobachtungen über elektrische Erscheinungen im Walde. VI. Absterben ganzer Baumgruppen durch den Blitz. Naturwiss Z Forst-Landwirtsch 3: 493-507

von Tubeuf C 1906 VII. Über sogenannte Blitzlöcher im Walde. Naturwiss Z Forst-Landwirtsch 4: 344-351

von Tubeuf C 1918 Über die Beziehungen der Baumphysiologie zur praktischen Harznutzung. Naturwiss Z Forst-Landwirtsch 16: 2-17

von Tubeuf C 1930a Das Problem der Knollenkiefer. Z Pflanzenkrank Pflanzenschutz 40: 225-251

von Tubeuf C 1930b Scheinmarkstrahlen im Holz der Kiefer. Z Pflanzenkrank Pflanzenschutz 40: 353-356

von Tubeuf C, Zehnder A 1903 Über die pathologische Wirkung künstlich erzeugter elektrischer Funkenströme auf Leben und Gesundheit der Nadelhözer. Naturwiss Z Forst-Landwirtsch 1: 448-461

Vozzo JA 1978 Radiographic terminology for biological research. U S For Serv Gen Tech Rep SO-18, 45 pp

Wade DD, Johansen RW 1986 Effects of fire on southern pine: Observations and recommendations. U S For Serv Gen Tech Rep SE-41, 14 pp

Wagenführ R 1961 Die Klassifizierung des Holz-Parenchyms - ein Beitrag zur Bestimmung der Holzarten (I & III). Holztechnologie 2: 110-118, 207-211

Wagg JWB 1948 The formation of resin canals in Douglas-fir as influenced by environment. J For 46: 768

Waisel Y, Fahn A 1965 A radiological method for the determination of cambial activity. Physiol Plant 18: 44-46

Waisel Y, Noah I, Fahn A 1966 Cambial activity in Eucalyptus camaldulensis Dehn. II. The production of phloem and xylem elements. New Phytol 65: 319-324

Walker NE, Cumbie BG 1968 Anticlinal divisions in the vascular cambium of Hibiscus. Trans MO Acad Sci 2: 66-72

Walker-Simmons M, Holländer-Czytko H, Andersen JK, Ryan CA 1984 Wound signals in plants: a systemic plant wound signal alters plasma membrane integrity. Proc Natl Acad Sci USA 81: 3737-3741

Wardrop AB 1949 The influence of pressure on the cell wall organization of conifer tracheids. Proc Leeds Philos Soc 5(pt 2): 128-135

Wardrop AB 1952 Formation of new cell walls in cell division. Nature 170: 329

Wardrop AB 1954 The mechanism of surface growth involved in the differentiation of fibres and tracheids. Aust J Bot 2: 165

Wardrop AB, Dadswell HE 1950 The nature of reaction wood. II. The cell wall organization of compression wood tracheids. Aust J Sci Res Ser B 3: 1-13

Wardrop AB, Dadswell HE 1952 III. Cell division and cell wall formation in conifer stems. Aust J Sci Res Ser B 5: 385-398

Wardrop AB, Dadswell HE 1957 Variations in the cell wall organization of tracheids and fibres. Holzforschung 11: 33-41

Wareing PF 1958 The physiology of cambial activity. J Inst Wood Sci 1: 34-42

Wareing PF, Roberts DL 1956 Photoperiodic control of cambial activity in Robinia pseudacacia L. New Phytol 55: 356-366

Wargo PH, Skutt HR 1975 Resistance to pulsed electric current: an indicator of stress in forest trees. Can J For Res 5: 557-571

Warren Wilson J 1978 The position of regenerating cambia: auxin/sucrose ratio and the gradient induction hypothesis. Proc R Soc Lond Ser B 203: 153-176

Warren Wilson J, Grange RI 1984 Regeneration of tissues in wounded stems: a quantitative study. Ann Bot 53: 515-525

Warren Wilson J, Warren Wilson PM 1961 The position of regenerating cambia - a new hypothesis. New Phytol 60: 63-73

Warren Wilson J, Warren Wilson PM 1981 The position of cambia regenerating in grafts between stems and abnormally-oriented petioles. Ann Bot 47: 473-484

Warren Wilson J, Warren Wilson PM 1984 Control of tissue patterns in normal development and in regeneration. In: Barlow PB, Carr DJ (eds) Positional controls in plant development. Cambridge University Press, 225-280

Weatherhead I, Barnett JR 1986 Development and structure of unusual xylem elements during graft union formation in Picea sitchensis L. Ann Bot 57: 593-598

Webb CD 1967 Interlocked grain in Liquidambar styraciflua L. XIV IUFRO Cong Munich Sect 22/41: 398-411

Webber IE 1934 Bearing of upright ray cells in the wood of Hibiscus mutabilis on usage of the term tile cell. Trop Woods 37: 9-13

Webber IE 1938 Intercellular cavities in the rays of dicotyledonous woods. Lilloa 2: 465-469

Weinstein AI 1926 Summary of literature relating to the volume, distribution, and effects of medullary rays in wood. J For 24: 915-925

Wenham MW, Cusick F 1975 The growth of secondary wood fibres. New Phytol 74: 247-261

Werker E 1970 The secretory cells of Pinus halepensis Mill. Israel J Bot 19: 542-557

Werker E, Baas P 1981 Trabeculae of Sanio in secondary tissues of Inula viscosa (L.) and Salvia fruticosa Mill. IAWA Bull 2: 69-76

Werker E, Fahn A 1969 Resin ducts of Pinus halepensis Mill. - their structure, development and pattern of arrangement. Bot J Linn Soc 62: 379-411

Westing AH 1964 The longevity and aging of trees. Gerontologist 4: 10-15

Wetzel S, Demmers C, Greenwood JS 1989 Spherical organelles, analogous to seed protein bodies, fluctuate seasonally in parenchymatous cells of hardwoods. Can J Bot 67: 3439-3445

Whalley BE 1950 Increase in girth of the cambium in Thuja occidentalis L. Can J Res Sec C 28: 331-340

Wheat DW 1977 Successive cambia in the stem of Phytolacca dioica. Am J Bot 64: 1209-1217

Wheeler EY, Zobel BJ, Weeks DL 1966 Tracheid length and diameter variation in the bole of loblolly pine. Tappi 49: 484-490

White DJB 1962 Tension wood in a branch of sassafras. J Inst Wood Sci 2: 74-80

White DJB, Robards AW 1966 Some effects of radial growth upon the rays of certain ring-porous hardwoods. J Inst Wood Sci 17: 45-52

White PR, Millington WF 1954a The distribution and possible importance of a woody tumor on trees of the white spruce, Picea glauca. Cancer Res 14: 128-134

White PR, Millington WF 1954b The structure and development of a woody tumor affecting Picea glauca. Am J Bot 41: 353-361

White PR, Tsoumis G, Hyland F 1967 Some aspects of the growth of tumors on the white spruce, Picea glauca. Can J Bot 45: 2229-2232

Whitmore FW, Zahner R 1966 Development of the xylem ring in stems of young red pine trees. For Sci 12: 198-210

Whitmore TC 1962 Studies in systematic bark morphology II. General features of bark construction in Diptocarpaceae. New Phytol 61: 208-220

Wieler A 1886a Ist das Markstrahlcambium ein Folgemeristem? Ber Deutsch Bot Ges 4: 73-77

Wieler A 1886b Ueber den Funktionswechsel der Markstrahlinitiale bei Holzgewächsen. Ber Deutsch Bot Ges 4: 259-266

Wieler A 1887 Beiträge zur Kenntniss der Jahresringebildung und des Dickenwachsthums. Jahrb Wiss Bot 18: 70-132

Wieler A 1898 Ueber die jährliche Periodicität im Dickenwachsthum des Holzkörpers der Bäume. Thar For Jahrb 48: 39-139

Wiesner J 1872 Untersuchung einiger Treibhölzer aus dem nördlichen Eismeere. Sitzungsber Math Naturwiss Cl Akad Wiss Wien 65 (Bd 1 Abth): 96-104

Wigand A 1856 Einige Beispiele anomaler Bildung des Holzkörpers. Flora 39: 673-681

Wigand A 1861 Beleuchtung von Schacht's Behandlung der Frage über die Intercellularsubstanz und die Cuticula. Flora 44: 81-94, 97-103

Wight W 1933 Radial growth of the xylem and the starch reserves of Pinus sylvestris: a preliminary survey. New Phytol 32: 77-96

Wilcox H, Czabator FJ, Girolami G, Moreland DE, Smith RF 1956 Chemical debarking of some pulpwood species. NY State Coll For Syracuse Tech Publ 77, 34 pp

Wilhelmi T 1969 Elektrische Potentialdifferenzen im Waldbäumen, ihre Abhängigkeit von meteorologischen Faktoren und ihre Beziehung zum Dickenwachstum. Schweiz Z Forstwes 120: 284-299

Wilkes J 1986 Anatomy of zones of ring shake in Eucalyptus maculata. IAWA Bull 7: 3-11

Wilkinson E, Musselman LJ 1979 The parasitic gymnosperm, Parasitaxus ustus. In: Musselman LJ, Worsham AD, Eplee RE (eds) Proc 2nd Int Symp on Parasitic Weeds N C State Univ Raleigh Suppl, 48 pp

Wilson BF 1963 Increase in cell wall surface area during enlargement of cambial derivatives in Abies concolor. Am J Bot 50: 95-102

Wilson BF 1964 A model for cell production by the cambium of conifers. In: Zimmermann MH (ed) The formation of wood in forest trees. Academic Press, New York, 19-36

Wilson BF 1966 Mitotic activity in the cambial zone of Pinus strobus. Am J Bot 53: 364-372

Wilson BF 1968 Effect of girdling on cambial activity in white pine. Can J Bot 46: 141-146

Wilson BF 1973 A diffusion model for tracheid production and enlargement in conifers. Bot Gaz 134: 189-196

Wilson BF, Howard RA 1968 A computer model for cambial activity. For Sci 14: 77-90

Wilson, BF, Wodzicki TJ, Zahner R 1966 Differentiation of cambial derivatives: proposed terminology. For Sci 12: 438-440

Wilson LF 1977 A guide to insect injury of conifers in the Lake States. U S For Serv Agric Handb 501, 218 pp

Wilton OC 1938 Correlation of cambial activity with flowering and regeneration. Bot Gaz 99: 854-864

Wilton OC, Roberts RH 1936 Anatomical structure of stems in relation to the production of flowers. Bot Gaz 98: 45-64

Winkler C 1872 Zur Anatomie von Araucaria brasiliensis. Bot Ztg 30: 581-585; 597-609

Wisniewski M, Bogle AL, Wilson CL 1984 Histopathology of canker development in peach trees after inoculations with Cytospora leucostoma. Can J Bot 62: 2804-2813

Wisniewski M, Bogle AL, Wilson CL 1985 Seasonal variation in cambial electrical resistance and its relation to growth in two cultivars of peach. Can J Plant Sci 65: 345-350

Wloch W 1976 Cell events in cambium, connected with the formation and existence of a whorled cell arrangement. Acta Soc Bot Pol 45: 313-326

Wloch W 1981 Nonparallelism of cambium cells in neighboring rows. Acta Soc Bot Pol 50: 625-636

Wloch W 1985 Time-variable frequency of events in domains of Tilia cambium. Acta Soc Bot Pol 54: 29-40

Wloch W 1987 Transition areas in the domain patterns of storeyed cambium of Tilia cordata Mill. Acta Soc Bot Pol 56: 645-665

Wloch W 1988 Chiralne zdarzenia komórkowe i wzór domenowy w kambium lipy (Tilia cordata Mill.). (Chiral cell events and domain pattern in the cambium of lime (Tilia cordata Mill.). Thesis Univ Slqski Katowice, 56 pp

Wloch W, Bilczewska E 1987 Fibrillation of events in the cambial domains of Tilia cordata Mill. Acta Soc Bot Pol 56: 19-35

Wloch W, Szendera W 1989 The storeyed and non-storeyed arrangement of rays in the storeyed cambium of Tilia cordata Mill. Acta Soc Bot Pol 58: 211-228

Wloch W, Wawrzyniak S 1990 The configuration of events and cell growth activity in the storeyed cambium of the linden (Tilia cordata Mill.). Acta Soc Bot Pol 59: 25-43

Wloch W, Zagórska-Marek B 1982 Reconstruction of storeyed cambium in the linden. Acta Soc Bot Pol 51: 215-228

Wodzicki TJ 1962 A contribution to the rate of xylem cell formation and differentiation of tracheids in Pinus silvestris L. Acta Soc Bot Pol 31: 219-227

Wodzicki TJ 1971 Mechanism of xylem differentiation in Pinus silvestris L. J Exp Bot 22: 670-687
Wodzicki TJ, Brown CL 1973 Cellular differentiation of the cambium in the Pinaceae. Bot Gaz 134: 139-146
Wodzicki TJ, Peda T 1963 Investigation on the annual ring of wood formation in European silver fir (Abies pectinata D.C.). Acta Soc Bot Pol 32: 609-618
Wodzicki TJ, Zajaczkowski S 1970 Methodical problems in studies on seasonal production of cambial xylem derivatives. Acta Soc Bot Pol 39: 509-520
Wodzicki TJ, Zajaczkowski S 1983 Variation of seasonal cambial activity and xylem differentiation in a selected population of Pinus silvestris L. Folia For Pol Ser A 25: 5-23
Wodzicki TJ, Abe H, Wodzicki AB, Pharis RP, Cohen JD 1987 Investigations on the nature of the auxin-wave in the cambial region of pine stems. Plant Physiol 84: 135-143
Wolkinger F 1969 Morphologie und systematische Verbreitung der lebenden Holzfasern bei Sträuchern und Bäumen. I. Zur Morphologie und Zytologie. Holzforschung 23: 135-144
Wolkinger F 1970 II. Zur Histologie. Holzforschung 24: 141-151
Wolkinger F 1971 III. Systematische Verbreitung. Holzforschung 25: 29-30
Wolter KE 1968 A new method for marking xylem growth. For Sci 14: 102-104
Woodwell GM, Miller LN 1963 Chronic gamma radiation affects the distribution of radial increment in Pinus rigida stems. Science 139: 222-223
Worbes M 1985 Structural and other adaptations to long-term flooding by trees in Central America. Amazoniana 9: 459-484
Yamamoto F 1992 Effects of depth of flooding on growth and anatomy of stems and knee roots of Taxodium distichum. IAWA Bull 13: 93-104
Yamamoto F, Kozlowski TT 1987a Effects of flooding of soil on growth, stem anatomy, and ethylene production of Thuja orientalis seedlings. IAWA Bull 8: 21-29
Yamamoto F, Kozlowski TT 1987b Effect of flooding of soil on growth, stem anatomy, and ethylene production of Cryptomeria japonica seedlings. Scand J For Res 2: 45-58
Yamamoto F, Kozlowski TT, Wolter KE 1987 Effect of flooding on growth, stem anatomy, and ethylene production of Pinus halepensis seedlings. Can J For Res 17: 69-79
Yamanaka K 1984 (Normal and traumatic resin canals in the secondary phloem of conifers). J Jap Wood Res Soc 30: 347-353
Yamanaka K 1985 Abnormal tissue in swollen stemwood of Chamaecyparis obtusa Sieb. & Zucc. IAWA Bull 6: 53-60
Yamanaka K 1989 Formation of traumatic phloem resin canals in Chamaecyparis obtusa. IAWA Bull 10: 384-394
Yang KC, Benson CA, Wong JK 1986 Distribution of juvenile wood in two stems of Larix laricina. Can J For Res 16: 1041-1049
Yoshimura K, Hayashi S, Itoh T, Shimaji K 1981a Studies on the improvement of the pinning method for marking xylem growth I. Minute examination of pin marks in taeda pine and other species. Wood Res Kyoto Univ No 67: 1-16
Yoshimura K, Itoh T, Shimaji K 1981b II. Pursuit of the time sequence of abnormal tissue formation in loblolly pine. J Jap Wood Res Soc 27: 755-760
Yoshimura R 1979 Anatomical structure of wood in a leaning stem of Yachi-damo, Fraxinus mandshurica var. japonica, IV. Parenchyma-strand length variation within the annual ring formed before and after artificial bending. Mokuzai Gakkaishi 25: 449-454

Yoshimura R, Fukazawa K, Ishida S 1984 Fiber structure variation within an annual ring of Yachi-doma (Fraxinus mandshurica Rupr. var. Japonica Maxim.). In: Sudo S (ed) Proc Pac Reg Wood Anat Conf Oct 1-7, 1984 Ibaraki Jpn, 121-123

Yoshizawa N, Matsumoto S, Idei T 1985 Morphological features of tracheid tips associated with compression wood formation in Larix leptolepis Gord. IAWA Bull 6: 245-253

Yoshizawa N, Tanaka Y, Idei T 1986 Development of vascular cambium and compression wood formation in the shoot of young spruce (Picea jezoensis var Hondoensis). IAWA Bull 7: 21-30

Yoshizawa N, Kiyomiya M, Idei T 1987 Variations in tracheid length and morphological changes in tracheid tips associated with the development of compression wood. Wood Sci Tech 21: 1-10

Yumoto M 1984 The trabeculae and its related structures. Res Bull Coll Expt For Hokkaido Univ 41: 205-259

Yumoto M, Ishida S, Fukazawa K 1982 Studies on the formation and structure of the compression wood cells induced by artificial inclination in young trees of Picea glauca. I. Time course of the compression wood formation. Res Bull Hokkaido Univ For 39(1): 137-162

Yunus M, Yunus D 1976 Ray and fusiform initials in the vascular cambium of some tropical trees. Geobios 3: 103-104

Yunus M, Yunus D, Iqbal M 1978 On the cambial structure of some industrially important tropical trees. Flora 167: 159-163

Zagórska-Marek B 1975 Growth activity of fusiform initials in storeyed cambium. Acta Soc Bot Pol 44: 537-552

Zagórska-Marek B 1984 Pseudotransverse divisions and intrusive elongation of fusiform initials in storeyed cambium of Tilia. Can J Bot 62: 20-27

Zagórska-Marek B, Hejnowicz Z 1980 Discontinuous lines on the radial face of wavy-grained xylem as a manifestation of morphogenetic waves in the cambium. Acta Soc Bot Pol 49: 49-62

Zagórska-Marek B, Little CHA 1986 Control of fusiform initial orientation in the vascular cambium of Abies balsamea stems by indol-3-ylacetic acid. Can J Bot 64: 1120-1128

Zahur MS 1959 A comparative study of secondary phloem of 423 species of woody dicotyledons belonging to 85 families. Cornell Univ Agr Exp Sta Ithaca Mem 358, 160 pp

Zajaczkowski S, Dodd RS 1987 Significance of supra-cellular organisation in development of cambium. IAWA Bull 8: 322

Zajaczkowski S, Wodzicki TJ, Bruinsma J 1983 A possible mechanism for whole-plant morphogenesis. Physiol Plant 57: 306-310

Zajaczkowski S, Wodzicki TJ, Romberger JA 1984 Auxin waves and plant morphogenesis. In: Scott TK (ed) Hormonal regulation of development II. The functions of hormones from the level of the cell to the whole plant. Encylopedia of Plant Physiology New Series Vol 10. Springer, Berlin, Heidelberg, New York, 244-262

Zamski E 1972 Temperature and photoperiodic effects on xylem and vertical resin duct formation in Pinus halepensis Mill. Israel J Bot 21: 99-107

Zasada JC, Zahner R 1969 Vessel element development in the earlywood of red oak (Quercus rubra). Can J Bot 47: 1965-1971

Zee S-Y 1968 Ontogeny of cambium and phloem in the epicotyl of Pisum sativuum. Aust J Bot 16: 419-426

Zelawski W 1980 Aging of the plant organism (A survey of experiments and hypotheses). Sov Plant Physiol (Transl) 27(4): 658-665

Zhengli L, Keming C 1988 Differentiation of secondary xylem after girdling. IAWA Bull 9: 375-383

Zhuravleva MV 1972 (A procedure for determining cambium activity in growing spruce trees). Lesnoi Zhurnal No. 1: 140-141. Transl Environ Can No. 2489, 5pp, 1984

Ziegenspeck H 1931 Über die Zwischenzustände bei der Bildung verholzter Membranen. Ber Deutsch Bot Ges 49: 381-386

Ziegler H 1964 Storage, mobilization and distribution of reserve material in trees. In: Zimmermann MH (ed) The formation of wood in forest trees. Academic Press, New York, 303-320

Ziegler H, Merz W 1961 Der "Hasel"-wuchs. Über Beziehungen zwischen unregelmässigem Dickenwachstum und Markstrahlverteilung. Holz Roh- Werkst 19: 1-8

Zijlstra K 1908 Die Gestalt der Markstrahlen im sekundären Holze. Rec Trav Bot Neerl 5: 17-50

Ziller WG, Stirling D 1961 Sapsucker damage in coastal British Columbia. For Chron 37: 331-335

Zirkle C 1931 Nucleoli of the root tip and cambium of Pinus strobus. Cytologia 2: 85-105

Zobel BJ, van Buijtenen JP 1989 Wood variation. Its causes and control. Springer, Berlin, Heidelberg, New York, 363 pp

Subject Index

Springer-Verlag and the Environment

We at Springer-Verlag firmly believe that an international science publisher has a special obligation to the environment, and our corporate policies consistently reflect this conviction.

We also expect our business partners – paper mills, printers, packaging manufacturers, etc. – to commit themselves to using environmentally friendly materials and production processes.

The paper in this book is made from low- or no-chlorine pulp and is acid free, in conformance with international standards for paper permanency.

MIX
Papier aus verantwortungsvollen Quellen
Paper from responsible sources
FSC® C105338

If you have any concerns about our products,
you can contact us on
ProductSafety@springernature.com

In case Publisher is established outside the EU,
the EU authorized representative is:
Springer Nature Customer Service Center GmbH
Europaplatz 3, 69115 Heidelberg, Germany

Printed by Libri Plureos GmbH
in Hamburg, Germany